THE LIGHT FANTASTIC

The Light Fantastic

A Modern Introduction to Classical and Quantum Optics

Second Edition

I. R. Kenyon

School of Physics and Astronomy
University of Birmingham

OXFORD

UNIVERSITY PRESS

OXFORD
UNIVERSITY PRESS

Great Clarendon Street, Oxford OX2 6DP

Oxford University Press is a department of the University of Oxford.
It furthers the University's objective of excellence in research, scholarship,
and education by publishing worldwide in

Oxford New York

Auckland Cape Town Dar es Salaam Hong Kong Karachi
Kuala Lumpur Madrid Melbourne Mexico City Nairobi
New Delhi Shanghai Taipei Toronto

With offices in

Argentina Austria Brazil Chile Czech Republic France Greece
Guatemala Hungary Italy Japan Poland Portugal Singapore
South Korea Switzerland Thailand Turkey Ukraine Vietnam

Oxford is a registered trade mark of Oxford University Press
in the UK and in certain other countries

Published in the United States
by Oxford University Press Inc., New York

First published 2008
Second Edition published 2011

British Library Cataloguing in Publication Data
Data available

Library of Congress Cataloging in Publication Data
Data available

Printed in Great Britain
on acid-free paper by
CPI Antony Rowe, Chippenham, Wiltshire

ISBN 978–0–19–958461–1 (Hbk)
ISBN 978–0–19–958460–4 (Pbk)

10 9 8 7 6 5 4 3 2 1

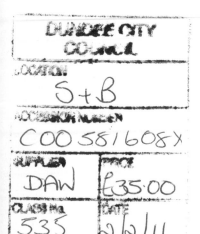

Preface

For the second edition chapters have been added to cover two areas that have seen interesting developments recently. These developments are the construction and applications of photonic crystals and quantum dots. The inclusion of photonic crystals has required an update of the analysis of interference filters. A full quantum calculation of atomic transition rates has been added, which helps when interpreting the relationship between absorption, spontaneous and stimulated emission. Sections have also been added which describe the operation of EM-CCDs and show how their use improves the resolution of ground-based telescopes. Finally a description of digital micromirror projectors has been added. Minor improvements and corrections have been made elsewhere. The opportunity has been taken to bring the classical section of the book in line with the quantum section by making the complex form of a sinusoidal plane wave in space-time $\exp\left[i(\mathbf{k} \cdot \mathbf{r} - \omega t + \phi)\right]$ throughout.

This book deals primarily with the properties and uses of electromagnetic waves and photons of visible light. Other regions of the electromagnetic spectrum are only treated where appropriate: for example there is coverage of optical fibre communication using near infrared radiation. Modern quantum theory originated in the observations of the quantum behaviour of electromagnetic radiation, and now, a century later the quantum behaviour of light offers tantalizing possibilities for computing and encryption. During that period a deeper understanding of electromagnetic radiation in terms of waves and photons has made possible the invention of lasers, optical fibre communication, space-based telescopes, the world wide web and digital cameras. The optoelectronics industry, undreamt of even forty years ago, has grown to be a major employer of scientists and engineers. Even crude measures, such as the hundred million solid state lasers made annually, the millions of kilometres of optical fibre installed, and the widespread availability of megapixel digital cameras and of DVDs, give a sense of this industry's economic and cultural impact. Studies of the subtle features of quantum theory, such as entangled states, have been facilitated by research tools dependent on the technological advances in optoelectronics, which illustrates the truism that technology and pure science go forward hand-in-hand. Clearly there is a necessity for a wide range of scientists and technologists to possess an up-to-date understanding of waves and photons so that they can make use of the theoretical, experimental and technological tools now available. The main objective of this text is to provide that basic understanding, which will be important if the reader is to follow future developments in this rapidly expanding field.

The text is designed to be comprehensive and up-to-date so that students at universities and colleges of technology should find this volume useful throughout their degree programme. Following an introductory chapter providing basic concepts and facts, the book is divided into three sections: the first section (Chapters 2–4) covers ray optics, the second section (Chapters 5–11) wave optics, and the final section (Chapters 12–20) quantum optics.

Huygens' principle is used to derive laws of propagation at interfaces in Chapter 2. On this basis the geometric optics of mirrors and lenses is treated in Chapter 3. Then the principles and design of optical instruments including microscopes, telescopes and cameras are outlined in Chapter 4. Aberrations from the paraxial theory and the simpler techniques for reducing them to tolerable levels are also described in Chapters 3 and 4.

The section on wave optics starts in Chapter 5 with the superposition rule for electromagnetic waves and its application to interference effects such as those seen in Young's crucial two slit experiment and the Michelson interferometer. Coherence and the relation to atomic wavepackets are both introduced in these simple examples. Interferometers including the Fabry–Perot and Sagnac varieties are described with their applications. Diffraction effects, both near and far field, and spectrometers are considered in Chapter 6. Fourier transforms, of which diffraction patterns are an example, are treated formally in Chapter 7. This allows the connection between Michelson interferograms and the source spectrum to be exploited in extracting spectra with standard infrared Fourier transform (FTIR) spectrometers. Chapter 8 pulls together themes in optical instrument design in describing the design of optical mirror telescopes, adaptive optics and aperture synthesis. Interferometric arrays are discussed and their performance compared to that of radio telescopes. Electromagnetic wave theory rests on Maxwell's equations and Poynting's theorem for the energy in electromagnetic waves. The electromagnetic wave equation and the laws of propagation of light at interfaces (Fresnel's laws) are derived directly from classical electromagnetic theory in Chapter 9. Dielectric multilayer filters are analysed. Modes of the electromagnetic field are introduced and the density of these states deduced. Chapter 10 carries the description of polarization forward to include circular polarization which is revealingly the polarization state of individual photons. Birefringence and the effects of applied fields are described together with devices using them such as DVD readers and LCD screens. Electromagnetic interactions with matter in semiclassical terms are discussed in Chapter 11: dispersion, absorption and scattering are described and shown to be related. Propagation in metals is described and surface plasma waves are analysed. This completes two sections devoted to the purely classical behaviour of light.

An account of the fundamental experiments that underpin the quantum theory of electromagnetic radiation in Chapter 12 opens the section on quantum optics. In Chapter 13 the dual wave–particle nature of electromagnetic radiation and the Heisenberg uncertainty principle are examined at length. An outline of the necessary elements of quantum mechanics and its application to atomic structure are provided as groundwork for the topics that follow. The principles underlying laser operation, as well as gas, solid state and semiconductor lasers, and their applications, are treated in Chapter 14. Topics include mode-locking and chirped pulse amplification. Detectors of radiation in the visible and near infrared are described in Chapter 15: these include the CCD and CMOS arrays used in digital cameras, photomultipliers and night vision devices. Optical fibre based communication principles, devices and systems, as well as optical fibre sensors are described in Chapter 16. This treatment includes a survey of the modes carried by fibres, and an analysis of absorption, dispersion and noise in long single mode fibre connections. Chapter 17 contains an account of the properties of photonic crystals, which are responsible in nature for the iridescence of opals and butterfly wings. There are discussions of practical applications in optical interconnects, compact optical modulators and photonic crystal fibres. Chapter 18 introduces the semiclassical calculation of decay rates and the behaviour of atoms in the resonant and near resonant laser beams. Effects including Rabi oscillations, electromagnetically induced transparency and slow light are introduced. After this the developments leading to the fabrication of optical clocks are described. Chapter 19 starts by introducing the second quantization of the electromagnetic field, in which fields are now presented as quantum mechanical operators. This is followed by a description of the study of correlations between photons, first observed by Hanbury Brown and Twiss. Then the theory and experimental methods for generating entangled photons are described, followed by experimental studies of two photon correlations in interferometers showing delayed choice and quantum erasure. A calculation of the emission and absorption rates of two state atoms is made, quantizing both atom and electromagnetic field. Chapter 20 introduces quantum dots, which have discrete quantized energy levels like atoms but considerably larger electric dipole moments. They therefore couple more strongly to electric fields. The optical properties of quantum dots within optical cavities are analysed using cavity quantum electrodynamics. Lastly there is a description of a quantum dot diode with the potential for use in quantum key distribution.

The text has been designed so that subsets of chapters are self-contained and well suited to accompany focused optics courses, while the complete text provides compact coverage for courses that extend through three or four years. A suggested reduced course could include all the chapters and sections listed here:

- Introduction and ray optics: Chapters 1 and 2;
- Lenses without abberations: Sections 3.1 to 3.6.1;
- Optical instruments: Sections 4.1 to 4.5.2, and 4.8;
- Wave optics and interferometers: Sections 5.1 to 5.7.1, and 5.8 to 5.9;
- Diffraction and gratings: Sections 6.1 to 6.9;
- Astronomical telescopes: Sections 8.1 to 8.3;
- Electromagnetic theory and Fresnel's laws: Sections 9.1 and 9.4 to 9.6, and 9.8 to 9.8.1;
- Polarization phenomena: Sections 10.1 to 10.4, and 10.5, and 10.5.2 to 10.7.1, and 10.8 to 10.8.3;
- Light in matter: Sections 11.1 to 11.6.2;
- Quantum behaviour of light: Chapter 12; Sections 13.1 to 13.5.2, and 13.11 to 13.13;
- Lasers and detectors: Sections 14.1 to 14.4, and 14.4.3 to 14.6; Sections 15.1 to 15.3.1, and 15.7 to 15.9;
- Optical fibre communication: Sections 16.1 to 16.2, and 16.4 to 16.6, and 16.9 to 16.10.1, and 16.13 to 16.14;
- Photonic crystals: All except Sections 17.3 and 17.3.1;
- Quantum interactions: Sections 18.3 to 18.8;
- Quantized electromagnetic fields: Sections 19.6 to 19.8;
- Cavities and quantum dots: All except Sections 20.3 and 20.4.

Acknowledgements

I would like to thank two Heads of the School of Physics and Astronomy at Birmingham University, Professors John Nelson and Mike Gunn, and also Professor Peter Watkins, Head of the Elementary Particle Physics Group at Birmingham, for their support and encouragement during the lengthy preparation of both editions of this textbook. Dr Sonke Adlung, the senior science editor at Oxford University Press, has always been unfailingly helpful and courteous in dealing with the many aspects of the preparation, and my thanks go to him for making my path easier. I am also grateful to Hannah McGuffie, Victoria Mortimer and April Warman at Oxford University Press for the smooth management of copy editing, layout, production and publicity.

Many colleagues have been more than generous in finding time in busy lives to read and comment on material for which they have a particular interest and expertise. First I want to thank Professor Ken Strain of the Institute for Gravitational Research, Glasgow University and the GEO600 team for reading and commenting on the material on gravitational wave detection and Professor Peter Tuthill of the Astronomy Department, University of Sydney, for reading and commenting on the sections concerning aperture synthesis with telescopes. I am indebted to Professor Chris Haniff of the Astrophysics Group at the Cavendish Laboratory, Cambridge University, who read through the material on modern interferometry with telescopes and aperture synthesis, and made extensive valuable comments. Also I wish to thank Professor Helen Gleeson of the University of Manchester for reading and commenting on the polarization chapter, particularly the section on liquid crystals. Dr Peter Norreys, Group Leader at the Central Laser Facility, Rutherford Appleton Laboratory, helped me by checking the material relating to extreme energy lasers. I am indebted to Dr Peter Pool of e2v Technologies who patiently answered my many questions about CCD stucture and readout; and who read and commented on the material on EM-CCDs. Ian Bennion, Professor of Optoelectronics at the University of Aston, was kind enough to look over the material on optical fibres and made some very useful suggestions for improvement; I extend my thanks to him. I am particularly grateful to Lene Hau, Mallinckrodt Professor of Physics and Applied Physics at Harvard University, who made comments on the sections concerning electromagnetically induced transparency, and to Professor David Wineland of the Time and Frequency Division of the National Institute for Standards and Technology, Boulder, for looking over the section on optical clocks. My thanks also to Professor Peter Vukusic of Exeter University, Professor Thomas Krauss, leader of the Microphotonics and Photonic Crystals Group at St Andrews University and Mr Marcel Spurney, also of St Andrews University, for making valuable suggestions for improvements to the chapter on photonic crystals. I am extremely grateful to both Ulf Leonhardt, Professor in Theoretical Physics at St Andrews University, and Rodney Loudon, Research Professor at the University of Essex, who guided me around several difficulties in the theory of the quantized electromagnetic field, and to Professor Loudon for a very careful reading of the chapter on the quantization of the electromagnetic field. My thanks also to Professor A. J. Shields of Toshiba Research Europe and Professor T. M. Hsu of the National Central University, Taiwan, for their kindness in finding time to comment on material concerning their experiments. In addition I am very much indebted to Professor M. Skolnick of Sheffield University for reading the whole of the final chapter and suggesting several significant improvements. My warm thanks to Professor Craig Mackay of the Astrophysics Institute at Cambridge for his expert comments on material relating to lucky imaging and its continuing development.

Turning to my Birmingham colleagues, I want to first thank Dr John Griffith whose all-round, enthusiastic knowledge of optics was always freely available over the years before I started work on this book. He also read and commented on a very early draft of the classical optics component of the book. Next I wish to thank Professor Mike Gunn, who took on the task of reading the description of semiclassical interactions between radiation and atoms. His expert advice and comments were very valuable. I especially thank Dr Ken Elliott for producing the spectra of helium gas, nitrogen gas, a fluorescent lamp and the Sun which appear in figures 1.8–1.15. My thanks also go to Professor Ted Forgan who was kind enough to read the chapter dealing with scattering, absorption and dispersion: his guidance and suggestions on particular points were very helpful. I wish to thank Professor Yvonne Elsworth for taking the time to read the chapter presenting Fourier optics, and for her comments. I would like to thank Dr Somak Raychaudry for reading and commenting on the chapter relating to astronomical telescopes. Dr Alastair Rae read and made many useful comments on the content of the chapters which introduce quantum concepts: Alastair's advice was particularly helpful in balancing the material. Dr Garry Tungate was brave enough to read the long laser chapter and shared with me the fruits of his experience, for which I am very much indebted to him. Thanks too to Dr Chris Eyles who took on the reading and commenting on the chapter dealing with detectors; he, as well as Dr Peter Pool of e2v Technologies, patiently answered many questions about CCD stucture and readout. Thanks too to Dr Ray Jones for comments on the early version of the first two chapters. I am grateful to Dr David Evans for carefully checking chapters on classical electromagnetism, and scattering and absorption; and to Dr Goronwy Jones for lively and useful comments on the chapter on the development of quantum mechanics. I am grateful to Dr Costas Mylonas for patiently answering a string of questions on the analysis of electromagnetically induced transparency. I am indebted to Dr Martin Long who generously found time for patient explanations: his insights clarified quite a few difficult points. Dr Chris Mayhew kindly offered to read the whole of the final draft of the first edition and I want to thank him for his useful and helpful comments.

In addition I would like to heartily thank the reviewer of both editions, Dr Geoffrey Brooker of Oxford University, for his thorough and detailed reports. His incisive and enlightening comments proved extremely beneficial.

The input from all these colleagues removed misunderstandings on my part, helped me to clarify arguments and brought points to my attention that I would otherwise have missed. The responsibility for any remaining errors should be laid at my door.

My thanks also go to the authors and publishers who have allowed me to use published figures, or adaptations of figures, or tables, each acknowledged individually in the text. I am grateful to Taylor and Francis, owners of the publications of the CRC Publishing Company regarding Table 12.1 and figure 11.5; the American Physical Society regarding figures 12.4, 12.7, 12.12, 13.16, 14.12, 16.30, 19.10, 19.12, 19.14 and 20.13; the Institute of Physics regarding figures 20.8 and 20.9; the American Institute of Physics regarding figures 17.11, 20.16, 20.17 and 20.18; the Astronomical Society of the Pacific regarding figure 8.20; Elsevier Publishing regarding figure 19.1; Pearson Education Ltd. regarding figure 15.1; Schott Glas regarding figure 3.33; and the Institution of Engineering and Technology regarding figure 16.1. In addition thanks to Dr L. Gardner, program manager at Lambda Research Corporation, USA, for supplying a lens design analysis from the OSLO® system; Dr Bernnd Lingelbach of HTW, Aalen and Institut fuer Augenoptik, Abtsgemuend, Germany, for supplying a figure of a mediaeval aspheric lens; Dr Mike Rietveld of the EISCAT Scientific Association, N-9027 Ramfjordmoen, Norway, for supplying a figure of delay data from the EISCAT radiosonde; and Michael Douglass of Texas Instruments for figure 2.20 illustrating the operation of a Texas Instruments digital micromirror device.

In producing some 500 diagrams I made almost exclusive use of the ROOT package developed by Dr Rene Brun and Dr Fons Rademakers and described in *ROOT – An Object Oriented Data Analysis Framework* which appeared in the Proceedings of AIHENP'96 Workshop, Lausanne, Nuclear Instruments and Methods in Physics Research A389(1997)81-6. ROOT can be accessed at http://root.cern.ch. I also thank Dr Brun for help while learning to use this sophisticated tool.

To my wife Valerie, without whose support I would not have attempted, let alone completed this book.

Contents

Introduction

1.1 Aims and overview

An understanding of the properties and manipulation of electromagnetic radiation in the visible and near visible part of the spectrum is of great importance in the physical sciences and related branches of engineering. It was, of course, in the study of electromagnetic phenomena that quantum effects were first encountered, while today optical experiments continue to reveal subtleties of the quantum behaviour of radiation. On the practical side optical sources, devices and detectors are crucial to both research and industry. A few examples will illustrate their value, starting with applications in pure research: ground and space based telescopes look back into the remote past of the universe; kilometre sized interferometers are being used to search for gravitational waves; while laser cooling enables researchers to bring ions to rest so that their transitions provide the ultimate in stable reference frequencies for optical clocks. In the commercial field applications also abound and are of increasing importance: the millions of kilometres of optical fibre now installed provide the modern data highway; semiconductor lasers are used in transmitting this data over fibres, other lasers are used to weld retinas while kilowatt power lasers routinely drill holes in inch-thick steel plates. Among consumer goods DVD readers and LCD screens function thanks to our ability to manipulate polarized light. The techniques and interests of fundamental research and industry overlap a good deal. For example charged coupled devices (CCDs) with several million pixels on a single chip are the preferred detectors in astronomy and are also found in digital cameras. A more exotic example is the interest of solitons to both pure and applied research: these are waveforms which can be transmitted over thousands of kilometres of optical fibre without change of shape. Finally entangled states of photons are studied to investigate questions about the logical structure of quantum mechanics, and at the same time these states offer the possibility of a form of secure quantum key distribution for cryptography when exchanging sensitive information. It is to introduce students of science and engineering to this intellectually stimulating and industrially significant subject that the text aims.

Optics has three increasing levels of sophistication. Ray or geometric optics is covered in Chapters 2–4, wave optics is covered in Chapters 5–11, and quantum optics is covered in Chapters 12–20. Ray and wave optics make up classical optics. Quantum optics provides the more complete description of the behaviour of electromagnetic radiation so it naturally subsumes classical optics. However ray and wave optics encapsulate significant components of optics and are very often adequate and more convenient to use. The emphasis here is on visible radiation and the near infrared radiation used in optical fibre communications.

The optics of paraxial (near axis) light rays incident on plane and spherical mirrors and interfaces is presented in Chapters 2 and 3 and applied to prisms and lenses. Matrix methods of analysing ray propagation are included. Chapter 4 is used to explain the principles and design of optical instruments. These are the tools for image formation and for transmitting and directing light beams. Performance parameters such as numerical aperture and etendue are introduced. The aberrations of lenses and mirrors and corrective techniques are reviewed.

Wave optics begins in Chapter 5 with the discussion of the superposition of electromagnetic waves and interference effects. Young's two-slit experiment and the Michelson interferometer are taken as examples. Interference effects are shown to occur only when the beams involved are coherent. The wave packet is identified with the coherence volume. Interferometers such as the Fabry–Perot and Sagnac devices are analysed. Chapters 6 contains material on diffraction, which is interference due to any pattern of apertures. The emphasis is on Fraunhofer diffraction produced by circular apertures and by gratings, the latter being the basis for spectrometers. Matrix methods for determining paraxial propagation are extended to laser beams. Examples of near field Fresnel diffraction are also presented.

Fraunhofer diffraction is an example of a Fourier transform. Chapter 7 introduces Fourier analysis and describes its application in Fourier transform infrared (FTIR) spectrometery. The convolution theorem, Parseval's theorem, the Wiener–Khinchine theorem and the bandwidth theorem are deduced. Abbe's theory of image formation is outlined. Accounts of holography and information processing close the chapter. Optical telescopes such as the Hubble Space Telescope (HST) are described in Chapter 8. The use of adaptive optics and lucky imaging to counter atmospheric turbulence are described. Aperture synthesis and interferometry with telescopes are discussed. Gravitational wave detectors and gravitational imaging are treated briefly.

In Chapters 9 and 10 the classical treatment of radiation is developed from Maxwell's equations. Fresnel's laws are derived, yielding the fractions of radiation reflected and transmitted at interfaces. Evanescent waves are treated. There follows a description of interference filters and beam splitters. Modes of electromagnetic radiation and their propagation in waveguides are the final topics in Chapter 9. Polarization is the topic of Chapter 10. Effects such as dichroism, birefringence and the optical activity of materials under applied fields are all analysed. Devices for manipulating polarization are covered and their use in DVD readers and liquid crystal displays.

Dispersion, absorption and scattering in materials are discussed in Chapter 11 in terms of the underlying atomic and molecular response. Scattering from scatterers much smaller than the wavelength (Rayleigh scattering) and of comparable size (Mie scattering) are described. Absorption and dispersion are analysed at frequencies around that of the resonant response of electrons bound in atoms. The optical properties of metals and plasmas are treated using Drude's free electron model. Energy and information carried by electromagnetic radiation are shown to travel with the group velocity of the wavepackets. Surface plasma waves and their use in medicine are introduced.

The quantum optics section of the book commences in Chapter 12 with a review of experimental evidence used to show that electromagnetic radiation possesses particle as well as a wave properties. de Broglie's insight that material particles also have a wave–particle nature, and its experimental validation are described. Bohr's model of an atom with quantized electron orbits is described. The reconciliation of the wave and particle nature of electromagnetic radiation, known as wave-particle duality, is discussed in detail. Heisenberg's uncertainty principle and its implications for the precision of measurements are treated. Chapter 13 is used to introduce quantum mechanics. Schroedinger's non-relativistic equation for particle motion is solved for the electron in the hydrogen atom. Quantization rules for optical transitions are noted. The experiments revealing the spins of electrons and photons are described.

Three chapters on applications of quantum phenomena in research and industry follow. The topics are lasers, photodetectors and fibre optics, which together underpin the world's telecommunications. Chapter 14 contains descriptions of laser types and their applications. The treatment commences by deducing the relationships between absorption, and spontaneous and stimulated emission in equilibrium. This leads on to the prerequisites for lasing. The unprecedented coherence of laser radiation is its key property. Descriptions of laser modes and gain follow. Applications described in interferometry and spectroscopy depend on the narrow line width of lasers. Semiconductor lasers are convenient sources for injecting light into optical fibre and several types are discussed. Solid state lasers use a crystal such as Ti:sapphire as the active lasing material. Their applications, including chirped pulse amplification in laser fusion devices, are described. Non-linear effects in certain crystals produced by high intensity beams – harmonic generation, parametric amplification, Raman scattering and Brillouin scattering – are all treated.

An account of electronic detectors of visible and near infrared radiation is given in Chapter 15. One common detector type is the semi-conductor photodiode. Their thresholds, efficiency, sensitivity, linearity and noise are all discussed. Schottky and avalanche photodiodes are described. The features of CCD imaging arrays are explained. Descriptions of photomultipliers and hybrid image intensifiers end the chapter.

Single mode optical fibre has especially low dispersion and absorption in the near infrared. As a result faithful transmission of data at high rates over intercontinental distances on single mode fibre has become commonplace. Mode transmission and dispersion of optical fibre are analysed in depth in Chapter 16. Optical link techniques, devices, and power budget are discussed, including wavelength division multiplexing. Soliton propagation is presented. Representative examples of optical fibre sensors are outlined, including current transformers.

The remaining chapters introduce modern topics in which potential for future developments is concentrated. Chapter 17 contains an account of photonic crystals. These are dielectric structures in which the relative permittivity has a periodic variation. Photonic crystals possess band-gaps analogous to the electronic band-gaps of semiconductors, but in this case it is electromagnetic waves which cannot penetrate the crystal. The optical properties of photonic crystals are analysed using Bloch waves and Brillouin zones. Compact Mach–Zehnder interferometers and improvements in the efficiency of LED emission are described. Photonic crystal fibres are also discussed.

Chapter 18 provides a semi-classical analysis of the interaction of electromagnetic radiation with atoms. The rate of decay of hydrogen from the excited 2p to the 1s ground state is calculated as an example of an electric dipole transition. Selection rules for transitions are explained. When an atom is exposed to a laser tuned to a resonance frequency of the atom, this is pumped cyclically between the ground and excited state. Such Rabi oscillations are analysed. Electromagnetically induced transparency and the associated slowing of light to vehicle speeds is described. The laser cooling and trapping of ions make these Doppler free sources the ultimate frequency references. A description of an optical clock based on a single mercury ion closes Chapter 18.

The full quantum theory of matter and radiation has all fields represented by operators. Chapter 19 is used to present this second quantization of the electromagnetic field. Now basic operators create and annihilate individual photons. Correlations are re-expressed in terms of the expectation values of these operators. This approach provides a framework within which many surprising effects are explained. A start is made with photon bunching from an incoherent source observed by Hanbury Brown and Twiss. Entangled states of photons are quantum mechanical states with no classical equivalent. Experimental methods for creating entangled pairs of photons are described. Interferometry with entangled photons is presented exploring subtleties such as complementarity, delayed choice and quantum erasure. A final section contains a full quantum mechanical calculation of transition rates.

In Chapter 20 quantum dots in the form of nanometre sized pieces of semiconductor are introduced. These are embedded in a second semiconductor with a wider band-gap. Thus the dot forms a potential well in which electrons ocupy discrete energy levels. A dot is therefore like a huge atom with a correspondingly large electric dipole moment. The surrounding semiconductor is micron sized with reflective walls making it an optical microcavity. Cavity electrodynamics is used to analyse how such a dot interacts with the modes of the microcavity. Examples are given of this behaviour including the alteration of the rate of spontaneous emission. Finally a quantum dot diode providing single photons on demand is presented. The application of such sources to quantum key distribution is outlined.

The remainder of this present chapter puts in place elementary material needed later. It includes a sketch of electromagnetic wave theory; a review of the electromagnetic spectrum; basic properties of optical materials; radiation terminology; and the Doppler shift.

1.2 Electromagnetic waves

Electromagnetic waves have properties resembling those of the more familiar mechanical waves. Waves on a stretched string or on the surface of deep water show local movement and yet the string or the water is not carried forward with the waves. What is carried forward is energy, which in the case of a wave on the sea is readily apparent to someone in a ship tossed by waves. In the case of electromagnetic waves the electric and magnetic fields at every point in the wave's path are the things that oscillate. Energy carried by the electromagnetic wave is detected in ways that depend on its wavelength. For example as you sit at your desk you might feel the Sun's rays warming you, or use a radio receiver to convert modulation of radio waves to sound, or receive light waves from this sentence on this page so that you can share my thoughts.

The electromagnetic fields are vector quantities which generally point transverse to the direction of wave motion so that the electric field, the magnetic field and the direction of motion of the wave are mutually perpendicular. This assertion is exactly true for a uniform plane wave in free space; where free space is the vacuum uncluttered by material objects. This is similar to the transverse displacement of a string when a wave travels along it. Suppose that an electromagnetic wave travels in the z-direction and the electric field vector points in the x-direction, then the magnetic field vector points in the y-direction. The definition of the units rests on the measurement of the force exerted on a charge:

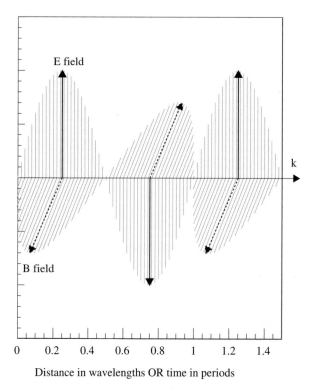

Fig. 1.1 The variation of the electric and magnetic fields (a) with position along the direction of wave motion for a plane electromagnetic wave with linear polarization; or (b) with time at a fixed point in space.

the force on a charge q coulombs (C) moving with velocity $\mathbf{v}$ metres per second $(\mathrm{m\,s^{-1}})$ is given by the Lorentz relation

$$\mathbf{F} = q\,[\,\mathbf{E} + \mathbf{v} \wedge \mathbf{B}\,], \tag{1.1}$$

where the force $\mathbf{F}$ is in newtons when the electric field $\mathbf{E}$ is expressed in volts per metre $(\mathrm{V\,m^{-1}})$ and the magnetic field $\mathbf{B}$ is in teslas (T). Figure 1.1 shows the electric and magnetic fields in a sinusoidal wave moving along the arrowed path, labelled $\mathbf{k}$. Representative field vectors are drawn at three points along the path, the remainder are indicated by shading. The diagram is a snapshot in time, and at a later moment the whole wave will have moved to the right. The distance between successive identical features (e.g. peaks) is the *wavelength* with the symbol λ. Alternatively we can regard figure 1.1 as showing the time variation of the wave at a fixed point in space. In this case the time between successive identical features is the *period* with the symbol τ. Then the number of peaks passing a fixed point per unit time is $1/\tau$, which is called the *frequency*, measured in hertz (Hz), and given the symbol f. Therefore, with f waves of length λ passing per second, the *wave velocity* is

$$v = f\lambda. \tag{1.2}$$

This is also known as the *phase velocity*. Electromagnetic waves can have any wavelength, with light being the electromagnetic waves detectable by the eye with wavelengths between 400 and 700 nm. The electromagnetic waves shown in figure 1.1 are called *plane polarized* or *linearly polarized* because each field vector remains in a plane as the wave propagates.

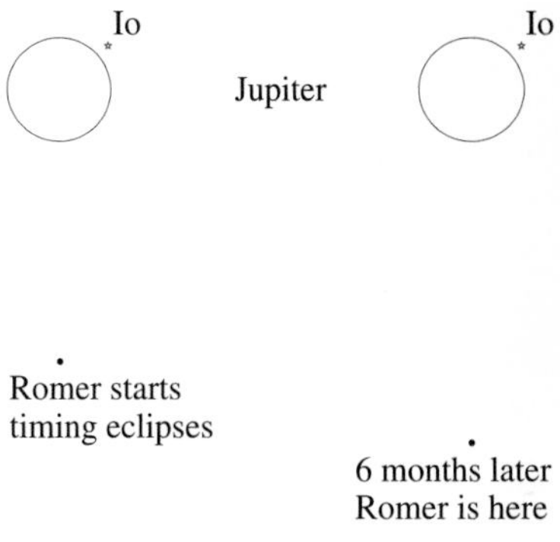

1.3   The velocity of light

Compared to the velocity of mechanical waves the velocity of light is extremely large; a distant lightning flash is seen well before the thunderclap is heard. However this simple observation does not allow any inference as to whether light arrives instantaneously, or whether its velocity is large but finite. Rømer in 1676 was the first to deduce that the velocity of light is finite: he had been timing a sequence of eclipses of the moon Io by its parent planet Jupiter. Jupiter orbits the Sun once every 11.8 years with Io in close attendance, Io orbiting Jupiter once every 1.77 days. The motion of the Earth around the Sun and of Io around Jupiter are so much more rapid than that of Jupiter that we can ignore the motion of Jupiter in following Rømer's argument. Rømer noted that, over a half-year during which the Earth–Jupiter separation was continuously increasing, the period of rotation of Io round Jupiter appeared to lengthen also. The eclipses of Io by Jupiter were finally twenty-two minutes later than was to be expected if Io's period of revolution around Jupiter was constant. Rømer correctly interpreted this delay as the time taken by light from Io to travel the distance that the Earth had moved away from Io over that half year. This effect is illustrated in figure 1.2.

Later Bradley used another astronomical method to obtain a value for the velocity of light close to the modern value, $3 \, 10^8 \, \mathrm{m \, s^{-1}}$. Bradley observed that distant stars appeared to change position during the year. Looking north for example, all the stars complete an annual circular orbit of 43 arcseconds diameter. (This is twenty times larger than the displacement observed in a year in the relative positions of the nearest stars with respect to the distant stars due to parallax.) Bradley correctly interpreted this *aberration of starlight* as being due to the motion of the Earth around the Sun. He argued that the direction which the starlight appears to come from is the vector difference between the velocity of the starlight and that of the Earth. A knowledge of the Earth's orbital velocity and the aberration yields the velocity of light. This is shown in figure 1.3.

Such astronomical methods were displaced by more precise Earth-based measurements in which the round trip time is measured for light to travel along a measured path to and from a mirror. If this is done in air then very long paths, of order tens of kilometres, can be used, but a small correction is needed to compensate for the difference between the velocity of light in vacuum and in air at the atmospheric conditions. The

Io   Io

Jupiter

Romer starts
timing eclipses

6 months later
Romer is here

Fig. 1.2 Rømer's method of determining the speed of light.

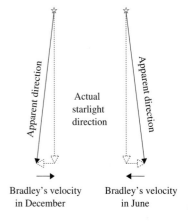

Apparent direction   Actual   Apparent direction
                        starlight
                        direction

Bradley's velocity   Bradley's velocity
in December   in June

Fig. 1.3 Aberration of starlight. The apparent direction of the same star is shown at times six months apart.

round trip time for a total path of 30 km is 0.1 ms so that a precision of 10^{-9} s (one nanosecond, 1 ns) in timing is required to get a precision of 1 part in 10^5 in the velocity determination. In order to acheive the necessary precision in timing electronically controlled shutters such as the Kerr cell described in Chapter 10 are used. This technique is commonly used to measure distances in surveying, in which case the velocity of light is the input and the distance the output.

In the 1960s a new approach to measuring the velocity of light took advantage of the relation given above: velocity equals wavelength times frequency. A source emitting a narrow range of frequencies is used, and both the frequency and wavelength of the radiation in vacuum are measured, and then multiplied together to give c. The velocity of electromagnetic waves in vacuum has been measured very precisely and is found to be a constant independent of the wavelength of the radiation.

Late in the nineteenth century Michelson discovered that the velocity of light is independent of the motion of the source and of the observer; his measurement will be discussed in Chapter 5. This result is quite different from the way the velocity of, for example, sound waves behaves. If an observer at rest measures the velocity of sound as v m s^{-1}, then on moving toward the source at a velocity u m s^{-1} the velocity will appear to rise to $(v + u)$ m s^{-1}, just as you would guess. For electromagnetic radiation the measured velocity is *constant* whatever the relative motion of source and observer! This experimental fact, the constancy of the velocity of light, is a fundamental feature of nature. Einstein made the constancy of the velocity of light in free space, whatever the motion of the source or observer, one of the two postulates on which he built the special theory of relativity in 1905. By 1984 the value of the velocity of light in vacuum determined from the product of wavelength and frequency measurements was:

$$c = 299\,792\,458 \text{m s}^{-1}.$$

The precision in the determination of c depended on the precision of the reference standards of length and time. In the case of the unit of time this was, and is, provided by atomic clocks based on the frequency of a microwave transition in caesium. Such clocks commonly agree to better than parts in 10^{12} so that atomic clocks are accepted as the primary standards of time (and frequency). The length of the second is defined to be 9 192 631 770 periods of the radiation emitted in a specified transition of ^{133}Cs. At that time the standard of length was defined in terms of a wavelength of krypton but with much poorer precision. Scientists therefore chose to define, once for all, the velocity of light in vacuum at its then best measured value, given above. This is an altogether reasonable approach because the velocity of light in vacuum is a constant of nature, whereas units of length and time are definitely not. This leaves the unit of length as something that has to be measured; the metre being the distance travelled by light in vacuum in (9 192 631 770/299 792 458) periods of the ^{133}Cs microwave transition.

1.4 A brief outline of electromagnetic wave theory

A *wave equation* is the equation of motion for the type of waves considered; sound, electromagnetic, water waves, etc. In the case of electromagnetic waves the wave equation is obtained from Maxwell's four equations which encapsulate the properties of the electromagnetic fields. The resulting wave equation in vacuum is extremely compact and applies to both electric and magnetic fields. In this section three key formulae from electromagnetic theory will be quoted: the wave equation, the equation for energy flow and the equation for energy storage in electromagnetic waves. A full derivation will be given later in Chapter 8. This introduction is restricted to electromagnetic waves travelling in free space: very conveniently it applies to a good approximation, and in many circumstances, to such waves travelling in air.

The simplest electromagnetic wave is a *plane wave* travelling in what we can choose to be the z-direction. With such a wave $\mathbf{E}$ has at any moment the same value over any plane surface perpendicular to the z-axis, and the same is true of $\mathbf{B}$. Taking $\mathbf{E}$ to lie along the x-direction, the wave equation for the electric field then reduces to

$$\partial^2 E_x/\partial z^2 = \varepsilon_0 \mu_0 \partial^2 E_x/\partial t^2, \tag{1.3}$$

where ε_0 is a scalar quantity called the permittivity of free space, and μ_0 is another scalar called the permeability of free space. The values of physical constants such as ε_0 and μ_0 are collected in Appendix A. We can try a sinusoidal plane wave solution, also known as an harmonic plane wave,

$$E_x = E_0 \cos{(2\pi z/\lambda - 2\pi ft + \phi)}, \tag{1.4}$$

where ϕ is a phase factor. The word *amplitude* is customarily defined to mean E_0, which is the maximum value that E_x takes, but it is also widely used for the instantaneous value of E_x in for example discussions of coherence. It will be clear from the context in which way the word is being used. Differentiating E_x twice with respect to z and t and inserting the results into the wave equation gives

$$-(4\pi^2/\lambda^2)E_x = -\varepsilon_0 \mu_0 (4\pi f^2)E_x, \tag{1.5}$$

so that the wave equation is satisfied provided that

$$f^2\lambda^2 = 1/\varepsilon_0\mu_0. \tag{1.6}$$

It was shown above that the velocity of sinusoidal waves equals the product of the frequency and wavelength, so the velocity of electromagnetic waves in free space must be

$$c = 1/\sqrt{\varepsilon_0\mu_0}. \tag{1.7}$$

On getting this result Maxwell substituted the then measured values of ε_0 and μ_0 into this equation in order to predict the wave velocity. He found that the predicted velocity agreed with the measured value of the velocity of light to within the precision with which c, ε_0 and μ_0 were known. Thus electromagnetic theory leads to a wave equation whose solutions, in free space, are waves travelling at exactly the measured speed of light. Maxwell's equations impose no restriction on the wavelengths or the frequencies possible; so the inference made was that *light is just one form of electromagnetic radiation.* Soon afterwards, in 1886, Hertz tested the idea that electromagnetic waves can be generated and detected well beyond the visible spectrum. He generated electromagnetic waves of frequencies $\sim$100 MHz from a spark gap in an oscillatory circuit, and he observed that an oscillatory current was produced in an identical circuit located several metres away. Early in the twentieth century Marconi successfully transmitted radio signals across the Atlantic; and a century later communication via electromagnetic waves has become all-pervasive.

The magnetic field obtained using Maxwell's equations is perpendicular to the direction of motion and to the electric field:

$$B_y = E_x/c. \tag{1.8}$$

An alternative way of writing the fields using the angular frequency $\omega = 2\pi f$ and the *wave number* or *propagation constant*[1] $k = 2\pi/\lambda$ is:

$$E_x = E_0 \cos\left(kz - \omega t + \phi\right), \tag{1.9}$$

and

$$B_y = \left(E_0/c\right) \cos\left(kz - \omega t + \phi\right). \tag{1.10}$$

Waves can of course travel in any direction. In the more general case that the wave travels in a direction given by the unit vector $\hat{\mathbf{k}}$ the wave at a point $(\mathbf{r}, t)$ in space-time is

$$E = E_0 \cos\left(\mathbf{k} \cdot \mathbf{r} - \omega t + \phi\right), \tag{1.11}$$

where $\mathbf{k} = k\hat{\mathbf{k}}$ is the *wave vector* and, as before, k is the wave number.

The shape of the wavefronts of light emitted from a point source would be spherical, but typical sources are generally in containers which restrict the angular range of the wavefront. In the ideal case of an unobscured sinusoidal wave radiating from a point source at the origin, the electric field at a point in space-time $(\mathbf{r}, t)$ would be

$$E = E_0 \cos\left(kr - \omega t + \phi\right). \tag{1.12}$$

Here again ω, k and ϕ are respectively the angular frequency, the wave number and the phase factor. As the spherical wavefront travels further and further from the source it approximates ever more closely to a plane

[1] Another definition used in spectroscopy for the wave number is $1/\lambda$.

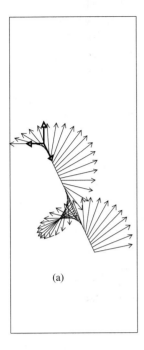
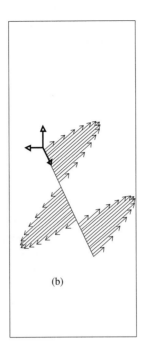

(a)   (b)

Fig. 1.4 The variation of the electric field with position along the direction of wave motion for (a) a circularly polarized electromagnetic wave and (b) a plane polarized wave of the same wavelength.

surface over any fixed area.

A second simple type of polarization is that known as *circular polarization*. In this case the electric field of a plane sinusoidal wave travelling in the z-direction is

$$\mathbf{E} = E_0 \left[\mathbf{e}_x \cos\left(kz - \omega t\right) + \mathbf{e}_y \sin\left(kz - \omega t\right) \right], \qquad (1.13)$$

where $\mathbf{e}_x$ and $\mathbf{e}_y$ are unit vectors along the x- and y-directions respectively. This electric field rotates with angular frequency ω in the xOy plane. The magnetic field rotates too, remaining at right angles to the electric field. The contrasting behaviour of the electric field in plane and circularly polarized waves is shown in figure 1.4. Light emitted by most sources is a mix of polarizations, changing from instant to instant. If the plane of polarization is entirely random then the source is said to be unpolarized. Lasers are the exception, in that lasers are generally designed to produce beams with plane polarization.

The electromagnetic waves carry energy and in vacuum the instantaneous energy density is

$$U = (\varepsilon_0 \mathbf{E}^2 + \mathbf{B}^2/\mu_0)/2, \tag{1.14}$$

where U is in joules/metre^3 ($\mathrm{J\,m^{-3}}$). In the case of a sinusoidal plane wave in free space, for which $B = E/c$, this reduces to

$$U = \varepsilon_0 E^2. \tag{1.15}$$

The time average of the energy density for a wave of the form given by eqns. 1.9 and 1.10 is obtained by taking the average over one cycle of oscillation $\tau = 2\pi/\omega$,

$$\overline{U} = \varepsilon_0 E_0^2 \int_0^\tau cos^2(kz - \omega t + \phi)\,\mathrm{d}t/\tau$$
$$= \varepsilon_0 E_0^2/2. \tag{1.16}$$

The flow of energy in an electromagnetic wave is the energy crossing unit area per unit time perpendicular to the wave direction. It is therefore a vector quantity, called the Poynting vector

$$\mathbf{N} = \mathbf{E} \wedge \mathbf{B}/\mu_0. \tag{1.17}$$

This *energy flux or power per unit area*, $\mathbf{N}$, is measured in watts/metre^2 ($\mathrm{W\,m^{-2}}$), and points in the direction the wave is travelling. Its magnitude is

$$N = E^2/(\mu_0 c) = E^2/Z_0. \tag{1.18}$$

where $Z_0 = \sqrt{\mu_0/\epsilon_0}$ is called the impedance of free space. The time average of the energy flow for a wave of the form given by eqns. 1.9 and 1.10 is

$$\overline{N}_z = \varepsilon_0 E_0^2 c/2, \tag{1.19}$$

which equals the product of the energy density in the field and the wave velocity.

Many discussions of light travelling through free space and simple materials use only a single vector field rather than the two vector fields, $\mathbf{E}$ and $\mathbf{B}$. This simplification is permissible because the magnitudes of the electric and magnetic fields are proportional, and no information is lost thereby. In analysing light propagating through optical systems with lenses and mirrors the simplification can be taken a step further to use a single scalar field when there is no dependence of the devices on polarization. Referring back to the expression for the Lorentz force eqn. 1.1 we see that the ratio of the electric/magnetic force is $E/[vB] = E/[v(E/c)] = c/v$, for an electromagnetic wave in free space. Evidently the magnetic force can be neglected unless the velocity of the electrons approaches the velocity of light, or if the material has a large magnetic permeability. Therefore in many cases where electromagnetic radiation interacts with matter the magnetic field may be neglected. It remains true that both fields are inextricably involved in carrying and storing energy in electromagnetic radiation.

1.4.1   More general waveforms

In the preceding attention was focused on the sinusoidal wave solutions to the wave equation. A more general travelling wave solution is

$$E_x = E_0 F(z \pm ct) = E_0 F(w) \qquad (1.20)$$

where F is any function whatever of the combination $w = (z \pm ct)$. This

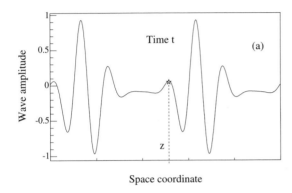

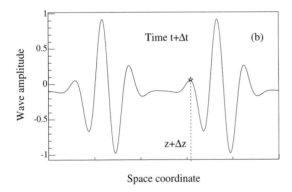

Fig. 1.5 An electromagnetic wave $F(z - ct)$ at two different moments t and $(t + \Delta t)$.

statement can be readily checked. Firstly, differentiating twice with respect to t gives

$$\partial^2 E_x / \partial t^2 = E_0 \, (\mathrm{d}^2 F / \mathrm{d}w^2)(\partial w / \partial t)^2 = c^2 E_0 \, (\mathrm{d}^2 F / \mathrm{d}w^2); \qquad (1.21)$$

then differentiate E_x twice with respect to z

$$\partial^2 E_x / \partial z^2 = E_0 \, (\mathrm{d}^2 F / \mathrm{d}w^2)(\partial w / \partial z)^2 = E_0 (\mathrm{d}^2 F / \mathrm{d}w^2). \qquad (1.22)$$

Hence the wave equation is satisfied. Figure 1.5 shows a wave of arbitrary shape at two instants t and $(t + \Delta t)$, where the star indicates a reference point (feature) on the wave which moves from z to $(z + \Delta z)$ in a time

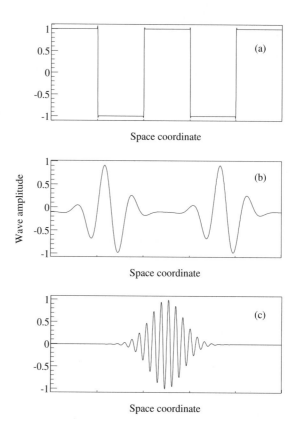

Fig. 1.6 Examples of electromagnetic waveforms: (a) a square repetitive wave; (b) an irregular repetitive wave; (c) a wavepacket or pulse.

interval Δt. The value of $F(z-ct)$ is the therefore the same at these two points in space-time, namely at (z,t) and $(z+\Delta z,\ t+\Delta t)$. Therefore

$$z - ct = (z + \Delta z) - c(t + \Delta t),$$
$$\text{and }\ \Delta z = +c\Delta t. \tag{1.23}$$

This means that the feature and hence the wave is moving rightward. The reader can check that $F(z+ct)$ represents a leftward moving wave. There are many possible waveforms of the type $F(z \pm ct)$ and some examples are shown in figure 1.6: (a) is a square wave, (b) is a wave of irregular shape but still a repetitive wave and (c) is a waveform called a *pulse* or *wavepacket* which is not repetitive. The reason we can concentrate on sinusoidal waves in the face of these and countless other possibilities is that whatever the waveform it can always be duplicated exactly by a sum of sinusoidal waves with their amplitudes and phase factors suitably chosen. There is a well defined procedure called Fourier analysis which is used to extract these harmonic components from any waveform. Fourier analysis is treated at length in Chapter 6, while here it will be sufficient to take note of this underlying simplicity: if results

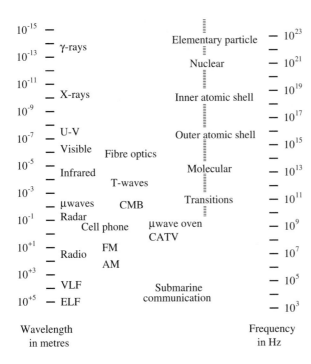

Fig. 1.7 The electromagnetic spectrum.

can be proved for harmonic waves then they must apply equally for any waveform.

1.5   The electromagnetic spectrum

Electromagnetic waves in free space can have any wavelength. At one extreme the wavelength of radiation at the mains frequency (EU 50 Hz, US 60 Hz) is huge ($6\,10^6$ m, $5\,10^6$ m); and at the other extreme the gamma rays emitted by a decaying π^0-meson have wavelength $2.9\,10^{-15}$ m. The way that electromagnetic waves interact with matter depends on their wavelength and this variation has affected how and when the different parts of the electromagnetic spectrum were first discovered, how they are named and in what ways the radiation in each part of the spectrum can be used.

On figure 1.7 the principal regions of the spectrum, the primary sources and important uses of electromagnetic radiation are all indicated. Radio and TV stations transmit waves of pre-assigned wavelengths which lie in the range from about 1 km to 1 m. Cable TV (CATV) and satellite TV use shorter wavelengths. Electromagnetic waves of very long wavelength (VLF and ELF) are used to communicate with submarines because radiation of shorter wavelengths is strongly absorbed by sea water. Microwaves are electromagnetic waves with wavelength from about 1 m to 0.1 mm. Radar, microwave transmitters for mobile phones as well as microwave ovens operate in this part of the spectrum. Microwave detectors have been used to observe the spectrum of the relic radiation from the early universe, the cosmic microwave background (CMB), whose intensity peaks at a wavelength around 2 mm. The region from 1 mm wavelength down to the red end of the visible spectrum is called the infrared, and overlaps the microwave region. An electric fire radiates most of its energy in the infrared, while the preferred wavelengths for telecom optical fibre links are in the near infrared (near, that is, to the visible spectrum). At these latter wavelengths the absorption of electromagnetic waves in glass fibre has a broad minimum. The visible part of the spectrum extends from 400 to 700 nm. Shorter wavelength radiation down to around 10 nm is called ultraviolet (UV) radiation. At yet shorter wavelengths the ultraviolet merges into the X-ray region, which extends from roughly 10 nm to 1 pm (10^{-12} m). Excited nuclei emit radiation of very short wavelength from 0.1 nm downward, and these waves are called γ rays. The shortest wavelength γ rays are emitted in the decays of elementary particles such as the π^0-meson. Apart from the visible spectrum, which is defined by the range of wavelengths to which the average eye is sensitive, none of the boundaries mentioned above is at all precise. For the present we should also note that visible radiation interacts very effectively with the atoms whose size is $\sim$0.1 nm.

Fig. 1.8 Emission spectrum of helium gas. This spectrum and the following emission spectra for nitrogen gas, a fluorescent lamp and the Sun's spectrum were recorded with a TV SPEC spectrometer with 600 lines per mm, made by Elliott Instruments Ltd., www.elliott-instruments.co.uk. Courtesy Dr K. H. Elliott.

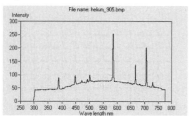

Fig. 1.9 Emission spectrum of helium gas, presented as a histogram of the intensity against wavelength. This spectrum was obtained automatically from the preceding spectrum.

Radiation of wavelength shorter than light can penetrate matter with increasing ease as the wavelength falls. Thus ultraviolet radiation penetrates the skin and can damage cells by ionizing the component DNA, and this penetrating power increases steadily as the wavelength decreases. The longer wavelength UV region is divided into UVA of wavelength from 320 to 400 nm, UVB of wavelength from 280 to 320 nm and UVC of wavelength shorter than 280 nm. UVB radiation damages living tissue at intensities not very much greater than those met on a sunny day in temperate climes. The more dangerous UVC radiation from the Sun is absorbed by oxygen and ozone; but the UVB is only absorbed by ozone, hence the concern about the depletion of the ozone layer in the upper atmosphere. The deep penetrating power of X-rays and γ-rays makes them useful tools in medical diagnosis and in material science when applied in a controlled manner.

1.5.1   Visible spectra

The reader will have seen the spectra produced when sunlight passes through a prism. Light of different colours is bent, *refracted*, through an angle which depends on the colour of the light. This variation of the bending with wavelength is known as *dispersion*. Light can also be dispersed into a spectrum when it is reflected from a grating of parallel, uniformly spaced grooves on a metal surface, the spacing being of order a few micrometres. Such *diffraction gratings* are described in Chapter 6.

Figures 1.8 and 1.9 show the same visible spectrum emitted by a gas of excited helium atoms. The source is a slit illuminated by a spectral lamp filled with pure helium gas, and through which an electric discharge is passed. The spectrum of light dispersed by the diffraction grating is shown in figure 1.8, and its intensity is histogrammed as a function of wavelength in figure 1.9. If the light were *monochromatic*, that is to say all of a single wavelength then the spectrum would be a single line in figure 1.8, at the image of the slit for that particular colour of light. In fact the spectrum contains many lines, each due to light of a particular narrow range of wavelengths. These narrow *spectral lines* are characteristic of the spectrum of a gas. Notice first that this spectrum and those in the following diagrams are truncated by the display software at 300 and 780 nm. Secondly note that for clarity the intensity scales of the histograms are offset by 40 units from zero.

The set of spectral lines is unique to the element emitting light in the spectral lamp and we shall see later in Chapter 12 that the spectral lines show regularities in the wavelengths for each element. Such regularities were a puzzle to the physicists who discovered them in the nineteenth century. The regularites arise from the structure of the atoms of each element and can only be understood using quantum theory. The interpretation of spectra, which was a key step in guiding the development of quantum mechanics, will be fully described in Chapters 12 and 13.

Figures 1.10 and 1.11 exhibit the emission spectrum of a spectral lamp containing nitrogen. Here the individual spectral lines are grouped into *bands* which are characteristic of molecular spectra, in this case diatomic nitrogen. The spectrum from a modern fluorescent lamp is more complex. This is shown in figures 1.12 and 1.13, where the spectral lines are superposed on a *continuum*. The spectral lines are emitted by the mercury gas within the discharge tube, and are characteristic of that element. Some of the UV radiation emitted by the mercury atoms is absorbed by a powder deposited on the inside wall of the glass envelope. This is a *fluorescent* material, meaning that its atoms absorb the UV radiation and then re-emit visible light with a delay that is less than a microsecond. Now condensed matter, such as the powder used to coat a discharge tube, emits over a range of wavelengths rather than producing line spectra. This broadening of the spectral lines into a continuum results from the proximity and strong mutual interaction of atoms in a solid or liquid. By using a fluorescent powder which absorbs the ultraviolet radiation emitted by the mercury atoms and re-emits visible light the light yield from the lamp is enhanced considerably. Consequently fluorescent lamps are among the most efficient lamps in converting wall plug power to visible light.

The final spectrum to be seen in figures 1.14 and 1.15 is the Sun's spectrum. This is a continuum marked by dark lines and corresponding

Fig. 1.10 Emission spectrum of nitrogen gas, Courtesy Dr K. H. Elliott.

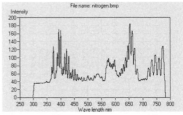

Fig. 1.11 Emission spectrum of nitrogen gas, presented as a histogram of the intensity against wavelength. This spectrum was obtained automatically from the preceding spectrum.

Fig. 1.12 Emission spectrum of a modern fluorescent lamp. Courtesy Dr K. H. Elliott.

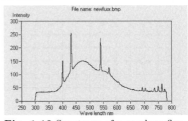

Fig. 1.13 Spectrum of a modern fluorescent lamp, presented as a histogram of the intensity against wavelength. This spectrum was obtained automatically from the preceding spectrum.

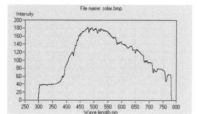

Fig. 1.14 Spectrum of the Sun. Courtesy Dr K. H. Elliott.

Fig. 1.15 Spectrum of the Sun, presented as a histogram of the intensity against wavelength.

notches in the histogram. This continuum is that of a hot body in thermal equilibrium, in this case the outer layers of the Sun at around 6000 K. The dark lines were interpreted by Fraunhofer as due to the absorption of light by cooler gases in the upper reaches of the Sun's atmosphere. These lines match in wavelength the emission lines of elements found on Earth, and this allows inferences to be made about the gases making up the Sun. The set of *Fraunhofer* lines due to absorption by helium in the Sun's outer atmosphere were initially a mystery because at that time helium had not been identified and was only later isolated on Earth. A comparison of the spectra of the fluorescent lamp and the Sun shows that the modern lamp mimics the shape of the Sun's spectrum quite well.

1.6 Absorption and dispersion

Electromagnetic waves passing through matter are partly absorbed by the atoms and molecules, and their velocity in matter is less than in vacuum, by a factor called the *refractive index*. The frequency of the wave is unchanged and hence from eqn. 1.2 it follows that the wavelength in matter is smaller than its value in free space by the value of the refractive index. Air has a refractive index of 1.0003 at standard temperature and pressure (STP: 293 K; 10^5 Pa) for visible light. Water has a refractive index of 1.333 at STP for visible light, but the value rises to around 9.0 for microwaves. This difference indicates that light and microwaves interact very differently with water; in fact water is quite transparent to visible light while it absorbs microwaves strongly enough to make cooking with microwaves practical. Materials transparent to light have refractive indices ranging from near unity for gases up to 2.47 for diamond, with those for common glasses lying in the range 1.5 and 2.0. The variation of the velocity of electromagnetic radiation with wavelength is known as *dispersion* and gives rise to effects such as the dispersion of light by a prism. For most materials that are transparent to light the refractive index falls smoothly with wavelength across the visible spectrum. A few materials show *anomalous* dispersion, with the refractive index rising with wavelength over a portion of the visible spectrum. *Transmittance* is the fraction of radiation transmitted by a material, and falls below unity because there is absorption in the body of material and reflection at the entry and exit surface. *Internal* transmittance is defined as the light intensity reaching the exit surface divided by that which entered the material. Figure 1.16 shows the variation of refractive index of several materials commonly used in making lenses or other transparent optical components: borosilicate crown glass and dense flint glass; fused silica which is useful in extending coverage into the UV; magnesium fluoride (MgF_2) which is useful in extending coverage far into the IR, as well as being used in coating lenses to reduce surface reflections. Crown glass is transparent from 350 to 2000 nm; flint glass from 420 to 2300 nm; fused silica from 260 to 2500 nm; and MgF_2 from 120 to 8000 nm. The normal

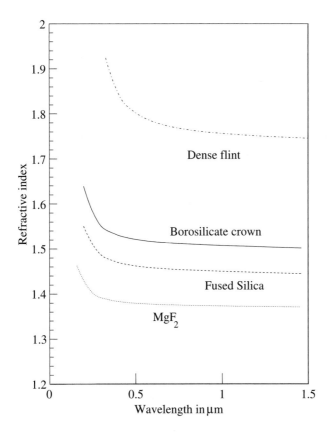

Fig. 1.16 The variation with wavelength of refractive index of optical glass, of fused silica and of MgF$_2$.

dispersion curves for glasses are well fitted by the Sellmeier empirical formula over the visible spectrum

$$n^2 = 1 + B_1/(\lambda^2 - C_1) + B_2/(\lambda^2 - C_2) + B_3/(\lambda^2 - C_3). \qquad (1.24)$$

The values of the constants B_i and C_i are often tabulated by manufacturers of optical materials. Expressions linking the wave number and the angular frequency of electromagnetic radiation in a material, or equally the wavelength and frequency, are called *dispersion relations*. The simplest is that for a vacuum, $\omega = kc$. Again eqn. 1.24 is another dispersion relation when we note that $n = ck/\omega$.

Figure 1.17 shows a correlation between optical properties which is crucial for us Earth dwellers. In figure 1.17(a) the absorption coefficient in water is plotted as a function of the wavelength. An absorption coefficient α is defined such that a beam of radiation will be reduced in intensity by a fraction $\alpha\, ds$ after passing through an infinitesimal layer

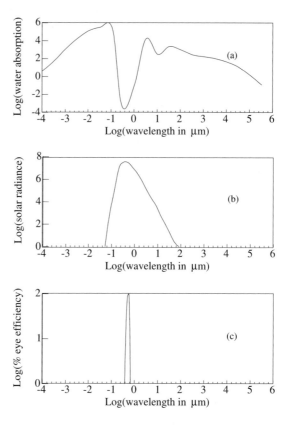

Fig. 1.17 Coincidences: (a) log of absorption coefficient for water in cm^{-1}; (b) log of the Sun's spectral irradiance in W m^{-2} μm^{-1} (c) log of the relative spectral efficiency of the human eye.

of thickness ds. That is

$$\mathrm{d}I(s)/\mathrm{d}s = -\alpha I(s).$$

Integration gives the dependence of the intensity with distance s to be

$$I(s) = I(0) \exp{(-\alpha s)}, \qquad (1.25)$$

which is known as *Beer's law*. The absorption in water is at a minimum at approximately 500 nm and rises very fast as the wavelength moves off that value. In figure 1.17(b) the spectrum of radiation emitted by the Sun is depicted. It is roughly the radiation spectrum of a black body at 6000 K, and peaks very close to the wavelength at which the absorption of water reaches a minimum. Finally figure 1.17(c) shows the relative spectral sensitivity of the retina of the human eye, with the peak sensitivity at 550 nm. Evidently the eye is well designed to use the available radiation illuminating the planet. If instead, the retina were designed to 'see' at wavelengths just a factor three away (180 nm or 1650 nm) the absorption by water (α now around 1 cm^{-1}) would be a severe restriction.

Radiation at 180 nm, in the ultraviolet, is in any case very damaging to tissue. Attenuation of the ultraviolet component of the Sun's radiation by water vapour, oxygen and ozone in the atmosphere is essential to protect our eyes and skin.

The main process by which the Sun's energy is converted into a form accessible to living things is plant photosynthesis, which involves photo-chemical transitions in complex molecules. Plant photosynthesis oper-ates at peak efficiency using light in the visible spectrum, which is just the wavelength range of the copious radiation penetrating to ground level. Much longer wavelength radiation cannot initiate these molecular processes, while shorter wavelength ultraviolet radiation would destroy the active molecules. The multiple coincidence presented in figure 1.17 is fundamental for life as we know it.

1.7   Radiation terminology

The time averaged energy flow of radiation across unit area of a surface is known as the *irradiance* and also as the *intensity*. Radiation is not generally directed in a beam but spread over a range of angles, hence a quantity, the *radiance*, is defined as the radiated energy per unit solid angle per unit area. The total energy radiated by a source or crossing a surface is called the *radiant flux*, ϕ_e.

How energy is distributed with wavelength is of importance because the physical effects of radiation depend strongly on the wavelength. The *spectral radiant flux* is the radiant flux per unit wavelength, $\phi_{e,\lambda}$, and naturally the total

$$\phi_e = \int_0^\infty \phi_{e,\lambda} \, d\lambda. \tag{1.26}$$

A significant property of radiation is whether it produces a visual effect. Therefore visual quantities are defined parallel to the energy/radiation quantities so far described. The fundamental factor used to relate visual and energy quantities is the *luminous efficacy* V_λ, so that the *spectral luminous flux* associated with the spectral radiant flux is

$$\phi_{v,\lambda} = V_\lambda \phi_{e,\lambda}, \tag{1.27}$$

where the subscripts e and v refer to energy and visual quantities. The variation of V_λ with wavelength was standardized by the Comité Inter-national d'Eclairage (CIE) so as to correspond to the response of the average human eye.

In daylight the sensitivity of the human eye is greatest at a wave-length of 555 nm. The unit of luminous flux, the lumen (lm), is defined by taking one watt of radiation at 555 nm to have a visual equivalent of 683 lm. The reason for such a strange conversion factor is that the use of the energy and visual units developed quite independently. Figure 1.18

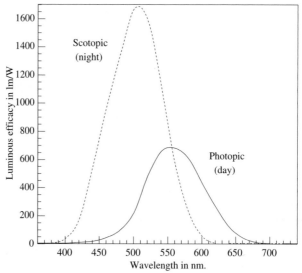

Fig. 1.18 The relative sensitivity of the average human eye as a function of the wavelength of light.

shows how the sensitivity of the eye varies with wavelength. The solid curve shows the variation of V_λ in daylight (photopic vision), while the broken line curve is the corresponding curve for night vision (scotopic vision). Daylight and night vision rely on different receptors in the retina: in daylight on the cones and at night on the rods. Cones come in three types which respond to blue, green and yellow light, respectively. The more sensitive rods all have the same spectral response so that night vision is monochrome. Their response peaks at 1700 lm for one watt of radiation at 507 nm wavelength.

The *luminous flux* from a source is

$$\phi_v = \int \phi_{v,\lambda}\,d\lambda = \int \phi_{e,\lambda} V_\lambda\,d\lambda, \qquad (1.28)$$

where the integral need only run over the visible spectrum from 400 to 700 nm.

The visual and energy quantities are shown in table 1.1 with corresponding pairs on the same line. Of the remaining quantites only the visual ones will be discussed because their units are the less obvious. The *luminous intensity* of a point source in a given direction is

$$I_v = d\phi_v/d\Omega, \qquad (1.29)$$

where $d\Omega$ is an element of solid angle, and I_v is measured in lumens per steradian or candelas (cd). If the point source is isotropic then, of

Table 1.1 Table of related visual and energy quantities and their units.

Energy parameter	Unit	Visual parameter	Unit
Radiant flux	W	Luminous flux	lm
Radiant intensity	$W\,sr^{-1}$	Luminous intensity	$lm\,sr^{-1}$
Irradiance/intensity	$W\,m^{-2}$	Illuminance	$lm\,m^{-2}$
Radiance	$W\,m^{-2}\,sr^{-1}$	Luminance	$lm\,m^{-2}\,sr^{-1}$

course, $\phi_v = 4\pi I_v$.

The *illuminance* E_v is the luminous flux per unit area, measured in lumens per square metre, or lux (lx). Finally the *luminance*, L_v, is the luminous flux per unit area per unit solid angle, measured in $cd\,m^{-2}$. Luminance is, in everyday speech, the *brightness* of an object. At luminances below $0.03\,cd\,m^{-2}$ human vision relies on the rods, and at luminances above $3\,cd\,m^{-2}$ on the cones. Between these extremes both types of receptor play a part. Typical luminances are $100\,cd\,m^{-2}$ for indoor lighting, $10^4\,cd\,m^{-2}$ in full sunlight and $10^{-3}\,cd\,m^{-2}$ in starlight. One $cd\,cm^{-2}$ may be called a stilb, while in describing the luminance of display panels one $cd\,m^{-2}$ is called a nit.

A *diffuse* source is one which looks equally bright in all directions. When viewed at an angle θ away from the normal the projected area is reduced by a factor $\cos\theta$. It follows that, in order to compensate, the luminance of a diffuse source should obey Lambert's law:

$$L_v(\theta) = L_v(0)\cos\theta. \qquad (1.30)$$

Integrating over the forward solid angle gives the total luminous flux from unit area of the source

$$
\begin{aligned}
E_v &= \int_{\phi=0}^{2\pi}\int_{\theta=0}^{\pi/2} L_v(\theta)\sin\theta\,d\theta\,d\phi \\
&= 2\pi\int_{\theta=0}^{\pi/2} L_v(\theta)\sin\theta\,d\theta \\
&= \pi L_v(0)\int_{\theta=0}^{\pi/2}\sin(2\theta)\,d\theta = \pi L_v(0). \qquad (1.31)
\end{aligned}
$$

A rough surface painted matt white is an excellent diffuse reflecting surface. Light from a compact source can be diffused by passing it through a ground glass screen.

It is surprising that the brightness of an extended source does not change with its distance from the observer, unlike a point source. This effect is noticable on a bright day: for example when viewing the white cliffs of Dover from the cross-Channel ferry. The area of cliff face lying within a fixed solid angle at the observer's eye pupil increases like the distance squared as the cliffs recede. At the same time the solid angle presented by the pupil to the cliff face falls off at the same rate. The two effects cancel and the luminance at the pupil remains the same. In turn this means that the illuminance of the image on the retina remains the same as the ferry moves away.

An *overall luminous efficacy* is commonly used when specifying the performance of lamps used in the home and workplace. It is defined as the rate of converting electrical power from the mains in watts to visible power expressed as lumens. This overall luminous efficacy

$$V = \phi_\text{v}/E, \tag{1.32}$$

where E is the electrical power drawn from the mains. Electrical energy may be lost before reaching the lamp: for example $\sim$15% of the electrical energy is consumed in the ballast resistors of fluorescent lamps and in the driver circuits of light emitting diodes.[2] Incandescent light bulbs produce around 15 lm for each watt of electric power drawn from the wall plug and have a luminous efficacy of $\sim$15 lm W^{-1}; linear fluorescent lamps, light emitting diodes and ceramic metal-halide lamps produce $\sim$80 lm W^{-1}; high pressure sodium lamps produce $\sim$120 lm W^{-1}. In the case of high pressure sodium lamps, emitting yellow light of wavelength 589 nm, the corresponding luminous efficacy taken from figure 1.18 is 500 lm W^{-1}. Hence the efficiency for the conversion of electrical to visible energy of even these lamps is only 120/500 or around 24%; the other 76% appears as heat.

[2]Sometimes this loss may be omitted, which makes a lamp appear more efficient than it really is.

When expressing the intensity ratio between the light entering and leaving an optical fibre link the decibel (dB) is often used. If the ratio of the light entering divided by the light emerging is R, then in decibels this becomes

$$n = 10 \log_{10} R. \tag{1.33}$$

Such a ratio unit is convenient when the intensity ratios are very large. If the loss along an optical fibre is n dB km^{-1} then over s km it is sn dB. Powers of, for example lasers, may be expressed in dBm in which the ratio is the laser power divided by 1 mW. Thus if the laser power is stated to be -5.0 dBm this means that the power P is given by

$$-5.0 = 10 \log_{10} [P/(1\,\text{mW})]. \tag{1.34}$$

Thus the laser power is $10^{-0.5} = 0.316$ mW.

Luminescence describes the emission of electromagnetic radiation in the visible and near visible parts of the spectrum following any excitation other than thermal excitation. Photoluminescence is luminescence produced by incident electromagnetic radiation, electroluminescence is that produced by an applied voltage and bioluminescence that emitted by living organisms such as fireflies.

1.8   Black body radiation

This is radiation which can be produced by simple apparatus and whose properties can be analysed simply too. Figure 1.19 shows the black body spectrum at three temperatures. As noted above, the Sun's emission

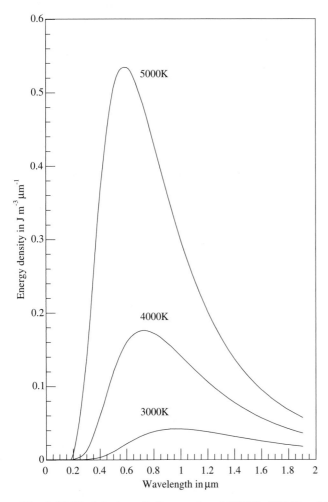

Fig. 1.19 Black body radiation spectra at 3000 K, 4000 K and 5000 K.

spectrum, before absorption in the Earth's atmosphere, is approximately that of a black body at 6000 K. Consider an enclosure whose walls are maintained at a constant and uniform temperature, and which is taken to be evacuated. Suppose a small body is suspended within this volume. It will reach thermal equilibrium at the temperature of the enclosure and in that condition it will radiate as much energy as it absorbs. This shows immediately that a good emitter at any wavelength must necessarily also be an equally good absorber of radiation at the same wavelength, and a poor emitter must be a poor absorber. It follows that the radiation in the enclosure at a given temperature is independent of the material in the walls of the cavity. The radiation in such an enclosure is known as *black body radiation*. If an aperture very much smaller than the linear dimensions of the enclosure is left open, then this acts as a source of black body radiation, and is a perfect emitter because it transmits all the radiation coming from inside. This aperture is also a perfect absorber of radiation falling on it because the radiation entering has a negligible chance of being reflected back out of the aperture. The *spectral emittance* ϵ_λ of any surface can be defined as the fraction of radiation emitted around wavelength λ compared to that emitted by a black body. Also the *spectral absorptance* a_λ can be defined as the fraction of incident radiation that a surface absorbs at the same wavelength. Then the result just deduced is that

$$\epsilon_\lambda = a_\lambda, \tag{1.35}$$

and is known as *Kirchhoff's law*. If a large absorptance could be allied to a smaller emittance, then a body made from this anomalous material suspended in the thermal enclosure would form a heat engine that violates the second law of thermodynamics. Practical black body sources are thermally insulated boxes, whose walls contain heaters under thermostatic control, the interior wall surfaces are also corrugated. Commercially available sources running typically at $1000\,\mathrm{K}$ have emittances of over 99% across the wavelength range 1 to $30\,\mu\mathrm{m}$.

1.9   Doppler shift

Everyone is aware that the tone of a single note siren of an emergency vehicle drops just as the vehicle passes by. When it is approaching the frequency is higher, and when it is receding the frequency is lower than that of the siren at rest. These frequency shifts constitute the *Doppler effect*, which is observed both for sound and for electromagnetic waves. The upper panel of figure 1.20 shows sound waves emitted by an approaching source. This source has velocity v_e; the sound has velocity v_s, frequency f and wavelength λ. In a short time Δt the source emits $f\Delta t$ waves and the leading one of these waves will travel a distance $v_s\Delta t$. In the same time the source moves a distance $v_e\Delta t$ so that the $f\Delta t$ waves are confined to a reduced distance $(v_s - v_e)\Delta t$. Thus the wavelength is compressed to

$$\begin{aligned}
\lambda' &= (v_s - v_e)\Delta t / f\Delta t. \\
&= \lambda(v_s - v_e)/v_s \\
&= \lambda(1 - v_e/v_s).
\end{aligned}$$

Hence the frequency heard is higher

$$f' = f/(1 - v_e/v_s). \tag{1.36}$$

The case that the observer is moving with velocity v_o and the source is at rest is shown in the lower panel of figure 1.20. This observer receives more waves per unit time than one at rest. In a time Δt the total number

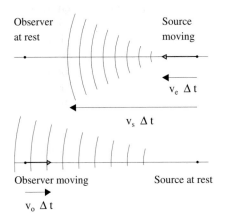

Fig. 1.20 Doppler shift of frequency when source or observer are moving.

of waves passing the observer is $(v_s + v_o)\Delta t$ compared to $v_s \Delta t$ for an observer at rest. Thus the frequency heard is

$$
\begin{aligned}
f' &= f(v_s + v_o)\Delta t / v_s \Delta t \\
&= f(1 + v_o/v_s).
\end{aligned}
\tag{1.37}
$$

In the above two equations the velocity of the source or observer needs to be replaced by the component of the velocity toward the observer or source respectively when the motion is not along the line joining them. So the first result becomes

$$
f' = f/(1 - v_r/v_s),
\tag{1.38}
$$

where v_r is the radial component of the source's velocity.

For electromagnetic radiation from a moving source in free space this analysis would give

$$
f' = f_0/(1 - v_r/c),
\tag{1.39}
$$

where c is the velocity of light and f_0 is the frequency of a moving source measured in the rest frame of the source. However there is an additional relativistic effect that must be taken into account known as *time dilation*. If the time between two events occuring in the rest frame of the moving source is measured in that frame, and also by the observer at rest then the time interval measured by the observer is longer by a factor $\gamma = 1/\sqrt{[1 - (v/c)^2]}$.

Although macroscopic objects on Earth do not travel at velocities close to that of light, some elementary particles do so and the result can be astonishing. μ-leptons created by primary cosmic rays in interactions with atmospheric atoms at heights of 200 km or so mostly survive to reach the Earth's surface. They are travelling close to the speed of light (relativistically) so the duration of this journey is $(200\,\text{km})/c \approx$ 0.67 ms or somewhat longer. The lifetime of a μ-lepton measured at rest in the laboratory is only $2\,\mu\text{s}$ and hence, viewed classically, few μ-leptons should survive to reach the Earth! However the time dilation factor increases the lifetime measured in the Earth's frame of reference by a factor of around 1000, and so most do reach us.

Thus the frequency of the moving source measured by the stationary observer is reduced compared to that measured in the rest frame of the source by an additional factor $\sqrt{[1 - (v/c)^2]}$,

$$f = f'\sqrt{[1 - (v/c)^2]}. \tag{1.40}$$

Note that this depends on v rather than its radial component. Combining eqns. 1.39 and 1.40 gives the frequency detected by the observer

$$f = f_0\sqrt{[1 - (v/c)^2]}/[1 - v_r/c]. \tag{1.41}$$

When the source moves directly toward the observer

$$f = f_0\sqrt{(1 + v/c)/(1 - v/c)}. \tag{1.42}$$

Turning to the situation that the observer is in motion it is found that the same eqns. 1.41 and 1.42 are obtained for the frequency of the electromagnetic radiation determined by the observer. This is but one example of the basic feature of the special theory of relativity that the relative velocity is the relevant parameter. In equations 1.41 and 1.42 we should therefore identify v as the *relative velocity of approach* of source and observer, and v_r as its radial component. When v is small compared to c eqn. 1.42 can be approximated by:

$$f = f_0(1 + v/c), \tag{1.43}$$

which, not surprisingly, is also the classical prediction.

1.10 Waveform conventions

Complex sinusoidal, plane waves are presented throughout in the form $\exp[i(\mathbf{k} \cdot \mathbf{r} - \omega t + \phi)]$. In some textbooks on classical optics the choice of $\exp[i(\omega t - \mathbf{k} \cdot \mathbf{r} - \phi)]$ for the complex waveform may be met. Calculations made with either choice yield identical results for all measurable quantities, but there can be a change of sign in intermediate quantities. Any such cases which might be of interest are indicated in notes.

Exercises

(1.1) The solar flux impinging on the Earth's atmosphere at the equator is $1.5\,\mathrm{kW\,m^{-2}}$. Calculate the magnitude of electric field there. What is the Sun's total energy output?

(1.2) Light from a laser of wavelength $633\,\mathrm{nm}$ in space is reflected from a comet and the returning light is found to be red-shifted by $10^{-1}\,\mathrm{nm}$. What is the relative velocity of the comet with respect to the observer?

(1.3) The Lyman α line in the atomic spectrum of hydrogen has wavelength $121.6\,\mathrm{nm}$ in the UV. The wavelength of this same line in the spectrum received from a quasar 01422+2309 is $561.79\,\mathrm{nm}$. Is this red shift conceivably due to the recession velocity of the quasar away from the Earth? What else could have stretched the wavelength by this factor?

(1.4) What are the periods, wavelengths, velocities and directions of these waves: (a) $A(x,t) = \cos\left[2\pi(15x + 3t)\right]$; (b) $B(x,t) = \exp i[2\pi(15x + 20y - 5t)]$?

(1.5) An electromagnetic wave $E(x,t) = 15.0\sin\left[2\pi(x/\lambda - ft - \phi)\right]$ has values $0\,\mathrm{Vm^{-1}}$ and $9.95\,\mathrm{Vm^{-1}}$ at locations $(t,\ x) = (0,\ 0)$ and $(0,\ 1500\,\mathrm{m})$ respectively. Calculate f, λ and ϕ, taking the longest wavelength solution.

(1.6) A laser emits a beam with $1\,\mathrm{kW}$ power in free space. What is the energy in a length of $1\,\mathrm{m}$ of the beam?

(1.7) $E_0\cos\left(kx - \omega t\right)$ and $E_0\cos\left(k'x - \omega't\right)$ are two electromagnetic waves in free space and $\omega' = \omega + \Delta\omega$ where $\Delta\omega$ is small compared to ω. Show that $k' = k - \Delta k$ where $\Delta k/k = \Delta\omega/\omega$.

(1.8) Mobile phones operate at a frequency of $1\,\mathrm{GHz}$. What is the wavelength? The power radiated by a digital low power mobile phone is $125\,\mathrm{mW}$. If the power is assumed to radiate isotropically what is the electric field at $2\,\mathrm{cm}$ from the antenna? The maximum limit recommended by the IEEE for RF power in head tissue is $1.6\,\mathrm{mW\,g^{-1}}$ over any $1\,\mathrm{g}$ of tissue, and is designed to limit any heating to a safe level.

(1.9) Microwave ranges operate at $2.45\,\mathrm{GHz}$. What is the corresponding wavelength?

(1.10) A beam of light of wavelength $500\,\mathrm{nm}$ falls perpendicularly on a screen with two holes. One beam travels in glass, the other in air. How long will the glass need to be to cause a delay of $1\,\mathrm{ns}$ between the beam in glass relative to that in air? Take the refractive indices of air and glass to be 1.0 and 1.5 respectively.

(1.11) Use figure 1.17 to calculate the fraction of radiation at $10\,\mathrm{cm}$ wavelength from the Sun that penetrates the atmosphere, assuming there is $10\,\mathrm{kg}$ of water in a column of $1\,\mathrm{m^2}$ area in the atmosphere.

Reflection and refraction at plane surfaces

2

2.1   Light rays and Huygens' principle

The crisp edges of shadows on a sunny day remind us that on the scale of everyday objects light travels in straight lines. The idea of a light ray indicating the path of light appears in depictions of the Sun during the reign of the monotheistic pharaoh Akhenaton (c. 1200 BC), as sketched in figure 2.1. Around 1000 AD Al Hazen made a simple and elegant experiment that validates the idea. He placed five lamps in one room and made a small hole in the partition separating it from an adjacent darkened room. Al Hazen saw distinct images of each flame on the wall of the darkened room, and he noted that he could remove each image simply by putting his hand in the appropriate ray path in the darkened room. In this and the following two chapters the properties of mirrors, lenses and optical instruments will be studied using ray optics. The first requirement is to understand how rays and waves are related.

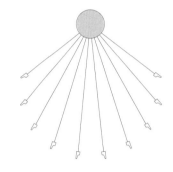

Fig. 2.1 Typical depiction of the Sun, Aten, during the reign of Akhenaten. Each ray ends in a pointing hand.

In the case of the plane wave, eqn. 1.11, $E = E_0 \cos(\mathbf{k} \cdot \mathbf{r} - \omega t + \phi)$, the rays follow the direction of the wave vector $\mathbf{k}$. Rays therefore point perpendicularly to the wavefronts. In the case of spherical waves given by eqn. 1.12, $E = E_0 \cos(kr - \omega t)$, the rays are outwardly directed radial lines perpendicular to the spherical wavefronts. At a boundary between two media, for example air and water, rays other than those exactly perpendicular to the surface change direction there: they undergo *refraction*. Refraction occurs because the velocity of light in matter varies from one material to another. The velocity also depends on the state of the material: the velocity being higher, for example, the less dense a gas is. Thus on a clear summer's day, when the air is hotter the closer it is to the tarmac of the road surface, light from the sky is refracted in this region so that it turns upward. A mirage is seen, an apparent pool of water reflecting the sky. The laws of reflection and refraction at surfaces can be deduced using *Huygens' principle* (1678), an early step in the development of the wave theory of light which is still useful. Huygens proposed a simple picture of how the 'disturbance' at one wavefront produces a disturbance at a later time. His principle states that all points on a wavefront can be treated as point sources of secondary spherical waves. Then at a later time the new position of the wavefront is the surface tangential to the forward going secondary

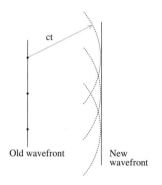

Fig. 2.2 The construction of a future wavefront, at a time t later. Three of the spherical waves used in Huygens' construction are shown.

waves. Figure 2.2 shows what happens in the case of a plane wave in free space. After a time t the spherical secondary wavefronts have radius ct. The new wavefront tangential to them is planar and a distance ct ahead; consistent with expectation.

Huygens' construction is adequate away from any obstruction in the wave path. However at the edge of an aperture the construction predicts that the wave spills round this edge. A more complete version of wave theory is needed in order to explain what happens at apertures. For the present it is sufficient to note that light of wavelength λ passing through an aperture of width a shows departures in angle of order λ/a from straight line propagation. When green light passes through an optical component of aperture $1\,\text{cm}$ this amounts to $5\,10^{-5}\,\text{rad}$, which we can safely neglect in this and the following chapter.

In the following two sections the laws of reflection and refraction will be deduced from Huygens' principle. These laws will then be used to study the imaging produced by plane mirrors, plane sheets of transparent material and simple triangular prisms. After that, total internal reflection (TIR) at a surface between two media is described: this is how light is guided along optical fibres so that the low losses required for telecom transmissions are achieved. Finally a digital micromirror device is described which is used in compact modern projectors. For simplicity the refractive index of air will be taken to be exactly unity in discussions of optical elements in air.

2.1.1　The laws of reflection

In figure 2.3 ABC is a plane wavefront which has just reached the mirror surface at A at a given moment. Huygens' construction of a wave at a later time begins with drawing spherical secondary waves from all points on this wavefront. Representative examples, originating from points A, B and C, are shown at a time t later when they have travelled out a distance ct. In Huygens' construction the wavefront at time t later

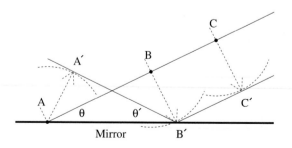

Fig. 2.3 Incident wavefront ABC and partially reflected wavefront A′B′C′ at a plane mirror. The arrow-headed broken lines indicate ray paths.

is the surface tangential to these secondary waves: A′B′C′. The part B′C′ is still an incoming wave while part A′B′ is reflected; and these movements are indicated by the arrows on the rays. In the triangles AA′B′ and B′BA:

$$AA' = BB' = ct; \qquad (2.1)$$

$$AB' \text{ is common;} \qquad (2.2)$$

$$\angle AA'B' = \angle ABB' = 90°. \qquad (2.3)$$

Thus the triangles are similar and we have the law of reflection:

$$\theta' = \theta. \qquad (2.4)$$

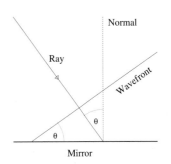

Fig. 2.4 The angles between the ray and the surface normal and between the wavefront and the mirror surface are equal.

Now the angle between the ray and the normal to the mirror is identical to the angle between the wavefront and the mirror surface, as shown in figure 2.4. Thus the incident and reflected rays make equal angles with that normal. The plane formed by the incident and reflected rays contains the normal to the surface at the point of reflection; it is called the *plane of incidence*.

2.1.2 Snell's law of refraction

The refraction of light can be handled similarly using Huygens' principle. In figure 2.5 light is incident on a plane interface separating a medium of refractive index n_1 from one of refractive index n_2; the velocity of light in the two media is thus $v_1 = c/n_1$ and $v_2 = c/n_2$. PP′ is perpendicular

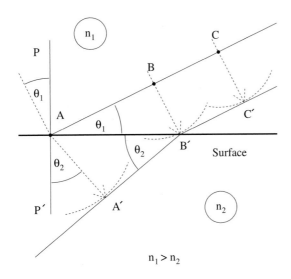

Fig. 2.5 Refraction at a plane surface separating media of refractive indices n_1 and n_2, with $n_1 > n_2$.

to the interface. The choice is made that $n_1 > n_2$: the first medium is said to be *optically denser* so that light travels more slowly in the first medium than in the second. ABC is an incoming wavefront that has just reached the interface at A. Spherical waves originating from A, B and C are shown at a time t later. Then:

$$\text{B}'\text{B} = ct/n_1 \text{ and } \text{AA}' = ct/n_2. \tag{2.5}$$

Now

$$\text{AB}' = \text{AA}'/\sin\theta_2 = \text{BB}'/\sin\theta_1. \tag{2.6}$$

Thus

$$ct/[n_2\sin\theta_2] = ct/[n_1\sin\theta_1], \tag{2.7}$$

so we obtain Snell's law

$$n_1\sin\theta_1 = n_2\sin\theta_2. \tag{2.8}$$

Here the plane of incidence contains the incident ray, the refracted ray, the surface normal and the reflected ray.

A different way of presenting Snell's law is informative and will also be useful later. The starting point is to note that the waves *match* at the interface: exactly at the boundary the wave peaks in the second medium are in precisely the same places as the wave peaks in the first medium. The wavefront A'B'C' in figure 2.5 illustrates this. The wavelength in the first medium, λ_1, can be expressed in terms of the free space wavelength, λ, thus

$$\lambda_1 = v_1/f = c/(f\ n_1) = \lambda/n_1, \tag{2.9}$$

with a similar expression for the second medium. Note that the frequency of electromagnetic radiation remains the same in going from one material to another because an electric field at frequency f arriving at the interface produces effects at the same frequency f in the medium on the far side of the surface. Now using the fact that the separation between peaks in the two media are equal along the interface, we have:

$$\lambda_1/\sin\theta_1 = \lambda_2/\sin\theta_2. \tag{2.10}$$

In terms of the wave number ($k = 2\pi/\lambda$), this becomes:

$$k_1\sin\theta_1 = k_2\sin\theta_2. \tag{2.11}$$

2.1.3   Fermat's principle

Yet another way can be used to prove Snell's law which illustrates a principle that was first enunciated in a clear form by Fermat. He proposed that the optical path taken between two points by light is the path which minimizes the travel time of light. The travel time is

$$T = \sum_i n_i\ell_i/c, \tag{2.12}$$

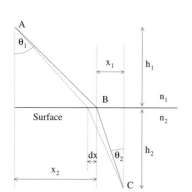

Fig. 2.6 ABC is the actual optical path of light rays between points A and C. The broken line is a nearby path.

where the sum is taken over all path elements of length ℓ_i and refractive index n_i. An *optical path length* is defined as

$$L = \sum_i n_i \ell_i. \tag{2.13}$$

Fermat's principle correctly predicts that rays in a uniform medium follow straight lines. Its success in the case of refraction is easily proved. The actual path of light, ABC, shown in figure 2.6 has an optical length

$$L = n_1 h_1 \sec \theta_1 + n_2 h_2 \sec \theta_2, \tag{2.14}$$

and the travel time is L/c. A small displacement, dx, of B to the left along the interface the results in a change in path length

$$dL = n_1 h_1 \sec \theta_1 \tan \theta_1 d\theta_1 + n_2 h_2 \sec \theta_2 \tan \theta_2 d\theta_2, \tag{2.15}$$

where $d\theta_1$ and $d\theta_2$ are the corresponding changes in the angles θ_1 and θ_2 respectively. From the diagram we have

$$x_1 = h_1 \tan \theta_1$$
$$x_2 = h_2 \tan \theta_2.$$

Now $dx = -dx_1 = dx_2$, so that

$$d\theta_1 = -\cos^2 \theta_1 dx/h_1,$$
$$d\theta_2 = +\cos^2 \theta_2 dx/h_2.$$

Substituting for $d\theta_1$ and $d\theta_2$ in eqn. 2.15 gives

$$dL = -n_1 \sin \theta_1 dx + n_2 \sin \theta_2 dx.$$

For a minimum of the path length we require $dL/dx = 0$, which immediately gives Snell's law. The reader may like to test Fermat's principle in the case of reflection. It is significant that from the simple idea that optical paths should be of *extremal* length it has been possible to reproduce ray optics. From the standpoint of classical physics Fermat's principle is simply the principle of least action applied to optics.

2.1.4 Simple imaging

Figure 2.7 shows how an image is formed by a plane mirror. A cone of rays diverging from the object A is reflected from the mirror between P and P′ and focused by the eye onto the retina at A″. The rays appear to diverge from the mirror image A′: AOA′ is perpendicular to the mirror and AO = OA′. Figure 2.8 shows another simple situation where the object viewed lies within a layer of a material of refractive index n. With the notation of the diagram:

$$n \sin r = \sin i. \tag{2.16}$$

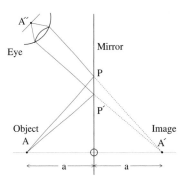

Fig. 2.7 Virtual image A′ of point A, distance a from mirror, and focused on the retina at A″.

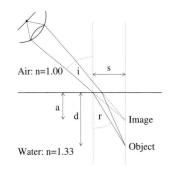

Fig. 2.8 Viewing an object in an optically denser medium.

For the case that the object is viewed close to the surface normal so that the angles i and r are small:

$$a = s/\tan i \approx s/\sin i; \; d = s/\tan r \approx s/\sin r. \qquad (2.17)$$

Then:

$$d/a = \sin i/\sin r = n. \qquad (2.18)$$

Hence the depth appears shallower by the ratio of the refractive indices. When an object is viewed through a block of glass the image is displaced sideways as in figure 2.9; and the displacement, d, changes as the block is rotated. Such a rotatable, parallel-sided glass plate provides simple lateral alignment in some optical instruments.

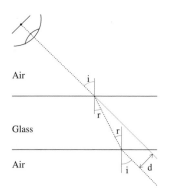

Fig. 2.9 Displacement of an image by a thick parallel-sided glass plate.

2.1.5 Deviation of light by a triangular prism

Simple prisms like that shown in figure 2.10 are used to disperse white light into the different colours. Light from a slit source perpendicular to the diagram is first *collimated* into a parallel beam. The incoming arrow indicates the direction in which this beam is incident on the prism. Light of a single wavelength will follow paths parallel to the arrowed line. Focussing beyond the prism produces a line image of the source again perpendicular to the diagram. Figure 2.11 shows typical spectra produced on the screen when the source slit is illuminated by discharge laboratory lamps each containing gas of one element. Each line in the spectrum is produced by light of a different wavelength emitted by the gas in the lamp. They are separated, *dispersed*, because the refractive index of glass varies across the visible spectrum. The angular deviation of the ray drawn in figure 2.10 is

$$\delta = i - r + e - s. \qquad (2.19)$$

Summing the angles of the triangle bounded by the ray and the prism edges, gives

$$(90° - r) + (90° - s) + \alpha = 180°,$$

so that,

$$r + s = \alpha. \qquad (2.20)$$

Then eqn. 2.19 becomes

$$\delta = i + e - \alpha. \qquad (2.21)$$

Applying Snell's law to the two refractions

$$\begin{aligned} e &= \sin^{-1}(n \sin s) \\ &= \sin^{-1}[n \sin(\alpha - r)] \\ &= \sin^{-1}(n \sin \alpha \cos r - n \cos \alpha \sin r). \end{aligned} \qquad (2.22)$$

Now

$$\sin r = (\sin i)/n, \;\; \text{and} \;\; \cos r = \sqrt{1 - (\sin^2 i)/n^2}.$$

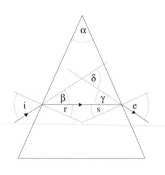

Fig. 2.10 Ray path through prism. The deviation δ is minimum in the symmetrical configuration where $i = e$.

Making these substitutions in eqn. 2.22 gives

$$e = \sin^{-1}\left[\sin\alpha\sqrt{(n^2 - \sin^2 i)} - \cos\alpha\sin i\right]. \qquad (2.23)$$

Then replacing e in 2.21,

$$\delta = i - \alpha + \sin^{-1}\left[\sin\alpha\sqrt{(n^2 - \sin^2 i)} - \cos\alpha\sin i\right]. \qquad (2.24)$$

The dependence of the deviation on the angle of incidence is plotted in figure 2.12 for a prism with an apex angle α equal to $30°$ and a refractive index of 1.5. There is a quite shallow minimum in the distribution of deviation against the angle of incidence. The minimum occurs in the symmetric arrangement where $e = i$, and $s = r$. At *minimum deviation* eqns. 2.20 and 2.21 become

$$r = \alpha/2, \text{ and } \delta_{\min} = 2i - \alpha. \qquad (2.25)$$

Applying Snell's law at minimum deviation at either surface, and using 2.25:

$$n = \sin i/\sin r = \sin\left[(\delta_{\min} + \alpha)/2\right]/\sin(\alpha/2). \qquad (2.26)$$

Now the refractive index of glass changes with wavelength. Thus minimum deviation occurs at different angles for the different colours, with the results shown in figure 2.11. In the case of borosilicate crown glass the refractive index changes from 1.51 to 1.54 between the red and blue ends of the spectrum. Of equal importance for spectroscopy, the refractive index changes monotonically with wavelength for any common type of glass so that the dispersion does not superpose colours. Consequently a prism gives a spectrum in which the wavelength increases smoothly from one end to the other. A measurement of the angle of minimum deviation easily determines the refractive index to one part in ten thousand.

When the prism angle is small such that $\sin\alpha \approx \alpha$ eqn. 2.26 becomes

$$n = (\delta_{\min} + \alpha)/\alpha, \qquad (2.27)$$

whence

$$\delta_{\min} = (n-1)\alpha. \qquad (2.28)$$

The deviation minimum is flatter for a narrow angle prism, so that this equation for the minimum deviation is a good approximation to the actual deviation over a wide range of angles of incidence around the symmetric arrangement.

2.2 Total internal reflection

When light is incident on a surface between two materials from the more optically dense material then at sufficiently large angles of incidence the refracted ray is suppressed. Rewriting Snell's law:

$$\sin\theta_2 = (n_1/n_2)\sin\theta_1, \qquad (2.29)$$

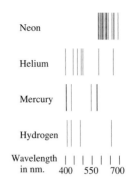

Fig. 2.11 Spectral lines observed with lamps containing gases of various elements at low presure.

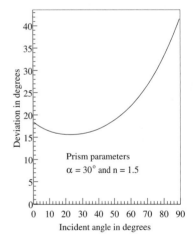

Fig. 2.12 The deviation of a ray passing through a prism as a function of the angle of incidence.

so that

$$\theta_2 > \theta_1. \tag{2.30}$$

There is therefore an angle of incidence called the *critical angle* θ_c, at which the angle of refraction will reach 90°:

$$\sin \theta_c = n_2/n_1. \tag{2.31}$$

At larger angles of incidence the light is totally reflected and there is

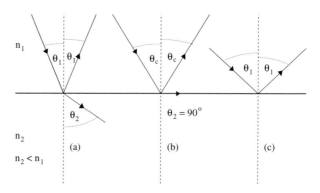

Fig. 2.13 Refraction at an optically less dense medium. Angle of incidence (a) less than the critical angle, (b) equal to the critical angle θ_c, and (c) greater than the critical angle.

no refracted ray travelling into the less optically dense material. Figure 2.13 shows the situation for increasing angles of incidence: for the angle of incidence less than the critical angle; equal to the critical angle of incidence when the refracted ray is parallel to the surface; and for a larger angle of incidence when the reflection is total. The property of *total internal reflection* (TIR), is used widely to guide light with low loss. These applications include the use of prisms in optical instruments to manipulate light beams, and transmission of electromagnetic radiation along optical fibres.

2.2.1   Constant deviation prism

Figure 2.14 shows a single compound prism which converts minimum deviation into a fixed deviation of 90°. One such minimum deviated ray is indicated on the diagram. The minimum deviation is produced not by a prism of apex angle 60° but effectively by two prisms, each of apex angle 30°. Between these a 45° prism is inserted to give a total internal reflection. The notional outlines and angles of these components are shown in the diagram. If the incident ray encounters the first half prism at minimum deviation, then it will leave this half prism perpendicular to the mid face. The TIR then deviates the ray exactly 90° so that it enters the second half prism perpendicular to its mid face also, and then completes the minimum deviation. The total deviation of the minimum deviated ray is just the 90° deviation produced by TIR: the deviations at the external surfaces of the half prisms are equal, but now in opposite senses, so they cancel. This prism is used in constant deviation spectrometers. The light is directed in a parallel beam at the prism by a collimator and observed by a telescope at right angles, all mounted in a fixed frame. Only the prism needs to be rotated to bring each part of the spectrum into view at minimum deviation.

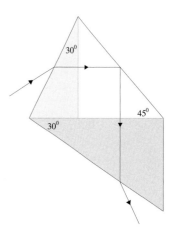

Fig. 2.14 Constant deviation prism. The dotted lines mark the boundaries of the hypothetical component prisms. The ray shown undergoes minimum deviation.

2.2.2 Porro prisms

Porro prisms have angles of 45°, 45° and 90°. A pair of them arranged as shown in figure 2.15 are used in binoculars to correct the inversion of the image produced by the lenses, giving an upright final image. There are four total internal reflections on the light path shown, and each reflection turns a right handed object into a left handed image, and vice versa. The overall effect is to preserve the original orientation, a right handed object remains right handed. Porro prisms used in binoculars

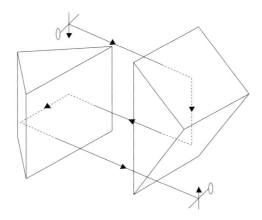

Fig. 2.15 This arrangement of two Porro prisms is used in binoculars to correct the inversion of the image produced by the lenses, giving a final upright image. The light path is shown for an object and its inverted image. Where the optical path lies inside the prisms the ray is indicated with a broken line.

have a narrow groove ground across the centre of the front face, whose purpose is to absorb scattered light. This is shown in figure 2.16. Light entering from outside the field of view is scattered at glancing angles from the front faces and would otherwise contribute a background haze. Another advantage accruing from the inclusion of a pair of Porro prisms in binocular designs is that the front lenses can be further apart than the eyes are; which gives improved depth discrimination. Again the use of the prisms allows the designer to fold the optical path, so making the instrument compact and easy to manipulate. Finally because the light is always incident nearly perpendicular to the prism faces there is no dispersion.

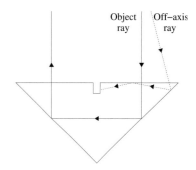

Fig. 2.16 Side view of one Porro prism. Rays potentially contributing a background of scattered light are trapped by the groove.

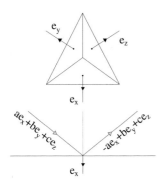

Fig. 2.17 The upper panel shows a corner cube and unit vectors perpendicular to the faces. The lower panel shows TIR from the *x*-face.

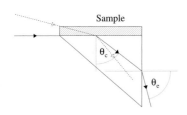

Fig. 2.18 The Pulfrich refractometer is illuminated by a focused beam. The critical ray's path carries the solid arrows.

2.2.3 Corner cube reflector

A corner cube reflector is, as the name implies, a corner cut from a glass cube. An example is shown in the upper panel of figure 2.17. The sloping face makes equal angles with the cube faces. When a ray enters through the sloping face it will undergo internal reflections in some order from each of the three perpendicular faces. It finally emerges accurately parallel to its original path but travelling in the opposite direction. To prove this assertion we consider an incident ray whose direction on entering the corner cube is along the direction given by the unit vector

$$a\mathbf{e}_x + b\mathbf{e}_y + c\mathbf{e}_z,$$

where the normals to the three mutually orthogonal surfaces of the corner cube are the unit vectors $\mathbf{e}_x$, $\mathbf{e}_y$ and $\mathbf{e}_z$. After reflection from the $\mathbf{e}_x$ surface the ray direction becomes

$$-a\mathbf{e}_x + b\mathbf{e}_y + c\mathbf{e}_z. \tag{2.32}$$

This is shown in the lower panel of figure 2.17. After reflections from all three orthogonal faces the exiting ray points along the direction

$$-a\mathbf{e}_x - b\mathbf{e}_y - c\mathbf{e}_z \tag{2.33}$$

which has reversed the incident ray. This property guarantees that whatever tilt the incoming beam has, it will always be reflected back whence it came. A square of 100 such corner cubes was left on the Moon during the Apollo XI mission and they provide a convenient mirror target for measurements of the Moon–Earth separation. Bicycle reflectors are arrays consisting of large numbers of small corner cubes moulded in clear plastic. They operate on the same principle and because the reflection at each facet is by TIR there is no need to metallize the back surface.

2.2.4 Pulfrich refractometer

Measurement of the critical angle of reflection can be used to obtain an accurate determination of refractive index for solid and liquid samples. Figure 2.18 shows a Pulfrich refractometer with the test material, refractive index n, sitting on a $45°/45°/90°$ reference prism of accurately known refractive index, $n_\mathrm{r} > n$. Light of the selected wavelength coming from the left is focused onto the boundary so that the beam covers an angular range down to glancing incidence at the interface. The angular range of the beam entering the prism has a sharp cut-off at the critical angle, θ_c, where $\sin\theta_\mathrm{c} = n/n_\mathrm{r}$. Then applying Snell's law to the critical ray at its exit from the prism

$$\begin{aligned}
\sin\theta_\mathrm{e} &= n_\mathrm{r}\sin(90° - \theta_\mathrm{c}) \\
&= n_\mathrm{r}\cos\theta_\mathrm{c} \\
&= n_\mathrm{r}\sqrt{1 - \sin^2\theta_\mathrm{c}} \\
&= \sqrt{n_\mathrm{r}^2 - n^2}.
\end{aligned} \tag{2.34}$$

Thus the specimen's refractive index is:

$$n = \sqrt{n_{\mathrm{r}}^2 - \sin^2 \theta_{\mathrm{e}}}. \tag{2.35}$$

All that is required to determine the sample's refractive index is the measurement of the angle θ_{e}.

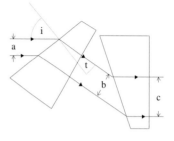

Example 2.1

Figure 2.19 shows a beam incident on two identical prisms. These are oriented so that the beam enters each at the same angle of incidence, and in the case shown leaves perpendicular to the exit surfaces. If the refractive index of the glass is n show that the beam width in the plane of the diagram is expanded by a factor

Fig. 2.19 An anamorphic prism pair.

$$c/a = [n^2 - \sin^2 i] / [n \cos i]^2.$$

If d is the width of the beam measured along the surface of the first prism

$$a = d \cos i; \quad b = d \cos t.$$

Thus

$$b/a = \cos t / \cos i.$$

Similarly

$$c/b = \cos t / \cos i.$$

Hence

$$\begin{aligned}
c/a &= \cos^2 t / \cos^2 i \\
&= [1 - \sin^2 t] / \cos^2 i \\
&= [1 - (\sin^2 i)/n^2] / \cos^2 i \\
&= [n^2 - \sin^2 i] / [n \cos i]^2. \tag{2.36}
\end{aligned}$$

The prisms do not affect the width of the beam in the perpendicular dimension. This arrangement of prisms is called an *anamorphic* pair and it is used to render circular the elliptically shaped profile of a laser diode beam. Anamorphic pairs are preferred over cylindrical lenses for *circularizing* laser diode beams because the prisms occupy less space and are cheaper. The degree of shaping can be altered as required by rotating the prisms.

2.3   Optical fibre

An optical fibre consists of a cylindrical glass core surrounded by concentric glass cladding of slightly lower refractive index. Modern optical fibre communication relies on total internal reflection to guide electromagnetic radiation along the core over paths extending to hundreds of kilometres. Near infrared radiation is used, this being the wavelength range for which absorption in the glass is at a minimum. Figure 2.20

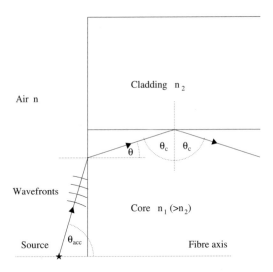

Fig. 2.20 Longitudinal cross-section through an optical fibre illuminated by a point source. The ray drawn is the one at the widest angle off axis to be totally internally reflected at the core–cladding interface.

shows a cross-section taken along the fibre length. Some radiation travels inside the core and is totally reflected at the core–cladding interface. For clarity the diameter of the core in the diagram is made overlarge. On the left is a point source and a ray from the source which meets the core–cladding interface at the critical angle. After the first reflection at this interface the ray will travel to the opposite side of the core where it will meet the core–cladding surface at the critical angle of incidence, just as in the first reflection. Evidently, once a ray is trapped in the core it generally remains in the core. The rays from the source which are retained inside the core lie inside the cone of semi-angle θ_{acc}. This acceptance angle will be a function of the refractive indices of the external medium, n, the core, n_{c} and the cladding, n_{e}. Snell's law can be applied to the ray where it enters the core:

$$n \sin \theta_{\mathrm{acc}} = n_1 \sin \theta. \tag{2.37}$$

At the core–cladding interface for the critical ray:

$$n_1 \sin \theta_{\mathrm{c}} = n_2 \sin 90° = n_2, \tag{2.38}$$

but $\theta = 90° - \theta_c$, so that:

$$n_1 \cos\theta = n_2. \tag{2.39}$$

Hence

$$\sin\theta = \sqrt{1 - n_2^2/n_1^2}. \tag{2.40}$$

Substituting for $\sin\theta$ in 2.37:

$$n \sin\theta_{\mathrm{acc}} = \sqrt{(n_1^2 - n_2^2)}. \tag{2.41}$$

The quantity $n \sin\theta_{\mathrm{acc}}$ is called the *numerical aperture* (NA) of the fibre. The square of the NA, defined in this way, is a standard measure of the light gathering power of an optical instrument, not simply of optical fibres. A fuller account of the properties and use of optical fibres is given in Chapter 16.

2.4   Micromirror projector

Digital micromirror devices (DMD), pioneered at Texas Instruments by L. J. Hornbeck, are now widely used in projectors[1] because of their compact size, high switching speed and robustness. The basic component is a rectangular array of typically 1–2 million square mirrors, with a pitch of $14\,\mu$m, the mirrors covering $\sim$90% of the total area. Each aluminium

[1]Digital Light Processing, $\mathrm{DLP}^{\mathrm{TM}}$, projectors.

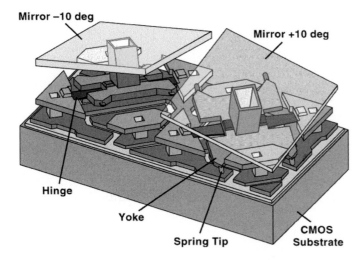

Fig. 2.21 A pair of landed mirrors from a digital micromirror array. Courtesy Texas Instruments.

micromirror swings on torsion hinges about a diagonal axis and is deflected electrostatically when voltage is applied to metal pads near its free corners. There are two stable orientations at $\pm 12°$ from the plane of the device, where the micromirror contacts mechanical stops. Figure

2.21 shows a pair of mirrors landed against the stops, in this design with ±10° rotation. The rotation from one stable orientation to the other takes only 15 μs. These orientations define the two digital states: in one position the micromirror deflects an incident light beam onto a screen and illuminates one small area (pixel) of that surface; in the other position the micromirror deflects the light away from the screen so that the pixel is dark. The image produced is seamless without visible gaps between pixels on the screen. Thanks to the full-on/full-off operation of the micromirrors the contrast between the illuminated and dark state of pixels on the screen is 2000:1 or better. The micromirror array is fabricated as the upper layer of a CMOS chip whose electronic cells below each micromirror apply the requisite timed deflection voltages. The final device is evidently a masterpiece of microtechnology.

Switching between colour in the simplest projectors is managed by directing white light through a rotating colour filter divided into 120° segments of red, green and blue. Figure 2.22 sketches one notional layout of a projector using a more sophisticated method for switching between colours. Three pure colour LEDs provide red, green and blue beams.

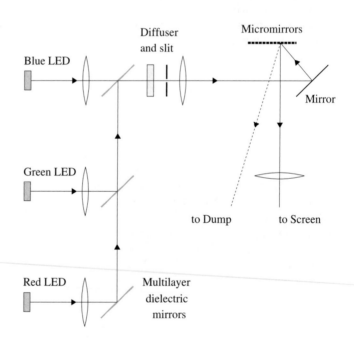

Fig. 2.22 Notional layout of a digital micromirror projector showing the light paths.

With a refresh rate of 60 frames a second the interval available for displaying one colour in each frame is 1/180 s or 5.6 ms. The LEDs are therefore active in turn for 5.6 ms intervals. Variations in image intensity are coped with by further subdividing this interval into unequal time segments. For example if a gray scale with 256 levels is required, then the time segments are 1/2, 1/4, 1/8.... and 1/256 of 5.6 ms. Each micromirror is selected 'on' or 'off' for each of these time segments by an eight bit control word input to the chip for that mirror. The actual optical switching begins when the mirror has moved so that light first enters the optics located between the micromirrors and the screen; and it ends when the mirror reflects all the light along that path. This optical switching time is only 2 μs, which is less than 10% of the shortest time segment.

Digital micromirror projectors have significant advantages over the previously dominant LCD projectors.[2] Their simple electromechanical components continue to function reliably for over 100,000 hours. Only the LEDs need replacement, and these too are long-lived. By comparison LCD projectors have lifetimes of a few thousand hours, with the colours gradually yellowing during that time. Finally digital micromirror projectors are more robust and compact, with some small enough to fit in one's pocket.

[2]LCD projectors are described below in Chapter 10.

A related application is the use of digital micromirror arrays in back-projection televisions, with screen diagonal lengths commonly around 70 inches. These televisions have correspondingly reduced power consumption, typically only 200 watts.

Exercises

(2.1) A thin sheet of plastic is placed on a microscope stage. The distance between the top and bottom surfaces as measured by the microscope is 300 μm. If the plastic has refractive index 1.55 what is its actual thickness?

(2.2) The angle of minimum deviation of a monochromatic beam of light is 20° as measured with a triangular prism with apex angle 25°. What is the refractive index of the prism for this wavelength of light?

(2.3) What is the critical angle for a diamond/air interface, and for a diamond/water interface? The refractive indices of water and diamond are 1.33 and 2.47 respectively.

(2.4) What is the NA of an optical fibre with core and cladding of refractive indices 1.505 and 1.50 respec-

tively at wavelength 1.3 μm?

(2.5) A glass thread in air would guide light in the same way that an optical fibre does. Why is this not a competitive solution for communications?

(2.6) Use Fermat's principle to prove the law of reflection.

(2.7) A light ray travels through a pile of clear accurately parallel faced glass sheets, each in contact with its neighbours above and below. The glass sheets can have different refractive indices. If the angle of incidence of the ray is 45° and it emerges from the final sheet what can you say about the direction of the emergent ray? Will it ever be the case that total internal reflection will prevent the ray traversing all the sheets?

Spherical mirrors and lenses

3.1   Introduction

A good proportion of the adult population need lenses, in the form of spectacles or contact lenses, to carry out day-to-day activities such as driving. Mirrors also abound: in the home concave mirrors provide a magnified close-up view of the face, and convex mirrors on the car driver's door give the driver a wider field of view, with the catch that the car following appears further off than it actually is! Everything seen live on TV comes through a lens system, often of sophisticated design. These lenses contain ten or more elements: they can zoom in without loss of focus, compensate for the shaking hand that holds them, and provide an image that faithfully reproduces the colour and proportions of the scene. Other lens systems are used by scientists and engineers to study the very small (microscopes) and the very distant (telescopes). On many production lines there are monitoring systems relying on lenses to give non-contact measurements of size, location and orientation of rapidly moving items. Lenses and mirrors can also produce an intense concentration of energy at the focal point: this ranges from a lens concentrating sunlight so that a piece of paper catches fire to the proposed ignition of nuclear fusion in deuterium/tritium pellets using laser beams.

The properties of single lenses, single mirrors and complex combinations can be predicted using the laws of reflection and refraction derived in Chapter 2. Most of the lenses and mirrors met in instruments have spherical surfaces. When the rays from an object are *paraxial*, that is to say they lie close to the axis of an optical system and make small angles with that axis, then the equations for calculating the image position for optical systems with spherical surfaces are relatively simple. This is the *paraxial approximation*, and the paraxial formulae for determining the image position in the cases of reflection and refraction will be derived first. Then the paraxial formulae for lenses and systems of lenses are derived and reformulated in the more convenient matrix form.

Rays from a point object that are either at wide angles to the optical axis of the lens/mirror, or travel far from the optical axis will not converge precisely at the paraxial image point. The deviations of these rays from the paraxial image are called *aberrations*. Because the refractive in-

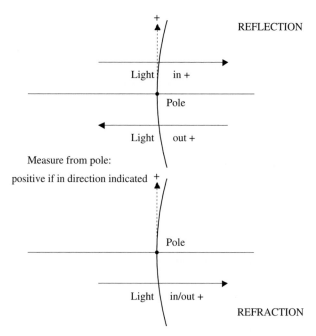

Fig. 3.1 Cartesian sign conventions.

dex changes with wavelength the position of the image formed by a lens will depend on wavelength, and this will be the case even for paraxial rays. The resulting deviations from perfect imaging are called *chromatic aberrations*. Mirrors and instruments that only use mirrors are evidently free from chromatic aberrations. In the latter half of the chapter aberrations and the techniques used to reduce them are described.

3.1.1   Cartesian sign convention

Figure 3.1 introduces the *Cartesian sign convention* that will be used systematically when analysing the paths of rays incident on mirrors, refracting surfaces and lenses. The point on the mirror or refracting surface lying on the optical axis is called the *pole* or *vertex*. In the case of that useful ideal, the thin lens having zero thickness, the pole is at the centre of the lens. This point is taken as the origin of Cartesian coordinates, and the object distance, image distance and radius of curvature are all measured from the pole. The positive direction along the optical axis is always the direction light is travelling. The sign of the object distance is determined with reference to the incident light's direction; the signs of the image distance and the radius of curvature are determined with reference to the direction of the outgoing light. Distances upward from the optical axis are positive. Most optical system consist of several elements. When calculating the effect of reflection or refraction at each element in turn coordinates relative to that element must be used.

These conventions lead to a relatively simple formalism: the paraxial lens and mirror formulae for imaging and magnification are identical; surfaces and lenses that converge (diverge) a parallel beam all have positive (negative) power; the transition to matrix manipulation is simple.

It can help in getting the signs correct to draw lines with arrowheads from the pole to the object, image, centre of curvature or focal point. Then in any mirror/lens formula the quantity inserted for the distance in question is positive/negative if the arrow points in/against the direction of the light. Equally when a length calculated using such a formula is positive/negative the arrow is drawn from the pole so that it points in/against the direction of the light.

Sometimes it is useful to distinguish *real* from *virtual* images or objects. In figure 2.7 the object is real but the image is virtual. The actual rays start from real objects and pass through real images. Objects and images are virtual when only the extensions of the actual rays pass through them.

3.2 Spherical mirrors

Ray paths after reflection from spherical mirrors are deduced from the reflection laws of Chapter 2. The reflected ray lies in the plane defined by the incident ray and the normal to the surface at the point of contact, and it makes an angle to the normal equal to that made by the incident ray. Figure 3.2 shows the ray construction used for a concave mirror

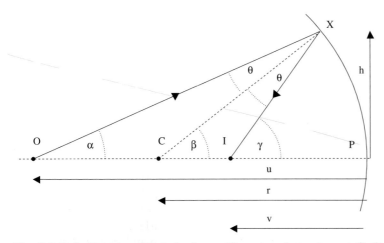

Fig. 3.2 Reflection at a spherical mirror with centre of curvature at C. O is the object and I the image. One ray from O is shown, reflected at X toward I.

forming the image of an object at O a distance u from the mirror. The mirror has radius of curvature r and its centre of curvature is at C. A representative ray is reflected from the mirror at X a height h above the optical axis and travels to cut the axis at I a distance v from the mirror. The angles are drawn large here for clarity: the rays considered are paraxial so that the angles are in fact very small. We will show that

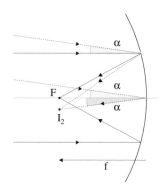

Fig. 3.3 F is the image of a distant point object lying on the optical axis. I_2 is the image of a similar distant point lying off-axis.

all paraxial rays reflected from the mirror converge on this same point, proving that I is the image of O.

Here the arrow for the object distance points against the direction that the incident light travels so that the object distance u is negative. On the other hand the arrows for the image distance v and to the centre of curvature point along the direction of the outgoing light, so that v and r are both positive. Lastly the point X is above the optic axis, making the height h positive. Then in the paraxial approximation

$$\alpha = h/(-u); \quad \beta = h/r; \quad \gamma = h/v. \tag{3.1}$$

Applying the law of reflection at X the angles marked θ are equal. From triangles OCX and OXI respectively

$$\beta = \alpha + \theta, \quad \text{and} \quad \gamma = \alpha + 2\theta. \tag{3.2}$$

Eliminating θ we obtain

$$2\beta = \gamma + \alpha. \tag{3.3}$$

Substituting for the angles and cancelling h gives an equation independent of the choice of ray

$$1/v = 1/u + 2/r, \tag{3.4}$$

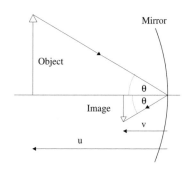

Fig. 3.4 Transverse or lateral magnification of a finite width object.

which is the paraxial imaging equation for a spherical mirror. Now consider light coming from a distant object lying on the optical axis. It arrives as a beam parallel to the axis. After reflection this beam converges at the *focus* or *focal point* F, as shown by the solid arrowed lines in figure 3.3. The plane perpendicular to the optic axis through the focus is called the *focal plane*. Substituting in the above equation shows that the distance from the mirror to the focal plane is $r/2$, which is called the *focal length* f, of the mirror. It locates the focus half way between the mirror and its centre of curvature. The focal length is thus positive for a converging mirror, that is one concave toward the incoming light: it is negative for a diverging, convex mirror. Then

$$1/v = 1/u + 1/f. \tag{3.5}$$

The *power* of the mirror is defined as $P \equiv 1/f$. The mirror equation, eqn. 3.5, applies equally to concave as well as convex mirrors and for all locations of the object; whether it is in front of the mirror, or virtual and behind the mirror.

Figure 3.3 shows images formed by light from two distant ojects, such as stars: one on the optical axis, the other off axis at an angle α. The incoming parallel rays from the on-axis star are shown as solid lines in the figure; while the incoming parallel rays from the second star are drawn as broken lines. Each set converges to an image point in the focal plane: for the on-axis star at F, and for the star off-axis at I_2. The angular separation α of the stars determines the separation of their images in the focal plane, and for paraxial rays this separation is simply,

$$s = f\alpha. \tag{3.6}$$

Images of extended objects produced by curved mirrors are generally different in size from the objects. As illustrated in figure 3.4 the *transverse* or *lateral magnification* is

$$m_{\mathrm{t}} = v/u. \tag{3.7}$$

Next consider a rod of length Δu placed *along* the optical axis at a distance u from the mirror. Differentiating the mirror equation, eqn. 3.5 relates the image length along the optical axis Δv to Δu,

$$(1/v^2)\Delta v = (1/u^2)\Delta u.$$

The *longitudinal magnification* is thus

$$m_{\mathrm{l}} = \Delta v/\Delta u = (v/u)^2. \tag{3.8}$$

If the object in figure 3.4 moves toward the mirror Δu is positive. Then Δv is also positive and the image moves away from the mirror.

3.2.1 Ray tracing for mirrors

Ray diagrams are helpful in visualizing the formation of images. The three most useful rays are shown in figures 3.5 and 3.6 for concave and convex mirrors respectively. Actual ray paths are drawn in solid lines, while the construction lines described below are broken.

- A ray from the object towards the focus in the case of figure 3.6 and away from the focus in figure 3.5. After reflection this ray travels parallel to the optical axis.
- A ray from the object going parallel to the axis. After reflection from the mirror it travels through the focal point, as in figure 3.5, or away from the focal point as in figure 3.6.
- A ray from the object pointing toward the centre of curvature as in figure 3.6 or directly away from the centre of curvature as in figure 3.5. After reflection this ray will travel back along its own path because it is travelling radially with respect to the mirror.

Having three (or more) rays permits a simple cross-check on the accuracy of the ray tracing: they must intersect at a single image point. All the rays from the object will, after reflection, either converge towards a real image; or in the case of a virtual image, as in figure 3.6, the reflected rays diverge from the image. The concave mirror example shows how an enlarged image is formed of a person's face placed nearer than the focal point. The convex mirror gives a wide field of view and a demagnified image, which are features useful for surveillance.

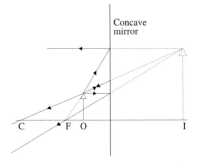

Fig. 3.5 Ray tracing for an object O in front of a concave mirror with centre of curvature C and focal point F. The image is formed at I. Rays are solid lines, construction lines are broken.

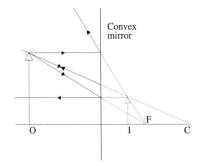

Fig. 3.6 Ray tracing for an object O in front of a convex mirror centre of curvature C and focal point F. The image is formed at I. Rays are solid lines, construction lines are broken.

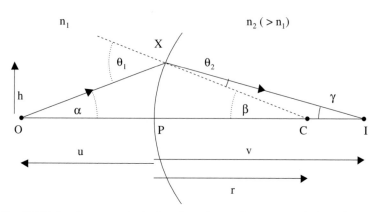

Fig. 3.7 Refraction at a spherical surface. A ray, shown arrowed, from the object O is refracted at X towards the image I. The line from O through the centre of curvature C is the optic axis.

3.3   Refraction at a spherical interface

In this section an equation is found for the position of the image when an object is viewed through a spherical refracting surface. Lenses rely for their focusing power on the refraction at a pair of spherical interfaces so the analysis presented here will prepare the ground for deriving the thin lens equation. As noted earlier we make the paraxial approximation: all rays are at small angles to the optical axis and any distances off-axis are also small. Figure 3.7 shows the path of a paraxial ray from an object, O, in a medium of refractive index n_1, which is refracted at a spherical interface into a second medium of refractive index n_2 at X. Then the ray cuts the optical axis at I. The surface has radius r and we choose $n_2 > n_1$. Applying Snell's law to the refraction at X, and remembering that all the angles are small, we have

$$n_1\theta_1 = n_2\theta_2. \tag{3.9}$$

For the angles shown, and recalling that according to the Cartesian sign convention u will have a negative value

$$\alpha = h/(-u); \quad \beta = h/r; \quad \gamma = h/v. \tag{3.10}$$

From triangles OXC and CXI respectively,

$$\theta_1 = \alpha + \beta, \text{ and } \theta_2 = \beta - \gamma. \tag{3.11}$$

Substituting for θ_1 and θ_2 in eqn. 3.9

$$n_1(\alpha + \beta) = n_2(\beta - \gamma).$$

Rearranging this gives

$$n_1\alpha + n_2\gamma = (n_2 - n_1)\beta.$$

Now substituting for the angles using eqn. 3.10 gives

$$n_2 h/v - n_1 h/u = (n_2 - n_1)h/r.$$

Cancelling h gives the imaging equation for a spherical refracting surface in the paraxial approximation

$$n_2/v - n_1/u = (n_2 - n_1)/r. \tag{3.12}$$

The quantity

$$P = (n_2 - n_1)/r \tag{3.13}$$

is known as the *bending power* or simply the *power* of the refracting surface.

3.4   Thin lens equation

Figure 3.8 shows a standard thin lens with spherical refracting surfaces. The lens is assumed to be immersed in one medium of refractive index n_1 while the lens itself has refractive index n_2. Applying eqn. 3.12 to

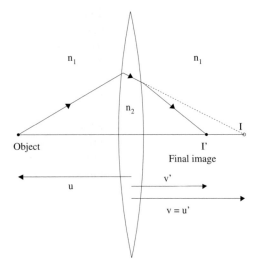

Fig. 3.8 Formation of an image of an object on the optical axis of a thin lens. The intermediate image produced by refraction at the first lens surface is I. This is imaged by refraction at the second surface at I'.

the refraction at the first surface

$$n_2/v - n_1/u = (n_2 - n_1)/r_1, \tag{3.14}$$

where r_1 is the radius of curvature of the first surface. Although the rays do travel a finite distance inside the lens this is neglected in what is called the *thin lens approximation*. Then the object distance of I from the second surface is just the image distance of I from the first surface

$$u' = v.$$

Applying eqn. 3.12 now to the refraction at the second surface

$$n_1/v' - n_2/u' = (n_1 - n_2)/r_2, \qquad (3.15)$$

where r_2 is the radius of the second lens' surface, and v' is the distance of the final image, I', from the lens.   Adding eqns. 3.14 and 3.15 gives

$$n_1/v' - n_1/u = (n_2 - n_1)/(1/r_1 - 1/r_2).$$

Here we drop the prime on the final image distance to obtain the paraxial *thin lens equation*

$$n_1/v - n_1/u = (n_2 - n_1)/(1/r_1 - 1/r_2) \equiv 1/f. \qquad (3.16)$$

If the lens is in air the thin lens equation becomes

$$1/v - 1/u = (n - 1)/(1/r_1 - 1/r_2) \equiv 1/f, \qquad (3.17)$$

where u is the object distance, v is the final image distance and n is the refractive index of the lens. Recalling the analysis of the mirror we can recognize that the quantity f defined above is the lens focal length. The bending power of a thin lens is the sum of the powers of the two surfaces

$$F = P_1 + P_2$$
$$= (n - 1)(1/r_1 - 1/r_2) \qquad (3.18)$$
$$\equiv 1/f. \qquad (3.19)$$

If an object is located at infinity then its image is at the focus of the lens and vice versa. The situation for the lens is symmetric: eqn. 3.16 gives the same focal length f whichever lens surface is facing the incoming light. Reversing the lens exchanges the radii, $r_1 \Rightarrow -r_2$ and $r_2 \Rightarrow -r_1$, and then $1/r_1 - 1/r_2$ is unchanged. Figure 3.9 shows how light from a distant point on axis gives an image at a distance f from the lens.

All the lenses discussed so far have positive focal length f and bring a parallel beam to a *real* focus, in the language of Section 3.1.1. Having two convex faces they are called *biconvex* lenses. Both surfaces are positive because each would alone converge a parallel beam: referring to eqn. 3.13 P is positive for both surfaces. By contrast in figure 3.10 a *biconcave* lens is drawn, each of whose surfaces would diverge a parallel beam and have negative power. Then eqn. 3.17 shows that the focal length f is negative. After passing through the lens an incident parallel beam diverges so that, as indicated by the dotted construction lines in figure 3.10, it appears to come from a focal point in the incident medium. The focal point is *virtual*. In the case that one surface is convex and the other concave the net focusing depends on the relative curvature of the surfaces. If the positive (convex) surface is more strongly curved then the lens is convergent and has positive focal length. It is called a *positive meniscus* lens. Conversely if the concave surface is more strongly curved then the lens is overall divergent: it is a *negative meniscus* lens. Some

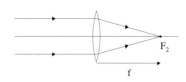

Fig. 3.9 Image formed by a positive lens with light incident from an object at infinity.

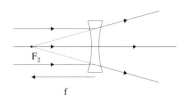

Fig. 3.10 Image formed by a negative lens with light incident from an object at infinity.

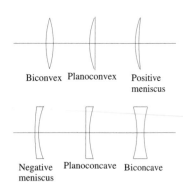

Fig. 3.11 The range of lens shapes. Lenses in the top row are all converging lenses and those in the bottom row are all diverging lenses.

examples of lens types are drawn in figure 3.11. Proceeding from left to right along the upper row, and then along the lower row the lens power steadily decreases. A plane sided sheet of glass marks the boundary between positive and negative lenses: it has infinite focal length and zero power.

3.4.1   Ray tracing for lenses

Three rays useful in visualizing the formation of images by a thin lens are shown in figure 3.12 for a positive lens and in figure 3.13 for a negative lens. The rays are:

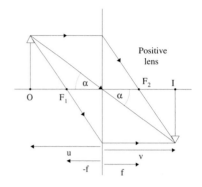

Fig. 3.12 Ray tracing through a positive lens. The object at O produces a real image at I.

- A ray from the object through the *pole* (centre) of the lens. This ray emerges undeviated because at the centre of the lens the faces are parallel. The displacement sideways is negligible because the lens is thin.
- A ray from the object travelling parallel to the optical axis. On leaving the lens it travels towards the second focal point F_2 in the case of a positive lens, and away from F_2 in the case of a negative lens.
- A ray through the first focal point F_1 for a positive lens or towards F_1 for a negative lens. After the lens it travels parallel to the optical axis.

Notice that for the negative lens the focal points (F_1 and F_2) have switched sides because the focal length is negative. The rays from the object in figure 3.12 converge after passing through the lens to a real image. In the case of the negative lens in figure 3.13 the rays diverge after passing through the lens so that they appear to come from a virtual image behind the lens. The distances from the lenses have been chosen to be $u = -2|f|$ in each case. Applying the thin lens equation 3.17 gives $v = 2f$ for the positive lens and $v = 2f/3$ for the negative lens example. The reader may like to construct examples where a positive lens produces a virtual image or a negative lens produces a real image.

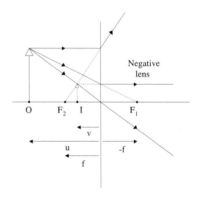

Fig. 3.13 Ray tracing through a negative lens. The object at O produces a virtual image at I.

The lateral magnification produced by a thin lens can be calculated by considering the ray through the lens' pole in either figure 3.12 or 3.13. The height of the object is $-u \tan \theta$ and of the image $-v \tan \theta$, where θ is the angle the ray makes with the optical axis. Hence the transverse magnification is

$$m_t = v/u. \tag{3.20}$$

When a short object is put along the optical axis the magnification can be obtained by differentiating eqn. 3.17:

$$-(1/v^2)\Delta v + (1/u^2)\Delta u = 0,$$

so that longitudinal magnification is

$$m_l = (v/u)^2. \tag{3.21}$$

If the object moves to the right, then the image moves right too.

Example 3.1

Two lenses are separated by 10 cm, the first of focal length +20 cm, the second of focal length −20 cm. An object is placed 40 cm in front of the positive lens. Where is the final image? What is its size? Is it upright or inverted? Use ray tracing to check the answer.

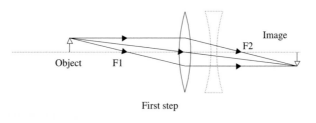

First step

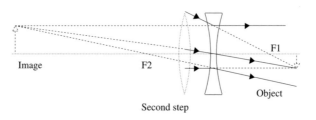

Second step

Fig. 3.14 Steps in dealing with image formation by a sequence of lenses. In the first step the second lens is ignored, and in the second step the first lens is ignored. The first and second focal points are labelled for each step.

The action of each lens is treated in its turn, ignoring the other one. These two steps are illustrated in figure 3.14, where the lens with the broken outline is the inactive one. For the first lens, in eqn. 3.17

$$u = -40, \ f = +20,$$

so that

$$1/v = -1/40 + 1/20 = 1/40.$$

Hence the image distance is +40 cm.

Now this image lies 30 cm to the right of the negative lens. The rays are refracted before they get to this image point by the negative lens, so this is a virtual object for the second lens. The values to be inserted in eqn. 3.17 when calculating the final image produced by the negative lens are

$$u = +30, \ f = -20,$$

so that

$$1/v = +1/30 - 1/20 = -1/60.$$

Hence the final image distance is located 60 cm to the left of the negative lens, and so it is virtual. Rays emerge from the negative lens pointing back to this location. The image magnification is the product of the individual lens magnifications. Thus

$$m = (v/u)_{+\text{lens}}(v/u)_{-\text{lens}} = [+40/(-40)][-60/(+30)] = +2,$$

hence the image is twice as big as the original object and upright.

Figure 3.14 shows the ray constructions required following the method given in Section 3.4.1. The upper diagram shows the construction for the positive lens; the lower for the negative lens. In the lower diagram the actual ray paths are drawn with full lines and their extensions to the object and image are drawn in broken lines. Notice that the rays used for the construction are generally different for the two lenses.

3.4.2   Magnifiers

A single positive lens or *loup* is used by electronic technicians to improve on the eye's ability to resolve detail when wiring circuit boards. In the lower half of figure 3.15 an object is placed at the *near point* which is the point closest to the eye at which one can focus comfortably. For young people the near point is about 25 cm from the eye. In the upper diagram a positive lens is placed in front of the eye, and now the eye is *relaxed*, that is to say it is focused at infinity. The factor increase in the image size produced by using the lens is

$$m = \tan\alpha/\tan\beta$$
$$= d_{\text{near}}/f_{\text{e}}, \tag{3.22}$$

where d_{near} is the distance to the near point. The magnification m cannot be increased indefinitely. A limit is reached when the lens' surfaces become sufficiently curved that departures from paraxial imaging become substantial. Magnification beyond a factor of about 10 with good quality imaging requires the use of a compound microscope containing two or more lenses.

3.5   Matrix methods for paraxial optics

The analysis of optical systems with several optical components is best handled using matrix methods. All refractions on the path of a ray are similar to that occuring at X in figure 3.16 for a surface of radius of curvature r. The paraxial approximation applies so that:

$$\phi = y/r, \tag{3.23}$$

and Snell's law gives

$$n_2\theta_2 = n_1\theta_1. \tag{3.24}$$

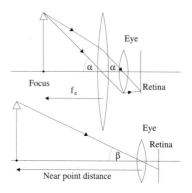

Fig. 3.15 In the lower diagram the eye views an object located at the near point. In the upper diagram the same object is viewed through a positive lens, of focal length f_{e}, with the eye relaxed.

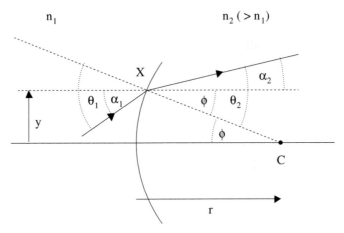

Fig. 3.16 Incident and refracted ray at an interface at a typical location in a compound lens system.

The θ angles are related to the tilts of the incoming and outgoing rays respectively

$$\theta_2 = \alpha_2 + y/r; \quad \theta_1 = \alpha_1 + y/r. \tag{3.25}$$

Substituting for θ_1 and θ_2 in eqn. 3.24 yields

$$n_2\alpha_2 + n_2 y/r = n_1\alpha_1 + n_1 y/r,$$

$$\text{so} \quad n_2\alpha_2 = n_1\alpha_1 - (n_2 - n_1)y/r,$$

$$\text{or} \quad n_2\alpha_2 = n_1\alpha_1 - Fy, \tag{3.26}$$

where F is the *power* of the surface $(n_2 - n_1)/r$. The lateral position after refraction is unchanged

$$y_2 = y_1. \tag{3.27}$$

Equations 3.26 and 3.27 tell us what happens to position and angle at an interface. The analogous equations for a ray travelling in one and the same material between two interfaces a distance l apart are $n_2 = n_1$ and $\alpha_2 = \alpha_1$ so that

$$n_2\alpha_2 = n_1\alpha_1, \tag{3.28}$$

$$\text{and} \quad y_2 = y_1 + n_1\alpha_1 t, \tag{3.29}$$

where $t = l/n_1$. This looks unnecessarily complicated, but brings the advantage that the effect of free propagation in a medium and refraction at a surface can each be expressed by a matrix operating on the same column vector

$$\begin{pmatrix} n_2\alpha_2 \\ y_2 \end{pmatrix} = \mathbf{M} \begin{pmatrix} n_1\alpha_1 \\ y_1 \end{pmatrix}. \tag{3.30}$$

In the case of refraction,

$$\mathbf{M}_r = \begin{pmatrix} 1 & -F \\ 0 & 1 \end{pmatrix}, \tag{3.31}$$

and for travel in one medium,

$$\mathbf{M}_t = \begin{pmatrix} 1 & 0 \\ t & 1 \end{pmatrix}. \tag{3.32}$$

In order that reflections can be included simply in the matrix analysis the light path must be unfolded at each reflection as shown in figure 3.17. This shows the actual path of a ray in the upper panel and the corresponding unfolded path in the lower panel. In the lower panel we see that each refractive surface crossed after the reflection is placed at its mirror image in a plane through the mirror's pole. Then the matrix for a reflection is

$$\mathbf{M}_m = \begin{pmatrix} 1 & -F \\ 0 & 1 \end{pmatrix}. \tag{3.33}$$

None of these three matrices depend on n_1 or y_1. The matrix that describes an optical system will be the product of a sequence of such matrices $\mathbf{M}_{t1}$, $\mathbf{M}_{r1}$, $\mathbf{M}_{t2}$ .... $\mathbf{M}_{tn}$

$$S = \mathbf{M}_{tn}....\mathbf{M}_{t2}\mathbf{M}_{r1}\mathbf{M}_{t1}. \tag{3.34}$$

$\mathbf{M}_r$, $\mathbf{M}_m$ and $\mathbf{M}_t$ have unit determinant, so any overall product matrix S will also have unit determinant $(ad - bc = 1)$:

$$S = \begin{pmatrix} a & b \\ c & d \end{pmatrix}. \tag{3.35}$$

We will show in the following section that $-b$ is in fact the power of the system.

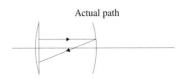

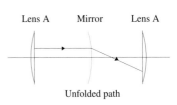

Fig. 3.17 Ray paths at reflection: in the upper panel the actual path, in the lower panel the unfolded path.

3.5.1   The equivalent thin lens

In the paraxial approximation any system of lenses can be replaced by an equivalent thin lens. The result is proved formally in Appendix B using matrix methods. Figure 3.18 shows the location of the *cardinal planes* of the equivalent thin lens: the cardinal planes are the *focal planes*, the *principal planes* and the *nodal planes*. The corresponding *cardinal points* are located where the optical axis cuts each such cardinal plane. P_1 and P_2 mark the two principal planes. In the case of a single thin lens the principal planes would simply coincide. The focal length, f, of the equivalent lens is measured from the principal plane to the focal plane. Ray tracing in figure 3.18 resembles that for a single thin lens. A ray striking the first principal plane at height h will emerge from the second principal plane at precisely the same height h. The two principal planes are therefore planes of *unit magnification*. When, as here, the same medium fills the object and image spaces the principal points are also points of unit angular magnification, that is they are nodal points.

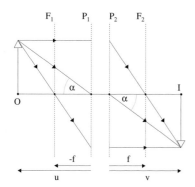

Fig. 3.18 Equivalent thin lens for a compound lens system. F_1 and F_2 mark the focal planes. P_1 and P_2 mark the principal planes, and also the nodal planes when the exit and entrance media are identical.

[1] As for example in eqns. 3.12 and 3.13 when u is set to $-\infty$.

A ray passing through the first nodal point at an angle α to the optical axis will emerge from the second nodal point at exactly the same angle α. If on the other hand the image and object media are different the principal and nodal planes do not generally coincide. It is at first sight surprising that the cardinal planes may lie inside or outside the region occupied by the lenses; the principal planes may even cross over, that is to say the first principal plane can lie to the right of the second principal plane. In this case the right hand half of the diagram in figure 3.18 would be moved bodily to the left so that it overlapped the left half. The prescriptions given for ray tracing remain valid in all such cases.

When the refractive index of the image region is n_{out} we can use Snell's law to show that the relation between the power and the focal length is $F = n_{\mathrm{out}}/f$.[1] Now consider a ray incident parallel to the optic axis and apply eqn. 3.35. This gives

$$n_{\mathrm{out}}\alpha_{\mathrm{out}} = by_{\mathrm{in}}.$$

If the outgoing ray, which passes through the focus, is extended back it strikes the principal plane at a height y_{in}. Thus

$$f = -y_{\mathrm{in}}/\alpha_{\mathrm{out}} = -n_{\mathrm{out}}/b,$$

whence we have $b = -F$. Take the example of two thin lenses, of powers F_1 and F_2 separated by a distance t in air. The overall matrix

$$
\begin{aligned}
S &= \begin{pmatrix} 1 & -F_2 \\ 0 & 1 \end{pmatrix} \begin{pmatrix} 1 & 0 \\ t & 1 \end{pmatrix} \begin{pmatrix} 1 & -F_1 \\ 0 & 1 \end{pmatrix} \\
&= \begin{pmatrix} 1 - tF_2 & -F_1 - F_2 + tF_1F_2 \\ t & 1 - tF_1 \end{pmatrix}.
\end{aligned}
\tag{3.36}
$$

The total power of the lenses is thus $F_1 + F_2 - tF_1F_2$. Rewriting this in terms of focal lengths

$$1/f = 1/f_1 + 1/f_2 - t/f_1f_2. \tag{3.37}$$

3.6   Aberrations

Monochromatic aberrations are the departures from paraxial imaging that appear in practical optical instruments: point objects are no longer imaged as point images. In addition there are *chromatic* aberrations which arise because the refractive index of glasses varies with the wavelength of the radiation, so that in turn the power of a surface given by eqn. 3.13 varies with wavelength. Thus a point object emitting, or illuminated by white light is imaged in the different colours at different points. Chromatic aberration effects occur for paraxial as well as non-paraxial rays, but they do not affect pure mirror systems. The

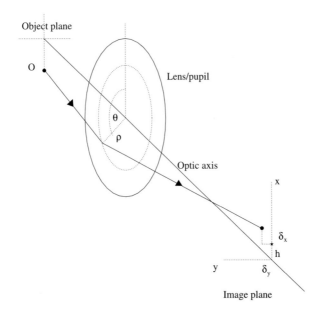

Fig. 3.19 Ray path from an object through the lens and continuing to the image plane.

chromatic and monochromatic aberrations for a typical 2.5 cm diameter crown glass lens of focal length 10 cm are similar in magnitude, ~ 1 mm at the focal plane, so that there is equal interest in reducing both types of aberration.

3.6.1   Monochromatic aberrations

Paraxial imaging is based on the approximation that *sines* of ray angles can be approximated by the angles. If the angle is large then more terms need to be included in the expansion of $\sin \theta$,

$$\sin \theta = \theta - \theta^3/3! + \theta^5/5! \ .... \tag{3.38}$$

Aberrations that arise from neglecting the second (third) term in the expansion are called third-order (fifth-order) aberrations. The third-order aberrations are dominant and are of five distinctive types. These third-order aberrations and their reduction by suitable lens combinations will be considered in the following sections. Figure 3.19 shows a schematic ray path from the object O which emerges from the lens (or exit pupil of a lens system) and intersects the image plane some distance from the paraxial image point (starred). Let the point on the lens have polar coordinates (ρ, θ), and the image point the Cartesian coordinates $(h + \delta x, \delta y)$, while the paraxial image point is $(h, 0)$. In the paraxial approximation the wavefront leaving the lens would be spherical with its centre at the image point. Actual wavefronts deviate from this shape and this is the origin of aberrations. The deviation of the actual from the ideal wavefront, which is for simplicity measured radially in the direction of the image, can only depend on two things: where the ray hits the lens relative to the optical axis, ρ; and where the image lies in the image plane relative to the optical axis, $\mathbf{h}$. Hence the deviation is some function of these vectors

$$\Delta R = \Delta R(\boldsymbol{\rho}, \mathbf{h}).$$

In the case of a system of lenses the lens aperture indicated in figure 3.19 is replaced by the exit pupil of the system, defined as follows. Each lens system has one aperture which is the most restrictive in limiting the angular size of the cone of rays from an object point which actually reach its image point. This limiting aperture is known as the *aperture stop*. Its image seen from the image space is called the *exit pupil*, and its image seen from the object space is called the *entrance pupil*. The ray passing through the centre of the aperture stop is called the *chief* or *principal ray*. This ray also passes through the centre of the entrance and exit pupils. Rays that pass through the edge of the aperture stop, and the entrance and exit pupils, are known as *marginal rays*.

If the system shown in figure 3.19 is rotated about the optical axis, then because the system is axially symmetric the aberration should not change. It follows that ΔR can only depend on rotational invariant quantities made up from $\boldsymbol{\rho}$ and $\mathbf{h}$. There are just three of these: ρ^2, h^2 and $\boldsymbol{\rho} \cdot \boldsymbol{h} = \rho h \cos\theta$. Then rewriting ΔR to include all possible quadratic and quartic terms made up from these quantities gives

$$\Delta R = a_1\rho^2 + a_2\rho h \cos\theta + a_3 h^2 + b_1\rho^4 + b_2\rho^3 h \cos\theta$$
$$+ b_3\rho^2 h^2 + b_4\rho^2 h^2 \cos^2\theta + b_5\rho h^3 \cos\theta + ... \qquad (3.39)$$

The quadratic terms with the a coefficients can be dropped, because they give overall movements of the image for paraxial and non-paraxial rays alike. Thus the term in ρ^2 gives focusing and moves the image plane for *all* rays; the term in $\rho h \cos\theta$ tilts *all* rays to make the image larger/smaller; while the term in h^2 is a constant over the whole aperture. The remaining quartic terms with b coefficients contain the third-order aberrations. It is straightforward to extract the image displacements from ΔR, the displacement of the wavefront from a sphere centred on the image point. The Cartesian coordinates at the exit pupil are $x = \rho \cos\theta$ and $y = \rho \sin\theta$. The tilts of the wavefront in the x- and y-directions are

$$(\partial\Delta R/\partial x) \quad \text{and} \quad (\partial\Delta R/\partial y).$$

Multiplying by the distance v from the lens (or exit pupil) converts these tilts into displacements at the image plane

$$\delta x = v\,(\partial\Delta R/\partial x)$$
$$\text{and} \quad \delta y = v\,(\partial\Delta R/\partial y). \qquad (3.40)$$

Then using $\partial\rho/\partial x = \cos\theta$, $\partial\rho/\partial y = \sin\theta$, $\partial\theta/\partial x = -\sin\theta/\rho$ and $\partial\theta/\partial y = \cos\theta/\rho$ these equations become

$$\delta x = v\cos\theta\,(\partial\Delta R/\partial\rho) - v\,(\sin\theta/\rho)\,(\partial\Delta R/\partial\theta), \quad \text{and}$$
$$\delta y = v\sin\theta\,(\partial\Delta R/\partial\rho) + v\,(\cos\theta/\rho)\,(\partial\Delta R/\partial\theta).$$

After some manipulation this yields

$$\delta x = c_1\rho^3\cos\theta + c_2(2 + \cos 2\theta)\rho^2 h + (3c_3 + c_4)\rho\,h^2\cos\theta + c_5 h^3, \quad (3.41)$$

$$\delta y = c_1\rho^3\sin\theta + c_2\rho^2 h\,\sin 2\theta + (c_3 + c_4)\,\rho\,h^2\sin\theta. \quad (3.42)$$

The five aberrations separated here with coefficients c_n are physically quite distinct and are called the *Seidel* aberrations:

- spherical aberration ($c_1 = 4vb_1$),
- coma ($c_2 = vb_2$),
- astigmatism ($c_3 = vb_4$),
- field curvature ($c_4 = v(2b_3 - b_4)$),
- distortion ($c_5 = vb_5$).

von Seidel developed the analytic study of aberrations in the 19th century after being engaged by the optical entrepreneur Zeiss to examine ways of improving lens performance.

3.6.2   Spherical aberration

The first terms in eqns. 3.41 and 3.42 are the only ones that contribute to aberrations of objects on axis,

$$\delta x = c_1\rho^3\cos\theta,$$
$$\delta y = c_1\rho^3\sin\theta.$$

These relations show that the intersections with the paraxial image plane of the rays passing through an annular section at radius ρ of the exit pupil lie on a circle. The radius of this circle increases with ρ so that the image becomes a circular blob. This *spherical aberration* is illustrated in figure 3.20 for an axial object at infinity. Positive lenses focus marginal rays more strongly than paraxial rays. The paraxial rays focus at F_P, while the marginal rays through the periphery of the exit pupil come to a focus at F_M. At the surface indicated by the dotted line lies the *circle of least confusion* where the spread of the rays is minimum. The distance $F_M F_P$ is the *longitudinal* spherical aberration, and the radius of the ray cone at the paraxial focus is the *transverse* spherical aberration. Figure 3.21 shows how a lens' surfaces can be *bent* while keeping the focal length fixed. The spherical aberration depends strongly on the lens shape and is smallest when the shape is close to being planoconvex, with the curved surface facing the light. In this configuration, with a parallel beam incident, the deviation of rays is shared equally between the two lens surfaces.

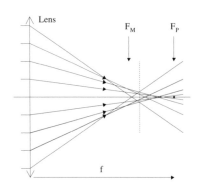

Fig. 3.20 Spherical aberration for a positive lens. The paraxial focus is F_P and the marginal ray focus is F_M. The broken line indicates the plane of the circle of least confusion.

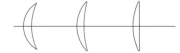

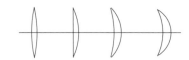

Fig. 3.21 Bending of the surface curvature of positive lenses, while keeping the focal length constant.

For certain combinations of object and image points spherical aberration is absent and these conjugate points are used in microscope design. In figure 3.22 P is a point on the surface of a planoconvex lens of refractive index n, such that its distance from the centre of curvature, C, of the convex face, radius r, is exactly r/n. Then we can show that rays from P passing through the lens at *any* angle diverge from a point image P′ where P′C is $n \cdot r$. P′CT and TCP are similar triangles because P′C $= n \cdot$ TC, CT $= n \cdot$ CP and the angle $\hat{PCT}$ is common. Hence

$$T\hat{P}'C = \alpha.$$

Also

$$P'\hat{T}C = \alpha'.$$

Applying the sine rule to $\Delta TP'C$

$$\sin(T\hat{P}'C)/r = \sin(P'\hat{T}C)/n \cdot r.$$

Multiplying by $n\,r$ we get

$$n \sin \alpha = \sin \alpha'.$$

This is exactly Snell's law so that TR is precisely the path of ray PT after refraction at T. This result holds for any ray from P which means that the image P′ is free of spherical aberration. P and P′ are known as *conjugate* points for the curved refracting surface. In high magnification microscopes the object is located at one conjugate point of the first lens so that spherical aberration is avoided while maximizing the numerical aperture.

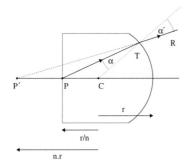

Fig. 3.22 Conjugate points (P and P′) for a spherical refracting surface whose centre lies at C. CP = r/n, and CP′ = $n \times r$, where r is the radius and n the refractive index.

3.6.3   Coma

Coma is familiar to anyone who has used a positive lens to focus the Sun's image on paper. When the lens is tilted the Sun's round image changes to a comet shaped flare. It is the following terms in eqns. 3.41 and 3.42 that cause coma

$$\delta x = c_2(2 + \cos 2\theta)\rho^2 h$$
$$\delta y = c_2\rho^2 h \sin 2\theta.$$

One term, $2c_2 \rho^2 h$, gives a simple radial displacement of the image point, whilst the sinusoidal terms cause the intersection of the rays with the paraxial image plane to travel round a circle as θ varies from 0 to π and again as θ increases from π to 2π. This is illustrated in figure 3.23 on which the chief ray follows the arrowed line. Rays travelling through an annular section of the exit pupil (fixed ρ) intersect the paraxial image plane in a ring as indicated by the letters a, b, c, d. The sizes of the rings and their radial offsets grow with the radius squared (ρ^2) of the annular section of the lens through which the rays pass. The circles overlap to give a comet shaped flare which may extend, as in figure 3.23, outward

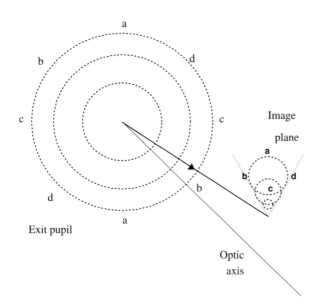

Fig. 3.23 Coma for an off-axis point object. Rays passing through a ring on the lens, like the ones shown, intersect the image plane on a ring. The total image is a comet shaped flare that starts at the paraxial image point and has a spread of 60°.

from the paraxial image, when it is called positive coma; or inward, which is known as negative coma. Coma also varies with lens shape. It is smallest for a positive lens close to planoconvex in shape, with the object far distant and the light incident on the curved face. This is very similar to the lens shape that also minimizes spherical aberration. A neat arrangement of two planoconvex lenses which eliminates coma when the object is not an infinity is shown in figure 3.24. The first lens produces an image at infinity, and the second refocuses the light and then the condition for reduced coma and spherical aberration holds for both lenses. This arrangement of biconvex lenses is often used in *condensers* which are optical systems for projecting light from a source onto the object viewed.

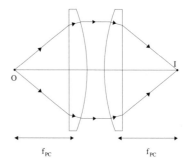

Fig. 3.24 An arrangement of two planoconvex lenses which minimizes coma and spherical aberration.

3.6.4 Astigmatism

Astigmatism is illustrated in figure 3.25, where the image of an off-axis point object consists of two line segments at right angles. The image nearer the exit pupil is a tangential line, the other is a radial line, while between them lies a circle of least confusion. The tangential line image is formed by the rays which lie in the plane containing the chief ray and the optical axis. These rays in it are called *tangential* or *meridional* rays. The radial line image is formed by the rays lying in the perpendicular *sagittal plane* which contains the chief ray. In figure 3.26 the object is shaped like a spoked wheel. The tangential image has a sharply defined

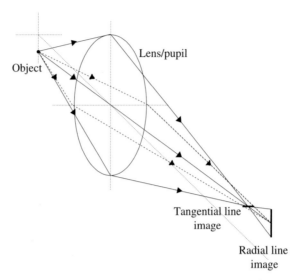

Fig. 3.25 Image formation with astigmatism. The solid lines are rays in the meridional plane containing the optical axis. The broken lines are sagittal rays forming a perpendicular plane. These give tangential and radial line images respectively.

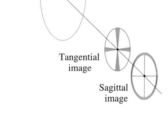

Fig. 3.26 Astigmatic image of a wheel-shaped object centred on the optical axis. At the tangential focus radial lines are blurred and at the sagittal focus the tangential lines are blurred.

rim while the spokes are fuzzy. Conversely the sagittal image has sharply defined spokes and a fuzzy rim.

Referring to eqns. 3.41 and 3.42, the components which cause astigmatism are

$$\delta x = 3c_3\rho\, h^2 \cos\theta,$$

$$\delta y = c_3\,\rho\, h^2 \sin\theta.$$

The tangential rays are those having $\theta = 0$ or π and have $\delta y = 0$. They form a point image in a tangential focal plane before reaching the paraxial image plane. On the other hand the sagittal rays with $\theta = \pi/2$ or $3\pi/2$ have $\delta x = 0$ and form a point image in a sagittal image plane one third as far from the paraxial image plane. The tangential rays when reaching the sagittal image plane still have $\delta y = 0$, and so form a *sagittal* or radial line image. Similarly the sagittal rays form a *tangential* line image in the tangential image plane.

An extreme degree of astigmatism is produced by cylindrical lenses, an example of which is shown in figure 3.27. There is focusing in the vertical plane only and the image of a distant point object is a horizontal line. Intermediate between cylindrical and spherical lenses are the *toric* lenses, which have different curvatures in perpendicular planes through the optical axis.

3.6.5   Field curvature

In eqns. 3.41 and 3.42 there is a second set of terms in $\rho\, h^2 \cos\theta$ and $\rho\, h^2 \sin\theta$ which resemble the terms responsible for astigmatism: namely those with the coefficient c_4.

$$\delta x = c_4 \rho h^2 \cos\theta,$$

$$\delta y = c_4 \rho h^2 \sin\theta.$$

Here however the coefficients c_4 are identical which makes all the difference. Following the argument made in the case of astigmatism, the
axial displacement of the tangential and sagittal images from the paraxial image plane due to these terms are equal, hence there is simply
a displacement of the image along the axial direction which increases
quadratically with h, the distance off axis. This aberration is called *field
curvature* and the curved image surface is called the *Petzval surface* Σ_P.
Any astigmatism gives further displacements, with the tangential image
surface (Σ_T) moving three times further from Σ_P than the sagittal image
surface (Σ_S). In figure 3.28 the displacements of the image planes are
shown as a function of the radial distance off axis for a biconvex lens.

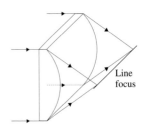

Fig. 3.27 Cylindrical lens.

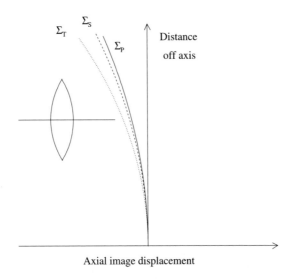

Fig. 3.28 Image surface curvature for a positive biconvex lens. See text for details.

In the case of a negative lens the field curvature is away from the lens
so that by a judicious combination of lenses the field curvature can be
reduced.

The total displacement of the Petzval surface in an image formed by
a series of thin lenses is

$$\Delta z \propto \sum_j (1/[n_j \cdot f_j]), \tag{3.43}$$

where n_j is the refractive index, and f_j the focal length of the jth lens.
With a single negative and positive lens made of the same glass the

field can be made flat by choosing $f_- = -f_+$. The overall focal length can be adjusted by the choice of f_+ and the lens separation: see eqn. 3.37. One widely used technique employed in lens systems to remove an unacceptable level of field curvature is to add a *field flattener*, which is a negative lens placed a small distance u from the final image plane of the system. The final image position is given by

$$v = uf/(f - u) \approx u,$$

so the image is otherwise unchanged in size.

3.6.6 Distortion

The fifth and final terms in eqns. 3.41 and 3.42 are

$$\delta x = c_5 \, h^3,$$
$$\delta y = 0.$$

They give a purely radial displacement of the image point that increases with the cube of the distance off-axis, h. If c_5 is negative (positive) then points on the image are displaced radially inward (outward), the displacement growing with the distance of the object off axis. Both types of distortion are shown in figure 3.29 for the object in the lower panel. The images are said to show *barrel* and *pincushion distortion*. The distortion produced by any lens is reversed if the lens is turned around, so a pair of lens arranged symmetrically will cancel each other's distortion.

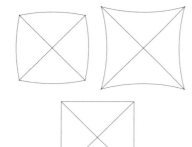

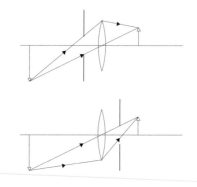

Fig. 3.29 Image distortion. The lower diagram shows a square object. The upper diagrams show images with barrel and pincushion distortion.

Fig. 3.30 The effect of aperture stop placement.

Aperture stop effects

The positioning of the aperture stop in a lens system significantly affects the aberrations. Thin positive lenses do not show distortion if the stop is placed against the lens. However if, as in figure 3.30, the stop is placed elsewhere distortion results: barrel distortion if the stop is in front of the lens and pincushion distortion if the stop is behind the lens. The essence of the aperture stop's influence, both in this example and in cases where its effect is favourable, is that it limits the range of rays that form the image. An intuitive way to view this effect is to note that in the upper diagram the distance rays travel to the lens is longer than that travelled by the ray through the lens' pole, so that the object distance is larger and the magnification smaller than for the paraxial rays. Consequently barrel distortion results. In the lower diagram the object distance is shorter for rays at wide angles, the magnification is correspondingly larger and pincushion distortion results.

In general the introduction of an aperture stop will only affect an aberration if one of the earlier aberrations in the sequence (spherical aberration, coma, astigmatism, field curvature, distortion) is present. The introduction of an aperture stop will affect neither the spherical

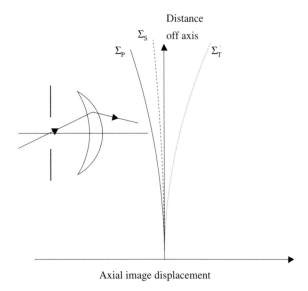

Fig. 3.31 Positive meniscus lens with an aperture stop and the resulting image surfaces. Σ_P, Σ_S and Σ_T are the Petzval, sagittal and tangential image surfaces respectively. The surface of least confusion is close to the ordinate axis.

aberration nor the Petzval field curvature. There is an optimal location of the stop for any given lens bending which minimizes the coma, and this is called its *natural* location.

Bending the lens surfaces moves the image field surface. The meniscus lens with a front aperture stop, shown in figure 3.31, gives astigmatic images, but the surface of least confusion is now flat. In addition the stop also reduces coma. This lens arrangement was often used in cheap cameras. It is not possible to choose a lens shape that will minimize simultaneously all the aberrations even if the aperture stop is located at the natural location. Combining a shape close to planoconvex and having the aperture stop at the natural location minimizes spherical aberration and distortion, but the level of astigmatism is then unacceptable. With a meniscus shape the aberrations are all similar but not particularly small. Combinations of lenses are needed to give images of overall high quality.

3.6.7 Chromatic aberration

Prisms made of glass disperse white light into a spectrum; and because any segment of a lens is prism-shaped it follows that the red and blue images produced by a single lens will not coincide. This *chromatic* aberration in the image of a point object on-axis is shown in figure 3.32. Blue light is focused more strongly than the red because the refractive index of glasses and most other transparent substances falls off with increasing wavelength. This is known as normal dispersion. The distance between

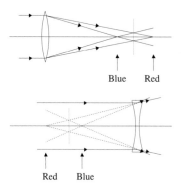

Fig. 3.32 Chromatic aberration produced by a positive and a negative lens of a distant point object on axis. The circle of least confusion lies in the plane indicated by the dotted line.

the red and blue images is the *longitudinal* chromatic aberration. At the waist in the ray envelope lies the circle of least confusion where the image spot is smallest. A commonly used measure of the dispersion is *dispersive power*

$$\nu = (n_F - n_C)/(n_D - 1), \tag{3.44}$$

where the refractive indices n_F, n_D and n_C are those of the glass measured at three wavelengths of Fraunhofer absorption lines in the Sun's spectrum. Historically these were selected to span the visible spectrum: the F-line is blue (486 nm), the D-line is yellow (589 nm) and the C-line is red (656 nm). Another measure of the dispersion is the *dispersive in-*

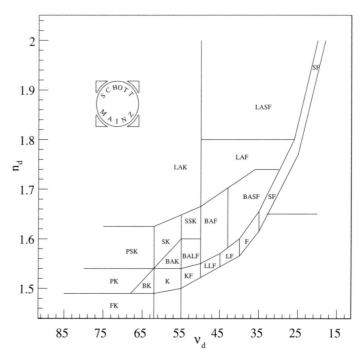

Fig. 3.33 A selection of Schott glasses: refractive index plotted against the dispersive index (Abbe number). The *less* dispersive crown glasses lie to the left of the centre line and the *more* dispersive flint glasses to the right. (With permission from *Schott Guide to Glass*: courtesy Wolfgang R. Wentzel, Schott AG, Hattenbergstr. 10, 55122 Mainz, Germany.)

dex, V, which is the inverse of the dispersive power. This is often the quantity specified by glass manufacturers: be aware that V is smaller if the dispersion is larger! Figure 3.33 plots the refractive index against the dispersive power for glass types produced by Schott AG.

There are two basic layouts of a pair of lenses with overall positive power that remove most of the chromatic aberration:

(1) A positive and negative lens made of glass with different dispersions which are in contact.

Table 3.1 Dispersion of two standard types of crown and flint glass. Crown glass is made of soda, lime and silica; flint glass of alkalis, lead oxide and silica. The code number of a glass is constructed, as indicated by the boldface characters from the D-line refractive index and the dispersive index.

Glass	486 nm F (blue)	589 nm D (yellow)	656 nm C (red)	Dispersive index V
BK7 517642 (crown)	1.5224	1.5168	1.5143	64.2
DF **620364** (flint)	1.6321	1.**6200**	1.6150	**36.4**

(2) A pair of positive lenses spaced apart by the mean of their focal lengths.

A lens *doublet*, that is a pair of lenses in contact, has focal length

$$1/f = 1/f_1 + 1/f_2.$$

Now for either lens

$$1/f_i = (n_i - 1)\rho_i,$$

where ρ_i is the geometric factor $1/r_1 - 1/r_2$ formed from the radii of curvature of the lens' surfaces, and n_i is the refractive index of the glass at 589 nm. Thus for each lens the difference in focal length between 486 nm and 656 nm is given by

$$\begin{aligned}\Delta(1/f_i) &= \Delta n_i \rho_i \\ &= \nu_i(n_i - 1)\rho_i \\ &= \nu_i/f_i.\end{aligned} \tag{3.45}$$

Using this result the change in the focal length of the system between 486 nm and 656 nm is given by

$$\Delta(1/f) = \nu_1/f_1 + \nu_2/f_2. \tag{3.46}$$

This chromatic difference in focal length vanishes if

$$\nu_1 f_2 = -\nu_2 f_1. \tag{3.47}$$

Dispersive power is positive for glass, so that in order to satisfy this equation one lens needs to be converging, the other diverging. A simple convergent colour corrected system can be made by cementing a double convex crown glass lens to a planoconcave flint glass lens as shown in figure 3.34. This is called a Fraunhofer achromat. Table 3.1 gives the optical parameters for two types of optical glass that are frequently used in making achromatic doublets. Crown glass has the smaller dispersive power, hence from eqn. 3.47 the positive (crown glass) lens will have shorter focal length, making the focal length of the combination positive. In the case of thick doublets there is a further difficulty. Although

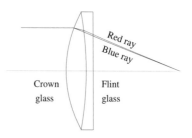

Crown glass   Flint glass

Red ray
Blue ray

Fig. 3.34 Achromat doublet lens.

the focal lengths for blue and red light have been made equal, the images do not coincide because the principal planes for the two colours are separated by a distance of order 10% of the lens thickness. What will remain true is that the magnification of objects is the same for both colours, which removes *lateral* chromatic aberration. In addition all achromatic doublets suffer residual chromatic aberration at other wavelengths away from 656 nm and 486 nm, a feature that is called *secondary colour*. The bending of the surfaces of an achromat remains at the disposal of the designer. Thus a lens shape for a given focal length can be chosen to minimize spherical aberration and coma, as well as chromatic aberration.

Combinations of thin lenses made of three different types of glass give colour correction at three wavelengths, and such lenses are called *apochromats*. The recent availability of expensive extra-low dispersion *ED* glass has made the correction of chromatic aberration that much easier, but with a cost penalty. An ED lens has lower dispersion across the visible spectrum than either an achromat or apochromat. ED lenses are essential for modern very long lens systems.

It is also possible to correct chromatic aberration at two wavelengths with lenses made of the same glass type. From eqn. 3.37 two lenses separated by a distance t have focal length

$$1/f = 1/f_1 + 1/f_2 - t/(f_1 f_2).$$

Then the difference between the focal lengths at 656 nm and 486 nm is given by

$$\Delta(1/f) = -\Delta f_1/f_1^2 - \Delta f_2/f_2^2 + t\Delta f_1/(f_1^2 f_2) + t\Delta f_2/(f_2^2 f_1)$$
$$= -\nu/f_1 - \nu/f_2 + 2t\nu/(f_1 f_2). \tag{3.48}$$

Thus in order to remove the chromatic aberration we must have

$$0 = [\nu/(f_1 f_2)] [-f_2 - f_1 + 2t],$$

that is

$$t = (f_1 + f_2)/2. \tag{3.49}$$

The lens separation must be half the sum of the two focal lengths.

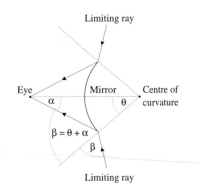

Fig. 3.35 Angular coverage with convex mirror.

3.7 Further reading

The seventh edition of *Principles of Optics* by M. Born and E. Wolf, published by Cambridge University Press (1999), contains a thorough mathematical acccount of aberrations and many other matters. The *Handbook of Optics*, volume 1, contains a broad discussion of aberrations in the context of fundamentals, techniques and design. The editor in chief is M. Bass and it was published in 1995 by McGraw-Hill, New York.

Exercises

(3.1) Two lenses are separated by 25 cm, the first of focal length -30 cm, the second of focal length $+40$ cm. An object is placed 10 cm in front of the negative lens. Where is the final image? What is its size? Is it upright or inverted? Use ray tracing to check the answer.

(3.2) Apply the results of Appendix B to locate the cardinal points of the system described in the previous question.

(3.3) A shopkeeper uses a convex mirror of radius of curvature 1 m and arc length across the diameter of 50 cm to view his shop. When he is 2 m from the mirror what angular coverage does this give him? What angular size would a customer's hand (10 cm) subtend at the shopkeeper's eye if the customer is 7 m from the mirror. Figure 3.35 shows the arrangement.

(3.4) An object is located 25 cm from a concave mirror whose radius of curvature is 40 cm. The object 0.1 cm in length lies along the optical axis. Where is its image located and how long is it? Use ray tracing to check the image location.

(3.5) What radius of curvature is required to make a planoconvex lens with focal length 30 cm from glass of refractive index 1.7? How would you arrange a pair of planoconvex lenses and an aperture stop in order to mimimize aberrations. What aberrations are reduced by this arrangement.

(3.6) Design an achromatic doublet from the glasses given in table 3.1 having a focal length of 45 cm. If only lenses of the same glass type are available how could you reduce the chromatic aberration?

(3.7) Show that the smallest separation of a *real* image and object is four times the focal length of the (positive) lens and that the lens is then midway between image and object.

(3.8) Show that if the object is a distance x_1 from the first focus and the image is a distance x_2 from the second focus of a lens of focal length f, then $x_1 x_2 = -f^2$. This can be done analytically or by using similar triangles in figure 3.12.

(3.9) A biconcave lens made from glass of refractive index 1.65 has surfaces with radii of curvature 25 cm and 45 cm. What is the focal length of the lens?

(3.10) Apply matrix methods to a lens of thickness T, refractive index n having surfaces with radii of curvature r_1 and r_2. What is the focal length and where are the principal planes of the lens?

(3.11) A lens of refractive index 1.73 has focal length 20 cm in air. What will its focal length become when immersed in water?

Optical instruments

4

4.1 Introduction

The refracting telescope and the microscope are the first instruments discussed in this chapter; the one used to magnify distant objects, the other to magnify tiny objects. Camera design and construction are examined next. Several special lens systems will be described, including the telephoto, zoom and telecentric lenses. The last is vital in process control in industry. Reflecting telscopes will be discussed in Chapter 8. In order to give bright images optical instruments need to have large lenses that collect non-paraxial as well as paraxial rays. Another fundamental reason that aperture size is important is that the smallest detail that can be seen clearly depends critically on the aperture size: even aberration-free optical systems produce extended images of point objects due to the wave nature of light. The plane beam of light incident on any aperture is *diffracted*, that is it spreads out on emerging from the aperture by an amount that depends on the ratio of the wavelength to the aperture size. Diffraction is discussed more fully in Chapter 6, where it will be shown that a circular aperture of diameter D limits the *angular resolution* with radiation of wavelength λ to at best

$$\Delta\theta = 1.22\lambda/D. \qquad (4.1)$$

The corresponding resolution in an image in the focal plane is then $f\Delta\theta$. Aberrations can only make the resolution worse; hence the methods for reducing aberrations need to be applied if the limit on resolution imposed by diffraction is to be attained.

The final sections of the chapter contain descriptions of widely used, non-standard lenses. Aspheric lenses are slightly non-spherical lenses and simplify the removal of aberrations in modern camera lens design. Graded index lenses rely on axial or radial refractive index variation, rather than surface curvature to give focusing. Lastly Fresnel lenses are discussed, which are flattened versions of normal lenses.

4.2 The refracting telescope

The simple refracting telescope design illustrated in figure 4.1, using a pair of lenses to give improved magnification over a single lens, was invented early in the seventeenth century. It uses two positive lenses, a

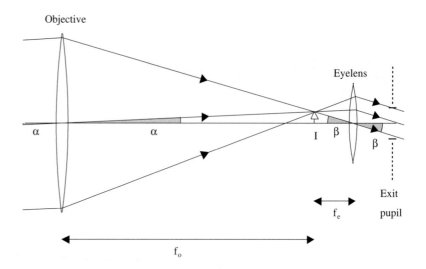

Fig. 4.1 Terrestrial telescope design as used in binoculars. The eye ring is located at the exit pupil, where the image of the objective is formed by the eyelens. The angular magnification is β/α.

large radius *objective* and a smaller *eyelens* spaced apart by the sum of their focal lengths. The objective produces an intermediate image I in its focal plane and the eyelens images this at infinity. A telescope produces angular magnification: parallel rays from a point on an object at infinity exit as parallel rays toward the image point at infinity. Such optical systems are called *afocal* and are unique in not possessing principal planes. The magnification achieved, M_θ, is the angular size of the image divided by the angular size of the same object seen without the telescope. Using the notation of figure 4.1

$$M_\theta = \beta/\alpha$$
$$\approx \tan\beta/\tan\alpha$$
$$\text{so that } M_\theta = f_\text{o}/f_\text{e}. \tag{4.2}$$

As mentioned in the previous chapter, one element in any optical instrument will restrict the amount of light from the object reaching the image more than the others. This element is known as the *aperture stop*. In order to determine which element is the aperture stop the location and size of the image of each element in the object space is calculated.[1] Then the element whose image subtends the smallest angle, as seen from the object, is the aperture stop. Its image is called the *entrance pupil*. Similarly the image of the aperture stop seen from the image side is called the *exit pupil*. In the case of a telescope the aperture stop is the objective lens and, because it is the first element, it is also the entrance pupil. The exit pupil of the telescope, that is the image of the objective produced by the eyelens, is at a distance v_p behind the eyelens where

$$1/v_\text{p} = -1/(f_\text{o} + f_\text{e}) + 1/f_\text{e}$$

[1]This image is produced by all the optical elements between the element in question and the object space.

$$= f_\text{o}/[f_\text{e}(f_\text{o} + f_\text{e})].$$

Then

$$v_\text{p} = f_\text{e}(f_\text{o} + f_\text{e})/f_\text{o}. \tag{4.3}$$

Now $f_\text{o} \gg f_\text{e}$ so the image is close to the focal plane of the eyelens. If D is the objective diameter the exit pupil diameter is

$$D' = D v_\text{p}/(f_\text{o} + f_\text{e})$$
$$= D/M_\theta. \tag{4.4}$$

Using eqn. 4.1 the equivalent angular resolution of the image due to diffraction is

$$\Delta\theta_\text{image} = \Delta\theta\, M_\theta$$
$$= (1.22\lambda/D)\,(D/D')$$
$$= 1.22\lambda/D'. \tag{4.5}$$

When used with the eye the light collected by the instrument should all enter the observer's eye. The eye pupil is placed at the exit pupil, which is therefore called the *eyering* for visual observation, and ideally the eye and exit pupil sizes should be the same. Both the angular resolution and light gathering power of the eye–telescope combination are limited by whichever of the eye pupil and the telescope's exit pupil is the smaller. If the eye pupil is the smaller then, when viewed by eye, some of the light collected by the telescope and some of its potential resolution are wasted. Alternatively if the eye pupil is the larger the light-gathering and resolution of the eye are not fully utilized. Matching the position and size of the exit and entrance pupils of coupled optical systems has to be a general design goal: in that way no component throws away the light transmitted by the others, and equally no component degrades the resolution of the others. The eye pupil diameter depends on the light intensity; from a value of 2 mm in bright light to 8 mm in near darkness. In full daylight the pupil diameter is 2.5 mm so the eye resolution at 550 nm wavelength light (green light) is $\sim 3\,10^{-4}$ rad, or 1 minute of arc. Thus the eye can comfortably resolve points 0.1 mm apart on an object placed 25 cm away at the near point.

In Chapter 2 numerical aperture was introduced to quantify the light collecting power of an optical fibre. Here we generalize the definition:

$$\text{NA} = n \sin\theta, \tag{4.6}$$

where θ is the semi-angle subtended by the entrance pupil at the object lying in a medium of refractive index n.

4.2.1   Field of view

Although the eyelens in a telescope does not restrict the illumination of the image, it does restrict the angular range over which objects can be seen. This range is called the *field of view*, and the eyelens is called the *field stop*. Its image in the object space is called the *entrance window* of the system, and the *exit window* is the eyelens itself. Similar definitions will apply to other optical instruments, the field stop being the aperture whose image in the object space restricts the field of view from the entrance pupil. Suppose the eyelens has diameter d, then the *apparent* field of view seen when the eye is placed at the exit pupil is

$$\alpha' = d/f_e. \tag{4.7}$$

The *actual* angular field of view is the angle subtended by the eyelens at the objective:

$$\alpha = d/(f_o + f_e). \tag{4.8}$$

Combining the last two equations

$$\alpha' = (f_o + f_e)\alpha/f_e. \tag{4.9}$$

Then using eqn. 4.2

$$\alpha' = (M_\theta + 1)\alpha. \tag{4.10}$$

The distance from the eyering to eyelens is called the *eye relief* and this needs to be at least 1.5 cm for comfortable use. The full field is only seen from the centre of the exit pupil; from other points across the exit pupil a smaller angular range is visible, so the outer part of the field of view is less well illuminated. This effect is called *vignetting*. In order to remove the outer poorly illuminated part of the view an aperture can be placed at the intermediate image in a telescope or microscope; and this aperture then becomes the field stop. The image produced by the telescope is inverted and to correct this either a further lens is needed or an inverting prism. The Porro prism described in Chapter 2 is one choice.

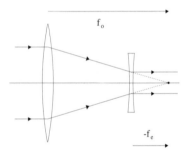

Fig. 4.2 Design suitable for either a Galilean telescope or a beam expander. The arrows on the rays would be reversed when used as a beam expander.

Another type of telescope shown in figure 4.2 uses a negative power eyelens located so that the lenses are separated by the sum of their focal lengths. This design, the Galilean telescope, produces an upright image so that this design is used in opera glasses. Although not invented by Galileo he improved the performance to ×30 power by experiment and skilful lens grinding. He was then able to observe four of the moons circulating round Jupiter, the mountainous surface structure of the Moon, and found that Venus shows phases like the Moon. These observations dealt the death blow to the old cosmology in which the Sun, Moon, planets and stars were perfect spheres and themselves lay on transparent spheres that rotated around a static Earth. The lens arrangement of the Galilean telescope is often used to expand the diameter of laser beams which are inherently narrow. For this application the Galilean design has the advantage over a design using two positive lenses that there is no intermediate focus (see figure 4.1). This avoids the local heating that occurs at the focus of an intense laser beam.

Example 4.1

A binocular has an objective of focal length 16 cm and diameter 48 mm. What eyelens' properties will give a magnification of 8 times and a 5° field of view? What will the eye relief and eyering diameter be? What will be the apparent field of view?

The overall angular magnification is given by eqn. 4.2 so the focal length of the eyelens is

$$f_e = 16/8 = 2\,\text{cm}.$$

If the field of view is 5° then using eqn. 4.8 the eyelens diameter is

$$d = 18\pi(5/180) = 1.57\,\text{cm}.$$

The eye relief is the distance of the eyering from the eyelens, given by eqn. 4.3

$$v_p = 2 \times 18/16 = 2.25\,\text{cm}.$$

In addition the eyering diameter is given by eqn. 4.4

$$D' = 48/8 = 6\,\text{mm}.$$

The apparent field of view is given by eqn. 4.10

$$\alpha' = 9 \times 5 = 45°.$$

Aberrations would be large for the system just outlined and the single objective and eyelens are replaced by lens combinations; the eyelens replacement being called an *eyepiece*.

4.2.2   Etendue

In a complex optical system consisting of several components, the numerical apertures and fields of view of the components should match from component to component along the chain. That is to say the exit pupil (window) of any component should coincide in position and size with the entrance pupil (window) of the following component. If just one component has a significantly smaller pupil (window) than its neighbours the additional light gathering power (field of view) of the other components is simply wasted.

Numerical aperture and field of view are not however independent properties, and this permits more flexibility than the bare statements above imply. It may be possible to trade field of view for light gathering power and vice versa by inserting some combination of lenses between the unmatched components. A quantity which expresses the overall light gathering power, called the *etendue*, determines whether components may be matched in this way or not. The etendue is first defined and then the way the trading occurs is illustrated with a single lens.

The solid angle Ω subtended by the entrance pupil at the object measures the light gathering power, that is the fan of rays received from any point on the object. Equally the unobstructed area A of the object visible in the image defines the field of view. The product of these two quantities and the refractive index squared

$$\mathcal{T} = n^2 \Omega A \tag{4.11}$$

is variously known as the *throughput*, *luminosity* or *etendue*. Another useful expression for the etendue can be obtained by re-expressing the solid angle in terms of the numerical aperture, $\mathrm{NA} = n \sin \theta$. Taking the semi-angle θ subtended by the entrance pupil to be sufficiently small, we have to a good approximation,

$$\Omega = 2\pi(1 - \cos \theta) = \pi \theta^2, \tag{4.12}$$

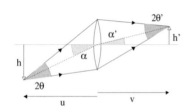

Fig. 4.3 Etendue through one optical component.

and then

$$\mathcal{T} = \pi (\mathrm{NA})^2 A. \tag{4.13}$$

Taking the product of the etendue and the radiance gives the total radiant flux in the beam, so the etendue gives a measure of the light that an optical system can transmit. Suppose the incident radiance on an optical system over the wavelength interval $\Delta\lambda$ at wavelength λ is $I(\lambda)\Delta\lambda$, then the flux of radiation through the system in this wavelength interval is

$$F = \mathcal{T} I(\lambda) \Delta\lambda. \tag{4.14}$$

Figure 4.3 shows a single lens placed between media of refractive indices n and n'. The object and image heights are h and h', while their areas are h^2 and $(h')^2$ respectively. The cone of rays accepted by the entrance pupil from any point on the object has semi-angle θ, and these rays converge in a cone of semi-angle θ' at the image point. In the paraxial approximation the ratios are

$$h'/h = v\,\alpha'/(u\,\alpha) = v\,n/(u\,n'),$$
$$\text{and } \theta'/\theta = -u/v.$$

Therefore

$$(n')^2 \Omega' A' = (n')^2\,\pi(\theta')^2\,(h')^2 = n^2 \pi \theta^2\,h^2 = n^2\,\Omega\,A.$$

Thus at each intermediate image through an optical system the image size and the angular spread of the fan of rays illuminating a point on the image will both change but the product, the etendue, is seen to be invariant. More precisely, the etendue of the light that makes its way through the whole of an optical system is invariant at each step: we cannot include light cut off by an aperture at some intermediate surface. It is always possible in principle to match two optical components with the same etendue by using intermediate lenses. However this may be impractical if the difference in pupil sizes is very large.

Some related conclusions can be drawn about the image brightness. Here brightness is always used to mean the radiance or the luminance, as defined in Section 1.7. Sticking to energy parameters, we see that the invariance of the etendue means that radiance or brightness is preserved. This result is known as the law of *conservation of radiance*. There is one special case: that of a point source, for example a distant unresolved star. With such an object there is no area so that the etendue cannot even be defined. Clearly the bigger the telescope aperture stop (the objective) used to collect light from a point source the better.

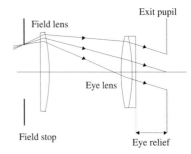

Fig. 4.4 Huygens eyepiece.

4.3 Telescope objectives and eyepieces

Generally the field of view needed in a telescope is narrow so that an achromatic doublet or apochromatic triplet is adequate as the objective. Among the aberrations spherical aberration and coma can be kept small by appropriate lens shaping. Astigmatism, field curvature and distortion are less important over narrow fields of view. In the case that a wide field is needed, one of the camera lenses described below would be used. Eyepieces on the other hand are required to give good image quality over the apparent field of view, which is wider in angle than the field of view by a factor equal to the angular magnification of the telescope. Three popular examples of eyepieces are shown in figures 4.4, 4.5 and 4.6. Each has two components: the *field lens* and the *eyelens*. As its name implies the field lens enlarges the field of view accessible by the eyelens. To do this the field lens is placed near the focal plane of the objective so that it does not change the image location appreciably but does pull in ray bundles from the edge of the field of view. The field lens is not placed too close to the image of the objective; if it were, then any imperfections of the field lens would be seen superposed on the final image. Many optical instruments use a field lenses to extend the field of view. *Huygens* eyepiece is the oldest, cheapest and least satisfactory. Two planoconvex lenses are spaced apart by half the sum of their focal lengths so that spherical aberration, coma and lateral chromatic aberration are small, but there is considerable longitudinal chromatic aberration and pincushion distortion. One great drawback is the small eye relief of only ~4 mm. The Ramsden eyepiece (not shown) has facing planoconvex lenses and an eye relief of over 10 mm. The *Kellner* eyepiece is a Ramsden eyepiece with an achromatic doublet for the eyelens. This gives a well corrected wide field of view. Another significant advantage which it has over the Huygens' eyepiece is that the image formed by the objective lies in front of the eyepiee. A calibration scale can be mounted there and because

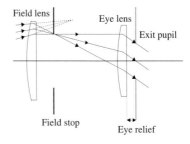

Fig. 4.5 Kellner eyepiece.

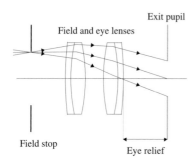

Exit pupil

Field and eye lenses

Field stop

Eye relief

Fig. 4.6 Ploessl eyepiece.

this is viewed through the complete eyepiece it will be colour corrected.

The *field stop* is no longer the edge of the single lens eyepiece, as it was for the schematic telescope of Section 4.2, but is an aperture placed at the location of the image formed by the objective. Finally the Ploessl eyepiece uses a symmetric layout of achromatic doublets which gives superior imaging. Its flat field is particularly important for use with media such as film and CCDs which cannot adapt, like the eye, to a curved field. Typical eyepieces have focal length 10 mm or 25 mm. Kellners give a 40°, and Ploessls give a 50° field of view.

The eyepieces described here are used in both microscopes and in telescopes. Eyepieces are matched to the objective. Take for example a 2.5 m focal length objective with a diameter of 25 cm used visually. Suppose a Ploessl eyepiece is used to give a total magnification of ×100, then it would require a focal length of 25 mm. If this Ploessl eyepiece has a field of view of 50°, then the actual field of view given by eqn. 4.10 would be 0.5°.

4.4   The microscope

Microscopes are designed to give high magnification and sufficient resolution to distinguish features as small as a few wavelengths of light. A

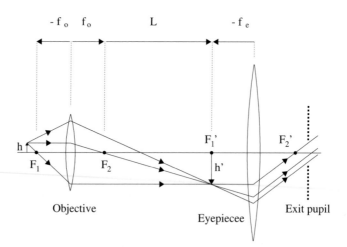

Fig. 4.7 Schematic structure of a microscope. The objective forms an image in the focal plane of the eyepiece, so that the final image is at infinity.

microscope is depicted schematically in figure 4.7. The objective gives an image at the focal plane of the eyepiece and the final image at infinity is viewed with the eye relaxed. Suppose the objective has focal length f_o, the eyepiece has focal length f_e, and that the intermediate image lies a distance L beyond the focus of the objective. Then the magnification produced by the objective shown in the diagram is

$$M_o = h'/h = -L/f_o. \qquad (4.15)$$

The eyepiece magnification is given by eqn. 3.22, $m = d_{near}/f_e$, so that the overall magnification of the microscope is

$$M = -(L\,d_{near})/(f_e\,f_o). \qquad (4.16)$$

The objective must accept a wide cone of rays from the object in order to give a well resolved and bright image, and it must also be aberration free. A typical high power oil immersion objective might have focal length 2.0 mm, giving $M_o = 100$ and NA 1.4 (and would be labelled $100\times$ NA 1.4 2 mm). The first stages of such an objective are shown in figure 4.8 and illustrate the use of conjugate object and image points to give perfect images (see Section 3.6.2). Note that the lens is effectively extended to the object (slide) by filling the gap between them with oil of the same refractive index as the lens. The object P and its image $P_{1/2}$ are conjugate points for the lens surface labelled 1. $P_{1/2}$ also lies at the centre of curvature of surface 2 so that the rays enter the second lens undeviated. Finally $P_{1/2}$ and its image, P_3, are conjugate points for the surface 3. The wide cone of rays captured from P emerge in a much tighter cone from P_3, so that standard lenses can be used to handle the beam thereafter. In this way a NA as large as 1.4 is achievable. More details of the microscope are shown in figure 4.9, where two achromats or apochromats can be seen to complete the objective. The image is corrected for spherical aberration, coma and chromatic aberration; and the field is flat. The eyepieces are similar to those mentioned for telescopes, with a Ploessl eyepiece being shown in figure 4.9. The size of a microscope is dictated by the average person's reach, and one common choice is to make the distance between the object and intermediate image equal to 195 mm.[2]

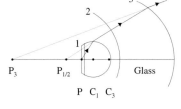

Fig. 4.8 Oil immersion objective showing conjugate points.

The illumination of the object is required to fill the angular acceptance of the objective in order that this acceptance is not wasted. Figure 4.10 shows a source and condenser lenses below the microscope stage providing what is known as *Koehler* illumination. This arrangement has a number of simple advantages. Firstly the area illuminated, the numerical aperture and the brightness can be adjusted independently. The left hand diaphragm (field stop) is imaged at the object and controls the field of view. Then the right hand diaphragm (aperture stop) is used to match the numerical aperture of the condenser to that of the microscope: if it is smaller then the resolution of microscope is degraded; if it is larger light that does not enter the objective directly can be scattered in, thereby causing a background haze. Finally the intensity can be changed by altering the source voltage or with *neutral density* filters that absorb equally across the spectrum. A further advantage is that

---

[2]This is the DIN (Deutsche Industrie Norm) standard.

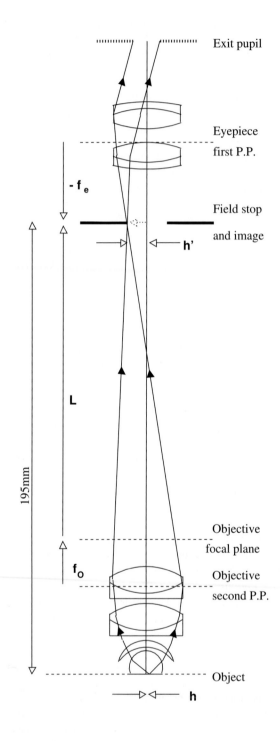

Fig. 4.9 Microscope construction showing the ray cone collected from an off-axis object point.

light from each point on the source forms a parallel beam at the object. This ensures that the illumination is coherent and also that it is uniform whatever the variations of brightness across the source. This topic will recur in Chapter 7.

The high intensity of illumination needed in a microscope leads to considerable scattering of light from the region around the point being viewed into the microscope. This significant background can be eliminated by the widely used technique of *confocal* illumination. The field stops in the illumination system and in the microscope are replaced by small apertures that limit the region on the object illuminated and viewed. This region is typically a circle of diameter $1\,\mu$m. Background scattered light is thus eliminated. The image intensity is sampled by an electronic detector behind the eyepiece and an image of the whole object is obtained by moving the microscope stage carrying the slide so that the illluminated spot is raster scanned across the whole slide.

4.5 Cameras

Cameras are used not only for standard photography, but also to record images produced by optical instruments such as microscopes. For all these uses film has been mainly supplanted by electronic detectors, notably the *charge coupled device* (CCD). A CCD consists of silicon photodiodes arranged in a rectangular array. When light is absorbed on a photodiode a charge proportional to the product of the light intensity and its duration is produced and stored. This process will be described in detail in Chapter 15. The exposure time is controlled by an electromechanical or electronic shutter. After exposure the charge on each diode is amplified, digitized and stored in memory; and from this data an image of the scene can be reconstructed. A typical CCD *format* for a compact digital camera ('point and shoot') is $6.6\,$mm$\times8.8\,$mm, with 3.6 million *pixels* each of area $4\,\mu$m$\times4\,\mu$m. Light falling on each pixel can be focused onto its photodiode by means of a microlens array. Larger CCDs are used in the digital version (DSLR) of the single lens reflex cameras described below. Such CCDs will usually be $16\,$mm$\times24\,$mm (APS-C format) in area, or $24\,$mm$\times36\,$mm (full frame) matching the film size of SLR cameras. In the latter case the CCDs would typically have 12 million $8\,\mu$m$\times8\,\mu$m pixels.

Film in traditional SLR cameras contains photosensitive crystals of silver halide which are in the range 0.1–$2\,\mu$m across. The *grain* distribution is not uniform and an equivalent pixel size, that contains roughly equal numbers of grains, is about $3\,\mu$m. The pre-existing SLR image format was $24\,$mm$\times36\,$mm and the focal length of the standard lenses was about $50\,$mm, giving a field of view $40°\times27°$. Initially, users of compact digital cameras expected a similar field of view, and so the lenses of such cameras have focal lengths of order $12\,$mm. Diffraction

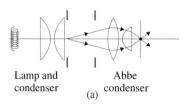
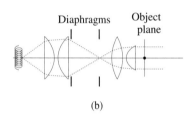

Fig. 4.10 Koehler illumination. (a) shows the imaging of the left hand diaphragm (field stop) at the object. The other diaphragm is the aperture stop. (b) shows how the source plus condenser give uniform illumination at the object.

imposes an irreducible limit on the angular resolution achievable of

$$\Delta\theta = 1.22\lambda/D,$$

where D is the lens diameter. Thus the resolution in lateral distance at the focal plane is

$$\Delta t = f\Delta\theta = 1.22\lambda f/D, \qquad (4.17)$$

which can be re-expressed either in terms of the numerical aperture

$$\Delta t = 0.61\lambda/\mathrm{NA}, \qquad (4.18)$$

or using the $f/\#$, which is the ratio of a lens' focal length to its diameter,

$$\Delta t = 1.22\lambda(f/\#). \qquad (4.19)$$

Equations 4.17, 4.18 and 4.19 are equally valid for optical systems other than a camera lens. At a typical aperture of f/8, Δt is 5 μm which is comparable to the resolution inherent in the pixel or film granularity. Lens apertures in SLR cameras can be as large as f/1.2 giving a lens diameter of around 40 mm. One factor that bears on the image resolution is the detail which the human eye can resolve, which was found above to be $\sim 3\,10^{-4}$ rad. When a photograph is viewed from a distance of 30 cm the corresponding spatial resolution will be 90 μm. Of course the eye looks at a final image of say 20 cm width rather than the CCD or 125 film. In order to reach that final image size the magnification required from a 24 mm×36 m film or CCD image is about a factor of 6. The resolution obtained on the final image with a DSLR camera having 8 μm pixels is therefore 50 μm, with the resolution possible with fine grain film being a few times better.

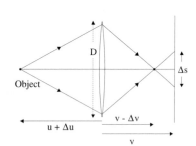

Fig. 4.11 Depth of field. A point object is displaced a distance Δu from the location at which its image is in focus. The image becomes a circle of diameter Δs.

The depth of field is defined to be the distance along the optical axis that the scene remains effectively *in focus*, that is to say the image of a point object remains smaller than one pixel. Figure 4.11 shows a camera in which the film/CCD plane is offset from the image plane. From eqn. 3.8 the relation between the axial image and object displacements is

$$\Delta u = \Delta v(u/v)^2.$$

In the case of a pinhole camera the depth of field is virtually unlimited: with a 0.5 mm hole placed 25 cm from a screen the equivalent $f/\#$ is 500.

If Δs is the pixel size then the maximum permitted image offset, Δv, in figure 4.11 is given by

$$\Delta v/v = \Delta s/D,$$

where D is the lens diameter. Combining the last two equations yields

$$\Delta u = (\Delta s/D)(u^2/v). \qquad (4.20)$$

The image plane is quite close to the second focal plane, so that to a good approximation v may be replaced by f. Then the depth of field is

$$\Delta u = (f/\#)(u/f)^2\Delta s. \qquad (4.21)$$

Taking as an example a DSLR camera ($f = 50\,\text{mm}$, $\Delta s = 8\,\mu\text{m}$) with $u = 3\,\text{m}$, and $f/\# = f/8$ gives $\Delta u = 23\,\text{cm}$. The eye tolerates a far larger defocusing than a single pixel.

Both CCDs and panchromatic film record the full visible spectrum. In the case of CCD a filter is placed over each pixel to restrict the sensitivity to either the red, green or blue parts of the spectrum. One arrangement of filters is to repeat a basic four-pixel pattern across the CCD: $\begin{smallmatrix} R & G \\ G & B \end{smallmatrix}$.[3] The sensitivity of film is denoted by an ISO number (*speed*) proportional to the film *density* (darkening) for a given exposure. At very low intensities of illumination of the scene photographed both film and CCDs record no image, and equally at very high intensities they saturate. CCDs are linear over a larger range of intensity than film. Standard film has the appellation ISO200, and a *faster* film with twice the sensitivity film has ISO400. The equivalent *speed* of a CCD depends not only on the area of the pixels but also on the level of amplification of the charge deposited. This amplification can be adjusted electronically and automatically to match light level, aperture and exposure time. The ISO range of DSLR cameras with large $8\,\mu\text{m} \times 8\,\mu\text{m}$ pixels is typically 100 to 1600. Some CCDs have arrays consisting of alternate small and large pixels, and this provides even more flexibility.

[3] This is called the Bayer colour filter array after its inventor.

Digital cameras offer the huge advantage that the image can be inspected immediately, and then retained or deleted at will. In addition facilities for electronic storage, manipulation and transmission become available once the image is transferred to a PC memory. On many digital cameras the pixel charge is digitized with 8 bits giving a scale of intensity from 1 to 2^8, that is from 1 to 256. When the data from the CCD is stored electronically further bits are needed for labelling. Thus for one million pixels of order 8 million bits (8Mb) are required, that is one million bytes (1MB). With the larger 12–20 million pixel CCDs with 12 bit resolution the storage required per picture is correspondingly larger. The full raw data from the CCD may be stored for later processing, but generally the memory requirement is reduced by factors up to 16 by preserving the data in the JPEG (Joint Photographic Expert Group) format. The content of blocks of pixels are Fourier transformed and the number of coefficients which are kept will depend on the degree to which the data is to be compressed. The rate of taking pictures with a digital camera depends on the rate at which the charges on the CCD are read out, digitized and transfered to memory. With a typical $40\,\text{MHz}$ *clock speed* the readout from a 1 million pixels takes of order 25 ms, allowing continuous shooting.

4.5.1 Camera lens design

An early lens design that has persisted in cheap cameras is the meniscus lens shown in figure 3.31. The field stop is in the natural location reduc-

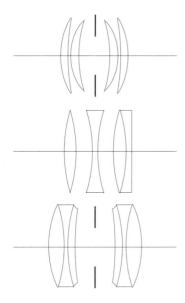

Fig. 4.12 Modern camera lens families. Reading from the top these are: Double Gaussian, Air-spaced Triplet, and Double Anastigmat

ing coma and spherical aberration and flattening the field. Distortion and astigmatism are not corrected. With a symmetric pair of lenses facing a central stop, coma, distortion and lateral chromatic aberration cancel partly or completely. Thus a step forward was to have a near-symmetric pair of achromatic doublets, retaining the meniscus shape. A near-symmetric pair of achromatic doublets of meniscus shape facing a central stop was the basis of the rapid rectilinear camera lens. Correcting astigmatism/field curvature as well as chromatic aberrations required the invention of high refractive index low dispersion crown glass and low refractive index high dispersion flint glass. The resulting cemented triplets of a flint biconcave lens sandwiched between two crown biconvexes are called *anastigmats*, that is, flat field achromats. A triplet of air spaced lenses, again with a biconcave between two biconvexes, has enough flexibility with six surfaces to permit designers to effectively remove all the aberrations simultaneously. This design is the basis a group of lenses such as the Tessar, a modern variant of which is shown in the central panel in figure 4.12 One further development of symmetric lenses overcomes higher (fifth) order spherical aberration. This is the double Gaussian triplet shown in the upper panel in figure 4.12; the individual lenses will often be achromatic doublets of meniscus shape. The final modern design shown in the lowest panel of figure 4.12 uses a symmetric pair of anastigmats.

Lens designers with a particular aim can search for an existing design in a database, and if necessary, proceed from there by iteration. Current computer based packages perform analytic calculation of lens system properties beyond the simple paraxial behaviour. These packages also perform ray tracing starting from an object point anywhere in the field of view. The rays are distributed in direction so that the points where they cross the aperture stop cover this surface in a uniform fine grid. The plot of intersections of the rays across the selected *image* plane then gives an accurate measure of the quality of the image. This data is displayed and also stored for analysis and comparison with alternative lens designs. All the parameters of the lenses (glass properties, surface curvatures, thickness, etc.) can be varied before a ray tracing sequence. The effects of including aspheric surfaces, mirrors and diffractive elements can also be modelled exactly. Many modern lens designs include one or more aspheric surfaces, and their inclusion gives a flexibility that usually results in a design with fewer lenses for the same image quality. As an example of the detailed analysis feasible with the OSLO® design suite,[4] figure 4.13 shows the final design and residual aberrations for a near-symmetric camera lens.

[4]Lambda Research Corp, 80 Taylor St., PO Box 1400, Littleton MA 01460-4400, USA.

4.5.2   SLR camera features

Single lens reflex cameras are the commonest film cameras. A sketch of the components is shown in figure 4.14. The lens has a focal length of typically 50 mm and an aperture of f/# 1.2, that shown being a double

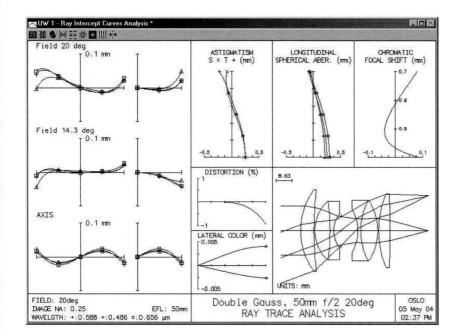

Fig. 4.13 Example of a double Gaussian lens design and residual aberrations; supplied by Dr L. Gardner, OSLO® program manager, Lambda Research Corp.

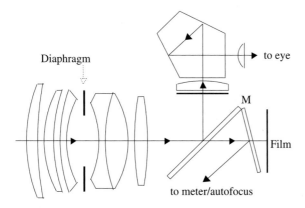

Fig. 4.14 The main optical components of a single lens reflex camera (SLR).

Gaussian design. The viewfinding, focusing and the light metering are all 'through the lens' (TTL) by means of mirrors, hinged at M, which snap out of the way just before the photograph is taken. The shutter that delimits the length of the exposure can be located close to the iris diaphragm at the field stop, and has a similar iris structure. Alternatively the shutter can be a focal plane shutter just in front of the film, in which case it is in the form of a blind with a slit that is dragged rapidly past the film. The front mirror has a central portion that is partly silvered so that some light travels to a second mirror and is reflected there to a light meter and autofocus system. Light reflected by the first mirror goes through an eyepiece and a pentaprism that restores the orientation[5] of the scene. In a DSLR camera a CCD replaces the film and the image is reproduced on a visual display for the user. Figure 4.15 shows the components of a simple autofocus system. The light passes through two well separated off axis apertures in front of a lens which forms an image on the equivalent of the film plane. Beyond this the rays fall on a transparent screen of *lenslets* (small lenses). These refocus the two cones of rays onto an array of photodiodes. If the image of the scene is in focus at the film, the photodiodes illuminated are exactly a distance t apart. When the image is not in focus on the film the separation is greater or smaller than t. The distance t is sensed electronically and the appropriate lens movement performed to regain focus.

[5] The top right hand surface of the pentaprism is roof-shaped, the ridge running parallel to the line drawn. Light travelling upward in the pentaprism first strikes the far roof surface, is then reflected towards the reader to the near roof surface. From there it follows the arrowed path element, and is reflected horizontally to exit through the rear wall. These four reflections restore the image orientation.

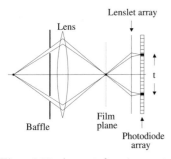

Fig. 4.15 An autofocusing system. Two ray pencils are shown which form images on the photodiode array.

4.5.3   Telecentric lenses

Telecentric lenses provide a projected rather than a perspective view of small objects and this makes them of great value in machine vision and for the monitoring of production lines. The characteristic feature of the telecentric lens design is to locate the aperture stop at the second focal plane, as shown in figure 4.16 for a single lens. I is the point image formed of the point O on some object. The rays that reach I leave O in a cone of rays that has the principal ray as its axis, and because the aperture stop is located in the focal plane the principal ray is parallel to the optical axis in the object space. Then, if the object is displaced in either direction parallel to the optical axis the path of the principal ray is unaltered. Although this movement of the object causes the image at I to go out of focus gradually, the image remains centred on I. This property is necessary in checking alignment of pins on PC boards, or of laser drilled holes. The lens aperture must be kept small enough that the depth of field over which objects remain well defined is adequate for the task. Evidently the front component of a telecentric lens has to be as wide as the area viewed at one time. A typical telecentric lens has a 5 cm diameter objective, a focal length of 5 cm and a working distance of 10–20 cm. The fully corrected telescope is some 30 cm in length; the image is formed on a CCD and the data is processed electronically. Typically the image scale varies by less than 1% for axial displacements of ± 10 mm in such instruments.

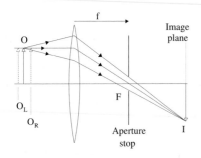

Fig. 4.16 A telecentric lens with the aperture stop located in the second focal plane.

4.5.4   Telephoto lenses

The image size of a distant object is simply its angular size multiplied by the lens focal length. In telephoto lenses a long focal length is achieved in a compact format by moving the second principal plane well out in front of the lenses. Figure 4.17 shows the basic design consisting of a positive lens placed in front of a negative lens. The positive lens focuses a distant point on the optical axis at F_+, and this image is refocused by the negative lens at I. From eqn. 3.37 we take the expression for the equivalent focal length of a pair of lenses separated by a distance t:

$$1/f = 1/f_1 + 1/f_2 - t/f_1 f_2.$$

A simple choice is to make $f_2 = -f_1$, so that:

$$f = f_1^2/t.$$

In figure 4.17 the equivalent thin lens, with its second principal plane at $P_2 P_2$, would focus the incident ray shown there along the broken line. The distance between the focal plane and the last lens surface is called the *back focal length*. Taking an infinitely distant object and using eqn. 3.17 for each lens gives this distance:

$$s_{2f} = f_1^2/t - f_1.$$

Telephoto lenses have relatively long focal lengths, so that they are prone to blurred images arising from camera-shake. In order to eliminate this weakness makers build in sensors to detect yaw (rotation about a vertical axis) and pitch (rotation about a horizontal axis perpendicular to the optical axis). Motors displace the lens elements, or the CCD, laterally to compensate for these motions with a response time of milliseconds. It also follows from eqn. 4.21 that the long focal length of the telephoto lens implies a shallow depth of field.

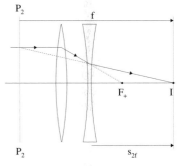

Fig. 4.17 A simple telephoto lens.

4.5.5   Zoom lenses

The zoom lens is used to achieve very dramatic effects. In a zoom the overall focal length changes and hence the magnification, while the object viewed remains continuously in sharp focus. Figure 4.18 shows the basic components of a zoom lens, some of which will in practice be achromats, multiple achromats or have aspheric surfaces. The front lens is essentially the objective that focuses the object. Behind this is a negative lens that moves axially during the zoom. Finally there are a pair of relay lenses that bring the image to a focus on film or an electronic detector. Placing the field stop between the relay lenses maintains a constant f/# during the zoom. The position of the negative lens is shown at the start (1) and end of the zoom (2), together with representative rays from the object showing that the image plane does not move. The second principal plane of the equivalent thin lens is shown before, $P_1 P_1$,

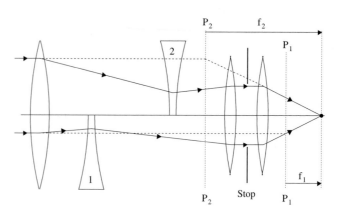

Fig. 4.18 Zoom lens showing the position of the mobile lens, the principal plane and ray paths: before the zoom below the optical axis (1), and after the zoom above the optical axis (2).

and after the zoom, P_2P_2. After the zoom the equivalent focal length is several times larger so the image is correspondingly enlarged. A focal length zoom from 200 to 1200 mm is commonly available, giving a 6:1 change in the magnification.

Zoom lenses represent the state of the art in lens design. The number of lens elements required to remove all aberrations adequately over the full zoom is rather large, 20 elements in some cases. At each air/glass or glass/air interface the fraction of incident light reflected is, as we shall show later, for light incident perpendicular to the surface

$$R = (n-1)^2/(n+1)^2, \tag{4.22}$$

where the glass has refractive index n. The resulting reflection coefficient is 0.04 (0.09) for each glass/air interface where the glass has refractive index 1.5 (1.9). Thus with 40 surfaces the fraction of the incident light transmitted would be less than $0.96^{40} \approx 0.20$ with usual lens materials. An associated problem arises from the internal reflections between lens surfaces of light from bright objects in the scene photographed. If, as is not uncommon, the brightness across the view photographed varies by a factor one hundred, then multiple reflections of the bright objects produce blobs and a background haze that dominate the picture. Therefore, where there are many surfaces in a lens system, each surface must be given anti-reflection coatings. Such coatings, which are discussed in Chapter 9, can reduce the reflection coefficient to a fraction of a percent per surface. Digital cameras have smaller focal lengths than film cameras so the difficulties of manufacturing zoom lenses are much reduced due to the lenses being physically smaller, and widespread use is now made of aspheric lenses. Nowadays excellent zoom lenses are fitted as standard on many digital cameras.

4.6   Graded index lenses

Graded index or GRIN lenses focus by refraction inside the lens, rather than by refraction at the surfaces of a conventional lenses. The glass composition and hence the refractive index changes within the lens. In a similar way, on a hot day, the differentially heated air above tarmac produces a mirage. There are two approaches: one is to use a radial refractive index gradient, the other an axial refractive index gradient. The first approach is employed in lenses which couple laser diodes to optical fibres or fibres to fibres and these lenses are usually a couple of millimetres in diameter. Because the focusing is internal the faces of graded index lenses may be simply flat, parallel surfaces.

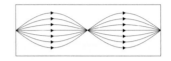

Fig. 4.19 A GRIN lens with radial refractive index gradient, of pitch 1.0. Several meridional ray paths are shown.

Figure 4.19 shows the path of rays through a GRIN lens having a radial refractive index gradient given by

$$n(r) = n(0) \sqrt{[1 - (g\,r)^2]}, \qquad (4.23)$$

where g is called the gradient factor. Meridional rays, that is to say rays lying in a plane containing the optical axis, follow sinusoidal paths along the lens with a wavelength $2\pi/g$.

From Snell's law we have

$$n(r) \cos \alpha(r) = n(r - dr) \cos \alpha(r - dr)$$
$$= .... = n(0) \cos \alpha(0), \qquad (4.24)$$

where $\alpha(r)$ is the angle the ray makes with the optical axis at radius r. At the furthest point from the axis, $r = R$, the ray is parallel to the axis and hence $n(R) = n(0) \cos \alpha(0)$. Then using eqn. 4.23, $R = \sin \alpha(0)/g$. The optical path length in one complete cycle is straightforward but tedious to evaluate

$$l = 4 \int_0^R n(r)\,dr/\sin \alpha(r)$$
$$= n(0)\,\pi\,[1 + \cos^2 \alpha(0)]\,/g, \qquad (4.25)$$

which for paraxial rays reduces to $l = 2\,n(0)\,\pi\,/g$. The optical length axially is $l = n(0)\,p$, where p is the wavelength of the sinusoids shown in figure 4.19. Hence $p = 2\,\pi/g$.

Representative parameters for a GRIN lens are: 1.8 mm diameter, n_0 around 1.6, g around 0.33/mm. A quantity called the *pitch* is defined to be the length of the lens in wavelengths. The lens shown in figure 4.19 has a pitch of 1.0 and produces at one end an image of an object placed at the other end. Similarly a lens with pitch 0.25 will focus a parallel beam. Medical endoscopes have integral pitch and can be 30 cm or more long. Some radial graded index lenses provide magnification of order ×10.

In the second class of graded index lenses the index varies linearly along the axial direction. The variation in refractive index is generally much greater than in the radially graded GRIN lenses, being as much as 0.15. Shaping the lens surfaces gives new effects. Suppose that the lens is planoconvex and that the refractive index is increasing along the optical axis, then a ray near the axis will travel through glass of on average higher refractive index than rays through the edges of the lens. This feature provides a useful way to compensate for spherical aberration.

4.7   Aspheric lenses

An alternative way to correct aberrations is to give the lens an *aspheric* (departing from spherical) profile. Such lenses have a long history and an example from the Gotland Museum in Sweden is shown in figure 4.20. Aspheric lenses are now widely made directly using computer controlled surface grinding techniques, or by casting from moulds prepared that way. Aspheric lenses capture a very wide cone of rays from a source while their surface shaping removes any spherical aberration. Planoconvex lenses of focal length 1 to 5 cm with NA as high as 0.75 are very useful as the first stages of condensers. Fully corrected lens systems can be made with fewer elements when aspheric lenses are included in the design.

4.8   Fresnel lenses

Compared with the lenses discussed above the Fresnel lens has relatively poor optical quality, but it is cheap and can be made with a very large area. Sections through a normal lens and its equivalent Fresnel lens are shown in figure 4.21. At the right of the figure one half of the Fresnel lens is seen as viewed from along the optical axis. Each annular section of the Fresnel lens has the same surface curvature as the corresponding piece of the normal lens. However the core of the normal lens has been removed from the Fresnel lens, making the latter flat in profile. Each ring of the Fresnel lens focuses in the same way as the corresponding ring in the normal lens which means that the Fresnel lens produces a similar image. However the image quality is poorer because of the light loss and the scattering at the steps between rings. Large area Fresnel lenses can be conveniently cast in plastic. They are often intended to be used at a specific object distance, and are often aspheric. Each overhead projector (OHP) uses a largish area Fresnel lens to focus the light from a substage condenser through the transparency being projected. The transparency sits directly on the Fresnel lens which concentrates the light incident into the projection lens. This makes it possible to have uniform illumination over a wide area. Fresnel lenses are used anywhere that a large area beam of moderate quality is needed, while wallet sized plastic Fresnel lenses make practical magnifiers for reading fine print, and are much lighter than glass lenses.

Fig. 4.20 Mediaeval rock crystal lens (c. AD1000) from a Viking hoard found in Sweden. Courtesy Dr. B. Lingelbach, Aalen Technical University.

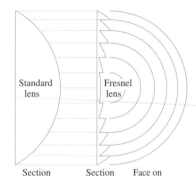

Fig. 4.21 A normal lens and the equivalent Fresnel lens. On the left are sections which contain the optical axis in each case; on the right, one half of the Fresnel lens is seen as viewed along the optical axis.

Exercises

(4.1) Show that the Galilean telescope of figure 4.2 produces angular magnification $M_\theta = f_o/f_e$. What is the factor by which this telescope expands a laser beam? In a Galilean telescope the objective has focal length 16 cm and diameter 44 mm, and the eyelens −2 cm focal length and diameter 10 mm. What is the angular magnification? What is the position and diameter of the image of the objective which is formed by the eyelens? In the complete optical instrument made up of this Galilean telescope and the observer's eye what is the aperture stop? Assume the eye pupil has diameter 5 mm and is 10 mm from the eyelens.

(4.2) A microscope has an objective of 2 mm diameter and focal length 10 mm, an eyelens of diameter 15 mm and focal length 25 mm, with tube length 160 mm. Calculate the magnification of each stage of the microscope. Determine the position and size of the exit pupil. What is the diameter of the field of view?

(4.3) Whereas a normal eye when fully relaxed brings objects at infinity into focus, a nearsighted person's fully relaxed eye brings objects much closer into focus. Suppose this *far point* is at 3 m, what focal length lens is needed to correct this nearsightedness?

(4.4) A person with long sight has a near point at 1 m rather than the usual 25 cm. What focal length lens will correct this?

(4.5) In an optical illusion two thin identically curved concave mirrors are placed on a table one on top of the other, the lower one facing up and the upper facing down with their rims in contact. The mirrors are some 20 cm across and the upper one has a central hole 2 cm in diameter. Viewers see a coin resting on the top of the upper mirror at its centre and attempt to pick it up. In fact the coin is resting on the centre of the lower mirror. How is this illusion achieved?

(4.6) What arrangement of two identical positive lenses will invert an image while leaving its linear size unchanged?

(4.7) Using the same approach as in Section 4.5 to show that if a camera is focused at infinity then the nearest point in focus is at $u = Df/\Delta s$, where Δs is the pixel size of the detector, f the focal length of the lens and D the lens diameter. For a camera with f/# of 1.4, focal length 12 mm and pixel size 4 μm, what is u?

(4.8) An eyepiece is constructed from two identical planoconvex lenses with curved surfaces facing and separated by $2f/3$ where f is the focal length of either lens. What is the focal length of the eyepiece and where are the principal planes? Which lens is the field stop?

(4.9) A perfect oil immersion objective of the type shown in figure 4.8 accepts rays out to 90° off axis. The refractive index of the spherical lens and the oil is 1.4. What is the semi-angle of the cone of rays that the next lens must be designed to accept.

(4.10) Show that the NA of a GRIN lens of radius R is $gRn(0)$ using the notation of Section 4.6.

Interference effects and interferometers

5

5.1 Introduction

The development and applications of the classical wave theory of light form the content of this and the succeeding six chapters. That light waves exist became accepted following Young's observation, early in the nineteenth century, of interference between light emerging from a pair of slits both illuminated by the same monochromatic source. As mentioned in the first chapter the wave theory of electromagnetic radiation, including light, was put on a sound theoretical basis around 1864 by Maxwell. It emerged that ray optics works in everyday situations because the wavelength of light is very small compared to everyday objects.

Electromagnetic fields are vectors and add together in the same simple way as force vectors. This fundamental property is called the *superposition principle*. It will be used in this chapter to explain interference effects in which a wave is divided up, the two parts travel different paths, and these are then superposed. Young's double slit experiment and Michelson's interferometer illustrate the two possibile procedures: respectively the division of the wavefront by apertures, and division of the wave amplitude by partially reflecting mirrors. The analysis of interference patterns using complex amplitudes and *phasor* diagrams are introduced at this point. A standard simplification made in this and the following chapters is to drop unnecessary constants when calculating intensities where it is the variation in intensity which is of interest: for example when the electric field is E the intensity is taken to be E^2.

Whether interference effects are seen or not with a given apparatus depends on the *coherence* of the wavetrains being superposed. Coherence is treated at length in this chapter, and much use is made of wavepackets in discussing coherence.

Practical applications of interference in devices used to measure wavelengths, distances, velocities and angular velocities are described in this chapter. Multiple beam interference produced with Fabry–Perot etalons is introduced, and is shown to give enhanced resolution and precision in wavelength measurement.

5.2   The superposition principle

If electromagnetic radiation from several sources is incident on any given point, the total electric field there is simply the vector sum of the electric fields produced at that point by each source acting alone

$$\mathbf{E} = \mathbf{E_1} + \mathbf{E_2} + \text{....} \qquad (5.1)$$

Equally, adding the individual magnetic fields vectorially gives the overall magnetic field. These statements go by the name of the superposition principle. When the point of interest is located in matter (rather than in free space) there can be differences betweeen the physical effect produced by the fields of each source alone and by the total field. For example two lasers illuminating a metal surface may melt the metal, whereas the individual beams leave it intact. Again we might choose a laser whose light can cause atoms in a gas to be excited to a level A and a second laser whose light can cause a transition of atoms from level A to a higher level B. Neither laser alone could access atomic state B but together they can do so.

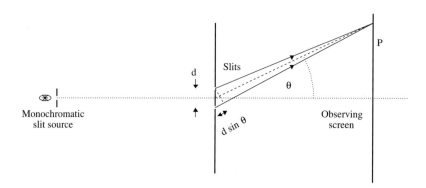

Fig. 5.1 Young's two slit interference experiment. The distances from the slits to the source and from the slits to the screen are very large compared to the slit separation. A typical slit separation might be 1 mm.

5.3   Young's two slit experiment

Young's apparatus is shown in figure 5.1. A monochromatic slit source illuminates a pair of slits separated by a small distance d, and beyond the slits is a screen. These slits are located symmetrically about the axis so the waves arriving at each of them are in phase. Young observed bright and dark fringes on the screen, and he realized that these fringes were due to *interference* between light from the two slits.

Figure 5.2 shows an analogous situation in which waves are produced on a water tank by a pair of closely spaced plungers which move at the same frequency and in phase. Stars locate the plungers on the figure. In some directions the waves from the sources are in step and give large amplitude waves which show up as the alternating white peaks and black troughs: the waves *interfere constructively*. In other directions the waves arrive out of step and interfere *destructively* leaving the surface undisturbed, which then appears grey in the figure. Following Young we now appreciate that the electromagnetic fields in light from the two slits in figure 5.1 interfere in a similar way.

The fields in a light wave oscillate at frequencies of $\sim 10^{14}$ Hz; frequencies which neither the eye nor any detector can follow. Rather they respond to the time average of the intensity, which is proportional to the square of the electric field. Thus the detected intensity of light whose electric fields are like those displayed in figure 5.2 would appear as shown in figure 5.3. The regions of constructive interference, shown white, would be brightly illuminated: and the regions of destructive interference, shown black, would be in darkness. A screen placed anywhere in front of the slits would be covered in bright and dark interference fringes, so that these are called *non-localized* fringes.

5.3.1 Fresnel's analysis

Fresnel, shortly after Young's observations, used the superposition principle to add the Huygens' waves from apertures in an obstructed light beam. The secondary waves are spherical and have an electric field

$$E = V_0 \cos{(kr - \omega t)}/r, \qquad (5.2)$$

where the angular frequency ω and wave vector amplitude k are the same as for the incident wave, and the distance and time are measured from the origin of the secondary wave on the apertures. The factor V_0 depends on the incident wave amplitude, and the factor $1/r$ ensures that the total power radiated remains constant with the distance from the source. At any point beyond the apertures the total electric field is the sum of the secondary wave electric fields. When the path lengths of secondary waves to the point of observation are different there is a phase difference and this is the origin of interference effects such as that seen by Young. Adding Huygens wavelets from a long slit produces a cylindrical wave: because the wavefronts are cylindrical, path lengths from the slit must be measured in the plane of figure 5.1. In the present case, as in may others, the path lengths of the interfering waves are almost equal, so that the effect of the factor $1/r$ in eqn. 5.2 is an overall constant multiplying the total amplitude.

Each slit in figure 5.1 is a source of Huygens waves and their waves arriving at the point P will have travelled distances differing by $d \sin \theta$. When this distance is an integral number of wavelengths m, the waves

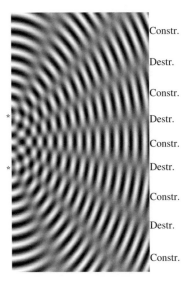

Fig. 5.2 Surface water waves produced by sources at the two stars. The bands of constructive and destructive interference are labelled.

Fig. 5.3 Time averaged intensity pattern computed for electromagnetic waves with wavelength identical to those in figure 5.2. White indicates brightness, black darkness.

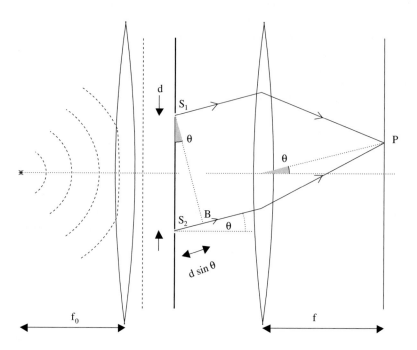

Fig. 5.4 A compact arrangement of Young's two slit interference experiment. The point source is at the focal point of one converging lens. The observing plane is in the focal plane of a second converging lens. Incoming wavefronts are shown with broken lines.

arrive in phase at P and interfere constructively,

$$d\sin\theta = m\lambda, \tag{5.3}$$

giving a central bright fringe at which m is zero. Destructive interference occurs when

$$d\sin\theta = (m + 1/2)\lambda. \tag{5.4}$$

At small angles the separation between adjacent bright fringes is

$$\Delta\theta = \lambda/d, \tag{5.5}$$

hence the ratio d/λ must be kept small enough so that the fringes can be seen by eye. The above simple analysis applies provided that the apertures are all narrow and the distance of the slits from source and screen are large. If the source slit is so wide that the path length from different points on the source to a slit varies by more than a fraction of a wavelength the interference pattern will be blurred. If the two slits are themselves broad, then the pattern of interference becomes more complicated; something which is investigated in the following chapter.

Figure 5.4 shows a compact setup with the source and screen lying at the focal planes of positive lenses. The light from the source arrives as a plane wave normally incident at the slits, and hence in phase. In addition light leaving the two slits in a direction making an angle θ to the optical axis will arrive at the point P having travelled paths differing by $d \sin \theta$. This conclusion is based on the useful fact that all points on the inclined plane wavefront at $S_1 B$ are the same *optical* distance from the image point P in the focal plane. P lies a distance $f \tan \theta$ off axis where f is the focal length of the second lens. Then the electric fields produced at P by radiation from the two slits are

$$E_1 = E_0 \cos (-\omega t); \quad E_2 = E_0 \cos (-\omega t + \phi).$$

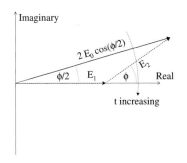

Fig. 5.5 Phasor diagram for Young's two slit experiment.

Light must leave S_2 earlier than light leaves S_1 in order that they both reach P at the same time, making ϕ positive. Another general point to recall here is that a path difference of one wavelength produces a phase difference of 2π, so that

$$\phi = 2\pi d \sin \theta / \lambda = k\, d \sin \theta.$$

In order to manipulate the mathematics pictorially the complex forms of the field are used

$$E_1 = E_0 \exp [-i\omega t]; \quad E_2 = E_0 \exp [-i(\omega t - \phi)].$$

These complex amplitudes are shown on a *phasor* diagram in figure 5.5; that is an Argand diagram taken at the moment when E_1 lies along the real axis. The superposition principle requires we add the fields vectorially giving a resultant, which at time $t = 0$ is

$$E = 2E_0 \cos (\phi/2) \exp (i\phi/2); \tag{5.6}$$

or at any time

$$E = 2E_0 \cos (\phi/2) \exp [-i(\omega t - \phi/2)],$$

with real part

$$E = 2E_0 \cos (\phi/2) \cos (\omega t - \phi/2). \tag{5.7}$$

Hence the intensity

$$I = 4E_0^2 \cos^2 (\phi/2) \cos^2 (\omega t - \phi/2). \tag{5.8}$$

Detectors record the average intensity taken over many full cycles of total duration T. Their response is thus[1]

$$\overline{I} = 4E_0^2 \int_0^T \cos^2 (\phi/2) \cos^2 (\omega t - \phi/2)\, \mathrm{d}t/T$$
$$= 2E_0^2 \cos^2 (\phi/2) = 2E_0^2 \cos^2 (\pi d \sin \theta / \lambda). \tag{5.9}$$

---

[1] In detail

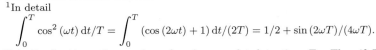

$$\int_0^T \cos^2 (\omega t)\, \mathrm{d}t/T = \int_0^T (\cos (2\omega t) + 1)\, \mathrm{d}t/(2T) = 1/2 + \sin (2\omega T)/(4\omega T).$$

$4\omega T$ is the 8π times the number of cycles completed in time T. Thus if T is the response time of a detector ($> 1\,\mathrm{ns}$) and ω is an optical angular frequency ($\sim 10^{15}\mathrm{rad\,s}^{-1}$) the second term is negligible, making the integral equal to $1/2$.

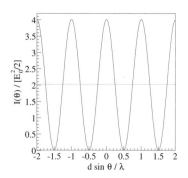

Fig. 5.6 Intensity in Young's two slit experiment as a function of the direction off axis θ, expressed as a multiple of the single slit intensity. The dotted line shows the mean intensity. λ is the wavelength and d the slit separation.

This variation is shown in figure 5.6 as a function of the angle off axis. The peak intensity is four times the time averaged intensity ($E_0^2/2$) produced by a single slit. Averaging over a complete spatial cycle of the fringe pattern gives

$$I_{\text{average}} = 2 \int_0^{2\pi} E_0^2 \cos^2(\phi/2)\, \mathrm{d}\phi/(2\pi) = E_0^2,$$

so the average intensity over the fringe pattern is the same as the sum of the intensities for two independent slits. No light is lost or gained, it is simply redistributed.

Time averaging for monochromatic waves

It will often be useful to use complex waves, making the replacement

$$F = f \cos(kz - \omega t + \phi) \Rightarrow F_c = f \exp\left[i(kz - \omega t + \phi)\right], \qquad (5.10)$$

where F is any actual electric or magnetic field. The only measurable quantities in optics are intensities and fluxes, both being products of two fields FG averaged over the time of response of the detector. In the case of monochromatic waves

$$FG = f \cos(kz - \omega t + \phi) g \cos(kz - \omega t + \psi)$$
$$= (fg/2)\left[\cos(2kz - 2\omega t + \phi + \psi) + \cos\alpha\right],$$

where $\alpha = \phi - \psi$. The time average over many optical periods is

$$\overline{FG} = (fg/2)\cos\alpha.$$

Compare this result with the product of the *instantaneous* complex fields

$$F_c G_c^* = fg \exp i\alpha,$$

and it is seen that

$$\overline{FG} = \mathcal{R}e\left[F_c G_c^*\right]/2. \qquad (5.11)$$

Hence for intensities

$$\overline{E^2} = \mathcal{R}e\left[E_c E_c^*\right]/2 = E_c E_c^*/2, \qquad (5.12)$$

while for energy flow the time average of the Poynting vector is

$$\overline{\mathbf{E} \wedge \mathbf{H}} = \mathcal{R}e\left[\mathbf{E}_c \wedge \mathbf{H}_c^*\right]/2. \qquad (5.13)$$

Visibility

For simplicity the amplitudes due to the two slits have been assumed to be equal. If they are unequal the cancellation when the interference is destructive will not be complete. Suppose the amplitudes are E_{01} and E_{02} then the total electric field is

$$E = E_{01} \cos\omega t + E_{02} \cos(\omega t - \phi),$$

and the instantaneous intensity is

$$I = E_{01}^2 \cos^2 \omega t + E_{02}^2 \cos^2(\omega t - \phi) + 2E_{02}E_{01}\cos\omega t \cos(\omega t - \phi)$$
$$= E_{01}^2 \cos^2 \omega t + E_{02}^2 \cos^2(\omega t - \phi) + E_{02}E_{01}[\cos(2\omega t + \phi) + \cos\phi].$$

The time average of this intensity is

$$\bar{I} = E_{01}^2/2 + E_{02}^2/2 + E_{02}E_{01}\cos\phi,$$

with a minimum that does not fall to zero. A quantity called the *visibility* is defined which expresses the degree of cancellation

$$V = (I_{\max} - I_{\min})/(I_{\max} + I_{\min}), \qquad (5.14)$$

where $I_{\max}$ and $I_{\min}$ are the observed maximum and minimum intensities respectively. In the case considered

$$V = \frac{2E_{01}E_{02}}{E_{01}^2 + E_{02}^2}.$$

The visibility defined in this way has the overall intensity normalized away so that its value is limited to the range between 0 to 1. Figure 5.7 shows two sets of fringes with visibilities 1.0 amd 0.5. Fringes with unit visibility are clearcut, while low visibility fringes are harder to detect because of the poor contrast between the maxima and minima of intensity.

5.3.2   Interference by amplitude division

In Young's two slit experiment *wavefront* division produces interfering beams. The alternative process is *amplitude* division in which a partially reflecting surface divides the light into reflected and transmitted beams which are subsequently both directed onto a surface where the interference pattern is observed. First consider the simple case of monochromatic illumination. Figure 5.8 shows an arrangement in which part of the incident light is reflected by a pair of partially reflecting surfaces inclined at a small angle to one another. Interference will only be seen if the two beams can be focused on the same region of the viewer's retina. This requires that the eye is focused as shown in the diagram, at a specific depth below the mirrors. Therefore the fringes are said to be *localized*, in contrast to Young's fringes which can be detected everywhere in front of the slits. Of course fringes can only be seen if light is incident from the direction that reflects into the viewer's eye. A broad diffuse source is needed if fringes are to be seen over a broad angular range. In the familiar case of white light from the sky falling on an oil slick the direction of the bright fringes will be different for different wavelengths so that coloured bands are seen apparently lying on the oil.

When viewing the surfaces near normal incidence bright fringes of constructive interference are visible if the optical path difference between

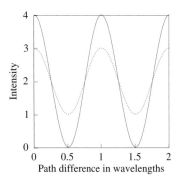

Fig. 5.7 Intensity distributions for fringes with visibility 1.0 (solid line) and visibility 0.5 (broken line).

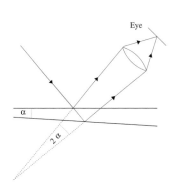

Fig. 5.8 Construction to show the localization of the interference between reflections from a pair of plane reflecting surfaces inclined at a small angle α.

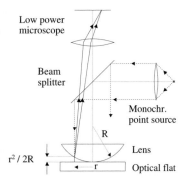

Fig. 5.9 Setup for observing Newton's rings. Reflections for the same incident ray path from the lens lower surface and optical flat upper surface are drawn as solid lines.

reflections is an integral number of vacuum wavelengths ($m\lambda$). Between consecutive fringes the path difference changes by λ and the spacing of the reflecting surfaces by $\lambda/2$. Such fringes form a contour map of regions of equal thickness. They are called fringes of *constant thickness* and also *Fizeau* fringes. *Newton's rings* are circular fringes of equal thickness seen when a lens is placed on top of an optical flat: interference occurs between the reflections from the adjacent surfaces of the lens and the optical flat. The experimental setup is illustrated in figure 5.9 with a planoconvex lens whose curved face has a radius of curvature R. At a radial distance r from the point of contact, the sagitta of the lens is $r^2/2R$. Light incident there parallel to the optical axis will be reflected either at the lens or at the flat and then arrive at the microscope; the two paths differ by an amount r^2/R. An unexpected observation is that the centre of the pattern where the physical path difference is zero appears dark rather than bright. This comes about because a reflection from an optically denser medium (here the air/optical flat interface) at normal incidence produces a phase change differing by π from the phase change for a reflection from a less dense medium (here the lens/air interface). Thus bright fringes are seen when

$$r^2/R = (m + 1/2)\lambda. \tag{5.15}$$

The quadratic dependence on r means that the fringes get more tightly packed the further one moves from the optical axis. Departures from fringe circularity indicate an imperfection whose importance can be estimated using the above equation. Viewing Newton's rings provides a quick practical test of lens quality.

When two flat reflecting surfaces are aligned parallel as sketched in figure 5.10 the fringes are now found to be localized at infinity: with the eye relaxed the two reflections arrive at the same point on the retina. AC is drawn perpendicular to the reflected rays and the difference in path length between the two reflections is A$'$C. The two interfaces are now taken to be identical so that the phase changes are the same at each reflection. A bright fringe is seen when the path difference is a whole number of wavelengths

$$2nd\cos\theta = m\lambda, \tag{5.16}$$

θ being the angle between the direction the viewer is looking and the perpendicular to the surfaces and n is the refractive index of the medium. If an extended diffuse source is used then each bright fringe seen by the viewer will extend to form a circle subtending a semi-angle θ with respect to the surface normal. These fringes are called fringes of *equal inclination* or *Haidinger* fringes, and they are *localized* at infinity. A positive lens can be used to project the fringes onto a screen or photodetectors. Notice that the fringe order m is very large for even a 1 mm spacing of the mirrors and that m is largest at the *centre* of the pattern.

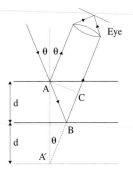

Fig. 5.10 Fringes localized at infinity from interference between *similar* parallel reflecting surfaces.

If the source in figure 5.10 were pointlike its two images would, like Young's slits, produce non-localized fringes. It is worth asking what it is about a broad source that suppresses these non-localized fringes. First note that different regions of the source produce fringe patterns which are displaced from one another. Then note that light from these different regions is incoherent, hence what is seen is the sum of the intensities of the fringes from all regions of the source. The overall effect with a broad source is therefore uniform illumination – with one exception. The one exception is illustrated in figure 5.10 where the eye is focused at an infinite distance. In that case each area of the source gives a bright fringe in the direction satisfying eqn. 5.16 and so the fringes from all areas match exactly.

5.4 Michelson's interferometer

This interferometer pictured in figure 5.11 was designed by Michelson to produce equal inclination fringes. With it he and Morley made measurements that underpinned the theory of special relativity. Modern versions of this interferometer are widely used in spectroscopy, especially in the infrared. The source must be both broad and diffuse; a lamp positioned

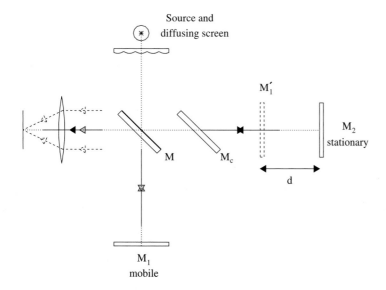

Fig. 5.11 Michelson interferometer. The two paths are indicated by grey and black arrowheads. d is the distance of M_1' (M_1's image in M) from M_2.

behind a ground glass screen is normally adequate. Light from the broad diffuse source undergoes amplitude division at a beam splitting mirror M which reflects 50%, and transmits 50% of the incident light. One beam is reflected from the movable mirror M_1, and the other from the fixed mirror M_2. The beams returning after these reflections are recombined by M and focused by the positive lens onto a detector. The lens and detector could be an observer's eye. M_c is a glass plate cut from the same sheet as M; its inclusion equalizes the thickness of glass traversed by the two beams and hence eliminates any chromatic dispersion.

The detector receives light directly from M_2 and also from the image M_1' of M_1 formed by M. When M_1' and M_2 are parallel (M_1 and M_2 perpendicular) there is a circular pattern of fringes of equal inclination illustrated in figure 5.12; the semi-angle θ of any bright fringe satisfying eqn. 5.16 with d now being the separation of M_1' from M_2. One useful feature of the Michelson is that having a virtual mirror M_1' the path difference can be set to zero. A significant advantage of the Michelson interferometer is that light from all points on the broad source contributes to the fringe formation, a point explained in the last paragraph in the last section. This makes for fringes far brighter, and easier to use, than those obtained with wavefront division.

When the mirror M_1 is moved to increase the path difference eqn. 5.16 tells us that a fringe corresponding to given value of m will expand (θ increases) and fresh fringes enter at the centre of the pattern. Between the appearances of successive fringes the mirror M_1 moves exactly one half wavelength. This simple fact is the basis for measuring mechanical movement directly in terms of wavelengths of atomic transitions, which themselves are determined solely by the laws and constants of atomic physics. Macroscopic lengths can therefore be expressed in fundamental, reproducible units. This is of course intellectually satisfying, but is now *essential* to many modern industries.

The alignment of M_2 parallel to M_1' is made using monochromatic light. A piece of wire is hooked over the ground glass diffuser and M_2 is rotated until the two reflected images of this wire coincide. At the moment they do so the fringes appear and are generally straight fringes of equal thickness because there is some small remaining tilt between the mirrors. Further delicate rotation of M_2 will remove the tilt and give circular fringes. Using eqn. 5.16 the reader may like to show that the angular spacing between adjacent fringes increases as the gap between M_1' and M_2 is reduced; which suggests a way of bringing M_1' into coincidence with M_2. When M_1' approaches coincidence with M_2 the fringes become very broad and any departures from flatness in the mirrors causes the fringes to lose their circular shape. At zero separation light across the whole spectrum is in step and around this setting a white light source produces a few brightly coloured fringes.

Fig. 5.12 Fringes of equal inclination seen with Michelson interferometer when the mirror M_2 and virtual mirror M_1' are parallel.

5.4.1   The constancy of c

In the late nineteenth century scientists imagined that light travelled as waves on an otherwise undetectable aether that existed everywhere and through which matter moved without much affecting the aether. In modern terms the aether would be an absolute frame with respect to which light would have a constant velocity. If the Earth's velocity relative to the aether were $\mathbf{v}$ the measured velocity of light on Earth would be changed by $-\mathbf{v}$. Michelson and Morley attempted to detect evidence for the Earth's motion through the aether using the Michelson interferometer.

Suppose such an interferometer with arms of equal length L moves with velocity v with respect to the aether in a direction parallel to the length of the M_2 mirror arm. As shown in figure 5.13 the length of the return path starting from mirror M via M_2 is $(L+D)+(L-D)=2L$, where $2D$ is the total displacement of M_2 through the aether during the time the light is travelling to and fro. The return path via M_1 is $2\sqrt{L^2+D^2}$ which is approximately $2L+D^2/L$ because v is much smaller than c. The difference between the two paths is thus D^2/L. This difference will reverse if the interferometer is rotated through $90°$ so that the motion is now along the direction of the M_1 mirror arm. From this argument it would follow that there would be a fringe shift on rotating the interferometer of

$$\Delta m = (2D^2/L)/\lambda. \qquad (5.17)$$

Now

$$D/L = v/c,$$

and substituting for D/L in the previous equation gives

$$\Delta m = (2L/\lambda)(v/c)^2. \qquad (5.18)$$

Taking the actual values used, $L = 10\,\text{m}$ and $\lambda = 500\,\text{nm}$, gives

$$\Delta m = 4\,10^7\,(v/c)^2.$$

At the time of the measurement it was understood that the Earth's orbital velocity was $30\,\text{km}\,\text{s}^{-1}$ so that Δm would have been 0.2, and if one uses instead the solar system's $600\,\text{km}\,\text{s}^{-1}$ velocity relative to the cosmic microwave background this shift should be twenty times larger. Michelson was able to exclude a fringe shift greater than 10^{-4} of a fringe. According to the first postulate of the special theory of relativity electromagnetic radiation travels in free space at the same velocity in all inertial frames, that is to say in frames that have constant relative velocity. This neatly explains the equality of travel times irrespective of how fast the source or observer move. The concept of an aether is therefore seen to be an error.

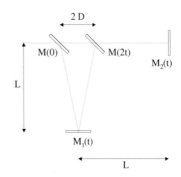

Fig. 5.13 Michelson–Morley experiment showing the position of the mirrors at the specified times in the conjectured aether.

5.5 Coherence and wavepackets

In the above introduction to interference effects pure sinusoidal, that is monochromatic waves with identical polarization were the rule; after division the wave trains acquired a phase difference dependent only on their path difference to the point where they interfere. When waves have identical wavelength and a fixed phase difference they are said to be *fully coherent*. This ideal situation is only well approximated by laser beams. If such a fully coherent source were available the interference pattern would be present however large the mirror separation in Michelson's interferometer, simply getting weaker as the light intensity from the far mirror faded away. In practice when a standard laboratory monochromatic source is employed the interference pattern disappears at a mirror separation of only a few centimetres. Such sources give *partially* coherent beams. A white light source is much less coherent than a laboratory monochromatic source: the fringes disappear if mirror M_1 moves only a wavelength or so from the null position. The coherence of a beam depends on the form of the wavetrain of radiation produced by the source and this connection will now be investigated.

The wavetrain from any source is made up of very large numbers of wavepackets emitted by individual atoms or molecules. The electric field distribution in a wavepacket will have a form similar to that shown in figure 1.6(c).[2] A one watt torch bulb emits around 10^{18} wavepackets per second. Laboratory reference sources containing a chemically pure gas at low pressure have a spectrum consisting of *spectral lines*, so that the radiation is confined to a few narrow wavelength intervals. The wavepackets corresponding to these spectral lines are typically of duration 10^{-10} s and the number of oscillations in a wavepacket is of order 10^5 rather than the handful shown in figure 1.6(c).

[2] When the quantum nature of electromagnetic radiation is met it will emerge that the wavepackets describe the position and motion of the quanta of radiation. Wavepackets are being presented here with an eye to this basic connection. A crude translation into quantum language would replace the word wavepacket by the word photon. However wavepackets can contain many photons, and in laser beams usually do so.

Over a time interval short compared to the wavepacket duration the electric field from the ith radiating atom would be approximately

$$E_i = E_0 \cos\left(\omega t + \phi_i\right), \tag{5.19}$$

where for simplicity the amplitude E_0, the angular frequency ω and the wavepacket duration are taken to be the same for all the atoms. Each atom radiates independently and so the phase ϕ_i is quite random. Summing the electric fields from all the radiating atoms gives

$$E = E_0 \sum_i \cos\left(\omega t + \phi_i\right)$$

$$= E_0 \left[\cos\omega t \sum_i \cos\phi_i - \sin\omega t \sum_i \sin\phi_i\right]$$

$$= E_0\,\xi\,\cos\left(\omega t + \beta\right).$$

In any other short interval different atoms would be emitting radiation and the factors ξ and β would have new values. Thus the electric field

$$E = E_0\,\xi(t)\cos\left[\omega t + \beta(t)\right] \tag{5.20}$$

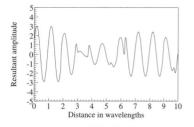

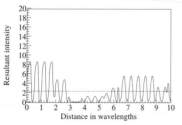

Fig. 5.14 The electric field and intensity produced when there are ten randomly phased wavepackets of the same wavelength and of unit magnitude, and each about five wavelengths long.

with the magnitude $\xi(t)$ and phase $\beta(t)$, varying with time. It will be the case that over an interval, taken at some arbitrary time t_1, and short compared to the duration of a wavepacket the wavetrain is approximately sinusoidal. If one then chooses another short section of the wavetrain at a much earlier or later time t_2, this wave section too is approximately sinusoidal. However the phase between the two sections is quite random because the time interval is much longer than the wavepacket length: the relative phase will not, except by chance, equal $\omega|t_1 - t_2|$. Figure 5.14 illustrates this, but the reader should bear in mind that typical laboratory sources emit wavepackets that are 10^5 waves long, rather than the few waves shown here. The overall electric field and the intensity are seen instantaneously over a test region ten wavelengths long produced by a wavetrain containing ten wavepackets at any given time, each wavepacket of length five wavelengths, and all having random phases with respect to each other. As time passes this pattern will continually change. The horizontal line on the intensity plot indicates the long term mean intensity. Segments of a wavetrain at instants separated by an interval short compared to the wavepacket duration are fully coherent. However segments separated by intervals long compared to the wavepacket duration have sometimes one phase, sometimes another. Thus the fringes produced by a Michelson interferometer have good contrast when the path difference between the arms is nearly zero but will fade at large path differences to leave finally a uniformly illuminated field of view.

The total instantaneous sum of the two beams at the detector in the interferometer is, using eqn. 5.20,

$$E = E_0 \cos[\omega t + \beta(t)] + E_0 \cos[\omega(t+s) + \beta(t+s)],$$

where s is the time delay between the arms and for simplicity, and without loss of generality in the result, $\xi(t)$ is taken to be unity. The instantaneous intensity is then

$$
\begin{aligned}
I &= E_0^2\{\cos^2[\omega t + \beta(t)] + \cos^2[\omega t + \omega s + \beta(t+s)] \\
&\quad + 2\cos[\omega t + \beta(t)]\cos[\omega t + \omega s + \beta(t+s)]\} \\
&= E_0^2\{\cos^2[\omega t + \beta(t)] + \cos^2[\omega t + \omega s + \beta(t+s)] \\
&\quad + \cos[2\omega t + \omega s + \beta(t) + \beta(t+s)] + \cos[\omega s + \beta(t+s) - \beta(t)]\}.
\end{aligned}
$$

The time average of this is

$$\bar{I} = E_0^2 + E_0^2 \int_0^T \cos[\omega s + \beta(t+s) - \beta(t)]\,\mathrm{d}t/T,$$

where $E_0^2/2$ is the mean intensity of either beam alone. If the beams are coherent so that $\beta(t) = \beta(t+s)$ then this becomes

$$\bar{I} = E_0^2 + E_0^2 \cos(\omega s),$$

which shows maximum interference as s is varied, going from 0 to $2E_0^2$. In the case of incoherent beams $\beta(t) - \beta(t+s)$ varies randomly and the integral vanishes leaving

$$\bar{I} = E_0^2,$$

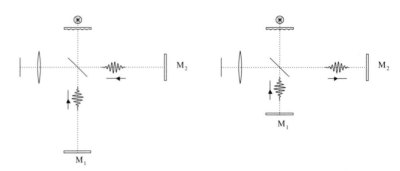

Fig. 5.15 Segments of one wavepacket travelling in the Michelson interferometer. In the left hand panel the arms are nearly equal in length, while in the right hand panel they are of very different length.

which is simply the sum of the individual intensities. These results bring out the key feature that for incoherent beams the time average intensity, which is what is detected, is just the sum of the individual intensities. However maximal interference is seen in the time average of coherent beams.

The above analysis releases us from having to tediously sum the electric fields for incoherent sources, calculate the intensity and time average – only to find that the result is always the sum of intensities. It justifies the simple rubric: with coherent sources add the amplitudes, but with incoherent sources add the intensities.

It helps in understanding the nature of coherence to consider individual wavepackets passing through a Michelson interferometer as shown in figure 5.15. In the left hand diagram the path difference d is made much smaller than the length of a wavepacket. After the reflection the segments of the wavepacket arrive in coincidence at the detector or observer's eye with a phase difference $2\pi d/\lambda$ at the centre of the field of view. This will be equally true for all wavepackets. Consequently they all contribute to an identical interference pattern and this is what is seen. The right hand diagram shows the opposite extreme in which the path difference is much greater than the wavepacket length. The two segments of any wavepacket arrive at the detector at different times so they do not overlap and so interference between them is impossible. When the wavepacket segment that took the longer path arrives at the detector it can coincide with part of an entirely different wavepacket that was emitted later by the source and took the shorter path. Such pairs of parts have quite random phase; some will interfere constructively and just as many will interfere destructively. Averaged over the response time of any detector the overall effect is a uniform illumination. The path difference over which interference can be observable is called the *coherence length* L_c for the radiation, and evidently this is simply the length of a wavepacket. The corresponding *coherence time* is $\tau_c = L_c/c$.

A useful indication of the degree of coherence of interfering beams is given by the visibility of the fringes observed. When the difference between the lengths of the arms of the Michelson interferometer is increased from zero the fringe visibility, defined by eqn. 5.14, declines from unity when the beams are fully coherent to zero when the path difference exceeds the coherence length and a uniform intensity is seen.

Two physically identical sources produce incoherent beams because the atoms in the two are emitting wavepackets quite randomly. This is equally true of the different parts of a single source. Therefore periods of time when the interference happens to be constructive between them will be matched by equal intervals of destructive interference. Therefore, averaging over the response time of any detector, a uniform intensity results. Intensities rather than amplitudes are added. Lasers are the exception to the rule that the atoms in a source emit photons with random phases. In Chapter 14 the lasing mechanism will be described in detail. For the moment it is sufficient to appreciate that in lasers the phase of the radiation emitted is locked to that of the existing radiation in the laser. This produces wavepackets that are almost pure sinusoidal waves over relatively long periods of time. Lasers are therefore capable of producing beams of radiation that are coherent over correspondingly long times; a coherence time as long as a millisecond is readily achieved. This is a big jump in coherence time compared to what other sources could offer in the infrared, visible and near ultraviolet. Many novel research and manufacturing possibilities were opened up by this increase in coherence time and length, and the accompanying increase in radiance.

5.5.1 The frequency content of wavepackets

Any wavepacket can be duplicated with a sum of monochromatic waves at frequencies around the mean frequency of the radiation. The technique of Fourier analysis which is used in resolving a wavepacket into these frequency components and in extracting their magnitudes is a key topic in Chapter 7. Here simple arguments will be made which relate the spread of frequencies, Δf, to the duration of a wavepacket, Δt. Suppose that the wavepacket is the sum of waves with frequencies $f = \omega/2\pi$ and wave vectors $k = \omega/c$

$$W(x,t) = \int A_f \cos(kx - 2\pi f t + \phi_f) \, \mathrm{d}f. \qquad (5.21)$$

Without any loss of generality we can take the peak of the wavepacket to be the origin ($x = 0$). At this point, the peak of the wavepacket, the contributing waves are all in phase so we can put $\phi_f = 0$ also. Then

$$W(0,t) = \int A_f \cos(2\pi f t) \, \mathrm{d}f. \qquad (5.22)$$

As time passes the contributing waves at the origin gradually slip out of phase with each other and the wavepacket electric field at the origin falls, i.e. the wavepacket moves elsewhere. If $f_{\max}$ and $f_{\min}$ are the maximum and minimum frequencies in the wavepacket then the waves at the origin cancel one another after a time $\Delta t/2$ such that

$$2\pi f_{\max}(\Delta t/2) = 2\pi f_{\min}(\Delta t/2) + \pi$$

i.e. $\Delta t\,(f_{\max} - f_{\min}) = 1,$

$$\text{or } \Delta t\,\Delta f = 1. \tag{5.23}$$

This relation relates the frequency spread of the constituent waves in a wavepacket to the wavepacket duration, and will be refined in Section 7.3.3. Now the time during which the waves from a source remain coherent is called the *coherence time*, τ_c, and is equal to the wavepacket duration Δt. Hence a relation between the coherence time of light from a source and the frequency spread of the light can be written down

$$\tau_c = 1/\Delta f. \tag{5.24}$$

Correspondingly the *coherence length* of the waves in free space from a source will be:

$$L_c = c\tau_c = c/\Delta f = \lambda^2/\Delta\lambda. \tag{5.25}$$

The coherence length so defined is more precisely the *longitudinal coherence length*, while the *transverse coherence length* specifies the lateral distance over which radiation from a source remains coherent. A typical source used to produce the line spectrum of an element is a low pressure gas discharge tube, and for such a source the intense lines in the atomic spectrum have wavepackets lasting of order 10^{-10} s. Sources once used as wavelength references have longer coherence times: nitrogen-cooled, low pressure krypton sources produce an orange-red line whose wavepackets are 0.75m in length.

5.5.2   Optical beats

Beating between optical beams is analogous to the beating of tuning forks of very similar frequencies. From what has been said above about the incoherence of one optical source with another it is evident that the observation of optical beats between sources must require very special experimental conditions. The analysis of how and when optical beats can occur helps to bring out the significance of the coherence time of the radiation and its interplay with the response time of the detector used. Suppose that in Young's two slit experiment the slits are illuminated by different beams whose polarizations are identical but whose angular frequencies are ω_1 and ω_2. Suppose also that the coherence times of both sources are longer than the response time of the detectors used. The electric field at point P in figure 5.1 would be the sum of two pure sinusoidal waves during an interval short compared to the coherence time

$$E_0 \cos\left(k_1 s_1 - \omega_1 t\right) + E_0 \cos\left(k_2 s_2 - \omega_2 t\right), \tag{5.26}$$

where $k_i = c/\omega_i$ and s_i is the path length from slit i. The sources could be beams from separate lasers. The intensity is then

$$
\begin{aligned}
I &= E_0^2 \left[\cos^2 \left(k_1 s_1 - \omega_1 t \right) + \cos^2 \left(k_2 s_2 - \omega_2 t \right) \right. \\
&\quad \left. + 2 \cos \left(k_1 s_1 - \omega_1 t \right) \cos \left(k_2 s_2 - \omega_2 t \right) \right] \\
&= E_0^2 [\cos^2 \left(k_1 s_1 - \omega_1 t \right) + \cos^2 \left(k_2 s_2 - \omega_2 t \right) \\
&\quad + \cos \left(\Sigma(ks) - \Sigma \omega t \right) + \cos \left(\Delta(ks) - \Delta \omega t \right)],
\end{aligned} \tag{5.27}
$$

where $\Delta \omega$ is the difference $(\omega_2 - \omega_1)$, $\Sigma \omega$ the sum $(\omega_2 + \omega_1)$ and so on. The detector response commences with the release of electric charge, and is followed by electronic processing. The duration of the response varies and it is this *spread* which constitutes the response time, t_d, of the detector. A representative value for photomultipliers and photodiodes is 10 ns. The first three terms in eqn. 5.27 oscillating at 10^{14} Hz would average out over the many full cycles in the detector response time to $E_0^2/2$, $E_0^2/2$ and 0 respectively. However the final term has an angular frequency $\Delta \omega$, which may be low enough that the detector can sample the waveform more than once in each cycle. This condition is fulfilled if

$$
t_d \Delta \omega < 2\pi.
$$

Then taking t_d to be 10 ns the critical frequency difference below which the detector can sense the variations in the intensity is

$$
\Delta f_d = \Delta \omega_d / 2\pi = 1/t_d = 100 \, \text{MHz}.
$$

A screen covered with detector pixels would record the transient fringe movements provided the frequencies of the two stable sources were closer that Δf_d. Then eqn. 5.27 reduces to

$$
I = E_0^2 [1 + \cos \left(\Delta(ks) - \Delta \omega t \right)]. \tag{5.28}
$$

The oscillations observed are *optical* beats analogous to audible beats. The fringes would be transitory, lasting for a coherence time and then reforming. If however the source frequencies differ by much more than Δf_d the response of the detector to the final term in eqn. 5.27 is also zero leaving an intensity $I = E_0^2$ everywhere across the screen. That is uniform illumination with an intensity equal to that produced by the two slits separately.

In summary, clear optical beats can be detected when the sources are highly coherent and close enough in frequency that the beat period is much longer than the response time of the detectors involved. The beating together of radio waves shows different features because detectors at radio frequencies respond linearly to the electric field unlike detectors of visible radiation which respond to the intensity.

Quasi-monochromatic sources

It follows from the preceding that if the spread in angular frequency of radiation from a source $\Delta \omega$, and the response time of the detectors

τ_d, are such that $\Delta\omega\tau_d \ll 2\pi$, then interference fringes produced by all the frequency components will superpose giving high visibility fringes. In such a cases the coherence time then greatly exceeds the response time and the source is said to be *quasi-monochromatic*. The equations given above for evaluating time averages of intensities and energy flux in radiation, eqns. 5.12 and 5.13, apply to such quasi-monochromatic radiation.

5.5.3 Coherence area

The finite width of a source imposes limitations on the coherence of a wavetrain transverse to the direction of travel. In figure 5.16 the source illuminating Young's slits is Δy wide and at a distance z from the slits, the angle subtended by the source at the slits is θ_s while the semi-angle subtended at the source by the slits is θ_d. The path difference between light arriving at the upper slit from the two edges of the source is

$$\Delta y \sin\theta_d = d\Delta y/2z,$$

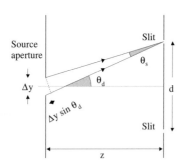

Fig. 5.16 Transverse coherence length due to finite source dimension.

and there is a similar spread in the case of the lower slit. This leads to a spread in the relative phase between the light arriving at the two slits of

$$kd\Delta y/2z = \pi d\Delta y/(z\lambda).$$

Thus the light arriving at the slits is coherent only if this phase difference is small. A corresponding *lateral coherence length* can be defined, within which the slits must lie from one another in order that there can be interference

$$d_c = z\lambda/\Delta y = \lambda/\theta_s. \tag{5.29}$$

For a circular source the corresponding *lateral coherence area* is then

$$A_c = \lambda^2/(\pi\theta_s^2). \tag{5.30}$$

Arranging this differently gives

$$A_c\left(\pi\theta_s^2\right) = \lambda^2. \tag{5.31}$$

Now referring to eqn. 4.11 we see that the left hand side in the equation here is exactly the etendue $\mathcal{T}_c$ of the beam from the source into the coherence area. This is a general result, the etendue from a source into its coherence area is exactly λ^2:

$$\mathcal{T}_c = \lambda^2. \tag{5.32}$$

The coherence area and the coherence length taken together define a *coherence volume*. At all points within this volume the waves have a constant phase relation and could be brought together in some apparatus to interfere. This might be in Young's two slit experiment or a Michelson interferometer. Each wavepacket in the beam will, at any given moment on its journey, determine a coherence volume.

5.6 Stokes' relations

There is a general principle, called *time reversal invariance*, stating that if time were reversed then physical processes could reverse exactly. The only known violations are observed in particular weak decays of elementary particles. In principle then, time reversal invariance can be used to relate the reflection and transmission coefficients for light incident on one or more plane parallel dielectric layers from its opposite ends. However if there is any absorption this is not reversed on the macroscopic scale we are interested in: absorption occurs whichever direction the light is incident.[3] We now apply time reversal invariance to relate the reflection and transmission coefficients at a single interface when there is negligible absorption.

[3]The individual steps in all the light matter interactions making up the absorption would have each to be reversed in order to produce a valid time-reversed process.

Figure 5.17 shows both the original and the time-reversed processes: r and t are the reflection and transmission amplitude coefficients for light incident from above, r' and t' are those for light incident from below. We take the point on the interface to be the origin of the spatial coordinates: there the incident, reflected and transmitted waves all have electric fields of the form $a \exp\left(-i\omega\tau\right)$ where ω is the angular frequency, τ is the time and $a = |a| \exp\left(i\phi\right)$. Any alteration of the time origin can be absorbed by a change in the phase ϕ. Therefore for consistency time reversal must change to ϕ to $-\phi$, as well as τ to $-\tau$. Time reversal therefore replaces the wave amplitude by its complex conjugate. Two time reversals evidently restore the original waves. Together the time reversed reflected and transmitted waves reproduce the incident beam, while the beam at the bottom left of (b) must vanish. Then

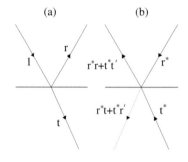

Fig. **5.17** Reflected and transmitted amplitudes: (a) for forward process and (b) for the time reversed process.

$$r^*r + t^*t' = 1, \qquad (5.33)$$
$$r^*t + t^*r' = 0. \qquad (5.34)$$

Now for a single surface, below the critical angle of incidence, the coefficients are real and the equations simplify to

$$r^2 + tt' = 1, \qquad (5.35)$$
$$r' = -r. \qquad (5.36)$$

These four equations are known as *Stokes' relations* after their 19th century discoverer. They apply equally for multiple layers of non-absorbing dielectric. Stokes' relations can be derived in an equivalent manner from the requirement of conservation of energy between the incident and outgoing light beams, which again excludes cases of absorption.

5.7 Interferometry

A vast range of research and industrial devices have been built which make use of interference effects. Applications include: measuring distances with precision of a fraction of a wavelength; stabilizing laser wavelengths and measuring wavelengths; testing and measuring the imperfections of optical components; inertial guidance in aircraft; measurement of refractive indices of gases; and revealing fluid structure and flow patterns in Tokomaks and wind tunnels. Predominantly these devices involve amplitude rather than wavefront division. The underlying reason is that wavefront division requires lateral coherence across the area covering the slits and this in turn implies a small area source, and in turn this means weak illumination. By contrast, taking an example of amplitude division, the Michelson interferometer has a broad source providing strong illumination. Amplitude division lends itself to flexible designs, with derivatives of the Michelson interferometer design being frequently met.

5.7.1   The Twyman–Green interferometer

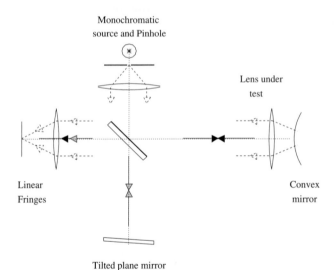

Fig. 5.18 Twyman–Green interferometer with a lens under test. The optical quality convex mirror is positioned to return the rays from the lens under test along their incident direction.

The Twyman–Green interferometer sketched in figure 5.18 is a variant of Michelson's inteferometer, which is used to examine optical component quality, in this case the quality of a lens. The combination of a point monochromatic source located at the focus of a positive lens provides a parallel coherent beam. An optical quality convex spherical mirror is positioned behind the lens being tested and is moved relative to the lens until the rays are retroreflected back along their incident path. In the other arm a plane mirror is tilted so that straight line fringes are seen by eye or imaged onto a detector array. If the lens were perfect the fringes would be equally spaced and straight. Any imperfections cause distortions which can be interpreted in order to define a grinding and polishing sequence which would correct the lens. Usually the image is captured on a CCD array and the analysis carried out by proprietary software.

In many industries standards of length are maintained using *gauge blocks*. Gauge blocks are simple cuboids with one pair of opposite calibration faces both optically flat and parallel. The material used has a low coefficient of thermal expansion, such as the metal Invar (64% Fe, 36% Ni) with coefficient $1.3\,10^{-7}\,\mathrm{K}^{-1}$ or the ceramic Zerodur with coefficient as low as $10^{-8}\mathrm{K}^{-1}$. The thickness of the gauge block between the calibration surfaces is first measured mechanically, giving a thickness between t_{lo} and t_{hi}, which differ by $\sim 10\,\mu\mathrm{m}$. Next the gauge block is wrung in contact with the mirror in one arm of an interferometer as shown in figure 5.19. Fringes are seen from the mirror surrounding the gauge block and from the gauge block. The lateral displacement between the fringes on mirror and gauge block, expressed as a fraction of a fringe spacing Δn_i, is measured for several wavelengths λ_i. Then the actual thickess of the block is

$$t = (n_i + \Delta n_i)\lambda_i, \tag{5.37}$$

where the values of the integers n_i are as yet unknown. Benoit's *method of exact fractions* provides a way to obtain these values and hence to determine t with a precision of tens of nanometres.

For one wavelength λ_1 the values of n_1 consistent with values of the thickness lying between t_{lo} and t_{hi} are each taken in turn and a thickness calculated. One of these thicknesses will be correct, but the question is which one. Using each such thickness in turn, $t(\mathrm{trial})$, the values of the interference order for each other wavelength are calculated,

$$n_i(\mathrm{trial}) + \Delta n_i(\mathrm{trial}) = t(\mathrm{trial})/\lambda_i. \tag{5.38}$$

For the correct choice of $t(\mathrm{trial})$, and only for that choice, all the trial fractions $\Delta n_i(\mathrm{trial})$ will match the corresponding measured values Δn_i. This match determines t. In practice three or more wavelengths are used to eliminate the possibility of accidental coincidences between the trial and measured values of the Δn_i's.

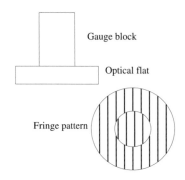

Fig. 5.19 Side view of gauge block on an optical flat, together with a notional fringe pattern seen with a Twyman–Green interferometer. The fringe modulation would be cos^2 in practice and thus less crisp than shown here.

5.7.2 The Fizeau interferometer

This interferometer, shown in figure 5.20, uses amplitude division within a single arm to produce interfering beams. It too is used for optical testing. An intense coherent beam from a laser is focused by a microscope objective onto a pinhole. This *spatial filter* produces a beam with lateral coherence over a wide angle. Such a design allows the interferometer to accomodate the very large area lenses and mirrors that are used in the space industry. A collimating lens directs the beam through an optically flat reference surface and then onto the component under test. In the example shown a mirror is being tested for flatness. Light returning

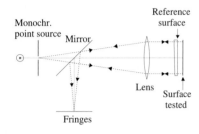

Fig. 5.20 Fizeau interferometer used to test an optical flat.

from the mirror passes again through the reference surface and collimating lens and finally is reflected by the partially reflecting surface onto an image plane, rather than travelling back to the pinhole. The reference surface is of high quality, flat to say $\lambda/50$, and the reference plate is a slightly wedge shaped so as to throw the reflections from its other surface out of the field of view. Interference fringes are formed between the waves reflected from the reference surface and the surface under test. If the test surface were perfect and tilted slightly with respect to the reference surface a pattern of linear equally spaced fringes would be seen. The capture of the actual fringe pattern, and its analysis, proceed as described above for the Twyman–Green interferometer.

5.7.3   The Mach–Zehnder interferometer

The Mach–Zehnder interferometer is shown in figure 5.21. Monochromatic light from an extended diffuse source is divided by the beam splitter BS1 and the separated beams travel via either mirror M1 or mirror M2 to meet again at the second beam splitter, BS2. Emerging together from BS2 the beams are focused by the lens onto an image plane where their interference fringes are observed. The design permits the two paths to be widely separated: one path can cross a multimetre sized volume such as a wind tunnel or a nuclear fusion device while the other path skirts that volume. Whenever a test chamber is placed in one arm, glass plates identical to those belonging to the test chamber are inserted in the other arm in order to cancel their optical effects on the fringe pattern. When the mirroring surfaces are all set at exactly $45°$ to the rectangular shape of the beam path the interference fringes are located at infinity and appear in the focal plane of the lens. By tilting BS2 from the $45°$ orientation the fringes are moved so that they are localized within the test volume and the image plane is moved correspondingly, as shown in the figure. Photographs of the fringes then show any mechanical structure in the test volume in focus together with the fringes. The structure might be an aircraft wing in a wind tunnel. In a dynamical situation involving gas compression and flow any variation in the refractive index of the gas in the test chamber will alter the optical path length and distort the fringe pattern. In this way the flow lines of gas around an aircraft model become observable and are available for interpretation. Interferometers used to observe the gas flow in fusion reactors employ CO_2 lasers of wavelength $10.6\,\mu$m and can detect variations in the plasma electron density of one part in a thousand ($10^{17}\,\mathrm{m}^{-3}$ in $10^{20}\,\mathrm{m}^{-3}$).

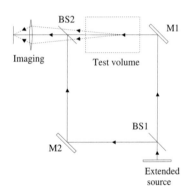

Fig. 5.21 Mach–Zehnder interferometer.

5.7.4   The Sagnac interferometer

The Sagnac interferometer is unique in its ability to allow the detection of rotation. Figure 5.22 shows that the device consists of a closed optical path around which coherent beams from the laser source circulate in opposite directions. This design is called *passive* because the laser is not integral with the ring. A beam splitter BS divides the laser beam to provide two equal amplitude counterrotating beams. After making a circuit round the four arms the beams exit through the same beam splitter and arrive superposed at the detector. An *active ring laser* is formed from a monolithic glass block with surface mirrors in place of the individual mirrors in figure 5.22. The glass is doped with lasing material so that it forms both the laser and the closed path. When the ring is rotated the path for one direction of travel is shortened and for the other lengthened; an effect that can be detected optically and used to measure the rate of rotation. A simple question but an awkward question is this: rotation with respect to what? The answer supplied by the general theory of relativity is that the rotation measured is that with respect to the local inertial frame, that is to say the frame of the fixed stars. It is not the rotation relative to the Earth or relative to the solar system that is measured. In analysing this effect we treat the ring as a circular evacuated loop of radius R. The rate of rotation with respect to the inertial frame is Ω. Then the angle rotated in the time it takes light to complete one circuit is

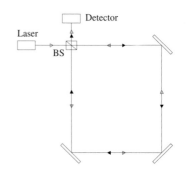

Fig. 5.22 Sagnac interferometer; passive ring gyro layout.

$$\Delta\theta = 2\pi R\Omega/c.$$

Thus the paths for the two senses of rotation differ in length by

$$\Delta s = 2R\Delta\theta = 4\pi R^2\Omega/c.$$

The corresponding phase difference induced between the beams after one rotation is then

$$\Delta\phi = 2\pi\Delta s/\lambda = 8\pi^2\Omega R^2/(c\lambda)$$
$$= 8\pi\Omega A/(c\lambda), \tag{5.39}$$

where A is the area enclosed by the loop. This result generalizes for the case of an arbitrarily shaped loop of vector area $\mathbf{A}$ having angular velocity $\boldsymbol{\Omega}$ to

$$\Delta\phi = 8\pi\boldsymbol{\Omega}\cdot\mathbf{A}/(c\lambda). \tag{5.40}$$

The sensitivity to rotation can be improved by using a multiturn loop of optical fibre to carry the light; the phase change is then multiplied up by the number of turns.

In an active ring with the laser in the loop, the complete orbits of the counter-rotating beams differ by Δs, and so the times for making one complete orbit differ by

$$\Delta\tau = \Delta s/c = 4\pi R^2\Omega/c^2,$$

between the counter-rotating beams. Thus the fractional difference between the orbital periods is

$$\Delta\tau/\tau = 2R\Omega/c$$

and the frequency difference between the counter-rotating beams from the one laser is

$$\Delta f = f\Delta\tau/\tau = 2R\Omega/\lambda.$$

This result generalizes to

$$\Delta f = 4\mathbf{\Omega} \cdot \mathbf{A}/(\lambda P), \tag{5.41}$$

where P is the length of the perimeter of the loop. These results, in particular eqns. 5.40 and 5.41, are valid whatever material fills the path provided that this is also rotating at the angular velocity $\mathbf{\Omega}$, and that the wavelength used is the wavelength in free space.[4]

A practical difficulty met with in using active Sagnac rings is that the frequencies of the two laser beams tend to lock together at a common frequency when the rotation frequency Ω is small. The coupling between the beams comes from back scattering at the mirrors in the optical path. In order to reduce this scattering to a low enough level that the coupling is removed it has been necessary to develop super efficient mirrors with reflection coefficients as high as 0.999 999! Compact active Sagnac interferometers arranged in a set of three with orthogonal axes are used to provide inertial guidance. The necessary dimensional stability is achieved by carving the frame containing the three units from a common block of Zerodur ceramic. These *ring gyroscopes* have no moving parts and offer superior precision to mechanical gyroscopes, so they have been fitted in high performance aircraft. More recently these gyroscopes have been replaced in many applications by the simpler fibre optic gyros described in Chapter 16.

5.8   Standing waves

An important application of the superposition principle is to the reflection of plane sinusoidal electromagnetic waves incident on the surface of good conductors. Suppose the *incident* electromagnetic wave travels in the direction of x decreasing, and its electric field is

$$E_i(x,t) = E_0 \cos(kx + \omega t) \tag{5.42}$$

parallel to the surface. At the surface, taken to be $x = 0$, the incident field is:

$$E_i(0,t) = E_0 \cos(\omega t). \tag{5.43}$$

Metals are good conductors, therefore the simplifying assumption is made that the free electrons within the conductor move instantaneously so keeping the electric field parallel to surface extremely small. This is close to the actual behaviour in the metals copper, silver and gold up

---

[4]See for example Chapter 1 of *The Fibre Optic Gyroscope* by H. Lefevre, published in 1993 by Airtech House, Boston.

to optical frequencies. The displaced electrons generate an electric field equal and opposite to the incident electric field at the surface:

$$E_{\mathrm{r}}(0, t) = -E_0 \cos{(\omega t)}. \tag{5.44}$$

Applying Huygens' principle, this disturbance produces a *reflected* plane

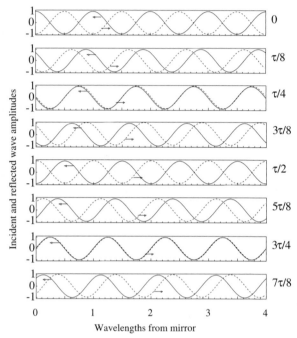

Fig. 5.23 Travelling waves from the right reflected by a perfect mirror; incident/reflected waves shown as full/broken lines: shown at intervals of one-eighth of a period.

wave propagating in the positive x-direction away from the surface:

$$E_{\mathrm{r}}(x, t) = -E_0 \cos{(-kx + \omega t)}. \tag{5.45}$$

The superposition principle determines the total electric field to be

$$\begin{aligned} E(x, t) &= E_i(x, t) + E_{\mathrm{r}}(x, t) \\ &= E_0 \cos{(kx + \omega t)} - E_0 \cos{(-kx + \omega t)} \\ &= -2E_0 \sin{(\omega t)} \sin{(kx)}. \end{aligned} \tag{5.46}$$

Figure 5.23 shows the incident wave (solid curves) and reflected wave (broken curves) at successive intervals of one eighth of a period ($\tau/8$). Figure 5.24 shows the corresponding total wave amplitude. This oscillates, but at certain locations called *nodes* the amplitude is always zero. This is therefore called a *standing* wave. Starting at the conductor the nodes are spaced at intervals of one half wavelength of the incident travelling waves. Midway between each pair of nodes lie the *antinodes* where the field variation is largest. Figure 5.25 shows the incident and reflected

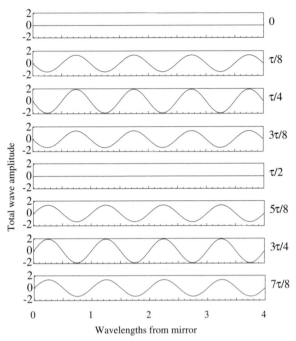

Fig. 5.24 The resultant standing waves from the incident and reflected waves shown on figure 5.23. Note that the ordinate range is twice that in figure 5.23.

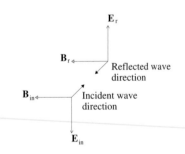

Fig. 5.25 The orientation of the electromagnetic fields and the direction of wave propagation of the incident and reflected waves at the mirror.

field vectors at the surface of a good conductor. The relative orientation of **E**, **B** and the wave direction **k** is preserved after reflection and so the magnetic field is not cancelled out at the surface. If the metal surface is rough the reflections from nearby points on the surface travel in different directions and a mirror smooth to a fraction of a wavelength is needed in order to produce simple standing waves. Everyday mirrors consist of metal-coated smooth glass surfaces, while at longer wavelengths a wire mesh is adequate for satellite dish antennae and radio-telescopes. This saves on cost and makes radio-telescopes easier to support in high winds.

Standing electromagnetic waves are readily demonstrated by directing microwaves onto a metal sheet. The standing wave pattern produced by a 10 cm wavelength source will have nodes spaced 5 cm apart, which are readily observed using a diode detector. By 1890 Wiener had photographed standing waves produced with visible light. As shown in figure 5.26 a glass plate coated with photographic emulsion rested with one edge on a plane front-metallized mirror and was tilted at a small angle α (10^{-3} rads) to the mirror. Plane monochromatic waves whose wavefronts were accurately parallel to the mirror surface illuminated the plate. The antinodes of the electric field are visible in figure 5.26, spaced a distance $\lambda/2$ apart perpendicular to the mirror's surface. This spacing is amplified by the tilt to $\lambda/(2\alpha) \approx 0.125$ mm along the emulsion's surface. When the photographic plate was processed the expected, equally spaced black stripes were seen where the light had activated the silver halide grains. At the edge of the emulsion in contact with the mirror, which is where the electric field is zero and the magnetic field is at its largest, the emulsion was clear. Hence the photochemical effect which leaves a developable image in the emulsion is caused by the action of the electric rather than the magnetic field.

In a laser two mirrors face one another across a region containing active material. Standing electromagnetic waves develop between the mirrors using energy drawn from the active material. These waves are analogous to those on a violin string when bowed: in each case there is a node at the two end points. The simplest *fundamental* violin string oscillation has an antinode midway along the string and a wavelength of twice the string length, $2L$. Vibrations are also possible with $2,3\ldots$ antinodes along the wire. Each distinct pattern of oscillation is called a *mode*, with the number of antinodes defining the *order*, n of a mode. Figure 5.27 shows modes with n equal to 1, 2, 3 and 30, where successive displacements of the string are indicated at time intervals of $\tau/16$. In the case of standing electromagnetic waves between mirrors the order of these modes is very much higher. For the nth mode the required wavelength λ is given by

$$n\lambda/2 = L, \text{ i.e. } \lambda = 2L/n. \tag{5.47}$$

An optical arrangement consisting of facing parallel mirrors of high reflectivity is called a *Fabry–Perot etalon* or *cavity*. The electric field in the Fabry–Perot etalon has to vanish at both mirrors and so has the form:

$$E = 2E_0 \sin(\omega t) \sin(kx), \tag{5.48}$$

where

$$k = 2\pi/\lambda = n\pi/L, \tag{5.49}$$

with n integral in order that the transverse electric field vanishes at $x = L$ as well as $x = 0$. In helium–neon gas lasers the active gas is contained in a 50 cm long Fabry–Perot cavity and the lasing wavelength is 633 nm, making n around 1.6 million.

5.9   The Fabry–Perot interferometer

The Fabry–Perot interferometer shown in figure 5.28 is a single arm instrument with a broad diffuse source and a lens to focus the Haidinger fringes formed by light repeatedly reflected between the glass plates. These glass plates are closely spaced with their inner faces being highly reflecting and optically flat. Together with their supporting frame the plates form a *Fabry–Perot etalon*. The inner surfaces of the plates can be aligned parallel by screws that hold the plates against a spacer of Zerodur or other material having a very low coefficient of thermal expansion. Interference is observed between the multiple beams produced by successive reflections from the highly reflecting surfaces of the etalon.

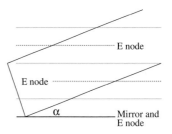

Fig. 5.26 Wiener's experiment producing standing waves of light at a mirror surface.

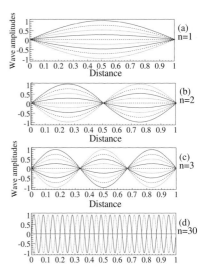

Fig. 5.27 Standing wave patterns for (a) the $n = 1$ mode, (b) the $n = 2$ mode, (c) the $n = 3$ mode, (d) the $n = 30$ mode. In (a), (b), and (c) the waves are shown at intervals of $\tau/16$, and at intervals of $\tau/2$ in (d).

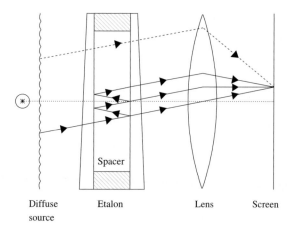

Diffuse   Etalon   Lens   Screen
source

Fig. 5.28 Fabry–Perot interferometer. The inner faces of the etalon have large reflection coefficients. A lens brings the parallel rays from multiple reflections to a common focus. One set of reflections is shown. Rays from other places on the source like the broken line ray give similar sets of reflections.

Some representative reflections of a ray from one point on the source are drawn in the diagram with solid lines. Parallel rays from all other points on the source, such as the one indicated by a broken line, will undergo similar sets of reflections and their fringes will coincide. It will emerge below that by using multiple reflections a chromatic resolving power is obtained which far exceeds that of prisms or diffraction gratings. A thin layer of aluminium on the etalon faces is commonly used to give a uniform reflectivity across the visible spectrum of 80%, while reflectivities as high as 99% are readily achievable with more complex coatings. The outer surface of each plate is flat, uncoated and makes a small angle to the inner surface so as to remove its own reflections from the field of view.

Suppose that by some means the reflection coefficient of the facing surfaces were simultaneously increased from the glass/air value of a few % to 90% while one watched. Initially a set of broad equal inclination circular fringes would be seen, just like those produced by the Michelson interferometer. When the reflection coefficient increased the bright fringes would get progressively narrower, as shown in figure 5.29, but would remain fixed in position. Finally when each surface reflected 90% of the incident light the total transmission would fall to $10\% \times 10\% = 1\%$, which would be concentrated in the narrowed bright fringes. Thus light from two closely spaced wavelengths whose fringes would overlap appreciably in a Michelson interferometer produce well separated fringes in a Fabry–Perot interferometer. This behaviour is now analysed quantitatively. Figure 5.30 shows the transmitted waves following multiple reflections in a Fabry–Perot etalon for an angle of incidence θ. The optical path difference between the successive transmitted rays drawn in the figure is given by the left hand side of eqn. 5.16, $2nd\cos\theta$, with d being the separation of the mirrored surfaces and n the refractive index of the material between them. Thus the phase delay between successive rays transmitted, in terms of the wavelength in free space and frequency is

$$\delta = 4\pi nd\cos\theta/\lambda = 4\pi nd\,f\cos\theta/c. \tag{5.50}$$

We ignore the weaker diverted reflections at the outer unsilvered surfaces of the etalon plates. Let the reflection coefficient for the wave amplitude at the air/glass interface be r and the transmission coefficient be t; while for the glass/air interface let these coefficients be r' and t' respectively. Assuming that there is no absorption at the surfaces we can use Stokes' relations between these coefficients given in eqns. 5.35 and 5.36. The superposition principle gives a total transmitted amplitude

$$\begin{aligned} E_t &= tt'[1 + r^2\exp(i\delta) + r^4\exp(2i\delta) + ...\\ &= tt'/[1 - r^2\exp(i\delta)]\\ &= (1-r^2)/[1 - r^2\exp(i\delta)] \end{aligned} \tag{5.51}$$

where we have use Stokes' relation eqn. 5.35 in the last line. For reference later the reflected amplitude is

$$\begin{aligned} E_r &= r' + tt'[r\exp(i\delta) + r^3\exp(2i\delta) + ...]\\ &= -r + (1-r^2)r\exp(i\delta)/[1 - r^2\exp(i\delta)], \end{aligned}$$

where both eqns. 5.35 and 5.36 have been used. Then

$$E_r = r[\exp(i\delta) - 1]/[1 - r^2\exp(i\delta)]. \tag{5.52}$$

Now putting $r^2 = R$ and $1 - r^2 = T$ in eqn. 5.51

$$E_t = T/[1 - R\exp(i\delta)].$$

The transmitted intensity at an angle θ is therefore

$$\begin{aligned} I(\theta) = E_t E_t^* &= T^2[(1 - R\cos\delta)^2 + R^2\sin^2\delta]^{-1}\\ &= T^2[1 - 2R\cos\delta + R^2]^{-1}\\ &= [T/(1-R)]^2\,[1 + 4R\sin^2(\delta/2)/(1-R)^2]^{-1}. \end{aligned} \tag{5.53}$$

Thus

$$I(\theta) = 1/[1 + 4R\sin^2(\delta/2)/(1-R)^2]. \tag{5.54}$$

This function is plotted against δ in figure 5.31 for a few values of R, and it shows that as R increases the bright fringes sharpen while retaining their peak intensities. The degree of sharpness is quantified by calculating the full width at half the maximum intensity (FWHM) of

Fig. 5.29 Appearance of Fabry–Perot fringes for the same mirror spacing as in figure 5.12 with a finesse of 15.

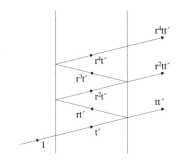

Fig. 5.30 Successive contributions to the wave amplitude transmitted by a Fabry–Perot etalon.

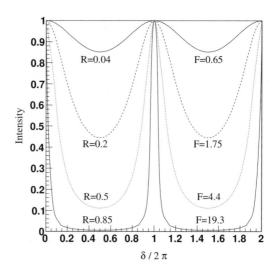

Fig. 5.31 Intensity distribution plotted against $\delta/2\pi = 2nd\cos\theta/\lambda$. The intensity is normalized to the incident intensity.

bright fringes. Suppose the value of δ at one peak of intensity is $2\pi m$ and that its value when the intensity has fallen off to half maximum is $2\pi m + \delta_{1/2}$, then

$$1 + \sin^2(\delta_{1/2}/2)\left[4R/(1-R)^2\right] = 2,$$

whence

$$\delta_{1/2} = 2\sin^{-1}\left[(1-R)/2\sqrt{R})\right]$$
$$\approx (1-R)/\sqrt{R},$$

because all the angles are small. Thus the FWHM of each fringe, $2\delta_{1/2}$, is $2(1-R)/\sqrt{R}$. The corresponding FWHM in frequency of each fringe is obtained using eqn. 5.50 with θ set to zero

$$2\Delta f_{1/2} = c\delta_{1/2}/(2\pi nd). \tag{5.55}$$

A convenient measure of the sharpness called the *finesse* is defined as the ratio between the phase change separating successive fringes and the phase change across the fringe at half maximum, $2\delta_{1/2}$. The finesse is thus

$$\mathcal{F} = \pi/\delta_{1/2} = \pi\sqrt{R}/(1-R). \tag{5.56}$$

When the reflection coefficient R is close to unity, it follows[5] that

$$R^{\mathcal{F}} \approx \exp(-\pi) \approx 0.04,$$

---

[5]$\ln R^{\mathcal{F}} = \ln(R)\pi\sqrt{R}/(1-R)$. Now putting $S = 1 - R$ we get $\ln R^{\mathcal{F}} = \ln(1-S)\pi\sqrt{(1-S)}/S$. Then because S is very small, $\ln R^{\mathcal{F}} \approx -\pi$.

which is similar in magnitude to the reflection coefficient at an uncoated glass/air interface. Thus $\mathcal{F}$ can be treated as the number of round trips within the etalon, beyond which the wave intensity becomes insignificant. Any lack of flatness in the etalon surfaces will be amplified by this factor so that the finesse expected with a given reflection coefficient can only be actually attained if the flatness is better than $\lambda/\mathcal{F}$. A reflection coefficient of 0.9 and a finesse of 30 represent standard values.

The narrowness of the Fabry–Perot fringes makes this interferometer a useful spectrometer. The *chromatic resolving power* of a spectrometer is defined to be the ratio of the wavelength to the smallest separation in wavelength that can be resolved with the apparatus. In the case of the Fabry–Perot instrument two wavelengths, λ and $\lambda - \Delta\lambda$, will just be resolvable when their mth order bright fringes are separated by $2\delta_{1/2}$. If, close to the centre of the pattern, the interference order for wavelength λ is m, then

$$m\lambda = (m + \delta_{1/2}/\pi)(\lambda - \Delta\lambda).$$

Then to a good approximation

$$\pi m\Delta\lambda = \lambda\,\delta_{1/2},$$

and the chromatic resolving power

$$\mathrm{CRP} = \lambda/\Delta\lambda = \pi m/\delta_{1/2}. \tag{5.57}$$

Now using the definition of finesse from eqn. 5.56

$$\mathrm{CRP} = \mathcal{F}m.$$

Using eqn. 5.16 to replace m, and recalling that near the centre of the pattern $\cos\theta \approx 1$, gives

$$\mathrm{CRP} = 2nd\mathcal{F}/\lambda. \tag{5.58}$$

Taking a 10 mm spacing in air and a finesse of 30 gives a resolving power of 10^6 at 633 nm wavelength. Now the difference in wavelength between individual spectral lines produced by most sources is large enough that adjacent circular fringes due to two spectral lines can correspond to very different values of m in eqn. 5.16; that is to say they are from different orders. The wavelength interval within which there is no overlap of orders and hence no ambiguity in assigning the order of a line of unknown wavelength is called the *free spectral range*. Clearly the increase in δ over the free spectral range is 2π, and the corresponding frequency change can then be obtained from eqn. 5.50

$$\Delta f_{\mathrm{fsr}} = c/(2nd). \tag{5.59}$$

Then

$$\Delta\lambda_{\mathrm{fsr}} = \lambda^2 \Delta f_{\mathrm{fsr}}/c = \lambda^2/(2nd). \tag{5.60}$$

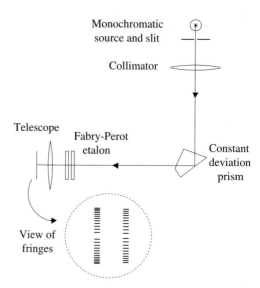

Fig. 5.32 Two stage spectrometer using a constant deviation prism and a Fabry–Perot etalon. In the insert the Fabry–Perot fringes across two spectral lines are shown.

With a 1 mm etalon spacing and a wavelength of 633 nm the free spectral range is only 0.2 nm which shows that one needs to preselect a narrow wavelength interval in order to determine an unknown wavelength. Preselection can be achieved in a two stage spectrometer, with the first stage being a grating or prism spectrometer or alternatively a bandpass filter of the sort to be described in Chapter 9. A two-stage spectrometer using a constant deviation prism of the type described in Section 2.2.1 is shown in figure 5.32. Light from a slit source is focused into a parallel beam by the collimator and falls on the constant deviation prism. This gives an outgoing beam that is observed by a telescope. For any given orientation of the constant deviation prism the telescope receives a narrow range of wavelengths centred on the wavelength for which that orientation produces minimum deviation. This window in wavelength can be scanned through the spectrum simply by rotating the constant deviation prism. The etalon is placed to intercept the beam from the prism and the resulting image produced by the telescope is observed by eye or let fall on a detector array. The inset in figure 5.32 shows the final image. With the Fabry–Perot etalon removed this would consist of repeat images of the source slit, one for each spectral line resolved by the prism. With the Fabry–Perot etalon in place one now sees through each slit image a diametral slice of the circular Fabry–Perot pattern for the spectral line involved. If the spectral line is in fact two close lines of wavelengths λ_1 and λ_2 which cannot be resolved by the prism, then that particular slit image will bear two interlaced sets of Fabry–Perot fringes satisfying $m_1\lambda_1 = 2d\cos\theta_1$ and $m_2\lambda_2 = 2d\cos\theta_2$.

Another arrangement for carrying out spectroscopy with a Fabry–Perot etalon is shown in figure 5.33. In this case the wavelength of the incident beam is limited to less than the free spectral range by an interference filter of the sort described in Section 9.7. The detector views the central spot and a scan through the free spectral range in wavelength is made by altering the product nd in eqn. 5.16. One method is to pump gas into the volume within the etalon; in principle this method requires a knowledge of the pressure and temperature in order to obtain the refractive index.

A neat way to avoid precise measurements of pressure and temperature is to make a pressure scan by introducing gas through a constant flow valve into an evacuated etalon. Then the pressure and hence the refractive index will vary linearly with time. Suppose the spectral line of interest has a wavelength $\lambda + \Delta\lambda$ where λ is the wavelength of a nearby reference line, so that $\Delta\lambda$ is the quantity to be measured. During the scan consecutive fringes due to the reference line will make their appearances at the centre when

$$m\lambda = 2n_m d, \tag{5.61}$$

$$(m+1)\lambda = 2n_{m+1}d. \tag{5.62}$$

Between these a fringe due to the line of interest will appear when

$$m(\lambda + \Delta\lambda) = 2n'_m d. \tag{5.63}$$

All fringe appearances are timed electronically. The above three equations can be used to give

$$\Delta\lambda = [\lambda/m](n'_m - n_m)/(n_{m+1} - n_m). \tag{5.64}$$

Now, because the scan is linear the ratio of differences in refractive index is equal to the ratio of the corresponding measured time intervals between fringe appearances. Finally m can be eliminated from this result by using eqn. 5.61

$$\Delta\lambda = [\lambda^2/2dn_m](n'_m - n_m)/(n_{m+1} - n_m), \tag{5.65}$$

in which n_m is the last unknown. During the experiment the gas remains at low pressure, keeping the refractive index very close to unity. Thus inserting a value for n_m based on a rough measurement of this pressure will yield a precise value of the required wavelength. A check is necessary on the linearity of the scan: the time intervals between consecutive appearances of the reference fringe with orders m, $m+1$ and $m+2$ should be equal.

An alternative scanning technique is to move one of the etalon plates while keeping it parallel to its stationary partner. Movements of a few

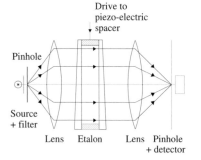

Fig. 5.33 Scanning Fabry–Perot spectrometer using a piezoelectric spacer.

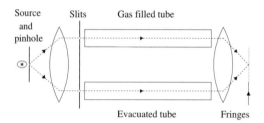

Fig. 5.34 The Rayleigh refractometer.

microns are adequate. The preferred scanning technique is to use piezo-electric crystals, which undergo a change in length in the direction of an applied electric field. Three such crystals placed at 120° intervals around the rim of the spacer provide sufficiently smooth parallel displacement of the moving plate.

The required size of the pinhole in front of the detector will now be estimated. Using this and the etalon size the etendue of the Fabry–Perot spectrometer is calculated. If the etalon separation is such that at wavelength λ a fringe sits at the exact centre of the pattern,

$$2nd = m\lambda.$$

The pinhole is made large enough to accept the half width of the central fringe. Suppose the angular radius of the pinhole as seen from the lens is θ, then

$$2nd\cos\theta = [m - \delta_{1/2}/(2\pi)]\lambda.$$

Subtracting this equation from the previous one and using an approximation adequate at these small angles, $\cos\theta = 1 - \theta^2/2$, gives

$$nd\,\theta^2 = \delta_{1/2}\lambda/(2\pi).$$

Using eqn. 5.57 to replace $\delta_{1/2}$,

$$\theta^2 = m\Delta\lambda/(2nd)$$
$$= \Delta\lambda/\lambda, \tag{5.66}$$

which is the inverse of the CRP. Hence the solid angle subtended by the pinhole at the lens is

$$\Omega = \pi\theta^2 = \pi\,\Delta\lambda/\lambda. \tag{5.67}$$

The etendue is then the product of this solid angle and the area of the beam at the lens

$$\mathcal{T} = (\pi D/2)^2(\Delta\lambda/\lambda). \tag{5.68}$$

where D is the diameter of the clear portion of the etalon. The reader will recall that the etendue is invariant along a beam, and may like to check that taking the product of the area of the pinhole times the solid angle of the beam at the pinhole gives an identical result.

In the following two chapters the chromatic resolving power and etendue of other spectrometers are also presented and remarks made about the relative merits of different devices.

Exercises

(5.1) In Young's experiment the slit spacing is 0.1 mm and monochromatic light of wavelength 633 nm is used. What is the fringe spacing if the image focusing lens has focal length 1.5 m?

(5.2) The first lens in Young's experiment shown in figure 5.4 has focal length 0.3 m. How narrow should the source aperture be in order to ensure that the slits are coherently illuminated?

(5.3) A source emits light of mean wavelength 500 nm. The wavepackets are 1 m long. What is the spread in frequency and in wavelength? What fraction is the frequency spread of the mean frequency? What is the corresponding fraction in wavelength?

(5.4) Two optically flat glass plates are viewed in light of wavelength 633 nm. Straight line fringes are seen, spaced apart at intervals of 1.5 cm across the surface. What is the angle between the plates?

(5.5) Show that at latitude 50°N the Earth's rotation causes a frequency splitting of 176 Hz between counter-rotating 633 nm laser beams in an active ring laser. The beam circuit can be assumed circular of radius 1 m.

(5.6) Suppose a highly coherent laser beam of wavelength 633 nm is interrupted by an electronic shutter so that random pulses each of duration 10 fs are produced. What is the coherence length and what are the spreads in wavelength and frequency of the beam?

(5.7) The accompanying figure 5.34 shows a Rayleigh refractometer. Light from coherently illuminated slits goes through separate hollow airtight arms of length L each with glass entry and exit windows.

The separate beams are brought together by a lens with fringes appearing in the image plane. Both arms are initially evacuated and gas is introduced slowly into one arm. This causes the fringes to move across the field of view; up or down in the diagram. If m fringes pass through the centre of the field while the gas is entering, show that the refractive index of the gas, n, is given by

$$n - 1 = m\lambda/L.$$

(5.8) In the Rayleigh refractometer the slits need to be well separated, which means the fringes are very close together. Suppose the wavelength of light is 500 nm, the slit separation is 2 cm and an imaging lens of 10 cm focal length is used to bring the beams together. What is the fringe separation and how would you view them?

(5.9) Calculate the finesse of a Fabry–Perot etalon whose plates have a reflection coefficient of 0.95 at 500 nm. If the plates are 3 mm apart in air what is the free spectral range? What is the potential chromatic resolving power obtainable at this finesse? What degree of optical flatness in the surfaces of the etalon is needed in order to exploit this potential? The etalon has a clear region of diameter 2.5 cm. Suppose the etalon is used as shown in figure 5.33 with the imaging lens having a focal length of 5 cm. What pinhole diameter is required on the imaging screen to cover the central fringe? What is the system etendue as defined in Section 4.2?

(5.10) Find an expression linking the free spectral range, finesse and FWHM of the fringes of a Fabry–Perot etalon.

Diffraction

6

6.1 Introduction

Diffraction is taken to mean any interference effect due to the interruption of a wavefront by apertures or obstacles, often disposed in regular arrays. The pattern of illumination is very different when the plane of observation is near to the diffracting surface and when it is a large distance away. Near the diffracting surface the pattern is the geometric shadow with fringes close to the shadow edges, and as the plane of observation moves further away this pattern changes smoothly into one that has lost any obvious resemblence to the geometric shape of the apertures and obstacles producing it. In the limit that the source and image plane are infinitely far from the diffracting surfaces, the pattern is called *Fraunhofer* diffraction. In practice this limiting condition is simple to produce: the source is placed at the focus of one positive lens and the plane of observation in the focal plane of another positive lens. Then the source and observing plane are effectively at an infinite distance from the diffracting apertures. Diffraction at finite distances is called *Fresnel* diffraction. Both Fresnel and Fraunhofer diffraction will be treated using the the Huygens–Fresnel picture of secondary waves introduced in the previous chapter. Fraunhofer diffraction is easier to analyse and is the basis of many research tools and technological applications; it will therefore receive more attention than Fresnel diffraction.

The first section below contains a discussion of the theoretical basis of the Huygens–Fresnel picture of diffraction. In the following sections the analysis of diffraction at a single long wide slit is presented, and then that for a rectangular slit. This is followed by a treatment of diffraction by arrays of equally spaced, identical slits. In the limit of very large numbers of slits these arrays are called *diffraction gratings*. Next the diffraction at a circular aperture is described, which leads to the formula previously used in Chapter 4 to calculate the resolving power of optical systems. Grating spectrometers are described and their performance compared to prism and Fabry–Perot instruments. After this Fresnel diffraction at a long broad slit is treated. It is shown how the diffraction pattern evolves, as the distance from the aperture increases: from a geometric shadow to a Fraunhofer pattern. Fresnel diffraction at a circular aperture is then treated, including a discussion of Fresnel zones and zone plates. Then a section is devoted to the schemes used in optical lithography to acheive the high resolution in electronic chip manufacture. After this some comments are made about the limitations of the theory underlying the analysis presented in this chapter when applied to regions very close to the apertures. In the final part of the chapter simple *Gaussian* laser beams and the role of diffraction in their evolution in optical systems are described.

6.2 Huygens–Fresnel analysis

The simple Huygens–Fresnel picture of interference between secondary waves presented in the previous chapter has a fundamental difficulty. Apparently the construction with secondary waves could equally well lead to a backward going wave as to a forward going wave. This difficulty was resolved by Kirchhoff's analysis of wave propagation at apertures, which is described in Appendix C. Kirchhoff obtained the following expression for the *spatial* part of the amplitude at P of a secondary wave originating from a point A in an aperture which is illuminated by a small monochromatic source at S of area dS as shown in figure 6.1

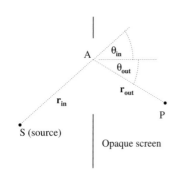

Fig. 6.1 Path lengths and angles used in Kirchhoff's analysis.

$$E = C \exp\left[ik(r_{\text{out}} + r_{\text{in}})\right] \left[(\cos\theta_{\text{in}} + \cos\theta_{\text{out}})/2\right] dS/(r_{\text{out}} r_{\text{in}}). \quad (6.1)$$

Here k is the wave number of the radiation, r_{in} is the length of the path from the source to A and θ_{in} the angle this makes with the normal to the aperture surface at A. r_{out} and θ_{out} are the corresponding parameters for the path from A to the point P where the amplitude is observed. C is a constant that depends on the source intensity. The *inclination factor* $(\cos\theta_{\text{in}} + \cos\theta_{\text{out}})/2$ is unity in the forward direction and falls to zero in the backward direction, and it is this which eliminates a backward propagating secondary wave. $1/r_{\text{in}}$ and $1/r_{\text{out}}$ are the range factors in spherical waves which ensure that the total flux remains constant as they expand. If the incident wave is planar then eqn. 6.1 reduces to

$$E = D \exp\left[ikr_{\text{out}}\right] \left[(1 + \cos\theta_{\text{out}})/2\right]/r_{\text{out}}, \quad (6.2)$$

where D is another constant that depends on the source intensity. Wherever the inclination and range factors change little over the aperture and image region they only affect the overall magnitude and not the pattern of interference. In such cases they can often be factored out of the analysis.

6.3 Single slit Fraunhofer diffraction

Figure 6.2 shows the standard experimental layout used to produce Fraunhofer diffraction. Incident plane waves are produced by placing a monochromatic source at the focus of a positive lens. These waves are incident normally on an opaque plane sheet pierced by a single long slit. Beyond this sheet another positive lens images the diffraction pattern onto a screen in its focal plane. The use of lenses brings the source and image plane in from infinity, so making a compact experimental setup with which to observe Fraunhofer diffraction.

The incident wave across the broad slit of width d gives rise to Huygens secondary waves whose resultant at the screen can be calculated. The

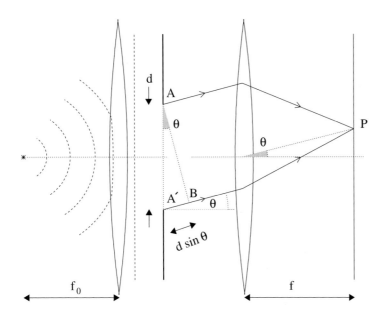

Fig. 6.2 Fraunhofer diffraction at a finite width slit.

path lengths to P will differ by the distance each slit element is from the line AB, drawn perpendicular to the rays travelling to P. For light from an element of the slit of width dx at a distance x from A the extra path length compared to light from A is $x \sin \theta$ and the corresponding phase delay is $\phi(x) = 2\pi x \sin \theta / \lambda = kx \sin \theta$. In the paraxial approximation the phase delay is kxx'/f where x' is the lateral coordinate of P: an expression symmetric in x and x' which will prove useful later. Thus the Huygens wave from an element (x, dx) of the slit gives a contribution to the wave at P

$$dE_{\mathrm{p}}(x) = E_0 \, dx \, \exp\left[i(ks - \omega t + \phi)\right],$$

where s is the optical path length from A to P, and E_0 is a constant expressing the contribution to the field per unit width of the slit. Summing these contributions gives the total amplitude $E_{\mathrm{p}}(\theta)$, from which we get the intensity at P

$$I(\theta) = E_{\mathrm{p}}^{*}(\theta)E_{\mathrm{p}}(\theta).$$

Evidently the common factor $\exp i(ks - \omega t)$ disappears in the intensity

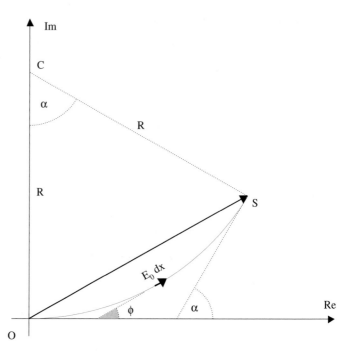

Fig. 6.3 Phasor diagram for Fraunhofer diffraction at a finite width slit. The short arrow is a representative phasor due to a short segment of the slit width. The resultant phasor amplitude for the whole slit is the chord length OS.

so it is only needful to add up the phasors $E_0 \mathrm{d}x \exp\left(i\phi\right)$. Then

$$
\begin{aligned}
E_\mathrm{p}(\theta) &= \int_0^d E_0 \exp\left(ikx\sin\theta\right)\mathrm{d}x &&(6.3)\\
&= E_0\left[1 - \exp\left(ikd\sin\theta\right)\right]/(ik\sin\theta)\\
&= E_0 d\,\exp\left[(ikd\sin\theta)/2\right]\,\mathrm{sinc}[(kd\sin\theta)/2],
\end{aligned}
$$

where as usual $\mathrm{sinc}(x) = \sin x/x$. Thus the intensity is

$$
I(\theta) = (E_0 d)^2\,\mathrm{sinc}^2[(kd\sin\theta)/2]. \tag{6.4}
$$

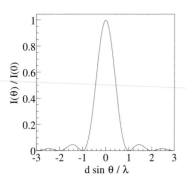

Fig. 6.4 Intensity distribution for Fraunhofer diffraction at a finite width slit.

This calculation of $E_\mathrm{p}(\theta)$ at P is expressed diagramatically in the phasor diagram, figure 6.3. The contribution to the amplitude at P from an element of the slit located at $(x,\mathrm{d}x)$ has magnitude $E_0 \mathrm{d}x$ and has phase angle $kx\sin\theta$. Thus its phasor has length $E_0 \mathrm{d}x$ and is inclined at an angle $\phi = kx\sin\theta$ to the real axis. Adding all the phasors vectorially gives a circular arc which turns through an angle $\alpha = kd\sin\theta$ and has radius $R = E_0 d/\alpha = E_0/k\sin\theta$. The resultant amplitude at P is the chord of this arc

$$
\begin{aligned}
E_\mathrm{p}(\theta) &= 2R\sin\left(\alpha/2\right)\\
&= E_0\,d\,\mathrm{sinc}[(kd\sin\theta)/2] \tag{6.5}
\end{aligned}
$$

as before. Whenever the arc makes one or more complete circles the resultant intensity is zero. For this to be the case

$$kd \sin \theta = 2n\pi,$$

which reduces to

$$d \sin \theta = n\lambda \tag{6.6}$$

where n is a non-zero integer. On approaching the forward direction $\sin \theta \to 0$, and applying l'Hôpital's rule

$$\text{Limit}_{\theta \to 0} \left[\sin \theta / \theta \right] = 1. \tag{6.7}$$

Thus there is a maximum in intensity in the forward direction for which eqn. 6.4 gives

$$I(0) = (E_0 \, d)^2. \tag{6.8}$$

Then eqn. 6.4 can be rewritten compactly

$$I(\theta) = I(0) \, \text{sinc}^2 [(kd \sin \theta)/2]. \tag{6.9}$$

Figure 6.4 shows this intensity distribution calculated for Fraunhofer diffraction at a single long slit. The outer bright fringes are half as wide as the central one. Their peaks lie nearly midway in angle between the minima and the first two have intensities only 0.047 and 0.017 of the forward intensity.

Actual diffracting screens have two-dimensional apertures and the analysis for one-dimensional slits is easily extended to two dimensions. The resulting patterns are simple only where there is some symmetry in the apertures, and of these cases the rectangular and circular apertures are treated below.

6.4   Diffraction at a rectangular aperture

Up to this point it has been assumed in calculating the effects of interference and diffraction that the slit length, L, is so large that λ/L is effectively zero. This excludes any diffraction in the direction of the slit length so that the diffraction pattern is two dimensional, lying in the plane perpendicular to the slit length. The calculation of diffraction will now be extended to a rectangular aperture with limited length and width. Monochromatic plane waves are incident normally on an aperture of width d_x in the x-direction and d_y in the y-direction. Let $\mathbf{e}_x$, $\mathbf{e}_y$ and $\mathbf{e}_z$ be the orthogonal unit vectors in and perpendicular to the aperture plane. Then consider the Huygens wave from an element of the aperture of area $\mathrm{d}x \, \mathrm{d}y$ located at

$$\mathbf{r} = x\mathbf{e}_x + y\mathbf{e}_y.$$

The phasor amplitude at a point P which lies in the direction given by the wave vector

$$\mathbf{k} = k_x \mathbf{e}_x + k_y \mathbf{e}_y + k_z \mathbf{e}_z$$

will be

$$dE(\mathbf{k}) = E_0 \exp\left(i\mathbf{k}\cdot\mathbf{r}\right)\mathrm{d}x\,\mathrm{d}y$$
$$= E_0\left[\exp\left(ik_x x\right)\mathrm{d}x\right]\left[\exp\left(ik_y y\right)\mathrm{d}y\right].$$

The total amplitude at P is obtained by integrating this expresion over x and y. These integrals are independent and have been evaluated in the previous section. Re-using eqn. 6.9 gives for the total intensity at P

$$I(\mathbf{k}) = I(0)\mathrm{sinc}^2(k_x d_x/2)\mathrm{sinc}^2(k_y d_y/2), \qquad (6.10)$$

where $I(0)$ is the intensity in the forward direction. The diffraction pattern is simply the product of those for the two dimensions separately. This distribution is shown in figure 6.5 for a rectangular aperture *twice as tall* as it is wide. Correspondingly the fringes are *twice as wide* as they are tall. The figure shows how the image would appear on a photographic negative or CCD that has been deliberately overexposed in order to enhance the weaker intensity peaks that lie both along and off the axes.

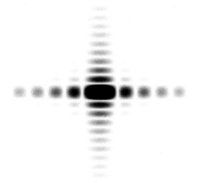

Fig. 6.5 Intensity pattern for Fraunhofer diffraction at a rectangular aperture.

6.5   Diffraction from multiple identical slits

The experimental arrangement shown in figure 6.2 is a template for producing Fraunhofer diffraction using an opaque sheet with any choice of apertures. A very useful arrangement is to have a row of identical slits, d wide, regularly spaced a distance a apart centre-to-centre, as shown in figure 6.6. Analysis of the diffraction produced by such an array will prepare the ground for the discussion of diffraction gratings and spectrometers. In the remainder of this section we suppose that the slits are sufficiently long that the variation in the diffraction pattern is all along a line perpendicular to the slit lengths. It is seen in figure 6.6 that the path length from a slit centre to P changes by $a\sin\theta$ between successive slits. Correspondingly the phase difference of the secondary waves arriving at P from the slit centres is

$$\beta = ka\sin\theta. \qquad (6.11)$$

The phasor addition of the amplitudes at P is illustrated for the case of three slits in figure 6.7. In the upper diagram the dotted line arcs show the phasor contributions of the elements in each slit; the resultants of individual slits are the phasors drawn with open arrowheads; and their resultant is the long phasor ending in a solid arrowhead. In the lower diagram the phasors of the three slits are drawn for four choices of β: namely 0, $2\pi/3$, π, $4\pi/3$ and 2π. If the amplitude at P with a single slit open is unity, then these four choices for β yield amplitudes 3, 0, 1, 0 and 3 respectively, while the corresponding intensities are 9, 0, 1, 0 and 9. As β changes between these configurations the intensity is either falling or rising monotonically. Some simple conclusions can be inferred from figure 6.7 about the pattern of fringes seen when there are N slits.

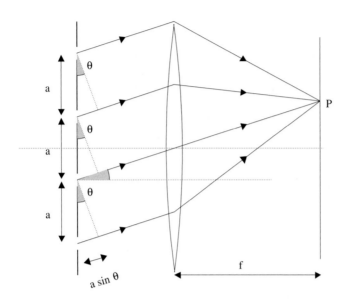

Fig. 6.6 Fraunhofer diffraction at an array of equally spaced identical slits.

- There are *principal* maxima whose intensity is N^2 *times* that of a single slit maximum.
- There are $N - 2$ much lower intensity *subsidiary* maxima between each pair of principal maxima.

The same results will now be obtained analytically. The wave at P due to the mth slit is

$$E_{\mathrm{p}}(\theta)\exp\left[\,i(m-1)ka\sin\theta\,\right],$$

where $E_{\mathrm{p}}(\theta)$ is the single slit contribution given in eqn. 6.5. Making use of eqn. 6.9 the intensity at P due to all N slits is

$$I_N(\theta) = (E_0 d)^2 \mathrm{sinc}^2[(kd\sin\theta)/2]X_N^* X_N, \qquad (6.12)$$

where

$$X_N = \sum_{m=1}^{N} \exp\left[\,i(m-1)ka\sin\theta\,\right] = \frac{1 - \exp\left(iNka\sin\theta\right)}{1 - \exp\left(ika\sin\theta\right)}$$

$$= \exp\left\{[i(N-1)ka\sin\theta]/2\right\}\left[\frac{\sin\left[(Nka\sin\theta)/2\right]}{\sin\left[(ka\sin\theta)/2\right]}\right],$$

where we have used the result

$$\sum_{m=1}^{N} x^{m-1} = (1 - x^N)/(1 - x).$$

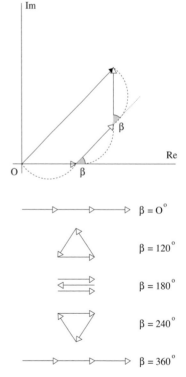

Fig. 6.7 Phasor diagrams for Fraunhofer diffraction at three equally spaced identical slits. The upper panel shows the slit phasors and their resultant. Below the alignments for intensity minima and maxima appear.

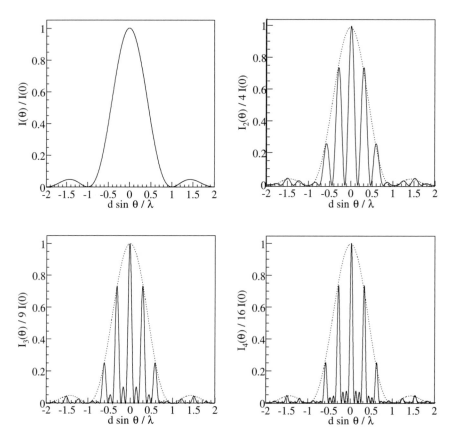

Fig. 6.8 Fraunhofer diffraction patterns for one, two, three and four slits. The slit widths and spacing are the same in each case. The single slit pattern is shown as a dotted line in the multislit plots.

Thus

$$X_N^* X_N = \left[\frac{\sin\left[(Nka\sin\theta)/2\right]}{\sin\left[(ka\sin\theta)/2\right]} \right]^2.$$

Substituting this result in eqn 6.12 gives the intensity at P

$$I_N(\theta) = (E_0 d)^2 \operatorname{sinc}^2[(kd\sin\theta)/2] \left[\frac{\sin\left[(Nka\sin\theta)/2\right]}{\sin\left[(ka\sin\theta)/2\right]} \right]^2.$$

Writing this more succinctly

$$I_N(\theta) = I(0) \operatorname{sinc}^2(\alpha/2) \left[\frac{\sin\left(N\beta/2\right)}{\sin\left(\beta/2\right)} \right]^2 \tag{6.13}$$

where $I(0) = (E_0 d)^2$ is the forward intensity due to a single slit, and where we repeat that

$$\alpha = kd\sin\theta; \quad \beta = ka\sin\theta,$$

d being the slit width and a the slit spacing centre to centre. We see that the intensity pattern for N slits contains a multiple slit pattern

$$M_N(\theta) = \sin^2(N\beta/2)/\sin^2(\beta/2)$$

which modulates the single slit intensity pattern

$$I(0)\,\mathrm{sinc}^2(\alpha/2).$$

$M_N(\theta)$ has zeroes wherever

$$N\beta/2 = m\pi,$$

where m is an integer, with the important *exception* that whenever m/N is also equal to an integer p,

$$\beta/2 = p\pi. \tag{6.14}$$

In this case l'Hôpital's rule gives

$$\sin^2(N\beta/2)/\sin^2(\beta/2) \to N^2, \tag{6.15}$$

with the result that the principal maxima are N^2 times brighter than the single slit maximum. These principal maxima occur at angles given by

$$a\sin\theta = p\lambda. \tag{6.16}$$

Diffraction patterns with one, two, three and four identical equally spaced slits are shown in figure 6.8, all with the same slit width and spacing. In the multislit diagrams the single slit pattern scaled up by a factor N^2 is shown as a dotted line. The intensities at the principal maxima touch this *envelope*, while the subsidiary maxima are much weaker.

Figure 6.9 shows phasor diagrams when there are a large number of slits, in this case 37 slits. Reading from the top panel down these diagrams relate to the forward principal maximum, the adjacent minimum, the first subsidiary maximum, and another principal maximum lying midway between principal maxima. In each case the resultant phasor amplitude is written in brackets. The resultant phasors are drawn with full arrowheads for the two subsidiary maxima. With each slit contributing unit intensity the first subsidiary maximum has intensity close to $D^2 = (2N/3\pi)^2$, which is 0.047 that of a principal maximum. Thereafter the subsidiary maxima decline in intensity with those midway between the principal maxima having about the same intensity as a single slit maximum.

6.6   Babinet's principle

The diffraction pattern produced by an opaque screen with any arrangement of apertures is related to that produced by the *complementary*

Principal maximum (N)

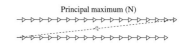

First minimum (0)

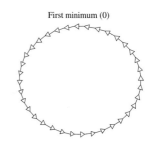

First subsidiary maximum (D)

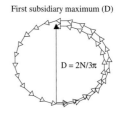

$D = 2N/3\pi$

Middle subsidiary maximum (1)

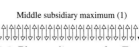

Fig. 6.9 Phasor diagrams for Fraunhofer diffraction produced by a large number of slits.

screen: this complementary screen is transparent wherever the first screen is opaque and opaque wherever the first screen is transparent. Suppose that at P, lying in the focal plane of the second lens in figure 6.2, the light amplitude with the first screen in place is A_1; and A_2 when the complementary screen replaces it. Also suppose the amplitude at P when both screens are removed is A_0. Then clearly

$$A_0 = A_1 + A_2.$$

However A_0 is only non-zero in the exact forward direction, that is when P lies on the optical axis. Elsewhere

$$A_2 = -A_1.$$

It follows that the intensities produced by complementary screens are identical except in the forward direction where the amplitudes add to give the unobstructed amplitude. This is known as *Babinet's principle*.

6.7 Fraunhofer diffraction at a circular hole

The circular aperture, of radius r, shown in figure 6.10 is used to produce Fraunhofer diffraction in the standard experimental arrangement. Light from a point $\mathbf{s}$ on the aperture with wave vector $\mathbf{k}$ travels to P in the image plane. There is a phase difference of $\mathbf{k} \cdot \mathbf{s}$ at P compared to light also travelling to P from the centre of the aperture. The vectors involved are

$$\mathbf{s} = s\cos\phi\,\mathbf{e}_x + s\sin\phi\,\mathbf{e}_y,$$

$$\mathbf{k} = k\cos\theta\,\mathbf{e}_z + k\sin\theta\,\mathbf{e}_x,$$

where $\mathbf{e}_x$, $\mathbf{e}_y$ and $\mathbf{e}_z$ are unit vectors along orthogonal axes drawn in the diagram. Thus

$$\mathbf{k} \cdot \mathbf{s} = ks\sin\theta\cos\phi,$$

and the wave at P is

$$E = E_0 \int_0^r \int_0^{2\pi} \exp{(iks\sin\theta\cos\phi)}s\,\mathrm{d}\phi\mathrm{d}s, \qquad (6.17)$$

where E_0 includes constants and also the complex exponents that vanish when the intensity is calculated. Integration over ϕ gives a Bessel function of order zero[1]

$$E = 2\pi E_0 \int_0^r J_0(ks\sin\theta)s\mathrm{d}s,$$

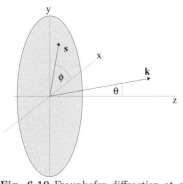

Fig. 6.10 Fraunhofer diffraction at a circular aperture. The vectors $\mathbf{k}$ and $\mathbf{s}$ point from the hole's centre toward a point on the image and to an element of the aperture respectively.

---

[1] *Table of Integrals, Series and Products* by I.S. Gradshteyn and I.M. Ryzhik, edited by A. Jeffrey, 5th edition 1994; published by Academic Press, London.

$$\int_0^{2\pi} \exp{(iz\cos\phi)}\mathrm{d}\phi = 2\pi J_0(z),$$

$$\int_0^r z J_0(z)\mathrm{d}z = r J_1(r).$$

and the integral over s yields a Bessel function of order one

$$E = 2\pi r^2 E_0 J_1(kr\sin\theta)/(kr\sin\theta).$$

Finally the intensity at P is

$$I(\theta) = 4\pi^2 r^4 E_0^2 [J_1(kr\sin\theta)/(kr\sin\theta)]^2.$$

Noting that in the forward direction this expression reduces to $\pi^2 r^4 E_0^2$, the intensity in the diffraction pattern of a circular hole can be rewritten as

$$I(\theta) = I(0)\left[\frac{2J_1(kr\sin\theta)}{(kr\sin\theta)}\right]^2. \tag{6.18}$$

This function is plotted in figure 6.11. The bright central spot is named the *Airy disk*, after the 19th century astronomer who was the first to calculate this distribution. 84% of the light falls within the Airy disk and its angular radius θ is given by

$$\sin\theta = 0.61\lambda/r = 1.22\lambda/D, \tag{6.19}$$

where D is the diameter of the hole. Lord Rayleigh proposed what is now the standard definition of the limit of the resolving power of a lens system: using an optical system with entrance pupil of diameter D, two *point* objects are resolvable if their angular separation exceeds $1.22\lambda/D$. At the limit the intensity maximum of one object's image would lie at the first minimum of the other object's image. This *Rayleigh criterion* was already used in Chapter 4 to evaluate the resolution of optical instruments.

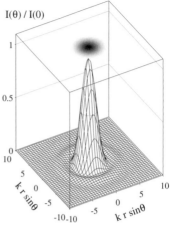

Fig. 6.11 The intensity distribution for diffraction at a circular hole. It is also shown in projection on the roof of the box as it might appear on a photographic negative with only the Airy disk visible.

6.8   Diffraction gratings

It follows from the analysis of Fraunhofer diffraction by multiple slits that when the number of slits becomes large the principal maxima become very narrow, while the subsidiary maxima are so weak as to be undetectable. Opaque screens with large numbers of identical equally spaced slits are known as diffraction *gratings* and are widely used in studying the spectra emitted by sources. The key point is that the principal maxima are so narrow that the principal maxima of spectral lines of closely similar wavelength are separate and distinct: they are resolved. The *chromatic resolving power* of spectroscopic devices is defined as

$$\text{CRP} = \lambda/\Delta\lambda, \tag{6.20}$$

where $\Delta\lambda$ is the smallest difference in wavelength at which it is possible to separate two spectral lines at wavelength λ. The practical limit occurs for gratings when the principal maximum for wavelength $\lambda + \Delta\lambda$ coincides in angle with the minimum adjacent to the same maximum for wavelength λ. In the case of the pth order maximum produced by light of wavelength λ on a grating with N slits

$$Np\lambda = Na\sin\theta,$$

and at the adjacent minimum

$$(Np + 1)\lambda = Na \sin(\theta + \Delta\theta). \tag{6.21}$$

If this minimum coincides with the pth order principal maximum for wavelength $\lambda + \Delta\lambda$ it follows that

$$Np(\lambda + \Delta\lambda) = Na \sin(\theta + \Delta\theta). \tag{6.22}$$

Subtracting eqn. 6.21 from eqn. 6.22 gives

$$Np\,\Delta\lambda - \lambda = 0,$$

whence

$$\lambda/\Delta\lambda = Np. \tag{6.23}$$

It is worth emphasizing that N is the number of grating lines illuminated by the source being studied. Lines out of the beam can play no part in diffracting the beam.

There can be confusion when fringes of one order overlap those of the adjacent order, which will happen if the range of wavelengths in the incident radiation is large. In order to avoid overlaps between the first and second order principal maxima the spread of wavelengths must be such that

$$2\lambda_{\mathrm{min}} > \lambda_{\mathrm{max}},$$

where λ_{min} and λ_{max} are the shortest and longest wavelengths. The widest permissible spread is called the *free spectral range* of the grating, which in terms of λ_{max} is

$$\lambda_{\mathrm{fsr}} = \lambda_{\mathrm{max}}/2 \text{ (first order)}. \tag{6.24}$$

For any higher order, p, overlaps with both the adjacent orders, $p - 1$ and $p + 1$, must be avoided. In this case

$$\lambda_{\mathrm{fsr}} = \lambda_{\mathrm{max}}/(p + 1) \text{ (higher order)}. \tag{6.25}$$

Gratings met in research and industry are nearly always reflection gratings because these are straightforward to manufacture with line densities up to thousands per millimetre. Figure 6.12 shows a parallel beam being reflected from the faces of the reflective elements, which take the place of slits in a reflection grating. The path difference between light reflected from adjacent elements is now $a(\sin\theta - \sin\alpha)$ where α is the angle of incidence and θ the angle of reflection. This is to be compared to a delay of $a\sin\theta$ in the case of a transmission grating with light incident normally. Therefore the sole change required in the analysis carried through for a transmission grating to make it applicable to a reflection grating is to replace $\sin\theta$ everywhere by $(\sin\theta - \sin\alpha)$. The equation giving the angular location of principal maxima becomes

$$a(\sin\theta - \sin\alpha) = p\lambda. \tag{6.26}$$

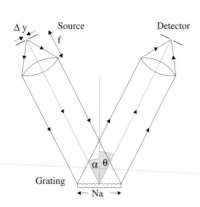

Fig. 6.12 Reflection grating.

6.9 Spectrometers and spectroscopes

An instrument used to view spectra by eye is called a *spectroscope*, while one employed with any sort of electronic detector is called a *spectrometer*. One simple design is shown in figure 6.13. A massive cylindrical frame supports a central rotating table designed to carry the dispersing element, which could either be a reflecting grating or a prism. Two arms protrude from the frame. One arm is fixed rigidly to the frame and carries a collimator with a variable width entry slit. Light from a source illuminating this slit emerges as a parallel beam onto the grating on the central table. The second arm carries a telescope which receives the light from the reflecting grating. It brings the light to a focus at a slit, behind which the detector is placed. This telescope arm can rotate independently about the central vertical axis, and carries a vernier that travels, as the arm moves, over a graduated angle scale that runs around the edge of the frame. A comparison will now be made between the chromatic resolving powers obtainable with such a spectrometer using in one case a grating and in the other a prism.

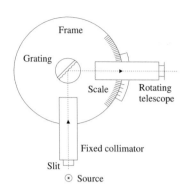

Fig. 6.13 Simple spectrometer or spectroscope design.

The chromatic resolving power of a prism is, unexpectedly, limited by diffraction. In figure 6.14 a parallel beam containing light of two nearby wavelengths is dispersed by a prism at minimum deviation: these wavelengths are λ and $\lambda + \Delta\lambda$. The prism has vertex angle α and it is assumed the whole face is illuminated down to the base, which has length s. BD and BE are wavefronts for the two wavelengths after dispersion, and w is the width of these beams. When working at minimum deviation the angle $\widehat{\text{CBE}} = (\alpha + \delta_{\min})/2 \equiv \theta$. Then eqn. 2.26 gives the refractive index

$$n = \sin\theta / \sin(\alpha/2). \tag{6.27}$$

Differentiating eqn. 6.27 with respect to wavelength gives

$$dn/d\lambda = \cos\theta [\, d\delta_{\min}/d\lambda\,] / [\, 2\sin(\alpha/2)\,]. \tag{6.28}$$

The two wavelengths will be resolvable provided that the change in the deviation in θ between them is greater than the angular width of the diffraction peak for a slit of width w. In this limit

$$\Delta\delta_{\min} = \lambda/w. \tag{6.29}$$

If $\Delta\lambda$ is the precise change in wavelength that produces a change in minimum deviation λ/w we have

$$\Delta\delta_{\min}/\Delta\lambda = \lambda/(w\Delta\lambda). \tag{6.30}$$

$\Delta\lambda$ is small so we can replace $d\delta_{\min}/d\lambda$ by $\Delta\delta_{\min}/\Delta\lambda$ in eqn. 6.28 giving

$$dn/d\lambda = \lambda\cos\theta / [\, 2w\Delta\lambda\sin(\alpha/2)\,] \tag{6.31}$$

Rearranging this equation gives the chromatic resolving power of the prism

$$\lambda/\Delta\lambda = [2w\sin(\alpha/2)/\cos\theta](dn/d\lambda)$$
$$= s(dn/d\lambda). \tag{6.32}$$

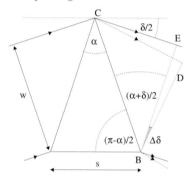

Fig. 6.14 Separation of wavelengths λ and $(\lambda + \Delta\lambda)$ by a prism at minimum deviation. The incoming and outgoing beams are then symmetric. For clarity $\delta_{\min}$ is abbreviated to δ on this diagram.

For reference a comparison is made between the chromatic resolving power of a 5 cm wide grating with 1200 lines/mm and that of a prism of base length 5 cm made of DF flint glass[2] with high dispersion $dn/d\lambda = 10^{-4}\,\mathrm{nm}^{-1}$. Then using eqn. 6.23 and the above equation, we have

$$(\lambda/\Delta\lambda)_{\mathrm{grating}} = 60\,000$$
$$(\lambda/\Delta\lambda)_{\mathrm{prism}} = 5000$$

Gratings are cheaper to make and simpler to handle so they are preferred for most applications. The chromatic resolving power of gratings falls far short of that obtained with a Fabry–Perot etalon. However the free spectral range of an etalon is so narrow that often a two stage instrument must be used to avoid confusion between orders: the first stage uses a prism or grating to select a narrow band of wavelengths which are made the input to the etalon.

6.9.1   Grating structure

The first useful gratings were produced by mechanically engraving lines on metal. Light is reflected off one face of the V-shaped grooves, which act as the slits. Defects, in the form of cyclic variations in the depth or spacing of grooves, were hard to avoid and resulted in low intensity satellites close to the principal maxima, known as *ghosts*. These could be confused with real but weak lines in the spectrum. Nowadays the position of the engraving tool is monitored and controlled using an interferometer which makes it possible to effectively eliminate such cyclic or random errors. The master metal gratings are used as moulds from which polymer replica gratings are cast, after which a film of aluminium is deposited on the replicas. The newer holographic gratings are produced photographically. First a polymer sheet coated with photoresist is exposed to an interference pattern formed by intersecting UV laser beams and the resist is broken down at locations of high intensity in the interference pattern. Afterwards the resist surface is chemically etched to remove the degraded material, leaving a rippled surface that forms the grating. Finally aluminium is deposited on the grating. This holographic process gives finer and more regular line spacing, and in addition the resulting gratings are freer of random defects than the ruled gratings. 3600 lines/mm is a standard line density achieved in holographic gratings compared to 1200 lines/mm with ruled gratings.

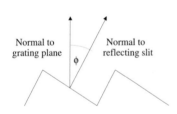

Normal to grating plane ϕ Normal to reflecting slit

Fig. 6.15 Blazed grating surface.

Most of the light incident on a grating will end up in the central zero order fringe which is at the same location for all wavelengths. The light going into this central *white* fringe is wasted as far as any spectroscopic study is concerned. The situation can be improved in the case of ruled gratings by shaping the scribing tool so that the groove cross-section has the appearance shown in figure 6.15. With this profile the normal to each reflecting facet is now inclined at an angle ϕ with respect to the normal to the plane of the grating and hence the direction of the peak of the single slit diffraction pattern is rotated. This process is known as *blazing* and ϕ is called the blaze angle. Blazing leaves the directions of principal maxima unchanged because the plane containing the slit centres is unchanged. The diffraction pattern (for ten slits) is shown both without and with blazing in figure 6.16. Evidently blazing at an appropriate angle can improve the brightness of the first principal maxima by a very big factor. The wavelength whose first principal maximum on one side lies exactly at the centre of the rotated single slit envelope is called the *blaze* wavelength, and is given at normal incidence by

$$\lambda_{\text{blaze}} = a \sin 2\phi. \tag{6.33}$$

Holographic gratings can be blazed to a limited extent by etching the grating with an ion beam.

6.9.2   Etendue

The etendue is calculated for the instrument shown in figure 6.12. Collimator and telescope are represented as single lenses with identical focal lengths f, the collimator slit width along the direction of dispersion is Δy, and its length is w. The grating is taken to have N lines with interval a and to be a square of side length $W = Na$. Etendue is invariant through the system so it can be calculated at any convenient aperture; here the beam at the collimator slit is considered. Only light passing through the area of the grating is useful, and its area projects to cover an area $W^2 \cos \alpha$ of the collimator lens. This active area at the lens subtends a solid angle at the collimator slit

$$\Omega = W^2 \cos \alpha / f^2. \tag{6.34}$$

Ideally the slit width should subtend an angular width at the grating equal to the FWHM of a principal maximum. If it is any larger the maxima would be smeared to a greater effective width; if it is any narrower it would restrict the light etendue unnecessarily. The next step is to determine this ideal slit width. Differentiating eqn. 6.26 with respect to wavelength gives

$$\text{d}\alpha = p \, \text{d}\lambda / (a \cos \alpha). \tag{6.35}$$

Now the lateral displacement at the collimator slit plane of $\text{d}y$ produces a change in the angle of incidence on the grating

$$\text{d}\alpha = \text{d}y / f. \tag{6.36}$$

Using the previous two equations to eliminate $\text{d}\alpha$ gives

$$\text{d}y = pf \, \text{d}\lambda / (a \cos \alpha).$$

Hence for a resolution $\Delta\lambda$ the slit width must be as small as

$$\Delta y = pf \, \Delta\lambda / (a \cos \alpha),$$

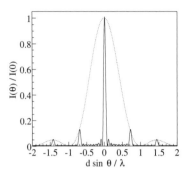

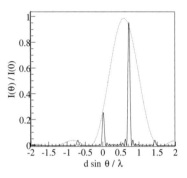

Fig. 6.16 The diffraction pattern seen with unblazed (upper panel) and blazed (lower panel) gratings.

and the slit area is then

$$A = wpf\,\Delta\lambda/(a\cos\alpha). \tag{6.37}$$

Using eqns. 6.34 and 6.37 to substitute for Ω and A in eqn. 4.11 gives the etendue of the spectrometer

$$\begin{aligned}
\mathcal{T} &= [W^2\cos\alpha/f^2]\,[wpf\,\Delta\lambda/(a\cos\alpha)]\\
&= W^2 wp\,\Delta\lambda/(fa).
\end{aligned} \tag{6.38}$$

Using eqn. 6.26 again to replace p, we get

$$\mathcal{T} = W^2\,(w/f)\,[\Delta\lambda/\lambda]\,(\sin\theta - \sin\alpha). \tag{6.39}$$

The luminous flux through the system is simply obtained by multiplying the etendue by the radiance incident on the input slit in the wavelength range $(\lambda,\Delta\lambda)$

$$\begin{aligned}
F &= \mathcal{T}\,I(\lambda)\Delta\lambda\\
&= I(\lambda)\,[\Delta\lambda^2/\lambda]\,W^2\,(w/f)\,(\sin\theta - \sin\alpha).
\end{aligned} \tag{6.40}$$

Once the chromatic resolving power and the incident luminance are chosen, the slit length and grating area should both be increased, and the f/# reduced as far as feasible. Of course the light from the source needs to illuminate the whole slit and its fan of rays should be wide enough to cover the whole grating; otherwise the potential etendue calculated above will not be available.

6.9.3   Czerny–Turner spectrometer

A very widely manufactured spectrometer design is based on the Czerny–Turner *mounting* shown in figure 6.17. The grating rotates about an axis through its centre and perpendicular to the plane of the diagram. Concave mirrors are used to collimate the light from the entry slit and focus the diffracted light onto the exit slit. The use of mirrors avoids any chromatic aberration that could arise with lenses, and the symmetric layout means that coma cancels between the two reflections. Folding the optical paths keeps the instrument relatively compact. When the grating is at a given orientation the first principal maximum for some wavelength determined by the slit location will pass through the exit slit onto the detector. Therefore the spectrum can be scanned across the detector by rotating the grating. The entry slit is often of fixed width, while the exit slit has a variable width which can be changed to alter the chromatic resolving power. In order to optimize the light entering from the source the source can be imaged with a lens so as to just fill the input slit. The cone of rays from the source should also be made just wide enough to fill the grating. Light passing alongside the grating is not only wasted but also gets reflected to give a background haze falling on the output slit.

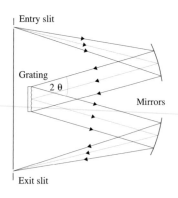

Fig. 6.17 Czerny–Turner spectrometer components.

In common with other spectrometers the Czerny–Turner can be used as a *monochromator*. In this role the source illuminating the input would ideally have a flat continuous spectrum ($I(\lambda)$ is constant). Then the output slit itself becomes a source of narrow bandwidth radiation that can be tuned in wavelength by rotating the grating.

6.9.4 Littrow mounting

This simple design for a grating spectrometer uses a single lens, or mirror, as shown in figure 6.18. The blazed grating is oriented so that the incoming light is incident at the blaze angle and then the reflected light returns almost parallel to the incident beam. The mirror is an off-axis section of a paraboloid, and by giving this a slight tilt the reflected beam is cast a little below the plane of the diagram so that the detector does not overlap the source. Littrow mountings suffer from astigmatism and coma, but are extremely compact. A grating in the Littrow mounting can be used as a monochromator to tune the wavelength of a dye laser, as shown later in figure 14.10.

6.9.5 Echelle grating

Michelson appreciated that the chromatic resolving power of a grating

$$\lambda/\Delta\lambda = pN$$

is proportional to the order p as well as to the number of lines N. He reasoned that it should therefore be possible to obtain a high chromatic resolving power with a coarse grating by using a high order of diffraction. Today the most practical grating whose design is based on this principle is the *echelle* grating, and this is illustrated in figure 6.19. It is an extreme form of a blazed grating with deep, coarse steps and is used in a Littrow mounting for which eqn. 6.26 becomes

$$2a \sin \phi = p\lambda, \qquad (6.41)$$

where ϕ is the blaze angle. Echelle gratings provide a very useful intermediate level in chromatic resolving power between the Fabry–Perot spectrometer and the grating spectrometer. Thus for example with the *R2* grating, which has a blaze angle, ϕ, such that $\tan \phi = 2$, and 316 lines/mm, the diffraction order at 500 nm wavelength is around 11 and the chromatic resolving power with a 10 cm width grating is thus 350 000.[3] Such wide line spacing is hard to achieve holographically and echelle gratings must be engine ruled under interferometer control.

An echelle alone has a small free spectral range: if illuminated by a wide range of wavelengths the higher order spectra from shorter wavelengths overlap the lower order spectra from longer wavelengths. Instruments designed for satellite surveys of the Earth use an echelle in series

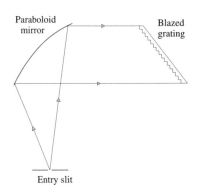

Fig. 6.18 Littrow mounting with a blazed grating with the light incident on the grating perpendicular to the reflective surfaces.

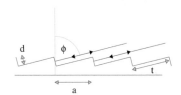

Fig. 6.19 Echelle grating showing incoming and outgoing beam directions in a Littrow mounting.

[3]A standard grating with similar resolving power in first order to this echelle would have 350,000 lines in a similar width, which would be difficult to make. It would be illuminated head on rather than edge on as in the case of the echelle, so that the optics would need much larger apertures.

with a prism, their directions of dispersion being at right-angles. This achieves good resolution across a wide wavelegth range. Sometimes the two elements are combined into a single component, called a *grism*, with the grating formed directly on the surface of the prism.

6.9.6   Automated spectrometers

Simple modern *spectrometers* contain a fixed grating and a CCD array to capture the whole spectrum simultaneously. Figure 6.20 illustrates the basic elements of such a spectrometer. The CCD array is two dimensional being made wide enough to cover the image of the slit length, with typically $10\,\mu\text{m} \times 10\,\mu\text{m}$ pixels. If a standard grating were used the image of the input slit would trace out a curved surface as the input wavelength changes. Instead special gratings are constructed in order to keep the whole image focused on the plane surface of the CCD array. These are holographic gratings in which the slit spacing changes uniformly across its length, and this feature gives some supplementary focusing which flattens the image plane. These spectrometers are made to be hand-held and interface to a PC with immediate display of the spectrum. Even so the resolution can be as good as $0.2\,\text{nm}$ over the visible spectrum.

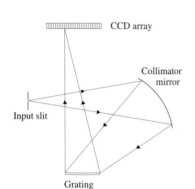

Fig. 6.20 Spectrometer with focusing grating and CCD detector array.

6.10   Fresnel and Fraunhofer diffraction

Figure 6.21 contrasts the conditions for observing Fresnel and Fraunhofer diffraction. When, as shown in the upper diagram, the incident and outgoing waves at the aperture have plane wavefronts the phase of the light arriving at P depends linearly on the position at which the Huygens wave originated across the slit and it is this feature that makes analysis straightforward. Linear dependence of phase on the position across the slit can be regarded as the *distinguishing* feature of Fraunhofer diffraction. Much more common is the situation shown in the central diagram, where the viewing plane is at a finite distance from the slit and the diffraction is known as Fresnel diffraction. P is the point at which the light is observed and is a perpendicular distance r from the plane of the slit, and a transverse distance ρ from a point S on the slit. Then the distance SP is

$$s = \sqrt{r^2 + \rho^2} \approx r + \rho^2/2r. \qquad (6.42)$$

If it is assumed that plane monochromatic waves are incident on the slit then the relative phase of light arriving at P from S is $k\,[r + \rho^2/2r]$, which varies *quadratically* with the position of the point of origin across the slit. This quadratic dependence is *characteristic* of Fresnel diffraction and makes it relatively more complicated to analyse than Fraunhofer diffraction. At large enough distances the wavefronts converging at P become sufficiently flat that the diffraction approaches the Fraunhofer limit. At this point the term $\rho^2/2r$ has become small compared to the wavelength. The explicit criterion for Fraunhofer diffraction to apply for a slit of width w is that

$$(w/2)^2/(2r) \leq \lambda/8,$$

or equivalently

$$r \geq w^2/\lambda = r_{\mathrm{R}}, \tag{6.43}$$

where r_{R} is known as the *Rayleigh distance*. Thus far the incoming waves at the slit were assumed to be plane. If the source is instead at a finite distance r_{s} then the condition for Fraunhofer diffraction to apply is that both

$$r \geq r_{\mathrm{R}} \text{ and } r_{\mathrm{s}} \geq r_{\mathrm{R}}, \tag{6.44}$$

so that the curvature of both incoming and outgoing waves is negligible. If either condition is violated the phase of the light arriving at P depends quadratically on ρ, giving Fresnel diffraction. In the case of a circular aperture of radius r

$$r_{\mathrm{R}} = \pi r^2/\lambda. \tag{6.45}$$

In the lower diagram in figure 6.21 the observation plane lies at the image plane of the source, and the aperture is located anywhere between source and image plane. According to the definitions given, the diffraction pattern observed with this experimental layout is also, surprisingly, an example of Fraunhofer diffraction. In order to understand this conclusion, first imagine that a negative lens is placed immediately before the aperture and that it has the correct focal length $-f$ to give a parallel beam. In addition imagine a second, positive lens of focal length f to be placed immediately after the aperture. With this new setup there will be Fraunhofer diffraction because the incident and emerging waves are planar at the aperture. The new setup differs from the original in having coincident lenses of equal and opposite powers at the aperture. Thus it is optically equivalent to the original setup: the image will be in in the same place and the same size as before. Hence the original setup in the lower diagram produces Fraunhofer diffraction.

To summarize: in all cases where observation is made in the image plane of the source Fraunhofer diffraction is observed, which includes the standard setup shown in figure 6.2. For all other arrangements there is Fresnel diffraction. In Fresnel/Fraunhofer diffraction the phase of light arriving at the observing plane is quadratically/linearly dependent on the distance across the aperture of the point at which the Huygens secondary wave originates.

6.11 Single slit Fresnel diffraction

The phasor diagram for Fresnel diffraction differs markedly from that for Fraunhofer diffraction shown in figure 6.3. In that plot the change in phase angle between phasors contributed by successive elements of the slit was constant because of the linear dependence of phase on position

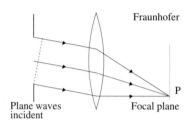

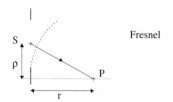

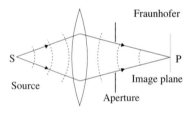

Fig. 6.21 Examples of Fraunhofer diffraction (upper and lower panels), and Fresnel diffraction (centre panel).

across the slit. By contrast with a quadratic dependence of the phase angle on position the arc formed by the phasors will curl up, and the result looks as shown in figure 6.22. This is called a *Cornu spiral*. The calculation of the intensity distribution for Fresnel diffraction at a linear slit shown in the middle panel of figure 6.21 starts from the expression for the Huygens wave arriving at P from S, an element of the slit of width $d\rho$ at a distance ρ across the slit

$$dE = \exp{(iks)}d\rho/s,$$

where the factor $1/s$ allows for the fall-off in amplitude as the wave spreads out with distance. Thus the total wave at P is

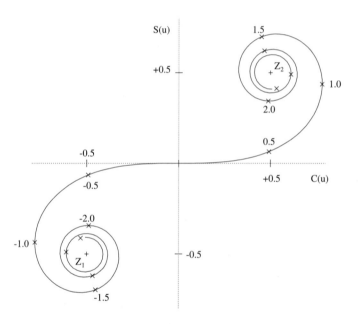

Fig. 6.22 Phasor plot for Fresnel diffraction at a slit. The resultant amplitude is the chord length between the points with values of u corresponding to the two edges of the slit. The tighter turns of the Cornu spiral, which are omitted here, would converge at the crosses. Points where $u = \pm0.5, \pm1.0, \pm1.5...$ are indicated by $\times$s.

$$E = \int \exp{(iks)}d\rho/s.$$

Using the approximation from eqn. 6.42 and taking constant factors outside the integral

$$E = [\exp{(ikr)}/r] \int \exp{(ik\rho^2/2r)}d\rho,$$

where the difference in the amplitude factor between $1/s$ and $1/r$ across the slit is ignored. Making a change of variable to

$$u = \sqrt{2/\lambda r}\, \rho \qquad (6.46)$$

and dropping the constant multipier gives

$$E(u_1, u_2) = \int_{u_1}^{u_2} \cos\left(\frac{\pi}{2}u^2\right) du + i \int_{u_1}^{u_2} \sin\left(\frac{\pi}{2}u^2\right) du, \qquad (6.47)$$

where u_1 and u_2 are the values of u at the two edges of the slit. This equation contains the *Fresnel* definite integrals

$$C(u) = \int_0^u \cos\left(\frac{\pi}{2}u^2\right) du; \qquad (6.48)$$

$$S(u) = \int_0^u \sin\left(\frac{\pi}{2}u^2\right) du. \qquad (6.49)$$

The amplitude given by eqn. 6.47 can be rewritten

$$E(u_1, u_2) = [\, C(u_2) - C(u_1)\,] + i\,[\,S(u_2) - S(u_1)\,]. \qquad (6.50)$$

Thus the intensity at P is, apart from constants,

$$I(u_1, u_2) = [\, C(u_2) - C(u_1)\,]^2 + [\,S(u_2) - S(u_1)\,]^2. \qquad (6.51)$$

These results become more approachable once they are displayed graphically. In figure 6.22 the trajectory of the function $C(u) + iS(u)$ as u runs from $-\infty$ to ∞ is drawn on an Argand diagram with representative values of u indicated along the path. The total phasor amplitude given by eqn. 6.47 is that part of the loop between the points where u is equal to u_2 and u_1. The magnitude of the resultant amplitude at P in figure 6.21 is then the length of the chord joining the two ends, while the light intensity at P is the square of this chord length.

The quadratic dependence of phase on position across the slit has the following effect: the angle which the curve in figure 6.22 makes with real axis increases quadratically with the parameter u, and hence the phasor curve coils up more and more tightly as $|u|$ increases. Only the first few loops are shown; asymptotically they spiral into the end points Z_1 and Z_2 in figure 6.22. The length $Z_1 Z_2$ represents the amplitude of the electric field at P when the obstructing screen is removed, namely the amplitude of a freely propagating wave. A line from the origin to either end point represents half this amplitude and corresponds to the situation in which one slit jaw is moved off to infinity so that one side of the incident wave is unobstructed, and the other half of the incident wave is blocked. P would then lie exactly at the edge of the geometric shadow. This gives an intensity at P exactly one-quarter what it would be in the freely propagating wave. If P moves laterally into the shadow then one end, u_2, of the phasor remains fixed at the upper end point Z_2 while the other, u_1 end of the phasor moves away from the origin

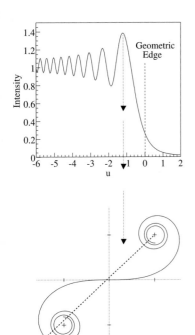

along the curve toward Z_2. Thus the amplitude (chord length) and intensity (chord length squared) diminish steadily as P moves deeper into the shadow. If P moves in the other direction away from the geometric shadow, the u_2 end of the phasor remains at Z_2, while the u_1 end now moves away from the origin along the curve toward the lower end point Z_1. In this case the phasor length oscillates and so will the intensity. As P moves steadily away from the shadow edge the u_1 end of the phasor moves round loops that grow tighter around Z_1 and the intensity oscillates with gradually diminishing swings. The intensity variation near the edge of the geometric shadow is shown in the upper panel of figure 6.23, normalized to the unobstructed wave intensity. Note that the illumination is highest just outside the geometric shadow, being around 40% brighter than in the unobstructed wave. The phasor producing maximum intensity is drawn onto the Cornu spiral in the lower panel of figure 6.23.

How Fresnel diffraction changes as the plane of observation moves away from the slit until it lies at the Rayleigh distance from the slit is illustrated in figure 6.24. On the left are shown three planes at selected distances from the slit, and on the right the light intensity distributions observed on those planes: the correspondence between each surface and its intensity curve is indicated by a shared line style. On the surface clos-

Fig. 6.23 Fresnel diffraction at a linear edge. In the upper panel the intensity distribution is shown around the edge of the geometric shadow. The phasor giving the highest intensity is drawn on the Cornu spiral in the lower panel.

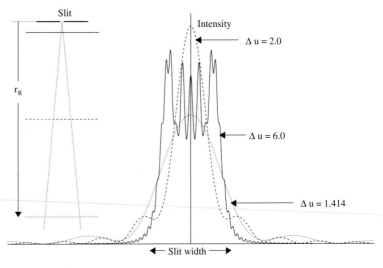

Fig. 6.24 Fresnel diffraction at a slit. On the left are shown the planes and on the right the patterns seen in those planes. Plane and plot are drawn with matching line styles.

est to the slit indicated by the full line in figure 6.24 $\Delta u = u_2 - u_1 = 6$, which is a long section of the Cornu spiral. As P moves across the illuminated region of the surface indicated by the solid line the length of the chord on the Cornu spiral oscillates strongly in length, and the intensity is shown by the full line curve. The pattern still has some resemblance to the geometric shadow. At a surface further off, indicated by the broken line, the arc length $\Delta u = 2$ is shorter and the chord length changes less violently as P moves through the illuminated region. The corresponding intensity curve is shown as a broken line. The third, dotted line surface is at the Rayleigh distance from the slit, and here $\Delta u = 1.414$. As the point of observation P moves across the illuminated region the dotted intensity curve is traced out. This has a shape approximating to the Fraunhofer diffraction pattern for a single slit. Notice that the minima of intensity are not zero, and will only fall to zero in the Fraunhofer limit.

6.11.1 Lunar occultation

When the Moon's surface passes across the line joining the observer on Earth to a star the Moon is said to occult the star. This alignment is shown in figure 6.25. At occultation the star's image at the Earth's surface will be diffracted. Although the distances from the Earth to Moon, r_m, and from Moon to star, r_s, are literally astronomical the diffraction is Fresnel diffraction. This is because the aperture is infinitely wide so that the Rayleigh distance given by eqn. 6.43 is also infinite. Thus the fringe pattern is that shown in the lower panel. From figure 6.22 we see that the first maximum of intensity occurs where $u \approx 1.2$, and the second where $u \approx 2.4$. Then using eqn. 6.46 the fringe spacing is

$$\Delta\rho \approx 1.2\sqrt{\lambda r_m/2}. \tag{6.52}$$

Taking the wavelength to be 500 nm gives a fringe spacing of about 12 m over the Earth's surface. The fringes travel at the speed of the Moon's shadow so that the signal produced at a light sensitive detector will be a series of pulses corresponding to the intensity maxima.

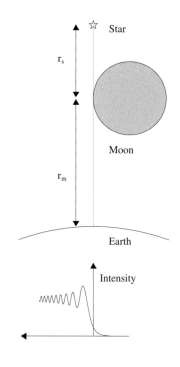

Fig. 6.25 Lunar occultation.

6.12 Fresnel diffraction at screens with circular symmetry

The prediction for the diffraction pattern produced by a circular disk provided a convincing early test of the wave theory of light. In 1818 Fresnel submitted a paper on wave theory in a competition judged by a panel appointed by the French Academy of Sciences. One panel member, Poisson, calculated from Fresnel's theory that when a circular obstacle is illuminated by a point source there should be a bright spot at the exact centre of the geometric shadow. This prediction appeared to be absurd and easy to refute: Arago, the panel chairman, did the experiment and then saw that the bright spot was really present. This spot became known as *Poisson's spot*. The analysis of Fresnel diffraction at a circular aperture is presented here in a more qualitative way than that used for a rectangular aperture. A more complete discusion would become complicated and not add any significant insights.

Figure 6.26 shows a screen illuminated by light from a point source shining through a circular hole. Huygens waves originating from an annulus of the wavefront at the hole will arrive at the point P on axis with equal phase delay $k(s_1+s_2)$ where s_1 is the distance from source to annulus and s_2 the distance from the annulus to P. With the approximation made in eqn. 6.42

$$s_1 + s_2 = r_1 + r_2 + \rho^2/2r_1 + \rho^2/2r_2.$$

Then the phase becomes

$$k(r_1 + r_2) + k\rho^2/2r_1 + k\rho^2/2r_2,$$

and dropping the piece common to all annuli, $k(r_1 + r_2)$, leaves the relative phase

$$\phi = k\rho^2(1/r_1 + 1/r_2)/2 = k\rho^2/2R, \tag{6.53}$$

where

$$R = r_1 r_2/(r_1 + r_2). \tag{6.54}$$

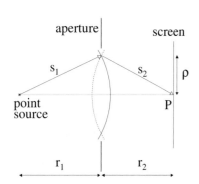

Fig. 6.26 Fresnel diffraction at a circular aperture.

Now imagine that the surface of the incoming wavefront is divided into annular zones such that between adjacent zones the path length to P changes by one half wavelength. Then the Huygens waves from adjacent zones will arrive at P with a phase difference of π, so they tend to cancel one another's contributions to the amplitude at P. Annular zones drawn in this manner are called *Fresnel zones* and the outer radius of the mth Fresnel zone is given by

$$k\rho_m^2/2R = m\pi,$$

whence

$$\rho_m^2 = mR\lambda. \tag{6.55}$$

The zones therefore have equal areas

$$\pi(\rho_m^2 - \rho_{m-1}^2) = \pi R\lambda$$

in this approximation. Consequently the cancellation of contributions to the amplitude from succesive Fresnel zones is quite precise. The total amplitude at P is

$$E = \int_0^{\rho_0} 2\pi \exp\left[i\rho^2 k/2R\right]\rho\,d\rho.$$

Making the substitution $\xi = \rho^2 k/2R$ gives

$$E = \lambda R \int_0^{\xi_0} \exp\left(i\xi\right) d\xi = i\lambda R\left[1 - \exp\left(i\xi_0\right)\right], \tag{6.56}$$

which indeed oscillates around zero as ρ_0 and ξ_0 increase.

When a more precise calculation is made, starting from Kirchhoff's expression for the secondary wave in eqn. C.8, the corrections to an individual zone almost cancel. On the one hand the zone area increases slightly with increasing m; on the other hand the amplitude at P falls off with m, both because the distance from the zone increases and because the inclination factor grows smaller. If there are n zones exposed in the aperture the total amplitude can be written as follows

$$E = 0.5E_1 + (0.5E_1 + E_2 + 0.5E_3)$$
$$+(0.5E_3 + E_4 + 0.5E_5)$$
$$\left\{ \begin{array}{ll} + \ .... + 0.5E_{n-1} + E_n & (n \text{ even}) \\ + \ .... + 0.5E_n & (n \text{ odd}) \end{array} \right\}.$$

Contributions from odd and even zones have opposite sign so that the contributions which are bracketed together cancel, leaving

$$E = 0.5(E_1 + E_n). \tag{6.57}$$

Now imagine that the aperture is made wide enough that the path to P from the whole spherical wavefront emitted by the source is unobstructed. Then the final, nth, zone will be in the backward direction and will have an inclination factor zero. This zone therefore makes zero contribution at P, so the previous equation reduces to

$$E = 0.5E_1,$$

showing that if the wavefront is completely unobstructed the wave amplitude on axis is only half as large as when a circular hole exposes just the central Fresnel zone. It is now possible to use this analysis to explain the origin of Poisson's spot.

An aperture exposing exactly n Fresnel zones has as its complement a disk that exactly covers n zones. Suppose that the electric field at the centre of the pattern with the disk in place is E_{disk}. Babinet's principle requires that the sum of the complementary wave amplitudes equals the amplitude for a freely propagating wave on axis. Thus

$$E_{\text{disk}} + 0.5(E_1 + E_n) = 0.5E_1.$$

It follows that $E_{\text{disk}} = -0.5E_n$, which is non-zero. Therefore there is always a bright spot at the centre of the geometric shadow of a circular disk illuminated by a point source. A clean ball bearing makes an excellent circular disk; but if the outline of the disk used departs from the circular by an area as small as a single Fresnel zone Poisson's spot is lost.

Fig. 6.27 Zone plate seen from beam direction.

6.12.1 Zone plates

A *zone plate* is a flat circular screen which has alternate transparent and opaque annular zones, the outer radius of the mth zone counting from the centre is $\rho_m \propto \sqrt{m}$. An example is shown in figure 6.27 with opaque even zones. If such a zone plate is placed between a point source and a screen, as in figure 6.26 such that eqn. 6.55 is satisfied,

$$\rho_m^2 = mR\lambda \tag{6.58}$$

then the zones on the zone plate precisely match the Fresnel zones. The contributions to the amplitude at P due to light passing through each clear zone are then all in phase. If there are $n/2$ transparent zones the central intensity is roughly $n^2/4$ larger than that due to the central zone alone and $n^2/2$ times larger than the intensity with the unobstructed incident radiation. In effect the zone plate focuses the light from the point source at P. Using eqn. 6.54 to replace R in eqn. 6.58 gives

$$\rho_m^2/m = \frac{r_1 r_2}{r_1 + r_2}\lambda,$$

so that

$$1/r_1 + 1/r_2 = m\lambda/\rho_m^2.$$

If the distances are measured with the Cartesian sign convention this becomes

$$1/r_2 - 1/r_1 = m\lambda/\rho_m^2,$$

which is identical to the thin lens formula with $\rho_m^2/m\lambda$ being the equivalent focal length. Moving off axis the focusing property is soon lost, and in addition the image obviously suffers chromatic aberration.

While the focusing property of a zone plate is mainly a curiosity for light its use is important in nearby regions of the electromagnetic spectrum for which lenses are difficult or impossible to construct; that is, for X-rays and short wavelength UV light. In addition zone plates are used to focus electron beams used in the commercial fabrication of electronic circuits on silicon wafers. The focal lengths are then about 1 mm and the zones' widths are about 1 nm.

6.13 Microprocessor lithography

The progress exhibited in Moore's law, that the number of transistors in processors doubles every 18 months, has been due in large measure to improvements in optical lithography. Each layer of circuitry on a silicon wafer requires the following sequence of operations that make up the lithographic process. First the wafer is coated with a photoresist which when irradiated changes to a form that can be chemically etched away. Then the features required for that layer are transferred from an optical mask (made by optical or electron beam lithography) onto the wafer by a lens system using Koehler illumination as shown in figure 6.28. The arrowed rays are the principal rays from the two edges of the mask and these are incident normally at the wafer surface. This telecentric arrangement has the advantage that the image will not be displaced laterally by height variations across the chip surface. After

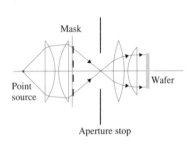

Fig. 6.28 Imaging a mask onto a wafer using Koehler illumination. Principal rays are drawn from the edges of the wafer.

exposure, etching and subsequent cleaning the wafer is ready for further processing which may involve deposition of material or removal of material in the regions exposed by the etching. The smallest achievable feature according to the Rayleigh criterion is

$$\Delta s = (1.22\lambda/D)v = 0.61\lambda/\text{NA}, \tag{6.59}$$

where v and D are the image distance and exit pupil diameter respectively. Ever finer features can be produced by reducing the wavelength λ: from 1997 the 248 nm KrF laser has been used and more recently the 193 nm ArF laser. As noted in discussing figure 1.16 the internal transmittance of fused silica falls steeply in this region so that further reduction in wavelength requires a change in approach to less well-studied materials and to mirrors. Shortening the wavelength also reduces the depth of field given by eqn. 4.21. Taking the image to lie in the focal plane, the depth of field obtainable with the above resolution is

$$\Delta v = 0.305\lambda/(\text{NA})^2. \tag{6.60}$$

At a wavelength of 248 nm and a NA of 0.6 the depth of field is only $0.15\,\mu$m, which is less than the surface height variation over a wafer. Consequently several exposures may be needed at different depth settings to adequately expose the resist across all the wafer. The telecentric illumination ensures that this does not lead to any lateral shift of the image. Several methods are used to circumvent the diffraction limit so that features of less than $\lambda/2$ are consistently realized, and simple examples of these techniques are now described.

One method used is to immerse the region between the final lens and wafer in a liquid of refractive index n and refocus to keep v unchanged. The wavelength in the liquid is lower by the factor n than the wavelength in air, and hence the resolution is similarly improved. Figure 6.29 shows an example of the technique known as phase shift masking (PSM). In the upper panel a section of a mask with opaque sections formed by chromium deposited on quartz is shown together with both the amplitude pattern and intensity distribution at the wafer. Through diffraction the structure of the mask has been totally obscured. In the lower panel, one clear region of the mask is treated to give a phase shift of π, which might be obtained by coating with molybdenum silicide or by etching away a layer of quartz. Now the structure of the mask is resolved in the intensity distribution on the wafer. An array of sophisticated types of PSM are deployed in practice. The effect of diffraction when features drop below one wavelength in size is to round corners, extend tracks and broaden long tracks. Therefore instead of making the mask exactly the same shape as that required on the wafer in the upper panel of figure 6.30, the mask is shaped as shown in the lower panel. This is called optical proximity correction. Detailed optical modelling is needed to optimize the shapes used.

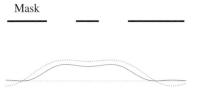

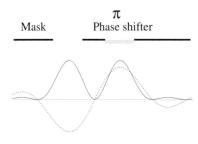

Fig. 6.29 Phase shift masking. The mask in the lower panel has a coating over one open region. The amplitudes and intensities at the wafer are shown with broken and full lines respectively.

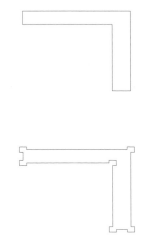

Fig. 6.30 A mask with optical proximity corrections is shown in the lower panel. This yields a satisfactory approximation to the shape shown above when projected onto the wafer.

6.14   Near field diffraction

The Huygens–Fresnel addition of secondary waves fails when used to determine the electric field amplitude of electromagnetic waves very close to the edges of obstacles. This is because the underlying theoretical justification for the Huygens–Fresnel analysis made by Kirchhoff no longer applies. Kirchhoff assumed that the electric field across any aperture is exactly what it would be in the absence of any obstacle, and that over the area of the obstacles it is zero. Reflections of waves from obstacles are thus neglected although they are important close to the edges of obstacles. Secondly Kirchhoff's analysis assumes that the electric field is a scalar quantity. This is adequate only if the points at which interference is observed are far enough from the aperture that the light from all points across the diffracting surface arrives nearly parallel. Then scalar addition yields a good approximation to the total vector field amplitude. However the waves from the two edges of a slit to a nearby point travel in very different directions. Their electric field vectors will not usually be parallel and should not be added as if they were scalars. Near to the diffracting surface it is necessary to solve Maxwell's equations with the appropriate boundary conditions imposed by the edges of the apertures; for example, one requirement is that at the surface of a good conductor the tangential component of the electric field vanishes.

6.15   Gaussian beams

A Fabry–Perot etalon with a large spacing between mirrors is called a cavity and may have mirrors which are flat or curved. Figure 6.31 shows a longitudinal section through a beam confined in a Fabry–Perot cavity with, in this case, mirrors of radius R spaced a distance R apart. This is called a *confocal* cavity because the focal points of the mirrors coincide, and *symmetric* because the radii are equal. The beam boundary does not have straight line edges, rather it has an outline that is hyperbolic with a waist. The standing waveforms that develop have curved wavefronts, shown as dotted lines in the figure, and their intensity falls off as the edge of the mirror is approached.

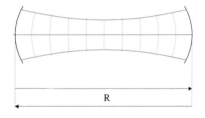

Fig. 6.31 Gaussian beam confined in a symmetric confocal Fabry–Perot cavity. The curved broken lines are wavefronts.

These standing waves are examples of solutions of Maxwell's wave equation subject to the requirement that there is a node of the electric field at the mirror surface and that outside the mirror area the intensity vanishes. Light within a laser is usually confined within a Fabry–Perot cavity, and consequently the solutions to Maxwell's equations for radiation confined in such a cavity have special interest. For the present we can note that in a laser the cavity contains an active material which establishes and maintains the standing waves of electromagnetic radiation in the cavity. Part of this wave escapes through one partially transmitting mirror to form the external laser beam. Figure 6.32 shows a longitudinal slice through such a laser beam after passing through a

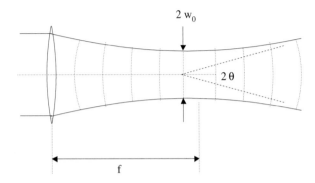

Fig. 6.32 Gaussian beam focused by an aberration-free lens. The curved dotted lines are wavefronts and the broken lines are asymptotic to the beam outline at a large distance along the axis.

perfect converging lens. Although there is no aberration the beam does not come to a point focus but has the same characteristic waist as the beam in the cavity. The solution that has been selected for analysis is the simplest of all the possible waveforms that can occur in a Fabry–Perot cavity. This waveform has a radial distribution which falls off from the optical axis with a Gaussian profile, and has no azimuthal variation. It is called the *Gaussian* or TEM_{00} mode. Here TEM indicates that electric and magnetic fields are transverse to the direction of the waves, while the subscript zeroes specify that the profile has no radial or azimuthal nodes. Because of its simple compact shape this is the preferred laser beam shape. Usually the mirror edges or some internal aperture restrict the area of the mode contained in the cavity, lightly clipping the tail of the Gaussian. However the other broader modes spill much more over the mirror edges and are not built up by repeated passes through the active medium.

The electric field distribution in a Gaussian beam of angular frequency ω and wave number k is $E = E_{00} \exp[-i\omega t]$ with[4]

$$E_{00} = (w_0/w) \exp(-i\phi) \exp(-r^2/w^2) \exp[ik(z + r^2/2R)], \quad (6.61)$$

where w, ϕ and R are complicated functions of the distance, z, along the beam axis; r is the radial distance off axis. This and any other paraxial solution to Maxwell's equations with boundary conditions has, built-in, the effects of diffraction caused by these same boundaries. Thus the radial confinement in the third term on the right hand side in eqn. 6.61 is accompanied by an angular divergence apparent in the final term of that equation. Imperfections of lenses or mirrors increase the divergence

---

[4]The details of the derivation can be found in Chapter 4 of the fourth edition of *Principles of Lasers* by O. Svelto, published by Plenum Press, New York (1998).

of the beam, while the aberration-free waveform of eqn. 6.61 is said to have *diffraction limited* divergence. The evolution of a Gaussian beam as it travels is more complex than that of a beam following ray optics because ray optics ignores diffraction. Rather than derive the Gaussian waveform an attempt is made here to interpret the terms in eqn. 6.61 in three steps.

First note that the radial distribution of the TEM$_{00}$ mode is a Gaussian:

$$E_{00}[\text{radial}] = \exp\left(-r^2/w^2\right), \tag{6.62}$$

where $w/\sqrt{2}$ is the root mean square radial width. This wave can be pictured as a plane wave launched at $z = 0$ with $w = w_0$. Thereafter the mode undergoes diffraction and at a distance along the axis much greater than the Rayleigh distance, $\pi w_0^2/\lambda$ the beam boundary asymptotically approaches a cone in shape. The semi-angular spread is then

$$\theta = \lambda/w_0\pi, \tag{6.63}$$

where λ is the wavelength. In detail the radius of the Gaussian beam evolves with distance as follows:

$$\begin{aligned} w^2 &= w_0^2 + z^2\theta^2 \\ &= w_0^2\left[1 + (z\lambda/\pi w_0^2)^2\right]. \end{aligned} \tag{6.64}$$

Secondly the spreading waveform tends to a spherical shape with a radius of curvature R. Referring to figure 6.33 the sagitta at a radius r off axis is $r^2/2R$. Thus the phase at the point (z,r) is the same as that at the point $(z + r^2/2R, 0)$. Consequently the wave dependence on distance is contained in eqn. 6.61 in a term

$$E_{00}[\text{motion}] = \exp\left[ik(z + r^2/2R)\right]. \tag{6.65}$$

Collecting the terms in eqn. 6.61 with an explict dependence on r gives

$$\exp\left\{[ikr^2/2][1/R + 2i/(kw^2)]\right\}. \tag{6.66}$$

Evidently the curvature of the wavefront is not exactly $1/R$ but has acquired this complex form[5]

$$\begin{aligned} 1/q &= 1/R + 2i/kw^2 \\ &= 1/R + i\lambda/\pi w^2. \end{aligned} \tag{6.67}$$

It follows that at the plane $z = 0$,

$$q_0 = -i\pi w_0^2/\lambda. \tag{6.68}$$

If the aperture at launch is made very small we would have $q = z$, and for finite apertures this becomes

$$q = z - i\pi w_0^2/\lambda. \tag{6.69}$$

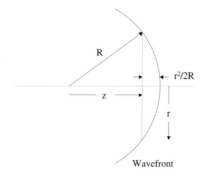

Fig. 6.33 Gaussian beam wavefront.

[5] Here the imaginary part would have the opposite sign if the choice of complex wave were $\exp\left[i(\omega t - kz)\right]$ rather than $\exp\left[i(kz - \omega t)\right]$.

Thirdly, and finally, the total energy in the wave must remain constant as it expands. This requires a normalization factor in the Gaussian waveform

$$E_{00}[\text{normalization}] = q_0/q = (w_0/w)\exp\left(-i\phi\right), \qquad (6.70)$$

where

$$\phi = \tan^{-1}\left(z\lambda/\pi w_0^2\right) = \tan^{-1}\left(z\theta/w_0\right). \qquad (6.71)$$

Only a single parameter is required in addition to the wavelength to characterize the Gaussian wave fully. The parameter conventionally chosen is the width of the beam w_0 at its waist which has been taken to lie at $z = 0$.

The alternative expressions for q, eqns. 6.67 and 6.69, provide a connection between R, z and w. Using this link we have

$$R = z\left[1 + (w_0/z\theta)^2\right] = z\left[1 + (\pi w_0^2/z\lambda)^2\right]. \qquad (6.72)$$

As required the wavefronts become flat at the *waist* at $z = 0$. Rearrangement of eqns. 6.64 and 6.72 gives the following useful expressions for w_0 and z in terms of R and w

$$w_0^2 = w^2/\left[1 + (\pi w^2/\lambda R)^2\right] \qquad (6.73)$$
$$z = R/\left[1 + (\lambda R/\pi w^2)^2\right]. \qquad (6.74)$$

These expressions can be used to calculate the location of the waist and its width when the width and curvature of the beam are known at an arbitrary location along the beam.

In Section 5.9 we saw that monochromatic light within a Fabry–Perot cavity whose wavelength was such that successive reflected waves are in phase can remain in the cavity for a number of reflections comparable to the finesse; which can easily reach a few hundred. If in addition, the waveform reproduces its radial and azimuthal distributions after a complete round trip, rather than spreading, conditions are then excellent for storing such a beam. The requirement that a Gaussian beam reproduces itself after each pass to and fro in the cavity is that

$$0 < (1 - L/R_1)(1 - L/R_2) < 1, \qquad (6.75)$$

where L is the mirrors' separation and $R_{1,2}$ are the mirror radii of curvature.[6] Otherwise the beam will diverge steadily at each pass. The region of stability is shown shaded in figure 6.34. Both the symmetric confocal and planar cavities are borderline cases, called *conditionally stable*, but are simple examples to analyse. In practice some displacement to within the shaded region ensures stability. There are plans to store light of 100 kW power in 4 km long Fabry–Perot cavities in the gravitational wave detectors described in Chapter 8.

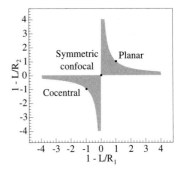

Fig. 6.34 Stability diagram for Fabry–Perot cavities.

[6]This condition is derived in exercise 6.13.

In the case of a symmetric confocal cavity the waist is midway between the mirrors and the mirrors are located at $z = \pm R/2$. Applying eqn. 6.72 to these mirror surfaces gives

$$R = \pm 2\pi \omega_0^2/\lambda, \tag{6.76}$$

so that the phases at the mirror given by eqn. 6.71 are $\pm \pi/4$. In order for the waves to duplicate their form after two traversals of the etalon we must have

$$2kL = 2p\pi + \pi, \tag{6.77}$$

where p is integral. All the more complex modes beyond the Gaussian TEM_{00} mode either share this condition or have

$$2kL = 2p\pi. \tag{6.78}$$

Equation 6.78 was the condition for standing waves found with the simpler analysis of a Fabry–Perot etalon carried through in the previous chapter. In that analysis it was implicitly assumed that the mirrors and the plane waves were of infinite extent laterally.

It is important to note that the modes of a confocal cavity are highly degenerate: modes with a particular value of p may be a mix of modes with different numbers of longitudinal nodes and different transverse distributions. The condition given by eqn. 6.78 is also that for the TEM_{00} mode in a plane mirror etalon.

We see that all the modes of the symmetric confocal cavity have wavenumbers which are integral multiples of $\pi/2L$, whatever their angular distribution. This simple distribution in the mode wavenumbers makes the symmetric confocal cavity particularly useful for spectroscopy: if a monochromatic beam is incident having a relatively broad angular range it will be able to couple efficiently to a combination of these equal frequency cavity modes. With the other stable cavity configurations this simplicity is lost. If for example the mirrors of a 10 cm symmetric confocal cavity are moved to be just 11 cm apart, moving inside the zone of stability in figure 6.34, the cavity modes essentially form a continuum. When therefore non-confocal cavities are used in spectroscopy a Gaussian beam is required. Scanning across a wavelength range can be made through changing the etalon spacing or by filling it with gas and altering the gas pressure.

As noted above, lasers are constructed with the active material inside a Fabry–Perot etalon. Generally a simple TEM_{00} mode is preferred because it is compact and, as we see below, its behaviour in optical systems can be calculated by matrix methods parallel to those used for paraxial ray beams. Consequently the confocal cavity is avoided because the higher order modes, having the identical frequencies, would so easily be excited with the desired Gaussian mode.

6.15.1   Matrix methods

The matrix formulation for tracing paraxial rays through an optical system that appears in Section 3.5 can be extended quite simply so as to apply to Gaussian beams. The matrices deduced in Section 3.5 which describe the action of optical components apply equally to Gaussian beams. A new complex parameter has emerged which is the analogue of the radius of curvature of a wavefront in standard ray optics

$$1/q = 1/R + i\lambda/(\pi w^2). \tag{6.79}$$

In the case of paraxial rays

$$R = y/\alpha,$$

where y is the distance the ray lies off axis and α is the ray slope at the same location. Referring back to Section 3.5 an optical element for which the matrix operation is

$$\mathbf{M} = \begin{pmatrix} a & b \\ c & d \end{pmatrix},$$

will produce a new radius of curvature

$$\begin{aligned} R' &= (c\alpha + dy)/(a\alpha + by) \\ &= (c + dR)/(a + bR). \end{aligned} \tag{6.80}$$

Correspondingly for a Gaussian wavefront the effect on the complex curvature, q, is

$$q' = (c + dq)/(a + bq). \tag{6.81}$$

A simple example of applying this formalism is now given. The Gaussian beam is incident on a lens of focal length f with the lens placed at the beam's waist where the wavefront is flat. Then using eqn. 6.79

$$1/q = i\lambda/(\pi r^2),$$

where r is the radius of the waist. The matrix describing the operation of the lens on the beam is

$$\mathbf{M} = \begin{pmatrix} 1 & -1/f \\ 0 & 1 \end{pmatrix},$$

so that the wave out of the lens has

$$q' = 1/\left[(i\lambda/\pi r^2) - 1/f\right].$$

Rearranging this result gives

$$1/q' = i\lambda/\pi r^2 - 1/f.$$

It follows that the radius of curvature of the emerging wavefront is $-f$, so that the beam is converging. In addition the radius of the beam leaving the lens is identical to that entering the lens, r. The lens focuses the Gaussian beam to a narrower waist rather than to a point focus, which would be the case for a plane wave. The location of the beam waist relative to the lens is obtained using eqn 6.74

$$-z = f/\left[1 + (\lambda f/\pi r^2)^2\right]$$

which shows the waist is closer to the lens than the ray focus would be. The beam waist is given by eqn. 6.73,

$$r_0^2 = r^2 / [\, 1 + (\pi r^2 / \lambda f)^2 \,].$$

Then if $\pi r^2 \gg \lambda f$ the waist radius is approximately $\lambda f / \pi r$.

Exercises

(6.1) Check the set of (bulleted) conclusions given in Section 6.5 by drawing phasor diagrams for the case of four identical slits.

(6.2) A diffraction grating 4 cm long has 1000 lines/mm. What is its potential chromatic resolving power? What blaze angle will make 500 nm the blaze wavelength for light incident perpendicular to the plane of the grating? What is the free spectral range of the grating?

(6.3) What aperture diameter would a telescope require to resolve stars that are 1.3 arcsec apart at a wavelength 500 nm?

(6.4) What is the frequency of the pulses that are received by a detector on Earth which is set up to observe lunar occultation of a distant star? The Earth's radius is 6.38 10^6 m.

(6.5) What width diffraction grating of 200 lines/mm is required to just resolve the two sodium D lines at 589.592 and 588.995 nm in first order diffraction? If a lens of 0.5 m focal length is used to image the fringes what will the physical separation of the lines be? What f/# lens is required?

(6.6) A grating has two slits each width d and separated centre-to-centre by a. How many maxima will lie under the first maximum of the corresponding single slit envelope?

(6.7) If a grating is immersed in water how does this change the angular position of the maxima?

(6.8) If alternate slits of a grating are covered with a layer of transparent material that has an optical thickness 0.5 wavelengths longer than that of the same thickness of air what will be the effect on the Fraunhofer diffraction pattern?

(6.9) All the ten zones of a zone plate are transparent. Alternate zones starting with the second from the centre are covered with the material mentioned in

the previous question. If a parallel monochromatic beam is incident normally on the zone plate what is the intensity at the bright central point at focal plane compared to that observed with a conventional zone plate of the same dimensions?

(6.10) A 5 cm square reflection grating with 1200 lines/mm is used as in figure 6.12 with lenses of focal length 1 m, and θ is fixed at 60°. The slit lengths are also 5 cm long. What value should α have in order that the detector intercepts the first order maximum for 632 nm? What is the potential chromatic resolving power achievable with the grating, and how narrow should the slits be in order to attain this? What is the etendue of the instrument?

(6.11) Show that the etendue of the TEM$_{00}$ Gaussian mode is λ^2.

(6.12) Using the expressions for the radius of the wavefront and the diameter of a Gaussian beam, show that $\lambda R / \pi w^2 = \pi w_0^2 / z \lambda$. Hence prove eqns. 6.73 and 6.74.

(6.13) Using eqns. 3.32 and 3.33 show that the transfer matrix for a complete round trip in a cavity with concave mirrors of radius R_1 and R_2 set a distance apart L, starting from the mirror with radius R_2, is

$$S = \begin{pmatrix} a & b \\ c & d \end{pmatrix},$$

with

$$a = 1 - 2L/R_1 - 4L/R_2 + 4L^2/R_1 R_2,$$
$$b = -2/R_1 + 4L/R_1 R_2 - 2/R_2,$$
$$c = 2L - 2L^2/R_1,$$
$$d = -2L/R_1 + 1.$$

Then show using eqn. 6.81 that if a Gaussian mode propagating in this cavity reproduces itself after one round trip, that is $q' = q$,

$$q = (d - a)/2 \pm \sqrt{[(d - a)^2/4 + bc]}.$$

Now use the fact that the q for a Gaussian mode has a non-zero imaginary part to prove that

$$(a + d)^2 < 4.$$

For this step you need to recall that for any transfer matrix

$$ad - bc = 1.$$

Finally show that eqn. 6.75 follows from these limits.

Fourier optics

7.1 Introduction

The technique of Fourier analysis will first be introduced in an elementary way. This relates the time distribution of any wavetrain to its frequency distribution in a precise and simple manner. Equally it provides the relationship between spatial and *spatial frequency* distributions. Diffraction of light at apertures acquires a new interpretation: the spatial distribution of the apertures in coordinate x is being in effect *Fourier transformed* into the equivalent distribution in spatial frequency, $k_x = k \sin \theta$, where θ is the outgoing direction of the light. Optical transforms, correlations and power spectra are then discussed.

The application of the Michelson interferometer to study the correlations between a light beam at two instants was discussed in Chapter 6. This *autocorrelation* will be used to give a quantitative measure of coherence. Fourier transforming an autocorrelation gives the frequency distribution; its application to spectroscopy is described here. Parameters for quantifying image resolution are explained.

The view of diffraction as a Fourier transformation process led Abbe to the modern theory of image formation, and hence to a better understanding of the resolution of optical instruments. Abbe's theory and its applications will therefore be described here. The use of acousto-optic cells to modulate and deflect light, and holography are also described, with applications to information processing.

7.2 Fourier analysis

Pure sinusoidal waves such as $\cos(kx)$ are mathematical abstractions because they would extend to infinity. They have simple mathematical properties, for example their integrals and differentials to all orders are well defined and simple. In the late eighteenth century Fourier showed that any repetitive wave (whether infinite in space or time) can be expanded as a sum of pure sinusoidal waves with numerical coefficients called a *Fourier series*. This analysis can be extended to wavetrains that are finite in extent.

Suppose the function $f(x)$ repeats its form at equal intervals λ in coordinate x, then Fourier showed that the following expansion is generally valid

$$f(x) = c_0/2 + c_1 \cos kx + c_2 \cos 2kx + ... + s_1 \sin kx + s_2 \sin 2kx + ..., \quad (7.1)$$

where $k = 2\pi/\lambda$. These discrete sinusoids are orthogonal over the range $-\lambda/2$ to $\lambda/2$, as we now demonstrate:

$$\int_{-\lambda/2}^{\lambda/2} \cos mkx \cos nkx \, dx$$

$$= \int_{-\lambda/2}^{\lambda/2} \cos (m+n)kx \, dx/2 + \int_{-\lambda/2}^{\lambda/2} \cos (m-n)kx \, dx/2$$

$$= \lambda/2 \text{ if } m = n; \ 0 \text{ if } m \neq n.$$

$$\int_{-\lambda/2}^{\lambda/2} \sin mkx \sin nkx \, dx$$

$$= \int_{-\lambda/2}^{\lambda/2} \cos (m-n)kx \, dx/2 - \int_{-\lambda/2}^{\lambda/2} \cos (m+n)kx \, dx/2$$

$$= \lambda/2 \text{ if } m = n; \ 0 \text{ if } m \neq n.$$

$$\int_{-\lambda/2}^{\lambda/2} \cos mkx \sin nkx \, dx$$

$$= \int_{-\lambda/2}^{\lambda/2} \sin (m+n)kx \, dx/2 - \int_{-\lambda/2}^{\lambda/2} \sin (m-n)kx \, dx/2 = 0.$$

Hence the Fourier series coefficients are easy to extract,

$$c_m = (2/\lambda) \int_{-\lambda/2}^{\lambda/2} f(x) \cos (mkx) \, dx, \quad (7.2)$$

$$s_m = (2/\lambda) \int_{-\lambda/2}^{\lambda/2} f(x) \sin (mkx) \, dx. \quad (7.3)$$

Notice that when $f(x)$ is an even (odd) function of x the sine (cosine) coefficients are all zero. One simple example is a periodic square wave with repeat distance λ, which is defined in the region between $-\lambda/2$ and $\lambda/2$ by

$$f(x) = 1 \text{ for } |x| < \lambda/4; \ f(x) = 0 \text{ for } \lambda/4 \leq |x| < \lambda/2.$$

This is a symmetric function of x so that the s_m's all vanish, while

$$c_m = (2/\lambda) \int_{-\lambda/4}^{\lambda/4} \cos (mkx) \, dx = (2/m\pi) \sin (m\pi/2),$$

giving

$$f(x) = 1/2 + (2/\pi)[\cos kx - \cos (3kx)/3 + \cos (5kx)/5 - ...]. \quad (7.4)$$

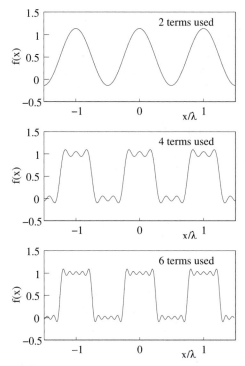

Fig. 7.1 Partial summations of the Fourier series for a square repetitive wave.

The contributions from the first two, four and six terms are shown in figure 7.1. A more compact way of writing the Fourier series is

$$f(x) = \sum_{-\infty}^{\infty} F_m \exp\left(imkx\right), \qquad (7.5)$$

with

$$F_m = (1/\lambda) \int_{-\lambda/2}^{\lambda/2} f(x) \exp\left(-imkx\right) \mathrm{d}x.$$

where $F_{\pm m} = (c_m \mp is_m)/2$, and $F_0 = c_0/2$. Waves from actual sources like the pulse shown in figure 1.6(c) have finite length and are called *non-periodic*. As the following example shows, they are the limiting forms of periodic waves. We rewrite the square repetitive wave as

$$f(x) = 1 \text{ for } |x| < \lambda/4M; \quad f(x) = 0 \text{ for } \lambda/4M \le |x| < \lambda/2$$

with $M = 1$. Now let λ and M increase by the same factor so that λ/M remains fixed at a finite value d. As shown in figure 7.2 this leaves the pulse centred on the origin unaffected, while increasing the pulse to pulse separation. If λ and M are increased to infinity while preserving their ratio then what remains is a single square pulse of width $d/2$ at the origin – which is a non-periodic wave. At the same time the change in the argument, $kx = 2\pi x/\lambda$, between successive terms in the Fourier

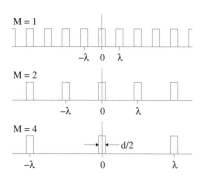

Fig. 7.2 Illustration of steps in the change from a repetitive waveform to an isolated wave.

series tends to zero. In this limit the sum of terms becomes an integral, where F_m is replaced by $F(k)dk$, $F(k)$ being a continuous function of k. It follows, after some manipulation, that

$$f(x) = \int_{-\infty}^{\infty} F(k) \exp\left(+ikx\right) \mathrm{d}k/2\pi, \tag{7.6}$$

with

$$F(k) = \int_{-\infty}^{\infty} f(x) \exp\left(-ikx\right) \mathrm{d}x. \tag{7.7}$$

This pair of equations are complementary, and the pair of variables linked, in this case x and k, are called *conjugate variables*. $F(k)$ is known as the *Fourier transform* of $f(x)$ and taking its Fourier transform, with the sign in front of i reversed, returns $f(x)$. For convenience the relationships will also be written in this way: $F(k) = \mathrm{FT}[f(x)]$ and $f(x) = \mathrm{FT}[F(k)]$. Rewriting the expression for $f(x)$

$$f(x) = \int_0^{\infty} F(k) \exp\left(+ikx\right) \mathrm{d}k/2\pi + \int_{-\infty}^0 F(k') \exp\left(+ik'x\right) \mathrm{d}k'/2\pi$$

$$= \int_0^{\infty} F(k) \exp\left(+ikx\right) \mathrm{d}k/2\pi$$

$$+ \int_0^{\infty} F(-k) \exp\left(-ikx\right) \mathrm{d}k/2\pi, \tag{7.8}$$

where the replacement $k' = -k$ has been made in the second integral. From this separation we see that if $f(x)$ is *real* the two quantities being integrated must be complex conjugates of one another, that is to say

$$F(-k) = F(k)^* \text{ and } |F(-k)| = |F(k)|. \tag{7.9}$$

After putting $F(k) = |F(k)| \exp\left[i\alpha(k)\right]$ eqn. 7.8 becomes *for real $f(x)$*

$$f(x) = \int_0^{\infty} |F(k)| \cos\left[kx + \alpha(k)\right] \mathrm{d}k/\pi. \tag{7.10}$$

Thus far Fourier analysis has been applied to spatial distributions. It is equally valid and valuable in relating distributions in time and frequency, which are the other pair of conjugate variables appearing in a general wavefunction with respect to time, t. For these we choose

$$f(t) = \int_{-\infty}^{\infty} F(\omega) \exp\left(-i\omega t\right) \mathrm{d}\omega/2\pi, \tag{7.11}$$

where

$$F(\omega) = \int_{-\infty}^{\infty} f(t) \exp\left(i\omega t\right) \mathrm{d}t, \tag{7.12}$$

and if *f(t) is real*

$$f(t) = \int_0^{\infty} |F(\omega)| \cos\left[\omega t - \alpha(\omega)\right] \mathrm{d}\omega/\pi, \tag{7.13}$$

It is important to note that the signs of the exponents in eqns. 7.6 and 7.7 could be reversed so that

$$f(x) = \int_{-\infty}^{\infty} F(k) \exp\left(-ikx\right) \mathrm{d}k/2\pi,$$

$$F(k) = \int_{-\infty}^{\infty} f(x) \exp\left(+ikx\right) \mathrm{d}x.$$

The combination of eqns. 7.6 and 7.7 is the appropriate choice if $\exp\left(ikx\right)$ is the complex spatial waveform. Equally the combination of equations presented in this note is required if the complex spatial waveform is $\exp\left(-ikx\right)$.

where $F(\omega) = |F(\omega)| \exp[i\alpha(\omega)]$. Equations 7.10 and 7.13 show that, as expected, the expansion of measurable, and therefore purely real quantities requires only contributions of positive frequency with real coefficients. Another useful result is that the Fourier transform of a *Gaussian* distribution is also a Gaussian

$$\mathrm{FT}[\exp(-x^2/2\sigma^2)/\sqrt{2\pi}\sigma] = \exp(-\sigma^2 k^2/2). \qquad (7.14)$$

The major consequence of the preceding analysis is that it becomes straightforward to calculate the results of any frequency dependent physical process which is caused by a finite wavetrain, provided that effects add linearly. First the wavetrain is resolved into its pure sinusoidal components with their numerical coeficients. Then the effects of the physical process are calculated for each frequency component. Finally these are summed to give the total effect.

7.2.1   Complex field usage

An electric field is of course a real quantity. However, as indicated earlier, it can sometimes be useful to extend it to become a complex quantity. Suppose the electric field at a given location is

$$E_{\mathrm{r}}(t) = \int_{-\infty}^{+\infty} e(\omega) \exp(-i\omega t) \, \mathrm{d}\omega/2\pi, \qquad (7.15)$$

where $e(\omega) = |e(\omega)| \exp[i\alpha(\omega)]$. Then

$$E_{\mathrm{r}}(t) = 2 \int_0^{+\infty} |e(\omega)| \cos[\alpha(\omega) - \omega t] \, \mathrm{d}\omega/2\pi. \qquad (7.16)$$

The appropriate imaginary part is got by replacing the cosine term *at all frequencies* by a corresponding sine term:

$$E_{\mathrm{i}}(t) = 2 \int_0^{+\infty} |e(\omega)| \sin[\alpha(\omega) - \omega t] \mathrm{d}\omega/2\pi. \qquad (7.17)$$

The complex field is thus

$$E(t) = E_{\mathrm{r}}(t) + iE_{\mathrm{i}}(t) = 2 \int_0^{+\infty} e(\omega) \exp(-i\omega t) \, \mathrm{d}\omega/2\pi. \qquad (7.18)$$

Evidently the Fourier transform of the complex field is $2e(\omega)$ when $\omega > 0$ and zero when $\omega < 0$.

The Dirac delta function

The use of this function, $\delta(x)$, of a variable x simplifies the Fourier analysis of repetitive pulses. It is defined such that

$$\int_{-\infty}^{\infty} \delta(x - a) f(x) \, \mathrm{d}x = f(a). \qquad (7.19)$$

Table 7.1 Table of Fourier transforms.

$f(x)$	$F(k)$
$f(ax)$	$F(k/a)/|a|$
$\delta(x)$	1
$\exp(-x),\ x > 0$	$(1 - ik)/(1 + k^2)$
$\exp(-x^2/2\sigma^2)/(\sqrt{2\pi}\sigma)$	$\exp(-k^2\sigma^2/2)$
$\Pi(x)$: $= 1$ for $|x| < 0.5$, $= 0$ for $|x| > 0.5$	$\mathrm{sinc}(k/2)$

The function $\delta(x-a)$ has unusual properties: it is zero everywhere except for a positive spike at $x = a$ which is infinitely narrow, while at the same time the area under this spike is unity. Two useful results follow

$$\delta(bx) = \delta(x)/|b| \ \text{ for } \ b \neq 0, \tag{7.20}$$

$$\int_{-\infty}^{\infty} \exp\left[\,ix(\alpha - \beta)\,\right] \mathrm{d}x = 2\pi\,\delta(\alpha - \beta). \tag{7.21}$$

Taking the Fourier transform of a function $f(x)$ twice, and using eqn. 7.21, gives

$$\int\int f(x') \exp\left(ik(x - x')\right) \mathrm{d}k\, \mathrm{d}x'/2\pi = f(x),$$

which confirms that the original function is obtained. The reader may now like to prove that, using the conventions of equations 7.6 and 7.7,

$$\mathrm{FT}[f(x - x_0)] = \exp(-ikx_0)\,\mathrm{FT}[f(x)],$$
$$\mathrm{FT}[F(k - k_0)] = \exp(+ik_0x)\,\mathrm{FT}[F(k)], \tag{7.22}$$

which are called the *shift theorems*.

7.2.2   Diffraction and convolution

If the reader refers back to eqn. 6.3 it is seen that the expression for the amplitude of the diffracted wave is simply the Fourier transform of the wave amplitude at the diffracting aperture, in that case for a slit of width d. The conjugate variables are the coordinate across the screen, x, and the lateral component of the wave vector $k_x = k \sin\theta$. With a plane wave incident normally on the screen the phase factor required for light travelling in a direction making an angle θ with the forward direction is $\exp(-ik_x x)$, and this factor constitutes the Fourier transforming factor.[1] Diffraction of the same incident beam by a screen whose transmission coefficient is $f(x)$ gives an amplitude in the direction θ which is its Fourier transform

$$E(k_x) = \int_{-\infty}^{\infty} f(x) \exp(-ik_x x)\, \mathrm{d}x.$$

[1]In eqn. 6.3 the sign was reversed: referring to figure 6.2 θ was being measured using one sense for the lateral coordinate (upward positive) and x with the opposite sense (downward positive).

For this reason the lens following the screen is called the *Fourier* lens and its focal plane is called the *Fourier transform* plane or simply the *transform* plane.

In Section 6.5 the intensity of the diffraction pattern for multiple broad slits turned out to be the direct product of the pattern for a single broad slit and the pattern for multiple (infinitely) thin slits. Figure 6.8 illustrates this. This type of simplification is very general. The transmission coefficient for multiple identical slits can be described using two functions. One function, $h(x)$, describes an individual slit centred at $x = 0$: $h(x) = 1$ for $|x| < a/2$ and zero elsewhere. The other function describes the placement of the slit centres at x_1, x_2,... : $g(x) = \sum_m \delta(x - x_m)$. The transmission coefficient is then explicitly

$$f(x) = \int_{-\infty}^{\infty} g(x')h(x - x')\,dx', \qquad (7.23)$$

an expression that is called a *convolution* and which is written in shorthand form as

$$f(x) = g(x) \otimes h(x) = h(x) \otimes g(x). \qquad (7.24)$$

With the example given of multiple identical slits

$$f(x) = \int \sum_m \left[\delta(x' - x_m)h(x - x')\right] dx' = \sum_m h(x - x_m).$$

This convolution is shown pictorially in figure 7.3. When the functions are more complicated than delta functions the results are less easy to visualize. The simplification in the diffraction pattern of an array of identical apertures emerges when the Fourier transform of the convolution is taken:

$$F(k) = \int_{-\infty}^{\infty} \int_{-\infty}^{\infty} g(x')h(x - x') \exp(-ikx)\,dx\,dx'.$$

Putting $y = x - x'$ gives

$$F(k) = \int_{-\infty}^{\infty} g(x') \exp(-ikx')\,dx' \int_{-\infty}^{\infty} h(y) \exp(-iky)\,dy = G(k)\,H(k),$$
$$(7.25)$$

where $G(k) = \mathrm{FT}[g(x)]$ and $H(k) = \mathrm{FT}[h(x)]$. Alternatively this result can be written

$$\mathrm{FT}\left[g(x) \otimes h(x)\right] = G(k)\,H(k), \qquad (7.26)$$

which is known as the *convolution theorem*. The factorization of the diffraction pattern of several broad slits into the broad single slit pattern, $H(k)$, and the multiple thin slit pattern, $G(k)$ seen in eqn. 6.13 gives one example of the operation of the convolution theorem. Two other examples are illustrated in figure 7.4. This figure shows the Fraunhofer diffraction patterns for two, and for five randomly placed circular apertures. The patterns seen are products of the pattern for a single circular aperture multiplied by the multiaperture pattern.

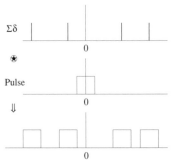

Fig. 7.3 The figure shows the convolution of a sum of delta functions with a square pulse. The resultant is a set of square pulses centred on the delta function locations.

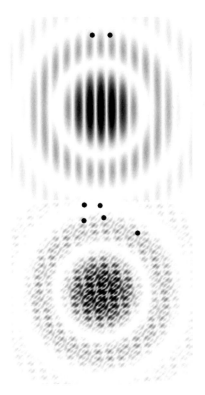

Fig. 7.4 The diffraction patterns for two and five identical circular apertures. The small black circles indicate the hole size and spacing for the two cases.

Optical transfer function

Any optical system produces a more or less blurred image of a point object due in part to aberrations and in part to diffraction. We suppose that the illumination is incoherent, as it would be for a camera viewing a landscape or a telescope viewing a star field. The spread of light intensity around the ideal image point is called the *point spread function* or PSF; and the ratio of the peak intensity to the peak intensity in the limit with no aberrations is called the *Strehl ratio*. When the Strehl ratio is of order 0.5 or larger it is given approximately by

$$S = 1 - [2\pi\sigma(\lambda)/\lambda]^2 \qquad (7.27)$$

where $\sigma(\lambda)$ is the rms *wavefront error*; that is the departure of the wavefront from an unaberrated shape. Excellent optical systems for which the Strehl ratio is larger than 0.8, corresponding to a wavefront error of $\lambda/14$ or less, are conventionally called *diffraction limited*.

The PSF determines how easy it will be to detect a point source against a background using an optical system and detector such as a CCD array. The observer would only be able to detect the source provided that the charge deposited in the pixels inside the image significantly exceeds the noise in adjacent pixels. The noise includes contributions from background and from the detector itself in the absence of any radiation. If the PSF is reduced by a factor n then the size of the image of the point source is similarly reduced. Its signal is then concentrated in n^2 fewer pixels so that the *signal to noise ratio* or SNR increases by the same factor.

If the PSF is the same over the whole image then the intensity, I, across the image is related to that across the object, O, through a convolution

$$I(x) = O(x) \otimes \text{PSF}(x) = \int \text{PSF}(x - mx')O(x')\mathrm{d}x', \qquad (7.28)$$

where m is the magnification. Fourier transforming the PSF resolves the smearing of the image as a function of spatial frequency, $k/2\pi = 1/\lambda$, so that the effects of smearing on small and large scale features in the image are separated. This Fourier transform is called the *optical transfer function* or OTF. Then from the convolution theorem we have

$$\text{FT}[I] = \text{OTF } \text{FT}[O]. \qquad (7.29)$$

The OTF is in general complex and can be expanded thus:

$$\text{OTF} = \text{MTF } \exp{(i\text{PTF})}, \qquad (7.30)$$

where MTF is called the *modulation transfer function*, and PTF the *phase transfer function*, both being functions of the spatial frequency and both real. If features of some spatial frequency k in the object have visibility (contrast) $V(k)$, this is degraded in the image to MTF$(k) \otimes$ $V(k)$. One simple method used to measure the MTF is to record an image of a test screen. This screen carries parallel lines, spaced λ apart, the intensity varying sinusoidally like $\cos(2\pi x/\lambda)$. The visibility of the image is then compared with that of the original for different choices of the spatial frequency $1/\lambda$. Figure 7.5 shows some representative plots of the MTF: one for a perfectly corrected optical system, labelled C; and others for two imperfect systems, labelled A and B. The MTF for C falls to zero at a spatial frequency determined by diffraction at the aperture stop. System A is superior for low spatial frequencies but would be the poorer choice for high spatial frequencies, so that B is better for recording fine detail. This comparison shows that the OTF gives a more complete representation of optical performance than a single number such as the resolving power or Strehl ratio.

When coherent illumination is used the amplitudes at the image plane are added, rather than the intensities. The interference effects exploited in microprocessor lithography and described in Section 6.13 are one example. These effects invalidate an analysis based simply on adding intensites.

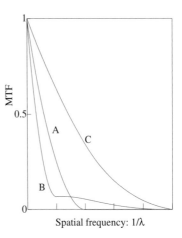

Fig. 7.5 Examples of the dependence of the modulation transfer function on spatial frequency for a well corrected lens (C) and for two imperfect lenses. Of these two B performs better at higher spatial frequencies.

7.3   Coherence and correlations

The coherence of light beams and sources, first introduced in Chapter 5, will now be given a quantitative basis in terms of correlations between beams or of correlations within a single light beam (autocorrelations). Fourier analysis provides the key by which the spectrum of a source can be extracted from the autocorrelations of the light emitted.

The instantaneous intensity produced by a pair of interfering beams of light with complex fields E_1 and E_2 is

$$I(t) = |E_1(t) + E_2(t)|^2 = |E_1(t)|^2 + |E_2(t)|^2 + 2\mathcal{R}e\,[\,E_1^*(t)E_2(t)\,].$$

Electronic detectors and our eyes measure an *instantaneous* intensity which is the average over a time T, very much longer that the period of the light oscillations. Thus the measured quantities are

$$\overline{I(t)} = \overline{I_1(t)} + \overline{I_2(t)} + 2\mathcal{R}e\,\overline{[\,E_1^*(t)E_2(t)\,]},$$

with the bar indicating a time average. The interference term contains what is called a *correlation* of E_1 and E_2, namely $\int_0^T E_1^*(t)E_2(t)\mathrm{d}t/T$. Usually beams of any interest have constant intensity so that

$$I = I_1 + I_2 + 2\mathcal{R}e\overline{(E_1^*(t)E_2(t))},$$

where I_1 and I_2 are constants.

It will be assumed that the beams involved are *stationary*, which means that their fluctuations (due to the random emission of wavepackets with random phases) are not changing in character with time. It follows that the correlations defined here do not depend on when they are measured. From here on, unless otherwise specified, all beams with constant intensity are assumed to be stationary.

The size of the correlation term depends not only on how similar the beams are, but also on how intense they are. In order to remove the dependence on intensity the correlation is divided through by $\sqrt{I_1 I_2}$ to give

$$g^{(1)}(1,2) = \overline{E_1^*(t)E_2(t)}/\sqrt{I_1 I_2} \tag{7.31}$$

which is called the *degree of first order coherence* between the beams. If the interference is between light from the same source at times separated by τ

$$I = 2I_0 + 2\mathcal{R}e\overline{(E^*(t)E(t+\tau))} \tag{7.32}$$

with the inteference term being an *autocorrelation* $\int_0^T E^*(t)E(t+\tau)\mathrm{d}t/T$. The degree of first order coherence is in this case

$$g^{(1)}(\tau) = \overline{(E^*(t)E(t+\tau))}/I_0. \tag{7.33}$$

For stationary beams $g^{(1)}(1,2)$ is independent of time and $g^{(1)}(\tau)$ depends only on the interval, τ, *between* measurements.

Bearing in mind that the maximum value of the real part of any complex quantity x is $|x|$ and its minimum value is $-|x|$, it follows that the maximum and minimum intensities are

$$I_{\mathrm{max/min}} = 2I_0 \pm 2|\overline{(E^*(t)E(t+\tau))}|.$$

Hence the visibilty is

$$V = |\overline{(E^*(t)E(t+\tau))}|/I_0$$
$$= |g^{(1)}(\tau)|, \tag{7.34}$$

so that the visibility provides a simple measure of the modulus of the degree of first order coherence. Figure 7.6 shows examples of how the intensity would vary across an interference pattern produced by beams of equal intensity. The x-coordinate could represent either the position across a Young's two slit pattern or the position of the mobile mirror in Michelson's interferometer. In the upper diagram there is full coherence and $|g^{(1)}|$ is unity, with $g^{(1)}$ having whatever phase shift there is between the wavetrains. Then in the second diagram there is partial coherence with $|g^{(1)}|$ being 0.5 and finally in the third diagram there is incoherence with $g^{(1)}$ being zero.

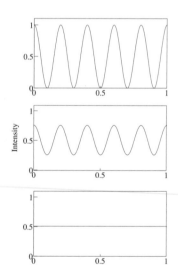

Fig. 7.6 Fringes with various visibilities produced by two souces: upper panel for fully coherent sources; centre panel for partially coherent sources; lower panel for incoherent sources.

7.3.1   Power spectra

The power per unit area of a light beam in free space is given by Poynting's formula eqn. 1.18

$$N(t) = E(t)^2/Z_0,$$

measured in $\mathrm{W\,m^{-2}}$, where $E(t)$ is the real electric field. The total energy radiated per unit area is obtained by integrating over time

$$\mathcal{E} = \int_{-\infty}^{\infty} E(t)^2 \mathrm{d}t/Z_0. \tag{7.35}$$

The electric field at any time can be re-expressed in terms of its Fourier transform $e(\omega)$ which is a function of the angular frequency ω. That is to say

$$E(t) = \int_{-\infty}^{\infty} e(\omega) \exp(-i\omega t)\mathrm{d}\omega/2\pi. \tag{7.36}$$

Thus the total energy radiated per unit area becomes

$$\mathcal{E} = \int\int\int_{-\infty}^{\infty} e^*(\omega)e(\omega') \exp(i[\omega - \omega']\,t)\,\mathrm{d}t\,\mathrm{d}\omega\,\mathrm{d}\omega'/(4\pi^2 Z_0)$$

$$= \int\int_{-\infty}^{\infty} e^*(\omega)e(\omega')\delta(\omega - \omega')\,\mathrm{d}\omega\,\mathrm{d}\omega'/(2\pi Z_0)$$

$$= \int_{-\infty}^{\infty} e^*(\omega)e(\omega)\,\mathrm{d}\omega/(2\pi Z_0), \tag{7.37}$$

where the fact that $E^* = E$ has been used in writing the first line. Comparing eqns. 7.35 and 7.37 we see that the total energy radiated can be expanded in an identical manner in terms of either the temporal distribution, $E(t)$, or of its Fourier transform, the frequency distribution, $e(\omega)$. This result is known as *Parseval's theorem* and provides the justification for treating $|e(\omega)|^2/Z_0$, measured in $\mathrm{W\,m^{-2}\,Hz^{-1}}$, as the actual distribution of electromagnetic energy in *frequency* (rather than angular frequency). It is therefore called the *spectral energy distribution*.

Using the fact that E is real and eqn. 7.9 we have $|e(-\omega)| = |e(\omega)|$, and then eqn. 7.37 becomes

$$\mathcal{E} = 2\int_0^{\infty} |e(\omega)|^2 \mathrm{d}\omega/(2\pi Z_0), \tag{7.38}$$

with, as seems reasonable, only positive frequency components contributing. The corresponding energy per unit *angular frequency* interval is then $|e(\omega)|^2/(\pi Z_0)$, while

$$P(\omega) = |e(\omega)|^2/(\pi Z_0 T) \tag{7.39}$$

is the mean power per unit area per unit angular frequency during the time T that the beam is on.

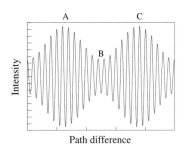

Fig. 7.7 Schematic of interferogram for the sodium D lines.

7.3.2   Fourier transform spectrometry

Michelson interferometers are widely employed to measure spectra in the visible and especially the infrared part of the electromagnetic spectrum using a technique called *Fourier transform spectrometry*. The apparatus is arranged as shown in figure 5.11 to study a source's spectrum. Mirrors M_1' and M_2 are set parallel in order to give circular fringes and a detector is placed behind a small circular hole centred on the focused fringes. The intensity seen by the detector is recorded continuously during a scan in which the mobile mirror is moved at constant speed. This recorded intensity pattern is called an *interferogram*, and we shall see that taking its Fourier transform yields the power spectrum of the source.

A simple example of an interferogram which illustrates how information on spectra can be extracted is shown schematically in figure 7.7. It is an interferogram recorded with the D lines of a sodium gas source at 588.995 nm ($\lambda - \Delta\lambda$) and 589.592 nm (λ), all other spectral lines being filtered out. Effectively there are two incoherent sources each giving a separate intensity fringe pattern at the detector. These two intensities add to give the total intensity detected and recorded as the interferogram. At points of maximum visibility (A,C) the fringe patterns are in step: their peaks in intensity coincide and their troughs in intensity coincide. Correspondingly at a minimum of visibility (B) they are exactly out of step. Between successive maxima of visibility the path difference must change by a certain number of wavelengths, n, for the longer wavelength and by $n+1$ wavelengths for the shorter wavelength. n is determined to the nearest integer simply by counting the number of waves under the envelope between A and C. Knowing this value

$$(n + 1)(\lambda - \Delta\lambda) = n\lambda,$$
$$\text{so} \quad \Delta\lambda = \lambda/n,$$

which provides a measurement of $\Delta\lambda$ if λ is already known.

The chromatic resolving power obtained with the Michelson interferometer in this way depends on the total change in path difference during the scan. It will be possible to just resolve two wavelengths provided that two adjacent maxima of intensity, A and C, lie within the full scan length (full range of the path difference), x_w. Then $x_w = n\lambda$ and the chromatic resolving power

$$\lambda/\Delta\lambda = n = x_w/\lambda. \tag{7.40}$$

The next step will be to consider the analysis of more typical spectra consisting of many lines, and for which Fourier analysis is essential.

The intensity at the detector, assuming that the beams from the mirrors have equal intensity, is given by eqn. 7.32

$$I = 2I_0 + 2\overline{(E(t)E(t + \tau))}$$

where E(t) and $E(t+\tau)$ are the values of the real electric field at times t and $(t+\tau)$ respectively during the scan. Only the autocorrelation term is of interest here. Writing the electric field in terms of its frequency components

$$E(t) = \int_{-\infty}^{\infty} e(\omega) \exp\left(-i\omega t\right) d\omega/2\pi,$$

and using $E^* = E$, the autocorrelation term is

$$\overline{E(t)E(t+\tau)}$$
$$= \int \int \int e^*(\omega)e(\omega') \exp\left(i\omega t\right) \exp\left[-i\omega'(t+\tau)\right] d\omega\, d\omega'\, dt/(4\pi^2 T),$$

where T is the duration of the measurement. Using the equality

$$\int \exp\left[i(\omega-\omega')t\right] dt = 2\pi\, \delta(\omega-\omega'),$$

the autocorrelation simplifies thus

$$\overline{E(t)E(t+\tau)} = \int_{-\infty}^{\infty} e(\omega)e^*(\omega) \exp\left(-i\omega\tau\right) d\omega/(2\pi T) \qquad (7.41)$$

We have seen in eqn. 7.39 that $|e(\omega)|^2/(\pi Z_0 T)$ is simply the power spectrum $P(\omega)$, so that taking the Fourier transform of eqn. 7.41 gives

$$P(\omega) = \text{FT}\left\{\overline{E(t)E(t+\tau)}\right\}/(\pi Z_0), \qquad (7.42)$$

where $Z_0 = \sqrt{\mu_0/\varepsilon_0}$ is the impedance of free space. This relation between the power spectrum and the Fourier transform of the autocorrelation provides the required link betwen the interferogram and the power spectrum. Equation 7.42 is one version of the *Wiener–Khinchine theorem*. It must be emphasized here that both the autocorrelation and the power spectrum are devoid of phase information. The same power spectrum can produce very different waveforms in time depending on the relative phases of its frequency components, but these different waveforms will share an identical autocorrelation function.

The expression for the spectral power distribution when the electric field is extended to the complex form defined by eqn. 7.18 is

$$P(\omega) = FT\left\{\overline{E^*(t)E(t+\tau)}\right\}/(4\pi Z_0), \qquad (7.43)$$

differing by a factor $1/4$ from eqn. 7.42. Its derivation is given as an exercise at the end of the chapter.

When analysing interferograms produced with a Michelson interferometer eqn. 7.42 can be re-expressed more conveniently in terms of the wave number $k = \omega/c$ and the path difference x

$$P(k) = P(\omega)(d\omega/dk) = (c/\pi Z_0)\text{FT}\left\{\overline{E(t)E(t+x/c)}\right\}. \qquad (7.44)$$

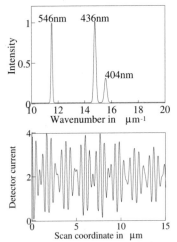

Fig. 7.8 In the upper panel the spectrum of a mercury lamp with broad lines at 404, 436 and 546 nm is shown, and in the lower panel the interferogram. We shall see later that the wave number is directly proportional to the energy change in a molecular/atomic transition, which enhances the usefulness of this display.

Figure 7.8 shows a reproduction of a typical spectrum and interferogram obtained in an undergraduate experiment using a mercury lamp.[2] There are limitations on the information about a spectrum that can be extracted from a Michelson interferogram. Firstly the scan length, x_w, restricts the interferogram to a finite *window*, and secondly the detector only samples the intensity at discrete steps, x_s apart, and not continuously. Figure 7.9 helps to illustrate this, showing in the upper panel that no useful information can be extracted for radiation of wavelength longer than the scan length x_w. The lower panels show that sampling a wavetrain of wavelength $2x_s/3$ can give an identical set of measurements to sampling one of wavelength $2x_s$. This effect is known as *aliasing*. It can be inferred that measurements every x_s along the interferogram only sample waves of wavelengths greater than $2x_s$ adequately. This inference is given precision by the Whittaker–Shannon sampling theorem: it states that sampling at intervals x_s over an infinite scan length would uniquely determine a spectrum containing only waves with wavelength longer than $2x_s$.

These limitations can be quantified when the Fourier transform is examined. What is being measured is not $I(x) = \overline{E(t)E(t + x/c)}$ but the product

$$I' = I\,C\,W = I(x) \sum_m \delta(x - mx_s)\,W(x)$$

where C is a *comb* function consisting of a sum of delta functions at intervals x_s, and W is a *window* function, a square pulse of unit area and width x_w. According to the convolution theorem the Fourier transform of I' is the convolution

$$P'(k) = \text{FT}(I) \otimes \text{FT}(C) \otimes \text{FT}(W). \tag{7.45}$$

The Fourier transform of the square window gives a broad distribution in k

$$\text{FT}[W] = \int_{-x_w/2}^{x_w/2} \exp\left(-ikx\right) \mathrm{d}x/x_w = \text{sinc}(kx_w/2), \tag{7.46}$$

and if the spectrum is monochromatic with $k = k_0$, then $\text{FT}(I) = \delta(k - k_0)$ and

$$\text{FT}[I] \otimes \text{FT}[W] = \text{sinc}\left[(k - k_0)x_w/2\right]. \tag{7.47}$$

Two spectral lines will be regarded as just resolved if the maximum of the sinc function of one falls at the first minimum of the sinc function of

---

[2] If the number of samplings is N, an $N \times N$ matrix inversion is needed to extract the Fourier transform of the interferogram. When, as is typically the case, N is of order 10 000 this process is very time consuming. J. W. Cooley and J. W. Tukey, *Math. Computing* **19**, p. 297 (1965), invented a fast Fourier transform technique that reduces the number of inversions to $N \ln(N)$. However the spectrum obtained is discrete, being determined at the wavelengths $x_w, x_w/2, x_w/4, \ldots, 2x_w/N$ only.

the other line. This happens when the separation in wave number, Δk, between the lines is such that

$$\Delta k = 2\pi/x_\mathrm{w}. \tag{7.48}$$

Then the chromatic resolving power is

$$\lambda/\Delta\lambda = k/\Delta k = x_\mathrm{w}/\lambda, \tag{7.49}$$

which refines the estimate given in eqn. 7.40.

By contrast the effect of the comb is to give aliases of any spectral line. The Poisson summation theorem gives

$$\sum_m \delta(x - mx_\mathrm{s}) = (1/x_\mathrm{s}) \sum_n \exp\left(ink_\mathrm{s}x\right), \tag{7.50}$$

where both summations run from $-\infty$ to $+\infty$ and $k_\mathrm{s} = 2\pi/x_\mathrm{s}$. This equality is a good approximation in the present case when the number of samples is several hundred. Then

$$\mathrm{FT}[C] = (1/x_\mathrm{s}) \int \sum_n \exp\left[-i(k - nk_\mathrm{s})x\right]\mathrm{d}x$$

$$= (2\pi/x_\mathrm{s}) \sum_n \delta(k - nk_\mathrm{s}). \tag{7.51}$$

Suppose again that the spectrum is monochromatic so that $\mathrm{FT}[I] = \delta(k - k_0)$. It follows that

$$\mathrm{FT}[I] \otimes \mathrm{FT}[C] = (2\pi/x_\mathrm{s}) \sum_n \delta(k - k_0 - nk_\mathrm{s}). \tag{7.52}$$

These aliases are displaced in wave number from the actual line by integral multiples of $2\pi/x_\mathrm{s}$. Aliases from radiation of wavelengths *below* $2x_\mathrm{s}$ which could simulate spectral lines above that wavelength must be eliminated by inserting an optical filter which removes radiation having wavelengths shorter than $2x_\mathrm{s}$.

The Michelson spectrometer has two important advantages over the grating spectrometer which make it the preferred instrument in many situations. The first, *Jacquinot* advantage, is that the etendue is about one hundred times larger than for a comparable grating spectrometer. This means that weaker sources can be studied and also weaker lines identified. The second, *Felgett* advantage, is that the Michelson detector receives all the wavelengths throughout the scan, whereas the detector in a grating spectrometer receives a restricted wavelength range. With N samples the time required to examine the same spectrum with a grating spectrometer in the same detail as with a Michelson is longer by a factor N, which is usually 1000 or more.

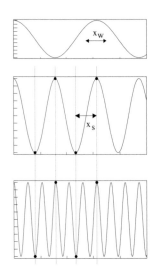

Fig. 7.9 In the upper panel the scan length, x_w, is much shorter than the wavelength. In the lower panels aliasing is shown between sine waves of wavelength $2x_\mathrm{s}$ and and $2x_\mathrm{s}/3$ for a sampling interval x_s. Samplings are indicated by dots.

When studying infrared spectra the radiation from any part of the apparatus falling on the detector can give a background illumination that overwhelms the signal of interest in the case of both grating and prism spectrometers. At room temperatures the spectrum of black body radiation peaks at a wavelength around $10\,\mu$m. The superior etendue of the Michelson is then a prime advantage. One simple technique used to remove this background is to pulse the radiation studied by placing a shutter in front of the source. Then the *pulsed* part of the radiation received on the detector is selected electronically using a *lock-in amplifier*. In addition, because the Michelson spectrometer measures correlations it is correspondingly less sensitive to a background which is constant with time.

7.3.3 Line width and bandwidth

The classical view of atomic and molecular transitions is that the electric field emitted by an individual, isolated atom undergoes damped harmonic oscillations

$$E(t) = E(0) \exp\left(-\gamma t/2\right) \exp\left(-i\omega_0 t\right), \text{ for } t \geq 0, \quad (7.53)$$

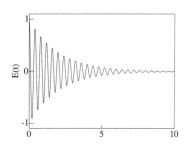

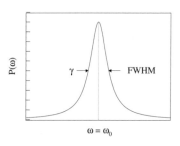

Fig. 7.10 Lorentzian spectrum. In the upper diagram the electric field distribution in time is shown; in the lower diagram the electromagnetic energy spectrum. Radiation at optical frequencies would typically have tens of thousands of oscillations under the decay curve.

in complex form. The real part of the electric field is shown in the upper panel of figure 7.10, and is zero for negative values of t. The intensity decreases with time like $\exp\left(-\gamma t\right)$ with the mean value of t being $1/\gamma$; this parameter is known as the *lifetime* of the state radiating. As might be expected, $\omega_0/2\pi$ will emerge as the central frequency of the radiation. We shall find in later chapters that quantum theory provides a refined interpretation of these quantities. The Fourier transform of this waveform is

$$e(\omega) = \int_0^{+\infty} E(0) \exp\left(-\gamma t/2\right) \exp\left[i(\omega - \omega_0)t\right] \mathrm{d}t$$
$$= E(0)/[\gamma/2 - i(\omega - \omega_0)].$$

Now the power spectrum is proportional to $|e(\omega)|^2$, so that the power spectrum normalized to unit total power is

$$P(\omega) = \frac{\gamma/2\pi}{[\gamma^2/4 + (\omega - \omega_0)^2]}. \quad (7.54)$$

This is called a *Breit–Wigner* or *Lorentzian* line shape, and this too is shown in figure 7.10. The power radiated peaks at an angular frequency ω_0 and has dropped to half its peak value at angular frequencies $\omega_0 \pm \gamma/2$, so that γ is known as the *line width*. Lifetimes of excited states of *isolated* atoms undergoing electric dipole transitions are typically around 10^{-8} s which gives *natural* line widths around $0.1\,$GHz. These natural line widths are appropriate to stationary, isolated atoms. Atoms in gases at sufficiently low temperature and low pressure approximate to this ideal. The atoms or molecules in a source will emit with random phases (apart from lasers) so that the total wavetrain will continuously change in phase and amplitude. This lack of coherence affects neither the power spectrum nor the autocorrelation function because these are insensitive to the relative phases of the frequency components.

Spectral lines emitted by laboratory gas discharge sources are further broadened by collisions of the radiating atoms with other atoms and by Doppler shifts arising from their own motion. When an atom which is emitting a wavetrain collides with another atom there will be an unpredictable change in phase between the wavetrain before and after collision. This broadens the frequency distribution while retaining a Lorentzian shape. In the atmosphere the interval between collisions, τ_{coll}, is around $5\,10^{-9}\,\text{s}$ which can be converted to a wavelength spread by using the relation given in eqn. 5.24: we get

$$\Delta\lambda = \lambda^2/c\tau_{\text{coll}} = 10^{-3}\,\text{nm}.$$

According to the kinetic theory of gases $\tau_{\text{coll}} \propto 1/(P\sqrt{T})$ where P is the pressure and T here the absolute temperature.

When the atom moves with velocity v towards the observer the angular frequency of the radiation emitted shifts from the rest value ω_0 to $\omega_0 + \omega_0 v/c$. The kinetic theory predicts the velocity distribution of atoms in a gas to be

$$F(v)\mathrm{d}v = \exp\left(-mv^2/2k_{\text{B}}T\right)\mathrm{d}v,$$

where k_{B} is the Boltzmann constant and m the atomic mass. Then the frequency distribution of the intensity of radiation is, apart from a constant factor,

$$P(\omega)\mathrm{d}\omega = \exp\left[-m(\omega - \omega_0)^2 c^2/(2\omega_0^2 k_{\text{B}}T)\right]\mathrm{d}\omega,$$

which has the *Gaussian* shape mentioned above. The constant factor can be chosen so that the integral is unity. We use the result for a Gaussian

$$\int_{-\infty}^{\infty} \exp\left(-x^2/2\beta^2\right)\mathrm{d}x = \sqrt{2\pi}\beta; \tag{7.55}$$

then the standard form having unit area under the curve is

$$P(\omega) = \exp\left[-(\omega - \omega_0)^2/2\sigma^2\right]/\sqrt{2\pi}\sigma, \tag{7.56}$$

with a width parameter σ equal to $\sqrt{\omega_0^2 k_{\text{B}}T/mc^2}$. For gas atoms in the atmosphere the width at $500\,\text{nm}$ wavelength is around $1\,\text{GHz}$ in frequency or $10^{-3}\,\text{nm}$ in wavelength. This is much larger than the natural line width.

Where the broadening is the same for all atoms or molecules in a source, whether it is due to collisions in a gas or interactions between nearby atoms in a uniform crystal the broadening is called *homogeneous*. This leads to a Lorentzian line shape. Where the broadening varies for different subsets of the atoms, as with Doppler broadening or when a

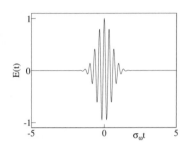

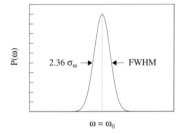

Fig. 7.11 Gaussian spectrum. It would contain radiation emitted by many atoms. In the upper diagram the electric field distribution in time is shown; in the lower diagram the em energy spectrum.

crystal has inhomogeneities, this is called *inhomogeneous* broadening. Inhomogeneous broadening produces a Gaussian line shape. This is the characteristic form when there are many uncorrelated individual contributions to the broadening. Mixtures of similar degrees of homogeneous and inhomogeneous broadening give an intermediate, *Voigt* line shape. There is further discussion of line widths from the quantum viewpoint in Chapter 13, and in Chapter 14 with reference to lasers.

Pulses with an electric field that follows a pure Gaussian can be produced by lasers, a topic developed later in Chapter 14. Such a waveform, shown in the upper panel of figure 7.11, has special properties. Its time distribution is

$$E(t) = \exp\left(-i\omega_0 t\right)\exp\left(-t^2/4\sigma_t^2\right) \tag{7.57}$$

where $\sqrt{2}\sigma_t$ is the root mean square deviation from the mean angular frequency at ω_0. The waveform's intensity distribution is

$$I(t) = \exp\left(-t^2/2\sigma_t^2\right). \tag{7.58}$$

We use the Wiener–Khinchine theorem to extract the power spectrum for this Gaussian pulse. After that there is a discussion of the theorem's application to incoherent as well as to coherent spectra. Starting with the autocorrelation we have

$$\overline{E^*(t)E(t+\tau)} \propto \int_{-\infty}^{\infty} \exp\left(-i\omega_0\tau\right)\exp\left\{-[t^2 + (t+\tau)^2]/4\sigma_t^2\right\}\mathrm{d}t$$

$$= \exp\left(-i\omega_0\tau\right)\exp\left(-\tau^2/8\sigma_t^2\right)$$
$$\int_{-\infty}^{\infty} \exp\left[-(\sqrt{2}t + \tau/\sqrt{2})^2/4\sigma_t^2\right]\mathrm{d}t$$
$$= \sqrt{2\pi}\sigma_t \exp\left(-i\omega_0\tau\right)\exp\left(-\tau^2/8\sigma_t^2\right)$$

The Wiener–Kinchine theorem, in the form presented as eqn. 7.43, gives

$$P(\omega) \propto \int_{-\infty}^{\infty} \exp\left[i(\omega - \omega_0)t)\right]\exp\left(-t^2/8\sigma_t^2\right)\mathrm{d}t$$
$$= 2\sqrt{2\pi}\sigma_t \exp\left[-2\sigma_t^2(\omega - \omega_0)^2\right], \tag{7.59}$$

where use has been made of eqn. 7.14. Normalizing to unit power and replacing σ_t by $1/2\sigma_\omega$ produces

$$P(\omega) = \exp\left[-(\omega - \omega_0)^2/2\sigma_\omega^2\right]/[\sqrt{2\pi}\sigma_\omega]. \tag{7.60}$$

This is shown in figure 7.11 together with the frequency distribution of the electric field. Notice that the Lorentzian and Gaussian shapes are quite different with the former having a longer tail.

The existence of a general relationship for a waveform between the spread in frequency and in time was broached earlier in Section 5.5.1. Equation 7.60 has shown that when the electric field of a pulse has a pure Gaussian shape like that shown in figure 7.11 the angular frequency distribution is also Gaussian. In addition the widths of the power distribution in angular frequency and the *intensity distribution* with time satisfy the relation

$$\sigma_t \, \sigma_\omega = 1/2 \qquad (7.61)$$

This result is known as the *bandwidth theorem*. Now the full width at half maximum (FWHM), $\Delta t = \sqrt{8 \ln(2)}\sigma_t = 2.35\sigma_t$, so the above equation when written in terms of the FWHM of frequency and time becomes

$$\Delta f \Delta t = 0.44. \qquad (7.62)$$

With any other pulse shape the product of the width of the power distribution in angular frequency and its intensity distribution in time is always larger than for the simple Gaussian shown in figure 7.11. Measurement errors broaden *both* the angular frequency and time distributions so that the product of the widths is also increased by measurement error. Therefore in all cases $\Delta f \Delta t \geq 0.44$. Whenever the product of the widths of a wavepacket with a Gaussian envelope approaches the limiting value given by eqn. 7.61 it is said to be *transform limited*.

It is important to appreciate that the sinusoidal waves making up the electric field waveform in figure 7.11 have a simple phase relationship: they are all in phase at the peak of the Gaussian. On the other hand waves from a source whose power spectrum is Doppler broadened into a Gaussian have random phase relationships. In the case of Doppler broadening the envelope of the electric field is therefore broad and irregular and not a simple Gaussian. Nonetheless, in this case the Wiener-Khinchine theorem guarantees that the autocorrelation function will also be a Gaussian because the two quantities it connects, the autocorrelation function and the power spectrum, are devoid of phase information.

The term *bandwidth* is also used in the analysis of the response of detector systems consisting of a detector of radiation and the electronics for amplifying and frequency filtering the current from the detector. Suppose that $h(t)$ is the current due to unit radiation intensity on the detector system, and that the Fourier transform is $H(\omega)$. Then the bandwidth of the system is defined as

$$B = \int_0^{+\infty} |H(\omega)|^2 \, d\omega / 2\pi, \qquad (7.63)$$

where the definition uses $|H(\omega)|^2$, rather than $H(\omega)$ because it is the *power* that detectors respond to.

As an example consider the case of a system which has a response time τ, so that it effectively integrates the input over a time τ. For this system we can put

$$h(t) = 1/\tau \quad \text{for} \quad -\tau/2 < t < +\tau/2.$$

A wavepacket can be formed with a Gaussian profile whose carrier wave under the envelope alters in frequency along the wavepacket. The example of chirped pulses will be met in Section 14.11.1: these are pulses in which the carrier wave frequency increases/decreases smoothly with time. Such wavepackets are not transform limited.

Expressing this response as a function of angular frequency gives

$$H(\omega) = \int_{-\tau/2}^{+\tau/2} \exp{(i\omega t)}\, \mathrm{d}t/\tau$$
$$= \mathrm{sinc}[\,\omega\tau/2\,]. \tag{7.64}$$

Therefore the bandwidth expressed in terms of frequency is

$$B = \int_0^\infty \mathrm{sinc}^2[\,\omega\tau/2\,]\, \mathrm{d}\omega/(2\pi).$$

Substituting g for $\omega\tau/2$ this becomes[3]

$$B = \int_0^\infty \mathrm{sinc}^2(g)\, \mathrm{d}g/(\pi\tau) = 1/(2\tau). \tag{7.65}$$

Following the same steps for a Lorentzian response, which for unit intensity input would give an output intensity $\exp{(-t/\tau)}/\tau$, the bandwidth is

$$B = 1/(4\tau). \tag{7.66}$$

Example 7.1

The etendue and chromatic resolving power are the parameters most useful when comparing different types of spectrometer. In order to determine the etendue we need to consider the area and angular spread of the beam falling on the detector. One essential restriction is that the circular aperture defining the area of the detector exposed should contain just the central fringe and no more. If there is a maximum of intensity at the exact centre of the fringe pattern for wavelength λ when the optical path difference between the two arms is x, then

$$x = n\lambda.$$

The adjacent minimum is at an angle θ given by

$$(n - 1/2)\lambda = x\cos\theta = x(1 - \theta^2/2 + ...).$$

Taking the difference of the last two equations gives, to an approximation adequate for the small angles involved,

$$\lambda = x\theta^2.$$

Referring to figure 7.12, the angular radius θ_{hole} of the aperture in front of the detector is made small enough to accept only the central fringe:

$$\theta_{\mathrm{hole}}^2 = \lambda/x.$$

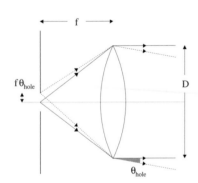

Fig. 7.12 Michelson spectrometer aperture.

---

[3]Integral 3.821/9 in the 5th edition of *Table of Integrals, Series and Products* by I. S. Gradshteyn and I. M. Ryzhik, edited by A. Jeffrey, and published by Academic Press, New York (1994). $\int_{-\infty}^{\infty} \{\sin^2{(ax)}/x^2\}\, \mathrm{d}x = a\pi,\ a > 0$.

Next let the focal length of the exit lens be f so that the area of the hole defining the area of the detector exposed is

$$A = \pi f^2 \theta_{\text{hole}}^2 = \pi f^2 \lambda / x.$$

The other parameter defining the etendue is Ω, the solid angle subtended at the hole by the exit lens. If D is the lens diameter

$$\Omega = \pi D^2 / 4f^2.$$

Using eqn. 4.11 the etendue is

$$\mathcal{T} = A\Omega = \pi^2 D^2 \lambda / 4x. \tag{7.67}$$

A typical instrument might have a scan length of 2 cm and a lens of 3 cm diameter. At $1\,\mu$m wavelength, for such an instrument, eqn. 7.49 gives a chromatic resolving power of $2\,10^4$. The etendue would be $0.11\,\text{mm}^2\,\text{sr}$, which is very much larger than the etendue of a grating spectrometer.[4]

[4]See exercise 6.10 in which the grating spectrometer has an etendue $\sim$100 times smaller.

7.4 Image formation and spatial transforms

The connection between diffraction and the resolution of detail seen in optical images was discovered in the mid 19th century by Abbe whilst he was trying to understand why larger but less well-corrected lenses used in microscopy gave more detail than smaller, better corrected lenses. Figure 7.13 shows a grating, AB, illuminated by monochromatic plane waves and imaged by a well-corrected lens. The illumination is therefore coherent and could be provided by the condenser arrangement for Koehler illumination illustrated in figure 4.10. For simplicity the object distance is taken to be twice the focal length, giving unit magnification. Parallel beams are shown emerging from the grating in the directions of the principal difraction maxima, and these produce bright lines in the transform plane, that is to say in the focal plane of the lens. Apart from these maxima the transform plane is dark. As they travel onward from the transform plane the beams spread out again to form the image of the grating, A′B′, at twice the focal length from the lens. Abbe pictured image formation as a two step process: in the first step the grating diffracts the incident light into parallel beams which form the Fourier transform of the grating; in the second step the lens performs a second Fourier transform to recover the image of the grating.

Abbe pointed out that even with a perfect lens the image cannot be exact because the lens only accepts the lower order (small angle) diffracted beams. If the lens is so small that it only captures the zeroth order diffracted beam then the image plane is of uniform brightness and all detail is lost. If instead the first and zeroth orders are captured the image is a grating of the correct spacing but with a sinusoidally varying

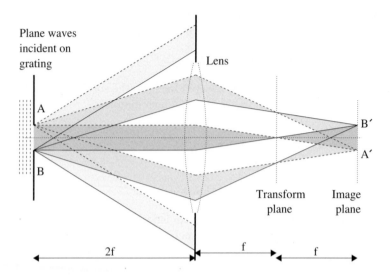

Fig. 7.13 Abbe's insight into image formation. The lower order diffractive beams from the grating AB are shown as shaded bands. A′B′ is the grating image. Rays travelling from A to A′ are drawn as broken lines; rays from B to B′ as solid lines.

amplitude across each element of the grating. We can see this by taking a distribution in the transverse component of the wave vector k_t which only contains these orders

$$A\delta(k_t) + B[\delta(k_t - k_0) + \delta(k_t + k_0)].$$

If θ is the semi-angle subtended by the lens at the object $k_0 = k\sin\theta$. Fourier transforming this expression gives a distribution in the lateral coordinate, x,

$$\int \exp(ik_t x)\{A\delta(k_t) + B[\delta(k_t - k_0) + \delta(k_t + k_0)]\}\,\mathrm{d}k_t$$
$$= A + B[\exp(+ik_0 x) + \exp(-ik_0 x)]$$
$$= A + 2B\cos(k_0 x).$$

The finest resolvable detail on an object will be detail for which the lens has just big enough diameter to capture the first order diffracted beam. If this limiting detail has linear dimension Δx then

$$k_0\Delta x = 2\pi,$$

and substituting for k_0 gives

$$\Delta x \sin\theta = \lambda. \tag{7.68}$$

This, Abbe's view of what is resolvable by an optical system, provides a complementary approach to the Rayleigh criterion given in Section 6.7.

Notice that the nth diffraction order for a grating of a given pitch coincides in angle with the first order for a grating with an n-times finer pitch, so we can say that higher order beams carry information about the fine detail of the grating. Put another way: higher diffraction orders correspond to features with higher spatial frequencies across the image and are needed to recover the sharp edges of the slits in the image. Of course most objects viewed with a microscope are more complex than gratings, but the conclusion remains valid that information on the fine detail in the image is carried by light diffracted at large angles. Although this wide angle light may be very weak it is essential for resolving fine detail. The fact that it makes a significant contribution despite its faintness was what had escaped notice before Abbe's investigations.

The diffractive orders are directly accessible in the transform plane and it is here that manipulations are most easily made to enhance the properties of images formed with a coherently illuminated object.

7.5   Spatial filtering

Any changes made to the high or low order spatial frequency components of an image by selectively obscuring regions of the transform plane is known as *spatial filtering*. Blocking the high order spatial frequencies will smooth the image. This will for example remove *noise* in the form of dust spots on a picture or the dot structure in pictures produced by half-tone printing. Conversely, removing low order spatial frequencies will enhance the outlines of images. Similar image editing facilities are provided by packages for manipulating graphics on PCs. The simple filtering shown in figure 7.14 produces a pure plane wavefront from a distorted wavefront, such as that from a diode laser. The incoming beam is focused on a pinhole located on axis in the focal plane of the first lens. Only the plane wave component in the incident beam passes through the pinhole; all the other components in the beam focus away from the pinhole. The waves emerging from the pinhole are spherical and the second lens converts these to a plane wave.

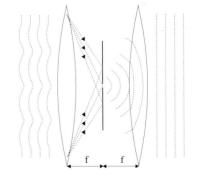

Fig. 7.14 Spatial filter consisting of a pinhole and lenses. Only the plane wave component from the incoming distorted wave focuses at, and passes through the pinhole.

Biological specimens studied with microscopes are mostly transparent and the structures of interest may only differ from the surrounding material in having a slightly different refractive index. One drastic approach is to dye the specimen with a material that is selectively absorbed by some structures; but this is not always feasible and it may damage the specimen. An alternative is to use *dark field* illumination which is shown in the lower diagram in figure 7.15. The left hand diaphragm forming the aperture stop in the condenser is perfectly opaque apart from a narrow clear annular aperture. In the plane conjugate to this diaphragm on the objective side an annular stop is placed which exactly covers the image of the annular aperture. This plane is the transform plane and the annular stop introduced blocks the zero order light. An observer

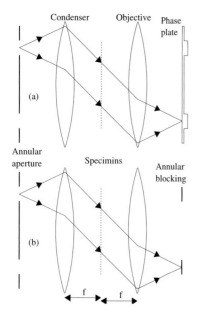

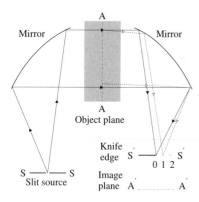

Fig. 7.15 The upper diagram shows phase contrast illumination, and the lower diagram shows dark field illumination.

Fig. 7.16 Schlieren photography. The diffraction orders at the plane of the knife edge are numbered 0, 1 and 2.

viewing a blank slide in the field of view would see a perfectly dark field. If a specimen slide is then inserted the higher spatial frequencies in its image, which are normally swamped by the zero order beam, can now produce an image. In this image the edges of structures are particularly clear.

A more subtle technique earned a Nobel prize for Zernike in 1953 and this is illustrated in the upper diagram of figure 7.15. The annular condenser aperture just described is again used, but now a transparent *phase plate* replaces the annular stop in the objective. This phase plate is made thicker over the same annular region which was opaque in dark field illumination to give a phase delay of $\pi/2$ for light travelling through the annulus relative to light missing it. Suppose the amplitude of light at the transform plane is $E_0 \sin(\omega t)$ with normal illumination and a clear slide. When a specimen is inserted the light passing through any given area undergoes a small relative phase shift ϵ, determined by how much the local refractive index deviates from the mean refractive index. The corresponding amplitude is

$$E_0 \sin(\omega t + \epsilon) = E_0 \sin(\omega t) + \epsilon E_0 \cos(\omega t),$$

and the intensity is $E_0^2(1 + \epsilon^2) \approx E_0^2$, showing that the weak image has been swamped. Things are quite different when the phase plate is inserted because the zero order component is phase shifted to $E_0 \cos(\omega t)$. Thus the total amplitude becomes $E_0(1 + \epsilon) \cos(\omega t)$ and the intensity becomes $E_0^2(1 + \epsilon)^2 \approx E_0^2(1 + 2\epsilon)$. This gives a detectably large change in intensity for even a small change in refractive index.

7.5.1 Schlieren photography

This technique is used to make visible the variations in density in a fluid in motion around large scale objects, and the apparatus used is sketched in figure 7.16. A monochromatic slit source is focused by a mirror so as to throw a parallel beam through the volume of interest, which is shown shaded in the figure: this might, for example, be a wind tunnel. Afterwards the light falls on a second mirror whose focal plane (the transform plane) is S'S' and which images the object AA at A'A'. A knife edge is placed at S'S' parallel to the slit source and positioned to block off part of the zero order as well as all the higher diffraction orders on one side. Regions of refractive index differing from the mean then appear as bright and dark streaks, just as in the phase contrast microscope. Mirrors are preferred to lenses for the focusing because the area of the field of view is comparatively large. The mirrors shown are off-axis paraboloids whose aberrations remain small over the required angular range.

7.5.2   Apodization

Attempts to observe faint objects near brighter ones are made difficult because the diffraction rings around the image of the brighter object can easily swamp the fainter one's image. The rings outside the central Airy disk can be suppressed by placing a filter over the telescope objective whose transmission coefficient falls with a Gaussian dependence on the radial distance off axis. The diffraction pattern is the Fourier transform of this shape, which is also a Gaussian. Figure 7.17 contrasts the diffraction patterns for a clear circular aperture and a Gaussian shaded aperture. While the Gaussian is broader its lack of any outer rings means that there is a much improved chance of detecting a target against the glare of a brighter companion. *Apodization* is the term used to describe this procedure.

7.6   Acousto-optic Bragg gratings

Ultrasound waves in a solid or liquid produce sinusoidal variations in density and hence similar changes in refractive index across the material. Incident light will diffract from such density gratings and at the same time undergo a minute Doppler shift in frequency because the grating is travelling at the speed of sound. Both the deflection and frequency shift have significant applications in information processing. The techniques, their analysis and some applications will be reviewed here.

The crystals used to excite the ultrasound waves such as quartz are chosen for their high piezo-electric coefficients at ultrasound frequencies of up to many gigahertz. Figure 7.18 shows an incident plane wave light beam being diffracted from a *Bragg cell*. The Bragg cell consists of a piezo-electric crystal which generates ultrasound waves, an acousto-optic material through which they travel, and finally an absorber that suppresses reflections. Choices for the acousto-optic material are extra-dense flint glass and lithium niobate. If the acoustic wave has sufficient width W the incident light will be diffracted entirely into the zero order and first order on one side. This can be understood with the help of the lower panel in figure 7.18. The directions of the waves being diffracted in first order from planes of maxima in refractive index are shown, θ' being the angle which the incident light beam makes with the acoustic wavefronts inside the cell. Diffraction maxima will satisfy the usual relation

$$p\lambda' = \Lambda(\sin\theta' + \sin\alpha'),$$

where θ' and α' define the incoming and outgoing wave directions, where Λ and λ' are the respective wavelengths of acoustic and of light waves in the crystal, and where p is integral. There is an additional requirement for observing a maximum when the acoustic wave is broad: light diffracted at all points along a given acoustic wavefront must be in phase, which means that the angle of reflection (α') equals the angle of inci-

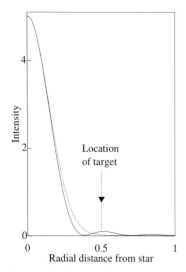

Fig. 7.17 The image intensity of a bright point source produced by a Gaussian shaded aperture (broken line) and that for a sharp aperture boundary (solid line).

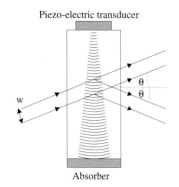

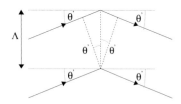

Fig. 7.18 The upper diagram shows an acousto-optic Bragg cell diffracting a laser beam. The lower diagram shows the scattering of light from a pair of acoustic waves inside the Bragg cell.

dence (θ'). Then the requirement for a diffraction maximum becomes

$$2\Lambda \sin\theta' = p\lambda'.$$

Expressing this in terms of the angle at which the light waves are incident in the air and their wavelength in air this becomes

$$2\Lambda \sin\theta = p\lambda.$$

Lastly the angle of incidence is made sufficiently small that the radiation is restricted to zero, or first order with

$$2\Lambda \sin\theta_{\mathrm{B}} = \lambda, \tag{7.69}$$

where θ_{B} is given the subscript B because this equation is also the Bragg condition met in X-ray diffraction. In order for any diffraction to occur the width of the optical beam, w, should span at least one acoustic wavelength: $\Lambda < w/\cos\theta \approx w$ because θ is small. Now

If the width W of the acoustic wave is sufficiently narrow then the angular separation between the diffractive orders, λ'/Λ, becomes smaller than the angular spread of the acoustic beam Λ/W. As a result diffraction of the optical beam in many orders can occur. This is called the Raman–Nath regime, but is not of any further interest to us here.

$$\Lambda = V/F, \tag{7.70}$$

where V and F are respectively the velocity and frequency of the acoustic waves. Therefore the requirement on Λ can be re-expreesed as a frequency limit which acoustic waves must exceed in order to produce any useful acousto-optic effect

$$F \geq V/w. \tag{7.71}$$

In flint glass with a light beam of $1\,\mathrm{mm}$ width $F \geq 3\,10^6\,\mathrm{Hz}$.

The light scattered from the acoustic wave is Doppler shifted from frequency f_0 to frequency $f_0 + F$. It is thus possible to make use of either the deflection or the frequency shift of light within a Bragg cell. These changes are determined by the acoustic wave frequency, which in turn is the frequency at which the piezo-electric crystal is driven. The frequency shift is used in applications requiring heterodyning of optical beams, while beam deflection is used in information processing.

7.6.1   Microwave spectrum analysis

Figure 7.19 shows the elements of a system for analysing microwave spectra, which is used to scan continuously for incoming radar signals. A laser beam is incident on the Bragg cell and the incoming microwave signal is applied to the piezo-electric transducer. Each microwave frequency produces a specific laser beam deflection and the deflected beams are projected by the Fourier lens onto a linear detector array such as a CCD. Using eqns. 7.69 and 7.70 we can relate the accessible range of microwave frequencies to the angular spread of the microwave beam

$$\Delta F = V\Delta(1/\Lambda) = 2V\cos\theta_{\mathrm{B}}\Delta\theta_{\mathrm{B}}/\lambda \approx 2V\Delta\theta_{\mathrm{B}}/\lambda. \tag{7.72}$$

This is maximized when $\Delta\theta_B$ fills the angular divergence of the acoustic beam.[5] The number of distinguishable frequencies across this bandwidth N_r is twice the angular divergence $\Delta\theta_B$ divided by the angular divergence of the optical beam, because the deflection of the optical beam is twice the Bragg angle. Thus

$$N_r = 2\Delta\theta_B/\Delta\theta_{laser} = 2\Delta\theta_B w/\lambda. \qquad (7.73)$$

Bandwidths of several gigahertz are obtained and of order a thousand distinguishable frequency steps. Of course the lens focal length, the CCD detector array pixel size and number of pixels need to be matched to this performance.

7.7 Holography

Holograms are familiar from the embossed images on credit cards and displays in various promotions. The first holograms were made in 1948 by Gabor, who received a Nobel prize in physics for this work. However it needed the introduction of the laser to make the technique of practical use in the optical domain. Prototype systems are manufactured which store data in holograms with fast access and readout. The principles and several applications of holography will be described in the following sections.

The images seen through holograms have depth and when the observer's viewpoint is changed the scene changes exactly as if the original scene were present. What is special about a hologram is that, unlike a normal photograph which records the light intensity falling on it, the hologram records the amplitude and phase of the light from the scene. Coherent illumination is necessary, and can readily be provided by a laser. The exposure of a hologram and the later image reconstruction are shown in figure 7.20. During the exposure the light from a laser is divided into two parts that remain coherent with one another. One beam, called the *reference* beam, falls directly on the film that will become the hologram, while the other beam illuminates the object to be recorded. Light scattered from this object forms an *object* beam which also falls on the film. The fringes on the film produced by interference between the object beam and the reference beam contain phase information from the object beam, and this pattern is stored as exposed and unexposed regions on the film. Processing preserves these fringes as dark and light areas.

Afterwards the hologram is replaced in its initial position with respect to the reference beam, but now the object is removed. The observer sees

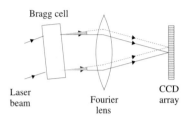

Fig. 7.19 Bragg cell diffracting a laser beam. Different microwave frequencies cause different beam angular deflections.

[5]The requirement on the relative angular spreads of the two beams is very different when the acoustic beam is used simply to Doppler shift the frequency of the optical beam. Then the need is to maximize the power transfer into the laser beam and so the angular spreads of the optical and acoustic beams are made equal.

a realistic three-dimensional virtual image of the object exactly where the object would have been. In addition there is a second real image

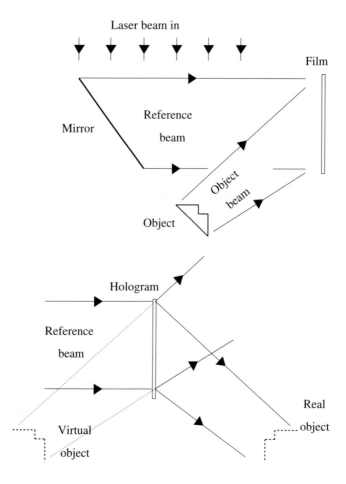

Fig. 7.20 The upper diagram shows the exposure to produce a hologram. The lower diagram shows the images produced when the hologram is illuminated with the reference beam again. The part of this reference beam that is not scattered by the hologram is omitted from the lower diagram.

which has its surface features inverted, so that, for example, a nose would appear to project into rather than out of a face. This is known as a *pseudoscopic* image. The observer sees the virtual image through a window formed by the hologram; and if the hologram is broken in two, either piece alone gives the same view – through a window that is now smaller. This property shows that the information about all the image is held over all the hologram; which is very different from how information is held on a normal photograph.

7.7.1 Principles of holography

An elementary example of holography is shown in figure 7.21 where a point object is placed in an incoming plane laser beam at a distance R from the recording film. The fringes produced by interference between the direct light and that scattered by the object are circular and the mth bright fringe has a radius ρ_m, such that

$$\sqrt{R^2 + \rho_m^2} - R = m\lambda,$$
$$\text{then} \quad \rho_m^2/2R = m\lambda,$$
$$\text{and} \quad \rho_m^2 = 2mR\lambda, \tag{7.74}$$

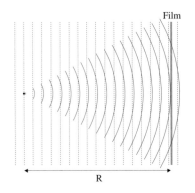

Fig. 7.21 The diagram shows a point object in a reference beam. The reference beam and the scattered light from a point object produce circular interference fringes on the film.

to an adequate approximation. This is exactly the same expression as that for the radii of the even order Fresnel zones given in eqn. 6.55. Therefore the pattern appearing in the developed *hologram* will be a zone plate. When this is illuminated by the reference beam the focusing property of the Fresnel zone plate discussed in Section 6.12.1 will produce a virtual image where the point object had been and a second real image a distance R in front of the hologram. For convenience these images are shown separately in figure 7.22. An extended object generates a more complex fringe pattern in the recording medium, each point on its surface that can *see* the hologram making its own contribution at each point on the hologram, which is the case we now consider.

Suppose that the reference beam in figure 7.20 is $\mathcal{E}_r = E_r \exp(-i\omega t)$; then the object beam will be

$$\mathcal{E}_o(x) = E_o(x) \exp(ikx \sin\alpha) \exp[i\phi(x)] \exp(-i\omega t),$$

where x is the coordinate running up the page and α is the angle the object beam makes with the reference beam. E_o and E_r are real. The phase, $\phi(x)$, and amplitude, $E_o(x)$, are determined by the detailed shape of the object. Using eqn. 5.12 the time averaged intensity of the light falling on the recording material at x is

$$
\begin{aligned}
T(x) &= \mathcal{R}e\{[\mathcal{E}_o(x) + \mathcal{E}_r]^*[\mathcal{E}_o(x) + \mathcal{E}_r]\}/2 \\
&= \{E_r^2 + E_o^2 + E_r E_o \exp[i(kx\sin\alpha + \phi(x))] \\
&\quad + E_r E_o \exp[-i(kx\sin\alpha + \phi(x))]\}/2.
\end{aligned}
\tag{7.75}
$$

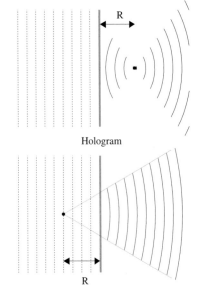

Fig. 7.22 The diagrams show the reconstruction of a hologram of a point object in the reference beam. The reference beam beyond the hologram is omitted for clarity. In the upper diagram the reconstruction of the real image is seen, and in the lower the reconstruction of the virtual image.

For simplicity we will take this intensity distribution to be reproduced in the transmission amplitude of the final processed hologram. If this is the case then when the hologram is illuminated by the reference beam (with the object removed) the transmitted light amplitude will be

$$
\begin{aligned}
E(x) &= E_r T(x) \\
&= E_r(E_r^2 + E_o^2) \exp(-i\omega t) \\
&\quad + E_r^2 E_o \exp[i(kx\sin\alpha + \phi(x))] \exp(-i\omega t) \\
&\quad + E_r^2 E_o \exp[-i(kx\sin\alpha + \phi(x))] \exp(-i\omega t).
\end{aligned}
\tag{7.76}
$$

Of these *three* terms the first is a forward beam aligned along the reference beam. Apart from a constant the second term is the object beam and produces the virtual image where the object had been. The final term describes a beam tilting downward at angle α to the reference beam and this produces the real image. Its amplitude is the complex conjugate of the object beam and results in a pseudoscopic image. The three emerging beams are also recognizable as the diffracted beams of order $+1$ (virtual), 0 (forward) and -1 (real).

7.7.2   Hologram preparation

The following are some general points that bear on the preparation of holograms in the laboratory. In order that the reference and object beams are coherent the path differences should be small compared to the coherence length of the laser, and the coherence area should extend well beyond the object being photographed. With beams inclined at an angle α it is evident that the fringe spacing is of order $\lambda/\sin\alpha$, so that there is a need to keep the whole apparatus stable during the exposure to the level of parts of a wavelength. This is less of a problem in laboratories in which an intense pulsed laser is available, for example a ruby laser giving $20\,\mathrm{ns}$ long pulses at $694\,\mathrm{nm}$. The emulsion needs to contain fine enough silver bromide grains (under $10\,\mu\mathrm{m}$ diameter) to record the fringes, but because fine grains require as much light to render them developable as coarse grains the film is relatively slow. In the object beam there are large local fluctuations of intensity due to interference effects and so a suitable arrangement is to have the reference beam about three times more intense than the object beam.

When the hologram is in the near/far field of the object it is known as a Fresnel/Fraunhofer hologram. If the object lies in the reference beam and this is incident perpendicular to the film the hologram produced is called an inline hologram. Other alignments give off-axis holograms, which are the simplest to produce and avoid the confusion of superposed images seen with an inline hologram. If the reference and object beams arrive from opposite sides this gives a reflection hologram.

When looking at a hologram in one's hand many large circular features catch the eye. These are due to dust and other imperfections; the interference fringes holding the image information are too small to be detected by eye. *Phase holograms* can be produced by treating the exposed film so as to rehalogenate the silver grains. These renewed silver halide grains then migrate to the nearby unexposed silver halide leaving regions of depleted optical density. Phase diffraction gratings have a considerable advantage over amplitude gratings because the fraction of light which they diffract into each first order image can be as large as 33%, as against 6% for amplitude gratings; phase holograms share this advantage. Modern materials such as photopolymers and photoresists are widely used to produce phase holograms. When the former are exposed to light the monomers polymerize and this produces a useful change in refractive index; when the latter are exposed to light they become soft and can be etched away with solvents leaving the image in relief.

Surfaces with any degree of roughness seen in laser light show many small bright *speckles* that change with any movement of the viewer.

Speckles are the result of interference between light diffracted from nearby parts of the surface and hence their angular size is simply the angular resolution of the camera or the eye doing the viewing. On the film plane this is $1.22\lambda f/D$, f and D being the respective focal length and entry pupil diameter. Speckles will appear on holograms, and one way to reduce their impact is to take multiple exposures with the beam *incident* on the object being displaced a minute amount between exposures.

7.7.3 Motion and vibration analysis

Holograms can be exposed in such a way as to reveal patterns of motion of musical instruments and living tissue. One approach is to freeze the motion by pulsing the laser source on for a few nanoseconds, and then to repeat the pulse after a short delay. The interference patterns from the two exposures will match over surfaces that have not moved or have moved a whole number of wavelengths during the delay. These regions will appear bright in the reconstructed image and form contours of equal, known movement. An alternative is to take a long exposure which will give a uniform intensity at moving points, while any nodes will appear very bright.

Holograms formed by double exposures are widely used to render visible in three dimensions the flow of fluids, as for example in a wind tunnel. This contrasts with the projected images obtained with Schlieren photography or with the Mach–Zehnder interferometery described in Section 5.7.3. In the case of a wind tunnel the first exposure would be made with the air at rest, and the second with the air flowing. When the scene is reconstructed from the hologram the contours of increased and reduced density appear as bright regions in three dimensions.

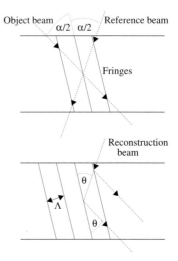

Fig. 7.23 The upper panel shows the formation of interference fringes in a thick emulsion by the object and reference beams. The lower panel shows Bragg scattering from the interference fringes after chemical development.

7.7.4 Thick holograms

The fringes recorded in a hologram in each small local area are inclined at equal angles to the reference and object beams as shown in figure 7.23 with spacing

$$\Lambda = \lambda/[\,2\sin{(\alpha/2)}\,],$$

where α is the angle between these beams. This separation is of order one wavelength so that if the recording medium is tens of microns thick each local region has many parallel fringes which act as Bragg planes in reconstruction. They are equivalent to the planes of high density in the acousto-optic modulator described in Section 7.6. There is constructive interference between light reflected from these layers provided that the angles of incidence and diffraction, θ, satisfy

$$\lambda = 2\Lambda\sin\theta,$$

where Λ is the layer spacing. This condition holds when the reconstruction beam points in exactly the direction of the reference beam, and

Bragg scattering occurs in the direction of the beam forming the virtual image. This of course strengthens the virtual image at the expense of the real, pseudoscopic image. It has however an even more important and useful *negative* effect.

When the beam used in reconstruction is incident at some angle different from the reference beam the Bragg condition is no longer satisfied and no image is seen. As a consequence the way is open to record multiple images in a thick hologram. These can be recorded and reconstructed with the reference beam rotated to a new angle for each image. This makes it feasible to produce a display hologram that changes as one walks past; a new hologram is revealed at each step the viewer takes. These holograms are usually phase holograms in order that the brightest possible virtual image is obtained. Photopolymer films used to make such thick holograms are typically $100\,\mu$m thick.

Thick reflection holograms produced by two reference beams incident normally on the recording material from opposite directions will produce planar interference fringes parallel to the surface and spaced at intervals of $\lambda/2$. This structure will reflect almost all the light at wavelengths close to λ and very little at other wavelengths. It constitutes what is called a *notch* filter. Such filters are useful for making safety glasses which reduce the intensity of a laser beam by many orders of magnitude while not reducing the ambient light at other wavelengths.

By using three laser beams simultaneously and panchromatic film (responsive to wavelengths across the visible spectrum) a colour hologram can be produced. The choices of the 476.5 nm Ar$^+$ ion laser line, the 532 nm Nd:YAG line and the He:Ne 632.8 nm line give good rendering of colour. Viewing would seem to require that the same three lasers are used: each wavelength has produced an interference pattern and this will generate an image of that colour when the corresponding laser illuminates the hologram. However there is crosstalk because each reconstruction beam is also diffracted by the fringes produced by the other lasers. Crosstalk will be suppressed if the recording medium is sufficiently thick, because a reconstruction beam of colour Y will only satisfy the Bragg condition for the fringes made with colour Y and for no others. As a bonus it is now possible to use a white light source in the reconstruction phase. Only those components in the white light beam whose wavelengths are extremely close to one of the laser wavelengths can satisfy the Bragg condition and these alone contribute to the image. Together these three Bragg selected wavelength bands form a full colour image. All other wavelengths in the reconstructing white light beam pass onward in the reference beam direction. The Bragg diffraction requirement for obtaining an image from a thick hologram effectively insulates the three colour images from each other. Figure 7.24 shows a simple layout invented by Denisyuk for producing and viewing full colour holograms. The three laser beams are spatially filtered by a pinhole to give pure

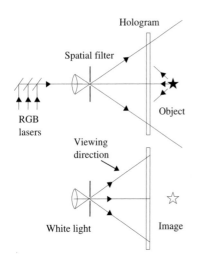

Fig. 7.24 The upper panel shows Denisyuk's method for recording a colour hologram. Viewing in white light is shown in the lower diagram.

spherical wavefronts. Each reference wave passes through the recording material with some portion being reflected back from the object in an object beam. Interference fringes will lie roughly parallel to the surface of the recording medium because the reference and object beam enter the recording layer from opposite directions. Several layers of nodes will therefore be formed through the depth of the recording medium so that the Bragg isolation of each colour is acheived. The viewing of the Denisyuk hologram in white light is also shown in figure 7.24.

7.8   Optical information processing

The techniques of spatial filtering and holography underlie methods that are used for processing information. Potential applications include the automatic matching of fingerprints and large scale data storage and retrieval. These techniques use coherent light and rely on the properties of Fourier transforms. The remaining sections of this chapter will be used to introduce a few representative developments in optical information processing.

7.8.1   The $4f$ architecture

Figure 7.25 shows one common arrangement for optical processing with coherent light, called the $4f$ geometry. Coherent plane waves arrive from the left, and might be provided by a laser plus beam expander with spatial filtering. The components shown – object plane, the first transform lens, the transform plane, a second transform lens and the image plane – are spaced at intervals of a focal length. For simplicity the focal lengths are taken to be equal, and the discussion is restricted to one transverse dimension. As an example of optical processing we consider how to compare a sequence of patterns, $f_i(x)$, with a fixed pattern $g(x)$, where x is the dimension perpendicular to the optical axis. The Fourier transform of $g(x)$ is generated in the form of a hologram as explained below and this, $G(k)$, is placed in the transform plane. Then in turn the patterns $f_i(x)$ are placed in the object plane. The light amplitude transmitted through the transform plane will therefore be the product $F_i(k)G(k)$. This is Fourier transformed by the second lens so that at the image plane the amplitude is

$$h(x) = \text{FT}[\, F_i(k)\, G(k)\,] = f_i(x) \otimes g(x), \qquad (7.77)$$

where the convolution theorem has been used. Thus if the patterns match so that $f_i(x) = g(x)$,

$$h(x) = \int g(x')g(x - x')\mathrm{d}x. \qquad (7.78)$$

Usually $g(x)$ is some fairly random pattern, so that if the patterns match $h(x)$ will be large in the forward direction, where $x = 0$, and small elsewhere. On the other hand, whenever $f_i(x)$ is different from $g(x)$, $h(x)$

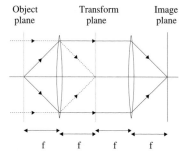

Object plane   Transform plane   Image plane

Fig. 7.25 $4f$ architecture for coherent optical processing.

will have a uniform distribution. A pattern match is therefore signalled by a bright spot appearing on the optical axis in the image plane.

A defect of this simple system is that with an inline arrangement the unscattered reference beam travels forward making it difficult to distinguish when a match is achieved. It is necessary to displace the signal indicating a match between target and test patterns to a location well away from the optical axis. Figure 7.26 shows the layout used to produce the hologram of the target pattern $G(k)$ which has this desirable property. Part of the broad planar laser beam passes through a transparency that carries the target pattern and is then focused onto the recording film. The remainder of the laser beam is deflected onto the same sensitive area. Its amplitude at a point x on the recording plane is $E_\mathrm{r} \exp\left(-ikx\sin\alpha\right)$, where α is the deflection and k is the wave number of the laser light. If $g(x_0)$ is the target pattern at x_0 in the object plane, then its diffraction pattern (Fourier transform) on the film plane at coordinate x is

$$G(kx/f) = \int g(x_0) \exp\left[ikx_0(x/f)\right] \mathrm{d}x_0,$$

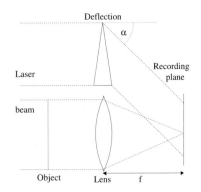

Fig. 7.26 Setup for producing a VanderLugt filter.

where the sine of the angle in which the light is diffracted is approximated by x/f. This arrangement produces a *Fourier transform* hologram in which the Fourier transform of the object interferes with the reference beam. The particular layout shown in figure 7.26 makes up a *Vander-Lugt filter*, named after its originator. The effect of using a reference beam tilted at an angle α is that the bright spot indicating a match in the comparison is displaced a corresponding distance $f\sin\alpha$ below the optical axis on the image plane in the detector arrangement shown in figure 7.25. There are practical difficulties in getting a positive match when the scale and orientation of the patterns being compared are not the same.

Both the target and test patterns can each be presented electronically using a *spatial light modulator*. This is usually a *liquid crystal display* (LCD) of the sort to be described in Chapter 10. Effectively the LCD forms a screen a few centimetres across having a million or so pixels, each of which can be independently set to transmit or absorb light. Referring to figure 7.25 the LCD screen placed in the transform plane carries the fixed pattern $G(k)$ while the screen placed in the object plane can be switched from one pattern to another, $f_i(x)$, in tens of microseconds, making possible comparisons at high rates.

7.8.2 Data storage and retrieval

The ability to store many distinguishable holograms each written with the reference beam incident at a different angle implies the possibility of high capacity data storage on holograms. Data densities of up to 100 bits/μm^2 have been achieved, which compares well with around 10 bits/μm^2 on DVDs but is still less than the 160 bits/μm^2 (100 Gbits per square inch) obtained with current longitudinal recording on magnetic disks. The parallelism of data storage on holograms can offer high access rates: all the information on an individual hologram or *page* is available simultaneously. An access time of 1 ms to each 1 Mbit hologram implies a data rate of 1 Gbs^{-1}, which is many times faster than the sequential read out rate from other media.

Exercises

(7.1) Calculate the Fourier expansion for a repetitive sawtooth wave, $f(x)$, having a repeat distance λ and for which $f(x) = x$ when $-\lambda/2 < x < \lambda/2$?

(7.2) What is the fringe count between successive maxima of visibility on an Michelson interferogram taken with a sodium source filtered to pass only the two D lines?

(7.3) (a) Show that convolution is commutative, that is that $f(x) \otimes g(x) = g(x) \otimes f(x)$. (b) In figure 7.3 imagine that one delta function $\delta(x - x_m)$ is replaced by $-3\,\delta(x - x_m)$. What difference does this make to the convolution?

(7.4) A radio-frequency beam with frequencies ranging from 40 to 60 MHz drives a lithium niobate acousto-optic Bragg cell. The optical beam is from a HeNe laser of wavelength 632.8 nm and for this wavelength lithium niobate has a refractive index 2.2. What is the microwave wavelength? What width of laser beam is required to permit 100 frequencies to be distinguished across microwave input frequency range? What angular spread of microwave beam is required to cover the bandwidth? The velocity of ultrasound in lithium niobate is 6600 ms^{-1}.

(7.5) A Cd state emits a 643.8 nm line which has a lifetime of 4.1 10^{-10} s. What is the line width in frequency and in wavelength?

(7.6) A thin transmission hologram is recorded with a HeNe laser of wavelength 632.8 nm, the object and reference beams being inclined at $\pm 30°$ to the normal to the surface respectively. What is the average fringe spacing in the emulsion?

(7.7) A camera is used to photograph the virtual image in the setup described in the previous question. As large a depth of field as possible is required, however the speckle size increases as the lens is stopped down. What would be a suitable aperture stop to keep the speckle size on film under 10 μm?

(7.8) (a) Calculate the degree of first order coherence of a source emitting a single spectral line with Lorentzian shape given by eqn. 7.53. (b) What would be the visibility of fringes seen with a Michelson interferometer using this source?

(7.9) (a) Calculate the degree of first order coherence of a source emitting a single spectral line with Gaussian shape given by eqn. 7.56. (b) What would be the visibility of fringes seen with a Michelson interferometer using this source?

(7.10) A monochromatic source of angular frequency ω_0 is pulsed on and off with a cycle time of τ_p. If the pulses are square between zero and unity what is the frequency distribution of the radiation?

(7.11) Show that for both the Lorentzian and Gaussian power distributions, with unit total power, the value at resonance is approximately the inverse of the FWHM.

(7.12) If $F(k)$ is the Fourier transform of $f(x)$, what are the Fourier transforms of $\mathrm{d}f/\mathrm{d}x$ and $\mathrm{d}^2 f/\mathrm{d}x^2$?

(7.13) Prove the version of the Wiener–Khinchine theorem for the complex extension of electric fields given in eqn. 7.43.

Astronomical telescopes

8.1 Introduction

The major telescopes now being used for astronomical observations in the visible and near visible regions of the spectrum are reflecting telescopes with primary mirrors of diameters ranging up to 10 m. These huge apertures make it possible to catch enough light to record distant sources active when the universe was less than 10% of its present age. Mirrors rather than lenses are the practical optical elements at this scale. Firstly it is essential to support such large pieces of glass over their whole area in order to prevent the physical sagging which would otherwise alter their optical properties significantly; and this can only be done for mirrors. Additionally mirrors are free of chromatic aberration, lose little light through absorption in their surfaces and function at wavelengths at which lenses would absorb strongly. In the following section the main design features of astronomical telescopes will be described. Details of some representative large telescopes are listed in table 8.1.

Atmospheric extinction of the radiation from astronomical sources is primarily due to water, carbon dioxide, ozone and oxygen molecules. Each species has absorption bands and they all scatter radiation at other wavelengths. The molecular scattering of light[1] is proportional to λ^{-4}. Absorption below 300 nm wavelength is mainly due to ozone and is almost complete. Starting at 300 nm, the *optical window*, with extinction of 10% or so in the visible, extends into the infrared with increasingly frequent absorption bands, some almost opaque, to around 15 μm. Most molecular densities fall off exponentially with height, and at a height of 4 km a telescope is above 95% of the atmospheric water vapour. However the major part of the ozone is above 20 km.

[1] This is called Rayleigh scattering and is discussed in Chapter 11.

The twinkling of the stars is an indication of atmospheric turbulence, and this turbulence reduces the resolving power of any telescope on the Earth's surface. Typically for telescopes located at 3000 m high sites in dry areas the best resolution achievable (the *seeing*) at 500 nm wavelength is around 0.4 arcsec, which is what could be achieved with a telescope above the atmosphere having an aperture diameter of only 25 cm; but the larger telescope retains its advantage in gathering more light. *Adaptive optics* is the term used to describe the ways and means by which the image distortions due to atmospheric turbulence are corrected in real time. A complementary technique called *lucky imaging* is

Table 8.1 A table of the parameters for some representative large telescopes.

Telescope	Design	Diameter/segments	Mounting	Location/height
Keck I,II	Ritchey–Chretien	10 m/36	Alt-az	Hawaii/4208 m
Hobby–Eberle	Spherical primary	9.2 m/91	Fixed elevation	Texas/2072 m
VLT1-4	Ritchey–Chretien	8 m/1	Alt-az	Chile/2635 m

also described. These techniques for recovering the full potential of large ground-based telescopes are described in the third part of the chapter. Adaptive optics has scientific applications elsewhere: in controlling the wavefronts of laser beams used in microscopy, in laser-induced fusion and in studies on human vision.

Individual telescopes are being used together in arrays in order to further extend the resolution in the visible and near visible part of the spectrum. Light from the telescopes is brought together to interfere and the fringe patterns, *interferograms*, used to give information on the source. At Mauna Kea the two 10 m Keck telescopes and four 1.8 m telescopes will be used as an interferometer with a largest separation of 135 m – giving a potential resolution at 1 μm wavelength of approximately $7.4\,10^{-9}$ rad or 1.5 mas (milliarcsec). Interferometry with multiple apertures is discussed in the fourth part of the chapter.

Several Michelson interferometers have been built with kilometre long arms to detect the tiny change in relative length between their arms which is expected when a gravitational wave strikes the Earth. Only rare, catastrophic events, such as supernova explosions in our galaxy are likely to give detectable signals. This very different application of interferometry in astrophysics is the topic of the fifth segment of the chapter. A final short segment of the chapter is used to describe the observations of gravitational lensing and includes a simple account of the origin of this general relativistic effect.

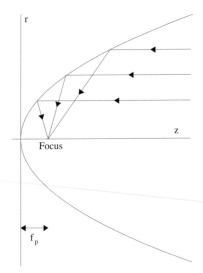

Fig. 8.1 Paraboloid mirror.

8.2   Telescope design

Telescopes need to have a large entry pupil in order to collect the maximum light and so to detect weak and distant astronomical sources; and also to resolve the details of the structure of sources. An *unresolved* star is one whose angular diameter is much smaller than the angular resolution of the telescope viewing it. The entry pupil of a telescope is the *primary* mirror, the one which first intercepts the light. Now the angular resolution due to diffraction alone, $\Delta\theta$, was calculated in Section 6.7,

$$\Delta\theta = 1.22\lambda/D, \qquad (8.1)$$

for light of wavelength λ at a circular aperture of diameter D. In the case of the Hubble Space Telescope, the HST, with a 2.4 m diameter primary mirror, the resolution limit imposed by diffraction is $2.54\,10^{-7}\,\mathrm{rad}$ or 52.4 mas. The aberrations in a mirror telescope are made comparably small by using mirrors whose shapes are paraboloids and hyperboloids of revolution.

Paraboloid mirrors have the useful property that all incident rays parallel to the axis pass, after reflection, through the geometric focus as shown in figure 8.1. A paraboloid has the equation $r^2 = 4f_{\mathrm{p}}z$ where z is the axial distance from the pole of the paraboloid and r is the distance off axis. f_{p} is the equivalent paraxial focal length of the mirror. However the aberration of a paraboloidal mirror rises rapidly when the target moves off axis: the angular size of comatic flare projected on the sky is

$$\alpha \approx \theta D^2/(4f_{\mathrm{p}}^2), \tag{8.2}$$

where θ is the angle off axis. Fortunately hyperboloid shaped mirrors also give a good axial focus and have off-axis aberrations that are opposite in sign to those of paraboloidal mirrors. The combination of one mirror of each shape has the potential for producing an image for which the off-axis aberrations largely cancel. Figure 8.2 shows how all the rays directed towards one geometric focus of a hyperboloid mirror will, after reflection, pass through the second focus. Using the coordinates shown the equation of the hyperboloid is $r^2 = (z^2 - a^2)(e^2 - 1)$ where e is called the eccentricity. The linear magnification is the ratio of the image to object distance $m = (e + 1)/(e - 1)$. If the hyperboloid mirror is placed so that its first geometric focus coincides with the focus of the paraboloid, then the final image will lie at its second focal point. The size of the image of a star of angular diameter θ produced by the two-mirror combination would be

$$D_* = mf_{\mathrm{p}}\theta. \tag{8.3}$$

As noted above, the off-axis aberrations are very much reduced if the curvatures of the mirrors are chosen suitably. The standard layout for the primary and secondary mirrors is shown in figure 8.3 with the final image being projected through a hole in the primary. The classical *Cassegrain* telescope shown in the upper panel has a convex secondary mirror, while the *Gregorian* telescope has a concave secondary mirror. A cross shaped frame of thin rods, called a *spider* supports the secondary from the telescope tube. Thus light travelling toward the primary is Fresnel diffracted by the secondary and its spider, and if the image of a star is overexposed it acquires thin cruciform arms. In both these telescope designs the primary is the aperture stop and also the entrance pupil, so that it is the diameter of the primary mirror which determines the telescope's potential light gathering power and angular resolution.

The Cassegrain is more compact than the Gregorian for the same focal length, which gives a decisive advantage when the telescope weighs tons and is several metres in length. In addition the Cassegrain secondary mirror and the hole in the primary can be made smaller. The

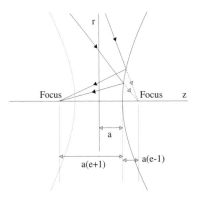

Fig. 8.2 Hyperboloid mirror. The dotted curve shows where the other half of the hyperboloid surface would lie.

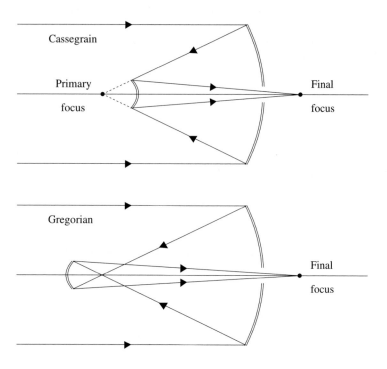

Fig. 8.3 Cassegrain and Gregorian telescopes.

final image surface is curved toward the secondary mirror in the classical Cassegrain and its variants. In one popular variant on the Cassegrain telescope the shapes of both mirrors have been altered slightly to make an aplanatic (free of coma as well as spherical aberration) telescope; the primary mirror also becomes hyperboloid in shape, but with an eccentricity only slightly different from that of a paraboloid mirror. This *Ritchey–Chretien* design was used for the HST, Keck and many other large telescopes. The principal residual aberration is then the astigmatism with angular size in the sky, given in arc seconds

$$\alpha = 0.5m\theta^2/(57.3f/\#), \tag{8.4}$$

where θ is the angle off axis in arc-minutes, m is the magnification produced by the secondary and $f/\#$ is the focal ratio for the complete telescope.

When observing a star or galaxy an astronomer needs to be able to hold the image at a fixed point in the field of view as the Earth rotates. Two mountings suitable for this purpose are widely used to support telescopes and are shown in figures 8.4 and 8.5. The first diagram illustrates the *alt-az* mounting in which one rotation axis is vertical and the other horizontal. In the alternative *equatorial* mounting one axis points to the

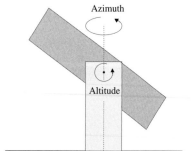

Fig. 8.4 Telescope in alt-az mounting.

pole star and the other is horizontal. The latter mounting has the advantage that once a target is in view the only motion required thereafter to retain the target in view is rotation around the polar axis. In the case of the alt-az mounting the tracking speed is variable and even when the target is tracked precisely its image rotates in the field of view. This latter feature requires correction using additional optical elements if a long exposure – which is often the case – is required. All these weaknesses of the alt-az mounting are less significant now that computer control can be complex, reliable and cheap. When the telescope weighs many tons, having a layout in which the centre of gravity can lie directly over the bearings is a decisive advantage for the alt-az mounting. It leads to a lighter, smaller frame and reduces the size of dome required to house the telescope. Multimetre telescopes mirrors are extremely difficult to cast, and those of over 5 m diameter are often segmented.[2]

Example 8.1

The HST is a Ritchey–Chretien telescope, with mirrors having outer diameters of 2.4 and 0.8 m. Their equivalent radii of curvature, on axis, are 11.04 and 1.36 m, so their paraxial focal lengths are $+5.52$ and -0.68 m respectively; the mirrors being placed 4.91 m apart. Thus the focus of the primary mirror, the *prime* focus, is 0.61 m beyond the secondary mirror. If the distance of the final focus from the secondary is v, then applying the mirror eqn. 3.5 gives

$$1/v = 1/0.61 - 1/0.68,$$

whence $v = +5.95$ m, putting the final focus 1.04 m behind the primary mirror. The overall focal length f is given by eqn. 3.37,

$$1/f = 1/f_p + 1/f_s - d/(f_p f_s),$$

where $f_{p,s}$ are the mirror focal lengths and d their separation and this yields a focal length 57.2 m. Then one arcsec in the sky projects onto 279 μm in the image plane: the *plate scale* is said to be 279 μm/arcsec. At this scale a pixel in the CCD arrays, which are typically 15 μm across, covers 0.05 arcsec. The diffraction limit at 500 nm wavelength given by eqn. 8.1 is 0.04 arcsec, so the pixel size is well matched to the attainable resolution. The long overall focal length relative to the short optical path within the telescope shows that the Ritchey–Chretien and classical Cassegrain telescopes are in fact mirror telephoto systems.

Initially the HST showed unexpected spherical aberration because the primary had been shaped to an incorrect profile, departing at the periphery by 2 μm from the correct profile. This was rectified by adding a pair of mirrors. A field mirror, M_1, placed a little ahead of the focal plane imaged the entry pupil onto a second mirror, M_2, so that points on M_2 were in one-to-one correspondence with points on the primary mirror. M_2 was shaped so as to cancel the aberration of the primary exactly.

[2]The twin 10 m Keck telescope primary mirrors are made up of 36 hexagonal segments cast from a ceramic material, Zerodur, of low thermal expansion coefficient ($\sim 10^{-7}$ K^{-1}). Each segment is 76 cm thick, 1.8 m across and weighs 0.4 t. A single Keck instrument plus supports weighs 300 t and occupies an eight-storey high spherical shaped dome.

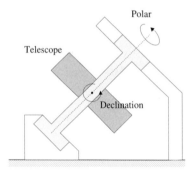

Fig. 8.5 Telescope in equatorial mounting.

8.2.1 Auxiliary equipment

The full range of spectrometers described in Chapters 6 and 7 are deployed in studying emission spectra of celestial sources in the visible and near visible parts of the spectrum. For high resolution work echelle gratings are often used. Combinations of crossed prisms and gratings permit the simultaneous examination of a wide spectral range. Where appropriate, optical fibres are used to transfer light from individual star images to corresponding individual points along the entry slit of a spectrometer. One end of each fibre peers through a precisely located hole in a plate covering the image plane, while the other end is placed at its chosen location along the spectrometer slit. In this way several hundred stellar spectra can be recorded simultaneously.

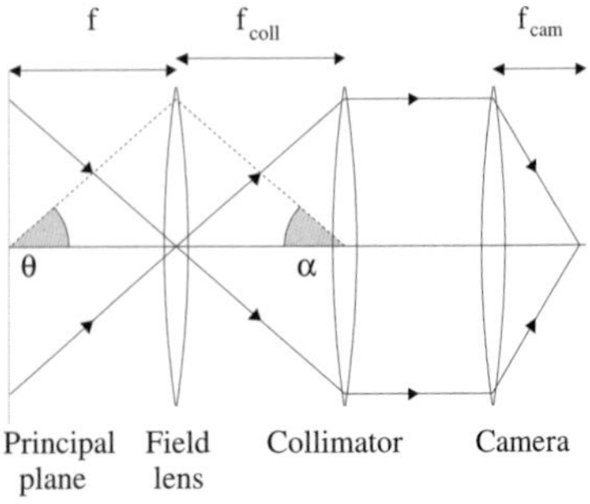

A very simple auxiliary component is the field lens used to produce a flat image plane to match the planar surface of detectors. This is positioned close to the secondary focus. More complex optics provides magnification or demagnification to match the image size or resolution to that of detectors. A focal reducer is shown in figure 8.6. This consists of a field lens near the telescope focus, a collimator and the detector camera. Rays from a star lying on the optical axis are shown as solid lines. The field lens images the exit pupil of the telescope to just fill the collimator aperture which is the entry pupil of the camera/collimator combination. If f_{col} and f_{cam} are their focal lengths the image size is changed by a factor $f_{\mathrm{cam}}/f_{\mathrm{col}}$. The field lens' area should match the field of view of the telescope, $f\theta$, where f is the telescope's focal length, and θ its angular field of view in the sky. In order to maintain the etendue along the optical chain we require $D_{\mathrm{col}}\alpha = D\theta$, where D_{col} is the collimator diameter, D the primary diameter and α is the angular field of view of the collimator. Additionally we see in figure 8.6 that $f\theta = f_{\mathrm{col}}\alpha$. Combining these two requirements determines the collimator diameter

$$D_{\mathrm{col}} = Df_{\mathrm{col}}/f.$$

The region of parallel beam between the collimator and camera in the focal reducer provides a natural location to insert dispersive elements such as prisms, gratings or Fabry–Perot etalons, or for simple filters.

Light auxiliary equipment can be mounted at the Cassegrain focus behind the primary mirror. Figure 8.7 shows an alternative in which a plane mirror diverts the light along the *altitude* axis to what is called the *Nasmyth* focus. In the case of the 10 m Keck telescopes heavy equipment at the Nasmyth focus is carried around on a horizontal frame which follows the telescope's azimuthal rotation. This avoids locating heavy gear at the Cassegrain focus where it would be carried around in three dimensions. Another alternative is the *Coude* focus, which requires further mirrors to put the image in a stationary position. This necessitates a long focus, rotating mirrors and a narrow field of view.

f   f_coll   f_cam

θ   α

Principal   Field   Collimator   Camera
plane   lens

Fig. 8.6 Image reducer for matching image area to detector area.

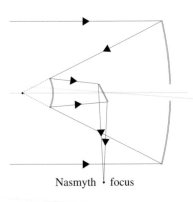

Nasmyth · focus

Fig. 8.7 Nasmyth focus along the altitude axis of a Cassegrain telescope in an alt-az mounting.

Light from stars is dispersed in travelling through the Earth's atmosphere, and when a star is not directly overhead (not at the *zenith*) the image becomes a coloured strip. This dispersion is removed by putting *Risley prisms* in the optical path. An individual Risley prism is made from two thin prisms glued face to face so that their deflections cancel. According to eqn. 2.28 a single prism of narrow angle α and refractive index n would provide a deflection $(n-1)\alpha$ and a dispersion $\alpha(\mathrm{d}n/\mathrm{d}\lambda)$. In a Risley prism the types of glass used for the component thin prisms are chosen so that

$$(n_1-1)\alpha_1 = (n_2-1)\alpha_2,$$

while at the same time there is a net dispersion. The compensator for atmospheric dispersion is a pair of Risley prisms in series. When viewing directly overhead they are rotated to be in opposition and their dispersions cancel; for viewing in other directions the angle between the pairs is changed so that their net dispersion cancels that of the atmosphere.

8.3 Schmidt camera

A radically different telescope design that gives a much wider field of view than the Ritchey–Chretien is the *Schmidt camera*. In 1929 Schmidt came to appreciate that if the aperture stop of a concave *spherical* mirror is placed at the centre of curvature then all directions of incidence through the centre of this stop to the mirror are radial and are equally good optical axes. Consequently with this layout the coma and astigmatism are eliminated. What remains is the spherical aberration, which Schmidt was able to reduce drastically by placing an aspheric glass corrector plate across the stop aperture. Figure 8.8 shows the principal components of Schmidt's design. The corrector plate cancels out spherical aberration by being convex at the centre (focusing) while it is concave at the periphery (defocusing). This surface figuring can only remove spherical aberration at a single wavelength so there is residual spherical aberration at other wavelengths. That remaining aberration can be further reduced if the corrector plate is made into an achromatic doublet like those discussed in Section 3.6.7. Two drawbacks of the Schmidt design are that the image surface is highly curved and that the image is difficult to access. A typical large aperture Schmidt is the UK Schmidt of 1.24 m diameter, focal length 3.07 m located at the Siding Spring observatory in Australia. It has a field of view of 6.6° over which the resolution is about 1 arcsec and has a plate scale of 67 arcsec/mm. Originally it was used to produce an atlas of the southern sky and is now also used for measuring the spectra of sources. Schmidts of much larger aperture are impractical because a larger thin corrector would sag significantly, and because the overall length, being twice the focal length, makes mounting difficult. A common variation on Schmidt's design, due to Maksutov, uses a thin meniscus lens to correct the spherical aberration.

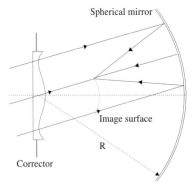

Spherical mirror

Image surface

R

Corrector

Fig. 8.8 Schmidt camera.

Stellar luminosities

The *apparent luminosity*, $\ell\,\mathrm{W\,m^{-2}}$, of a star is the actual power density arriving at the Earth. The measured *absolute luminosity*, L, of a star is the total radiated power in watts, which can be calculated from the apparent luminosity if the distance to the star, D, is known

$$L = 4\pi D^2 \ell.$$

Apparent magnitude, m, is a scale devised by Pogson in the nineteenth century to quantify the visual scale of intensity that went back in some form to classical Greece. Values of 1 and 6 correspond respectively to the brightest stars (excluding the Sun) and faintest stars visible to the eye. The apparent luminosity corresponding to an apparent magnitude zero is explicitly defined to be $2.52\ 10^{-8}\,\mathrm{W\,m^{-2}}$ in the V–band, that is using a broadband filter centred at 510 nm to capture all visible wavelengths. Then, because the eye responds logarithmically rather than linearly to light intensity the relationship between apparent magnitude and apparent luminosity is

$$m = -2.5 \log_{10}\left[\ell/(2.52\ 10^{-8}\mathrm{W\,m^{-2}})\right]. \tag{8.5}$$

Finally *absolute magnitude* is defined as the apparent magnitude which a star would have at a distance of 10 parsec from the Earth, i.e. at a distance of $3.086\ 10^{17}$ m. The Sun itself has absolute (V-band) magnitude 4.72 and an apparent magnitude -26.85. Very large telescopes can identify stars with apparent magnitudes as high (weak) as $+25$, for which the apparent luminosity is $2.52\ 10^{-18}\,\mathrm{W\,m^{-2}}$.

8.4   Atmospheric turbulence

Observatories are sited on remote mountains which enjoy clear skies and low levels of humanity's light pollution. However, unlike the HST, they cannot escape the effects of turbulence in the atmosphere. The energy from turbulent motion is dissipated as local heating in the atmosphere and this results in important variations locally in the refractive index of the atmosphere. These variations cause a star's image to *scintillate*, meaning to move and to change in intensity at a rapid rate. These movements are sufficiently small and rapid that all the eye registers is a twinkling. An incoming wavefront from a star is planar before entering the atmosphere but its surface becomes crumpled by the local variations in the refractive index of the atmosphere. The area over which the wavefront which arrives at the Earth is relatively flat (but not necessarily parallel to the incident wave above the atmosphere) is called the *Fried parameter r_0*. More precisely: over a distance r_0 the phase variation of the wavefront has a root mean square value of 1 rad. Diffraction at this equivalent aperture gives an angular resolution (point spread function) of approximately λ/r_0. At a mountain observatory site with good *seeing* r_0 around 25 cm at a wavelength of 500 nm and then the resolution is

0.4 arcsec. The shape of the wavefronts arriving at a telescope changes on a timescale τ_0 which can be 50 ms in conditions of good seeing at 500 nm. It follows that the resolving power of telescopes with mirrors of any larger diameter than r_0 is degraded to that of a mirror of diameter r_0, although their light gathering capacity is unaffected. The Fried parameter and τ_0 both vary strongly with wavelength

$$r_0 \propto \lambda^{6/5},$$

so that the extent of the point spread function, λ/r_0, is proportional to $\lambda^{-0.2}$. This gives a slow improvement in seeing as the wavelength increases.

The effect of this turbulence on the image seen depends on the size of the telescope's aperture. When an unresolved star is viewed through a telescope whose diameter d is less than the Fried parameter the image has an Airy disk of angular diameter λ/d. This image moves jerkily at intervals of average duration τ_0 with angular displacements of order λ/r_0. On the other hand if the telescope has diameter D much *larger* than the Fried parameter the image consists of speckles of angular size λ/D. These speckles, numbering roughly $(D/r_0)^2$, continually move, fade, coalesce and re-form with the same characteristic timescale τ_0. A long exposure produces a blurred image of angular size λ/r_0, and with Strehl ratio[3] $(r_0/D)^2$.

[3]See Section 1

8.5 Adaptive optics

The techniques described as adaptive optics are used to compensate for the effects of atmospheric turbulence and to recover the potential resolving power of large diameter telescopes. The first step is to sense the shape of the wavefront arriving from an unresolved star which is sufficiently close in direction to the target for their images to suffer essentially the same distortion; the reference star is called a *guide* star. This information on the wavefront shape is then used to deform a flexible mirror placed in the optics of the telescope in such a way that after reflection the wavefronts recover their planar shape. The control sequence has to respond on a timescale of milliseconds in order to compensate the changing distortion faithfully and in real time. Figure 8.9 shows schematically the adaptive optics which could be placed at the Nasmyth focus of a large telescope. After reflection from the deformable mirror the light encounters a beam splitter, with part being directed to a wavefront sensor and part going to the detectors that are recording data, which are called the *science instruments*. Electronic signals from the wavefront sensor are used as input to a processor that controls actuators which change the shape of the deformable mirror.

The commonest wavefront sensor is the *Shack–Hartman* sensor shown in figure 8.10. A planar, square array of identical lenslets focuses the

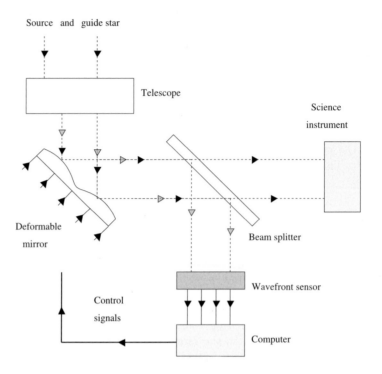

Fig. 8.9 Adaptive optics.

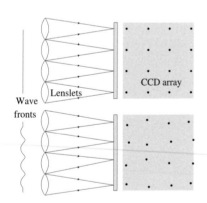

Fig. 8.10 Shack–Hartman wavefront sensor. In the upper panel an undistorted plane wave is incident and in the lower panel the incident wave is distorted. The pattern of image points on the detector array is shown on the right in each case.

incoming light from the guide star onto a CCD array having for example four pixels per lenslet. In the upper panel of the figure an undistorted plane wave is incident and the image on the CCD is a square array of dots, each dot lying on the optical axis of its lenslet. What is shown in the lower panel is the result when the incident wave is distorted. Each image spot is now displaced by a distance and in a direction determined by the local orientation of the wavefront across its particular lenslet. If the wavefont has a tilt of $\Delta\theta$ and the lenslet focal length is f, then the displacement is simply $f\Delta\theta$. A section through a distorted wavefront appears in figure 8.11 where the vertical lines separate the regions seen by individual lenslets. Once the direction and magnitude of tilt over each cell is known it is evident that the whole wavefront can be reconstructed with a precision set by the lenslet diameter. Note for future reference that the wavefront sensor will not detect an overall delay or advance of the wavefront due to a change in the refractive index common to the whole area of the telescope aperture. This *piston* component of the distortion becomes significant when light from two or more telescopes is brought together to interfere.

The most important part of wavefront correction is to remove the overall tip or tilt of the wavefront because this accounts for just under 90% of the image distortion. This correction can be performed using a rigid plane mirror that can be rotated about either of two orthogonal axes in its own plane as shown in figure 8.12. If this *tip/tilt* correction is insufficient a second, this time deformable mirror can be used to remove the remaining distortion. Deformable mirrors are available with up to several thousand actuators over the surface. One type of deformable mirror has a thin glass or ceramic skin which is bonded to actuators mounted on a flat rigid plate. The actuators are generally piezoelectric rods which expand or contract under an applied voltage and their movement flexes the mirror surface. The total correction sharpens the image of an unresolved star to the extent that in the case of Keck II the Strehl ratio improves from less than 0.01 to 0.35.

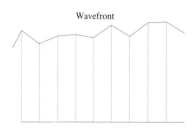

Fig. 8.11 Wavefront segments reconstructed by the Shack–Hartman wavefront sensor.

Adaptive optics on 8 m and the proposed 30 m telescopes in the visible spectrum, where turbulence is characterized by a physical and temporal scales of around 25 cm and 50 ms, imposes severe demands: the numerous actuators must be agile and the CCDs used to record the images from the separate Shack–Hartmann mirrors must be capable of fast noise-free readout. The recently developed EM-CCDs described in Section 15.7.4 are ideal for this role. They provide full frame readout at kHz rates with readout noise of less than one electron per pixel.

If the effective height of the turbulence in the atmosphere is $\overline{H}$ above ground level then the *isoplanatic angle* within which the wavefront distortion is uniform is given by

$$\theta_0 \sim r_0/\overline{H}. \tag{8.6}$$

$\overline{H}$ is typically 5 km so that with a Fried parameter of 25 cm the isoplanatic angle is only a few arc seconds. Unfortunately the isoplanatic regions around the available guide stars cover only a small fraction of the sky. A less satisfactory alternative is to use a narrow laser beam to generate an artificial guide star near to the target. The most effective method is to excite fluorescence at a wavelength 589 nm in sodium atoms concentrated in a layer lying around 90 km above the Earth. A different scheme is to back scatter a laser beam off the atmospheric molecules, a process that produces a guide star at 10 to 20 km above the Earth. Laser guide stars have an inherent drawback. On its way upward the laser beam is deflected by the turbulence and on its return it undergoes an almost equal and opposite deflection. The artificial guide star remains stationary and cannot give any information on the tip/tilt component of the atmospheric distortion. However any *faint* star lying within the isoplanatic angle of the target can be used to provide the pointing information that determines the tip/tilt correction, despite this star being too faint to help beyond that.

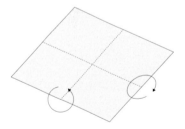

Fig. 8.12 Mirror used for tip/tilt correction.

Extreme care is taken to reduce variations in telescope performance arising from factors local to the telescope and its protective dome. Local convection, differential heating of the telescope components and radiative cooling of the mirrors during a night's observation are eliminated as far as possible. Measures such as cooling the dome interior to nighttime temperatures during the day, insulation of the dome from work areas below it, forcing a slight downdraft of air through the slit during viewing, and using baffles to reduce wind movement all contribute to maintaining a stable environment.

8.5.1   Lucky imaging

Lucky imaging is another technique used to avoid the degradation of astronomical images due to atmospheric turbulence. Its name belies the fact that it is a neat and systematic method which exploits the statistical nature of atmospheric turbulence. During any long observing period there are intervals during which the regions of near planar wavefronts from stars are much larger than r_0. In lucky imaging short exposures are recorded of duration ideally shorter than the average interval, τ_0, over which the local atmospheric distortion remains the same. Exposures are made at a rapid rate in order to utilize the available viewing time efficiently. Those images which have a high Strehl ratio for unresolved sources in the field of view are selected. These chosen frames are then superposed using unresolved sources to register the images with one another, a *shift and add* procedure.

Lucky imaging had been proposed by Hufnagel in 1966 and was brought to general notice by Fried in 1978. It only became a practical proposition with the development by e2v Technologies of the EM-CCDs described in Section 15.7.4. EM-CCDs can have a readout noise of well below one electron per pixel when read out at many megapixels per second: the corresponding frame read out rates can exceed 1 kHz. Cooling to liquid nitrogen temperatures renders negligible the noise generated by the dark current. Hence, cooled EM-CCDs provide the rapid noise-free sampling required to make lucky imaging feasible. As a comparison the readout noise per pixel of a standard CCD at megapixel rates is of order tens of electrons, enough to seriously degrade image quality.

Systems using adaptive optics, by contrast, are designed to correct the effects of atmospheric turbulence continuously. This is particularly difficult across the visible spectrum: for example at 500 nm wavelength r_0 is only 25 cm and τ_0 is of order 50 ms in conditions of excellent seeing. It was therefore lucky imaging that first produced diffraction limited images from a ground based 2.5 m telescopes at 810 nm wavelength, with resolution as good as that of the 2.5 m Hubble Space Telescope.[4] A comparison of images from the Hubble Space Telescope, lucky imaging at the Nordic optical telescope and the NAOMI adaptive optics system at the William Herschel telescope is shown in figure 8.13. At 2.2 μm wavelength both r_0 and τ_0 are about seven time larger than at 500 nm,

---

[4]See for example 'Diffraction limited 800 nm imaging with the 2.56 m Nordic Optical Telescope', by J. E. Baldwin, R. N. Tubbs, G. C. Cox, C. D. Mackay, R. W. Wilson and M. I. Andersen: *Astronomy and Astrophysics* 368, L1-4 (2001).

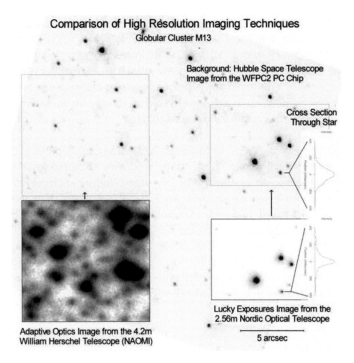

Comparison of High Resolution Imaging Techniques
Globular Cluster M13

Background: Hubble Space Telescope
Image from the WFPC2 PC Chip

Cross Section
Through Star

Lucky Exposures Image from the
2.56m Nordic Optical Telescope

Adaptive Optics Image from the 4.2m
William Herschel Telescope (NAOMI)

5 arcsec

Fig. 8.13 Comparison of lucky imaging with adaptive optics and the Hubble Space Telescope. The HST imaging at 870 nm has a resolution of 0.12–0.15 arcsec, NAOMI adaptive optics imaging at 660 nm has a resolution of 0.4 arcsec, the lucky imaging at 810 nm has a resolution of 0.10 arcsec. Image courtesy of Professor Mackay, Institute of Astronomy, University of Cambridge, UK.

and lucky imaging used alone loses much of its advantage.

Fried gave an estimate for the probability of obtaining a lucky image with a phase variation of less than one radian across the primary, corresponding to a Strehl ratio of greater than 0.37. For a telescope of diameter D his estimate was

$$P \simeq 5.6 \exp\left[-0.1557(D/r_0)^2\right]. \tag{8.7}$$

When $D = 6r_0$, which is the case for a 2.5 m diameter telescope imaging at around 800 nm, in conditions of good seeing, this probability is around 3%. In practice, in such conditions, when exposures are taken at around 20 Hz more than 10% of these prove useful.[5] Unfortunately Fried's estimate for the success rate at an 8 m diameter telescope is many thousand times less, so that lucky imaging alone becomes impractical.

The size of the isoplanatic patch is increased whenever a lucky image is recorded. Pursuing the example of a telescope whose diameter is 2.5 m,

---

[5] Private communication from Professor Craig Mackay.

the isoplanatic angle is increased from $r_0/\overline{H}$ to $D/\overline{H}$, becoming one arc minute. This vastly increases the chance of locating an unresolved star within the isoplanatic patch. In addition, because lucky imaging uses the whole of the primary aperture such stars can be proportionately weaker than those employed for adaptive optics. Overall, the area of sky containing guide stars is one hundred times larger for lucky imaging than for adaptive optics at visible wavelengths.

A combination of adaptive optics with lucky imaging deploys the advantages of both techniques. The adaptive optics compensates atmospherc turbulence to some level, with a corresponding increase in the effective size of the Fried parameter. An EM-CCD camera then records the improved images at high rate. Thanks to the increase in the Fried parameter the likelihood of recording lucky images is enhanced. This tandem approach has been applied at the Palomar 5 m telescope and a resolution of 35 milliarcsec was achieved, close to the diffraction limit for this telescope. It is currently, in 2010, the best resolution ever obtained in the visible or near infrared, whether from ground- or space-based telescopes.[6]

8.6   Michelson's *stellar* interferometer

The first successful measurement of a star's angular diameter was carried out by Michelson in 1921 using the stellar interferometer sketched in figure 8.14, making a type of measurement foreseen some 60 years earlier by Fizeau. Light from the star, Betelgeuse, was received by a pair of plane mirrors mounted on a long bar attached to the frame of the telescope[7] and reflected into the telescope by two further mirrors. These outer mirrors could be moved apart along the bar to a maximum separation of over 6 m. The resultant image at the telescope's focus was therefore crossed by fringes due to interference between the light following the separate mirror paths.

It is very important when thinking about astronomical interferometry to remember that the light from any point on a star's surface is incoherent with the light from any other point on that star, or equally on any other star. Consequently the intensity pattern seen by Michelson was the sum of the intensities of the interference patterns due to light from each region of the star.

The complex amplitude at the telescope's image plane caused by light from one edge A of the star would be

$$E = E_0[\exp(ikp_1) + \exp(ikp_2)]\exp(-i\omega t),$$

with light of angular frequency ω and wave number k. p_1 and p_2 are the

[7]This was the 2.5 m Hooker telescope located at a height of 1742 m on Mt Wilson in California. Betelgeuse is the red star in the right shoulder of Orion, α-Orionis. At a distance of 430 lightyears ($4.1\,10^{18}$ m) it is the nearest red supergiant star. The current accepted value for its angular diameter is 0.054 arcsec, making its geometric diameter about 650 times larger than that of the Sun.

---

[6]loc. cit.

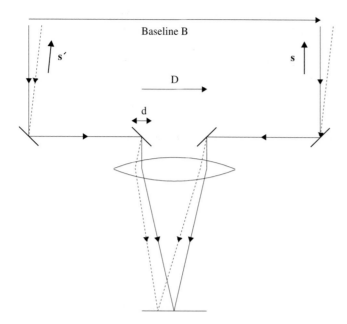

Fig. 8.14 Michelson's stellar interferometer. $\mathbf{s}$ and $\mathbf{s}'$ are unit vectors in the directions of the sources. $\mathbf{B}$ is the baseline vector length and $\mathbf{D}$ the vector separation of the inner mirrors.

lengths of the two different mirror paths and $\delta = p_1 - p_2$ their difference. Thus the time averaged intensity

$$I = EE^*/2 = E_0^2 \left[1 + \cos\left(k\delta\right)\right]. \qquad (8.8)$$

Let the *baseline*, that is the vector separation of the two outer mirrors, be $\mathbf{B}$; let $\mathbf{D}$ be the separation of the inner mirrors; and let the unit vector in the direction of A be $\mathbf{s}$. Then $\delta = \mathbf{s} \cdot (\mathbf{B} - \mathbf{D})$ at the geometric image point. The corresponding path difference for light arriving at the same image point from the opposite edge A' of the star is $\delta' = \mathbf{s}' \cdot \mathbf{B} - \mathbf{s} \cdot \mathbf{D}$. Thus there is a phase difference between the sets of fringes produced separately by the sources A and A'

$$\Delta = k(\delta - \delta') = k(\mathbf{s} - \mathbf{s}') \cdot \mathbf{B}. \qquad (8.9)$$

Now because $\mathbf{s}$ and $\mathbf{s}'$ are unit vectors and the star's angular size, $\Delta\theta$, is small we have $\Delta\theta = |\mathbf{s} - \mathbf{s}'|$. Then with $\mathbf{B}$ aligned parallel to $\mathbf{s} - \mathbf{s}'$ eqn. 8.9 reduces to

$$\Delta = kB\Delta\theta. \qquad (8.10)$$

In Michelson's apparatus Δ is varied by moving the outer mirrors apart symmetrically. When this happens B increases, Δ increases and the sets of fringes due to A and A' move apart. Eventually a separation of the outer mirrors is reached at which the phase difference Δ is exactly π. Then the fringes due to A and A' are exactly out of step and a uniform total intensity is produced across the star's image. Figure 8.15 illustrates

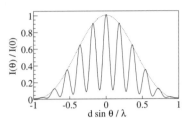

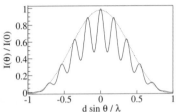

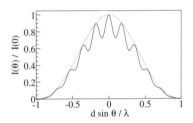

Fig. 8.15 Fringes with decreasing visibilities: 0.81, 0.33 and 0.14.

how the fringe visibility changes with increasing phase difference. In this figure the fringe envelope is the single slit diffraction pattern due to a mirror of width d acting as an entry aperture. The spacing of the fringes within the envelope is that for a point source viewed by the *inner* mirrors. There are therefore D/d fringes across the envelope, and what was very important for Michelson, their location would not change when the outriggers were moved in or out. At the setting giving zero visibility, where $\Delta = \pi$, eqn. 8.10 gives

$$\Delta\theta = \pi/kB = \lambda/(2B).$$

Michelson used this result to deduce the angular size of Betelgeuse, obtaining a value 0.047 arcsec. The essential feature of the measurement is that the angular resolution has been boosted to $\lambda/(2B)$ by using outriggers whose separation is B, whereas the telescope alone has a resolution of λ/D.

After Michelson's measurements the technique languished because long outriggers proved very unstable and because of atmospheric turbulence. Beginning in the 1940s measurements of stellar diameters were made with the new technique of *intensity* interferometry which requires less strict path equality. This technique, developed by Hanbury Brown, will be discussed in Chapter 19.

In more recent times the development of lasers, fast electronics and computers provided tools with which astronomers could carry stellar amplitude interferometry much further. Long baselines can be measured with a precision of tens of nanometres and altered sufficiently fast to compensate continuously for the piston component of atmospheric distortion. Interference patterns (interferograms) can now be recorded using light from separate telescopes or from an array of apertures placed in front of a single large telescope; and where Michelson only made a single measurement of the baseline length at which the fringe visibility went to zero, nowadays the visibility itself is measured for as many baseline lengths and orientations as possible. From this new information the images can be reconstructed of complex distant sources which might be too small for an individual telescope to resolve. This field of *aperture synthesis* interferometry will be described in the following sections.

8.7   Modern interferometers

Whatever the interferometer the difference between the lengths of the two optical paths has to be kept to less than the coherence length if interference is to occur. This poses a problem if the light from two separated telescopes is brought to interference as illustrated in figure 8.16. Each telescope collects light from a star, and this is focused into a parallel beam and guided by mirrors so that the beams from both telescopes are brought together at a common detector. The geometric paths from a star not lying directly overhead differ by much more than a coherence length. In order to cancel out this difference *delay lines* are incorpo-

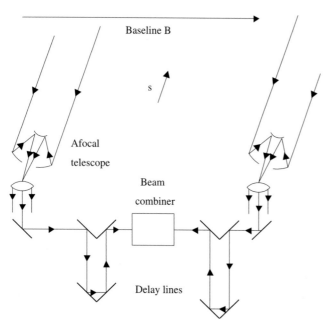

Fig. 8.16 Telescope interferometer with delay lines. The telescopes have auxiliary lenses to produce a parallel beam, making them afocal.

rated in the optical paths from each telescopes. The mobile mirrors in the delay lines are carried on trolleys moving along rails. The lengths of the complete optical paths are monitored continuously with laser-based heterodyne Michelson interferometers of the type described in Section 14.6.3.

Detectors, perhaps placed behind wavelength filters, respond to a range of wavelengths, which leads to modifications in the treatment given in the last section. Equation 8.8 for a single point object becomes

$$I = \int_{\lambda_0 - \Delta\lambda/2}^{\lambda_0 + \Delta\lambda/2} E_0^2 \left[1 + \cos(k\delta) \right] d\lambda, \qquad (8.11)$$

where the integral is taken over the range of wavelength detected. The integral reduces to

$$I = E_0^2 \, \Delta\lambda \left[1 + \mathrm{sinc}(\pi\delta/L_c) \, \cos(k_0\delta) \right], \qquad (8.12)$$

where $L_c = \lambda_0^2/\Delta\lambda$ is the coherence length of the radiation: see eqn. 5.25. Thus the length of the fringe train is of order $\lambda_0/\Delta\lambda$ fringes, which amounts to 10 fringes if the bandwidth is 10%. At the centre of the pattern the path difference is close to zero so that all wavelengths remain in phase giving an easily recognizable *white light* fringe; while at

For simplicity source intensity and detector efficiency are assumed constant over the range $(\lambda_0 \pm \Delta\lambda/2)$ and zero outside this range. $\ell = \lambda - \lambda_0$ is small and the approximation is made that

$$\cos(k\delta) = \cos\left[2\pi\delta/(\lambda_0 + \ell) \right]$$
$$= \cos\left[(2\pi\delta/\lambda_0)(1 - \ell/\lambda_0) \right]$$
$$= \cos(2\pi\delta/\lambda_0)\cos(2\pi\delta\,\ell/\lambda_0^2).$$

$$\int \cos(k\delta)\, d\lambda$$
$$= \cos(2\pi\delta/\lambda_0) \int_{-\Delta\lambda/2}^{\Delta\lambda/2} \cos(2\pi\delta\,\ell/\lambda_0^2)\, d\ell$$
$$= (\lambda_0^2/\pi\delta)\cos(k_0\delta)\sin(\pi\delta\,\Delta\lambda/\lambda_0^2)$$
$$= \cos(k_0\delta)\mathrm{sinc}(\pi\delta/L_c).$$

the edges of the pattern the fringes are coloured.

Atmospheric turbulence poses other difficulties for interferometry with telescopes because it causes the optical path lengths from source to each telescope to change rapidly and in an uncorrelated manner. One approach is to use telescopes or mirrors that have diameters of order of the Fried parameter, r_0, so that the wavefront distortion is at least uniform across each telescope and then to equip each telescope with tip/tilt correction using adaptive optics. However this leaves uncorrected the piston component of the distortion, which can be of order tens of microns in the visible spectrum and which is changing on a time scale of τ_0. The single slit envelope is unaffected by the piston component, but the two-slit fringes lying under it are affected and move relative to the envelope. Here lies a difficulty: only the two-slit fringes at around zero path difference are dispersion free and useful for measurement. With his relatively compact interferometer Michelson could hold these fringes in view using a glass block mounted in one of the two optical paths. This he would tilt by hand to centre the fringes on the envelope. In modern systems the fringes are tracked electronically and kept in view by altering the optical delay lines appropriately. Equality of the two total optical paths from the star is maintained to a precision of several nanometres for interferometers with baselines of order 100 m. Most observations have been made with interferometers using detectors sensitive to infrared radiation because τ_0, r_0 and the isoplanatic angle all improve (increase) with increasing wavelength.

A stellar interferometer in which, as in figure 8.14, the fringes fall on a CCD detector array in the image plane is termed an *image plane* interferometer. Figure 8.17 shows a second variant, called a *pupil plane* interferometer, in which the collimated beams impinge on a beam splitter and the light from each exit face falls on a single detector. These plane wavefronts interfere leaving the whole field of view at each detector uniformly illuminated: when one is dark the other is light – as we now show. Suppose A and A' are the complex amplitudes of the incoming beams, with $|A| = |A'|$, then the outputs falling on the two detectors are

$$A_1 = A + iA', \text{ and } A_2 = iA + A', \tag{8.13}$$

where the factor i is due to the phase difference between the reflected and transmitted beams from a beam splitter. Now if $A = -iA'$ then $A_1 = 0$ and $A_2 = 2A'$; thus when one exit pupil is dark the other is bright, and vice versa. Subsequently the electronic signals from the two detectors are subtracted to give a resultant

$$\Delta I = I_2 - I_1 = A_2 A_2^* - A_1 A_1^* = 4\Im m(A^* A'). \tag{8.14}$$

In the pupil plane interferometer the fringes are scanned across the detectors by dithering the position of a mirror in one optical delay line so that the path difference changes by many wavelengths. A single detector can be used rather than the detector array needed for image plane

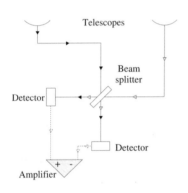

Fig. 8.17 Pupil plane interferometer. The full lines are light paths and the broken lines carry electrical signals.

interferometry.

8.8   Aperture synthesis

There is much more information contained in the fringe patterns than the single value of the angular radius of a star. By combining data from interferometers using baselines of different lengths and orientations it is possible to obtain images of celestial objects. The analysis techniques which are in use were for the most part developed earlier to extract images from analagous radiotelescope measurements. In this section the basic analysis will described, with later sections covering important experimental and analysis details.[8]

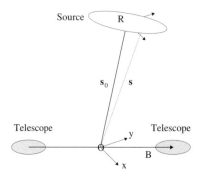

Fig. 8.18 Coordinate system for aperture synthesis.

Figure 8.18 shows an extended astrophysical source viewed by an interferometer. Two orthogonal axes, Ox and Oy, are indicated at the source and these are repeated at the telescopes, with the third orthogonal axis being directed along the unit vector $\mathbf{s_0}$ pointing from the centre of the baseline to a *reference* point on the source, R, called the *phase centre*. Let the source brightness be $I(\mathbf{s})$ where $\mathbf{s}$ is the unit vector pointing from the centre of the baseline to a representative point on the source. Then the intensity at the detector is, apart from constants,

$$P(\mathbf{s_0}, \mathbf{B}) = \int I(\mathbf{s}) \left[1 + \cos\left(k\delta \right) \right] \mathrm{d}A, \qquad (8.15)$$

where $\mathrm{d}A$ is an element of area of the source centred in a direction with the unit vector $\mathbf{s}$, and where $\mathbf{k}$ points along $\mathbf{s}$. The difference in length between the two optical paths from the elements of the source at $\mathbf{s}$ to the detector is

$$\delta = \mathbf{s} \cdot \mathbf{B} + p, \qquad (8.16)$$

where p is the difference between the optical paths from telescope entrance pupil to detector – including the delay lines. Now the detector responds to a range of wavelengths, usually restricted by a filter, for example the K-band from 2 to 2.4 μm. The overall path difference between the two paths must be held less than the coherence length which in this case is $\sim$10 μm. The choice is made to set the delay line length difference so the path difference is zero for the phase centre

$$\mathbf{s_0} \cdot \mathbf{B} + p = 0. \qquad (8.17)$$

Thus

$$\delta = \left[\mathbf{s} - \mathbf{s_0} \right] \cdot \mathbf{B} = \mathbf{\Delta s} \cdot \mathbf{B}. \qquad (8.18)$$

Now the angle that $\mathbf{\Delta s}$ subtends on the sky is $\Delta s / s_0$, which is simply Δs because s_0 is a unit vector. $\mathbf{\Delta s}$ is small and can therefore be resolved

---

[8]This section has been adapted from slides of Professor Haniff's talk 'Optical Interferometry – A Gentle Introduction' at the 2003 Michelson Interferometry Summer School at the California Institute of Technology, Pasadena, CA.

into component angles ℓ in the ORx plane and m in the ORy plane with $\Delta s^2 = \ell^2 + m^2$. In order to simplify the notation $k\mathbf{B}$ is similarly resolved into components u and v along the Ox and Oy axes respectively. Then the relative phase between light arriving along the two paths becomes

$$k\delta = \ell u + mv + \phi, \tag{8.19}$$

where ϕ is any small phase introduced by deliberately altering the difference between the lengths of the delay lines from the condition of zero path difference at the phase centre. Thus the intensity at the detector becomes

$$P(u, v; \phi) = \int \int I(\ell, m) \left[1 + \cos\left(\ell u + mv + \phi\right)\right] d\ell\, dm$$

$$= P_0 + \cos\phi \int \int I(\ell, m) \cos\left(\ell u + mv\right) d\ell\, dm$$

$$- \sin\phi \int \int I(\ell, m) \sin\left(\ell u + mv\right) d\ell\, dm, \tag{8.20}$$

where $P_0 = \int\int I(\ell, m)\, d\ell\, dm$ is the total intensity. The other two integrals in this equation are components of the Fourier transform

$$V(u, v) = \int \int I(\ell, m) \exp\left[i(\ell u + mv)\right] d\ell\, dm. \tag{8.21}$$

This transform is special: it converts the source brightness angular distribution into a distribution in u and v, that is into an intensity distribution in spatial frequency along the Ox and Oy axes respectively. Rewriting the detected intensity in terms of $V(u, v)$ gives

$$P(u, v; \phi) = P_0\{1 + \mathcal{R}e\left[V(u, v) \exp\left(i\phi\right)\right]\}. \tag{8.22}$$

In the case of the pupil plane interferometer $\mathcal{R}e\left[V(u, v) \exp\left(i\phi\right)\right]$ is the modulation observed when the delay line length is dithered. The very same modulation is seen across the image plane of the image plane interferometer. If the detected power is measured separately for ϕ set to zero in one case and ϕ set to $-\pi/2$ in the other, the respective power values will be $P_0(1 + \mathcal{R}eV)$ and $P_0(1 + \mathcal{I}mV)$. This gives enough information to be able extract the phase and magnitude of V. The maximum and minimum of intensity occur when

$$\mathcal{R}e\left[V(u, v) \exp\left(i\phi\right)\right] = \pm|V(u, v)|$$

and hence the visibility of fringes

$$\frac{(P_{\max} - P_{\min})}{(P_{\max} + P_{\min})} = |V(u, v)|$$

which shows that $|V(u, v)|$ is simply the visibility defined in eqn. 5.14. Therefore $V(u, v)$ is given the name *complex visibility*. It is important to keep firmly in mind that the visibility amplitude is not the amplitude of the fringes but rather the amplitude of their *contrast*. Putting eqn. 8.21

into words: the complex visibility of the fringe pattern is the Fourier transform of the source brightness distribution.[9]

We have seen that the visibility $V(u, v)$ is the Fourier transform of the source brightness $I(\ell, m)$ at a single point in the the (u,v) plane. What is needed to reconstruct the source distribution fully is a series of measurements of the visibility for different baseline lengths and orientations, giving a range of values of $k\mathbf{B}$ and hence of (u,v). Then if the distribution of these measurements is dense enough in the (u,v) space the following approximation can be made

$$
\begin{aligned}
I(\ell, m) &= \int \int V(u, v) \exp i(\ell u + m v) \, \mathrm{d}u \, \mathrm{d}v \\
&\approx \sum V(u, v) \exp i(\ell u + m v) \Delta u \Delta v,
\end{aligned}
\tag{8.23}
$$

where the sum is taken over all the measurements and where $\Delta u, \Delta v$ are the spacings between the measurements. Evidently both the phase and magnitude of the visibility have to be well determined if the image reconstruction is to be reliable.

For clarity in presentation of the analysis of stellar interferometers it has been assumed, thus far, that the starlight is monochromatic. In practice, in order that the fringes are bright enough to give reliable measurements, the spread of wavelengths used can be 10% of the mean wavelength. Thus the fringes of the different colours will only be superposed around the location where the optical path difference is zero, which is located at image of the phase centre on the target. Only here will the fringe contrast accurately reflect the true visibility. It is essential therefore to make measurements on the visibility amplitude close to the central white fringe. Also the displacement of the white fringe from the image of the phase centre determines the phase of the visibility: a displacement of f fringe widths gives a phase of $2\pi f$. Note that the number of clear fringes is much fewer than the number, B/d, expected with a monochromatic source: for a bandwidth of 10% there will be roughly 10 such fringes.

When the source is centrosymmetric so that $I(\ell, m)$ is symmetric around $\ell = 0$ and $m = 0$, examination of eqn. 8.21 shows that the complex visibility is always real. In this simple case all that is required are measurements of the visibility amplitude. In addition when the source is centrosymmetric the measurements only need to be made for one baseline orientation to fix the source intensity distribution. A simple example is a circular source of uniform intensity for which the calculation of Section 6.7 can be re-used. With the notation of figure 8.19 an

---

[9]This statement is one version of the *van Cittert–Zernike theorem*. See for example M. Born and E. Wolf, *Principles of Optics*, seventh edition, published by Cambridge University Press (1999).

element of the source at (ρ, α) covers a solid angle

$$\mathrm{d}\ell\,\mathrm{d}m = \sin\rho\,\mathrm{d}\rho\,\mathrm{d}\alpha = \rho\,\mathrm{d}\alpha\,\mathrm{d}\rho$$

to a good approximation, and $\mathbf{B}\cdot\mathbf{\Delta s} = B\rho\cos\alpha$. Then

$$V = \Delta I \int_0^\theta \int_0^{2\pi} \exp\left(ikB\rho\cos\alpha\right)\rho\,\mathrm{d}\rho\,\mathrm{d}\alpha,$$

where ΔI is the intensity per unit solid angle of the source and θ is its angular radius. Thus

$$\begin{aligned}
V &= 2\pi\,\Delta I \int_0^\theta J_0(kB\rho)\,\rho\,\mathrm{d}\rho \\
&= 2\pi\,\theta^2\,\Delta I\, J_1(kB\theta)/(kB\theta) \\
&= 2I\,J_1(kB\theta)/(kB\theta),
\end{aligned} \tag{8.24}$$

where I is the total intensity of the source. Figure 8.19 also shows this distribution for a source of 20 mas angular diameter observed at a wavelength of $2.2\,\mu$m. When the source is more complex in shape the measurement of the phase of the visibility is also required.

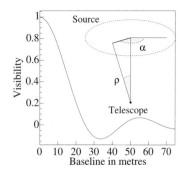

Fig. 8.19 Visibility versus baseline for a 20 mas source observed at a wavelength of $2.2\,\mu$m.

The effect of the piston component of the atmospheric distortion has been mentioned above. If the interferometer consists of a pair of telescopes there is the lag/lead in phase between them produced by changes in refractive index across the whole area of either telescope. Electronic systems are used to track these fringe movements and provide compensating delay line correction. However it is not possible to separate the displacement of the white fringe due to the visibility phase from the much larger effect due to the path length changes in the atmosphere. Thus each determination of the phase of the visibility contains an unknown error, larger than 2π. The phase can be recovered if a suitable centrosymmetric reference star lies within the isoplanatic region around the target by relating the positions of the target fringe to the reference star's image. In other cases phase recovery requires the techniques described in the sections following this.

8.9   Aperture arrays

The apertures used to collect the interfering beams can be mirrors themselves in alt-az or equatorial mountings which return the light to a fixed telescope, as in the COAST array at Cambridge in the UK, or they can be individual telescopes. In the case of COAST the mirrors were moved to various stations along three tracks forming a Y-shape of about 65 m extent. In this way baselines with a variety of lengths and orientations can be readily obtained. When the telescopes are stationary the rotation of the Earth provides the means to change the orientation of the baselines.

An alternative approach in interferometer imaging has been widely used. A plate with many identical circular holes across its surface is placed at the exit pupil plane of a large telescope. The individual holes function as separate telescopes and interference between all these pairs of *telescopes* is then simultaneously present in the image plane. If the orientations and lengths of the baselines formed by the pairs of apertures are all different then fringe patterns for each aperture pair have a unique and known fringe spacing and orientation. This permits the fringe patterns to be disentangled using Fourier analysis. With the aperture diameters being smaller than r_0 and the exposures being shorter than τ_0 a uniform distortion is frozen over each aperture. Frequent exposures are made to increase the overall signal relative to noise from electronics and scattered light. Figure 8.20 shows the aperture pattern used by Tuthill and colleagues[10] with the segmented Keck I telescope, projected back onto the primary mirror. The 21 apertures are placed so as to avoid any of the 210 baselines being the same as any other baseline, which makes these 210 fringe patterns distinguishable from one another because each has a known unique fringe orientation and fringe spacing.

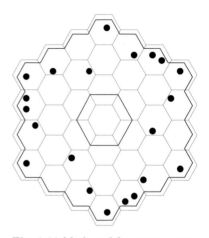

Fig. 8.20 Mask used for aperture synthesis on the Keck I telescope projected onto the primary mirror. The hexagons are the 36 segments from which the 10 m diameter mirror is made.

As noted earlier, it is essential to measure the phase accurately as well as the amplitude of the complex visibility in order to be able to reconstruct the source distribution. Of these the phase measurement is the most troublesome because the path difference is continually changing by tens of microns thanks to the piston component of the atmospheric distortion. If there is a bright unresolved star within the isoplanatic cone around the target of interest its white fringe can be used to give a phase reference. When no reference is available the following less complete solution is adopted. Visibilities are measured for each baseline provided by a triangle of apertures. The measured phases of the visibility function are then

$$\phi_{12} = \phi_{12}^0 + \Delta\phi_1 - \Delta\phi_2,$$
$$\phi_{23} = \phi_{23}^0 + \Delta\phi_2 - \Delta\phi_3,$$
$$\phi_{31} = \phi_{31}^0 + \Delta\phi_3 - \Delta\phi_1.$$

Here ϕ_{ij}^0 is the undistorted phase for the pair of apertures (i,j). The measured phase, ϕ_{ij}, is this true phase altered by the distortions at the two apertures involved: $\Delta\phi_i$ and $\Delta\phi_j$. Adding these three equations gives the *phase closure* relation

$$\phi_{12} + \phi_{23} + \phi_{31} = \phi_{12}^0 + \phi_{23}^0 + \phi_{31}^0. \tag{8.25}$$

This sum recovers the sum of the three actual phases. When there are many apertures the phase closure relations from all independent sets of three apertures provide almost as many equations as there are phases. Only a small amount of external information is then required to fully

[10]P.G. Tuthill, J.D. Monnier, W.C. Danchi, E.H. Wishnow and C.A. Haniff: *Publications of the Astronomical Society of the Pacific* 112, 555 (2000).

reconstruct the image. Measurements on each short exposure taken will yield different values of the visibility phases but give, within errors, the same *closure* phases. On the other hand the magnitude of the visibility will be the same within experimental errors from one exposure to the next. It is important to appreciate that, in general, the visibility phases and the image can be simultaneously reconstructed from the closure phases and visibility amplitudes.

8.10   Image recovery

Once the visibilities are available from enough baselines an attempt can be made to recover the image using eqn. 8.23. However, as is now shown, this image is imprinted with the distribution of the baselines in (u,v) space. Suppose $W(u,v)$ is the Fourier transform of the true source distribution $I(\ell, m)$. Then the image recovered is

$$R(\ell, m) = \int \int W(u, v) g(u, v) \exp\left[i(\ell u + m v)\right] \mathrm{d}u \, \mathrm{d}v, \qquad (8.26)$$

where $g(u, v)$ is unity at each point a measurement was made and zero elsewhere. Thus

$$R(\ell, m) = I(\ell, m) \otimes G(\ell, m), \qquad (8.27)$$

where

$$G(\ell, m) = \int \int g(u, v) \exp\left[i(\ell u + m v)\right] \mathrm{d}u \, \mathrm{d}v$$

is the Fourier transform of the distribution of the baselines. $G(\ell, m)$ is the diffraction pattern produced on the (l, m) plane by the sampling in the (u, v) plane. It has complicated lobes which smear the actual source distribution into the reconstructed image $R(\ell, m)$.

Various algorithms have been invented to try to extract the actual source distribution $I(\ell, m)$ from $R(\ell, m)$. A commonly used example called CLEAN will be described here.[11] The procedure is to pick out the brightest cell in the image located, let us suppose, at (ℓ_1, m_1) with intensity R_1. Then a distribution $\gamma G(\ell - \ell_1, m - m_1) R_1$ is subtracted from the image brightness distribution where γ is a constant of around 0.5 called the *loop gain*. This step removes what was the brightest region of the source taking account of the known smearing of the aperture pattern. This step is then repeated to remove the next brightest cell, and so on. Eventually the residual image after these successive subtractions will be of near-uniform low intensity containing only background light and detector noise. Finally in order to recover the best estimate of the actual image the 'sources' at (ℓ_i, m_i) with intensities $\gamma R(\ell_i, m_i)$ are convoluted with the point spread function (PSF) of the full telescope aperture. In practice γ needs to be tuned for each set of observations. Images obtained by Tuthill and colleagues from the Keck I telescope

---

[11] J.A. Hogbom, *Astronomy and Astrophysics Supplement* 15, 417 (1947).

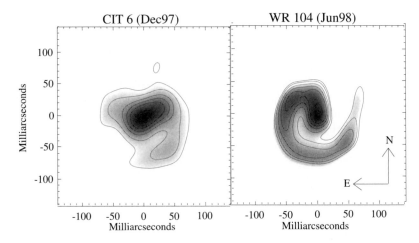

Fig. 8.21 Images of an evolved carbon star CIT6 and a Wolf–Rayet star WR104, obtained with aperture synthesis at the Keck I telescope. This figure originally appeared in the Publications of the Astronomical Society of the Pacific (P.G Tuthill, J.D Monnier, W.C Danchi, E.H Wishnow and C.A Haniff, *PASP* 112, 555 (2000)). Copyright 2000, the Astronomical Society of the Pacific; reproduced with the permission of the editors, and by courtesy of Professor Tuthill.

with a 15 hole mask using a related analysis technique are shown in figure 8.21.

It is perhaps surprising that the masking technique is so successful in producing a diffraction limited image after throwing away 90% or more of the incident light. However by restricting each aperture to the size of the Fried parameter and by making short exposures the temporal and spatial variation of the atmospheric distortion is eliminated from each exposure, and as a result fringes with good signal to noise are obtained. From such fringes obtained with multiple apertures the analysis described here can simultaneously eliminate the atmospheric distortion and reconstruct the image. Exposures with the full aperture simply superpose the images from all the area elements of the lens with their different and time varying distortions.[12]

8.11 Comparisons with radioastronomy

The resolution obtained with an interferometer of baseline length B at wavelength λ is λ/B. By chance the best resolution recently achieved with optical interferometers is very similar to that of current radiotelescopes. The difference in wavelengths ($\sim 1\,\mu$m and ~ 10 cm) is compensated by the difference in baseline lengths (~ 100 m and $\sim 10^4$ km). However radio interferometry and optical interferometry differ critically in the way radio and light waves are detected. Detectors of light re-

---

[12]I am indebted to Professors Tuthill and Haniff for clarification on a number of issues related to masking.

spond to the intensity of the radiation, which is proportional to the electric field squared EE^*. On the other hand detectors of radio waves are electric circuits in which the current is proportional to the electric field. The radio signal from one telescope can therefore be detected, amplified and even recorded without losing any of its phase content. This amplified signal (with a universal time marker) is then transmitted from its parent radiotelescope site to a distant location where it can interfere with another telescope's signal. In the contrasting case of optical interferometers coherent amplification with a phase reference is feasible but would degrade the signal to noise appreciably. The light from both telescopes must be transmitted directly along optical paths and brought to interfere. In addition the lengths of these paths must be measured to a fraction of the much smaller optical wavelength.

The development of the technology to measure distances of order $100\,\mathrm{m}$ to a precisions of $10\,\mathrm{nm}$, and to move mirrors to a comparable precision at speeds of $\mathrm{m\,s}^{-1}$ is relatively recent. The lasers required in the measuring process have themselves to be stable to an equivalent precision, namely better than one part in 10^{10}. Radio interferometry was implemented earlier from a simpler technological base and has accumulated vastly more data. The problems of phase and image recovery in the presence of atmospheric distortion are common to both types of interferometry. Consequently many analysis techniques, such as phase closure, which were originally developed for the interpretation of radio data have been re-used in analysing optical data.

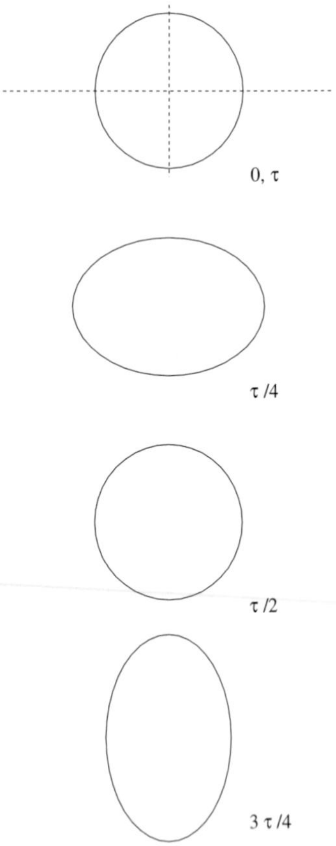

8.12   Gravitational wave detectors

Einstein, on the basis of his general theory of relativity, predicted the existence of gravitational waves. Any non-symmetric accelerating mass will emit gravitational radiation, but a perfectly spherical star collapsing radially would not radiate. The only experimental evidence is as yet indirect: this comes from measurements of the slow changes in the orbital period of a binary pulsar PSR1913+16. The period is changing at just the rate expected if the system is radiating energy in the form of gravitational waves, and for the measurement of this effect Hulse and Taylor were awarded the 1993 Nobel prize in physics.

Gravitational waves travel at the speed of light and distort space-time as they travel though it. The simplest gravitational wave excites quadrapole oscillations of space-time, and this motion is illustrated in figure 8.22. A circle of free masses, initially at rest, is shown at time intervals of one quarter period as a plane gravitational wave passes. In one half cycle the ring is stretched in one direction and squashed at right angles; in the next half cycle the distortion is reversed. If the arms of a Michelson interferometer are aligned along the dotted lines then the relative lengths of the arms would change at the frequency of the

$0, \tau$

$\tau/4$

$\tau/2$

$3\tau/4$

Fig. 8.22 Distortion of a circle of test masses produced by gravitational waves travelling perpendicular to the diagram.

gravitational wave, and the interference fringes would oscillate to and
fro in synchronism. Of course if the wavefronts are not parallel to the
plane of the interferometer there would be a reduced effect. The effects
of gravitational waves are all weak because the gravitational coupling
is itself weak compared to electromagnetism: for example two protons
repel one another electrostatically with a force 10^{36} times stronger than
their gravitational attraction. The amplitude of a gravitational wave is
expressed as the strain or change in length per unit length of the fabric of space-time. A supernova explosion occuring at the edge of our
galaxy is expected to give a burst of gravitational waves arriving at the
Earth lasting for a few milliseconds, with a frequency around 1 kHz, and
with a strain of around 10^{-18}. A pair of merging neutron stars would
give a chirp of radiation in which the frequency would rise from 40 Hz
to 1 kHz over several seconds with a strain of order 10^{-21}. A strain of
10^{-18} amounts to 10^{-6} nm in 1 km, and the measurement of such a small
quantity is a formidable challenge. This is the ultimate test of interferometry where displacements equal to 10^{-6} of a fringe must be detected.
Nonetheless the current rate of improvement in the sensitivity of huge
Michelson-type interferometers makes it likely that the direct detection
of gravitational waves will occur within a decade.

It might appear that this precision is unattainable because the displacements are much smaller than the diameter of an atom. However
an electromagnetic wave reflected from a surface is reflected from all the
atoms on that surface, so it is the *average* position of the atoms that
determines the phase of the reflected wave.

A sketch of the principal features of the current large gravitational
wave detectors is shown in figure 8.23. The volume of the optical paths is
enclosed and evacuated in order to eliminate refractive index variations
and convection currents. The laser preferred is a stabilized Nd:YAG
laser producing a beam of about 10 W at wavelength 1064 nm. This
feeds the beam splitter of a Michelson interferometer whose arms are as
long as feasible in order to maximize the displacement when a gravitational wave arrives. In the case of the larger LIGO detectors in the USA
each arm consists of a 4 km long Fabry–Perot cavity resonant with the
laser wavelength, while the GEO detector in Germany has arms 600 m
long. Throughout the interferometer the light beam is in the TEM_{00}
mode with its simple Gaussian profile. This makes it possible to use
small diameter mirrors despite the great distances involved. The mirrors and beam splitter are freely suspended so that the mirrors act as
the test masses for the gravitational wave. They have reflection coefficients close to unity so that the light makes many round trips within the
cavity. This feature effectively increases the cavity length by a similar
factor and hence too the sensitivity. By suitably positioning the beam
splitter it is arranged that the returning beams from the cavities interfere in such a way that all the light emerges toward the laser. Toward the
photodiode the beams interfere destructively, which is called the *dark*

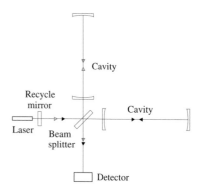

Fig. 8.23 Layout of a Michelson interferometer for gravitational wave detection.

fringe condition. This arrangement has two advantages. Firstly it is far easier to detect a small oscillation in light level at the photodiode if its quiescent state is dark, than if it were carrying a large current whose fluctuations can mask the signal. Secondly the light emerging toward the laser is reflected back into the Michelson by a recycling mirror in front of the laser; this *power recycling* technique boosts the light level in the cavities by an order of magnitude. Efficient recycling requires that the recycling mirror and the Michelson form another resonant cavity.

The difference between the travel times for N passes in the two arms of an interferometer when the gravitational strain is equal to $+S$ and $-S$ along the arms is

$$\Delta t = 4NLS/c,$$

where L is the rest length of each arm. The phase difference between light emerging after N passes along the arms is

$$\Delta\phi_s = 2\pi c \Delta t/\lambda = 8N\pi LS/\lambda = 4\pi Sc\tau_s/\lambda \qquad (8.28)$$

where τ_s is the storage time. In obtaining this simple result two effects were neglected. Firstly the frequency f_g of the gravitational wave may be high enough that the period of one oscillation is shorter that the storage time. In this case there can be a reversal of the strains during the time the light is in the arms which reduces the phase difference: in the limit of very high frequency gravitational waves there would be no phase difference. Secondly, all the light has been assumed to stay in the cavities for the full storage time, whereas it continually arrives and leaves through the entry mirror. τ_s must be redefined as the time it takes the intensity of radiation in the cavity to fall by a factor e when the laser is turned off. When both effects are taken into account the phase difference becomes

$$\Delta\phi = \Delta\phi_s/\sqrt{1 + (2\pi f_g \tau_s)^2}. \qquad (8.29)$$

At very high frequencies such that $f_g\tau_s \gg 1$ this reduces to

$$\Delta\phi = 2cS/(\lambda f_g),$$

and this is called the *storage time limit*. Current experiments use kilometre long cavities with finesses of order a few hundred, which means the storage times approach one millisecond. This implies that detectors are only likely to be sensitive to gravitational waves of frequencies up to a few kHz.

8.12.1   Laser-cavity locking

It has emerged in the last section that the operation of the interferometer requires that several cavities should resonate with the laser wavelength. The crucial process of searching for and maintaining this condition is called *locking* the cavity and the laser. In order to understand how this is achieved we first go back to figure 5.31, which shows the intensity transmitted through a Fabry–Perot etalon. As a result the reflected intensity must have the shape around resonance that is shown in figure 8.24. In this figure the variation of reflected intensity is plotted against the frequency of the light or equivalently the cavity length. Above resonance an increase in frequency causes a rise in the reflected intensity while below resonance an increase in frequency causes a fall in intensity. This difference is the basis of the *Pound–Drever* method for locking the cavity and laser. It is an illustration of how optics and electronics can be integrated to achieve sophisticated and delicate control systems.

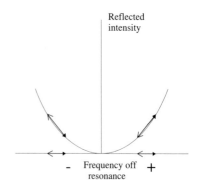

Fig. 8.24 The sign of the change in the reflected intensity from a Fabry–Perot cavity when the the cavity length is increased reverses at resonance. This is equally true when changes are made instead in the laser frequency.

Figure 8.25 shows a simplified outline of the equipment used. Light from the laser passes through an optoelectronic modulator across which a voltage is applied from a radio-frequency oscillator at tens of MHz. The effect of applying an oscillatory voltage across the modulator is to cause a small amplitude oscillation in its refractive index. Consequently the light emerging from the modulator has a synchronous oscillating phase. Optical modulators are discussed in more detail in Chapter 11. The electric field of light which has passed through such a modulator is

$$E = E_0 \exp\left[-i(\omega t + \beta \sin(\Omega t))\right], \qquad (8.30)$$

where ω is the angular frequency of the incident light, Ω is the angular frequency of the applied radio frequency voltage, and β is the very small amplitude of the induced oscillation in phase. This is known as *phase modulation*. With β small eqn. 8.30 can be approximated as follows

$$E = E_0 \exp(-i\omega t)\left[1 - i\beta \sin(\Omega t)\right]$$
$$= E_0 \exp(-i\omega t)\{1 - (\beta/2)\exp(i\Omega t) + (\beta/2)\exp(-i\Omega t)\}. (8.31)$$

Thus the beam emerging from the modulator now contains three different frequencies. The waves at angular frequency ω are called the *carrier* waves; and the waves at angular frequencies $\omega \pm \Omega$ are called the upper and lower *sidebands*. Now suppose the wave described by eqn. 8.31 is reflected *from a cavity* which is near resonance for the carrier. The sidebands will be well off resonance so their reflection coefficients are close to unity and change little with the wavelength. Thus the amplitude of the electric field of the reflected wave is

$$E_r = E_0 \exp(-i\omega t)\{r_c - i\beta \sin(\Omega t)/2\}, \qquad (8.32)$$

where r_c is the reflection coefficient of the cavity for the carrier at angular frequency ω. Hence the reflected intensity is

$$I = E_r E_r^* = A + B\sin^2(\Omega t) - (E_0^2\beta/2)\Im m[r_c - r_c^*]\sin(\Omega t), \quad (8.33)$$

where A and B need not be calculated in detail. As shown in figure 8.25 the reflected light falls on a photodiode detector which produces an electrical signal proportional to the light intensity. This signal is

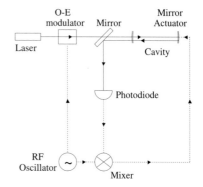

Fig. 8.25 Pound–Drever stabilization scheme. This locks the Fabry–Perot cavity length to an integral multiple of half wavelengths of a laser. The solid lines show the light paths while the broken lines carry electronic signals.

taken to the mixer where it is multiplied by a signal $K \sin(\Omega t)$ coming directly from the same radio-frequency oscillator that provides the beam modulation; finally the mixer output is time averaged. As a result only the term in eqn. 8.33 which has the $\sin(\Omega t)$ variation contributes to this time average. Apart from a multiplicative constant, the final output is

$$V = \Im m[\, r_{\mathrm{c}} - r_{\mathrm{c}}^* \,]. \qquad (8.34)$$

The reflection coefficient for a cavity is available in eqn. 5.52

$$r_{\mathrm{c}} = r\,[\, \exp i\delta - 1\,]\,/\,[\, 1 - r^2 \exp i\delta\,],$$

where r is the reflection coefficient at either mirror of the etalon and δ is the phase change accumulated by light in going to and fro once in the cavity. Close to resonance $\delta = 2n\pi + \epsilon$ with ϵ being small so that the reflection coefficient of the cavity can be approximated as follows

$$r_{\mathrm{c}} = i\epsilon r/(1 - r^2) = -r_{\mathrm{c}}^*.$$

Substituting this value for r_{c} into eqn. 8.34 gives

$$V = 2r\epsilon/(1 - r^2). \qquad (8.35)$$

This output has the desired property that it changes sign with ϵ at resonance and can be used as a control signal to reach resonance. This signal drives an actuator to move one of the cavity mirrors (as shown in figure 8.25), or is used to tune the laser frequency.

8.12.2   Noise sources

Three sources of noise restrict the sensitivity of the interferometers used in gravitational wave searches: seismic, thermal and shot noise. How important these are and how the detectors are designed to reduce the overall noise are now described. There is some variation between the importance of different noise sources, depending on the detector designs.

Seismic noise encompasses ground motion due to Earth tremors, to the wind blowing over the surface and to human activity. This last source can involve movements of tens of microns due to vehicles moving at hundreds of metres distance. The intensity of seismic noise increases as the frequency falls and dominates over other sources below approximately 10 Hz. The mirrors and beam splitter are all suspended from spring loaded supports by pairs of thin wires like pendulums. The supports absorb vertical disturbances and pendulum bobs are insensitive to high frequency horizontal movements of their supports. Fine control of a mirror's position and damping is made magnetically.

Thermal noise is due to the minute thermal vibrations of all the mechanical components in the optical chain. It dominates other noise sources over the mid-frequency range around 100 Hz. Each mechanical object has its own natural frequency of vibration and at these frequencies the thermal vibrations can stimulate motion that amounts to a significant noise as far as gravitational wave detection is concerned. The pendulums formed by the mirrors and their suspensions swing at around one hertz and the wires vibrate like violin strings at a few hundred hertz. These are sharp resonances and only affect a narrow frequency range. The mirrors themselves can vibrate and are made of fused quartz for which the resonances are also sharp. However at frequencies away from the sharp resonances the mirrors when locked are highly stable.

The final noise source of importance arises from the natural fluctuations in the intensity of the laser beam.[13] These fluctuations can both mask true signals and also affect the stabilization and they form the dominant noise source at high frequencies. In summary the noise is equivalent to an uncertainty in the arm length

$$\Delta L = \sqrt{\hbar c\lambda/(8\pi\tau P)}, \qquad (8.36)$$

where $\hbar$ is Planck's constant$/2\pi$, P is the optical power in the cavity and τ the duration of the measurement. Over a 4 km path and with a laser of 1 μm wavelength the detectable strain is

$$S = 2.80\,10^{-19}/\sqrt{P\tau} \qquad (8.37)$$

with P in watts and τ in ms. Evidently P must be made as large as possible. This is partly achieved by locking the cavities on resonance with the laser so that the energy in the cavities is larger than the laser beam intensity by a factor roughly equal to the cavity finesse. In addition the power recycling increases the light stored in the Fabry–Perot cavities by a further large factor. The aim is to maintain powers of up to 100 kW in the future. Mirror surfaces have therefore to be smooth enough to avoid scattering or absorbing energy from the beams. The mirrors through which the light enters the cavities have to be highly homogeneous to reduce internal scattering and the consequent heating which changes the refractive index of the glass. The power cannot be increased indefinitely because eventually practical problems arise through the heating of optical components.

The gain in sensitivity brought about by increasing the light stored in the Fabry–Perot cavities has two equivalent interpretations. From one viewpoint we can see that the longer the storage time is, the longer will be the path over which the light travels during the passage of a gravitational wave and hence the higher the sensitivity. From another perspective the increase in light stored reduces the fractional fluctuation in light intensity and this also increases sensitivity.

The noise levels achievable with current detectors are notionally indicated in figure 8.26. Gravitational signals need to produce strains larger

[13]This is a quantum effect caused by the variation of the number of photons at any one time. The topic is discussed in Chapters 14 and 15. The mean number of photons detected is $n = \lambda P\tau/hc$, where hc/λ is the energy of each photon. The fluctuations in the photon count are Poissonian so that $\Delta n = \sqrt{n}$. From eqn. 19.82 the corresponding minimum uncertainty with which the electromangetic wave's phase can be determined is $\Delta\phi = 1/(2\Delta n)$. Therefore the minimum uncertainty in the path length is $\Delta L = \lambda(\Delta\phi/2\pi)$, that is $\sqrt{\hbar\lambda c/(8\pi P\tau)}$.

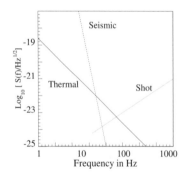

Fig. 8.26 Noise sources of modern kilometre long gravitational wave detectors.

than the noise levels shown there in order to be detectable. The value of S is determined by multiplying the ordinate from the graph by the square root of the bandwidth of the signal. In broad terms the gravitational wave detectors are optimized for detection of waves in the audio frequency range 10–5000 Hz.

8.13   Gravitational imaging

The first critical test for the general theory of relativity was provided by observations to find out whether light passing near to the Sun would be deflected by the distortion of space-time due to the sun's mass. A simple analogy is provided by rolling a ball bearing across a flat, taut, horizontal sheet of rubber. Its path would normally follow a straight line. On the other hand if a heavy weight is placed on the sheet the sheet is bowed down in that region so that the ball bearing would be deflected from a straight line path. In 1919 the deviation of light from Mercury as it approached extinction behind the Sun was measured and found to agree with the predicted 1.750 arcsec. Quantatively a ray of light passing at a minimum distance r from a mass M is deflected through an angle

$$\alpha = 4GM/(c^2 r) = 2R_\text{s}/r, \qquad (8.38)$$

Mercury's image was only visible because there was, by chance, an eclipse at the same time it passed into the Sun's 'shadow'. A comparison was made of the position of Mercury against the star field on photographs recorded during the eclipse and at another time when the Sun's image was far off. The measurement was unfortunately only accurate to around 0.5 arcsec. Radiotelescopes can follow a planet's radio image very close to the Sun without the help of an eclipse, and they confirm the general relativistic prediction within an experimental error of 0.01 arcsec.

called the *Einstein* angle. G is the gravitational constant $(6.67\,10^{-11}$ kg^{-1}m^{3}s$^{-2})$ and R_s $(2GM/c^2)$ is the *Schwarzschild* radius for a mass M. In the case of the Sun (with mass $1.99\,10^{36}$ kg and radius $6.69\,10^5$ km)

Fig. 8.27 Partial Einstein rings from several sources at varying distances produced by the galaxy Abell 2218. The image was made with Hubble Space Telescope; courtesy Professor R.Ellis (California Institute of Technology), Professor W. Crouch (University of New South Wales) and the NASA Space Telescope.

this Schwarzschild radius is only 1.4 km. When source and mass are axially symmetric around a line through the observer and all are separated from one another by very large distances the observer may see a circular *Einstein ring* image of the source extending all round the deflecting mass. Sections of Einstein rings can be seen in figure 8.27 recorded by the HST. The deflecting mass is the galaxy Abell 2218 and the arcs are images of sources at various large distances beyond this galaxy. The image does not need to be in colour because the gravitational deflection is achromatic!

Other effects are also observed when there is good alignment of source, the deflecting mass and the Earth. A few quasars have been detected whose high redshifts are sufficiently large that they must be far enough from the Earth that they ought to be undetectably faint. What happens in such cases is that an intervening galaxy is focusing a cone of rays toward the Earth and in so doing making the quasar appear brighter than it really is. This same focusing effect is relied on to assist in the search for brown dwarf stars. Such stars may contribute to the dark matter in our galaxy, and being dark they are not easy to detect. The technique to detect them is, paradoxically, to monitor continuously the intensities of a large number of stars in a small satellite galaxy of our own galaxy called the Large Magellanic Cloud (LMC). Whenever a brown dwarf in our galaxy passes in front of a star in the LMC the focusing effect can cause the apparent brightness of the star in the LMC to rise by a big factor and then to fall back to its original value after the 'eclipse'. Similar *microlensing* events involving a brief change in intensity can be mimicked by astrophysical processes that cause the star in the LMC to heat up temporarily. However such processes affect the distribution of the star's energy output across the spectrum. This alternative cause of star brightening can therefore be excluded if the star shows an identical and simultaneous rise and fall in intensity for blue and red light separately. Only a few brown dwarf stars have so far been detected by this method so that it is unlikely that collectively they are a major contributor to dark matter.

Exercises

(8.1) An unresolved star of apparent magnitude +1.6 is observed with a telescope whose primary has diameter 2.4 m. The image falls on a detector which has a response of $1 \, A \, W^{-1}$ of luminous intensity. What charge accumulates on the detector in 1 ms?

(8.2) Suppose, in conditions of seeing where the Fried parameter is 20 cm, that the distortion is entirely tip/tilt. Calculate the rms variation in the separation between the images of a laser guide star and a real guide star in the image plane of a telescope of focal length 30 m. The two stars can be assumed to lie in the same isoplanatic cone.

(8.3) Two unresolved stars of equal intensity have an-

gular separation 0.003 arcsec. What separation is required for a pair of two telescopes used as an interferometer in order to resolve the stars in light of wavelength 500 nm? In what way is the orientation of the baseline important?

(8.4) Show that if an astronomical source is centrosymmetric the complex visibility is always real.

(8.5) What power level is required in a 4 km long cavity in the arm of a gravitational wave interferometer working at an optical wavelength of $1 \, \mu m$ in order to make it feasible to detect gravitational waves with strain 10^{-21} and frequency 1 kHz? Estimate the storage time if the cavities have a finesse of 500.

(8.6) Calculate the complex visibility of two equal intensity unresolved sources an angular distance 2θ apart.

(8.7) According to the Sparrow criterion practically all the information on the detailed features of a source can be retrieved if the detector pixel size is half the resolution limit of the telescope. Hence relate the optimal $f/\#$ of the telescope to the pixel size, p, and the mean wavelength, λ, of the radiation detected. Can you suggest any reason for caution in interpreting the features detected in the image when the samplings across the image are spaced more closely than this limit (oversampling)?

(8.8) A classical Cassegrain telescope has a primary mirror with focal length 8 m and a secondary for which a is 4 m and e is 1.2. Calculate the separation of the mirrors, the location of the final focus and the plate scale.

(8.9) What is the angular field of view over which a paraboloidal mirror of 2.4 m diameter and focal length 5 m produces an image with less than 0.5 arcsec distortion? What is the value of the corresponding field of view for the Hubble Space Telescope design?

(8.10) The Fabry–Perot cavities in a gravitational wave detector have confocal mirrors a distance L apart and the radiation stored has wavelength λ. What is the minimum width, w, of the TEM_{00} mode at the mirrors? If the mirrors are 4 km apart and the light has wavelength 500 nm what is the value of the corresponding width parameter w_0?

Classical electromagnetic theory

9

9.1 Introduction

In 1864 Maxwell presented a unified theory of electric and magnetic fields, which is now encapsulated into four fundamental equations that take his name. These, together with Lorentz's formula for the force on any charge in an electromagnetic field, are the basic elements of classical (that is non-quantum mechanical) electromagnetic theory. In this chapter Maxwell's equations will first be introduced and used to infer, as Maxwell did, that electromagnetic waves exist and that light is simply one form of electromagnetic radiation. Next, the energy content and energy flow in electromagnetic waves will be discussed. The expression for the energy flow in an electromagnetic field will be deduced by applying the law of conservation of energy to the electromagnetic field. This completes the formal basis of the classical theory of electromagnetism and provides the starting point for discussing the behaviour of light passing through matter. That discussion takes up the two following chapters and covers polarization, absorption and dispersion effects.

Many optically useful materials are dielectrics, that is poor conductors of electricity, and are also only weakly magnetic. The emphasis in this chapter will be on homogeneous and isotropic dielectrics when the fields in the electromagnetic radiation are sufficiently small that the response of matter is linear, and the dielectric is stationary. The label used here for such materials in such fields is HIL. In Chapter 9 the discussion will be extended to anisotropic dielectrics, materials with which light can be manipulated through its polarization. The high electromagnetic fields in laser beams produce very striking non-linear effects in materials and these will be discussed after introducing lasers in Chapter 14. Propagation of electromagnetic radiation in metals, insofar as it affects optics, is dealt with in Chapter 11.

When the integral versions of Maxwell's equations are applied at an interface between different media they impose simple relationships between the fields on either side of the surface. In the central portion of this chapter the consequences of these *boundary conditions* are followed through for electromagnetic waves impinging on interfaces between dielectrics. This analysis not only proves the laws of reflection and refraction from first principles but also determines the amplitudes and intensities of the reflected and refracted waves. The reflected and refracted amplitudes are found to depend strongly on the polarization of the incoming waves. Another product of the analysis presented here is an explanation of wave behaviour in total internal reflection and frustrated total internal reflection, a topic broached in Chapter 2.

Anti-reflection coatings are essential for the optics of modern cameras because the number of lenses employed is such that most of the light would otherwise be lost in reflections. On the other hand, laser safety goggles are required to reflect essentially all the incident radiation over a restricted wavelength range. An account is given below of the use of thin multiple layers of dielectrics to form *interference filters* which either selectively reflect or transmit a range of wavelengths.

The last portion of the chapter is used to discuss modes of the electromagnetic field and their propagation in simple waveguides.

9.2   Maxwell's equations

The electric and magnetic fields, $\mathbf{E}$ and $\mathbf{B}$, can be determined directly by measuring the Lorentz force on a charge q travelling with velocity $\mathbf{v}$, namely

$$\mathbf{F} = q(\mathbf{E} + \mathbf{v} \wedge \mathbf{B}), \qquad (9.1)$$

and hence they are reasonably regarded as the physical em fields. The effects these fields have on matter will now be outlined in order to put Maxwell's equations in context.

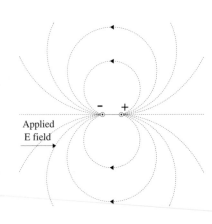

Fig. 9.1 Electric field lines around an electric dipole.

The action of an applied electric field on an atom or molecule is to pull apart the positive charges (nuclei) and negative charges (electrons) so that an electric dipole, $\mathbf{p} = \mathbf{d}q$ is produced, where q is the total charge of either sign and $\mathbf{d}$ is the vector separation of the centre of gravity of the positive charge from the centre of gravity of the negative charge. Electric fields within an atom are usually much larger than the applied fields so that the separation of the centres of the positive and negative charges produced by the applied electromagnetic field is small in comparison to the size of the atom/molecule. Applying Coulomb's law, the electric field due to a proton or electron at a distance equal to an atomic diameter, $0.1\,\mathrm{nm}$, is predicted to be of order $10^{11}\,\mathrm{V\,m^{-1}}$. For comparison the electric field within a continuous $1\,\mathrm{kW}$ laser beam is only around $10^{6}\,\mathrm{Vm^{-1}}$, but it can be large enough to disrupt atoms in very high power pulsed lasers. In some *polar* materials the molecules have an intrinsic electric dipole moment and these dipoles usually point in random directions in the absence of any applied electric field. In such polar materials an electric field exerts a torque $(\mathbf{p} \wedge \mathbf{E})$ on each molecular dipole. In liquids and gases, where these dipoles are free to rotate, this torque tends to align them parallel to the electric field direction. The alignment of the molecular/atomic dipoles, whether in polar or non-polar materials, caused by an applied electric field is called *polarization*. Figure 9.1 shows the electric field produced by a dipole which indicates that the electric field produced by the aligned dipoles in a dielectric opposes the applied field. An auxiliary field $\mathbf{D}$ called the *electric displacement* is therefore introduced

$$\mathbf{D} = \varepsilon_0\mathbf{E} + \mathbf{P}. \tag{9.2}$$

$\mathbf{P}$ is the *polarization* of the material, defined as the electric dipole moment per unit volume $N\mathbf{p}$, N being the number density of the electric dipoles. The materials considered in this chapter are homogeneous, isotropic and at sufficiently low intensity of the radiation so that $\mathbf{P}$ is linearly proportional to $\mathbf{E}$, namely

$$\mathbf{P} = \varepsilon_0\chi\mathbf{E}. \tag{9.3}$$

Thus

$$\mathbf{D} = \varepsilon_0(1 + \chi)\mathbf{E} = \varepsilon_0\,\varepsilon_{\mathrm{r}}\,\mathbf{E}, \tag{9.4}$$

where χ and ε_{r} are scalar constants for a given HIL material, and are respectively the *electric susceptibility* and the *relative permittivity*.[1] An *atomic/molecular susceptibility* is also defined

$$\alpha = p/(\varepsilon_0 E) = P/(N\varepsilon_0 E). \tag{9.5}$$

In general magnetic dipoles are induced in any material by an applied magnetic field. There is thus an auxiliary magnetic field defined by

$$\mathbf{H} = \mathbf{B}/\mu_{\mathrm{r}}\mu_0, \tag{9.6}$$

where μ_{r} is called the *relative permeability* of the material. Equations 9.4 and 9.6 are called the *constitutive* relations for a material. For the most part the effects of the magnetic dipoles are unimportant at optical frequencies so we can generally take $\mu_{\mathrm{r}} = 1$ in all that follows.

Maxwell's equations encapsulate the understanding of electromagnetism that had been achieved by Faraday and other experimenters in the early part of the 19th century. The integral forms are

$$\int_S \mathbf{D} \cdot d\mathbf{S} = \int_V \rho\,dV, \tag{9.7}$$

$$\int_S \mathbf{B} \cdot d\mathbf{S} = 0, \tag{9.8}$$

$$\oint \mathbf{E} \cdot d\mathbf{l} = -\int_S (\partial\mathbf{B}/\partial t) \cdot d\mathbf{S}, \tag{9.9}$$

$$\oint \mathbf{H} \cdot d\mathbf{l} = \int_S \mathbf{j} \cdot d\mathbf{S} + \int_S (\partial\mathbf{D}/\partial t) \cdot d\mathbf{S}, \tag{9.10}$$

where ρ and $\mathbf{j}$ are the free charge density per unit volume and free current density per unit area respectively. In the first two equations the surface

[1] The question of what $\mathbf{D}$ physically represents is not spelled out anywhere that I have looked. Feynman who possessed a very penetrating insight concluded that it is simply a useful tool.

integrals are over a closed surface and the volume integrals are over the volume enclosed. In the last two equations the line integrals are over a closed path and the surface integrals over any surface spanning that closed loop. The first of Maxwell's equations is known as Gauss' law of electrostatics and relates the flux of the displacement current through a closed surface to the total charge inside the volume. It is equivalent to Coulomb's law. The second equation is its magnetic counterpart, and the zero on the right hand side expresses the fact that free magnetic poles (monopoles) have never been detected. The last two equations are Faraday's law and the Ampere–Maxwell law. Faraday's law relates the *effective* electric potential around a closed loop to the rate of change of magnetic flux through that loop, written here for the case that the loop is stationary. Analogous to this the Ampere–Maxwell law relates the magnetic field integrated round a closed loop to the sum of the free and *displacement* currents through that loop. The differential forms of the equations are obtained by applying two theorems of vector calculus which hold for any vector **Z**. These are Gauss' theorem and Stokes' theorem, respectively

$$\int_S \mathbf{Z} \cdot d\mathbf{S} = \int_V \nabla \cdot \mathbf{Z} \, dV, \tag{9.11}$$

$$\oint \mathbf{Z} \cdot d\mathbf{l} = \int_V (\nabla \wedge \mathbf{Z}) \cdot d\mathbf{S}. \tag{9.12}$$

For example applying the latter to Faraday's law gives

$$\int (\nabla \wedge \mathbf{E}) \cdot d\mathbf{S} = - \int_S (\partial \mathbf{B}/\partial t) \cdot d\mathbf{S},$$

which is true for any surface so we have

$$\nabla \wedge \mathbf{E} = -\partial \mathbf{B}/\partial t,$$

which relates fields at a single place. The differential forms of Maxwell's equations[2] are therefore

$$\nabla \cdot \mathbf{D} = \rho \tag{9.13}$$
$$\nabla \cdot \mathbf{B} = 0 \tag{9.14}$$
$$\nabla \wedge \mathbf{E} = -\partial \mathbf{B}/\partial t \tag{9.15}$$
$$\nabla \wedge \mathbf{H} = \mathbf{j} + \partial \mathbf{D}/\partial t. \tag{9.16}$$

In free space in the absence of any charges Maxwell's equations reduce to

$$\nabla \cdot \mathbf{E} = 0 \tag{9.17}$$
$$\nabla \cdot \mathbf{B} = 0 \tag{9.18}$$
$$\nabla \wedge \mathbf{E} = -\partial \mathbf{B}/\partial t \tag{9.19}$$
$$\nabla \wedge \mathbf{B} = \mu_0 \, \varepsilon_0 \, \partial \mathbf{E}/\partial t. \tag{9.20}$$

[2] Maxwell's equations are consistent with the special theory of relativity. When the observer shifts to a different inertial frame the Lorentz coordinate transformations convert Maxwell's equations to an identical set of equations where the fields are now those measured in the new frame. It had seemed before Einstein's discovery of the special theory an odd quirk that the corresponding Galilean transformations of Newtonian mechanics did not reproduce Maxwell's equations in the new inertial frame. These matters are discussed in Chapter 11 of the third edition of J.D. Jackson's *Classical Electrodynamics* published by John Wiley and Sons, New York (1998).

Materials, mainly metals, which contain large numbers of electrons that are not bound to individual atoms, but are free to carry current when an external electromagnetic field is applied, are good electrical conductors, while those materials which contain very few free electrons are called *dielectrics*. Electrical conductivity determines the current density through the relation

$$\mathbf{j} = \sigma \mathbf{E}, \tag{9.21}$$

where the range of values that the conductivity σ in $\Omega^{-1}\,m^{-1}$ can take is huge: copper, an excellent conductor, has a conductivity of $6.45\,10^7\,\Omega^{-1}\,m^{-1}$ while a dielectric such as glass has a conductivity $10^{-12}\,\Omega^{-1}\,m^{-1}$. In the right hand side of eqn. 9.16 the second term is the *displacement current* whose size will depend on the polarization of the material and, because of the time differential, on the frequency. There is such a big difference between the relative importance of the conduction and displacement currents in conductors on the one hand, and in dielectrics (and free space) on the other hand, that it is helpful to separate the discussion of these two classes. The discussion of the behaviour of waves in dielectrics will begin in this chapter and the discussion of the behaviour of electromagnetic waves passing through conductors is postponed until Chapter 11.

Boundary conditions

It is essential to be able to connect the values of the electromagnetic fields in one medium to those in another at any interface. Making the connection requires the use of the integral equations and not the differential forms because the latter refer to fields at the same place. Figure 9.2 shows projections of a plane surface between materials 1 and 2. The broken lines enclose a volume of pillbox shape which straddles the surface, the flat faces being of unit area and lying a negligible distance from the interface. The actual interface is supposed to carry a surface charge of $Q\,C\,m^{-2}$. Then applying Gauss' law to this volume gives

$$D_{2n} - D_{1n} = Q, \tag{9.22}$$

where D_{2n} (D_{1n}) is the component of $\mathbf{D}$ perpendicular to the surface in material 2(1). Similarly Gauss' magnetic law gives

$$B_{2n} - B_{1n} = 0, \tag{9.23}$$

relating the normal components of the magnetic field. Next consider the broken line closed path shown straddling the interface in figure 9.3. Here there is assumed to be a surface current density of $j_s\,A\,m^{-1}$ travelling along the interface, perpendicular to, and into the plane of the diagram. The long arms are of unit length and they are a negligible distance from the interface. Applying the Maxwell–Ampere law to this path gives

$$H_{1t} - H_{2t} = j_s, \tag{9.24}$$

where H_{1t} (H_{2t}) is the tangential component of the magnetic field $\mathbf{H}_1$ $(\mathbf{H}_2)$ at the interface. Applying Faraday's law to the same circuit gives

$$E_{1t} - E_{2t} = 0. \tag{9.25}$$

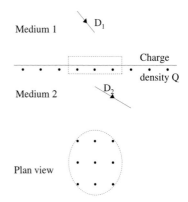

Fig. 9.2 Integration pillbox volume straddling an interface between media.

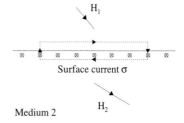

Fig. 9.3 Integration loop straddling an interface between media.

Consequently when, as is usually the case for dielectrics, there is no surface charge or current, the tangential components of $\mathbf{E}$ and $\mathbf{H}$, and the normal components of $\mathbf{D}$ and $\mathbf{B}$ are all *continuous* at such an interface.

9.3   The wave equation

Maxwell's equations lead simply to wave equations for electric and magnetic fields. Consider the example of free space first. Taking the curl of eqn. 9.19, and then using eqn. 9.20 gives

$$\nabla \wedge (\nabla \wedge \mathbf{E}) = -\partial/\partial t(\nabla \wedge \mathbf{B}) = -\mu_0 \, \varepsilon_0 \, \partial^2 \mathbf{E}/\partial t^2. \tag{9.26}$$

The identity that is valid for any vector field $\mathbf{X}$,

$$\nabla \wedge (\nabla \wedge \mathbf{X}) = \nabla(\nabla \cdot \mathbf{X}) - \nabla^2 \mathbf{X}, \tag{9.27}$$

when applied to $\mathbf{E}$ gives

$$\nabla \wedge (\nabla \wedge \mathbf{E}) = -\nabla^2 \mathbf{E},$$

because of eqn. 9.17. Then eqn. 9.26 can be rewritten

$$\nabla^2 \mathbf{E} = \mu_0 \, \varepsilon_0 \, \partial^2 \mathbf{E}/\partial t^2. \tag{9.28}$$

Starting from eqn. 9.20 a similar set of steps gives

$$\nabla^2 \mathbf{B} = \mu_0 \, \varepsilon_0 \, \partial^2 \mathbf{B}/\partial t^2. \tag{9.29}$$

These last two equations are wave equations; and it was shown in Chapter 1 that there are plane wave solutions. A suitable sinusoidal plane wave solution is

$$\mathbf{E} = \mathbf{E}_0 \exp\left[i(\mathbf{k} \cdot \mathbf{r} - \omega t)\right], \tag{9.30}$$

whose real part is the actual electric field, with $\mathbf{E}_0$ having Cartesian components (E_{0x}, E_{0y}, E_{0z}). One reason for using a complex form is that many mathematical manipulations will be simpler. Thus

$$
\begin{aligned}
\nabla \cdot \mathbf{E} &= \partial E_x/\partial x + \partial E_y/\partial y + \partial E_z/\partial z \\
&= \{E_{0x}\partial/\partial x + E_{0y}\partial/\partial y + E_{0z}\partial/\partial z\}\{\exp\left[i(+k_x x + k_y y + k_z z - \omega t)\right]\} \\
&= \{ik_x E_{0x} + ik_y E_{0y} + ik_z E_{0z}\}\{\exp\left[i(k_x x + k_y y + k_z z - \omega t)\right]\} \\
&= +i\mathbf{k} \cdot \mathbf{E}.
\end{aligned}
$$

Similarly

$$\nabla \wedge \mathbf{E} = +i\mathbf{k} \wedge \mathbf{E}, \quad \nabla^2 \mathbf{E} = -k^2 \mathbf{E},$$

while

$$\partial \mathbf{E}/\partial t = -i\omega \mathbf{E}, \quad \partial^2 \mathbf{E}/\partial t^2 = -\omega^2 \mathbf{E}.$$

Substituting the calculated differentials into eqn. 9.28 or eqn. 9.29 gives

$$k^2 = \mu_o \, \varepsilon_0 \, \omega^2.$$

Consequently both the wave equations will be satisfied provided the wave velocity

$$c = \omega/k = 1/\sqrt{\mu_0\,\varepsilon_0}. \tag{9.31}$$

It was pointed out in Chapter 1 that the right hand side of this equation was found to equal the velocity of light to within the experimental error. This equality established at a stroke that light is one form of electromagnetic radiation. According to the special theory of relativity the velocity of electromagnetic radiation in free space is constant irrespective of the motion of source or observer. The value of c was therefore fixed in 1984, by convention, at the experimental value it had at that time

$$c \equiv 299\,792\,458\ \mathrm{m\,s^{-1}}. \tag{9.32}$$

This leaves the units of length and time to be defined in a way consistent with this requirement.

Substituting the complex solution 9.30 into eqns. 9.17, 9.18 and 9.19 gives

$$i\mathbf{k} \cdot \mathbf{E} = 0, \tag{9.33}$$
$$i\mathbf{k} \cdot \mathbf{B} = 0, \tag{9.34}$$
$$\mathbf{B} = \mathbf{k} \wedge \mathbf{E}/\omega = \hat{\mathbf{k}} \wedge \mathbf{E}/c, \tag{9.35}$$

where $\hat{\mathbf{k}}$ is the unit vector $\mathbf{k}/k$. These results justify the statements made in Chapter 1 that $\mathbf{E}$, $\mathbf{B}$ and $\mathbf{k}$ form a right-handed set of orthogonal vectors for uniform plane electromagnetic waves travelling in free space: the waves are transversely polarized.

The preceding analysis requires very few changes when it is extended to study electromagnetic waves in HIL dielectrics. These are materials with high electrical resistance so there are essentially no free charges or currents, but they can be polarized by an applied electric field. Equations 9.13–9.16 with $\rho = \mathbf{j} = 0$ and $\mu_\mathrm{r} = 1.0$ yield a wave equation

$$\nabla^2\mathbf{E} = \mu_0\,\varepsilon_0\,\varepsilon_\mathrm{r}\,\partial^2\mathbf{E}/\partial t^2, \tag{9.36}$$

with a parallel equation for the magnetic field. A sinusoidal plane wave solution is

$$\mathbf{E} = \mathbf{E}_0\cos\left[\mathbf{k} \cdot \mathbf{r} - \omega t\right] = \mathcal{R}e\,\mathbf{E}_0\exp\left[i(\mathbf{k} \cdot \mathbf{r} - \omega t)\right]. \tag{9.37}$$

and the wave velocity is

$$v = \omega/k = 1/\sqrt{\mu_0\,\varepsilon_0\,\varepsilon_\mathrm{r}} = c/\sqrt{\varepsilon_\mathrm{r}}. \tag{9.38}$$

The magnetic field is

$$\mathbf{B} = \mathbf{k} \wedge \mathbf{E}/\omega = \hat{\mathbf{k}} \wedge \mathbf{E}/v. \tag{9.39}$$

Substituting the plane wave solution into the first two of Maxwell's equations shows that $\mathbf{E}$, $\mathbf{B}$ and $\mathbf{k}$ again form an orthogonal set. We can calculate the refractive index, n, from the wave velocity

$$n = c/v = \sqrt{\varepsilon_{\mathrm{r}}}. \tag{9.40}$$

It is worth noting that the relative permittivity can change a great deal with the wavelength of electromagnetic waves. For example water has a relative permittivity of 80 at low frequencies, while at optical wavelengths the refractive index of water is 1.33. Water molecules have an intrinsic dipole moment, so evidently the time required to get these molecular dipoles to align with the applied field is much longer than the period of light oscillations. This variation of wave velocity with wavelength is called *dispersion*. Glass and other materials used in passive optical components are all HIL dielectrics. A *characteristic impedance* is defined for any material as the ratio

$$Z = E/H, \tag{9.41}$$

which gives in the case of dielectrics

$$Z = \sqrt{(\mu_0/\varepsilon_0\,\varepsilon_{\mathrm{r}})} = \mu_0 c/n. \tag{9.42}$$

[3]This quantity was previously mentioned in Chapter 1.

In the case of free space the characteristic impedance[3] $Z_0 = 377\,\Omega$.

When the values of the fields in a plane wave are inserted into the expression for the Lorentz force eqn. 9.1, the electric and magnetic contributions are of magnitude qE and qEv/c respectively. Electron velocities in matter are much less than c so that it is generally adequate to ignore the magnetic force in a dielectric.

The relative importance of conduction and displacement currents for a sinusoidal plane wave can be inferred by comparing $\mathbf{j} = \sigma\mathbf{E}$ and $\partial\mathbf{D}/\partial t = -i\omega\varepsilon_0\varepsilon_{\mathrm{r}}\mathbf{E}$. The ratio of their magnitudes is

$$R_{\mathrm{c/d}} = \sigma/\omega\varepsilon_0\varepsilon_{\mathrm{r}}. \tag{9.43}$$

For glass the ratio, $0.113/\varepsilon_{\mathrm{r}}\omega$, is small at even low frequencies so it is appropriate to ignore the conduction current for dielectrics. However, in the case of copper this ratio becomes $0.728\,10^{19}/\omega$ which remains large up to optical frequencies.

9.3.1   Energy storage and energy flow

The energy stored in a capacitor at constant voltage and in an inductance carrying constant current provide simple examples of energy storage in electric and magnetic fields respectively. These cases will be considered here and the general expression for the total energy stored in any electromagnetic field will then be (plausibly) inferred.

The energy stored in a capacitor of capacitance C at a voltage V is $CV^2/2$. Assuming the capacitor has parallel plates, each of area A and a distance d apart, then ignoring edge effects $C = \varepsilon_0 \varepsilon_r A/d$. Also the electric field $E = V/d$. Hence the energy stored in the capacitor is

$$(\varepsilon_0 \varepsilon_r A/d)(Ed)^2/2 = \varepsilon_0 \varepsilon_r (Ad)E^2/2 = \mathbf{E} \cdot \mathbf{D}(Ad)/2,$$

and the energy stored per unit volume is $\mathbf{E} \cdot \mathbf{D}/2$. The energy stored in the magnetic field within a solenoid of inductance L carrying a current I is $LI^2/2$. Suppose the solenoid has area of cross-section A, length d and carries m turns per unit length. Then $L = \mu_0 m^2 Ad$, while $B = \mu_0 mI$. Consequently the energy stored in the solenoid's magnetic field is, ignoring end effects,

$$[B^2/\mu_0] (Ad)/2 = \mathbf{B} \cdot \mathbf{H}(Ad)/2,$$

so that the energy density in the magnetic field is $\mathbf{B} \cdot \mathbf{H}/2$. Thus the total energy density in an electromagnetic field is

$$U = (\mathbf{E} \cdot \mathbf{D} + \mathbf{B} \cdot \mathbf{H})/2. \tag{9.44}$$

Energy flow in electromagnetic radiation is a vector quantity $\mathbf{N}$ with units $\mathrm{W\,m^{-2}}$: in other words it is the power crossing unit surface area. Maxwell's equations do not in themselves determine this flow. Poynting realized that the extra ingredient was to apply conservation of energy to a volume of electromagnetic field. Take a volume V with surface area S. The energy within V has three components whose changes must balance out to zero. Firstly there is the change in stored energy per unit time $\partial/\partial t(\int_V U dV)$. Secondly there is the outward flow of energy per unit time through the whole surface $\int_S \mathbf{N} \cdot d\mathbf{S}$. This can be converted to a volume integral using Gauss' divergence theorem, giving $\int_V \nabla \cdot \mathbf{N} dV$. Thirdly there is the work done per unit time on whatever charges are enclosed in the volume $\int_V \mathbf{E} \cdot \mathbf{j} dV$, where $\mathbf{j}$ is the current density. In order that energy is conserved the total of these three contributions should be zero. Thus

$$\partial/\partial t \left(\int_V U dV \right) + \int_V \nabla \cdot \mathbf{N} dV + \int_V \mathbf{E} \cdot \mathbf{j} dV = 0.$$

This balance must be true for any volume so that

$$\partial U/\partial t + \nabla \cdot \mathbf{N} + \mathbf{E} \cdot \mathbf{j} = 0. \tag{9.45}$$

We now need to use Maxwell's equations for media. The scalar product of $\mathbf{E}$ with eqn. 9.16 gives

$$\mathbf{E} \cdot \mathbf{j} = \mathbf{E} \cdot (\nabla \wedge \mathbf{H} - \partial \mathbf{D}/\partial t). \tag{9.46}$$

Next we apply an identity valid for any pair of vectors to $\mathbf{E}$ and $\mathbf{H}$

$$\mathbf{E} \cdot (\nabla \wedge \mathbf{H}) = \nabla \cdot (\mathbf{H} \wedge \mathbf{E}) + \mathbf{H} \cdot (\nabla \wedge \mathbf{E})$$
$$= \nabla \cdot (\mathbf{H} \wedge \mathbf{E}) - \mathbf{H} \cdot \partial \mathbf{B}/\partial t,$$

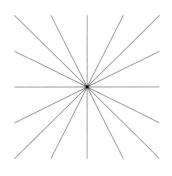

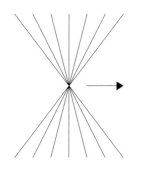

Fig. 9.4 Electric field lines: in the upper panel around a charge at rest and in the lower panel around a charge moving at constant velocity.

where eqn. 9.15 was used in replacing the second term on the right hand side. Substituting this result in eqn. 9.46 gives

$$\mathbf{E} \cdot \mathbf{j} = \nabla \cdot (\mathbf{H} \wedge \mathbf{E}) - \mathbf{E} \cdot (\partial \mathbf{D}/\partial t) - \mathbf{H} \cdot (\partial \mathbf{B}/\partial t).$$

Rearranging this equation with the help of eqn. 9.44

$$\partial U/\partial t + \nabla \cdot (\mathbf{E} \wedge \mathbf{H}) + \mathbf{E} \cdot \mathbf{j} = 0. \tag{9.47}$$

Comparing eqns. 9.47 with 9.45 allows us to identify the energy flow vector as

$$\mathbf{N} = \mathbf{E} \wedge \mathbf{H}, \tag{9.48}$$

which is known as the *Poynting vector*. In the case of a sinusoidal plane wave the actual (real) fields have Cartesian components

$$E_x = E_0 \cos(kz - \omega t)$$
$$B_y = (E_0/v) \cos(kz - \omega t)$$

which travel at velocity v and the magnitude of the Poynting vector is

$$N = [E_0^2/(\mu_0 v)] \cos^2(kz - \omega t),$$

along the z-direction. The time average of the Poynting vector taken over many cycles of the electromagnetic wave is

$$\overline{N} = E_0^2/(2\mu_0 v) = E_0^2/2Z. \tag{9.49}$$

Similarly the energy density is

$$U = \frac{1}{2}(\varepsilon_0 \varepsilon_{\mathrm{r}} E_0^2 + E_0^2/\mu_0 v^2) \cos^2(kz - \omega t)$$
$$= (E_0^2/\mu_0 v^2) \cos^2(kz - \omega t), \tag{9.50}$$

with a time average

$$\overline{U} = E_0^2/(2\mu_0 v^2). \tag{9.51}$$

The time averaged energy density and energy flow are necessarily very closely related; from eqns. 9.49 and 9.51

$$\overline{N} = v\overline{U}. \tag{9.52}$$

Electromagnetic fields also carry momentum as well as energy and it is the momentum of the radiation from the Sun that deflects comets' tails so that they point radially away from the Sun. This momentum can be determined by considering the reflection of a plane electromagnetic wave at normal incidence from the flat surface of a perfect conductor. The argument is only sketched here, but a rigorous proof can be found in Bleaney and Bleaney.[4] The magnetic field $\mathbf{H}$ is parallel to the surface in the dielectric but vanishes in the conductor so that the boundary

---

[4]Chapter 8 of the fifth edition of *Electricity and Magnetism* by B. and B. I. Bleaney, published by Oxford University Press (1983).

condition eqn. 9.24 reduces to $H = j_s$. Using eqn. 9.1 the *radiation pressure* on the surface is

$$\mathbf{P} = \mathbf{j}_s \wedge \mathbf{B} = BH\hat{\mathbf{k}}, \tag{9.53}$$

where $\hat{\mathbf{k}}$ is a unit vector perpendicular to the surface. Now the time average of the energy density in the incident waves is

$$\overline{U} = \overline{BH} = \overline{ED}.$$

Thus the relation between the radiation pressure and the energy flux in a plane electromagnetic wave is

$$\overline{\mathbf{P}} = \overline{\mathbf{N}}/v. \tag{9.54}$$

9.4 Electromagnetic radiation

The electric field of a plane electromagnetic wave, being transverse, is very different from the electric fields met in electrostatics. The upper panel of figure 9.4 shows the radial electric field of an isolated static electric charge. If the same charge is moving with constant velocity the field lines are compressed in the direction of motion as shown in the lower panel; but they remain radial. What is needed to generate travelling waves with transverse fields is for the charge to accelerate. Figure 9.5 pictures a simple example: the electric field lines are drawn for a charge that is at rest at A up till time t_0, it then moves to B and is again at rest there from time t_1. Before t_0 the lines point back to A, and after t_1 they point back to B. Between these times there are kinks in the field lines which give the electric field a transverse component. As time passes the kinks move continuously away from the charge with velocity c in free space. If the charge is instead made to oscillate between A and B the field's transverse component will alternate in direction giving waves with transverse fields oscillating at the frequency of the charge's motion. Over an area of a wavefront whose dimensions are small compared to its distance from the source the wavefront approximates to a plane wave.

The simplest source of radiation is an oscillating electric dipole. The motion is simple harmonic with the charges oscillating about their common centre. Referring to figure 9.1 the charges exchange positions every half cycle of the oscillation. As mentioned above an applied electromagnetic field acting on matter induces and aligns atomic and molecular electric dipoles. An excited atom can be classically pictured as one which has absorbed electromagnetic radiation and become a dipole in which the electron is oscillating about the much heavier nucleus. Such atomic excitations have frequencies ranging from the near infrared to the ultraviolet. Molecular excitations, which involve the vibration and oscillation of the nuclei, have frequencies in the near to far infrared region of the spectrum.The radiation from the dipole carries off energy and so the oscillation dies away: the radiation damps the dipole motion. In this

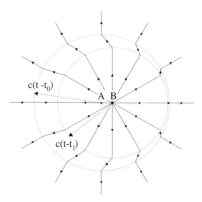

Fig. 9.5 Electric field lines from a charge at rest at A until time t_0; it then moves rapidly and comes to rest at time t_1 at B. The lines are drawn as they appear at a later time t.

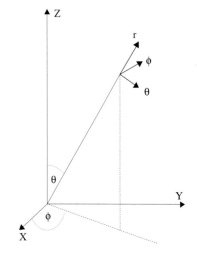

Fig. 9.6 Local spherical polar coordinates with the dipole axis as the polar axis $\theta = 0$. The r, θ and ϕ local axes point in the directions of increasing r, θ and ϕ respectively.

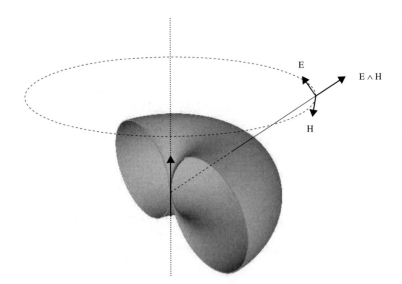

Fig. 9.7 A polar plot of the amplitude distribution of radiation from a dipole in the radiation zone. The dipole direction is shown by the central arrow. The directions of the fields and the Poynting vector are also shown at a representative point. Lines of **H** are tangential to the dotted line circle.

classical view the excited atom loses its excitation energy by radiating at its natural frequency.

At points very close to an electric dipole the field pattern is quite complex, while in the *radiation zone* at distances large compared to the wavelength the fields have a simple form. Using the local axes shown in figure 9.6 the field components in the radiation zone are

$$E_\theta = \frac{-\omega^2 p_0 \sin\theta}{4\pi\varepsilon_0 c^2 r} \cos\left(kr - \omega t\right), \tag{9.55}$$

$$B_\phi = -E_\theta/c, \tag{9.56}$$

where $p_0 \cos\omega t$ is the oscillating dipole moment and $\omega/k = c$. The power radiated is given by eqn. 9.48. The energy crossing an element **dS** of a sphere centred on the dipole per unit time is

$$\mathbf{N}\cdot\mathbf{dS} = \mathbf{E}\wedge\mathbf{H}\cdot\mathbf{dS} \propto \sin^2\theta dS. \tag{9.57}$$

This angular distribution of the radiated power is displayed in a polar diagram in figure 9.7, where the distance from the origin to the shaded surface is proportional to $\sin^2\theta$. The intensity of radiation falls off toward the axis and is zero exactly along the axis, which agrees with what is seen in figure 9.5: that there is no transverse component of the electric field along the axis through AB. Integrating over all directions gives the total power radiated:

$$
\begin{aligned}
W = \int \mathbf{N} \cdot \mathbf{dS} &= \int_0^{2\pi} \int_0^{\pi} E_\theta H_\phi r^2 \sin\theta \, \mathrm{d}\theta \, \mathrm{d}\phi \\
&= [(\omega^4 p_0^2)/(16\pi^2 \varepsilon_0 c^3)] \cos^2(kr - \omega t) \int_0^{2\pi} \mathrm{d}\phi \int_0^{\pi} \sin^3 \theta \, \mathrm{d}\theta \\
&= [(\omega^4 p_0^2)/(6\pi \varepsilon_0 c^3)] \cos^2(kr - \omega t).
\end{aligned}
$$

Averaging over time gives

$$
\overline{W} = \omega^4 p_0^2 / (12\pi \varepsilon_0 c^3) \tag{9.58}
$$

per dipole. More complicated charge distributions than dipoles are possible in an atom, such as a quadrupole which consists of a pair of dipoles oppositely aligned. Quadrupole oscillations occur and of course radiate. However the radiation from quadrupoles and other multipoles is weak compared to dipole radiation when the radiating structure is much smaller than the wavelength of the radiation – which is the case for atoms radiating light. The dominance of dipole over other more complex radiation persists in quantum theory and will be discussed later. On a larger scale the antennae of radio stations are true classical electric dipoles: they are in the form of conducting wires in which alternating currents flow.

9.5   Reflection and refraction

Electromagnetic theory is now applied to the behaviour of electromagnetic waves at interfaces between HIL dielectrics of different refractive indices n_1 and n_2. Figure 9.8 shows the wave vectors ($\mathbf{k}$) and fields at a plane interface between two dielectrics when a plane sinusoidal electromagnetic wave is incident with its electric field transverse to the plane of incidence. This is called a *transverse electric* (TE) wave, or alternatively an *s-polarized* wave. The subscripts i, r and t refer to the incident, reflected and transmitted plane waves respectively. Both the materials are HIL so that the electric field in the reflected and refracted waves will also be transverse to the plane of incidence. There are no free charges or currents. The alternative case when the magnetic fields are perpendicular to the plane of incidence is shown in figure 9.9 and bears the labels *transverse magnetic* (TM), or *p-polarization*. All the results obtained in the present section apply equally to p- as well as to s-polarized light. In the following section the case of s-polarization is calculated in detail, while for the case of p-polarization only results are stated.

The axes are oriented as shown in figure 9.8 with the surface being the plane $z = 0$, and with $x = 0$ being the plane of incidence. Along the y-axis, which lies in the surface, the plane waves reduce to the form $E = E_0 \exp[i(k_y y - \omega t)]$, where (k_x, k_y, k_z) are the components of the wave vector along the axes. The boundary conditions derived above state

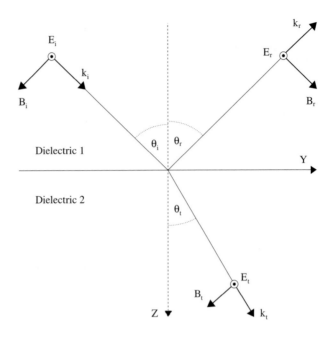

Fig. 9.8 Incident, reflected and transmitted wave vectors at a plane interface between two dielectrics for the case that the electric field is perpendicular to the plane of incidence: s-polarization.

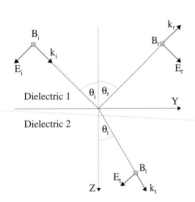

Fig. 9.9 Incident, reflected and transmitted wave vectors at a plane interface between two dielectrics for the case that the electric field lies in the plane of incidence: p-polarization.

that, provided there are no surface charges or currents, the components of **E** and **H** transverse to the surface and the components of **D** and **B** perpendicular to the surface are all continuous at the surface. This means, for example, that the transverse component of **E** just above the surface in the first medium is identical to the transverse component of **E** just below the surface in the second medium. This continuity of the transverse component of the electric field requires that

$$E_{0i} \exp\left[i(k_{iy}y - \omega_i t)\right] + E_{0r} \exp\left[i(k_{ry}y - \omega_r t)\right]$$
$$= E_{0t} \exp\left[i(k_{ty}y - \omega_t t)\right]. \tag{9.59}$$

If the fields are to match in this way at the surface for all times, it follows that

$$\omega_r = \omega_i = \omega_t,$$

which for simplicity we write ω. The above equation can be rewritten

$$v_1 k_r = v_1 k_i = v_2 k_t, \tag{9.60}$$

where v_1 and v_2 are the velocities of light in the two dielectrics. The first equality simply tells that the reflected wave has the same wave number and wavelength as the incident wave. The second equality can be rewritten again as

$$n_1 \lambda_i = n_2 \lambda_t, \tag{9.61}$$

where n_1 and n_2 are the two refractive indices. Thus it is the wavelength that changes from material to material while the frequency remains the same. The field values must equally match across the surface any point, that is at each and every value of y, hence

$$k_{ry} = k_{iy} = k_{ty}, \qquad (9.62)$$

which we write simply as k_y. Expanding the results in eqn. 9.62 gives

$$k_r \sin \theta_r = k_i \sin \theta_i,$$
$$k_t \sin \theta_t = k_i \sin \theta_i. \qquad (9.63)$$

Remembering that $k_r = k_i$ the first of these two equalities is just the law of reflection $\theta_r = \theta_i$. The second equality simplifies to Snell's law $n_1 \sin \theta_i = n_2 \sin \theta_t$.

It is worth emphasizing these laws are the consequence of the invariance of the component of the wave vector parallel to the surface under reflection or refraction.[5]

This analysis also provides an explanation of what happens in the process of total internal reflection (TIR) met earlier in Section 2.2. Then $n_1 > n_2$ and we choose an angle of incidence greater than the critical angle θ_c, so that $\sin \theta_i > n_2/n_1$. Thus eqn. 9.62 gives

$$k_y = k_i \sin \theta_i$$
$$= (n_1/n_2) k_t \sin \theta_i > k_t.$$

It follows that the component of the refracted wave vector *perpendicular* to the surface, k_{tz}, is imaginary,

$$k_{tz}^2 = k_t^2 - k_{ty}^2 < 0. \qquad (9.64)$$

Hence $k_{tz} = \pm i\kappa$ where κ is real and positive. Here the positive sign must be taken because the negative sign gives a wave whose amplitude increases exponentially in the less dense medium, which is physically unreasonable. The transmitted wave is thus

$$E_t = E_{0t} \exp(-\kappa z) \exp[i(k_y y - \omega t)]. \qquad (9.65)$$

This is an *evanescent* wave: it travels parallel to the interface, while its amplitude and intensity fall off exponentially with the distance from the surface. Its intensity

$$N = N_0 \exp(-2\kappa z) \qquad (9.66)$$

drops by a factor e over a distance $1/2\kappa$. If the incident wave is s-polarized then the magnetic field

$$\mathbf{B} = \mathbf{k} \wedge \mathbf{E}/\omega = +i\kappa E_t \mathbf{e}_y/\omega - k_y E_t \mathbf{e}_z/\omega,$$

which has a component along the y-direction and hence the evanescent wave is *not* a transverse wave. Similarly when the incident wave is p-polarized the electric field of the evanescent wave has a component along

[5] When either dielectric is inhomogeneous or anisotropic the calculation of light paths starts from this invariance property, but may lead to modifications of the laws of reflection and refraction. Optical paths in anisotropic dielectrics are calculated in the following chapter, while the photonic crystal fibres met in Chapter 16 are examples of inhomogeneous dielectrics.

the direction of travel.

In optical fibre an evanescent wave travels parallel to the fibre axis within the optically less dense cladding which surrounds the optically denser core. We can investigate qualitatively the fraction of total power travelling in the evanescent wave in single mode fibres. For values of θ_i above the critical angle

$$\kappa^2 = k_y^2 - k_t^2 = k_t^2[(n_1/n_2)^2 \sin^2 \theta_i - 1], \tag{9.67}$$

and if θ is close to $90°$

$$\kappa^2 \approx k_t^2(n_1^2 - n_2^2)/n_2^2.$$

Inserting into this equation the parameters for Corning SMF-28™ monomode optical fibre, namely $n_{core} = 1.4677$, $n_{cladding} = 1.4624$, at a wavelength $\lambda = 1.310\,\mu m$, gives

$$\kappa^2 \approx 0.3572\,\mu m^{-2}.$$

Then κ is $0.6\,\mu m^{-1}$ and the intensity falls by a factor e^2 in a distance of around $1.67\,\mu m$. Now the core radius is only $4.1\,\mu m$, whence it follows that the cross-sectional area of the cladding within this $1/e^2$ zone is $\pi(5.77^2 - 4.1^2)$, that is $51.8\,\mu m^2$, which is comparable to the area of the core itself, $52.8\,\mu m^2$. Consequently a significant fraction of the electromagnetic radiation through a monomode fibre travels within the cladding.

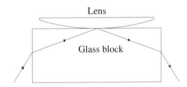

Fig. 9.10 Arrangement for observing the effects of frustrated total internal reflection.

Figure 9.10 illustrates a simple way in which to observe the effect of frustrated total internal reflection. A planoconvex lens of radius of curvature 50 to 100 cm is placed on an optical flat. Viewing as shown in the diagram, at an angle such that the angle of incidence at the glass block/air interface is significantly larger than the critical angle, a dark patch will be visible at the point of contact. Light in the evanescent wave penetrates into the gap and is then reflected at the lens surface. Then because there is a phase difference of π between the reflections from air/glass and a glass/air interfaces the two reflections interfere destructively at the point of contact. Applying eqn. 9.67 for light of 500 nm incident at $45°$ on a glass/air interface, the penetration depth is only 80 nm. Thus the penetration depth is much less than the separation between the lens and the glass block at which the first of Newton's rings appear and therefore none will be seen.

At X-ray wavelengths the refractive index of metals is slightly less than unity, which means that X-rays incident in free space at close to grazing incidence on a metal are totally internally reflected. The design of the XMM-Newton X-ray telescope, launched into space in 1999, makes use of this property. It consists of cylindrically symmetric thin alumininium mirrors with a 250 nm gold coating on their internal surfaces. An axial section of one mirror pair is pictured in figure 9.11; the first mirror is

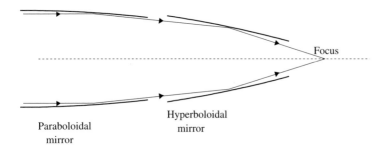

Fig. 9.11 XMM-Newton X-ray telescope using TIR. The degree of focusing is exaggerated: the actual focal length is about ten times the combined length of the mirrors.

a shallow paraboloid, and the second a shallow hyperboloid. Together they reflect and focus the X-rays from any source located close to the mirrors' axis. One such mirror pair would not collect many X-rays so 58 coaxial mirror pairs of graded diameters are arranged concentrically around one another. The outermost mirror has diameter 70 cm, and the innermost 31 cm diameter; the total length of the mirror assembly is 60 cm and the focal length measured from the paraboloid/hyperboloid junction is 7.5 m.[6]

9.6   Fresnel's equations

The analysis of reflection and refraction using electromagnetic theory is now continued in order to obtain the complex amplitudes of the reflected and refracted waves at the interface considered in figure 9.8. The exponential factors in equation 9.59 were shown to be equal so that the condition for continuity at the interface reduces to

$$E_{0i} + E_{0r} = E_{0t}. \tag{9.68}$$

Correspondingly, continuity of the tangential component of the magnetic field across the interface in figure 9.8 for s-polarization requires that

$$H_{0i} \cos \theta_i - H_{0r} \cos \theta_i = H_{0t} \cos \theta_t. \tag{9.69}$$

Making the substitution $H = En/(\mu_0 c)$ from eqn. 9.42 this becomes

$$n_1 E_{0i} \cos \theta_i - n_1 E_{0r} \cos \theta_i = n_2 E_{0t} \cos \theta_t. \tag{9.70}$$

Equations 9.68 and 9.70 can be solved simultaneously to give

$$r_s = E_{0r}/E_{0i} = (n_1 \cos \theta_i - n_2 \cos \theta_t)/(n_1 \cos \theta_i + n_2 \cos \theta_t), \tag{9.71}$$

$$t_s = E_{0t}/E_{0i} = 2n_1 \cos \theta_i/(n_1 \cos \theta_i + n_2 \cos \theta_t). \tag{9.72}$$

These results are called *Fresnel's equations* and apply for s-polarization. Fresnel's equations for p-polarization can be obtained in a similar man-

[6]In the following chapter the situation in which the refractive index can be less than unity will be discussed further. For the present take note that this does not imply that electromagnetic radiation or information can ever travel faster than *c*. Briefly the reason is that the distance electromagnetic waves can penetrate the metal is negligible.

ner and turn out to be significantly different:

$$r_p = E_{0r}/E_{0i} = (n_2 \cos \theta_i - n_1 \cos \theta_t)/(n_1 \cos \theta_t + n_2 \cos \theta_i), \quad (9.73)$$
$$t_p = E_{0t}/E_{0i} = 2n_1 \cos \theta_i/(n_1 \cos \theta_t + n_2 \cos \theta_i). \quad (9.74)$$

Figure 9.12 shows the reflected and transmitted amplitudes for both types of polarization when light is incident in air on glass of refractive index 1.5.

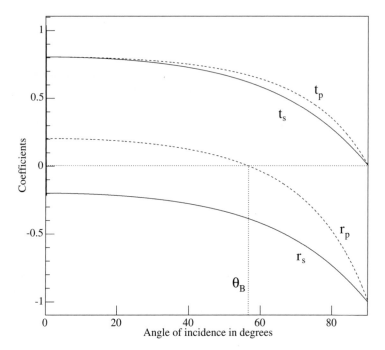

Fig. 9.12 Variation of the reflection and transmission amplitude coefficients for s/TE and p/TM polarized light as a function of the angle of incidence at a plane boundary between dielectrics in the case that light is incident in the optically less dense medium. The two refractive indices are 1.0 and 1.5.

Several comments can be made about the signs of the coefficients given by Fresnel's equations which illustrate the importance of associating a diagram of the field vectors with Fresnel's equations. First note that the sign of r_p can be reversed if in figure 9.9 the direction chosen as positive for the reflected electric field is reversed; that is the arrow labelled $\mathbf{E}_r$ is reversed. The effect of the choice of this direction becomes obvious at normal incidence, which is illustrated in figure 9.13. In the case of s-polarization the $\mathbf{E}_i$ and $\mathbf{E}_r$ are parallel, but point in opposite directions for p-polarization. This seems strange because at normal incidence the s- and p-polarizations are indistinguishable. However if we refer to figure 9.12 we find that at normal incidence $r_p = -r_s$, so that the reflected electric fields do in fact point in the same direction for s- and p-polarization at normal incidence.[7]

The relative intensities of the reflected and transmitted beams are determined by the fluxes in the beams over unit area of the interface. The time averaged absolute fluxes per unit *surface* area of the interface are given by

$$F = \overline{N} \cos\theta = E_0^2 \cos\theta / 2Z. \qquad (9.75)$$

where Z is the impedance of the material, which in a dielectric is given by eqn. 9.42. The *reflectance* and *transmittance* are defined as ratios of the flux of radiation leaving a surface area divided by the flux incident over the same surface area, thus

$$R_{p/s} = (F_r/F_i)_{p/s} = |r_{p/s}|^2, \qquad (9.76)$$
$$T_{p/s} = (F_t/F_i)_{p/s} = |t_{p/s}|^2 (Z_1 \cos\theta_t / Z_2 \cos\theta_i). \qquad (9.77)$$

Figure 9.14 shows the reflectances for the air/glass interface. At the angle of incidence θ_B, at which the reflected amplitude for p-polarized waves changes sign the reflected intensity vanishes. This angle is called *Brewster's angle*. From eqn. 9.73 it follows that

$$n_2 \cos\theta_B - n_1 \cos\theta_t = 0.$$

After using Snell's law to replace θ_t, this equation reduces to $\sin^2\theta_B = n_2^2/(n_1^2 + n_2^2)$. Then comparing this result with the identity $\sin^2\theta_B = 1/(1 + \cot^2\theta_B)$ gives

$$\tan\theta_B = n_2/n_1. \qquad (9.78)$$

If light is incident on a pile of glass plates at Brewster's angle then each plate reflects a fraction of the s-polarized light, while the light with p-polarization is transmitted. After passing through about ten plates the remaining s-polarized light is very weak. Therefore the beam reflected at Brewster's angle by a stack of ten plates is of course s-polarized, and additionally the transmitted beam is p-polarized.

At normal incidence the reflectance and transmittance are relatively easy to calculate because all the angles are $\pi/2$. For both polarizations

$$R_0 = (n_1 - n_2)^2/(n_1 + n_2)^2. \qquad (9.79)$$

If there is little absorption the transmittance is

$$T_0 = 1 - R_0 = 4n_1 n_2/(n_1 + n_2)^2. \qquad (9.80)$$

A useful reference quantity is the reflectance for a glass/air or air/glass interface at normal incidence. This is 4% for glass of refractive index 1.5, and consequently a sheet of glass reflects close to 8% of light incident normally.

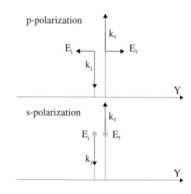

Fig. 9.13 Electric vectors in the incident and reflected waves at normal incidence. Upper panel for p-polarization and lower panel for s-polarization.

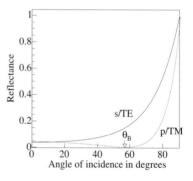

Fig. 9.14 Reflectance for both s- and p-polarization when light is incident in air on a plane glass surface. The glass has refractive index 1.5.

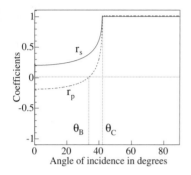

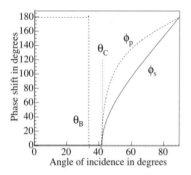

Fig. 9.15 The variation of the reflection amplitude coefficient with the angle of incidence on a plane interface between two dielectrics for both s- and p-polarization in the case that the light is incident in the optically denser dielectric. The refractive indices are chosen to be 1.5 and 1.0. In the upper panel the magnitudes of the coefficients are shown and in the lower panel their phases relative to the incident radiation. θ_B is Brewster's angle and θ_c is the critical angle.

When $\sin \theta_t$ is greater than unity (TIR) the reflection and transmission coefficients become complex, which has interesting physical consequences. Then $\cos \theta_t$ is purely imaginary so we set

$$\cos \theta_t = -i\chi, \qquad (9.81)$$

where χ is real and positive, and the negative sign is chosen to be consistent with eqn. 9.65. With the use of Snell's law this gives

$$\chi = \sqrt{(n_1/n_2)^2 \sin^2 \theta_i - 1}. \qquad (9.82)$$

Then substituting for $\cos \theta_t$ in Fresnel's equations

$$r_s = (n_1 \cos \theta_i + in_2\chi)/(n_1 \cos \theta_i - in_2\chi) \qquad (9.83)$$
$$t_s = 2n_1 \cos \theta_i/(n_1 \cos \theta_i - in_2\chi) \qquad (9.84)$$
$$r_p = (n_2 \cos \theta_i + in_1\chi)/(n_2 \cos \theta_i - in_1\chi) \qquad (9.85)$$
$$t_p = 2n_1 \cos \theta_i/(n_2 \cos \theta_i - in_1\chi). \qquad (9.86)$$

Both reflection coefficients have the form $\exp(i\alpha)/\exp(-i\alpha)$, that is $\exp(2i\alpha)$, which indicates that the incident light is fully reflected with a phase shift of 2α. This result confirms what was stated earlier: the transmitted waves only penetrate a short distance into the less optically dense material and there is no continuous flow of energy perpendicular to the surface. In figure 9.15 the amplitude reflection coefficients are shown in the case of a glass/air interface with $n_1 = 1.5$. In the upper panel the magnitudes appear, and in the lower panel the phases. The reflected p-polarized wave (TM) amplitude, of course, disappears when the light is incident at Brewster's angle.

On comparing figures 9.12 and 9.15 it is apparent that the signs of r_s and r_p for glass/air at normal incidence are opposite to those for air/glass at normal incidence. This is significant in experimental situations where, for example in observing Newton's rings in reflection, there is glass touching glass in air. The two reflections, glass/air and air/glass, at the point of contact have opposite phase and interfere destructively.

Fresnel's equations are consistent with Stokes' relations obtained in Section 5.6 by applying time reversal invariance: eqns. 5.33 and 5.34 are valid in general. In the more restricted case that the coefficients are real, which excludes total internal reflection, eqns. 5.35 and 5.36 also hold true. The Stokes' relations apply equally when the interface is made up of one or more thin surface layers on for example a glass block. This fact was already taken advantage of in analysing the function of the Fabry–Perot etalon, and can be useful in situations where the components of the surface are not known or where it would be tedious to calculate directly using Fresnel's laws. Reflection from dielectric–metal surfaces involves absorption and then the Fresnel coefficients are again complex. This case is considered in Chapter 11.

9.7 Interference filters

In some modern optical applications surfaces which effectively reflect very little light are essential, and in other applications it is equally important to have a reflectance as near 100% as possible. It was noted in Section 4.5.5 that because the number of surfaces in modern camera lenses is so large these surfaces are coated to reduce the reflectance to a fraction of a percent per surface. Safety goggles for those people working with lasers must reflect effectively all the radiation around the laser wavelength and yet transmit enough light at other wavelengths to provide useful vision. High power lamps can be fitted with *cold* mirrors that are designed to reflect visible light, yet transmit the accompanying longer wavelength infrared radiation. All these effects are achieved by coating the surfaces of the optical elements with one or more layers of dielectric. The basic principle is that the individual reflections should interfere constructively/destructively to enhance/cancel their reflection amplitudes. For this reason the optical thickness of layers is often made exactly one quarter or exactly one half wavelength These coatings are known as *interference filters*. Nowadays more than one hundred layers, alternating between two dielectrics, are routinely laid down on a single optical surface. Interference filters are used widely in optoelectronics for making mirrors that selectively reflect over a narrow wavelength interval, particularly in diode laser structures, and are then called *distributed Bragg reflectors*.

The range of wavelengths of radiation which cannot propagate within the structure of the multilayer filter is called a *photonic band-gap*, analogous to the electronic band-gaps in solids. Multilayer dielectric filters will be analysed here in such a way that the ground is prepared for the discussion of band-gaps in two- and three-dimensional photonic crystals in Chapter 17.

A simple method used to reduce reflections in optical devices is to coat each glass surface with a dielectric of optical thickness one quarter wavelength and refractive index lying between that of glass and air. The exact refractive index required is obtained using Fresnel's laws. Reflection at the air–dielectric and dielectric–glass interfaces at normal incidence have amplitudes $(n_{\mathrm{diel}} - n_{\mathrm{air}})/(n_{\mathrm{diel}} + n_{\mathrm{air}})$ and $(n_{\mathrm{glass}} - n_{\mathrm{diel}})/(n_{\mathrm{glass}} + n_{\mathrm{diel}})$ respectively. In addition the light reflected at the dielectric–glass interface travels half a wavelength further than that reflected at the air-dielectric interface, giving it a phase delay of π. The total reflected amplitude is thus

$$(n_{\mathrm{diel}} - n_{\mathrm{air}})/(n_{\mathrm{diel}} + n_{\mathrm{air}}) - (n_{\mathrm{glass}} - n_{\mathrm{diel}})/(n_{\mathrm{glass}} + n_{\mathrm{diel}}).$$

If the two reflections are required to cancel then

$$(n_{\mathrm{glass}} + n_{\mathrm{diel}})(n_{\mathrm{diel}} - n_{\mathrm{air}}) = (n_{\mathrm{diel}} + n_{\mathrm{air}})(n_{\mathrm{glass}} - n_{\mathrm{diel}}),$$

which gives $n_{\mathrm{diel}}^2 = n_{\mathrm{air}} n_{\mathrm{glass}}$. If the refractive index of the glass is 1.5

then cancellation of the reflections requires a dielectic of refractive index 1.224. In order to be practical a dielectric for optical coatings must be easy to evaporate onto glass, durable and should not absorb moisture. Among the suitable dielectrics magnesium fluoride has a refractive index 1.38 quite close to the value desired in a single anti-reflection coating. A quarter wave layer of magnesium fluoride achieves an amplitude reflection coefficient of -0.12 and a reflectance of 1.4% for light at normal incidence, compared to 4% in the absence of any coating.

A deeper analysis is necessary if we wish to calculate the performance of multiple layers for light incident at any angle, taking account of the multiple reflections.

9.7.1   Analysis of multiple parallel plane layers

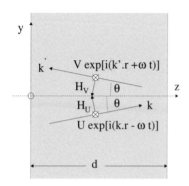

Fig. 9.16 Forward and backward travelling total waves within one layer of a filter; the filter can consist of many layers of different dielectrics. The electric fields point into the paper, with the magnetic fields pointing as shown.

In an individual layer, like that shown in figure 9.16, the backward going wave is built up from waves reflected in all later layers. This suggests that the analysis becomes very complicated with large numbers of layers. However things are made simpler once it is recalled that boundary conditions at a surface apply to the total light amplitudes arriving from all possible paths. We therefore concentrate on the total electric fields of the forward (rightward) and backward (leftward) going waves in any given layer. It will be assumed that there is negligible absorption, that is to say energy is conserved and time-reversal invariance will hold. Taking these electric fields at the left hand surface to be U and V respectively, then the corresponding electric fields at the right-hand surface are $U' = U \exp\left(+i\phi\right)$ and $V' = V \exp\left(-i\phi\right)$. The phase change across the layer is

$$\phi = 2\pi n d \cos\theta / \lambda, \tag{9.87}$$

where n is the refractive index of the dielectric, d its thickness, λ the free space wavelength and θ the angle the right-moving light beam makes with the normal to the layer's parallel faces. The relationships can be put into the general matrix form

$$\begin{bmatrix} U' \\ V' \end{bmatrix} = \mathbf{M} \begin{bmatrix} U \\ V \end{bmatrix}. \tag{9.88}$$

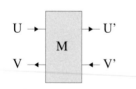

In the present case this *wave-transfer* matrix is

$$\mathbf{M} = \begin{bmatrix} \exp\left(+i\phi\right) & 0 \\ 0 & \exp\left(-i\phi\right) \end{bmatrix}. \tag{9.89}$$

The wave-transfer matrix for the effect of passage through an interface between two dielectrics will be deduced later. It is then straightforward to express the overall effect of any number of parallel layers of dielectric by an overall wave-transfer matrix

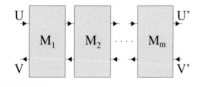

Fig. 9.17 In the upper panel the transfer matrix for crossing an interface or travelling within one layer is shown; in the lower panel the matrices for traversing a sequence of layers.

$$\mathbf{M} = \mathbf{M}_m.....\mathbf{M}_2\mathbf{M}_1, \tag{9.90}$$

where the individual matrices $\mathbf{M}_1$, $\mathbf{M}_2....\mathbf{M}_m$ describe the effect of the layers and interfaces in the order that the incident light traverses them.

The wave-transfer matrix for a single step and all steps is illustrated in figure 9.17. Travel through one layer of dielectric counts as one step and crossing an interface also counts as one step.

The effect of any number of layers and interfaces can be expressed as follows

$$U' = t_{12}U + r_{21}V', \tag{9.91}$$
$$V = t_{21}V' + r_{12}U, \tag{9.92}$$

where t and r specify the respective amplitude transmission and reflection coefficients, while the subscript $12(21)$ indicates light incident from the left (right).[8] Rearranging these equations to express U' and V' in terms of U and V gives

$$U' = (t_{12} - r_{12}r_{21}/t_{21})U + (r_{21}/t_{21})V, \tag{9.93}$$
$$V' = -(r_{12}/t_{21})U + V/t_{21}. \tag{9.94}$$

Correspondingly the general form of the wave-transfer matrix is

$$\mathbf{M} = \begin{bmatrix} t_{12}t_{21} - r_{12}r_{21} & r_{21} \\ -r_{12} & 1 \end{bmatrix} /t_{21}. \tag{9.95}$$

Stokes' relations hold

$$r_{12}^*r_{12} + t_{12}^*t_{21} = 1 \text{ and} \tag{9.96}$$
$$r_{12}^*/t_{12}^* = -r_{21}/t_{12}. \tag{9.97}$$

In addition if the incident and exit media are the same, the conservation of energy takes the following simple form because the incident and exit angles are now equal

$$r_{12}^*r_{12} + t_{12}^*t_{12} = 1. \tag{9.98}$$

It follows from comparing eqns. 9.96 and 9.98 that

$$t_{21} = t_{12}. \tag{9.99}$$

Then with the help of the last three equations eqn. 9.95 simplifies to

$$\mathbf{M} = \begin{bmatrix} 1/t_{12}^* & -r_{12}^*/t_{12}^* \\ -r_{12}/t_{12} & 1/t_{12} \end{bmatrix} \tag{9.100}$$

for such a lossless structure embedded in one medium.[9] This matrix has unit determinant. An important relationship involving t_{12} which we need later is that

$$|\text{Trace}[\mathbf{M}]|/2 = |\mathcal{R}e(1/t_{12})| = |\cos{(\alpha_{12})}|/|t_{12}|, \tag{9.101}$$

where $t_{12} = |t_{12}| \exp{(i\alpha_{12})}$.

The wave-transfer matrix for a single interface can be obtained by using Fresnel's laws to substitute for the amplitudes in eqn. 9.95. We

[8]In the case that the path is confined within one layer $t_{12} = t_{21} = \exp{(i\phi)}$, while $r_{12} = r_{21} = 0$.

[9]If the layers are left–right symmetric $r_{12} = r_{21}$, $r_{12}/t_{12} = -r_{12}^*/t_{12}^*$ and

$$\mathbf{M}_{\text{sym}} = \begin{bmatrix} 1/t^* & r/t \\ r^*/t^* & 1/t \end{bmatrix}.$$

Note that a single interface is not symmetric.

cannot use eqn. 9.100 because the incident and exit media at a single surface are necessarily different. Inserting these coefficients for TE (s-polarized radiation) gives

$$\mathbf{M} = \begin{bmatrix} u_2 + u_1 & u_2 - u_1 \\ u_2 - u_1 & u_2 + u_1 \end{bmatrix} /(2au_2). \tag{9.102}$$

Here $u_1 = n_1 \cos\theta_1$, $u_2 = n_2 \cos\theta_2$ and $a = 1$; n_1 and n_2 are the refractive indices, while θ_1 and θ_2 are the angles of incidence and refraction. In the case of TM (p-)polarized light $u_1 = n_1/\cos\theta_1$, $u_2 = n_2/\cos\theta_2$ and $a = \cos\theta_2/\cos\theta_1$. Note that it is easy to determine the angle of refraction after many layers of dielectric by repeated application of Snell's law:

$$n_j \sin\theta_j = n_{j-1} \sin\theta_{j-1} = \ldots\ldots = n_1 \sin\theta_1. \tag{9.103}$$

A structure of alternate layers of two dielectrics, each layer being one-quarter wavelength in optical thickness is commonly used, and is illustrated in figure 9.18. This particular structure produces the maximal interference because travel to and fro in each layer induces a phase change of exactly π at normal incidence. The wave-transfer matrix for propagation from A to B at normal incidence is obtained using eqns. 9.89 and 9.102,

$$
\begin{aligned}
\mathbf{M}_{\mathrm{AB}} &= \begin{bmatrix} i & 0 \\ 0 & -i \end{bmatrix} \begin{bmatrix} n_1 + n_{\mathrm{in}} & (n_1 - n_{\mathrm{in}}) \\ (n_1 - n_{\mathrm{in}}) & (n_1 + n_{\mathrm{in}}) \end{bmatrix} /2n_1 \\
&= \begin{bmatrix} i(n_1 + n_{\mathrm{in}}) & i(n_1 - n_{\mathrm{in}}) \\ -i(n_1 - n_{\mathrm{in}}) & -i(n_1 + n_{\mathrm{in}}) \end{bmatrix} /2n_1,
\end{aligned} \tag{9.104}
$$

where n_1, n_2 and n_{in} are the refractive indices of the two dielectrics and the incident medium respectively. The wave-transfer matrix for the segment B to C is similar,

$$\mathbf{M}_{\mathrm{BC}} = \begin{bmatrix} i(n_1 + n_2) & i(n_2 - n_1) \\ -i(n_2 - n_1) & -i(n_1 + n_2) \end{bmatrix} /2n_2, \tag{9.105}$$

while that for the final interface C to D is

$$\mathbf{M}_{\mathrm{CD}} = \begin{bmatrix} (n_{\mathrm{out}} + n_2) & (n_{\mathrm{out}} - n_2) \\ (n_{\mathrm{out}} - n_2) & (n_{\mathrm{out}} + n_2) \end{bmatrix} /2n_{\mathrm{out}}, \tag{9.106}$$

where n_{out} is the refractive index of the exit medium. The overall transfer matrix is

$$
\begin{aligned}
\mathbf{M} &= \mathbf{M}_{\mathrm{CD}}\mathbf{M}_{\mathrm{BC}}\mathbf{M}_{\mathrm{AB}} \\
&= \begin{bmatrix} (-n_{\mathrm{out}}n_1^2 - n_2^2 n_{\mathrm{in}}) & (-n_{\mathrm{out}}n_1^2 + n_2^2 n_{\mathrm{in}}) \\ (-n_{\mathrm{out}}n_1^2 + n_2^2 n_{\mathrm{in}}) & (-n_{\mathrm{out}}n_1^2 - n_2^2 n_{\mathrm{in}}) \end{bmatrix} \\
&\quad /(2n_1 n_2 n_{\mathrm{out}}).
\end{aligned} \tag{9.107}
$$

Comparing this result with eqn. 9.95 we can extract the overall amplitude reflection coefficient

$$r = -M_{21}/M_{22} = [(n_2/n_1)^2 - n_{\mathrm{out}}/n_{\mathrm{in}}]/[(n_2/n_1)^2 + n_{\mathrm{out}}/n_{\mathrm{in}}]. \tag{9.108}$$

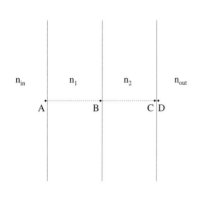

Fig. 9.18 Two layer dielectric filter. The wave paths are described in the text.

In the case of glass in air r will be reduced to zero if $n_2 = n_1\sqrt{n_{\text{glass}}/n_{\text{air}}}$. This requirement allows a flexibility in the choice of coating materials that is not possible with a single layer.

If the pair of layers is repeated s times the reflection coefficient becomes

$$r = [(n_2/n_1)^{2s} - n_{\text{out}}/n_{\text{in}}]/[(n_2/n_1)^{2s} + n_{\text{out}}/n_{\text{in}}], \qquad (9.109)$$

It is thus possible by simultaneously making the ratio (n_2/n_1) large and by using many layers to go to the other extreme and produce a coating with a reflection coefficient extremely close to unity. Magnesium fluoride and silicon dioxide with refractive indices 1.38 and 1.46 are suitable low refractive index materials, while titanium oxide and zinc sulphide are suitable materials with high refractive indices, 2.35 and 2.32 respectively.

In many applications it is important that high reflectance is maintained over a range of wavelengths and angles of incidence, and this question of range is considered next. The previous example of a lossless structure will be used again, having alternate layers of two dielectrics of refractive index n_1 and n_2. It will emerge that the crucial quantity in determining whether an incident wave is fully reflected from the stack is the trace of the transfer matrix. We first determine the wave-transfer matrix for the path AB indicated in figure 9.19 through a pair of layers at any wavelength. Using eqns. 9.89 and 9.102 we obtain, for either polarization,

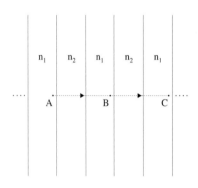

Fig. 9.19 Path through the basic cell consisting of two complete layers in a dielectric stack.

$$\mathbf{M} = \begin{bmatrix} \exp[+i\phi_1](u_2+u_1) & \exp[+i\phi_1](u_1-u_2) \\ \exp[-i\phi_1](u_1-u_2) & \exp[-i\phi_1](u_2+u_1) \end{bmatrix}$$
$$\begin{bmatrix} \exp[+i\phi_2](u_2+u_1) & \exp[+i\phi_2](u_2-u_1) \\ \exp[-i\phi_2](u_2-u_1) & \exp[-i\phi_2](u_2+u_1) \end{bmatrix}/(4u_1u_2),$$
$$(9.110)$$

where $\phi_1 = 2\pi n_1 d_1/\lambda$ and $\phi_2 = 2\pi n_2 d_2/\lambda$. In order to get the trace we need to evaluate the two diagonal elements. These are

$$M_{11} = \{+\exp[+i(\phi_1+\phi_2)](u_2+u_1)^2$$
$$- \exp[+i(\phi_1-\phi_2)](u_2-u_1)^2\}/(4u_1u_2),$$

and $M_{22} = M_{11}^*$. Then, after some manipulation, we obtain the trace

$$\text{Trace}[\mathbf{M}] = M_{11} + M_{11}^* = 2c_1c_2 - (u_1/u_2 + u_2/u_1)s_1s_2, \quad (9.111)$$

where $c_i = \cos\phi_i$ and $s_i = \sin\phi_i$. Using eqn. 9.101

$$\mathcal{R}e(1/t_{12}) = |\text{Trace}[\mathbf{M}]|/2 = |c_1c_2 - (u_1/u_2 + u_2/u_1)s_1s_2/2|. \quad (9.112)$$

Now if this quantity is greater than unity something unexpected happens. The transmission through a stack of such layers falls to zero; the wave is fully reflected and only an evanescent wave exists in the stack.

In order to demonstrate the behaviour claimed we first consider light travelling through two pairs of layers, along the path AC in figure 9.19. The wave-transfer matrix is the square of that for a single pair of layers, namely $\mathbf{M}^2$. Now $\mathbf{M}$ itself has the form specified by eqn. 9.100, so it follows that[10]

[10]$\mathbf{M}$ has the structure

$$\mathbf{M} = \begin{bmatrix} A & B \\ B^* & A^* \end{bmatrix},$$

with unit determinant $AA^* + BB^* = 1$. It follows that

$$\mathrm{Trace}[\mathbf{M}^2] = A^*A^* + 2AA^* + AA - 2$$
$$= (\mathrm{Trace}[\mathbf{M}])^2 - 2.$$

$$\mathrm{Trace}[\mathbf{M}^2]/2 = 2(\mathrm{Trace}[\mathbf{M}]/2)^2 - 1. \qquad (9.113)$$

Thus if $|\mathrm{Trace}[\mathbf{M}]|/2$ exceeds unity, then $|\mathrm{Trace}[\mathbf{M}^2]|/2$ exceeds unity by much more. With successively four, eight, sixteen,... layer pairs this modulus rapidly becomes very much greater than unity. Correspondingly the transmittance is negligible after sufficient layer pairs.

Taking the example of magnesium fluoride and titanium oxide as the two dielectrics, the maximum value of the quantity written in eqn. 9.112 is 1.15, and this maximum value increases as the *dielectric contrast* between the dielectrics increases. The reflectance of a filter made up of sixteen pairs of magnesium fluoride and titanium oxide layers all of equal optical thickness is shown in figure 9.20. Light is incident normally and λ_0 is the wavelength at which the optical thickness of the layers is exactly one-quarter wavelength. Using eqn. 9.112 the range of wavelengths $\pm\Delta\lambda$ over which radiation is totally reflected by the stack is given by

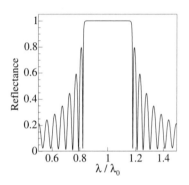

Fig. 9.20 The reflectance at normal incidence of an air/glass surface coated with a stack of sixteen pairs of magnesium fluoride and titanium oxide layers. Each layer has an optical thickness of one-quarter wavelength at wavelength λ_0. The reflectance is plotted against the wavelength divided by λ_0.

$$\Delta\lambda/\lambda_0 = (2/\pi)\sin^{-1}\left[\,(|u_2 - u_1|)/(u_2 + u_1)\,\right]. \qquad (9.114)$$

With alternate magnesium fluoride and titanium dioxide layers the range over which the reflection is total is $\pm 0.165\lambda_0$.

The range of wavelengths for which there is no transmission through a filter is an example of a *photonic band-gap*. Such band-gaps will be discussed more fully in Chapter 17, and are analogous to the electronic band-gaps described in Chapter 14.

The lobes outside the high reflectance band seen in figure 9.20 have to be suppressed in applications where a filter with a very sharp cut-off is required. Examples of this requirement will be met later in telecommunications, where data is transmitted along a single optical fibre using many laser beams of closely spaced wavelengths, maybe 0.4 nm apart. Each laser beam carries a separate stream of information from the others and has to be cleanly separated from them. The reflectance is *apodized*, that is to say the the lobes are removed, by making the refractive index variation across the filter follow the Gaussian modulation shown in the upper left hand panel of figure 9.21. The resultant reflectance as a function of wavelength is shown in the upper right hand panel. For comparison the lower panels show the refractive index variation and the reflectance variation without apodization.

Interference filters can be designed to produce a number of other subtle effects. For example a narrow transmission window can be produced

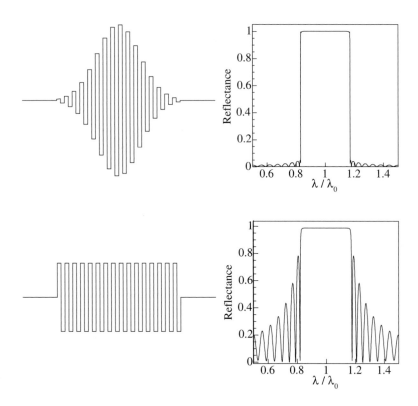

Fig. 9.21 Apodization of filters. The panels show the refractive index variation across the filter and the reflectance. In the upper panels the filter is apodized and in the lower the refractive index steps are constant.

within a broad wavelength range over which the transmission is effectively zero. Two reflective filters of the type just discussed are separated by a layer of optical thickness one half wavelength of the optically denser dielectric: the structure is symbolically $(HL)^m HH (LH)^m$, where L/H signifies a quarter wavelength of dielectric of low/high refractive index. This structure is a Fabry–Perot etalon with the two H layers forming the gap; its transmittance has very narrow width peaks at wavelengths such that $m\lambda_H = 4n_H d$, where m is an integer and n_H is the refractive index of an H layer. Thus the transmission maximum for $m = 1$ lies in the centre of the broad region of high reflectance. A filter with these characteristics is useful in picking out a narrow wavelength range while deleting all nearby spectral lines.

9.7.2 Beam splitters

Simple beam splitters are constructed by putting a multilayer coating on a glass plate, and the coating can be designed so that the reflected and transmitted beam intensities have a particular ratio, irrespective of whether the light has s- or p-polarization. One method is to sandwich a multilayer coating between the two halves of a glass cube cut diagonally as shown in figure 9.22. Useful ratios between the percentage reflected and transmitted beam intensities, for example 50/50, can be selected through the choice of a suitable sandwich design. Generally the outer faces of the beam splitter cube require an anti-reflection coating to suppress unwanted reflections.

Another possibility for splitting a beam is to deposit an iconel metal coating on a glass sheet, but this brings a bigger absorption loss. For example the reflected and transmitted intensities might both be 32%. Yet another beam splitter type consists of a coated polymer pellicle only a few microns thick supported in a circular metal frame. A 50:50 split is typical with the light incident at 45°. As a compensation for its inherent fragility a pellicle beam splitter introduces no optical aberrations. However there is interference between the reflections from the surfaces, just as in a Fabry–Perot etalon. Consequently the fraction of light transmitted/reflected oscillates as the wavelength changes by typically 5%.

The example analysed here will be for a *symmetric* dielectric coating between between glass prisms, where the angles of incidence at the coating are all 45°, and there is negligible absorption. In this case especially simple relationships hold between the reflected and transmitted amplitudes and their phases. Suppose beams are normally incident on the faces 1 and 2 with electric fields E_1 and E_2 respectively in figure 9.22. Then the symmetry of the interface requires that the emerging beams from the faces labelled 3 and 4 have electric fields respectively

$$E_3 = tE_1 + rE_2, \tag{9.115}$$
$$E_4 = rE_1 + tE_2, \tag{9.116}$$

where $r = |r| \exp(i\phi_r)$ and $t = |t| \exp(i\phi_t)$ are the respective reflection and transmission amplitude coefficients. Stokes' relation eqn. 5.33 applied to this symmetric interface yields

$$|r|^2 + |t|^2 = 1,$$

while Stokes' relation, eqn. 5.34, gives

$$rt^* + tr^* = 0.$$

In order that these terms cancel they must have opposite phase so

$$(\phi_r - \phi_t) = (\phi_t - \phi_r) \pm \pi,$$

thus finally

$$|\phi_r - \phi_t| = \pi/2. \tag{9.117}$$

Choosing $\phi_t = 0$, $\phi_r = \pi/2$, we have for this symmetric 50:50 beam splitter

$$t = 1/\sqrt{2}, \quad r = i/\sqrt{2}. \tag{9.118}$$

This illustrates the general property for symmetric beam splitters that the reflected and transmitted waves are out of phase by $\pi/2$.

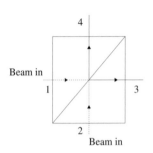

Fig. 9.22 Beam splitter using a multilayer coating between two 45° prisms.

9.8   Modes of the electromagnetic field

Mechanical systems have a number of degrees of freedom which define the number of independent types of motion that a system may undergo, and which we call *modes*. For example a set of n point masses has $3n$ degrees of freedom and each mode is simply the motion of one mass in one of the three orthogonal directions. Within electromagnetic wave theory we can identify corresponding modes of the electromagnetic field which are analogous to mechanical modes.

Modes of the electromagnetic field are solution of Maxwell's equations subject to whatever boundary conditions are imposed by optical elements. For instance, just outside a perfectly conducting mirror the tangential electric field would be zero. When the space is unbounded suitable modes with simple mathematical properties are sinusoidal plane waves, which can have any wavelength, any direction and any transverse polarization. Sinusoidal plane waves have simple mathematical properties and as was demonstrated in Chapter 7 any wavepacket with plane wavefronts can always be duplicated by a superposition of sinusoidal plane waves. These component waves will be grouped in wavelength around the mean wavelength of the wavepacket.

When there are boundary conditions there will no longer be modes at all wavelengths and travelling in all possible directions. Instead their distribution in wavelength and angle becomes *discrete*. The most interesting and useful optical arrangements are those for which the boundary conditions have some symmetry, such as the cylindrical symmetry of optical fibre or a Fabry–Perot cavity having circular mirrors. In the latter case the modes have wavelengths such that there are nodes of the electric field distribution at the mirror surfaces. If the cavity has small, well separated mirrors the paraxial approximation can be made. Waves are restricted to have the form

$$E(\mathbf{r}, t) = A(\mathbf{r}) \exp\left[i(kz - \omega t)\right] \qquad (9.119)$$

where A varies slowly along the beam (z) direction, so that $\partial A/\partial z \ll k$. Then the wave equation 9.36 simplifies to the paraxial Helmholtz equation

$$(\partial^2/\partial x^2 + \partial^2/\partial y^2)A + 2ik\,\partial A/\partial z = 0. \qquad (9.120)$$

The solutions are Gauss–Hermite functions[11] of which the simplest is the TEM$_{00}$ mode with a Gaussian profile and cylindrical symmetry, described in Section 6.15. The other modes have electric field distributions which have broader, more complex shapes, with one or more radial and azimuthal nodes across the mirror planes. The Gaussian mode has the most compact distribution of energy around the optical axis, so that

---

[11]See for example the fifth edition of *Optical Electronics in Modern Communications* by Amnon Yariv, published by Oxford University Press (1997).

spillage around the mirrors at each reflection preferentially depletes the other modes. Any wave trapped in a Fabry–Perot cavity can be resolved into a linear superposition of these *cavity* modes. More generally when there are two modes with the same transverse form but different wavelengths they are referred to as different *longitudinal* modes. Correspondingly modes of different transverse form are referred to as different *transverse* modes: the Gauss–Hermite modes are good examples. A wavepacket from the source will almost certainly not all enter a cavity, some being lost around its edges and some reflected from it. That part of the wavepacket that is captured is a new wavepacket made up of cavity modes whose wavelengths match those of the incident waves from the source.

Two important physical properties of modes can be usefully expressed in mathematical language. The first property is that each mode is *orthogonal* to all the other modes, meaning that it cannot be decomposed into a superposition of the other modes. Put another way they do not overlap: an integral over all space of the product of any pair of modes is always zero. The second property is that any waveform that is consistent with the boundary conditions, that is to say one which describes light confined within an optical structure, can always be replicated by a superposition of the modes for that structure. In a sense the modes are like unit vectors in a space in which each vector corresponds to a possible waveform consistent with the boundary conditions. These properties were explicitly proved in presenting Fourier analysis in Chapter 7 for sinusoidal waves, which are modes for electromagnetic waves in free unobstructed space.

If the material through which the wavepacket travels has some dispersion, then modes of different wavelength will travel at different speeds and hence the shape of the wavepacket changes. Individual modes travel without any change of shape; that is to say they propagate freely. We shall find later that optical fibre for long haul usage is constructed specifically so that only one mode propagates freely along it.

9.8.1   Mode counting

The modes of classical mechanical systems have equal time-averaged energies when they are in a state of thermal equilibrium. Thus it came about in the late 19th century that scientists anticipated that electromagnetic modes should, by analogy, also have equal energies in thermal equilibrium. The black body radiation spectrum was therefore confidently calculated – giving a prediction which diverged to infinity at short wavelengths! Experiment revealed instead that the spectrum peaked at a wavelength characteristic of the temperature of the body. The resolution of this difficulty led to the development of quantum mechanics and a profound change in the understanding of electromagnetic radiation.

The topic of black body radiation will therefore open the third section of this book where quantum phenomena are introduced.

In preparation we evaluate the number of modes available in unit volume of free space using sinusoidal waves as the modes. A simple approach is enough: the modes are counted inside a cubical box with perfectly conducting walls, shown in figure 9.23. This seems a very restrictive simplification, but it turns out that the result is quite independent of the shape or surface of the box and hence is universally valid. The tangential component of the electric field must vanish at the walls which lie at $x = 0$ and L, $y = 0$ and L and $z = 0$ and L. The electric field of a mode is a standing wave with components

$$E_x = A_x \cos(k_x x) \sin(k_y y) \sin(k_z z),$$
$$E_y = A_y \sin(k_x x) \cos(k_y y) \sin(k_z z),$$
$$E_z = A_z \sin(k_x x) \sin(k_y y) \cos(k_z z),$$

where $A_{x,y,z}$ are amplitudes, which must depend on the polarization. In terms of unit vectors oriented parallel to the edges, the wave vector is

$$\mathbf{k} = k_x \mathbf{e}_x + k_y \mathbf{e}_y + k_z \mathbf{e}_z.$$

The tangential components vanish automatically at the planes through the origin. These tangential components also vanish at the other faces provided that

$$k_x L = n_x \pi, \quad k_y L = n_y \pi, \quad k_z L = n_z \pi, \tag{9.121}$$

with n_x, n_y and n_z all being positive integers. Hence

$$n_x^2 + n_y^2 + n_z^2 = k^2 L^2 / \pi^2. \tag{9.122}$$

The numbers n_x, n_y and n_z are now used as the coordinates along orthogonal axes which define a new three-dimensional space. The possible choices of n_x, n_y and n_z define points which form the grid shown in figure 9.24 and are contained within the positive octant of a sphere of radius kL/π.[12] The number of modes is simply the volume of the octant $k^3 L^3 / 6\pi^2$, so that the number per unit volume of physical space is $k^3 / 6\pi^2$. Now applying eqn. 9.17 here gives

$$\partial E_x / \partial x + \partial E_y / \partial y + \partial E_z / \partial z = 0, \tag{9.123}$$

so that

$$A_x k_x + A_y k_y + A_z k_z = 0. \tag{9.124}$$

Thus only two of the amplitudes can be chosen freely, or put another way there are two independent polarizations. Then the total number of modes per unit volume is

$$\mathcal{N} = k^3 / 3\pi^2.$$

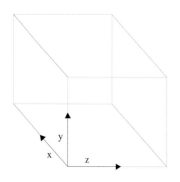

Fig. 9.23 Cubical box used for counting modes.

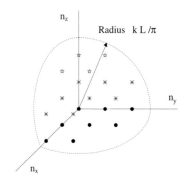

Fig. 9.24 The space displaying the electromagnetic modes in a reference volume. The labelling is explained in the text.

[12] Any mode with any negative values for n_x, n_y or n_z duplicates a mode with all positive values.

Finally the *density* of modes around wave number k is

$$\rho_k(k)\mathrm{d}k = \mathrm{d}\mathcal{N} = k^2\mathrm{d}k/\pi^2. \tag{9.125}$$

This can be re-expressed in terms of frequency using the equality

$$\rho_f(f)\mathrm{d}f = \rho_k(k)\mathrm{d}k.$$

Thus

$$\rho_f(f)\mathrm{d}f = 8\pi f^2\mathrm{d}f/c^3, \tag{9.126}$$

or in terms of the angular frequency

$$\rho_\omega(\omega)\mathrm{d}\omega = \omega^2\mathrm{d}\omega/(\pi^2 c^3). \tag{9.127}$$

If the medium is material of refractive index n, rather than free space, this result changes to

$$\rho_\omega(\omega)\mathrm{d}\omega = n^3\omega^2\mathrm{d}\omega/(\pi^2 c^3). \tag{9.128}$$

Once the density of modes is known a useful relationship between the mode count and the etendue of an optical system can be inferred. Because the modes are uniformly distributed in direction the fraction directed within the solid angle Ω, defined by the beam at the entrance pupil is $\Omega/4\pi$. If A is the pupil area then the light crossing this area in one second originates within a distance c/n from the pupil. The corresponding volume from which the light comes is therefore Ac/n, where n is the refractive index of the material preceding the pupil. Thus the number of modes which pass through the pupil per unit time is

$$\begin{aligned} N &= \rho(\omega)\mathrm{d}\omega(\Omega/4\pi)(Ac/n) \\ &= 2n^2\Omega A\mathrm{d}f/\lambda^2, \end{aligned} \tag{9.129}$$

where λ is the free space wavelength. Now using eqn. 4.11 we know that $n^2\Omega A$ is simply the etendue $\mathcal{T}$ of the optical system, so we have

$$N = 2\mathcal{T}\mathrm{d}f/\lambda^2. \tag{9.130}$$

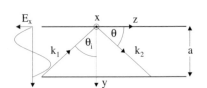

Fig. 9.25 A planar waveguide formed by plane parallel conducting surfaces. Wave vectors are drawn for the component plane waves making up the total wave that propagates within the gap. The electric field amplitude for one mode is drawn on the left.

Now it was proved earlier, see eqn. 5.32, that the etendue into a coherence area is λ^2. Therefore it is plausible to take the factor $\mathcal{T}/\lambda^2$ to count the number of modes per unit frequency interval, while the remaining factor two simply counts the number of independent polarization states. The overall product is the number of modes with different combinations of transverse profiles, frequency and polarization. Further insight into this result is obtained later by applying the uncertainty relation in Section 12.10.1.

9.9　Planar waveguides

Total internal reflection is used to trap electromagnetic waves in planar structures as well as in optical fibres. The analysis here of these simpler waveguides will provides a link to the analysis of optical fibres treated in Chapter 16. Figure 9.25 shows the wave vectors of a plane sinusoidal wave reflected between plane parallel conducting mirrors that lie a distance a apart. The incoming plane wave has its electric field in the x-direction, perpendicular to the diagram and parallel to the surfaces; while the wave vector direction makes an angle θ with the z-direction. The reflected wave is directed at $-\theta$. The incident electric field is

$$E_x = E_0 \exp\left[i(-\omega t - ky\sin\theta + kz\cos\theta)\right]$$

and the reflected wave which cancels this at the upper surface, $y = 0$, is

$$E_x = -E_0 \exp\left[i(-\omega t + ky\sin\theta + kz\cos\theta)\right].$$

Thus the total electric field is

$$\begin{aligned}
E_x &= E_0 \exp\left[i(-\omega t - ky\sin\theta + kz\cos\theta)\right] \\
&\quad - E_0 \exp\left[i(-\omega t + ky\sin\theta + kz\cos\theta)\right] \\
&= -2i\,E_0 \exp\left[i(-\omega t + kz\cos\theta)\right]\sin\left(ky\sin\theta\right).
\end{aligned}$$

This field will also vanish at the lower surface provided

$$k_y a = ka\sin\theta = m\pi, \tag{9.131}$$

where m is an integer. On the left of the figure the electric field amplitude is plotted for the case that $m = 2$. Waves satisfying eqn. 9.131 propagate without alteration and are the *modes* of the waveguide. The mth mode travels along the z-direction with wave vector

$$k_z^2 = k^2 - (m\pi/a)^2 \tag{9.132}$$

and with group velocity $c\cos\theta$. Although the electric field is transverse to the wave's direction of travel there is a component of magnetic field along the z-direction: such modes are therefore called TE modes. TM modes with a component of electric field along the direction of travel can be constructed in an analogous manner. Waves whose wavelengths are greater that $2a$ cannot satisfy eqn. 9.131 and are reflected back at the entry to the waveguide. In general an incident wave will therefore be made up of components which can propagate freely and other components which get reflected.

Similar modes occur when the radiation is confined by TIR within a layer of glass between two layers of glass of lower refractive index. In this case the requirement that a wave travels unaltered is now more complicated. At each reflection the wave penetrates into the lower refractive index medium as an evanescent wave and this leads to a phase delay. The reflection coefficient for a TE mode is given by eqn. 9.83 and the phase shift is $2\alpha_s$, where

$$\tan\alpha_s = n_2\chi/(n_1\cos\theta_i). \tag{9.133}$$

Then using eqn. 9.82 to replace χ

$$\tan \alpha_s = \sqrt{(n_1^2 \sin^2 \theta_i - n_2^2)}/n_1 \cos \theta_i$$
$$= \sqrt{(n_1^2 \cos^2 \theta - n_2^2)}/n_1 \sin \theta,$$

where we have used the fact that the angle θ is the complement of the angle of incidence θ_i in figure 9.25. In place of eqn. 9.131 we therefore have

$$k_y a = m\pi + \alpha_{sa} + \alpha_{sb}, \qquad (9.134)$$

where $2\alpha_{sa}$ and $2\alpha_{sb}$ are the phase shifts for reflections at the two surfaces bounding the guide layer. Any wave propagating within the layer would be a superposition of waves satisfying the boundary condition given by eqn. 9.134. In terms of rays the phase shift at total internal reflection indicates that the reflected ray leaves the interface at a point displaced along the surface from where the incident ray struck it. This *Goos–Haenchen shift* has been directly observed and it occurs because the ray penetrates a little into the less dense medium as an evanescent ray.

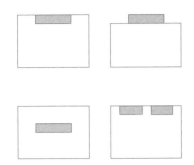

Fig. 9.26 Cross-sections of two-dimensional waveguides. The material of higher refractive index is shaded.

Cross-sections of waveguides with boundaries in two dimensions are shown in figure 9.26. In each case the external materials have lower refractive index, and in some cases one waveguide surface is bounded by air. At near infrared wavelengths the refractive indices of silicon and silica are respectively 3.5 and 1.46. Therefore such radiation is easily guided along silicon waveguides within silica, though with considerable absorption. A mode must satisfy a duplicate of eqn. 9.134, so we have for a guide of cross-section a_x by a_y embedded as shown in the left hand lower panel

$$k_x a_x = m\pi + 2\alpha_s, \qquad (9.135)$$
$$k_y a_y = m\pi + 2\alpha_s. \qquad (9.136)$$

Waveguides positioned as in the right hand lower panel can be sufficiently close so that the evanescent wave of one overlaps the other waveguide. There will therefore be coupling of the waves within the two waveguides, a property whose use in optoelectronics is described in Chapter 16.

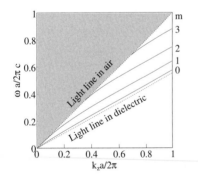

Fig. 9.27 Allowed modes of electromagnetic radiation in a dielectric sheet of thickness a in air. ω is the angular frequency and k_z is the wave vector component along the wave direction in the plane.

A simple example with which to illustrate the modes satisfying eqn. 9.134 is is provided by an infinite area thin plane sheet of glass with refractive index 1.8 and thickness a, suspended in air. The coordinates are as shown in figure 9.25: the z-direction lies in the direction the wave travels, the y-direction is perpendicular to the surface, and the electric field points in the x-direction. Then $\alpha_{sa} = \alpha_{sb} = \alpha_s$. Figure 9.27 illustrates the allowed modes on a plot of $\omega a/2\pi c$ against $k_z a/2\pi$, where ω is the angular frequency and k_z the wave vector component along the surface. There is a continuous distribution of modes in the shaded region; these are modes that can travel in glass or air because they arrive in glass at the air interface with an incident angle less than the critical angle. Points on the broken line boundary to the shaded region correspond to light travelling in air parallel to the surface. This boundary in called the *light line* in air and has the equation $\omega = k_z c$. Light at any smaller angle of incidence θ in air would lie at a point somewhere on a straight line through the origin with the equation $\omega = k_z c / \sin\theta$, above the light line in air. Light would propogate freely in unconfined dielectric anywhere above the light line in dielectric, which corresponds to light travelling parallel to the z-axis. Free propogation within the glass sheet is restricted to the modes trapped there by total internal reflection. These modes are shown in figure 9.27 as full lines called *dispersion curves*: each is labelled with the appropriate value of m. Light in such modes is said to be *index-guided* within the glass sheet: only the evanescent tails of the modes penetrate into the air on either side of the glass.

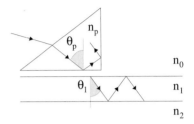

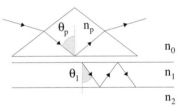

9.9.1   The prism coupler

Light can be transferred efficiently into a planar waveguide by the use of prism couplers as shown in figure 9.28. The prism is separated by a thin air gap from a planar waveguide of thickness d, whose refractive index is greater than that of the substrate on which it rests. Quantities relating to the prism, air, waveguide and substrate are labelled p, 0, 1 and 2 respectively. Monochromatic light is directed as shown so that when it arrives at the prism base its angle of incidence is greater than the critical angle for this glass/air interface. By making the air gap about half a wavelength thick it is arranged that the evanescent wave penetrates through the air gap into the planar waveguide below. The angle of incidence in the prism determines the angle of refraction in the waveguide: using Snell's law

Fig. 9.28 Two prism coupler arrangements for launching light into a planar waveguide.

$$\sin\theta_1 = (n_p / n_1) \sin\theta_p.$$

Therefore the *transverse component* of the wave vector in the waveguide is

$$
\begin{aligned}
k_y &= k_1 \cos\theta_1 \\
&= k_1 \sqrt{1 - (n_p/n_1)\sin^2\theta_p} \\
&= (2\pi/\lambda)\sqrt{n_1^2 - n_p^2 \sin^2\theta_p},
\end{aligned}
\tag{9.137}
$$

where λ is the free space wavelength. This wave entering the waveguide will occupy a trapped mode provided that it satisfies eqn. 9.134

$$k_y d = m\pi + \alpha_{10} + \alpha_{12}, \tag{9.138}$$

where m is an integer, while $2\alpha_{10}$ and $2\alpha_{12}$ are the phase shifts for reflections at the waveguide/air and waveguide/substrate interfaces respectively. Either the angle of incidence or the wavelength can be varied

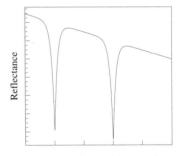

Wavelength or Incidence angle

Fig. 9.29 Variation of the reflection coefficient of a prism coupler with either wavelength or angle of incidence θ_p. The sharp dips occur when the light is coupled into a mode of the waveguide.

until both eqns. 9.137 and 9.138 are satisfied, and then light is *coupled* into the waveguide and travels along the waveguide. When this happens the intensity of the light reflected from the prism base shows a marked drop as illustrated in figure 9.29. Whenever the reflection coefficient dips in this way it is sure that the light is coupled into the waveguide.

The layouts shown in figure 9.28 are used to measure the properties of thin layers of dielectric that make up optoelectronic components and electronic chips. If the layer is thick enough to support two or more modes at a given wavelength then there is sufficient information available from measuring the angles of incidence θ_p at which coupling into the dielectric layer occurs to extract both the thickness and refractive index of this layer using eqn. 9.138. If the layer is so thin that only one mode is carried, the refractive index is needed in order to calculate the thickness. The prism coupling technique is complementary to that described in Chapter 11, which uses ellipsometry to study thin film properties.

Exercises

(9.1) Titanium oxide has refractive index 2.35. Calculate the velocity of light in this material and its characteristic impedance. What does the wavelength of light from a laser of wavelength 633 nm change to when it enters the oxide?

(9.2) An electric dipole 10^{-9} C m oscillates at a frequency 10 GHz. Calculate the electric and magnetic fields at 1 km distance in directions making angles of $90°$, $45°$ and $0°$ with the dipole axis. What is the time average energy flux in these three directions?

(9.3) Calculate the critical angle for a glass/air interface the glass having refractive index 1.5 at a wavelength 633 nm. What is the depth in air over which the light intensity falls by a factor e when light is incident at $42°$ on a glass/air interface, the glass having refractive index 1.5? Repeat the calculation for an angle of incidence $41.82°$.

(9.4) Show that if θ_B and θ'_B are the Brewster angles for light incident in opposite directions on a plane interface between dielectrics, their sum is precisely $90°$.

(9.5) Starting from Fresnel's equations prove the Stokes' relations $tt' = 1 - r^2$ and $r' = -r$. r and t are the amplitude reflection coefficients and r' and t' the corresponding coefficients when the same radiation follows the reverse path.

(9.6) Calculate the amplitude reflection and transmission coefficients using Fresnel's equations for light incident at $30°$ on an air/glass surface.

(9.7) Light is incident at Brewster's angle on a pile of n thin glass plates. The reflectance for the air/glass interface at Brewster's angle is R for s-polarized light. Calculate the transmittance of the complete stack for s-polarized light. You may neglect multiple reflections in this example.

(9.8) Express the laws of reflection and refraction in vector form. You can use $\mathbf{k}_i$, $\mathbf{k}_r$ and $\mathbf{k}_t$ to represent the incident, reflected and transmitted wave vectors, and $\mathbf{n}$ to represent a vector normal to the surface. In each case the magnitude $k = 2\pi/\lambda$ where λ is the wavelength in the material.

(9.9) Calculate the reflectance at normal incidence of glass of refractive index 1.5 in air which is coated with one double layer of magnesium fluoride and titanium oxide, each layer being of optical thickness one-quarter wavelength. The titanium oxide is in contact with the glass. Repeat the calculation for four double layers. What is the finesse of a Fabry–Perot cavity made of such mirrors? The refractive index of magnesium fluoride is 1.38 and that of titanium oxide is 2.35.

(9.10) Suppose the refractive index of the XMM-Newton mirror gold surfaces is 0.9999 at X-ray wavelengths.

At what angles of incidence will the X-rays be totally internally reflected in air?

(9.11) Is the relationship implicit in eqns. 9.115 and 9.116 that $r = r'$ inconsistent with Fresnel's Laws which require $r' = -r$?

Polarization

10.1 Introduction

The ability to produce and manipulate light in one state of polarization underlies the function of many modern devices such as liquid crystal displays (LCDs) and DVD reader/writers. This technology requires the use of materials that have an anisotropic response to light, which means using non-HIL materials. These include *dichroic* materials which absorb light in one state of polarization more than in the orthogonal state and so provide polarization filters.[1] *Birefringent* crystals on the other hand have a refractive index that varies with the orientation of the electric field of the light with respect to the crystal's axes. This property provides the means, for example, to rotate the plane of polarization of light. Similar anisotropic behaviour can be induced in certain materials when an external electric or magnetic field is applied. These field induced effects are of especial interest because they can be used to switch and to modulate light electronically at the high rates needed for optical communications.

[1] The word dichroic is sometimes also used with a different meaning: to describe a material which splits light into beams of different colours.

A first step in this chapter will be to establish the connection between states of circular polarization, introduced briefly in Chapter 1, and states of plane polarization. Dichroic and birefringent materials and their uses will be described in the next section. The Jones matrices, which provide a compact way of characterizing coherent states of polarization, are also introduced and explained. After this the optical effects of applied electric and magnetic fields and their applications are discussed. Birefringent liquid crystals are the essential element in a considerable fraction of consumer product displays; their properties and use will form the final topic in the present chapter.

10.2 States of polarization

Monochromatic electromagnetic radiation travelling in the z-direction in an HIL material can be in two independent states of plane polarization. One of these could have its electric field pointing along the x-direction, which in complex form is

$$\mathbf{E}_1 = E_0 \, \mathbf{e}_x \exp i(kz - \omega t),$$
$$\mathbf{B}_1 = (E_0/v) \, \mathbf{e}_y \exp i(kz - \omega t), \tag{10.1}$$

where $v = \omega/k$ is the wave velocity; the other would have fields orthogonal to these

$$\mathbf{E}_2 = E_0\,\mathbf{e}_y \exp i(kz - \omega t),$$
$$\mathbf{B}_2 = -(E_0/v)\,\mathbf{e}_x \exp i(kz - \omega t). \tag{10.2}$$

The time average of the intensity of the light waves formed by superposing these polarization states contains no interference terms

$$I = (\mathbf{E}_1 + \mathbf{E}_2)\cdot(\mathbf{E}_1 + \mathbf{E}_2)^*/2Z = (E_1 E_1^* + E_2 E_2^*)/2Z,$$

where the impedance $Z = \sqrt{\varepsilon_0\varepsilon_r/\mu_0}$. Thus these two polarization states can be used as *basis states*, in a manner analogous to the use of orthogonal vectors, to build any other plane polarized state. For example, by superposing them in the proportions $\cos\theta$ and $\sin\theta$ we get light travelling in the z-direction with its plane of polarization inclined at an angle θ to the x-direction. Of course one basis state cannot be formed from the other. The choice of the basis states is not unique: they could be chosen to have their electric fields at angles θ and $\theta + \pi/2$ to the x-direction.

Adding the chosen basis states with a phase difference of $\pi/2$ produces an electromagnetic wave with new properties. The electric field is

$$\mathbf{E}_\ell = (E_0/\sqrt{2})\left\{\,\mathbf{e}_x \exp\left[i(kz - \omega t)\right] + \mathbf{e}_y \exp\left[i(kz - \omega t + \pi/2)\right]\right\}, \tag{10.3}$$

of which the real part, the actual electric field, is

$$\mathbf{E}_\ell = (E_0/\sqrt{2})\left[\,\mathbf{e}_x \cos\left(kz - \omega t\right) - \mathbf{e}_y \sin\left(kz - \omega t\right)\right], \tag{10.4}$$

and this is shown at the location $z = 0$ in figure 10.1. A common convention is employed here when describing states of polarization: *the viewer is assumed to be looking towards the oncoming light.* Then in the present case the electric vector rotates anticlockwise around the z-axis with angular velocity ω, which is known as a state of *left circular* polarization. There is a second state of circular polarization in which the electric field vector rotates clockwise around the z-axis. This, *right circularly* polarized state, is obtained by adding the plane polarized states with the y-component lagging by $\pi/2$,

$$\mathbf{E}_r = (E_0/\sqrt{2})\left\{\,\mathbf{e}_x \exp\left[i(kz - \omega t)\right] + \mathbf{e}_y \exp\left[i(kz - \omega t - \pi/2)\right]\right\}, \tag{10.5}$$

and the resulting real part is

$$\mathbf{E}_r = (E_0/\sqrt{2})\left[\,\mathbf{e}_x \cos\left(kz - \omega t\right) + \mathbf{e}_y \sin\left(kz - \omega t\right)\right]. \tag{10.6}$$

When unequal contributions of the orthogonal plane polarized states are added with an arbitrary phase lead ϕ, the resultant electric field is elliptically polarized. The real electric field has components

$$E_x = E_a \cos\chi; \quad E_y = E_b \cos\left(\chi + \phi\right) = E_b(\cos\chi\cos\phi - \sin\chi\sin\phi) \tag{10.7}$$

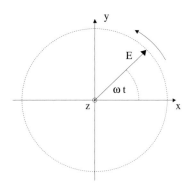

Fig. 10.1 Electric field of a left circularly polarized wave as seen looking toward the oncoming light. It rotates anticlockwise. The electric field of a right circularly polarized wave rotates clockwise.

where χ is $kz - \omega t$. Rearranging these equations gives

$$\sin \chi = \pm\sqrt{1 - (E_x/E_a)^2}, \quad \text{and}$$
$$E_y/E_b - (E_x/E_a)\cos\phi = -\sin\chi\sin\phi = \mp\sin\phi\sqrt{1 - (E_x/E_a)^2}.$$

Squaring this last line gives

$$(E_y/E_b)^2 + (E_x/E_a)^2 - 2(E_x E_y/E_a E_b)\cos\phi = \sin^2\phi, \qquad (10.8)$$

which is the equation of an ellipse tilted with respect to the x-axis at an angle α, which depends on ϕ. The dependence of α on ϕ will be determined next.

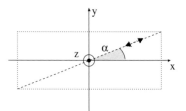

Let E_u and E_v be the components of the electric field refered to axes aligned on the ellipse's major and minor axes.

$$E_x = E_u\cos\alpha - E_v\sin\alpha,$$
$$E_y = E_u\sin\alpha + E_v\cos\alpha. \qquad (10.9)$$

Then when eqn. 10.8 is expanded in terms of E_u and E_v the coefficient of $E_u E_v$ will vanish; thus

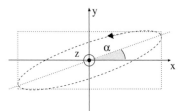

$$2\cos\alpha\sin\alpha(1/E_b^2 - 1/E_a^2) + 2(\sin^2\alpha - \cos^2\alpha)\cos\phi/(E_a E_b) = 0,$$

whence the tilt angle is given by

$$\tan 2\alpha = 2E_a E_b\cos\phi/(E_a^2 - E_b^2). \qquad (10.10)$$

Taking $E_a/E_b > 1.0$ $|\alpha|$ is less than $45°$, with α having the same sign as $\cos\phi$. In figure 10.2 the broken lines indicate the path followed by the end point of the electric field vector in the xOy plane for various choices of α. The phase lead, ϕ, increases from zero in the topmost panel to $\pi/2$ in the lowest panel. Notice that the largest excursions in E_x and E_y are respectively $\pm E_a$ and $\pm E_b$ whatever the phase lead.

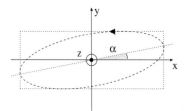

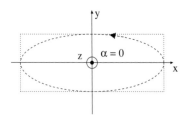

The product $\mathbf{E}_\ell \cdot \mathbf{E}_r^*$ vanishes and hence the time averaged intensity of any superposition of left and right circularly polarized waves

$$I = (a\mathbf{E}_\ell + b\mathbf{E}_r) \cdot (a\mathbf{E}_\ell + b\mathbf{E}_r)^*/2Z$$
$$= [aa^* E_\ell E_\ell^* + bb^* E_r E_r^*]/2Z \qquad (10.11)$$

Fig. 10.2 The paths of the end point of the electric field vector for elliptically polarized waves. The panels show from the top downward cases of increasing phase lead of the y-component field with respect to the x-component; starting from zero and ending with a lead of $\pi/2$.

contains no interference terms. Therefore the right and left circularly polarized states are orthogonal to one another. They make an equally good pair of basis polarization states, meaning that any plane, circular or elliptically polarized state can be reproduced by a linear superposition of right and left circularly polarized states. For example

$$\mathbf{E}_1 = (1/\sqrt{2})(\mathbf{E}_\ell + \mathbf{E}_r); \qquad \mathbf{E}_2 = (i/\sqrt{2})(\mathbf{E}_r - \mathbf{E}_\ell), \qquad (10.12)$$

reproduces the states of plane polarization that were introduced at the start of this section.

When a wave is reflected at a perfect mirror the electric field of the reflected wave cancels that of the incident wave at the surface of the mirror. Hence a linearly polarized wave incident normally is reflected with its polarization unchanged. When circularly polarized light is incident normally the electric field of the reflected wave rotates in the same sense as the electric field of the incident wave in order that they cancel at the surface. Now because the reflected wave is travelling in the opposite direction to the incident wave, it has the opposite circular polarization: right (left) circular polarization if the incident wave has left (right) circular polarization.

In Chapter 5 it was explained how light from actual sources consists of finite length wavepackets which have random phases. It is necessary to add that the polarizations of the wavepackets from sources are random. For example light from lamps met in the home contains wavepackets whose planes of polarization are uniformly distributed around the ray direction. The interference effects discussed in earlier chapters are only observed if the waves are not only coherent but also have the same polarization. If the beams being superposed have orthogonal polarizations then, as we have just seen, there is no interference. Interference is totally suppressed in Young's two slit experiment if polarizers which have orthogonal polarization, are placed one over each slit. We now discuss such polarizers.

10.3   Dichroism and Malus' law

Materials which absorb light of one plane of polarization more strongly than light of the orthogonal polarization are called *dichroic*. The most well known example is that of Polaroid sheet. Its basic constituent is a plastic material, polyvinyl alcohol, which is first stretched into a continuous sheet so that the long polymer molecules are lined up parallel in the stretch direction. Then iodine ions are deposited from vapour and these attach themselves along the length of the polymer molecules turning the latter into conductors. When light with its electric field parallel to the stretch direction falls on the Polaroid sheet a current flows along these conductive paths and energy is absorbed from the light beam. The process is so efficient that the beam is almost totally extinguished by a sheet only a few hundred microns thick. Light with its electric field transverse to the stretch axis cannot produce a current because there is no conductive path and hence light with this polarization is not absorbed. Commercial Polaroid sheets are available whose transmittance across the whole visible spectrum for one sense of polarization exceeds 50%, while being less than 10^{-4} for the orthogonal polarization. More robust dichroic filters made of glass are needed because high power laser beams would simply melt a sheet of polymer. In one type of glass filter the surface is covered with equally spaced, parallel metallic strips whose pitch is less than one wavelength. Alternatively the volume of the glass is loaded with parallel aligned, nanometre sized silver particles.

Figure 10.3 shows a beam of initially unpolarized light incident passing through a pair of polarizing sheets whose transmission axes are inclined at an angle θ. Assuming that there is full transmission for one sense of polarization and full absorption for the orthogonal sense, then the electric field of light emerging from the first sheet, the *polarizer*, is

$$\mathbf{E}_{\mathrm{p}} = E_0 \cos{(kz - \omega t)} \left[\mathbf{e}_x \cos\theta + \mathbf{e}_y \sin\theta \right],$$

while that emerging from the second sheet, the *analyser*, is

$$\mathbf{E}_{\mathrm{a}} = E_0 \cos{(kz - \omega t)} \mathbf{e}_x \cos\theta.$$

Hence the time averaged intensity

$$I(\theta) = [E_0^2/2Z] \cos^2\theta = I(0) \cos^2\theta, \tag{10.13}$$

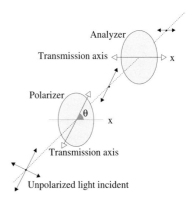

Fig. 10.3 Transmission of unpolarized incident light through a pair of Polaroids with their transmission axes making an angle θ. The arrows with solid heads indicate the electric vectors at each step.

a result known as *Malus's law*, where $I(0)$ is the intensity incident on the analyser. Thus the intensity transmitted by a polarizer when the light incident is unpolarized would be one-half the incident intensity. With *crossed* polarizer and analyser, that is with $\theta = \pi/2$, no light emerges from the analyser. A practical way to determine of the quality of a polarizer is to direct polarized light onto one face and then to measure the minimum and maximum transmitted intensities as the polarizer is rotated in its own plane about the beam as axis. The ratio of these intensities is called the *extinction ratio* and is 10^{-3} or less for simple sheet polarizers.

If a beam contains a mix of plane polarized light and unpolarized light the fraction of the polarized light can be determined by continuously measuring the transmittance through a polarizer as this is rotated through an angle of 90° about the beam axis. At maximum all the polarized and half the unpolarized light is transmitted, while at the minimum simply half the unpolarized light is transmitted. Thus the fraction of plane polarized light, the *degree of polarization* is

$$P = (I_{\mathrm{max}} - I_{\mathrm{min}})/(I_{\mathrm{max}} + I_{\mathrm{min}}). \tag{10.14}$$

10.4 Birefringence

The physical structure of many crystalline materials is anisotropic so that the ease with which an electric field can displace the electron clouds within atoms depends on the direction of the electric field relative to the crystal's axes. As a result the relative permittivity and refractive index also depend on the orientation of the electric field: this effect is called *birefringence*. There can be no such effect when electromagnetic waves travel either in amorphous materials, or in most liquids or in crystals

Table 10.1 The refractive indices of several uniaxial crystals, and the wavelength range over which their transmittance is high.

	Wavelength	n_o	n_e	High transmittance
Calcite	589 nm	1.658	1.486	350–4000 nm
Quartz	589 nm	1.544	1.553	200–2300 nm
LiNbO$_3$	633 nm	2.286	2.202	400–5000 nm
LiNbO$_3$	1300 nm	2.220	2.146	400–5000 nm
YVO$_4$	633 nm	1.993	2.215	400–4000 nm

such as cubic crystals which exhibit a high degree of internal symmetry. Among liquids the class of liquid crystals show birefringence and these provide the essential component of liquid crystal displays in TV and PC monitors; they are discussed in the final part of this chapter.

A birefringent crystal belonging to the *uniaxial* class is symmetric under rotations around its *optic axis*. Calcite (CaCO$_3$), crystalline silica (quartz, SiO$_2$) and lithium niobate (LiNbO$_3$) are commonly met uniaxial crystals. In the case of calcite the carbonate groups (CO$_3$) form parallel planes and the optic axis lies perpendicular to these planes. It is much easier to move electrons within rather than perpendicular to these planes and this is the origin of the birefringence. Figure 10.4 illustrates what happens when a narrow beam (ray) of light is incident perpendicular to a sheet of calcite whose optic axis lies in the plane of the diagram. The ray splits: one component obeys Snell's law, which in this case means it travels undeviated through the sheet; the other component is deviated on entering the calcite. When the calcite sheet is rotated about the incident ray the deviated ray rotates at the same rate. Anyone looking through such a calcite sheet sees not one, but two images of whatever lies behind the calcite. The undeviated image is formed by light with its electric field vector in the plane perpendicular to the optic axis, and this is called *ordinary* polarization. Light producing the other, rotating image has what is called *extraordinary* polarization: its electric vector points perpendicular to that of the light with ordinary polarization. This ability to split light according to its polarization using birefringent materials provides many useful tools in research and industrial applications. In calcite the electric field of light with ordinary polarization is acting in the plane in which the electrons are easy to move and as a result the velocity of the ordinary rays is less than that of the extraordinary ray. Materials like calcite, in which the extraordinary rays travel faster than the ordinary rays, are called *negative* uniaxial materials. Conversely in *positive* uniaxial materials the ordinary rays travel faster. Table 10.1 lists the refractive indices of several uniaxial crystals and the wavelength ranges over which their transmittance is high. As will be explained below, the extraordinary index varies with the direction of the ray. What is given in table 10.1, written n_e, is the extremal value of this index which is reached when the extraordinary ray travels exactly perpendicular to

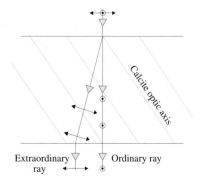

Fig. 10.4 Separation of ordinary and extraordinary rays in passing through a sheet of calcite.

the optic axis of the crystal. The ordinary refractive index is the same, n_o, whatever the ray direction.

10.4.1   Analysis of birefringence

In order to take account of the anisotropy of birefringent materials we shall retrace parts of the analysis of electromagnetic waves given in the previous chapter. Recall that Maxwell's equations were used to give an equation valid for either the electric or magnetic field: the wave equation, eqn. 9.28. Then a plane wave solution was attempted. This was found to satisfy the wave equation provided that the wave velocity satisfied the relation

$$v \equiv \omega/k = 1/\sqrt{\mu_0 \varepsilon_0 \mu_r \varepsilon_r}, \tag{10.15}$$

which in the case of a non-magnetic dielectric amounts to having a refractive index $n = \sqrt{\varepsilon_r}$. Here, we consider a sinusoidal plane wave travelling in a uniaxial crystal whose fields in complex form are

$$\mathbf{E} = \mathbf{E}_0 \exp\left[i(\mathbf{k} \cdot \mathbf{r} - \omega t)\right],$$
$$\mathbf{B} = \mathbf{B}_0 \exp\left[i(\mathbf{k} \cdot \mathbf{r} - \omega t)\right].$$

Then Maxwell's equations, 9.13, 9.14, 9.15 and 9.16, in the absence of free charge, reduce to

$$\mathbf{k} \cdot \mathbf{D} = 0; \quad \mathbf{k} \cdot \mathbf{H} = 0;$$
$$\mathbf{k} \wedge \mathbf{E} = \omega \mu_0 \mathbf{H}; \quad \mathbf{k} \wedge \mathbf{H} = -\omega \mathbf{D}. \tag{10.16}$$

These equations demonstrate that $\mathbf{k}$, $\mathbf{E}$ and $\mathbf{D}$ are all perpendicular to $\mathbf{H}$, and therefore coplanar, just as for isotropic materials. Taking the vector product of $\mathbf{k}$ with the third of the four equations gives

$$\mathbf{k} \wedge (\mathbf{k} \wedge \mathbf{E}) = \omega \mu_0 \mathbf{k} \wedge \mathbf{H} = -\omega^2 \mu_0 \mathbf{D}. \tag{10.17}$$

The relative permittivity is no longer a scalar quantity because the electric polarization of the material depends on the direction of the electric field with respect to the crystal axes. Fortunately if the material's absorption is negligible, which is often the case, the constitutive relation becomes relatively simple when the coordinate axes coincide with the crystal's principal axes

$$\begin{pmatrix} D_x \\ D_y \\ D_z \end{pmatrix} = \varepsilon_0 \begin{pmatrix} \varepsilon_x & 0 & 0 \\ 0 & \varepsilon_y & 0 \\ 0 & 0 & \varepsilon_z \end{pmatrix} \begin{pmatrix} E_x \\ E_y \\ E_z \end{pmatrix}. \tag{10.18}$$

Put more succinctly

$$\mathbf{D} = \varepsilon_0 \, \boldsymbol{\varepsilon} \cdot \mathbf{E},$$

where the diagonal elements of the matrix $\boldsymbol{\varepsilon}$ are ε_x, etc. In uniaxial crystals there is symmetry around the optic axis, which is taken here to be the z-axis. Then $\varepsilon_y = \varepsilon_x = \varepsilon_1$. Replacing $\mathbf{D}$ in eqn. 10.17 gives

$$\mathbf{k} \wedge (\mathbf{k} \wedge \mathbf{E}) = -(\omega^2/c^2)\boldsymbol{\varepsilon} \cdot \mathbf{E}.$$

In some crystalline materials the three relative permittivities appearing in eqn. 10.18 are all different. Such crystals are known as *biaxial* and have two optic axes. Their optical properties are more complex than those of uniaxial crystals and will not be discussed here in any detail. A full account of biaxial materials is given in *Polarization of Light* by S. Huard, published by John Wiley and Sons, New York (1990). Crystals of cubic symmetry are all isotropic; crystals with tetragonal, trigonal and hexagonal symmetry are all uniaxial; crystals with orthorhombic, monoclinic and triclinic symmetry are all biaxial.

Using the identity of eqn. 9.27 this becomes

$$(\mathbf{k} \cdot \mathbf{E})\mathbf{k} - k^2\mathbf{E} + (\omega^2/c^2)\varepsilon \cdot \mathbf{E} = 0. \tag{10.19}$$

This is the desired wave equation whose solution will yield the wave velocities, and the refractive indices, of the ordinary and extraordinary waves. Without losing generality, $\mathbf{k}$ can be taken to lie in the xOz plane inclined at an angle θ to the optic axis. Then eqn. 10.19 can be expanded to read

$$\begin{pmatrix} (\omega^2/c^2)\varepsilon_1 - k_z^2 & 0 & k_x k_z \\ 0 & (\omega^2/c^2)\varepsilon_1 - k^2 & 0 \\ k_x k_z & 0 & (\omega^2/c^2)\varepsilon_3 - k_x^2 \end{pmatrix} \begin{pmatrix} E_x \\ E_y \\ E_z \end{pmatrix} = 0. \tag{10.20}$$

There is one simple solution with the electric field pointing in the y-direction, which requires

$$(\omega^2/c^2)\varepsilon_1 - k^2 = 0,$$

that is

$$v \equiv \omega/k = c/\sqrt{\varepsilon_1}.$$

For this orientation of the electric field, in the xOy plane, the constitutive relation, eqn. 10.18, collapses to $\mathbf{D} = \varepsilon_0\varepsilon_1\mathbf{E}$. Thus $\mathbf{D}$ is parallel to $\mathbf{E}$ just as for isotropic materials. Light with this alignment of the electric field has *ordinary* polarization with refractive index $n_o = \sqrt{\varepsilon_1}$ independent of the direction of $\mathbf{k}$. Light with ordinary polarization obeys Snell's law whenever it enters or leaves the birefringent material.

The other independent solution involves the remaining two coupled equations (first and third lines) of eqn. 10.20 and has *extraordinary* polarization. In order for these equations to be consistent the determinant of the 2×2 matrix must vanish:

$$[(\omega/c)^2\varepsilon_1 - k_z^2][(\omega/c)^2\varepsilon_3 - k_x^2] - k_x^2 k_z^2 = 0. \tag{10.21}$$

After some manipulation this can be re-expressed as a requirement on the refractive index $n(\theta)$ of the extraordinary wave, when the wave vector makes an angle θ with the z-axis:

$$n(\theta) = kc/\omega. \tag{10.22}$$

This refractive index has to satisfy

$$1/n^2(\theta) = \cos^2\theta/n_o^2 + \sin^2\theta/n_e^2, \tag{10.23}$$

with $n_e = \sqrt{\varepsilon_3}$, and as before $n_o = \sqrt{\varepsilon_1}$. The result shows that the refractive index of the extraordinary wave varies with the angle the wave vector makes to the crystal's optic axis. At one extreme, when the wave travels along the optic axis the refractive index of the extraordinary wave is the same as that of the ordinary wave. At the other extreme when the wave vector is perpendicular to the optic axis the refractive index is n_{e} (examples given in Table 10.1). Substituting this value of n back into eqn. 10.20 gives the field vectors which, apart from a constant factor, are

$$\mathbf{E}_{\mathrm{e}} = -\mathbf{e}_x n_{\mathrm{e}}^2 \cos\theta + \mathbf{e}_z n_{\mathrm{o}}^2 \sin\theta, \qquad (10.24)$$

$$\mathbf{D}_{\mathrm{e}} = \varepsilon_0 n_{\mathrm{o}}^2 n_{\mathrm{e}}^2 (-\mathbf{e}_x \cos\theta + \mathbf{e}_z \sin\theta). \qquad (10.25)$$

These fields point in *different* directions. Both are perpendicular to the electric field of the ordinary wave, which confirms that the ordinary and extraordinary waves are independent solutions of eqn. 10.20.

10.4.2   The index ellipsoid

It is customary when visualizing the electric field and displacement vectors for the extraordinary waves to make use of the expression given for the energy contained in the electric field in eqn. 9.44

$$U_{\mathrm{e}} = \mathbf{E} \cdot \mathbf{D}/2.$$

We choose a value of the magnitude of $\mathbf{E}$ such that $U_{\mathrm{e}} = 1/2$, which makes the notation simpler and crucially does not affect the relative orientations of the field vectors. With the coordinate axes along the

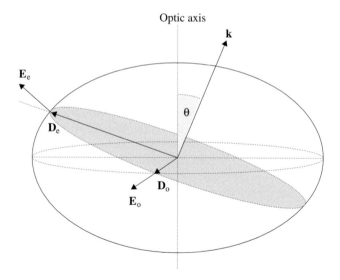

Fig. 10.5 Index ellipsoid for a negative uniaxial crystal. The section perpendicular to the wave vector is shaded, and the circular section perpendicular to the crystal's optic axis is also outlined. The electromagnetic field vectors for a wave with extraordinary (ordinary) polarization bear the subscript e (o). The angle between $\mathbf{E}_{\mathrm{e}}$ and $\mathbf{D}_{\mathrm{e}}$ is the angle labelled α in figure 10.6.

principal axes, the energy equation becomes

$$D_x^2/(\varepsilon_0\varepsilon_1) + D_y^2/(\varepsilon_0\varepsilon_1) + D_z^2/(\varepsilon_0\varepsilon_z) = 1. \qquad (10.26)$$

The surface described by this equation is the surface traced out by the endpoint of **D** and is drawn in figure 10.5. This surface is called the *index ellipsoid* and is drawn here for a negative uniaxial crystal. That for a positive uniaxial crystal would be an ellipsoid with the semi-axis along the optic (z-)axis longer than those in the orthogonal directions. A theorem from solid geometry states that at any point on the ellipsoid surface described by eqn. 10.26 the normal to the surface is the vector with components $(D_x/(\varepsilon_0\varepsilon_1),\ D_y/(\varepsilon_0\varepsilon_1),\ D_z/(\varepsilon_0\varepsilon_z))$. Evidently this is the vector **E**, so that **E** points normal to the surface of the index ellipsoid.

In figure 10.5 a wave vector **k** is drawn in an arbitrary direction. According to eqn. 10.16 **D** is perpendicular to **k**. Let us consider the case of ordinary polarization first: the electric displacement, $\mathbf{D}_\text{o}$ is perpendicular to the optic axis as shown in figure 10.5. It was proved in the last section that the displacement vector of the extraordinary waves is perpendicular to that of the ordinary waves, and is shown labelled $\mathbf{D}_\text{e}$ in the figure. The corresponding electric fields $\mathbf{E}_\text{e}$ and $\mathbf{E}_\text{o}$ are the normals to the index ellipsoid at $\mathbf{D}_\text{e}$ and $\mathbf{D}_\text{o}$ respectively. As expected, $\mathbf{E}_\text{o}$ is collinear with $\mathbf{D}_\text{o}$, but $\mathbf{E}_\text{e}$ and $\mathbf{D}_\text{e}$ are not collinear.

For certain directions of the wave vector the orientations of the field vectors of the extraordinary wave are simple to describe. When **k** points anywhere in the plane perpendicular to the crystal's optic axis $\mathbf{D}_\text{e}$ lies along the crystal's optic axis. In this case the constitutive relation for the extraordinary wave simplifies to $\mathbf{D}_\text{e} = \varepsilon_0\varepsilon_3\mathbf{E}_\text{e}$ so that its electric and displacement vectors are parallel and the refractive index is $n_\text{e} = \sqrt{\varepsilon_3}$. At the same time $\mathbf{D}_\text{o}$ and $\mathbf{E}_\text{o}$ lie perpendicular both to the crystal's optic axis and to **k**. The second simple case is when **k** points exactly along the crystal's optic axis. In this limit there is no distinction between the behaviour of the extraordinary and ordinary waves. Both have their electric fields in the plane perpendicular to the optic axis; for both polarizations **D** is parallel to **E**, and both refractive indices are n_o. When **k** points in a direction intermediate between the optic axis and the xOy plane then eqn. 10.23 shows that the refractive index for the extraordinary wave takes a value intermediate between n_o and n_e.

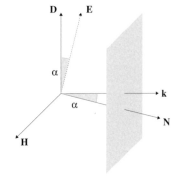

Fig. 10.6 The field vectors of an electromagnetic wave with extraordinary polarization travelling in a uniaxial crystal. The electric displacement, the electric field, the wave vector, and the Poynting vector are coplanar. The shaded surface is a wavefront.

10.4.3   Energy flow and rays

In general the electric displacement **D** of the extraordinary wave is not parallel to the electric field **E**, and therefore its Poynting vector ($\mathbf{N} = \mathbf{E} \wedge \mathbf{H}$) is generally not parallel to the wave vector. In other words the energy flow is not perpendicular to the wavefronts! The field vectors and a wavefront for an extraordinary wave are drawn in figure 10.6 exhibiting the general requirement that **D**, **E** and **k** are coplanar. It is seen that the velocity of the wavefront measured along the wave vector, v_p, is less than its velocity measured along the Poynting vector, v_r:

$$v_\text{r} = v_\text{p}/\cos\alpha, \tag{10.27}$$

where $\cos\alpha = \mathbf{D} \cdot \mathbf{E}/(D\,E)$. v_p is the usual wave/phase velocity and v_r is called the *ray velocity*.

A new surface, called the *ray surface*, is shown in figure 10.7 for the extraordinary and ordinary rays in the case of a negative uniaxial crystal. This surface is defined as the surface reached after unit time by light from a point source in the uniaxial material and is obtained by rewriting and solving eqn. 10.19 in terms of the vectors $\mathbf{N}$ and $\mathbf{D}$. The Poynting vector is radial and the wave vector is normal to the surface shown. Thus the ray surface for the ordinary wave is simply a sphere of radius c/n_o. In the case of extraordinary polarization the ray surface is the locus of the velocity vector $\mathbf{v}_\mathrm{r}$ satisfying the relation

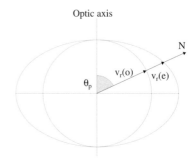

Fig. 10.7 Ray surfaces for extraordinary and ordinary waves travelling in a negative uniaxial crystal. The ray velocities are drawn for a particular choice of the direction of the Poynting vector $\mathbf{N}$.

$$1/v_\mathrm{r}^2 = \sin^2\theta_\mathrm{p}/v_\mathrm{e}^2 + \cos^2\theta_\mathrm{p}/v_\mathrm{o}^2, \tag{10.28}$$

where θ_p is the angle between the Poynting vector and the optic axis, $v_\mathrm{e} = c/n_\mathrm{e}$ and $v_\mathrm{o} = c/n_\mathrm{o}$. The relation connecting the various angles is

$$\theta_\mathrm{p} = \theta \pm \alpha, \tag{10.29}$$

where the positive/negative sign applies for negative/positive uniaxial crystals. The angles are always measured from the nearer direction of the optic axis.

10.4.4   Huygens' construction

Huygens' construction for wavefronts can now be applied to the propagation of light in uniaxial birefringent materials, with the ray surface providing the shape of the secondary wavelets. Figure 10.8 shows the propagation of a plane wave of infinite extent, having extraordinary polarization, over a time interval t. Each wavelet is a surface that reaches a distance $v_\mathrm{r}t$ from its origin in the direction θ_p with respect to the optic axis. A new wavefront at time t is drawn in figure 10.8 tangential to these wavelets. Although the wave vector $\mathbf{k}$ is perpendicular to the wavefronts, the Poynting vector $\mathbf{N}$ is not: it points from the wavelet origin to where the wavefront touches the wavelet. We note again that while $\mathbf{E}$ is normal to $\mathbf{N}$, $\mathbf{k}$ is normal to $\mathbf{D}$ and α is the angle between them.

It will generally be the case that a wave falling on a birefringent crystal has its electric field pointing in a direction other than one which would make it precisely an ordinary or an extraordinary wave when it enters the crystal. Such a wave is then a linear superposition of components with ordinary and extraordinary polarization, and as it progresses through the crystal the phase relationship between these components changes. In most cases the paths of the component extraordinary and ordinary waves also diverge, and unless the light is travelling along the optic axis the state of polarization changes. Consequently the pure extraordinary and the ordinary waves provide the appropriate basis states of polarization

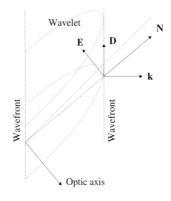

Fig. 10.8 Huygens' construction for a plane wave travelling in a negative uniaxial crystal. These ellipses are ray surfaces like those appearing in figure 10.7.

in birefringent materials. They alone retain the same polarization as they travel through a birefringent material.

10.5 Wave plates

Birefringent materials are used to modify the polarization of beams, turning plane polarized light into circularly polarized light or rotating the plane of polarization. They are also used to physically separate a beam into two orthogonally polarized components. Examples of these techniques and their applications are the topics of this and the following sections.

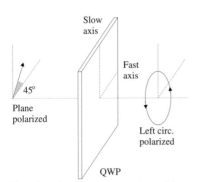

Fig. 10.9 A quarter-wave plate with an incident wave plane polarized at an angle of 45° measured in an anticlockwise sense from the fast axis. After passing the QWP the light is left circularly polarized.

[2]A piece of clingfilm makes quite a good quarter-wave plate, and a piece of clear Sellotape is a good approximation to a half-wave plate.

The difference between the refractive indices of light with ordinary and extraordinary polarization within a birefringent material means that a phase lag develops between these waves as they travel. A *quarter-wave plate* or QWP, shown in figure 10.9, is a slice of a uniaxial crystal cut with the optic axis lying parallel to its faces, and of such a thickness that there is a phase difference of $\pi/2$ between the ordinary and extraordinary waves of a particular wavelength after they have travelled through the plate. A plate of double the thickness is called a *half-wave plate* or HWP.[2] In quartz the extraordinary waves with electric field along the optic axis will travel slower than waves with the orthogonal ordinary polarization. These directions of polarization are therefore called the *slow* and *fast* axes, respectively, of the plate. The plate thickness, d, required to give a quarter-wave delay is such that

$$\lambda/4 = (n_e - n_o)d.$$

Using the values given in Table 10.1 shows that a quartz QWP has a thickness of $13.9\,\mu$m for light of wavelength 633 nm in air. Consider next that a plane polarized beam is incident on the QWP. If its plane of polarization makes an angle of 45° (anticlockwise as seen looking toward the oncoming light) with the fast axis it can be resolved into equal components with polarization along the fast and slow axes,

$$E_f(\text{in}) = (E_0/\sqrt{2})\cos(kz - \omega t),$$
$$E_s(\text{in}) = (E_0/\sqrt{2})\cos(kz - \omega t).$$

When these emerge from the QWP their phase difference is $\pi/2$,

$$E_f(\text{out}) = (E_0/\sqrt{2})\cos(kz - \omega t),$$
$$E_s(\text{out}) = (E_0/\sqrt{2})\cos(kz - \omega t + \pi/2) = -(E_0/\sqrt{2})\sin(kz - \omega t).$$

Evidently the emerging light is left circularly polarized. If the incident plane of polarization is instead at $-45°$ to the fast axis the emerging light will be right circularly polarized. With other orientations of the incident plane of polarization the emerging waves are elliptically polarized. This conversion process can evidently be put in reverse: when a circularly polarized beam is incident the emergent beam is plane polarized. Figure

10.10 shows a polarizing filter that is widely used to suppress reflections of ambient light. Only the vertically polarized component of the incident ambient light emerges through the first polarizer. This is converted to left circularly polarized light by the QWP. On reflection from the screen the direction of rotation of the polarization vector is unchanged but the direction of travel is reversed, so this reflected beam is right circularly polarized. Passage for a second time through the QWP yields horizontally plane polarized light – which is blocked by the polarizer.

A plane polarized wave passing through a HWP emerges with its polarization reflected in the fast axis of the HWP, as shown in figure 10.11. Suppose the incident electric field has components along the fast and slow axes

$$E_f(\text{in}) = E_0 \cos\theta \cos\phi; \quad E_s(\text{in}) = E_0 \sin\theta \cos\phi,$$

where $\phi = (kz - \omega t)$. Then the electric fields at exit are

$$E_f(\text{out}) = E_0 \cos\theta \cos\phi;$$
$$E_s(\text{out}) = E_0 \sin\theta \cos(\phi + \pi) = -E_0 \sin\theta \cos\phi.$$

Thus the component of electric field along the slow axis is reversed in travelling through the HWP.

10.5.1 Jones vectors and matrices

Jones vectors describe the polarization content of fully polarized beams and the related Jones matrices describe the action of optical components on the polarization of light passing through them. The restriction to fully polarized states still permits a useful range of practical applications, including those with laser beams. Plane polarized beams travelling along the z-direction and with their electric field pointing along the horizontal x-axis and the vertical y-axis are represented respectively by two-dimensional vectors

$$h = \begin{pmatrix} 1 \\ 0 \end{pmatrix} \exp i\phi; \quad v = \begin{pmatrix} 0 \\ 1 \end{pmatrix} \exp i\phi, \tag{10.30}$$

where $\phi = kz - \omega t$. Plane polarizers which only transmit light polarized along these axes are represented respectively by the matrices

$$H = \begin{pmatrix} 1 & 0 \\ 0 & 0 \end{pmatrix}; \quad V = \begin{pmatrix} 0 & 0 \\ 0 & 1 \end{pmatrix}. \tag{10.31}$$

The effects of plane polarizers on polarized beams can be summarized as follows:

$$Hh = h; \quad Vv = v; \quad Hv = Vh = 0.$$

Left and right circularly polarized beams are represented by the vectors

$$\ell = \begin{pmatrix} 1 \\ +i \end{pmatrix} \exp(i\phi)/\sqrt{2}; \quad r = \begin{pmatrix} 1 \\ -i \end{pmatrix} \exp(i\phi)/\sqrt{2}. \tag{10.32}$$

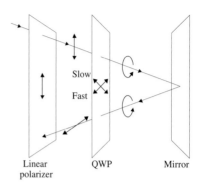

Fig. 10.10 Linear polarizer and quarter-wave plate used to suppress the reflections of ambient light from a screen. The screen is treated as a mirror. The polarization state of the light at each step is shown.

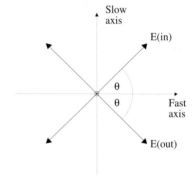

Fig. 10.11 A half-wave plate looking toward the oncoming light. The polarization state of a plane wave is shown before and after passing through the HWP.

Correspondingly the action of quarter-wave plates whose fast axes are respectively horizontal and vertical can be expressed by matrices

$$Q_\ell = \begin{pmatrix} 1 & 0 \\ 0 & +i \end{pmatrix}; \quad Q_r = \begin{pmatrix} 1 & 0 \\ 0 & -i \end{pmatrix}. \tag{10.33}$$

The effect of these QWPs on waves plane polarized at 45° to the fast axis is

$$Q_\ell \begin{pmatrix} 1/\sqrt{2} \\ 1/\sqrt{2} \end{pmatrix} \exp{(i\phi)} = \begin{pmatrix} 1 \\ +i \end{pmatrix} \exp{(i\phi)}/\sqrt{2} = \ell, \tag{10.34}$$

$$Q_r \begin{pmatrix} 1/\sqrt{2} \\ 1/\sqrt{2} \end{pmatrix} \exp{(i\phi)} = \begin{pmatrix} 1 \\ -i \end{pmatrix} \exp{(i\phi)}/\sqrt{2} = r, \tag{10.35}$$

The values of some Jones vectors and matrices depend on whether the complex wave form associated with a real electric field $\cos{(kz - \omega t)}$ is chosen to be, as here, $\exp{[i(kz - \omega t)]}$, or whether $\exp{[i(\omega t - kz)]}$ is the choice. If the second choice were made then all the imaginary terms in r, ℓ, Q_ℓ and Q_r etc. would reverse their signs.

agreeing with the earlier analysis. The effect of a sequence of dichroic and birefringent elements with Jones matrices J_1, J_2 ...J_n can be represented by the product matrix J_n....$J_2 J_1$. In particular the effect of a half-wave plate with its fast axis horizontal is

$$H_W = Q_\ell Q_\ell = \begin{pmatrix} 1 & 0 \\ 0 & -1 \end{pmatrix}. \tag{10.36}$$

Overall phase factors which occur in the matrices and vectors do not affect the state of polarization, so they can be dropped in any calculation used to predict the polarization state alone. If a plane polarized state described by a Jones vector $S(0)$ is rotated by an angle θ anticlockwise the new state has a Jones vector

$$S(\theta) = R(\theta)S(0), \tag{10.37}$$

where the rotation matrix $R(\theta)$ is given by

$$R(\theta) = \begin{pmatrix} \cos\theta & -\sin\theta \\ \sin\theta & \cos\theta \end{pmatrix}. \tag{10.38}$$

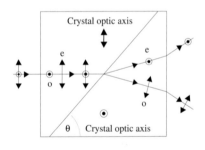

Fig. 10.12 Wollaston prism with unpolarized beam incident. The ray polarizations are indicated by the labels o and e.

Also if a polarizer or a waveplate is rotated anticlockwise in its own plane through an angle θ about the beam axis the Jones matrix is modified as follows

$$P(\theta) = R(\theta)P(0)R(-\theta). \tag{10.39}$$

10.5.2   Prism separators

These devices take as input an unpolarized beam and ideally output two orthogonally plane polarized beams travelling in well separated directions. With some devices only one beam is in a pure state of polarization, the other beam less so. Figure 10.12 shows a *Wollaston prism*, in which the component prisms are cut from a uniaxial crystal, in this case calcite, so that their optic axes lie at right angles to one another, as indicated. On entry to the first prism the Huygens wavelets for the extraordinary component are half ellipsoids whose axes of symmetry point upward in the diagram. Thus the ordinary and extraordinary wavefronts remain parallel to the entry surface and travel undeviated in the first prism. Thereafter their paths are shown in the figure labelled o and e. At the interface between the prisms the polarization components of the beam exchange their identities: the ordinary ray in the first region becomes the extraordinary ray in the second region and vice-versa. Consequently these rays are refracted at the interface away from the incident ray in opposite senses. When the angle labelled θ is set to $45°$ an angular separation of $20°$ is obtained at wavelength $589\,\text{nm}$. Calcite is widely used because it has high birefringence, transparency over a wide range of wavelengths, stability and is available cheaply. The contamination of the alternative polarization in either beam can be as low as 10^{-5} in standard calcite Wollaston prisms.

Separation of one pure plane polarized component is accomplished in *Glan prisms* by taking advantage of the difference in the critical angles for TIR of ordinary and extraordinary rays in uniaxial crystals. Two examples using calcite are drawn here: a *Glan–air* prism is shown in figure 10.13, and a *Glan–Thompson* prism in figure 10.14. In the Glan–air prism the optic axes of the component prisms lie perpendicular to the plane of the diagram and they are separated by an air gap. The prism is cut with one hypotenuse angle around $38.5°$ so that the angle of incidence of light arriving at the calcite/air gap interface is midway between the critical angle for the ordinary ray ($37.1°$) and that for the extraordinary ray ($42.3°$). The ordinary ray undergoes TIR while the extraordinary ray is partially transmitted and forms the required pure plane polarized, and in addition undeviated, beam.[3] The range in angle of incidence over which the extraordinary ray is transmitted and the ordinary ray undergoes TIR, is called the *acceptance* angle and amounts to $7.8°$ for a calcite Glan–air prism. The upper surface of the prism through which the ordinary ray would exit can be blackened to absorb it, or polished to give good transmission, depending on the application.

A much larger acceptance cone, of up to $30°$ full angle, is obtained with the Glan–Thompson prism. The gap between the component prisms is now filled with a transparent material like Canada balsam whose refractive index (1.55) lies midway between that for an ordinary ray and an extraordinary ray in calcite. With this modification the ordinary ray undergoes TIR at the calcite/balsam interface at angles of incidence greater than $69.2°$, while the extraordinary ray is partially transmitted at all angles of incidence. Therefore the aspect ratio of the Glan–Thompson prisms is made large, 2.5–3.0, so as to obtain the necessary large angles of incidence at the interface. The sandwich material in a Glan–Thompson prism can only withstand beam fluxes of up to $1\,\text{W}\,\text{cm}^{-2}$ without being damaged. On the other hand a Glan–air prism can cope with $100\,\text{W}\,\text{cm}^{-2}$. Both types of Glan prism provide only a single pure plane polarized beam travelling forward.

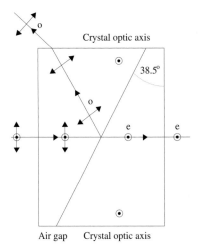

Fig. 10.13 Calcite Glan–air prism with unpolarized light incident. The s-polarized ray incident at the interface is the extraordinary ray; the p-polarized ray is the ordinary ray. The optic axes of the prism segments are also indicated.

[3]The gap in any Glan prism is many times the wavelength of light so that there is negligible frustrated TIR.

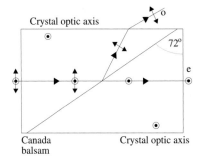

Fig. 10.14 Calcite Glan–Thompson prism with unpolarized light incident.

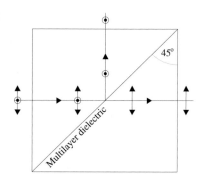

Fig. 10.15 Polarizing beam splitter made from glass prisms with a multilayer dielectric coating on the hypotenuse face.

10.5.3   Polarizing beam splitters and DVD readers

A beam-splitting prism designed to separate an unpolarized incident beam into pure orthogonally polarized beams emerging at 90° to one another is drawn in figure 10.15. The prisms are made of identical glass, either crown or flint. Recall that when a beam is incident at Brewster's angle on a stack of glass plates the p-polarized component is fully transmitted, and the s-polarized component partially reflected at each surface. With enough such surfaces the s-polarized component is almost completely reflected. In a polarizing beam splitter the function of the multiple layers of glass is performed by multiple dielectric layers on the hypotenuse face of one prism. Over a wavelength range of order 200 nm the transmittance of p-polarized light remains above 95% while the reflectance of the s-polarized light is close to 100%; cross contamination of the other polarization component is typically 0.1% and 2% respectively. All other surfaces in the beam paths are given antireflection coatings.

Figure 10.16 shows the optical layout for reading a DVD disk using a diode laser, a polarizing beam splitter (PBS), a quarter-wave plate (QWP) and a photodiode detector. The laser wavelength lies in the range 635 to 650 nm. The information on the disk is carried in the form of pits etched one quarter wavelength deep in the smooth reflective surface. The laser beam is focused to an image spot on the reflective surface, and the information is retrieved by the detector in the light reflected from the DVD surface. When the spot falls on the clear surface the detector receives full intensity, but when the spot covers a pit there is destructive interference, in the direction of the detector optics, between the reflection from the pit and the lands surrounding it. The pits follow a spiral track on the DVD with a pitch of 0.74 μm and lengths between 0.4 μm and 2.0 μm. The reader shown is mounted on an arm which maintains the image spot on the spiral track as the DVD rotates. The intervals of strong and weak reflections and the transitions are electronically converted to strings of binary zeroes and ones.

The laser light is polarized and its state of polarization at each stage along its path is indicated in the figure. Its initial polarization state is such that it is transmitted rather than reflected by the PBS. During the trip to and from the DVD the sequence already described for figure 10.10 is repeated with the result that the beam is orthogonally polarized on its return to the PBS and is reflected entirely into the detector. This arrangement prevents light reflected by the DVD from returning to the laser. If this were permitted then one of the laser mirrors could form a Fabry–Perot cavity with the DVD surface and hence affect the laser's operation. This deliberate decoupling makes it possible to have a compact unit in which the incident and reflected light share some common components.

In a read/write DVD the reflective layer is an alloy such as Ge-Sb-

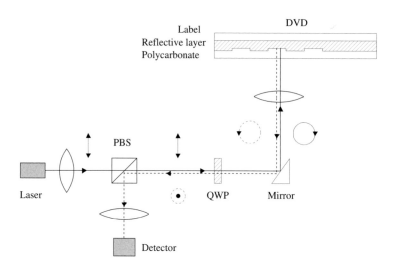

Fig. 10.16 DVD optical readout with the polarization of the laser beam indicated. For light travelling to the DVD the symbols are drawn with full lines, and for the light returning from the DVD with broken lines. PBS signifies the polarizing beam splitter.

Te whose reflection coefficient depends on its phase state. When this material is heated to 200°C it cools slowly to a crystalline state which has high reflectance; but when it is heated to 500–700°C it cools rapidly to an amorphous state with low reflectance. During the writing phase the power of the laser is raised so that it can melt the alloy: then bursts of high power and lower power produce non-reflective and reflective track segments respectively.

10.6   Optical activity

Some materials contain molecules or have a crystal structure with definite handedness, *chirality*, so that their refractive indices for right and left circularly polarized light differ. This property is called *circular birefringence*. Now a plane polarized beam can be resolved into right and left circularly polarized components as in eqn. 10.12. Consequently when a plane polarized beam travels through a circularly birefringent material a phase difference develops between its circularly polarized components. This causes the plane of polarization of the incident light beam to rotate as it passes through such materials, a property known as *optical activity*. Substances which cause the plane of polarization to rotate to the right include the natural sugars and most amino acids, and are called *d(extro)-rotatory*. Antibiotics are among the complementary class of *l(aevo)-rotatory* molecules. Quartz can function either way because it can crystallize in mirror image forms: in one form the silicon–oxygen chains follow a right handed helix, in the other form a left handed helix. Molecules whose structure has such a handedness are called *chiral*. They come in two forms, which are the mirror images of one another.

The handedness of a molecule is the same viewed from either end. Thus a set of randomly oriented molecules of one handedness produce a net effect on light. Optical activity can therefore be observed even in amorphous materials and liquids, provided their molecules are chiral. If the refractive indices for circularly polarized light travelling in an optically active material are n_ℓ and n_r, then the phase lead accumulated by the left circularly polarized component in travelling a distance z is[4]

$$\phi = (2\pi z/\lambda)(n_\ell - n_r). \tag{10.40}$$

[4]With the phase choice $(kz - \omega t) = \omega(nz/c-t)$ the phase lead of the left circularly polarized light is $(2\pi z/\lambda)(n_\ell - n_r)$.

Using Jones vectors the incident beam is

$$\begin{pmatrix} 1 \\ 0 \end{pmatrix} = \frac{1}{2}\begin{pmatrix} 1 \\ i \end{pmatrix} + \frac{1}{2}\begin{pmatrix} 1 \\ -i \end{pmatrix},$$

and the beam emerging has the Jones vector

$$\frac{1}{2}\begin{pmatrix} 1 \\ i \end{pmatrix} + \frac{1}{2}\exp\left(-i\phi\right)\begin{pmatrix} 1 \\ -i \end{pmatrix}$$

$$= [\exp\left(-i\phi/2\right)/2]\begin{pmatrix} \exp\left(i\phi/2\right) + \exp\left(-i\phi/2\right) \\ i[\exp\left(i\phi/2\right) - \exp\left(-i\phi/2\right)] \end{pmatrix}$$

$$= \exp\left(-i\phi/2\right)\begin{pmatrix} \cos\left(\phi/2\right) \\ -\sin\left(\phi/2\right) \end{pmatrix}. \tag{10.41}$$

Thus we see that the outgoing beam has its plane of polarization rotated clockwise through an angle equal to half the phase lead of the left over the right circularly polarized component

$$\beta = (\pi z/\lambda)(n_\ell - n_r). \tag{10.42}$$

In quartz the *specific rotatory power*, β/z is $21.7°/\text{mm}$.

Photoelasticity

When an object made of Perspex is placed between crossed polarizers and viewed in white light the field is dark. However, if the object is stressed, coloured bands appear across the area it covers, indicating that the stress has induced some birefringence. Regions that form a continuous band of one colour mark regions of equal stress. Models of large scale structures constructed from clear plastic can be studied in this way in order to assess the stresses that are likely to be present in the full scale object.

10.7   Effects of applied electromagnetic fields

Externally applied electric and magnetic fields exert forces on the electron clouds in atoms and molecules, which in turn cause changes to their configuration. This leads to a change in the relative permittivities appearing in the equation for the index ellipsoid, eqn. 10.26. The refractive index which at low values of the applied field is n_0, becomes

$$n = n_0 + \Delta n,$$

where Δn depends on the orientation of the applied fields $\mathbf{E}$ or $\mathbf{B}$, and of the polarization of the light with respect to the crystal's optic axis. Only the simpler cases will be of interest here, and to cover these cases the change is written conventionally

$$\Delta n = n^3(rE + sE^2 + r_B B)/2, \tag{10.43}$$

so that

$$-\Delta(1/n^2) = rE + sE^2 + r_B B. \tag{10.44}$$

Here r, s and r_B are coefficients which depend on the material, and on the orientations mentioned above. The corresponding optical effects are called respectively the Pockels effect, the Kerr effect and the Faraday effect and will be described in the following sections. Evidently r, s and r_B are tensors whose components depend on the orientation of the applied field and the polarization direction of the light. However the crystals are cut, and the fields applied, in such a way that usually only one component of the relevant tensor is of importance in each device.

The changes in optical behaviour of crystals produced by applied electric fields have provided a useful interface between electronics and optics. Electronic signals can be used to switch light on and off, or to modulate light with an analogue signal at rates up to tens of gigahertz. Optoelectronic devices exploiting this capability are frequently used in research and telecoms. One application has already been met in Section 8.12.1: an electro-optic modulator was used to lock together the frequencies of a cavity and a laser.

10.7.1 Pockels effect and modulators

The *Pockels effect* is the birefringence induced by an applied external electric field, and is linearly proportional to the field strength. When the applied electric field is reversed the effect is reversed, and we will show that for this reason no Pockels effect is observed in certain symmetric crystals or in liquids.

Suppose that the basic cell of a crystal contains atoms at coordinates $\mathbf{r}_i$ and that it is a *centrosymmetric crystal*: meaning that its basic cell is symmetric under the transformation: $\mathbf{r}_i \rightarrow -\mathbf{r}_i$. In such a material equal and opposite applied electric fields $\mathbf{E}$ and $-\mathbf{E}$ would produce the same change in the electron structure of the cell. This means that the change in refractive index will be the same for these two fields. Hence the Pockels effect must vanish in centrosymmetric crystalline materials and equally, using the same argument, in any isotropic liquid. In other less

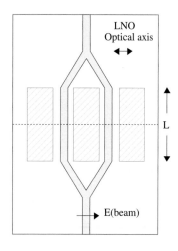

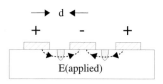

Fig. 10.17 Mach–Zehnder interferometer on a lithium niobate crystal. The lower diagram is a cross-section taken at the broken line. The indiffused titanium waveguides are shaded, and the gold electrodes are cross-hatched.

symmetric crystals it is the dominant electro-optic effect. The alteration in the refractive index produced by an applied electric field **E** is

$$\Delta n = n^3 r \cdot E/2, \tag{10.45}$$

where the Pockels coefficients depend on the relative orientation of the crystal axes, the plane of polarization and the direction of the applied electric field.

Crystals of the positive uniaxial material lithium niobate (LiNbO$_3$, or LNO) show a strong Pockels effect and are transparent for wavelengths from 400 to 5000 nm. Lithium niobate is *ferroelectric*: a crystal can have a very large permanent electric dipole moment in the absence of any applied electric field and this moment can be reversed by applying a sufficiently strong electric field. The mechanism involved is the formation of *domains* within which the dipole moments of the individual crystal cells line up parallel. In LNO alternate domains are parallel and antiparallel. An applied field causes the polarization of all the domains to align with itself so that a very large electric susceptibility and relative permittivity result. LNO possesses not only a strong Pockels effect, but also strong piezoelectric and acousto-optical responses. Potassium dihydrogen phosphate, KDP, is a similarly versatile crystalline material.[5] With no voltage applied the cell does not affect the polarization. Rapid switching in 1 ns or less is standard.

Figure 10.17 shows a device of a type used in telecoms to modulate the intensity of electromagnetic radiation in the visible or near infrared at frequencies of order 10 GHz. It consists of a LNO crystal a few centimetres long in which *waveguides* are formed to make the arms of a Mach–Zehnder interferometer. These waveguides are produced by diffusing vaporized titanium into the crystal. Their cross-section is a few microns across, similar to that of the core of a single mode optical fibre, and their refractive index is a little larger than that of the undoped LNO crystal. Thus light injected into one end of the waveguide is confined by TIR to travel along the waveguide just as in an optical fibre. Polarized light from a laser source enters via an optical fibre at one end of the waveguide, and exits into another optical fibre at the other end of the waveguide. Gold electrodes are deposited alongside the arms of the Mach–Zehnder and an electric potential is applied between them with the polarity shown in the figure. This voltage produces equal and opposite electric fields in the two arms and hence opposite Pockels effects. A simple alignment is shown here: the crystal optic axis, the direction of polarization and the applied electric field are all parallel. The refractive indices in the arms of the Mach–Zehnder are modified from their value

---

[5]The Pockels coefficients for these important complex crystals, LNO and KDP, are discussed and tabulated in *Polarization of Light* by S. Huard, published by John Wiley and Sons, New York (1990).

n_e in the absence of an applied electric field to become

$$n = n_e \pm r_{33}n_e^3 E/2,$$

where E is the applied field and r_{33} is $30.8\,10^{-12}$ m V^{-1} is the appropriate Pockels coefficient for LNO at 633 nm wavelength. Thus the phase difference between light in the two arms at the point they reunite is

$$\phi = (2\pi/\lambda)r_{33}n_e^3 EL,$$

λ being the vacuum wavelength of the light and L being the length of the electrodes. This phase difference can be re-expressed in terms of the electrode separation d and the applied voltage V as

$$\phi = (2\pi/\lambda)r_{33}n_e^3(LV/d). \tag{10.46}$$

If we define the electric field of the light entering each arm to be

$$E_{in} = E_0 \exp(-i\omega t),$$

then that emerging from the device is

$$E_{out} = E_0[\exp(-i\omega t) + \exp[-i(\omega t + \phi)]],$$

whose time averaged intensity is

$$I = E_{out}E_{out}^* = 2E_0^2[1 + \cos\phi]. \tag{10.47}$$

This behaviour is shown in figure 10.18 as a function of the applied voltage.

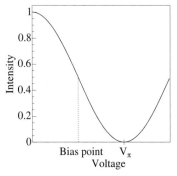

Fig. 10.18 The intensity transmitted through the Mach–Zehnder modulator versus the applied voltage.

The modulator is biased with a fixed voltage between the electrodes which is large enough to bring the operating point on the intensity versus voltage curve into the region where the slope is linear. Any small amplitude signal voltage applied on top of this fixed *bias* will be replicated in the variation of the intensity of the light transmitted by the modulator into the output fibre. The voltage that cuts the light off completely, V_π, is called the *half-wave voltage*. Evidently

$$V_\pi = \lambda d/(2n_e^3 r_{33}L). \tag{10.48}$$

Taking the wavelength to be 633 nm, the length of the electrodes to be 1 cm and their separation to be 10 μm, then the half-wave voltage for the LNO modulator being considered here would be 1 V. This value lies neatly within the range of voltages employed in modern electronics.

The time taken for light to pass between the two ends of the electrodes imposes an upper limit on the frequency of signals that can be used to modulate the light. This *transit time*, τ, equals $n_e L/c$ and only signals of frequencies well below $1/\tau$ will be reproduced without distortion in the light intensity output from the modulator. This limit can be raised appreciably by designing the electrodes so that they carry a *travelling wave* which propagates along the length of the electrodes parallel to the light in the waveguide.

10.7.2   Kerr effect

In liquids and centrosymmetric crystals the residual electro-optic effect is the *Kerr effect*, which is proportional to the square of the electric field. The applied electric field induces birefringence with the electrical field direction becoming the optic axis, and the material therefore behaves as a uniaxial crystal. If the applied field is E then the difference between the refractive index for light polarized parallel (extraordinary polarization) and perpendicular (ordinary polarization) to the applied electric field is expressed as

$$n_e - n_o = \kappa \lambda E^2, \tag{10.49}$$

where κ is called the Kerr coefficient. A material with a positive Kerr coefficient therefore has polarization properties similar to those of a positive uniaxial crystal. One liquid that has often been used, nitrobenzene,[6] has a very large Kerr coefficient, $2.4\,10^{-12}\,\mathrm{V}^{-2}\,\mathrm{m}$, while that for water is only $4.4\,10^{-14}\,\mathrm{V}^{-2}\,\mathrm{m}$. It follows also that

$$\Delta n = n_e - n = n_2 I, \tag{10.50}$$

where n is the refractive index in the absence of an applied field and I is the light intensity.

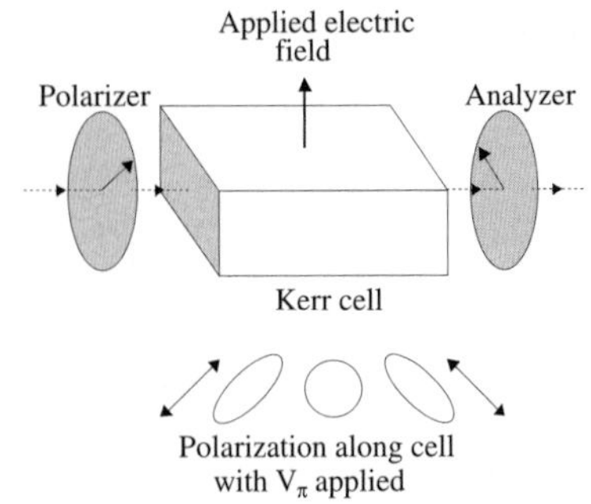

Figure 10.19 shows a fast optical shutter consisting of a Kerr cell located between a crossed polarizer and analyser. The electric field applied across the cell is oriented at $45°$ to the transmission axes of both polarizer and analyser. In the absence of any external electric field there is no transmission. However when an electric field is applied a phase delay is produced between light polarized along and perpendicular to the applied field direction. This delay can be inferred using eqn. 10.49 to be

$$\Delta\phi = 2\pi\kappa L(V/d)^2, \tag{10.51}$$

where L is the length and d the width of the cell containing the active material and V is the applied electric potential. When the phase difference is exactly π the Kerr cell is equivalent to a *half-wave plate* with, in the case of a material with a positive Kerr coeffiecient, the slow axis pointing along the direction of the applied electric field. The plane of polarization of the light therefore rotates by exactly $90°$ in passing through the Kerr cell and so the beam is transmitted by the analyser. For example, a cell of width $1\,\mathrm{cm}$ and length $4\,\mathrm{cm}$ filled with nitrobenzene requires a voltage of $2260\,\mathrm{V}$ to produce a phase difference of π. Switching is extremely fast, capable of following signals of frequencies up to $10\,\mathrm{GHz}$. This high switching speed has been exploited in the determination of the speed of light made by measuring the travel time of short pulses of light over a known path distance to and from a distant mirror. Kerr cell shutters can readily be constructed with apertures adequate to cover standard camera lenses and have been extensively used in high speed photography.

Fig. 10.19 Kerr cell sandwiched between crossed linear polarizer and analyser. The polarization state of the beam travelling through the Kerr cell is indicated for the case that the applied voltage makes the cell equivalent to a half-wave plate.

---

[6]Unfortunately this is a corrosive and unstable material.

10.7.3 Faraday effect

Faraday, in 1845, made the first observation of an effect of an applied electromagnetic field on light: he observed that the plane of polarization of light travelling in glass was rotated when a magnetic field was applied along the path of the light. In the *Faraday effect* a difference is produced between the refractive index of right and left circularly polarized light travelling along the direction of the magnetic field

$$\Delta n = n^3 r_B B, \tag{10.52}$$

where B is the applied field and r_B an optical constant that depends on the material. Over a path length L a phase difference $(2\pi/\lambda)\Delta n L$ develops between the two states of circular polarization. If instead the incident light is plane polarized we can use a result proved in Section 10.6: that the plane of polarization will rotate through an angle equal to half this phase difference,

$$\theta = (\pi/\lambda)\Delta n L = (\pi/\lambda)n^3 r_B L B,$$

where eqn. 10.52 was used to replace Δn. This result is generally contracted to read

$$\theta = \mathcal{V} L B, \tag{10.53}$$

where $\mathcal{V}$ is called the *Verdet constant*. By convention this constant is taken to be positive if the rotation is left handed when the light travels in the direction of the applied magnetic field. Materials containing paramagnetic ions have relatively large Verdet constants. The Verdet constant, expressing the rotation angle in radians, is typically $90\,\text{T}^{-1}\text{m}^{-1}$ for glass doped with terbium ions, but only $3.8\,\text{T}^{-1}\text{m}^{-1}$ for water, all measured at a wavelength in air of 633 nm. Many older texts quote the Verdet constant in the unit arc-min/amp: this value is obtained by dividing the constant in $\text{rad}\,\text{T}^{-1}\text{m}^{-1}$ by 235.1.

Unlike the Kerr effect, the rotation of the polarization induced by an external magnetic field is in one sense (say left handed as seen looking toward the oncoming light) when the light travels in the direction of the field and in the reverse sense when it travels in the direction opposite to the field (right handed when viewed looking toward the oncoming light). Thus the Faraday rotation is in the same sense when it is observed in a fixed direction, for example looking along the magnetic field direction. This feature makes it possible to construct *optical isolators* which permit light to travel through them in only one direction.

Figure 10.20 shows such an optical isolator consisting of a Faraday rotator sandwiched between a polarizer and analyser whose transmission axes are inclined at 45° to each other. A typical rotator is a terbium doped glass rod located inside a powerful permanent magnet. The paths and the corresponding polarizations of light entering from both the left and the right are also indicated in figure 10.20. In each case the plane

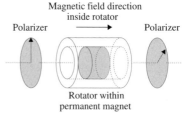

Fig. 10.20 Optical isolator consisting of a Faraday rotator sandwiched between polarizers with their transmission directions inclined at 45°. The polarization states for light travelling in either direction are shown below the device as seen from the left. The Faraday rotator is taken to have a positive Verdet constant.

of polarization is rotated clockwise through 45° by the Faraday rotator, as seen from the left hand end. Right moving light is transmitted but left travelling light is blocked.

Isolators are widely used when it is necessary to prevent light from a laser being reflected back into the laser. Such a reflected beam would be effectively confined in a cavity formed by laser mirror and the external reflecting surface, and this can easily disturb the laser's operation. When very high powers are involved the polarizers used would be Glan–air prisms rather than a sheet of Polaroid. A large Verdet constant in the visible part of the spectrum is associated with strong dispersion coming from atomic resonances in the ultraviolet. The factor λ^{-1} in the Verdet constant produces an additional increase in its value as the wavelength falls. Consequently a Faraday rotator needs to be tuned in order to function at different wavelengths by moving the glass rod further into or out from the magnet air core.

10.8   Liquid crystals

The liquid crystal display, LCD, is now ubiquitous, providing the displays in hand calculators, laptops and mobile phones. It has also almost entirely displaced the electron beam monitors due to a combination of advantages that are explained below. Liquid crystals are liquids which contain anisotropic molecules and which possess some degree of internal order. The molecules diffuse just like those in any liquid but they retain either some degree of alignment among the molecules (*nematic ordering*) or some degree of alignment and positional ordering of the molecules (*smectic ordering*). Thanks to this ordering liquid crystals are highly birefringent and optically active, properties which make them ideal elements for electronically driven displays.

When a crystalline substance melts to become a liquid crystal the latent heat is almost as large as that evolved in a transition from a crystal to a pure liquid, and when the liquid crystal is further heated so that it changes to a pure liquid the latent heat from this phase transition is much less. This shows that the ordering in a liquid crystal is much weaker than in a crystalline solid. The liquid crystals of interest here are called *thermotropic*, that is to say there is a limited temperature range over which they are stable as liquid crystals. Their molecules have a stiff rod-like centre section with more flexible extensions and there are strong attractive forces between neighbours.[7] Figure 10.21 contrasts the purely orientational ordering in a nematic, N, liquid crystal and the orientational ordering and layering in smectic liquid crystals. In smectic SmA liquid crystals the molecules align normal to the layers, while in

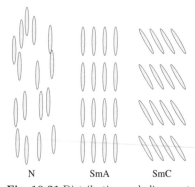

N   SmA   SmC

Fig. 10.21 Distribution and alignment of molecules in nematic, smectic A and smectic C liquid crystals.

---

[7]There are other liquid crystals that are formed when a material is dissolved in a solvent, such as soap in water. These *lyotropic* liquid crystals are not of importance in the applications discussed here.

smectic SmC liquid crystals the molecules are tilted. The average direction of alignment over a local region containing many molecules is called the *director*. The nematic liquid crystals are uniaxial and the director direction defines the local optic axis of the liquid crystal. Although the discovery of liquid crystals goes back to the 19th century the technological exploitation of their optical properties came only later, after George Gray and colleagues had in 1973 successfully synthesized liquid crystals which were stable over a useful temperature range around room temperature.[8] The structure of liquid crystals can be chemically engineered so that the difference between the extraordinary and ordinary refractive indices, Δn, has a value appropriate to the application, usually around 0.1 for LCD usage.

[8] G. W. Gray, K. J. Harrison and J. A. Nash: *Electronics Letters* 9, 130 (1973).

10.8.1   The twisted nematic LCD

Figure 10.22 shows the basic components of a simple *twisted nematic* (TN) liquid crystal display invented in 1977 by Schadt and Helfrich. A layer of uniaxial nematic liquid crystal, about $5\,\mu$m thick, is sandwiched between two glass sheets. On the inner glass surfaces a layer of polyimide is deposited, typically $100\,$nm thick, whose surface has been textured by gently rubbing in a fixed direction. This causes the liquid crystal molecules close to the surface to align nearly parallel to the rubbing direction. By arranging the rubbing directions on the two facing polyimide surfaces to be orthogonal, the director is forced to twist through 90° between these two surfaces. Polarizing sheets are glued to

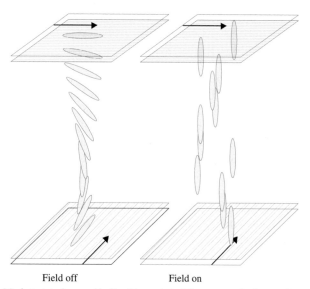

Field off   Field on

Fig. 10.22 A twisted nematic liquid crystal. The arrows indicate the transmission directions of the polarizers, and the light lines the rubbing direction on the polyimide layers. On the left transmitting light with no field applied; on the right blocking light when an electric field is applied.

the outer surfaces of the glass sheets with their transmission axes aligned parallel to the rubbing direction of the polyimide surface, which means the polarizers are crossed. When the LCD is used in transmission the panel is illuminated from below by a broad white light source.

The left hand panel in figure 10.22 shows the 'off' state with no voltage applied to the cell. Light entering from below is polarized parallel to the rubbing direction on entering the liquid crystal. In the liquid crystal the optic axis, coincident with the local director direction, rotates through 90° across the cell. Evidently left and right circularly polarized waves will have slightly different refractive indices because of the rotation of the optic axis across the liquid crystal. From the analysis made in Section 10.6 it follows that the plane of polarization of the light incident will rotate as it progresses. With the correct choice of the liquid thickness and the liquid crystal's birefringence the plane of polarization of the incident light can be made to follow the rotation of the optic axis through 90° and it will then be transmitted by the upper polarizer. The requirement for this *adiabatic* following is that the phase difference between the extraordinary and ordinary rays is much greater than the angular twist across the liquid crystal thickness[9]

$$(2\pi/\lambda)\Delta n\, z \gg \pi/2,$$

where Δn is this difference in refractive indices, and z the cell thickness. That is

$$(4\Delta n\, z/\lambda) \gg 1.$$

Thus if Δn is around 0.1 and the wavelength is 0.5 μm, then a thickness of 5–10 μm is adequate to produce a rotation through 90° which can be adiabatically followed by the electric field. As a result the light is transmitted by the upper polarizer.

In the 'on' state shown in the right hand panel an electric field is applied perpendicular to the cell. The molecules become polarized and, while those very close to the surfaces remain pointing along the rubbing axis, those throughout the body of the liquid crystal tip so that the director points along the field direction. With this new alignment the incident light travels along the optic axis of the liquid crystal, the plane of polarization of the incident light is no longer guided, and the light is blocked by the upper polarizer. What is shown in figure 10.22 is known as a normally white (NW) display. A normally black (NB) display is obtained if the transmission axis of the upper polarizer is set parallel to that of the lower polarizer.

Further details of the structure of a twisted nematic LCD are shown in figure 10.23. A white backlight is provided by either a fluorescent lamp or LEDs, plus diffusers. The electric field across the liquid crystal is applied via transparent conductive 100 nm thick indium tin oxide electrodes deposited on the glass ahead of the polyimide layer. It requires

[9]The ratio of the phase lag to twice the twist is called the *Mauguin parameter*.

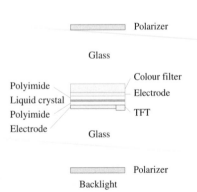

Fig. 10.23 Component layers of a pixel of a twisted nematic LCD.

only a few volts across the thin layer of liquid crystal to attain the high electric field needed to switch the molecular alignment. However the switching is slow because the molecules have to be turned against the viscous drag: a full cycle off/on/off takes at least 15 ms to complete. LCDs used as PC monitors are divided into pixels so as to match standard formats, for example the SXGA format with 1280×1024 pixels. Each pixel contains three electrically independent subpixels, covered by respectively red, green and blue colour filters. These filters are in the form of continuous vertical bands so that the subpixels on an 18-inch screen are then 0.28 mm tall and 0.093 mm wide. Individual subpixel voltages are applied using an active matrix (AM) addressing as shown in figure 10.24. Thin film transistors (TFT) located at one corner of each pixel are used to turn the pixel voltage on and off. Control voltages are applied via electrical buses: one set of which runs between the rows and the other between the columns of pixels. A data voltage is applied in sequence to each column in turn and at each of these steps a gating voltage is applied to just those rows that contain pixels that are to be turned on. If a DC voltage were applied continuously to a liquid crystal, then impurity ions would be driven onto the polyimide surfaces where they would adhere and affect its ability to align the liquid crystal molecules. In order to avoid this effect the voltage on any pixel is alternated between frames.

LCD panels have an overall thickness of a few millimetres. The transmission coefficient is restricted to 5% at best, because of the absorption in the many layers. This is still sufficient to outclass the electron beam tubes in brightness when using only modest backlight intensities. The displays in watches and calculators are illuminated by ambient light only, so that the power consumption is reduced to a tolerable level. In these devices a reflector forms the back layer.

The immediate advantages of the LCD over the electron tube for use in comparable displays are the flat format, light weight, avoidance of high voltage and roughly halved power consumption. A significant problem with TN LCDs was the rapid fall off in the contrast and the fidelity of the colour as the viewing direction is moved away from the perpendicular to the display surface. The origin of the problem is that when the molecules align perpendicular to the display surface the birefringence varies strongly with the viewing angle, and this is made worse by the molecules not being fully aligned. To alleviate this problem thin surface layers of diffusing and birefringent material are added to extend the angular range of satisfactory contrast and colour fidelity.

10.8.2 In-plane switching

A more effective solution is illustrated in figure 10.25 with the electrodes now located on one glass surface only. The non-transmitting 'off' state is shown in the left hand panel in which light emerging from the lower

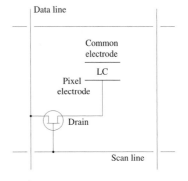

Fig. 10.24 Active matrix switching for LCD. The CMOS transistor obscures only a small fraction of the pixel area.

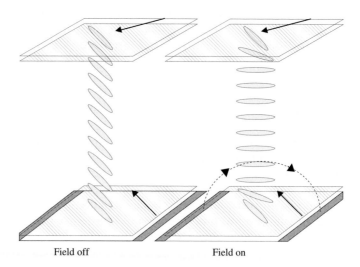

Field off   Field on

Fig. 10.25 In-plane switching LCD. The dark bands on either side of the lower surfaces are electrodes. The arrows mark the directions of the polarizers' transmission axes. In the right hand diagram with electric field applied the liquid crystal molecules set at 45° to the directions of the two transmision axes in the body of the liquid.

polarizer has its electric field along the director direction. Therefore the polarization is unchanged as the light crosses the liquid crystal and is blocked by the upper polarizer. In the 'on' state, shown in the right hand panel, the applied electric field is parallel to the glass surface rather than perpendicular to it as was the case for TN LCDs. Suppose the director of the liquid crystal, which defines its optic axis, sets at an angle θ to the direction of transmission of the lower polarizer as shown in figure 10.26. Then the electric field of the light entering the liquid has components parallel and perpendicular to the optic axis,

$$E_e = \cos\theta; \quad E_o = \sin\theta,$$

where all unnecessary factors have been removed. When these components emerge they have become

$$E_e = \cos\theta; \quad E_o = \sin\theta \exp(i\phi),$$

where the phase difference

$$\phi = (2\pi/\lambda)\Delta nd,$$

for a cell of thickness d and a difference in the refractive indices Δn; λ is the wavelength in free space. The component of the electric field along the direction of transmission of the analyser is therefore

$$E_A = \cos\theta\sin\theta - \cos\theta\sin\theta \exp(i\phi).$$

Thus the time averaged intensity of the light exiting the LCD is

$$I_A = E_A^* E_A = \sin^2(2\theta)\sin^2(\phi/2). \tag{10.54}$$

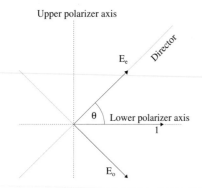

Fig. 10.26 Director and polarizer axes in twisted nematic cell with field off.

The intensity is maximized by having the director at 45° to the polarizers, and making the liquid thickness such that the phase lag ϕ is 2π. In order to keep the cell thickness large enough for cheap manufacture (several microns) Δn must be relatively small. Note that in switching between the two states shown in figure 10.25 the liquid crystal molecules remain parallel to the glass surfaces; this process is therefore called *in-plane switching* (IPS). Because the molecules' axes stay parallel to the glass the birefringence changes very little with the viewing angle. The contrast and colour fidelity remain excellent even when the LCD is viewed up to 80° off axis. Nowadays LCD pictures are generally bright ($\sim$500 cd m^{-2}) and of high static contrast ($\sim$2000:1) and can be used out of doors. The dynamic contrast attained by darkening the backlight in dark scenes can be several thousand. The other technology choice for TV monitors is to use plasma screens. These have a better inherent contrast, the pixels being black when off, and superior brightness when on. LCD as well as plasma response is now rapid enough to display fast sports without introducing any smearing.

Optical projectors are manufactured using LCDs.[10] The notional layout of such a projector is shown in figure 10.27 using three LCDs of around 20 mm lateral dimension and with a pixel pitch of about 15 μm. Mirrors with multiple layer dielectric coatings are used to selectively reflect the red, green or blue components of the incident white light onto the individual LCDs. The LCDs are of simple construction because no colour masks are needed and the beam is concentrated over a narrow angular range.

10.8.3   Polymer dispersed liquid crystals (PDLC)

Figure 10.28 shows the structure of a PDLC, in which bubbles of a nematic liquid crystal are dispersed randomly and densely within an isotropic polymer matrix. The bubbles are a few microns in diameter. Sheets of such structured material are now used to make windows which can be switched from clear to opaque in some tens of milliseconds. For this application the materials are selected to make the ordinary refractive index, n_{o}, of the uniaxial nematic liquid identical to the refractive index of the polymer. Both faces of the PDLC sheet are coated with transparent electrodes. As shown in the right hand panel, when a voltage is applied across the sheet it aligns the liquid crystal molecules perpendicular to the surfaces. Then light at normal incidence, irrespective of its polarization, enters a material with a single refractive index, n_{o}. The window is transparent. When the applied field is removed the molecules in each bubble take up orientations that are influenced by their interaction with molecules of the polymer surface and their optic axes become isotropically distributed. Thus the refractive index of the bubbles be-

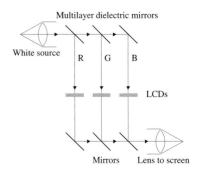

Fig. 10.27 LCD projector. The LCDs can be either simple twisted nematic or polymer dispersed LCDs because only a narrow beam is required.

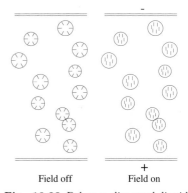

Fig. 10.28 Polymer dispersed liquid crystal. In the left hand panel the bubbles' refractive index is different from that of the polymer. In the right hand panel the refractive indices are the same for light incident normally.

---

[10]LCDs now share projection applications with micro-electromechanical devices. See Chapter 2.

comes n where

$$n^2 = (n_e^2 + 2n_o^2)/3,$$

which differs markedly from the polymer refractive index. In this state the densely packed bubbles in the PDLC window scatter the incident light so effectively that the surface looks uniformly dull.

10.8.4   Ferroelectric liquid crystals (FELC)

Nematic liquid crystals have a switching rate which, though more than adequate for visual displays, is not fast enough to be of much use in the spatial light modulators mentioned in Section 7.8.1. However another class of liquid crystals, SmC*, can be switched about one thousand times faster. Here the star indicates that the molecules have a definite handedness: they are *chiral*. The molecules possess an intrinsic electric dipole moment perpendicular to both the layer normal and the director direction. As shown in figure 10.29 the electric dipoles in each layer lie parallel and a layer forms a ferroelectric *domain*. In one layer the director makes a fixed angle with the layer normal, θ. However the chirality of the molecules causes the director direction to twist between layers so that the director direction for successive layers rotates slowly around the cone of semi-angle θ shown in figure 10.29. The director direction completes a full circuit around the cone after a few thousand layers.

A ferroelectric LCD cell is shown face on in figure 10.30. The thickness of the layer of liquid crystal is reduced to about $2\,\mu$m so that the directors are forced into one of the two orientations, OA and OB, 2θ apart. In this *surface stabilized* state all the molecules can be simultaneously aligned in one of these two alternative orientations by applying the electric field in the directions indicated. Switching is rapid because the applied field acts on the electric dipole moment of each domain, $\mathbf{P}$, giving a very high torque, $\mathbf{P} \wedge \mathbf{E}$. Another useful property of the surface stabilized regime is that once switched the alignment is stable and consequently power is only required to switch, and not to maintain the state of each pixel.

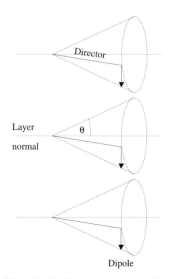

Fig. 10.29 Orientation of the director and electric dipole moment in one layer of a layered SmC* liquid crystal. Going rightward from one layer to the next the director rotates around the cone taking the dipole with it.

Crossed polarizers sandwich the FELC. In the upper panel of figure 10.30 the directors are aligned along OA so this is also the optic axis. The incident light has extraordinary polarization in the liquid crystal and is transmitted without change of polarization. It is therefore extinguished by the analyser. In the lower panel, with the electric feld reversed, the directors now align parallel to OB and this is now the optic axis. It follows that the incident light can be resolved into a component polarized along OB, and a component polarized at right angles along OC. The former has extraordinary polarization and the latter has ordinary polarization. Their electric fields on entering the liquid crystal are

$$E_e = \cos 2\theta; \quad E_o = \sin 2\theta.$$

After transmission, ignoring common factors, their values become

$$E_e = \cos 2\theta; \quad E_o = \sin 2\theta \exp(i\phi),$$

where ϕ is the phase difference arising from the difference between the ordinary and extraordinary refractive indices, Δn. We can now re-use the analysis that led to eqn. 10.54. This shows that the time averaged intensity passing through the analyser is

$$\overline{I}_A = \sin^2 4\theta \sin^2 \phi/2$$
$$= \sin^2 4\theta \sin^2 (\pi \Delta n d/\lambda),$$

where λ is the wavelength in air, Δn is the difference in the refractive indices and d is the thickness of the liquid crystal. The first factor in this equation is largest when θ is 22.5°, which is fortunately within the range of angles available with SmC* liquid crystals. The second factor is maximal when $d = \lambda/2\Delta n$. Although ideal in their role in spatial light modulators, the FELC but are not as yet competitors for the display market.

10.9 Further reading

Polarization of Light by S. Huard, published by John Wiley and Sons (1997). A comprehensive modern text on polarization including the Mueller/Stokes formalism and a detailed analysis of devices such as modulators.
Modern Optics by R. D. Guenther published by John Wiley and Sons (1990) contains a comprehensive and comprehensible description of the numerous ellipsoids used to describe the properties of birefringent materials.
'The history of liquid crystal displays' by H. Kawamoto in the *Proceedings of the IEEE*, Volume 90, number 4 (2002).

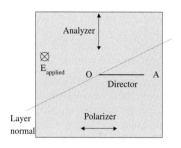

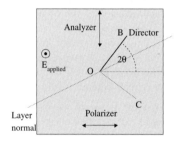

Fig. 10.30 Surface stabilized ferroelectric liquid crystal cell, seen face on. The alignments of the directors are shown for alternative signs of the applied electric field across the cell.

Exercises

(10.1) (a) What is the Jones matrix for a linear polarizer with its transmission axis at 45° to the x-axis? Can you interpret the result? (b) What is the Jones matrix for a layer of isotropic absorber which reduces the incident amplitude by a factor t?

(10.2) What is the instantaneous Poynting vector for the left circularly polarized wave whose electric field is given by eqn. 10.4?

(10.3) Consider a version of the Wollaston prism with the component prisms being cut with a very acute an-gle, 2.5°, instead of 45°. These two prism are also glued together. What is the phase delay between the two polarizations when the thicknesses of the two prisms where the beams cross them are d_1 and d_2? What useful property does this Babinet compensator possess?

(10.4) Show that the Wollaston prism in figure 10.12 with $\theta = 45°$ separates the two beams by 20°.

(10.5) An unpolarized light beam of intensity I_0 is incident perpendicularly on two Polaroid sheets in series. These are rotated in their own planes about the beam as axis. One rotates anticlockwise, the other clockwise, both at angular frquency ω. What is the intensity variation with time? At what frequency does the polarization vector of the light transmitted rotate?

(10.6) Determine over what range of input frequency the modulator described in Section 10.7.1 would be able to convert an electrical to an optical signal without distortion. What are the corresponding limits on the swing of the signal voltage around V_π such that the response remains linear to better than 0.3%?

(10.7) Calculate the magnetic field required to produce a $45°$ rotation of the plane of polarization of light of $633\,\text{nm}$ wavelength in a Faraday rotator made from a $2\,\text{cm}$ long terbium doped glass rod. If the rotator is then used with light of wavelength $533\,\text{nm}$ in the isolator shown in figure 10.20 what fraction of a reflected beam would penetrate the isolator?

(10.8) A laser beam of wavelength $589\,\text{nm}$ in free space is at normal incidence on a calcite plate whose optic axis is parallel to the surface. What are the wavelengths in calcite of the ordinary and extraordinary waves? What fraction of each polarization component enters the crystal?

(10.9) Resolve the elliptically polarized wave in eqn. 10.7, when ϕ is $\pi/2$, into a circularly and a plane polarized component. Write the result in terms of Jones vectors of this complex electric field.

(10.10) Show that the angle, α, between the electric field and electric displacement of the extraordinary wave is given by

$$\cos\alpha = [n_\text{e}^2 \cos^2\theta + n_\text{o}^2 \sin^2\theta]/\sqrt{n_\text{e}^4 \cos^2\theta + n_\text{o}^4 \sin^2\theta}.$$

This is also the angle between the Poynting vector and the wave vector, and is called the Poynting vector walk-off. Use eqns. 10.24 and 10.25; then $\cos\alpha = \mathbf{E}_\text{e} \cdot \mathbf{D}_\text{e}/(D_\text{e} E_\text{e})$.

(10.11) Prove eqn. 10.23 by writing the x- and z-components of eqn. 10.20 in terms of the refractive indices.

(10.12) Sketch the index ellipsoid for a positive uniaxial crystal. Hence check the sign change in eqn. 10.29 between negative and positive uniaxial crystals.

Scattering, absorption and dispersion

11.1 Introduction

Light travelling through matter interacts with the atoms and molecules, which results in three familiar effects: absorption, scattering and dispersion. These effects and the connections between them are described in this chapter and interpreted using the classical theory of how electromagnetic waves interact with matter. This approach provides many insights that, after reinterpretation, retain their value when quantum theory is developed in later chapters.

In the sections immediately below this one scattering will be discussed. Scattering from electrically polarizable particles with dimensions very much smaller than the wavelength of the radiation is called Rayleigh scattering. The blue colour of a clear sky is one consequence of Rayleigh scattering, in this case of sunlight scattered from molecules in the atmosphere. It will emerge that coherent Rayleigh scattering is the underlying process in slowing light down within transparent materials. Mie scattering is the name applied to scattering of light from larger particles whose size may be anywhere from about one tenth to a hundred times the wavelength. Interference effects between light scattered from different parts of a scatterer now come into play.

Short sections are used to introduce absorption and to point out correlations between absorption and dispersion that are due to atomic or molecular processes. Then the classical theory of dispersion and absorption in dielectrics based on atomic oscillators is outlined. Following this the optical properties of metals are interpreted using a classical model in which the conduction electrons are regarded as free within a metal. This model explains the shallow penetration of electromagnetic waves into metals at optical frequencies, and the accompanying high reflectance of metals. Free electrons undergo plasma oscillations at a frequency somewhere in the ultraviolet, which depends on the electron density in the metal. The effects of such oscillations on electromagnetic wave propagation are also interpreted using the model.

Difficulties over the definition of the velocity of electromagnetic waves, and the way these difficulties are resolved, are the topics considered in the next section of the chapter. For example, at wavelengths close to the centres of sharp spectral lines the velocity of an infinite sinusoidal plane wave can exceed c. However when the velocity of light is defined as the velocity of energy and information transfer by wavepackets rather than that of idealized infinite waves there is no violation of the postulates of the special theory of relativity. A final section describes the excitation of surface plasma waves at the boundary between a metal and a dielectric, and their recent use in biological sensors.

11.2   Rayleigh scattering

The scattering of sunlight from gas molecules in the upper atmosphere is responsible for the blue of the sky. In addition if the blue sky is viewed through a Polaroid, looking at right angles to the direction linking the observer to the Sun, the light is found to be strongly polarized. Scattering, as in this example, in which the particles doing the scattering are much smaller than the wavelength of the radiation, is called Rayleigh scattering. Whenever this is the case the electric field is uniform across each individual scatterer, taken here to be a gas molecule. An instantaneous dipole moment is induced in the molecule of magnitude

$$p(t) = \alpha \varepsilon_0 E_0 \exp(-i\omega t), \tag{11.1}$$

where $E = E_0 \exp(-i\omega t)$ is the electric field of the wave and α is the polarizability of the molecule. Each dipole, thus excited, radiates at the frequency of the incident light and in this way scatters light out of the incident beam. Using eqn. 9.58 the time averaged power scattered by one molecule is

$$U = \omega^4 \alpha^2 \varepsilon_0 E_0^2 / (12\pi c^3). \tag{11.2}$$

If there are N molecules per unit volume the power scattered out from unit area of the beam in a small distance $\mathrm{d}z$ along the beam is $\mathrm{d}W = NU\mathrm{d}z$. The incident power over unit area of the beam is $\varepsilon_0 c E_0^2/2$ so that the fractional power loss is

$$\begin{aligned} \mathrm{d}W/W &= -[N\omega^4 \alpha^2/(6\pi c^4)]\,\mathrm{d}z \\ &= -(8/3)\pi^3 \alpha^2 N\,\mathrm{d}z/\lambda^4. \end{aligned} \tag{11.3}$$

The strong dependence on wavelength causes blue light to be scattered about ten times more effectively than red light, which accounts for the blue of a clear sky. It also explains why light from the setting Sun, which has a long path through the atmosphere, should look red. Figure 11.1 has a viewer looking at the blue sky in a direction at right angles to the line passing through the Sun. Any molecule scattering light first absorbs the light, becoming polarized in essentially the same direction as the light absorbed, and then re-emits light. The angular distribution of the scattered light can then be inferred from figure 9.7. When the molecule is excited by light polarized perpendicular to the plane of scattering, as seen in the upper panel of figure 11.1, the observer is viewing

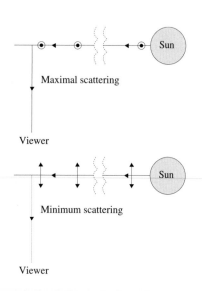

Fig. 11.1 Rayleigh scattering at $90°$. In the upper panel light is polarized perpendicular to the scattering plane, and in the lower panel it is polarized in the scattering plane.

in a direction for which the intensity is at a maximum. On the other hand, if the light is polarized in the plane of scattering, as illustrated in the lower panel of figure 11.1, then the molecular dipole points towards the observer and no light would be seen. However there can be some misalignment of the molecular dipole axis with the electric field inducing it due to asymmetry of the structure of the molecule, and then a little light would be seen. In any case the light received is strongly polarized perpendicular to the plane of scattering. Away from this viewing direction the polarization falls off rapidly.

The expression for the fractional power loss in eqn. 11.3 can be brought into a form that contains the refractive index instead of the molecular polarizability. Firstly using eqns. 9.4 and 9.5

$$\varepsilon_r = 1 + Np/\varepsilon_0 E = 1 + N\alpha. \tag{11.4}$$

Specializing to gases at low pressures, the relative permittivities are close to unity, so that to a good approximation

$$\varepsilon_r - 1 = 2(n - 1).$$

Replacing ε_r in the previous equation gives

$$\alpha = 2(n - 1)/N.$$

Finally replacing α in eqn. 11.3 yields

$$dW/W = -(32/3)\pi^3(n - 1)^2 \, dz/(N\lambda^4), \tag{11.5}$$

and after integrating over the path length z,

$$W(z) = W(0) \exp(-\beta z), \tag{11.6}$$

where $\beta = (32/3)\pi^3(n - 1)^2/(N\lambda^4)$. The distance, $1/\beta$, in which the intensity falls by a factor e is called the attenuation length. Light of 500 nm wavelength travelling in unpolluted air at sea level has an attenuation length of 65 km: that is a power loss in air of $12\,\text{Mm}^{-1}$ (parts per million in one metre).

The scattering from an individual dielectric sphere of radius a and refractive index n in air is calculated in exercise 11.10 below. A quantity called the *cross-section* is now defined as the total scattered flux divided by the incident flux per unit area of the incoming beam. It is thus the equivalent area from which light is removed by the scattering sphere:

$$\sigma = (8\pi/3)(2\pi/\lambda)^4 a^6 G^2, \tag{11.7}$$

where $G = (n^2 - 1)/(n^2 + 2)$. The intensity of Rayleigh scattering therefore increases with the cube of the geometric area of the scatterer, as well as falling off with the fourth power of the wavelength.

11.2.1   Coherent scattering

The scattering from the atoms or molecules in condensed matter is now examined and this will bring out the close connection between coherent scattering and refraction. Figure 11.2 shows a large area, very thin, flat sheet of some transparent dielectric of thickness s on which plane sinusoidal waves are arriving at normal incidence. The electric fields felt by all the molecules such as that labelled M are identical. Consequently their induced dipole moments are all in phase and can be written

$$p(t) = \alpha\varepsilon_0 E_0 \exp(-i\omega t).$$

These dipoles radiate at the same frequency as the incident radiation. In order to calculate the electric field at P due to scattered light the dielectric layer is divided into Fresnel zones centred on O. Then applying a result proved in Section 6.12, the total amplitude at P due to scattered radiation from the whole sheet is equal to half that produced by the first Fresnel zone. Now the field at P due to a dipole located on axis at O is given by eqn. 9.55

$$e_d(kz - \omega t) = \alpha\omega^2 E_0 \exp[i(kz - \omega t)]/(4\pi c^2 z)$$

pointing parallel to electric vector of the incident light. From eqn. 6.55 the surface area of the first zone is $\pi\lambda z$, and hence the number of dipoles within this zone is

$$N_d = \pi z\lambda(Ns) = 2\pi^2 z Ns/k,$$

where N is the number density of the molecules. In calculating the total electric field at P it is important to recall that the phasor formed by the electric fields in the first Fresnel zone turns through a half circle. The resultant electric field is therefore $2/\pi$ times the direct sum of that due to these dipoles, and is $\pi/2$ out of phase with the field produced by a dipole located on axis. After taking these factors into account the electric field at P due to scattered radiation is

$$\begin{aligned}
E_s &= (1/2)(2/\pi)N_d e_d(kz - \omega t + \pi/2) \\
&= (2\pi z Ns/k)\alpha\omega^2 E_0 \exp[i(kz - \omega t + \pi/2)]/(4\pi c^2 z) \\
&= (Nks\alpha/2)E_0 \exp[i(kz - \omega t + \pi/2)].
\end{aligned}$$

Thus the total electric field at P made up of the electric fields of the scattered and unscattered light is

$$\begin{aligned}
E_p &= E_0 \exp[i(kz - \omega t)] + E_s \\
&= E_0 \exp[i(kz - \omega t)][1 + iNks\alpha/2] \\
&\approx E_0 \exp[i(kz - \omega t)] \exp[+iNks\alpha/2], \qquad (11.8)
\end{aligned}$$

where the approximation is valid when s is sufficiently small. The effect of scattering is therefore equivalent to increasing the path length in the

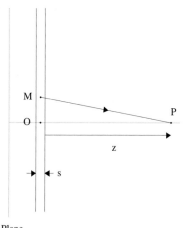

M

O

z

s

Plane
wavefront

Fig. 11.2 Coherent scattering from a thin plane layer of dielectric.

dielectric from s to $s + Ns\alpha/2$, that is by a factor which we can identify as the refractive index of the dielectric

$$n = 1 + N\alpha/2. \qquad (11.9)$$

In turn the relative permittivity is

$$\varepsilon_{\mathrm{r}} = (1 + N\alpha/2)^2 \approx 1 + N\alpha,$$

which agrees with eqn. 11.4.

What emerges from this analysis is that the refractive index of dielectrics has its origin in the interference of coherently scattered with unscattered light. The following picture is helpful. Light travels at its free space velocity c in the open spaces between atoms/molecules; scattering produces a wave of small amplitude which is delayed in phase by $\pi/2$ with respect to the direct wave; the resultant wave is little changed in magnitude compared to the incident wave, but suffers a phase shift proportional to the scattering amplitude.

Referring back to the previous section it can be seen that the scattering from a layer of gas was handled as if each molecule scattered incoherently from every other molecule. In direct contrast, in this section scattering from condensed matter has been treated as coherent. The reason for the difference in approach is that the scattering centres must be densely and uniformly distributed if the scattering is to be fully coherent. This is not the case in a gas, but it is nearly correct in a transparent solid like glass. If there were such a thing as a perfectly homogeneous material then the only consequence of Rayleigh scattering would be to give it a refractive index different from unity.

A simple argument is used here to make it plausible that the Rayleigh scattering from a gas can also be interpreted as scattering from the density fluctuations in the gas. Suppose that a unit volume of gas is divided into many equal volumes v which are small enough so that within each volume the molecules feel nearly the same electric field. Now the number of molecules per unit volume in a gas follows a Poissonian distribution, so that each cell would contain on average Nv molecules with an rms deviation $\sqrt{Nv}$. A simple view to take is that there are $1/v$ inhomogeneities, each containing $\sqrt{Nv}$ molecules, and that each inhomogeneity scatters incoherently of all the others. This yields a total intensity proportional to $(\sqrt{Nv})^2/v = N$, which is exactly the intensity expected from the scattering off N incoherent scatterers!

11.3 Mie scattering

When scattering takes place from larger particles whose size approaches the wavelength of light the analysis of scattering from a single particle becomes complicated. There is interference between light scattered from different parts of the same scatterer, and the phase delays between light travelling through different thicknesses of the scatterer must be taken into account. The first detailed study for spherical scatterers was made by Mie in 1908 and his analysis covered scattering from spheres of all diameters. The Rayleigh treatment is an adequate approximation for diameters up to roughly one-tenth of the wavelength of the radiation, while at diameters greater than one hundred times the wavelength of the radiation the ray theory is an adequate approximation. Scattering from spheres in the intermediate range where neither approximation works is thus generally called *Mie scattering*. Only the salient features of Mie scattering from dielectric spheres will now be discussed.

The interference between radiation scattered from different parts of the scatterer produces a diffractive angular distribution with lobes. As the diameter of the sphere increases, correspondingly the angular distribution gradually becomes more forward peaked with a lesser backward peak. When the sphere diameter is $\lambda/4$ (5λ) the ratio of the scattering intensity in the forward direction with respect to that in the backward direction is around 2.5:1 (2000:1). The total amount of scattering is expressed as an equivalent area from which the scattering sphere removes light from the beam: the *cross-section* per sphere. The rapid rise in cross-section with radius seen in Rayleigh scattering tails off and settles down to being proportional to the radius squared, and hence to the area of the scatterer, at radii a few times λ/n, where n is the refractive index of the scatterer. The strong dependence of the scattering intensity on the wavelength characteristic of Rayleigh scattering changes to a flatter dependence so that the light scattered from clouds, fog and aerosols, in which the droplet diameters are typically $10\,\mu$m, is white.

White paints contain a clear polymer matrix loaded with transparent particles of a very high refractive index material, titanium dioxide; these particles having dimensions of around half the wavelength of light. The Mie scattering is intense because of the big difference between the refractive indices of the polymer (~ 1.5) and titanium dioxide (~ 2.76 in one crystalline form), and is nearly constant across the visible spectrum. Thus the paint, or rather the backscattered ambient light, looks white. The scattering is sufficiently strong that a thin layer of paint can easily mask the colour of the underlying surface. A related effect was described in Section 10.8.3. Micron-sized liquid crystal droplets dispersed in a clear polymer matrix scatter light so effectively that light passing through is totally diffused. However when an electric field is applied the refractive index of the droplets becomes equal to that of the polymer and the panel is rendered transparent.

In a recent application the diameter of a sonoluminescent bubble has been tracked by Mie scattering during the rapid oscillations in which the diameter changes from ten microns to a fraction of a micron in one microsecond.[1] What was done was to measure the light scattered into a large solid angle centred on a scattering angle of 60° when the bubble was illuminated by a series of short pulses of light. The pulse duration was 0.2 ps and the repetition rate 76 MHz. In the angular range selected the integrated scattered intensity is proportional to the bubble area which provided a simple way to estimate the bubble diameter at each exposure.

11.4 Absorption

Light can be lost from a beam by absorption as well as scattering. Absorption of light takes place on electrons bound in atoms or molecules, and on free electrons in metals. When white light passes through a gas containing atoms of a single element, and the spectrum of the emerging light is viewed with a grating spectrometer, the spectrum is seen to be marked by multiple thin dark lines. These lines mark the individual narrow wavelength ranges at which absorption is strong and are characteristic of the atomic structure of the element.[2] The atoms may promptly re-emit the radiation or promptly emit radiation at a longer wavelength. This radiation is known as *fluorescence*, of which radiation at the same wavelength as the incident radiation is called *resonance fluorescence*. *Phosphorescence* is the term applied if there is a delay of greater than a microsecond before the secondary radiation emerges. In denser materials the absorption occurs over broad bands of wavelength rather than narrow lines, and of course some materials absorb all the light. The energy absorbed is generally converted through atomic processsses and collisions to heat. Scattering of light can therefore be interpreted as the absorption of light in which the absorption is followed by prompt re-radiation at the same wavelength.[3]

A beam of light traversing matter loses intensity through scattering out of the beam and through absorption. Coefficients of absorption, β_a, and scattering, β_s, can be defined such that in a thin layer of thickness dz the beam loses fractions of its intensity $\beta_a dz$ and $\beta_s dz$ through these two processes. Then the attenuation of a narrow beam over a distance z is given by

$$I(z) = I(0) \exp\left[-(\beta_s + \beta_a)z\right]. \tag{11.10}$$

With a broad beam the scattering from one part of the beam to another must be allowed for. At one extreme, in metals, the intensity of light falls by a factor e in a few nanometres while in normal window glass this distance is 0.3 m. Materials with an open structure such as plant leaves acquire their colour through selective absorption of the light which enters and is multiply scattered within the structure. Photosynthesis is the absorption process and what emerges is the unused green light. By contrast the colour of metals is due to the strong reflection from the surface. Gold appears reddish because it absorbs more strongly at the red end of the spectrum and reflects red light more effectively. How it comes about that strong reflection is associated with strong absorption

[1] K.R. Weninger, B.P. Barber and S.J. Putterman, *Physical Review Letters* 78, 1799 (1997).

[2] The fact that light restricted to a very narrow range of wavelengths forms a line in the image plane of a spectrometer has led to the light itself being called a spectral line.

[3] Raman and Brillouin scattering, which are processes that become significant at high intensities of illumination, will be considered later after quantum theory and lasers have been introduced.

in this case poses a quandary that is resolved in Section 11.6.

11.5   Dispersion and absorption

Dispersion and absorption are processes that share a common origin in atomic or molecular processes. Figure 11.3 shows schematically the

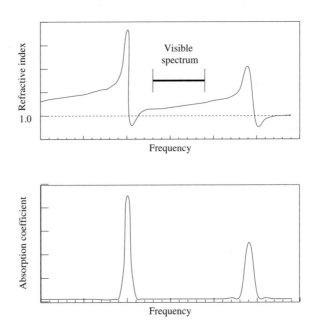

Fig. 11.3 Schematic plot of the refractive index and absorption coefficient for a material transparent across the visible spectrum.

variation of the refractive index and the absorption coefficient of a transparent material across the infrared to ultraviolet part of the spectrum. There are clear pairings between the absorption peaks and characteristically shaped oscillations in the value of the refractive index. In general many such pairs may be seen for a dielectric. Such sharp absorption peaks account for the lines seen in the absorption spectrum of a gas. The equivalent distributions for materials like glass or water show similar correlations, but with broader lines and additional features due to the mutual interaction of the densely packed atoms.

Each oscillation/peak pair appearing in figure 11.3 is due to a process in which energy is absorbed by atoms or molecules. Excitation of electrons in atoms happens through the absorption of ultraviolet and visible light, while molecular vibrations and rotations can be excited by the absorption of visible and infrared radiation down to microwaves. Transparent materials are transparent because for these materials the electronic absorptions occur in the ultraviolet and the molecular excitations in the infrared with none lying within the visible spectrum. Regions where the refractive index rises with increasing frequency (decreasing wavelength) are said to have *normal* dispersion and regions where the refractive index falls as the frequency increases are said to have *anomalous* dispersion. Common glass therefore has normal dispersion, as can be seen in figure 1.16.

A simple model of the atom as a classical oscillator will be used in the following sections to account for the shapes and correlations between the absorption and dispersion features illustrated in figure 11.3. The quantum interpretation presented later provides a deeper and quantitative explanation of atomic transitions. What the classical view provides are insights that do not lose their value when one uses quantum theory.

11.5.1   The atomic oscillator model

The process considered is the forced oscillation of an electron in an atom caused by the electric field of an electromagnetic wave. Electron velocities are tiny compared to c; thus the magnetic force $e\mathbf{v} \wedge \mathbf{B}$, of order evE/c, is negligible compared to the electric force eE. There are two other forces acting on an electron besides the force due to the electromagnetic wave. The first is the restoring electric force exerted by the stationary nucleus. Compared to the electric field of a nucleus at an electron, of order $10^{11}\,\mathrm{V\,m^{-1}}$, the electric field in any beam other than a high energy pulsed laser is very small. Thus the electron displacement is small compared to the atomic size and the restoring force will be linear in the displacement. The second force is an equivalent damping force used to represent the effect of all the processes which dissipate the energy absorbed by the electron. We shall see later that these include the re-radiation of the energy and processes such as atom–atom collisions in a gas. The equation of motion of the electron in this model is

$$m(\mathrm{d}^2x/\mathrm{d}t^2) = -eE_0 \exp\left(-i\omega t\right) - \xi(\mathrm{d}x/\mathrm{d}t) - \kappa x. \qquad (11.11)$$

Here $-e$ and m are the electron charge and mass respectively; the incident electromagnetic wave's electric field, $E = E_0 \exp\left(i\omega t\right)$, points in the x-direction; ξ is the damping constant and $-\kappa x$ is the restoring force. In the absence of any electromagnetic wave the electron would undergo damped harmonic motion with natural angular frequency, ω_0, given by

$$\omega_0^2 = \kappa/m.$$

Then, putting $\gamma = \xi/m$, the equation of motion can be rewritten

$$-eE_0 \exp\left(-i\omega t\right) = m[(\mathrm{d}^2x/\mathrm{d}t^2) + \gamma(\mathrm{d}x/\mathrm{d}t) + \omega_0^2 x]. \qquad (11.12)$$

The forced motion of the electron will have the same frequency as the driving electric field

$$x = x_0 \exp\left(-i\omega t\right), \qquad (11.13)$$

which when substituted in eqn. 11.12 gives

$$-eE_0 = m(-\omega^2 - i\gamma\omega + \omega_0^2)x_0,$$

after the common factor $\exp{(i\omega t)}$ has been cancelled. Whence

$$x_0 = (-eE_0/m)/(\omega_0^2 - \omega^2 - i\gamma\omega). \tag{11.14}$$

The displacement of the electron from its equilibrium position within the atom means that the atom acquires an electric dipole moment

$$p = -ex_0 \exp{(-i\omega t)},$$

and the *atomic polarizability* is then

$$\alpha = p/(\varepsilon_0 E) = (e^2/\varepsilon_0 m)/(\omega_0^2 - \omega^2 - i\gamma\omega). \tag{11.15}$$

The polarization of the material, that is the electric dipole moment per unit volume, is thus

$$P = N\alpha\varepsilon_0 E, \tag{11.16}$$

The previous analysis of Rayleigh scattering only applies well away from resonances and it also ignores damping. Then α is effectively $e^2/\varepsilon_0\kappa$.

where N is the number of atoms per unit volume. Using eqn. 11.4

$$\varepsilon_r = 1 + P/(\varepsilon_0 E) = 1 + N\alpha. \tag{11.17}$$

Substituting for α from eqn. 11.15 in the last line gives

$$\varepsilon_r = 1 + (Ne^2/m\varepsilon_0)/(\omega_0^2 - \omega^2 - i\gamma\omega)$$
$$= 1 + \omega_p^2/(\omega_0^2 - \omega^2 - i\gamma\omega), \tag{11.18}$$

where ω_p is called, in anticipation, the *plasma angular frequency*, and is given by

$$\omega_p^2 = Ne^2/(m\varepsilon_0). \tag{11.19}$$

This is a *resonant* response which is strongest when the incident radiation has a frequency close to the natural oscillation frequency of the electron in the atom. Close to an individual resonance we can take $\omega \approx \omega_0$. Then eqn. 11.18 simplifies to

$$\varepsilon_r = 1 + (\omega_p^2/\omega_0)/[2(\omega_0 - \omega) - i\gamma]. \tag{11.20}$$

The analysis is now continued for a gas, so that the relative permittivity is close to unity. Then the expression for the refractive index n is to a good approximation given by

$$n = \sqrt{\varepsilon_r} = 1 + (\omega_p^2/4\omega_0)/[\omega_0 - \omega - i\gamma/2]. \tag{11.21}$$

The refractive index therefore has a real and an imaginary part

$$n = n_r + in_i, \tag{11.22}$$

where n_i is both real and positive. Referring back to the analysis of TIR given in Chapter 9 it can be seen that if the refractive index has an imaginary part this means that the electromagnetic wave is attenuated.[4] After travelling a distance z the electric field becomes

$$E(z) = E(0) \exp\left[i\omega(nz/c - t)\right]$$
$$= E(0) \exp\left(-n_i z\omega/c\right) \exp\left[i\omega(n_r z/c - t)\right].$$

Therefore the intensity falls off exponentially,

$$I(z) = I(0) \exp\left(-2n_i\omega z/c\right), \qquad (11.23)$$

giving an absorption coefficient $\beta_a = 2n_i\omega/c$ proportional to n_i. The real part n_r must be identified as the standard refractive index which could be determined by measuring the phase delay of light passing through the dielectric. Separating the real and imaginary parts of eqn. 11.21 gives

$$n_r = 1 + \omega_p^2(\omega_0 - \omega)/(4\omega_0 R), \qquad (11.24)$$
$$n_i = \gamma\omega_p^2/(8\omega_0 R), \qquad (11.25)$$

where

$$R = (\omega_0 - \omega)^2 + \gamma^2/4.$$

This result confirms that the Lorentzian line shape introduced in Chapter 7 is the atomic line shape. Figure 11.4 shows this predicted variation of the real and imaginary parts of n as a function of frequency around the natural frequency of oscillation of the electron. The vertical scale shown is arbitrary and in the case of a gas the numbers shown for $n_r - 1$ and n_i would be scaled down by a factor of order 1000. This response reproduces both the shapes and correlations between the features present in figure 11.3: each absorption peak is accompanied by a characteristically shaped rapid variation of the refractive index involving a region of anomalous dispersion.

The responses of all the electrons in an atom have to be taken into account. Then the expression for the real and imaginary parts of the refractive index become

$$n_r = 1 + \omega_p^2 \sum_i [f_i(\omega_{0i} - \omega)/(4\omega_{0i} R_i)],$$
$$n_i = \omega_p^2 \sum_i [f_i\gamma_i/(8\omega_{0i} R_i)].$$

The coefficient f_i is called the *oscillator strength* of the ith electron. Such quantities can only be calculated using the quantum theory of atomic structure and its interaction with electromagnetic radiation.

---

[4]If the choice of complex waves were $\exp\left[i(\omega t - kz)\right]$ rather than $\exp\left[i(kz - \omega t)\right]$ then the sign in front of in_i in eqn. 11.22 would need to be negative. In addition $-i\gamma$ would be replaced by $+i\gamma$ in the preceding equations.

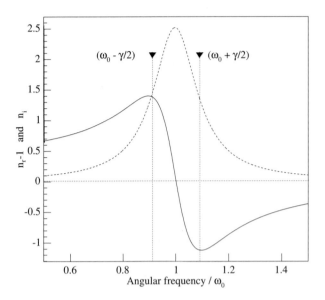

Fig. 11.4 Predictions for the real and the imaginary parts of the refractive index around resonance. The real part is indicated with a solid line and the imaginary part with a broken line. Note that $(n_r - 1)$ is plotted rather than n_r. The vertical scale is arbitrary. At $\omega = \omega_0 \pm \gamma/2$ the imaginary part is half the height at resonance, and the real part has its extremal values.

Consider the case when the damping is weak enough that $\gamma \ll \omega_0$. Then the peak absorption is at the angular frequency ω_0: the peak value of n_i is $\omega_p^2/(2\gamma\omega_0)$, and the width between the angular frequencies at which n_i falls to half its peak value is γ. Thus if the damping at the atomic level is weak the peak resonant absorption is large, the absorption line is narrow and the anomalous dispersion is more pronounced.

The assumption has been made implicitly in the preceding analysis that the electric field felt by the atoms is spatially uniform across the dielectric. In fact the local field felt by the atom will be the vector sum of this field and that due to the surrounding atoms. It can be shown that, when the dielectric is a non-polar liquid or a cubic crystalline material, the local electric field has a simple relationship to the applied field[5]

$$E_{\text{local}} = E + P/(3\varepsilon_0). \tag{11.26}$$

Then eqn. 11.16 becomes

$$P = N\,\alpha\,\varepsilon_0\,E_{\text{local}} = N\,\alpha\,\varepsilon_0\,[E + P/(3\varepsilon_0)],$$

---

[5]See for example Chapter 2 of the 7th edition of *Optics* by M. Born and E. Wolf, published by Cambridge University Press (1999). For other materials the numerical factor is different from 1/3.

and substituting $\chi\varepsilon_0 E$ for P gives the Clausius–Mossotti relation

$$N\alpha = \chi/(1 + \chi/3) = 3(n^2 - 1)/(n^2 + 2). \qquad (11.27)$$

With this replacement in eqns. 11.17 and 11.20 we get

$$(n^2 - 1)/(n^2 + 2) = (\omega_\mathrm{p}^2/3)\sum_i [(f_i/\omega_{0i})/(2(\omega_{0i} - \omega) - i\gamma_i)]. \qquad (11.28)$$

In the limit that the refractive index is close to unity the left hand side reduces to $(n-1)/3$, so that the equation collapses to that applicable to a gas. The mean separation of atoms is smaller by a factor ten in solids, liquids and high pressure gases than it is in gases at normal temperature and pressure (NTP, $20°\,\mathrm{C}$ and $10^5\,\mathrm{Pa}$). Thanks to this denser packing the interactions between atoms in liquids and in solids are stronger and change the absorption spectra a great deal. The narrow spectral lines observed with gases at NTP are replaced by broad bands. Excitation of molecules can take the form of vibrations in which the distance between the component nuclei oscillates, or of rotations of the molecule. The masses vibrating are now the nuclei so that the natural frequencies are correspondingly smaller and lie predominantly in the infrared/microwave region of the spectrum.

A further feature of many materials is the presence of molecular structures which possess permanent electric dipole moments. An applied electric field will cause these dipoles to align in the field direction, which makes another important contribution to the polarization. This motion is heavily damped in liquids and solids so that at optical frequencies this contribution to the relative permittivity is smoothly and slowly falling with increasing frequency. The glasses from which so many optical components are made are transparent to visible radiation because the atomic resonances lie in the ultraviolet and the molecular resonances in the infrared. Across the visible spectrum lying between these resonances the refractive indices of glasses shown in figure 1.16 therefore have normal dispersion and are well fitted by the various Sellmeier formulae, such as that given in eqn. 1.24. The poles of the terms, where their value diverges, give a good indication of the location in wavelength of the ultraviolet resonances.

There is an underlying physical connection between dispersion and absorption imposed by *causality*, which goes beyond the particular model used here. Causality is the basic requirement that effects must follow and not precede the forces producing them. The connection is expressed in the Kramers–Kronig relations linking the real and imaginary parts of the susceptibility. These are derived in Appendix I.

11.6   Metallic absorption and reflection

When an electromagnetic wave is incident on a metal it induces a current of free electrons. Energy transfered to the electrons is dissipated in collisions with the lattice of positive ions, and appears as heat. In this way the part of the wave entering the metal is attenuated within a short distance. It was pointed out in Section 9.3 that the ratio of the conduction current to the current arising from polarization in a good conductor like copper is large up to optical frequencies. When modelling the propagation of electromagnetic waves at these frequencies in metals it is enough to take account of the response of the free electrons only. A classical model of this type developed by Drude will be used here to analyse the propagation of electromagnetic waves in a metal.

In metals the least well bound electrons in each atom become detached from the parent atom and form a sea of *free* electrons. Travelling throughout the metal, their paths are punctuated by frequent, mostly inelastic, collisions with the lattice of positive ions, and these collisions serve to bring the ions and free electrons into thermal equilibrium. The effect of a collision is also to randomize the electron's direction after the collision.

An applied *constant* electric field E causes the electrons to acquire a drift velocity, v, in addition to their random motion; the total velocity is the vector sum of these two components. Newton's second law applied to the electron drift motion gives

$$m\mathrm{d}v/\mathrm{d}t = -eE,$$

$-e$ and m being the electron charge and mass respectively. However the electrons suffer collisions and after each collision the electron direction is random, which has the effect of reducing the mean drift velocity after collision to zero. A measure of the mean drift velocity can be obtained by integrating the last equation over the mean time, τ, between collisions

$$mv/\tau = -eE. \tag{11.29}$$

This result shows that the collisions with the lattice have the same effect as a damping force mv/τ opposing the electric force. At room temperatures in metals such as aluminium, copper and silver τ is around 10^{-14} s giving a drift velocity of order $0.02\,E\,\mathrm{m\,s}^{-1}$ where the field, E, is expressed in $\mathrm{V\,m}^{-1}$. Therefore drift velocities are many orders of magnitude less than the velocities of the free electrons. Equation 11.18 for the relative permittivity of a dielectric can be easily adapted to become applicable to a metal by making two simple changes. Firstly the term arising from the restoring force exerted by the parent nucleus is deleted, and secondly the damping force per unit momentum, γ is $1/\tau$. The equation obtained in this way for the relative permittivity of a metal is

$$\varepsilon_{\mathrm{r}} = 1 + \omega_{\mathrm{p}}^2/(-\omega^2 - i\omega/\tau). \tag{11.30}$$

Any contribution from the electrons that remain bound in the atoms has been ignored here. As before $\omega_{\mathrm{p}}^2 = Ne^2/m\varepsilon_0$, with N now being the number density of free electrons in the metal. When the frequency of the electromagnetic waves is well above $1/\tau$ there are many cycles of the waves between one electron–lattice collision and the next, which means that damping becomes less and less important as the wave frequency rises.

In the case of copper the number density of free electrons is around $8.5\,10^{28}\,\mathrm{m}^{-3}$ and the mean interval between collisions is approximately $2.4\,10^{-14}\,\mathrm{s}$. An estimate of the plasma frequency in copper, based on these values, is $2.6\,10^{15}\,\mathrm{Hz}$; the corresponding wavelength lies well into the ultraviolet. Using these values in eqn. 11.30 gives a value for the imaginary part of the refractive index in rough agreement with the observed value, but the real part is grossly underestimated. This indicates the limitations of the classical Drude model as far as quantitative predictions are concerned in the visible part of the spectrum. However the Drude model provides a consistent parametrization of the data at longer wavelengths. The intensity of light falls with distance in a metal like

$$I(z) = I(0)\exp\left(-\beta_a z\right), \tag{11.31}$$

where the absorption coefficient is given by

$$\beta_a = 2\omega n_i/c.$$

Thus the intensity falls off by a factor e in a distance called the *skin depth*

$$s = c/(2\omega n_i) = 0.08\lambda/n_i,$$

where λ is the wavelength in air. The skin depth is approximately $18\,\mathrm{nm}$ for light of wavelength $589\,\mathrm{nm}$ falling on copper. Of course metals reflect well – which at first sight seems hard to reconcile with the idea that they also absorb strongly. This point is now discussed.

The reflection coefficient for light falling at normal incidence on a dielectric is given in eqn. 9.79. In the case of reflection from metals where the refractive index can be complex this becomes

$$R_0 =\mid (n_1 - n_2)/(n_1 + n_2) \mid^2, \tag{11.32}$$

so that at an air/metal interface

$$R_0 = [(n_r - 1)^2 + n_i^2]/[(n_r + 1)^2 + n_i^2]. \tag{11.33}$$

It is clear that if the imaginary part of the refractive index is much larger than the real part, as is the case with metals, then the reflectance must be close to its upper limit of unity. The quandary mentioned above is resolved by noting that most of the incident light incident on a metal is reflected, while that small part that is transmitted is absorbed in a short distance within a metal. Figure 11.5 shows how the reflectances of copper, gold, silver and aluminium vary across the visible spectrum. Table 11.1 gives the values of the real and imaginary parts of the refractive index for several metals at a wavelength of $589\,\mathrm{nm}$. Thin layers of

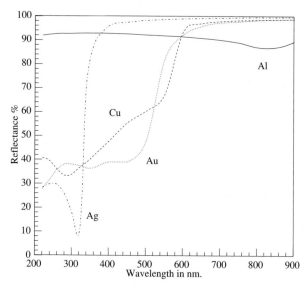

Fig. 11.5 Reflectance off fresh metal surfaces at normal incidence. From *Handbook of Optical Materials*, edited by M.J. Weber and published by the CRC Press, Boca Raton, 2003. Courtesy Taylor and Francis Group, and Professor Weber.

Table 11.1 Table of refractive indices of metals at 589 nm wavelength.

Metal	Real part	Imaginary part
Silver	0.18	3.64
Copper	0.62	2.63
Gold	0.47	2.83

silver and aluminium are deposited on glass to provide mirrors with high reflectance over a broad spectral range. Although multilayer dielectric coatings, which were described in Chapter 9, can give higher reflectance over comparable wavelength intervals these can only be deposited on items of small surface area. The primary mirrors of astronomical telescopes usually have an aluminium coating. Silver is not used despite its higher reflectance because it tarnishes quickly in air.

The upper panel of figure 11.6 shows the typical variation of the reflectance of a metal as a function of the angle of incidence. These curves bear a family resemblance to the corresponding ones for dielectrics shown in figure 9.12, but with larger values. Instead of the sharp change seen at Brewster's angle for purely dielectric interfaces, the phase difference between r_p and r_s falls steadily from π at normal incidence to zero at grazing incidence and passes through $\pi/2$ at what is called the *principal angle of incidence*, $\overline{\theta}_i$. The minimum of the amplitude of the p-polarized reflected wave typically occurs a couple of degrees below the principal angle of incidence for metals, while the angle at which the ratio of p- to s-polarized amplitudes is least is only a fraction of a degree below the principal angle of incidence. The lower panel in figure 11.6 shows the phase difference between reflected s- and p-polarized light.

The rapid extinction of light in a metal makes it impossible to measure the refractive index by using Snell's law, because only the reflected light is available for measurement. It is not adequate to simply measure the reflectances of s- and p-polarized light: their relative phase is needed as well in order to determine the complex refractive index. Suppose the incident light passes through a polarizer set at 45° to the plane of incidence. The light incident on the surface then has equal s- and p-polarized components which are in phase. In general, after reflection the components differ both in magnitude and in phase so that the reflected light is elliptically polarized. The ratio of the reflected amplitudes is

$$\rho = \tan\psi \exp(i\Delta), \qquad (11.34)$$

where $\tan\psi = |r_\mathrm{p}|/|r_\mathrm{s}|$ and Δ is their phase difference. The quantities $\tan\psi$ and Δ are measured using an *ellipsometer* of the sort shown in figure 11.7. A simple type of measurement involving *nulling* will be described. Monochromatic light is first passed through a polarizer set at 45° to the plane of incidence to give s- and p-polarization amplitudes which are equal and in phase. The compensator applies a phase difference between the s- and p-polarized waves which will compensate (null) the phase difference occuring at reflection. After reflection these waves are therefore in phase but their amplitudes are reduced by the respective reflection coefficients. Thus the light reaching the analyser is plane polarized but at an angle determined by the relative magnitude of the reflection coefficients for s- and p-polarization. This angle of polarization is determined by rotating the analyser so that the detector signal falls to zero. Δ and ψ can be calculated from the the compensator and analyser settings, and from these the reflection coefficients can be obtained by applying Fresnel's laws. The relationships are straightforward but lead to complicated expressions. Results are at their simplest when the angle of incidence is made equal to the principal angle of incidence, $\overline{\theta}_\mathrm{i}$. In this case the phase difference between the *s*- and *p*-polarized components after reflection is $\pi/2$, so that the compensator can be a QWP. If the plane of polarization after reflection is at the angle $\overline{\psi}$, then a good approximation for the refractive index of a metal surface is

$$n_\mathrm{r} = -\tan\overline{\theta}_\mathrm{i} \sin\overline{\theta}_\mathrm{i} \cos 2\overline{\psi},$$
$$n_\mathrm{i} = n_\mathrm{r} \tan 2\overline{\psi}.$$

Measurements are made of $\tan\psi$ and Δ at a range of angles of incidence to extract and check the complex refractive index. Variations on this technique are widely used in manufacturing and research to measure the thicknesses of single thin films and stacks of thin films of dielectrics where

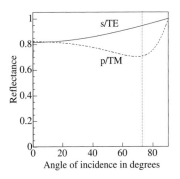

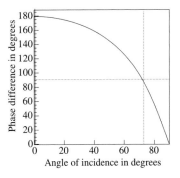

Fig. 11.6 The upper panel shows the reflectances from an air/gold interface for s- and p-polarized light of wavelength 589 nm. The lower panel shows the phase difference between the reflected p- and s-polarized light. The dotted lines in both panels mark the principal angle of incidence, $\overline{\theta}_\mathrm{i}$.

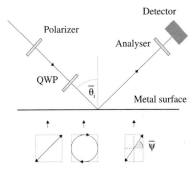

Fig. 11.7 Ellipsometer for determining the optical constants of metals. The successive polarization states are shown for a nulling measurement with light incident at the principal angle of incidence.

[6]For more details see *Spectroscopic Ellipsometry* by H. Fujiwara, published by John Wiley and Sons, New Jersey (2007).

the refractive indices are already known and used as inputs. Reflections from successive interfaces must be included in the analysis, following the lines described in Chapter 9. A modern development is spectroscopic ellipsometry in which the parameters ψ and Δ are measured for a range of wavelengths. Despite the complicated analysis required, ellipsometry gives results with very small errors because only angles are being measured, accurate to about $0.01°$, rather than relative intensities.[6]

11.6.1   Plasmas in metals

Plasmas are gases in which a proportion of the atoms or molecules are ionized into positive ions and electrons, and which are overall neutral. In its equilibrium state the spatial distributions of positive ions and electrons in a plasma are everywhere uniform and equal. One important example of a plasma is met in the upper atmosphere. The Sun's ionizing ultraviolet radiation maintains layers of plasma, known collectively as the *ionosphere*, at heights between 30 and 200 km. The free electrons in metals also qualify as a plasma with a number density of order $10^{28}\,\mathrm{m^{-3}}$ compared to the much lower values, around $10^{11}\,\mathrm{m^{-3}}$, observed in the ionosphere.

Any disturbance of the electric field causes the electrons to move at a much higher speed than the ions. This is because the electron mass is so much smaller, while the electrostatic force has the same magnitude for both an ion and an electron. Such a disturbance moves electrons out of a local region, while the positive ions are almost stationary. In this way the region acquires a net positive charge which attracts the displaced electrons. The displaced electrons then oscillate about their equilibrium position under this restoring force at the plasma frequency, ω_p, previously defined in eqn. 11.19. A simple analysis of plasma oscillations is made next before looking at their effects on electromagnetic wave propagation.

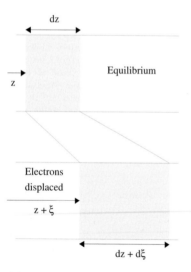

Fig. 11.8 Equilibrium and displaced positions of the free electrons in a plasma.

In figure 11.8 the upper panel shows a plasma in equilibrium and the lower panel the effect of displacing the electrons a distance ξ in the z-direction, where ξ varies in some as yet unspecified manner with z. The electron *number* density in the shaded region falls as a result of this movement by a factor $dz/(dz + d\xi)$. Thus the electron *charge* density within that region changes from $-Ne$ to $-Ne(1 - d\xi/dz)$, N being the equilibrium electron number density. The net charge density in that region, including both the stationary positive ions as well as the electrons, changes from zero in equilibrium to

$$\rho = Ne(d\xi/dz). \tag{11.35}$$

This induces an electric field which is given by Gauss' law

$$\nabla \cdot \mathbf{E} = \rho/\varepsilon_0.$$

In the case considered the field points in the z-direction so this last equation becomes

$$dE/dz = (Ne/\varepsilon_0)\, d\xi/dz,$$

which when integrated gives

$$E = Ne\xi/\varepsilon_0. \tag{11.36}$$

In turn this field acts on each electron, and the equation of motion is

$$md^2\xi/dt^2 = -Ne^2\xi/\varepsilon_0.$$

This is the differential equation for simple harmonic motion, and the long term solution is

$$\xi = \xi_0 \exp\left[-i\sqrt{(Ne^2/m\varepsilon_0)}t\right] = \xi_0 \exp\left[-i\omega_p t\right], \tag{11.37}$$

which justifies the name plasma frequency for $\sqrt{(Ne^2/m\varepsilon_0)}$.

Referring back to eqn. 11.30, it follows that at frequencies high enough that the electron/lattice collisions are unimportant

$$\varepsilon_r = 1 - \omega_p^2/\omega^2,$$

which can be rewritten as

$$n^2 = 1 - \omega_p^2/\omega^2. \tag{11.38}$$

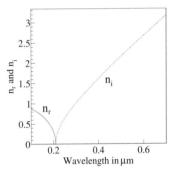

Fig. 11.9 Real and imaginary parts of the refractive index of sodium near the plasma frequency.

We therefore infer that at high enough frequencies the dominant effect of electromagnetic waves incident on a metal is to drive the electron plasma into forced oscillation. Figure 11.9 shows the behaviour of the real and imaginary parts of the refractive index of the alkali metal sodium around the plasma frequency predicted by eqn. 11.38. In the case of sodium the measured wavelength for the plasma oscillation, 210 nm, and that predicted using the free electron density, 209 nm, are in particularly good agreement and lie well in the ultraviolet. The onset of plasma oscillations has a profound effect on the transmission of electromagnetic waves. Below the plasma frequency n^2 is negative and hence the refractive index is purely imaginary. Then the reflection coefficient given in eqn. 11.33 simplifies:

$$R_0 \approx (n_i^2 + 1)/(n_i^2 + 1) = 1. \tag{11.39}$$

Reflection is therefore very strong. Above the plasma frequency n^2 is now positive and the refractive index is purely real and small, so that the reflection coefficient is also small. In the case of the plasmas in the upper atmosphere the plasma frequency is around 3 MHz. Below this frequency short range radio transmissions can travel large distances round the Earth thanks to reflections from the plasma. Above the plasma frequency the plasma becomes transparent and for this reason communication with satellites is at frequencies well above 3 MHz. When a spacecraft re-enters the Earth's atmosphere the ionization produced

In figure 11.5 only silver's reflectance shows any sign of a dip to mark the plasma frequency. Electrons are not free classical particles but inhabit quantized energy bands so that the predictions of the simple model need some modification. See for example Chapters 4 onward in *Optical Properties of Solids* by F. Wooten, published by Academic Press, New York (1972).

around it is so intense that communication at all frequencies is lost.

Attention must be drawn to two apparent violations of the theory of special relativity in the preceding analysis. First notice that above the plasma frequency the refractive index is *less* than unity. Secondly if we look back at figure 11.4 it is seen that in the regions of anomalous dispersion the refractive index can also become less than unity. It seems that in these cases electromagnetic waves would travel faster than *c*, their velocity in free space! The apparent violation of the special theory of relativity in regions of anomalous dispersion was already noted by contemporaries of Einstein and raised as a fundamental objection to the theory. In order to resolve these inconsistencies more careful consideration must be given to what is meant by the velocity of light.

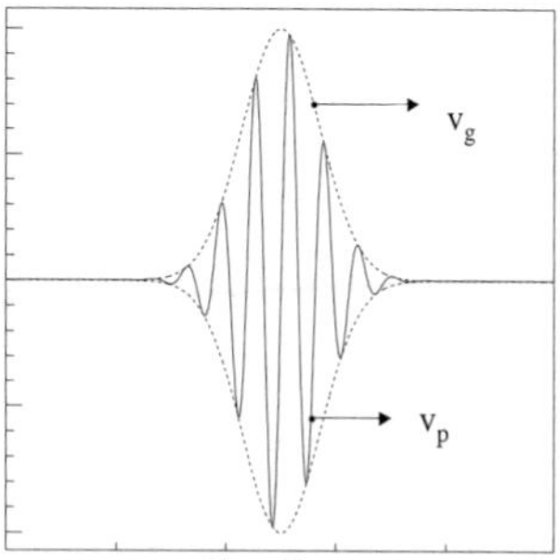

11.6.2   Group and signal velocity

Up to this point the definition used for the velocity of light has been

$$v_{\mathrm{p}} = \omega/k,$$

where ω and k are the angular frequency and the wave number of an infinite plane wave such as $\exp[i(kz - \omega t)]$. We shall call this velocity the *phase velocity* of light, as was done previously in Chapter 10. Infinite waves which span all space and time and have a unique frequency are however never met in nature. Rather electromagnetic radiation consists of wavepackets which are of finite duration and extent and which can be resolved into a superposition of infinite plane waves. There is even a problem in principle with measuring the velocity of a perfect sinusoidal wave. In order to recognize a place on the wavetrain for the purposes of timing its departure and arrival it must be marked in some way – which would change the wave from being a pure sinusoid. It therefore makes sense to consider the velocity of wavepackets as the true measure of the speed of light. The wavepacket peak in figure 11.10 indeed defines where the fields are large and where the Poynting vector is large. The form of the wavepacket is taken to have an electric field

$$E(z, t) = \int E(\omega) \exp[i(kz - \omega t)]\mathrm{d}\omega,$$

centred around wave number k_0 and angular frequency ω_0, where $E(\omega)$ varies slowly with ω. The phase velocity ω/k varies with frequency or equivalently with the wave number. Putting $\Delta\omega = \omega - \omega_0$ and $\Delta k = k - k_0 = \Delta\omega(\mathrm{d}k/\mathrm{d}\omega)$, this can be rewritten

$$E(z, t) = \exp[i(k_0 z - \omega_0 t)] \int E(\omega) \exp[i\Delta\omega(z\,\mathrm{d}k/\mathrm{d}\omega - t)]\,\mathrm{d}\omega, \quad (11.40)$$

where the first term describes the rapidly oscillating waves in figure 11.10 and the second term the envelope. Each wave in the envelope is a function of $(\mathrm{d}k/\mathrm{d}\omega)z - t$, so the velocity of the envelope is

$$v_{\mathrm{g}} = \mathrm{d}\omega/\mathrm{d}k, \quad (11.41)$$

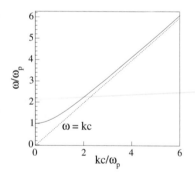

Fig. 11.10 Group and phase velocities of a wavepacket. There would be many thousands of individual wave oscillations inside a wavepacket from an actual light source.

Fig. 11.11 Angular frequency variation with the wave number above the plasma frequency ω_{p}. The dispersion relation for free space is also shown.

which is called the *group velocity*. Surprisingly a wavepacket does not necessarily travel at a velocity close to the mean velocity of its infinite plane wave components. In figure 11.10 the waves within the wavepacket travel at the phase velocity while the wavepacket itself, that is the envelope of the waves, travels at the group velocity. If one could travel parallel to and alongside the wavepacket at its velocity and somehow detect the internal waves, then they would appear to travel through the wavepacket. If the wave velocity exceeds the group velocity the waves would appear to flow forward inside the envelope. Energy and information travel at the group velocity so, provided the group velocity remains less than c, it is of little consequence if the phase velocity exceeds c.

Other useful expressions for the group velocity can be derived from eqn. 11.41

$$1/v_\mathrm{g} = 1/v_\mathrm{p} + (\omega/c)(\mathrm{d}n/\mathrm{d}\omega); \qquad (11.42)$$

$$1/v_\mathrm{g} = 1/v_\mathrm{p} - \lambda\,\mathrm{d}(1/v_\mathrm{p})/\mathrm{d}\lambda, \qquad (11.43)$$

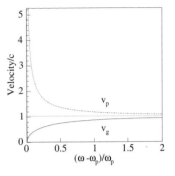

Fig. 11.12 Angular frequency variation with the group and phase velocities above the plasma frequency ω_p.

where λ is, as usual, the free space wavelength. Evidently in free space where there is no dispersion $v_\mathrm{g} = v_\mathrm{p} = c$. An example that brings out the importance of the group velocity is that of electromagnetic waves travelling in the ionosphere. Rewriting eqn. 11.38 gives

$$\omega^2 = n^2\omega^2 + \omega_\mathrm{p}^2 = k^2c^2 + \omega_\mathrm{p}^2, \qquad (11.44)$$

where k is the wave number in the plasma. This behaviour is drawn as a solid line in figure 11.11 for waves with frequencies above the plasma frequency, and asymptotically approaches the form for electromagnetic waves in free space $\omega = kc$. The corresponding phase and group velocities are displayed versus the angular frequency in figure 11.12. Close to the plasma frequency the phase velocity is very large, while the group velocity is close to zero. At high angular frequencies both velocities converge on c. Figure 11.13 shows an ionogram recording of the apparent heights at which short pulses of radio waves are reflected from the ionosphere. The ordinate is directly proportional to the delay between sending and receiving the signal back after reflection off the ionosphere. The abscissa is the frequency of the electromagnetic waves used. The peaks in the ionogram indicate reflection from individual ionized layers at increasing heights with correspondingly increasing electron densities. As the frequency of the transmitter is increased toward the plasma frequency of a particular layer the delay climbs steeply. Evidently the velocity of the signal penetrating the plasma layer slows down, reaching a minimum at the plasma frequency for that layer. This matches the behaviour expected if the velocity of the signal is indeed the group velocity.

Some further care is required in defining the velocity at which information can travel. Consider electromagnetic radiation with wavelength in the region of anomalous dispersion shown in figure 11.4. We can use the expression for the group velocity given in eqn. 11.42. When

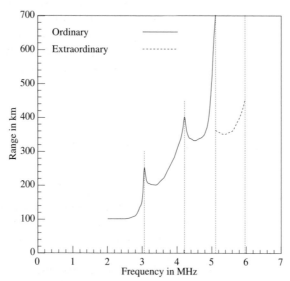

Fig. 11.13 Ionogram showing the apparent height at which pulses of radiation are reflected from the ionosphere versus the frequency. Courtesy Dr M. Rietveld of the EISCAT Scientific Association, N-9027 Ramfjordmoen, Norway.

$\mathrm{d}n/\mathrm{d}\omega$ is large and negative, which it certainly will be across a narrow spectral line, v_{g} can also become larger than c. A narrow line has a correspondingly sharp absorption peak so that radiation entering the medium is strongly absorbed. Consequently the frequency components in the wavepacket closest to the centre of the spectral line are preferentially absorbed, while the components of the wavepacket with frequencies above and below the spectral line are less attenuated. In such cases the energy and information all travel at a velocity less than c. This velocity at which information travels is termed the signal velocity .

K.E Oughstum and N.A. Cartwright in the *Journal of Optics* A4 (2002) S125, conclude that 'Superluminal energy and information transfer is not physically possible within the framework of the Maxwell–Lorentz theory in linear, causally dispersive systems.'

A different aspect of the distinction between group and phase velocity is met with when electromagnetic waves enter birefringent materials. Figure 10.8 shows how the Poynting vector of an electromagnetic wave with extraordinary polarization generally points in a different direction from the wave vector normal to the wavefronts. Energy travels at the ray velocity in the direction of the Poynting vector, while the wavefronts travel at the wave velocity. Now a wavepacket in a beam consists of infinite plane waves whose directions are distributed over a range of angles as well as wavelengths. These intefere constructively and travel with the group velocity in a direction determined by their Poynting vectors.

A surprising conclusion can be made from the analysis in this and the preceding chapter. Not only will the group velocity of light differ in magnitude from the phase velocities of the infinite sinusoidal plane waves making up the wavepacket in a dispersive medium, but also in a birefringent material a wavepacket of extraordinary waves will generally not even travel in the direction normal to the wavefronts of the constituent infinite plane waves.

11.6.3   Surface plasma waves

In addition to plasma waves that fill the volume of a metal, and which are for this reason called *bulk* plasma waves, *surface* plasma waves (SPW) also occur. These waves travel along the interface between a metal and a dielectric. One device first used by Kretschmann and Raether in 1968[7] for exciting SPWs is shown in the upper panel of figure 11.14. The prism is made from a dielectric with relatively large refractive index, $n_p = \sqrt{\varepsilon_p}$, for example quartz. A thin layer of metal is deposited on the prism base and the lower surface of this metal film is put in contact with a dielectric whose refractive index is much less than that of the prism. Because the metal layer is so thin, there can be total internal reflection at what is effectively a prism/dielectric interface. The evanescent wave can excite a surface plasma wave travelling along the metal/dielectric interface provided the phase velocity of the incident light matches that of a surface plasma wave. If this match is achieved the incident light feeds the surface plasma wave, with the result that the reflectance drops well below the 100% which normally signals TIR. The intensity distribution of the surface plasma wave as a function of distance from the metal/dielectric interface is illustrated in the lower panel in figure 11.14. It typically extends of order one wavelength into the dielectric, but less far into the metal. The upper panel of the figure also illustrates how to detect the presence of a surface plasma wave when using a white light source. After reflection the light emerges from the prism and falls on a diffraction grating. The resulting spectrum will show a gap at any wavelength for which TIR is suppressed when coupling to a surface plasma wave has occured. Because the surface plasma wave is sensitive to the physical and chemical content of a layer around one wavelength thick at the dielectric surface the device described is currently used as a biosensor. This application will be described after analysing the production of surface plasma waves.

If the material on which the metal is placed is non-magnetic then the surface plasma waves are p-polarized, and this case will be analysed now. The coordinate system is indicated in the upper panel of figure 11.14 with the plane $x = 0$ located on the metal/dielectric boundary. Then the magnetic field of the surface plasma wave points in the y-direction and has values

$$B_D = B_{D0} \exp\left(-\kappa_D x\right) \exp\left[i(k_z z - \omega t)\right], \tag{11.45}$$

$$B_M = B_{M0} \exp\left(\kappa_M x\right) \exp\left[i(k_z z - \omega t)\right], \tag{11.46}$$

where the subscripts D and M refer to the dielectric and metal respec-

[7]See *Surface Plasmons* by H. Raether, published by Springer-Verlag, Berlin (1988).

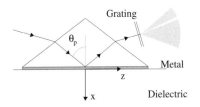

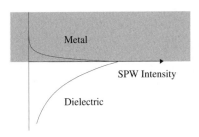

Fig. 11.14 The Kretschmann–Raether scheme for exciting surface plasma waves is shown in the upper panel. The lower panel shows the exponential decline in the intensity of the surface plasma wave with distance from the contact surface. The vertical scale is magnified in the lower panel.

tively. The wave vector k_D is given by

$$k_D^2 = k_z^2 - \kappa_D^2.$$

Also

$$k_D^2 = \omega^2/v_D^2 = \varepsilon_D \omega^2/c^2$$

where v_D is the wave velocity in the dielectric and ε_D its relative permittivity. Eliminating k_D from the last two equations gives

$$\kappa_D^2 = -\varepsilon_D \omega^2/c^2 + k_z^2, \tag{11.47}$$

with a similar relation for the wave vector in the metal

$$\kappa_M^2 = -\varepsilon_M \omega^2/c^2 + k_z^2. \tag{11.48}$$

Next we apply Maxwell's equation, eqn. 9.16, to the surface plasma wave in the dielectric and keep only the z-component

$$\kappa_D B_D = -i\varepsilon_D \omega E_{Dz}/c^2,$$

where E_{Dz} is the component of electric field in the z-direction. Rearranging this last equation gives

$$E_{Dz} = +i(\kappa_D c^2/\omega \varepsilon_D) B_D. \tag{11.49}$$

Similary for the wave in the metal we get

$$E_{Mz} = -i(\kappa_M c^2/\omega \varepsilon_M) B_M. \tag{11.50}$$

Finally we can impose the twin requirements that the z-component of the electric field is continuous at the surface, and that the magnetic field is continuous at the surface: $B_D = B_M$ and $E_{Dz} = E_{Mz}$. Then dividing eqn. 11.49 by eqn. 11.50 gives

$$\kappa_D/\varepsilon_D = -\kappa_M/\varepsilon_M. \tag{11.51}$$

This condition can only be satisfied if ε_M is negative; which is the situation in the case of a metal when the frequency of the incident electromagnetic radiation is less than the plasma frequency. When this is the case there can be a p-polarized surface plasma wave. Note also that its electric field is not transverse to the direction of propagation but lies in the zOx plane.

Proceeding further by squaring the last equation gives

$$\kappa_D^2/\varepsilon_D^2 = \kappa_M^2/\varepsilon_M^2.$$

Using eqns. 11.47 and 11.48 to substitute for κ_D and κ_M yields

$$\varepsilon_M^2/\varepsilon_D^2 = [k_z^2 - \varepsilon_M(\omega/c)^2]/[k_z^2 - \varepsilon_D(\omega/c)^2],$$

from which it follows that

$$k_z^2 = (\omega/c)^2 \varepsilon_M \varepsilon_D / (\varepsilon_M + \varepsilon_D), \qquad (11.52)$$

while

$$\kappa_{D/M}^2 = -(\omega/c)^2 \varepsilon_{D/M}^2 / (\varepsilon_M + \varepsilon_D). \qquad (11.53)$$

At high frequencies, but still below the plasma frequency, $\varepsilon_M = 1 - \omega_p^2/\omega^2$, and substituting this value for ε_M in eqn. 11.52 gives

$$k_z^2 = (\omega/c)^2 \varepsilon_D (1 - \omega_p^2/\omega^2) / (\varepsilon_D + 1 - \omega_p^2/\omega^2). \qquad (11.54)$$

Equation 11.54 is the surface plasma wave's *dispersion relation* and is drawn as a curved line in figure 11.15. As k_z tends to infinity ω tends to the limit $\omega_p/\sqrt{(1 + \varepsilon_D)}$, while near the origin it tends to $ck_z/\sqrt{\varepsilon_D}$. In the same figure the steeper, straight broken line follows the dispersion relation for light in free space incident at an angle θ_p, $\omega = k_z c/\sin\theta_p$. Evidently light in free space cannot excite a surface plasma wave because its phase velocity always exceeds that of the surface plasma wave; put another way their dispersion curves never cross away from the origin. However the phase velocity along the z-direction of light incident in the prism is $c/(\sqrt{\varepsilon_p}\sin\theta_p)$, and this flatter dispersion relation can intersect the dispersion curve for the surface plasma wave. Intersection is guaranteed if the slope is less than that of the dispersion curve for the SPW at the origin: $\sqrt{\varepsilon_p}\sin\theta_p > \sqrt{\varepsilon_D}$. The conditions for exciting a surface plasma wave can be produced either by varying the angle of incidence with monochromatic light, or by scanning in wavelength at a fixed angle of incidence, as is illustrated in figure 11.14. A sharp drop in the reflectance is the required signal. Using eqn. 11.53, and taking ε_D to be 2.0 and ε_M to be -10, the intensity of the surface plasma wave is predicted to fall off by a factor e in a distance of 0.11 (free space) wavelengths in the dielectric and 0.024 wavelengths in the metal. Gold is the metal preferred for the prism coating because it does not tarnish and also because the dips in reflectance are very narrow and deep. A gold layer which is around 50 nm thick gives an optimally deep and narrow dip.

When used in a biosensor the metal surface is coated with an appropriate agent to which the target (for example a bacterium such as *E. coli*) will bind. The fluid under test is then made to flow past the prepared surface. If this fluid contains any targets these are trapped by the surface agent and the accumulation of such material changes the relative permittivity ε_D of the surface layer, with a corresponding change in the frequency of the surface plasma wave. The resulting change in the wavelength at which the dip in reflectance occurs is large enough that a few picograms of target biomaterial per square millimetre are detectable. This approach has the flexibility of separating functions: the detector is sensitive yet biologically non-specific, while the biological agent is designed to trap a specific biotarget.

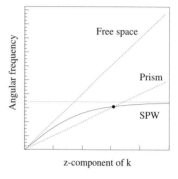

Fig. 11.15 Dispersion relations for a surface plasma wave, for an electromagnetic wave in free space and for an electromagnetic wave in the prism. The condition for which the wave in the prism and surface plasma wave match in frequency and wave number is marked by a spot.

From eqn. 11.52 it is apparent that when the metal and dielectric electric permittivities are close to cancelling the wavelength of the plasma wave along the surface direction becomes extremely short. This provides a practical means of coupling optical waves into nanostructures, thus linking opto-electronics and nanotechnology. At present this development is in its infancy.

11.7 Further reading

Optical Physics, third edition, by S. G. Lipson, H. S. Lipson and D. S. Tannhauser, published by Cambridge University Press (1995). A very imaginative book which concentrates firmly on understanding. It should prove useful in several areas besides the present chapter. 'How light interacts with matter' by V. F. Weisskopf in *Scientific American*, August 1968, pages 60–71 is also very informative.

Exercises

(11.1) Show that for a plasma $v_g = c^2/v_p$.

(11.2) Calculate the plasma frequency in a layer of the ionosphere where the electron number density is $10^{12}\,\mathrm{m}^{-3}$. What will the wave and group velocities be at a frequency 10 MHz?

(11.3) The power in a beam of light falls by a factor e in passing through 3 m of glass. What is this loss in $\mathrm{dB\,km}^{-1}$?

(11.4) Red ink is allowed to dry to form a thin solid layer on a glass sheet. It now appears in reflected light to be greenish. Why is this?

(11.5) Calculate the reduction of intensity of light of wavelength 589 nm after travelling through 200 km of air at NTP. The refractive index of air is 1.000292 and the number density of molecules is $3\,10^{25}\,\mathrm{m}^{-3}$.

(11.6) One empirical formula introduced to fit narrow resonances seen in the refractive index is

$$\varepsilon = 1 + \sum [A_i \lambda^2/(\lambda^2 - \lambda_i^2)].$$

Express the two constants for a given spectral line (A_i and λ_i) in terms of the electron natural frequency and the plasma frequency.

(11.7) Calculate the reflectance from copper of light of wavelength 589 nm using the data given in table 11.1.

(11.8) Monochromatic unpolarized light travelling in free space in the x-direction is scattered from a sphere of refractive index n, and radius a, this radius being much less than the wavelength of the light λ in free space. (a) Deduce using eqn. 11.27 that the electric dipole moment of the sphere is $p = p_0 \cos(\omega t) = 4\pi a^3 G \varepsilon_0 E$, where $E = E_0 \cos(\omega t)$ is the electric field of the light and $G = (n^2 - 1)/(n^2 + 2)$.

(b) Hence show that the time averaged flux scattered in solid angle $d\Omega$ in a direction making an angle θ_y with the dipole axis when dipole points along the y-axis is

$$dN = d\Omega(\omega^4 p_0^2 \sin^2 \theta_y)/(32\pi^2 c^3 \varepsilon_0).$$

(c) Write the corresponding expression if the dipole points along the z-axis.

(d) Show that

$$\sin^2(\theta_y) + \sin^2(\theta_z) = 1 + \cos^2 \theta_x,$$

where (θ_x, ϕ_x) are the polar angles with respect to the x-axis etc. The solid angle between the cones of semi-angle θ_x and $\theta_x + d\theta_x$ around the x-axis is $2\pi \sin(\theta_x)d\theta_x$.

(e) Hence show that the flux of radiation scattered into an angular interval between θ_x and $\theta_x + d\theta_x$ is

$$dN = \sin \theta_x d\theta_x [\omega^4 p_0^2 (1 + \cos^2 \theta_x)]/[32\pi c^3 \varepsilon_o].$$

(f) If the differential cross-section for scattering is defined as (dropping the subscript x)

$$d\sigma/d\theta = (dN/d\theta)/F,$$

where F is the incident flux per unit area, then show that the differential cross-section for Rayleigh scattering of a sphere is

$$d\sigma/d\theta = \pi k^4 a^6 G^2 (1 + \cos^2 \theta) \sin \theta.$$

(g) Hence show that the total Rayleigh cross-section from a single sphere is

$$\sigma = (8\pi/3)k^4 a^6 G^2,$$

where $k = 2\pi/\lambda$.

(11.9) Why is it that no plasma surface wave is excited at the metal/prism interface in figure 11.14?

(11.10) Prove eqn. 11.42.

The quantum nature of light and matter

12

12.1 Introduction

Just a century ago experiments with electromagnetic radiation led to a totally new insight, that not only matter comes in discrete packets but so too does electromagnetic radiation. These packets of radiation, first proposed by Planck and Einstein, are called *photons*, and electromagnetic radiation of frequency f Hz consists of photons each carrying a *quantum* of energy hf, where h is *Planck's constant* with a value $6.626\,10^{-34}$ J s. Just how small this is can be appreciated by noting that a one watt torch emits $\sim 10^{18}$ photons per second. Three pieces of experimental evidence were crucial to the acceptance of the concept of the quantization of radiation. The first came from the measurement of the spectrum of the radiation from a perfect absorber and emitter of radiation – a so-called black body. The second piece of evidence was provided by Lenard's and Millikan's studies of the photoelectric effect in which electromagnetic radiation liberates electrons from a metal surface. A third piece of evidence was obtained when Compton measured the change in wavelength of X-rays scattered by electrons. Somewhat later in 1924 de Broglie proposed that if electromagnetic waves had a particle nature, then for consistency matter should possess wave properties. Shortly thereafter interference effects were observed with electrons, and more recently they have been observed with other particle types including atoms and molecules. The experiments which demonstrated the particle nature of electromagnetic radiation and the wave nature of electrons are described in the first section of the chapter.

In 1911 Rutherford discovered the basic features of atomic structure. It became clear that an atom contains a nucleus that carries most of the mass and all the positive charge, but whose diameter is only one hundred-thousandth of the atom's diameter. Around this nucleus the much lighter, point-like, negatively charged electrons circulate at distances of order 10^{-10} m under the electrical attraction of the nucleus. Their total charge is equal and opposite to that of the nucleus. It was originally suggested that the electrons travelled in orbits like planets around the Sun, but this picture of the atom had a serious flaw. The electrons in orbits would be accelerating radially and hence they would radiate continuously. As a result they would lose energy and spiral rapidly into the nucleus. Bohr got around this difficulty by arbitrarily *quantizing* the orbits. He postulated that electrons only occupy certain orbits for which the angular momentum is an integral multiple of $h/2\pi$ and that while in these orbits the electrons do not radiate. In addition he postulated that an electron can move from one allowed orbit to another in a transition which is instantaneous and which is accompanied by the emission or absorption of a photon carrying the energy difference between the two orbits. Bohr's model was able to account for the existence of lines in atomic spectra and to explain quantitatively the striking regularities in the frequencies of the spectral lines of hydrogen and hydrogen-like atoms. These features had previously been incomprehensible, so this new understanding showed that here were the beginnings of a theory based on quantization. The successes, and crucial weaknesses, of the Bohr model will be described in the second part of the chapter. Today the Bohr model still provides a useful conceptual step towards the comprehensive theory of quantum phenomena, known as quantum mechanics.

If electromagnetic radiation, and matter, show both wave and particle properties then the obvious question to ask is how these properties can be integrated into a coherent conceptual and mathematical framework. In the third part of this chapter the interpretation of the *wave–particle duality* is discussed. Put very briefly, the wave intensity over a region of space takes on a new meaning as the probability of finding the associated particle in that region. However with probability comes uncertainty, and this is illustrated with examples. Each degree of freedom has an associated pair of *conjugate* variables, as for example in the case of a single particle the position and momentum in each dimension. These pairs can only be measured with limited precision, even with ideal measuring instruments. This limitation is quantified by Heisenberg's uncertainty principle for pairs of conjugate variables. Complementarity can also apply to pieces of information other than kinematic variables. In all cases, kinematic or not, exact knowledge of one piece of information implies total uncertainty about the other. As an example, in Young's double slit experiment one may either know which slit the photon passes through or one may observe the two slit interference pattern, but not both. If the paths are indistinguishable then interference is seen, but if the path is known (*welcher Weg information*) interference no longer occurs. The final section of the chapter is used to bring out the connections between photons, wavepackets, coherence volumes and modes of the electromagnetic field.

12.2   The black body spectrum

Black body radiation is electromagnetic radiation contained within an enclosure whose walls are maintained at a constant uniform temperature and, as discussed in Section 1.8, its spectrum is independent of the material of the enclosure. The practical form of a black body was described in Section 1.8. Figure 12.1 shows the spectra of black body radiation at three temperatures, like those that had been measured before 1900 by Lummer and Pringsheim, and by others. Their method was to disperse the black body spectrum with a diffraction grating and then to measure the heating effect of each wavelength segment of the spectrum.

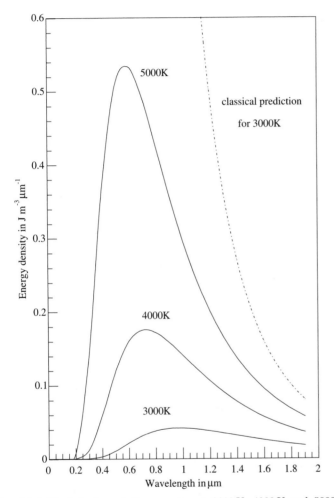

Fig. 12.1 Black body radiation spectra at 3000 K, 4000 K and 5000 K. Planck's quantum predictions which are indicated with full lines fit the data. The classical prediction for 3000 K is shown with a broken line.

It is straightfoward to make a prediction of the spectrum using classical theory, but the result is spectacularly wrong! The density of modes of electromagnetic radiation has been calculated in Section 9.8.1 to be

$$\rho(f)\mathrm{d}f = 8\pi f^2 \mathrm{d}f/c^3, \qquad (12.1)$$

where f is the frequency. In arriving at this value electromagnetic radiation is regarded as having two degrees of freedom, corresponding to the two polarizations. Classical thermodynamics then predicts that the energy should be partitioned equally between the modes with each mode having a mean energy in thermal equilibrium of $k_{\mathrm{B}}T$, where $T\,\mathrm{K}$ is the temperature and k_{B} is Boltzmann's constant, $1.381\,10^{-28}\,\mathrm{J\,K^{-1}}$. Thus the energy spectrum predicted is

$$W(f)\mathrm{d}f = 8\pi k_{\mathrm{B}} T f^2 \mathrm{d}f/c^3, \tag{12.2}$$

with $W(f)$ measured in $\mathrm{J\,m^{-3}\,Hz^{-1}}$. The equivalent distribution in wavelength is given by

$$W_\lambda(\lambda)\mathrm{d}\lambda = W(f)\mathrm{d}f,$$

hence

$$W_\lambda(\lambda) = W(f)\mathrm{d}f/\mathrm{d}\lambda = 8\pi k_{\mathrm{B}} T/\lambda^4, \tag{12.3}$$

and is indicated for a temperature of $3000\,\mathrm{K}$ by the broken line in figure 12.1. At high frequencies (short wavelengths) this prediction is wildly in error, a feature known as the *ultraviolet catastrophe*. Each of the rapidly increasing number of modes at high frequency has the same energy, which implies an infinite energy in the electromagnetic field! In 1901 Planck discovered what he took to be a temporary mathematical fix which brought the prediction into agreement with the data, and he imagined that his procedure would somehow be incorporated into classical theory. In fact this was the first step in revealing the quantum basis of nature.

What Planck proposed was that radiation is absorbed and emitted by the walls of the container in packets or *quanta* of energy,

$$E = hf = \hbar\omega, \tag{12.4}$$

where ω is the angular frequency of the radiation and h is known as Planck's constant; $\hbar$ is simply $h/2\pi$. If there are ℓ quanta in the mode its energy is $\ell h f$. Then the probability of there being ℓ quanta in any mode within the enclosure is given by the Boltzmann distribution for systems in thermal equilibrium

$$P(\ell) \propto \exp\left(-\mathrm{Energy}/k_{\mathrm{B}}T\right) = \exp\left(-\ell h f/k_{\mathrm{B}}T\right).$$

Putting $x = \exp\left(-hf/k_{\mathrm{B}}T\right)$ and normalizing so that the probabilities add up to unity gives

$$P(\ell) = x^\ell / \sum_\ell x^\ell = x^\ell(1-x).$$

Hence the mean number of photons in a mode is

$$\bar{\ell} = \sum_\ell \ell P(\ell) = (1-x)\sum_\ell \ell x^\ell = (x - x^2)\frac{\mathrm{d}}{\mathrm{d}x}\sum_\ell x^\ell$$

$$= (x - x^2)\frac{\mathrm{d}}{\mathrm{d}x}[1/(1-x)] = x/(1-x) = 1/(x^{-1} - 1).$$

Therefore

$$\bar{\ell} = 1/\left[\exp\left(hf/k_{B}T\right) - 1\right].\qquad(12.5)$$

Using the expression for the density of modes given above in eqn. 12.1, the energy spectrum of black body radiation was predicted by Planck to be

$$\begin{aligned}W(f)\mathrm{d}f &= \bar{\ell}hf\rho(f)\mathrm{d}f\\ &= (8\pi hf^{3}/c^{3})\mathrm{d}f/(\exp\left(hf/k_{B}T\right)-1),\qquad(12.6)\end{aligned}$$

with $W(f)$ again in $\mathrm{J\,m^{-3}\,Hz^{-1}}$. In a material of refractive index n there would be an additional factor n^{3} in the numerator

$$W(f)\mathrm{d}f = (8\pi hf^{3}n^{3}/c^{3})\mathrm{d}f/(\exp\left(hf/k_{B}T\right)-1).\qquad(12.7)$$

Planck found that this expression would fit the the observed black body spectra at all temperatures for which it was measured with the same value of h in each case. Quantization had solved the inherent weakness of classical wave theory and provided an excellent fit to the data. Thus quantization had to be taken seriously! The current measured value of Planck's constant is $6.626\,10^{-34}\,\mathrm{J\,s}$ making $\hbar$ equal to $1.0546\,10^{-34}\,\mathrm{J\,s}$. Note that in the limit of low frequencies and high temperatures, that is when $hf/k_{B}T$ is very small, Planck's formula reduces to the classical expression.

Several simple properties of the black body spectrum follow on from this analysis. The wavelength at which the spectrum peaks, λ_{peak}, is obtained by differentiating eqn. 12.6 and setting the result to zero. This gives

$$\lambda_{\mathrm{peak}} = 2.898\,10^{-3}/T,\qquad(12.8)$$

which was discovered experimentally in 1893 by Wien, and is known as *Wien's law*. The energy density in electromagnetic radiation in equilibrium at a temperature T can be obtained by integrating eqn. 12.6 over frequency. This gives

$$\mathcal{W} = \int_{0}^{\infty} W(f)\mathrm{d}f = 8\pi^{5}k_{B}^{4}T^{4}/(15c^{3}h^{3}),\qquad(12.9)$$

where we use the result that the definite integral $\int_{0}^{\infty} x^{3}\mathrm{d}x/(\exp x - 1)$ equals $\pi^{4}/15$. The units of $\mathcal{W}$ are $\mathrm{J\,m^{-3}}$

The flux of energy per unit time across unit area in the enclosure in one sense is of interest. This quantity is the irradiance or intensity of black body radiation measured in $\mathrm{W\,m^{-2}}$. Figure 12.2 shows a hemisphere drawn on a selected surface. Those modes whose normals lie between the two cones of semi-angles θ and $\theta + \mathrm{d}\theta$ make up a fraction $\sin\theta\mathrm{d}\theta/2$ of all the modes. Thus the total energy in these modes crossing unit area per unit time is

$$F = (\sin\theta\mathrm{d}\theta/2)(\mathcal{W}c\cos\theta).$$

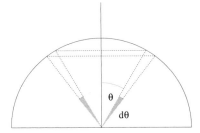

Fig. 12.2 Notional surface drawn within an enclosure containing black body radiation.

Integrating this result over all angles gives the irradiance or intensity of black body radiation

$$\mathcal{F} = (\mathcal{W}c/2) \int_0^{\pi/2} \sin\theta \cos\theta \, d\theta = \mathcal{W}c/4. \qquad (12.10)$$

Thus the intensity is

$$\mathcal{F} = 2\pi^5 k_B^4 T^4 / (15c^2 h^3) = \sigma T^4. \qquad (12.11)$$

This variation as the fourth power of the absolute temperature is known as *Stefan's law*, and the *Stefan–Boltzmann* constant, σ, is $5.76\,10^{-8}$ W m^{-2} K^{-4}.

The spectrum of the cosmic microwave background radiation which fills the universe has been measured with microwave dishes on board the COBE and WMAP satellites. The measurements reveal that this cosmic microwave background has a spectrum that deviates by only parts in one hundred thousand from the black body radiation spectrum at a temperature of 2.75 K. In fact it is the most perfect black body radiation spectrum ever observed.

12.3   The photoelectric effect

The *photoelectric effect* occurs when visible light or ultraviolet light falls on a metal or an alkali metal causing electrons to be emitted from the surface irradiated. By 1902 Lenard had shown that for each such metal there exists a threshold frequency below which no photoelectrons are produced, and this threshold frequency does not change however high the intensity of the incoming radiation. This behaviour is impossible to explain with electromagnetic wave theory alone. According to wave theory an electron in a metal would be forced to oscillate at the incoming wave frequency, this oscillation would build up in amplitude over time and eventually the electron would break free from the surface. Crucially, this sequence predicted by classical wave theory should proceed whatever the frequency of the incident electromagnetic radiation. Lenard also observed that the maximum energy of the emerging electrons depends solely on the frequency of the radiation and not on its intensity: facts which are again inconsistent with classical wave theory. Above threshold the number of electrons emitted was, as expected, proportional to the intensity of the radiation.

In 1905 Einstein proposed that in the photoelectric effect a single electron absorbs one of Planck's energy quanta from the incident radiation and escapes from the surface. Supposing it requires an energy ϕ to release an electron lying exactly at the surface of the metal, then all electrons will emerge with kinetic energy less than or equal to

$$\mathrm{KE}_{\mathrm{max}} = hf - \phi. \qquad (12.12)$$

Table 12.1 Cut-off wavelengths, frequencies and work functions for several metals from *CRC Handbook of Chemistry and Physics*, 77th edition, published by the Chemical Rubber Company Press, Boca Raton, FL, (1997). Courtesy Taylor and Francis Group.

Metal	Wavelength in nm	Frequency in THz	Work function in eV
Sodium	451	665	2.75
Lithium	428	701	2.90
Cesium	580	517	2.14
Copper	267	1120	4.65
Nickel	241	1240	5.15

ϕ depends on the metal and is called the *work function* of the metal. Electrons originating deeper in the metal will lose energy through collisions on their way out, so that KE_{max} is the maximum electron energy. Einstein's proposal introduces the required cut-off frequency $f_{co} = \phi/h$; radiation below this frequency cannot cause any photoemission of electrons.

Millikan, who was highly sceptical, set out to test Einstein's predictions in a series of experiments that extended Lenard's studies, and on which he, Millikan, continued working for over a decade. An outline of his later apparatus is shown in figure 12.3. A monochromator is used to select light from a strong line in the spectrum of a mercury lamp, this light then travels through a small window to fall finally on the metal surface being studied. The interior of the outer container is evacuated so that electrons do not lose energy in collisions with gas molecules, and so that the metal surface is not contaminated either. Because surface impurities change the threshold frequency so much, each metal surface was scraped immediately before being irradiated. Electrons ejected from the metal surface travelled to a copper Faraday cup and the resulting current was measured. A net negative voltage was applied to the cup relative to the metal and this voltage was increased until the current vanished. If this cut-off voltage is V and e is the electron's charge, then the maximum electron kinetic energy would be eV. One advantage accruing from the use of a copper cup is that copper has a high threshold frequency with the result that there was negligible photoemission caused by light reflected onto the cup. Millikan confirmed that there exists a threshold frequency for each metal he tested, and that with radiation of any lower frequency no electrons are emitted, however intense the radiation. Some modern determinations of the cut-offs for polycrystalline pure metal surfaces are given in Table 12.1. Secondly Millikan showed that when the kinetic energy of the highest energy *photoelectrons* was plotted against the frequency of the radiation the dependence fitted Einstein's predicted linear relation

$$KE_{max} = hf - \phi = h(f - f_{co}). \qquad (12.13)$$

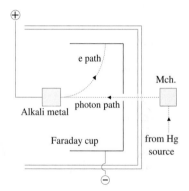

Fig. 12.3 Millikan's apparatus to study the photoelectric effect. Mch is a monochromator. The light passes through a window in the container.

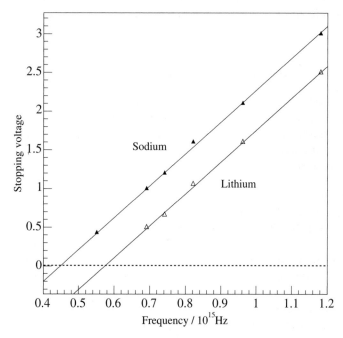

Fig. 12.4 Millikan's measurements of the stopping voltage against the radiation frequency for sodium and lithium. The voltages are corrected for the contact potentials that are present at contacts between different metals in the circuit. Adapted from R. A. Millikan, *Physical Review* 7, 355 (1916), by courtesy of the Amercan Physical Society.

This dependence is evident in the data collected by Millikan and shown in figure 12.4. From the slope of the lines fitted to the data for each metal Millikan extracted values for h which agreed with each other and with the value found by Planck. Millikan also confirmed that the number of photoelectrons is proportional to the intensity of the radiation. Later in 1927 Lawrence and Beams measured the delay between the moment that the radiation first arrives at the surface after the source is switched on and the moment at which electron emission commences. Classically the wave energy is spread over the whole surface of the wavefront and it would take some time before any individual electron could accumulate enough energy to escape from the surface. The delay should therefore increase as the intensity diminishes. By contrast quanta are localized so that emission should commence instantaneously when the light is turned on. Lawrence and Beams found that the delay was less than the resolution of their timing methods and gave an upper limit of $3\,10^{-9}$ s for the delay at low light intensities. A tighter limit for the delay was obtained by Forrester, Gudmundson and Johnson in 1955. They studied the photoelectric effect using light modulated at high frequencies and observed that the detector signal followed the modulation faithfully, showing that any delay was much less than the modulation period. Their conclusion was that the delay is significantly less than 10^{-10} s. The absence of any delay makes implausible a semiclassical picture, in which electromagnetic radiation is absorbed and emitted as quanta but travels purely as waves.

12.4   The Compton effect

Compton, following through the consequences of the quantization of radiation, appreciated that when X-rays scatter from matter the underlying process is the scattering of an individual quantum of electromagnetic radiation from a single electron which is initially at rest. This process is pictured in figure 12.5.

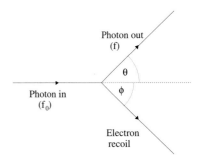

Fig. 12.5 Compton's particle interpretation of X-ray scattering from an electron.

The particle aspect of the quantum of energy is now explicit with electromagnetic radiation being pictured as consisting of particles called *photons* each carrying Planck's quantum of energy.

If the incoming photons had frequency f_0 Compton assigned to each an energy hf_0. He also assigned a momentum hf_0/c to the photons, which takes the relationship expressed in eqn. 9.54 between wave energies and momenta and applies it to an individual photon travelling in free space. The special theory of relativity gives the same relationship, $E = pc$. Suppose now that the scattered photon has frequency f, that the electron recoils with velocity βc, and that the scattering angles are as shown in figure 12.5. The electron must be treated relativistically because the kinetic energy of the recoiling electron is comparable to its rest mass energy. Initially its energy before being struck is the rest mass energy mc^2, while its energy and momentum afterwards are $m\gamma c^2$ and $m\beta\gamma c$ respectively, where $\gamma = 1/\sqrt{(1-\beta^2)}$. Energy is conserved in the scatter so that

$$hf_0 + mc^2 = hf + mc^2\gamma. \tag{12.14}$$

The components of momentum along and at right angles to the incident photon's direction are also conserved so that

$$hf_0/c - hf\cos\theta/c = m\gamma\beta c\cos\phi, \tag{12.15}$$
$$hf\sin\theta/c = m\gamma\beta c\sin\phi, \tag{12.16}$$

where θ and ϕ are the respective scattering angles of the photon and the electron relative to the incident photon's direction. Rearranging and squaring the energy equation, eqn. 12.14, gives

$$h^2(f_0^2 + f^2 - 2ff_0) = m^2c^4(\gamma - 1)^2.$$

Squaring the momentum equations eqns. 12.15 and 12.16 and adding the results gives

$$h^2(f_0^2 + f^2 - 2f f_0\cos\theta) = m^2\gamma^2\beta^2c^4.$$

Taking the difference between the last two equations yields

$$2h^2 f f_0(1 - \cos\theta) = 2m^2c^4(\gamma - 1).$$

The energy equation, eqn. 12.14, can be used to replace the right hand side of the last equation and this gives

$$2h^2 f\, f_0(1 - \cos\theta) = 2h(f_0 - f)mc^2,$$

which when divided by $2hf f_0 mc^2$ leaves

$$(h/mc^2)(1 - \cos\theta) = (1/f - 1/f_0).$$

Re-expressing this equation in terms of the wavelengths produces a simpler form

$$\lambda - \lambda_0 = (h/mc)(1 - \cos\theta). \tag{12.17}$$

This change in wavelength of the scattered X-rays is called the *Compton effect*. It depends only on physical constants and the scattering angle, and not at all on the nature of the scattering material. Numerically the term that quantifies the wavelength shift, h/mc, is 0.0024 nm which shows that it is necessary to use short wavelength radiation. If the experiment were attempted with visible light the change in wavelength would, even now, be hard to detect let alone measure precisely. Compton's X-ray photons had energies very much larger than the kinetic energy and binding energy of the electrons in the atoms which he used, and hence the assumption that the target electrons are free and at rest is fully justified.

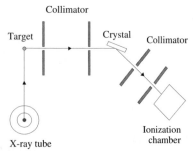

Fig. 12.6 Compton's apparatus for studying the wavelength change in X-ray scattering from various light elements.

Compton's apparatus is sketched in figure 12.6. Radiation from the molybdenum X-ray source is scattered off a graphite block. Collimators are used to select narrow beams of X-rays, a collimator being made of a pair of lead plates each pierced by a single small hole. The scattering angle off graphite is selected by positioning a first collimator appropriately. After emerging from this collimator the X-rays are Bragg scattered off a calcite crystal, with a second collimator set so that the angle of reflection is equal to the angle of incidence. Finally the Bragg scattered X-rays enter an ionization chamber filled with a gas whose atoms are readily ionized by X-rays. Under an applied voltage a current flows proportional to the X-ray flux. Figure 12.6 is drawn for the case that the scattering angle between the incident and scattered X-rays at the carbon target is 90°. At each such setting of the scattering angle from graphite Compton varied the Bragg scattering angle at the calcite crystal in steps and at each step recorded the current. The resulting angular distribution of the scattered X-ray intensity was then converted to a distribution in wavelength using Bragg's law. Figure 12.7 shows the spectra observed at three different scattering angles off a carbon target. The wavelength of the incident radiation is indicated by the broken line in each panel and the right hand peak contains the photons scattered from electrons. Compton found that the shift in wavelength of this right hand peak fitted his prediction precisely in magnitude, in its dependence on the scattering angle, and in its independence of the target material. In 1949 Hofstadter and McIntyre detected both the photon and the recoil electron, checking that they emerge in coincidence, to within 10^{-9} s.

The fit achieved to the black body spectrum, the explanation of the features of the photoelectric effect and the prediction and measurement of the Compton effect were the key elements in establishing that electromagnetic radiation has a particle nature.

12.5   de Broglie's hypothesis

In 1924 de Broglie pointed out that if electromagnetic waves possess particle properties, then it might be reasonable to suppose that material particles such as electrons should possess wave properties. He proposed that the relations connecting the wave and particle properties of electromagnetic radiation should also apply to material particles. Thus the frequency of the wave f associated with a particle of total energy E would be given by Planck's relation

$$E = hf. \tag{12.18}$$

The parallel relation for the momentum of a photon is

$$p = E/c = h/\lambda.$$

de Broglie therefore proposed a similar relationship for material particles

$$p = h/\lambda. \tag{12.19}$$

This is called the *de Broglie* relation, and λ is known as the *de Broglie wavelength* of material particles.

It is important to note that the energy concerned is the total energy of the material particle given by an expression from the special theory of relativity

$$E^2 = p^2c^2 + m^2c^4, \tag{12.20}$$

where m is the rest mass. This collapses to $E = pc$ for the massless photon and is the expression applied by Compton. At the other extreme, when velocities are small compared to c, $p \ll mc$ and

$$E = mc^2(1 + p^2/m^2c^2)^{1/2}$$
$$\approx mc^2 + p^2/2m,$$

where the binomial expansion is used. $p^2/2m$ is the usual kinetic energy of Newtonian mechanics, while mc^2 is the rest mass energy.

de Broglie's ideas were confirmed when in 1926 Davisson and Germer, and simultaneously G. P. Thomson, demonstrated the wave properties of electrons. Davisson and Germer accelerated electrons through potentials of tens of volts and then scattered them from a nickel crystal. They observed diffraction from the regular atomic layers in the crystal. Thomson, on the other hand, passed electrons through thin films of randomly

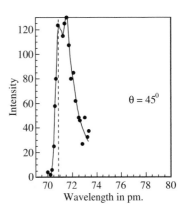

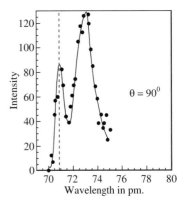

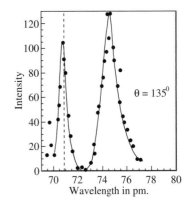

Fig. 12.7 Wavelength distribution of X-rays scattered from graphite. The vertical line indicates the wavelength of the direct beam (the K_α line of Molybdenum.). Taken from A. H. Compton, *Physical Review* 22, 411 (1923), by courtesy of the American Physical Society.

oriented microcrystals, and observed sharp circular diffraction rings at angles satisfying the usual Bragg condition

$$n\lambda = 2d \sin\theta, \tag{12.21}$$

where n is an integer, d is the crystal plane spacing, λ the electron wavelength and θ the angle between the electron path and the crystal surface.

If tiny material particles like electrons have wave properties then what implications does this have for macroscopic objects? A mass of one gram moving at $1\,\mathrm{m\,s^{-1}}$ has a de Broglie wavelength of $6.6\,10^{-31}\mathrm{m}$ so that we cannot expect to see diffraction of everyday objects. However because the de Broglie wavelength of electrons of around $2\,\mathrm{eV}$ is $1\,\mathrm{nm}$ diffraction effects will eventually impose a lower limit on the size of the gates in field effect transistors, and hence on the ultimate density of components in electronic processors. Nowadays the wave nature of material particles is exploited when using electron microscopes and neutron diffraction to explore the structure of matter.

12.6   The Bohr model of the atom

In parallel to the discovery of the quantum properties of electromagnetic radiation, experiments on atomic structure led to the appreciation that quantization was crucial in understanding the atom and atomic spectra. Rutherford in 1911 used α-particles (bare $^4\mathrm{He}$ nuclei, each with charge $+2e$) to bombard thin metal foils and had observed that substantial numbers were scattered into the backward hemisphere and some almost straight backward. His observations could only be consistently explained if the object within the atom which was scattering the α-particles carries most of the atomic mass, positive charge and is very much smaller than the atom. This scatterer is the nucleus which consists of neutral neutrons and charged protons and is typically $10^{-15}\mathrm{m}$ across. An electron has 1/2000th the mass of the proton or neutron and carries an equal and opposite charge to the proton. The electrons in an atom circulate around the nucleus in orbits extending to $10^{-10}\mathrm{m}$ in diameter.

Classically the electrons in such an atom must radiate continuously because they are accelerating radially. As a result the electrons would be expected to spiral rapidly into the nucleus while radiating over a broad frequency range. By contrast isolated atoms radiate at discrete wavelengths, the spectral lines, and their wavelengths form patterns that invite explanation. Hydrogen has a particularly simple atomic structure, with one electron orbiting a single proton nucleus. The emission spectrum of hydrogen gas excited by an electric discharge was already known at the time to consist of several series of spectral lines that could be fitted with one overall empirical formula

$$1/\lambda = R_{\mathrm{H}}(1/n^2 - 1/p^2), \tag{12.22}$$

where R_H is a constant known as the *Rydberg constant*. n and p are positive integers with the restriction that $p > n$. The currently accepted value of the Rydberg constant is $1.09678\,10^7 \mathrm{m}^{-1}$. Thus the spectrum consists of *spectral series*, each with a fixed value of n while p runs through the sequence $n + 1$, $n + 2$, $n + 3$, etc. The *Lyman* series is generated by the combinations $n = 1$, $p = 2, 3, 4, ...$ It starts with the *Lyman α-line* at $121.6\,\mathrm{nm}$ ($n = 1$, $p = 2$) and terminates at $91.3\,\mathrm{nm}$ ($n = 1$, $p = \infty$). Among the other series, also named for their discoverers, are: the Balmer series ($n = 2$) lying in the visible and ultraviolet; the Paschen series ($n = 3$), the Brackett series ($n = 4$) and the Pfund series ($n = 5$), all lying in the infrared part of the spectrum. An explanation of these spectral features was achieved by Bohr in 1913 using a simple quantized model of the atom.

   In the Bohr model the electrons in atoms are pictured as travelling in stable circular orbits which satisfy the requirement that the angular momentum is exactly an integral multiple of $\hbar$. The transition from an orbit of higher energy to one of lower energy is instantaneous and is accompanied by the emission of a photon which carries off the energy difference. Similarly an electron in a lower energy orbit can absorb a photon and instantaneously jump to a higher energy orbit provided the photon energy exactly matches the difference between the electron's energy in the two orbits. The photon frequency, f, is given by

$$hf = \Delta E, \tag{12.23}$$

where ΔE is the difference in energy between the two atomic states. It is now shown how the pattern of spectral lines for hydrogen emerges naturally from these simple postulates.

   Suppose that in an atom containing a single electron the radius of the electron's orbit is r, its speed is v and m is its mass. The quantization condition is

$$mvr = n\hbar, \text{ or } v = n\hbar/mr. \tag{12.24}$$

A justification for Bohr's quantization condition is revealed when the de Broglie relation from eqn. 12.19 is used to replace the momentum in eqn. 12.24. This gives

$$n\hbar = pr = hr/\lambda,$$

thus

$$n\lambda = 2\pi r. \tag{12.25}$$

Therefore the quantization condition requires that one complete orbit should contain an integral number of electron wavelengths. If the orbit length did not satisfy this condition then the electron wave, after travelling many times around the orbit, would interfere destructively with itself.

The electron is maintained in the stable orbit by the Coulomb attraction of the nuclear charge. Thus its radial equation of motion is

$$Ze^2/(4\pi\varepsilon_0 r^2) = mv^2/r, \tag{12.26}$$

where $-e$ is the electron charge and Ze is the nuclear charge, Z being unity for hydrogen. Using eqn. 12.24 to replace v in eqn. 12.26 gives

$$Ze^2/4\pi\varepsilon_0 = n^2\hbar^2/mr,$$

so that

$$r = 4\pi\varepsilon_0(n^2\hbar^2/mZe^2). \tag{12.27}$$

The radius of the first orbit in a hydrogen atom

$$a_\infty = 4\pi\varepsilon_0(\hbar^2/me^2) = 0.05292\,\text{nm} \tag{12.28}$$

is called the *Bohr radius* and characterizes the linear size of atoms; other orbits have radii n^2 larger. This prediction for the atomic size is consistent with the measured spacing of atoms in condensed matter if the outer electron orbits of one atom touch or slightly overlap those of an adjacent atom. A quantity needed later is the angular frequency of rotation of an electron in the nth Bohr orbit

$$\Omega_n = v/r = n\hbar/mr^2 = (Ze^2/4\pi\varepsilon_0)^2(m/n^3\hbar^3). \tag{12.29}$$

Finally the total energy of the electron in the nth orbit, made up of the kinetic and the potential energy, is

$$E_n = mv^2/2 - Ze^2/(4\pi\varepsilon_0 r).$$

The term mv^2 can be replaced using eqn. 12.26 to give

$$\begin{aligned} E_n &= -Ze^2/(8\pi\varepsilon_0 r) \\ &= -(Ze^2/4\pi\varepsilon_0)^2(m/2n^2\hbar^2). \end{aligned} \tag{12.30}$$

From this result the photon energy emitted/absorbed in a transition between the pth and nth orbit can be calculated,

$$\Delta E = E_p - E_n = (Ze^2/4\pi\varepsilon_0)^2(m/2\hbar^2)[1/n^2 - 1/p^2], \tag{12.31}$$

where $p > n$ is required for this to be positive. The wavelength of the photon emitted in such a transition is given by

$$1/\lambda = \Delta E/hc = R_\infty Z^2[1/n^2 - 1/p^2], \tag{12.32}$$

where

$$R_\infty = (1/4\pi\varepsilon_0)^2(me^4/4\pi\hbar^3 c). \tag{12.33}$$

This last expression needs some correction because the nucleus has been assumed to be immobile (of infinite mass). The corrected value in the

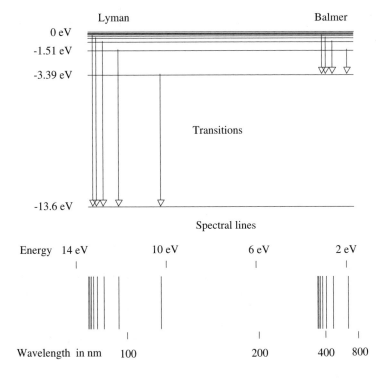

Fig. 12.8 The energy levels in the hydrogen atom are displayed together with transitions producing the Lyman and Balmer series. For clarity only the first few transitions are drawn. The spectral lines are shown below. Each series converges to a limit: in the case of upward transitions the limit is reached when the electron just escapes from the atom with zero kinetic energy.

case that the nucleus has mass M is

$$R = [mM/(M + m)](R_\infty/m) = \mu(R_\infty/m), \qquad (12.34)$$

where μ is called the reduced mass of the electron. Similarly the Bohr radius must be corrected

$$a_0 = 4\pi\varepsilon_0\hbar^2/\mu e^2. \qquad (12.35)$$

The prediction for the Rydberg constant from eqn. 12.33 using the accepted values for the constants agreed with the measured value.

Thus the Bohr model predicts the experimentally observed spectral series and the Rydberg constant with precision.[1] This is very convincing evidence for the quantization of atomic orbits. Figure 12.8 shows the energy levels of the hydrogen atom and some of the transitions. As the quantum number n in eqn. 12.30 increases the discrete energy levels

---

[1]Nowadays the success of quantum theory means that eqn. 12.33 is accepted: the measurement of the Rydberg constant is therefore one among several experimental measurements from which the constants appearing in eqn. 12.33 are determined.

pack more and more tightly together, converging on zero binding energy. Above this energy electrons are free and can have any positive energy. It follows that an electron in an atom can absorb any photon whose energy exceeds the electron's binding energy and emerge as a free photoelectron.

Only the Lyman series is seen in the absorption spectrum of hydrogen gas at low temperatures. The reason lies in the distribution of electrons between the energy levels in a collection of hydrogen atoms in thermal equilibrium. In thermal equilibrium at a temperature T K the number of hydrogen atoms with an electron in an excited state of energy ΔE above the lowest energy or *ground* state is given by the Boltzmann distribution

$$N(\Delta E) = N(0)\exp\left[-\Delta E / k_\mathrm{B}T\right], \tag{12.36}$$

where $N(0)$ is the number in the ground state. Therefore whenever the temperature is low enough that $k_\mathrm{B}T$ is very much smaller than the excitation energy of the first excited state the ground state is almost the only one occupied. Thus at a temperature of $20\,°\mathrm{C}$, for which $k_\mathrm{B}T$ is $0.025\,\mathrm{eV}$, absorptions in hydrogen gas will, almost entirely, involve Lyman line transitions from the ground state.

It must be the case that at the large scale the quantum description becomes equivalent to the classical description because at the macroscopic scale classical mechanics does work well. Bohr therefore proposed a *correspondence principle* which states that at very high quantum numbers the quantum description merges with the classical description of phenomena. For example the angular frequency of a photon emitted in a transition between adjacent levels given by eqn. 12.31 approaches, in the limit of very large quantum numbers n,

$$\omega_{n,n+1} = (Ze^2/4\pi\varepsilon_0)^2 m/(n\hbar)^3. \tag{12.37}$$

This is precisely the angular frequency of rotation of the electron in the nth orbit met with in eqn. 12.29. Now according to classical wave theory the angular frequency of the radiation emitted by a rotating electron is just the electron's rotation frequency, so we see that in a transition at high enough quantum numbers the quantum prediction has converged on the classical prediction.

12.6.1   Beyond hydrogen

The Bohr model fails to explain the spectrum of neutral helium, the next simplest element beyond hydrogen, and it was only following the development of quantum mechanics that a comprehensive explanation of all atomic spectra became possible. The Bohr model works moderately well for the alkali metals, whose atomic structure has a single electron outside a core of electrons tightly bound to the nucleus. Therefore the singleton electron feels the net electric field produced by the nucleus and the core of electrons, which is much like the electric field felt by the electron in a hydrogen atom.

A result useful in interpreting atomic structure within the Bohr model was obtained by Moseley in 1913. He compared the wavelengths of the sharp lines appearing in the X-ray spectra produced when a range of elements are bombarded by high energy electrons. These lines are emitted when the projectile electron ejects an electron from one orbit and an electron in a higher energy orbit drops into the orbit vacated. For the dominant high energy X-ray line, the K_α line, Moseley found that the wavelengths fitted a single expression

$$1/\lambda = 0.830\,10^7\,(Z-a)^2\,\mathrm{m}^{-1}, \tag{12.38}$$

where Z is the atomic number and a is around unity. This result is consistent with the interpretation that the K_α line is emitted in a transition between the second and first Bohr orbits. The prediction given by eqn. 12.32 for the wavelengths in this case is that

$$1/\lambda = R[1/1^2 - 1/2^2](Z')^2 = 0.823\,10^7\,(Z')^2\,\mathrm{m}^{-1},$$

where $Z'e$ is the effective charge acting on the active electron due to the nucleus and the other electrons. Only the single remaining electron in the innermost Bohr orbit lies inside the orbit of the active electron so that $Z'e$ is equal to $(Z-1)e$ where Ze is the nuclear charge. The close agreement of Moseley's fit to data with the prediction of the Bohr model is convincing evidence that the atomic number is identical to the nuclear charge divided by e. In turn this makes the atomic number equal to the number of electrons circulating the nucleus. That the atomic number determines the position of an element in the periodic table is now revealed to reflect the fact that the number of electrons in an atom determines its chemical properties.

12.6.2   Weaknesses of the Bohr model

It has been already noted that the Bohr model fails to explain the main features of spectra of elements beyond hydrogen. Its hybrid nature is also very unsatisfying: first a quantization condition is imposed for closed orbits and after that classical mechanics is used. As a result there is no explanation as to why transitions occur and hence no way of calculating rates at which transitions occur. Furthermore in many physical situations there is no obvious periodic condition that can be imposed comparable to the condition on the angular momentum of an electron orbiting a nucleus. The resolution of all these difficulties only came with the development of quantum mechanics.

12.7   Wave–particle duality

The acceptance of the existence of wave and particle behaviour of both electromagnetic radiation and matter led to a search for ways to reconcile these two types of behaviour. A closely connected issue was the replacement of Bohr's interim solution of grafting quantum conditions onto classical mechanics by a new coherent mathematical structure. In this section the interpretation of the *wave–particle duality* is discussed. The following chapter is used to describe the new quantum mechanics and its application to atomic structure.

The connection between the particle and wave properties of light is statistical, and this is equally true for electrons or other material particles. To be explicit, the probability of finding a photon in a given volume dV is simply determined by the instantaneous energy density, I, of the electromagnetic wave over the same volume

$$P dV = I dV / \int I dV, \qquad (12.39)$$

where P is called the *probability density*. The integral is taken over the whole of space to ensure that the total probability of the photon being found somewhere is unity.

Young's two slit experiment, met first in Chapter 5, provides a simple application of this statistical picture. Suppose the observation screen is a pixelated detector with granularity much finer than the fringe widths, and that all the pixels are equally efficient in detecting photons. Further, suppose that an extremely low intensity, monochromatic source is used, sufficiently weak so that at any given moment there is only ever a single photon within the volume between the source slit and the detector screen in figure 5.1. Figure 12.9 shows typical histograms of the photon distribution across the detector after 10, 1000 and 20 000 photons have been detected. For comparison the intensity calculated earlier in Chapter 5 is superposed in each case. An individual photon may hit anywhere across the screen, apart from locations where the wave intensity is precisely zero. Only the probability for arriving at each pixel is known, and the probabilities of reaching a given pixel are identical for each and every photon emerging from the source slit. The distribution observed when the number of photons is small is extremely ragged and does not resemble the wave intensity very closely. As the number increases the resemblance becomes ever closer, a convergence that is purely statistical. Recall that if the number of photons expected to strike a pixel is n, with n being large, the statistical uncertainty in this number is $\sqrt{n}$. Thus the fractional uncertainty on the number arriving at a pixel is $1/\sqrt{n}$, and this value falls as n rises. In a standard laboratory demonstration of Young's two slit experiment there are such large numbers of photons arriving at the screen per second that your eye cannot detect any statistical fluctuation.

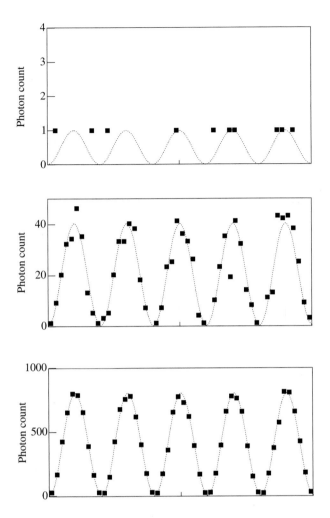

Fig. 12.9 Distribution of photons in the detection plane in Young's two slit experiment for 10, 1000 and 20 000 photons. The broken curves indicate the classical interference pattern.

A two slit experiment with a very low source intensity was first carried out in 1906 by Taylor and has been repeated many times. He used a photographic plate as the detector in an exposure lasting several weeks, after which the plate was processed and the anticipated fringes revealed. In this experiment and numerous similar ones there is at most only a single photon within the interference apparatus at any given moment. This has to mean that a photon is interfering with itself. The interference pattern does not change when multiple photons are in the apparatus so that we can conclude that there too each individual photon interferes with itself only.

It is when photons have high energies that their particle behaviour is most noticeable. For example a single optical photon can produce an electron via the photoelectric effect and this electron can be multiplied to give a detectable current pulse in a photomultiplier or avalanche photodiode. Particle behaviour is also evident in atomic, nuclear and molecular transitions. It is harder to associate particle behaviour with low frequency radio waves because each photon makes only a miniscule contribution to the total electric field and each such contribution cannot be individually detected. However the statistical fluctuation in the number of photons will be recognizable at low signal levels.

12.8   The uncertainty principle

Uncertainty is inherent in quantum theory but this uncertainty is quantifiable. We start by considering a thought (*gedanken*) experiment which Heisenberg used to illustrate the uncertainty that occurs when simultaneous measurements are made of position and momentum. Figure 12.10 shows a microscope used to detect ultraviolet photons scattered from an electron. The precision in the measurement of the electron's position is fixed by the resolution of the microscope. Taking Abbe's result, eqn. 7.68, the resolution in the electron's position is

$$\Delta y = \lambda / \sin \theta,$$

where λ is the wavelength of the radiation, and θ is the semi-angle subtended at the electron by the objective. Now the angle at which the photon enters the objective can be anywhere within the angular range of θ from the lens axis. Thus its momentum in the y-direction is uncertain by an amount

$$\Delta p_y = p \sin \theta,$$

where p is the momentum of the incoming photon. Momentum conservation then requires that this is also the uncertainty in the electron momentum. Multiplying together the uncertainties in the electron momentum and position gives

$$\Delta p_y \Delta y = \lambda p = h,$$

where de Broglie's relation has been used in making the second equality. The tightest limit on the product of uncertainties is obtained when the distributions in position and momentum have Gaussian rather than flat distributions, in which case the standard deviations satisfy

$$\Delta p_y \Delta y = \hbar/2.$$

If there are any instrumental errors in the measurement they can only increase these uncertainties. Taking into account that non-Gaussian distributions are possible and that there can be additional instrumental errors leaves only a limit on the product

$$\Delta p_y \Delta y \geq \hbar/2, \tag{12.40}$$

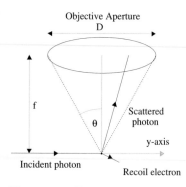

Fig. 12.10 Heisenberg's microscope being used to determine the location of an electron.

which is one expression of Heisenberg's *uncertainty principle*. This principle brings into sharp relief a fundamental difference between classical and quantum theory. In the classical view it is imagined that the momentum of the light beam probing the electron can be indefinitely reduced to make Δp_y as small as required. However the reduction in Δp_y can only be achieved in practice by increasing the photon wavelength, which increases Δy, but leaves the product of uncertainties unchanged. The uncertainty principle applies to the measurement in each of the three dimensions, from which it follows that simultaneous measurements of the vector position $\mathbf{r}$ and vector momentum $\mathbf{p}$ have uncertainties that satisfy

$$\Delta p_x \Delta p_y \Delta p_z \Delta x \Delta y \Delta z \geq \hbar^3/8. \qquad (12.41)$$

The product on the left hand side of this equation can be pictured as a volume element in a six-dimensional *phase space*. Three of the coordinates are the spatial coordinates, while the other three are the coordinates in momentum space. The inequality in the equation expresses the requirement that there is a limiting precision within which the kinematics of any particle can be known. The shape of this volume is dictated by the circumstances of the measurements, and is not necessarily cubical or spherical.

Spatial coordinates and time are treated in a unified way within the special theory of relativity: they merge to form a single space-time location (x, y, z, ct). Similarly energy and momentum form an energy–momentum vector $(p_x, p_y, p_z, E/c)$. One implication of this unification is that there must exist an energy–time uncertainty relation of the form

$$\Delta E \Delta t \geq \hbar/2. \qquad (12.42)$$

The interpretation of this uncertainty relation differs from that of the previous example because time is not a quantity being measured. Rather Δt is the time taken to make the measurement of the energy, and ΔE is the resulting uncertainty in the energy measured. Thus if a measurement is made of the energy of an excited state of an atom which decays with a lifetime τ the measurement of its energy has an uncertainty of at least $\hbar/2\tau$. The consequent uncertainty in the angular frequency of the light emitted in transitions from this state to the ground state is $\Delta\omega \geq 1/(2\tau)$. This can be recognized as the natural line width described in Section 7.3.1. In that section the origin of the natural linewidth was ascribed to a damping process, which is the classical equivalent of the exponential decay of excited states. Two other processes were described which broaden the spectral lines emitted by gases much more than their natural width: these are collisions and the Doppler effect. The effect of collisions is to reduce the time available for measurement, while the Doppler effect arises from the thermal motion of the atoms which emit the photons.

Quantum uncertainty relations bear a resemblance to the bandwidth theorems proved for classical wavepackets in Chapter 7. In that chapter a simple Gaussian pulse was introduced with a waveform

$$\psi(t) \sim \exp\left(-i\omega t\right) \exp\left[-(t-t_0)^2/4\sigma_t^2\right], \qquad (12.43)$$

and its Fourier transform, a waveform in angular frequency ω, was obtained. The corresponding intensity distributions in time and angular frequency were found to be Gaussians too with standard deviations σ_t and $\sigma_\omega = 1/(2\sigma_t)$. Their product $\sigma_t\sigma_\omega$ is equal to $1/2$. For all other distributions this product is larger, so we had the bandwidth theorem, eqn. 7.61,

$$\sigma_t\sigma_\omega \geq 1/2.$$

The corresponding relation linking position and wave vector uncertainties is

$$\sigma_x\sigma_k \geq 1/2.$$

If these expressions are multiplied by $\hbar$ the outcomes resemble the quantum uncertainty relations! There is an important distinction which this direct conversion obscures. In the classical view the shape of a wavepacket could be measured precisely to give the shape parameters σ_x and σ_k. In the quantum picture σ_x and σ_k are the uncertainties in the measurements on an individual photon.

The laws of classical physics are deterministic and statistical analysis is used as a practical tool only in dealing with systems containing very large numbers of particles. A good example is to be found in the kinetic theory of gases. By contrast, statistical behaviour is fundamental to even the simplest quantum systems.

It is informative to view diffraction as an artefact of the uncertainty principle. For example consider the diffraction pattern produced by a slit of width d on which light of wavelength λ is incident. The uncertainty in the lateral position of any photon going through the slit is d and using the uncertainty principle this implies an uncertainty in its transverse momentum of at least

$$\Delta p_y = \hbar/2d.$$

Consequently the angular spread of the photons is at least

$$\Delta\theta = \Delta p_y/p = \hbar/(2dp) = \lambda/(4\pi d),$$

which indeed reproduces the proportionality of the angular size of the first bright fringe to the quotient (wavelength)/(slit width).

12.9   Which path information

It goes without saying that to produce a two slit interference pattern the electromagnetic waves should pass through both slits. On the other hand a photon, a particle, ought to pass through one slit or the other. It is only *because* there is no information about which slit the photon passed through that an interference pattern is observed. With a symmetric pair of slits the wave amplitudes at each slit are equal and the probability of the photon being at either slit is 0.5. That is all we can know when we observe the interference pattern with its visibility of 100%. If on the other hand the photon is tagged in some way to identify its path the two slit interference pattern is inevitably destroyed. In such cases the wave amplitude at the slit where the photon is not detected must be zero, and the wave at the other slit naturally produces a single slit diffraction pattern.

Einstein suggested that it might be feasible to determine which slit the photon had passed through by measuring the recoil of the board in which the slits are cut, and still allow an interference pattern to be seen. Bohr demonstrated that this method fails because of the uncertainty principle. Figure 12.11 shows the arrangement. The momentum transfer to the board will depend on whether the photon is deflected leftward by the right hand slit or rightward by the left hand slit. The difference in momentum transfer between the two cases is pd/D, where p is the photon momentum, d is the slit spacing and D is the slit/screen separation. In order that the measurement made on the board's lateral momentum can be sensitive to this difference the precision of measurement should satisfy

$$\Delta p_y < pd/D.$$

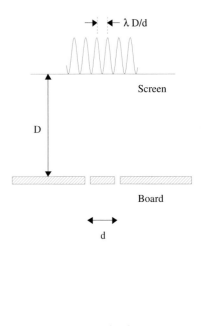

Next consider the effect of the board's movement sideways on the fringe pattern. Unless the position is known to better than the fringe spacing the fringe pattern will be smeared out and lost. Thus the lateral position of the board must be known to a precision

Fig. 12.11 Young's two slit experiment.

$$\Delta y < \lambda D/d.$$

Multiplying these two requirements together gives

$$\Delta p_y \Delta y < p\lambda.$$

The right hand side of the equation is just h. Therefore the proposed measurement to label the slit through which the photon passes *and* preserve the interference fringes would violate the uncertainty principle. One can either observe which slit the photon travels through or observe the two slit interference fringes, but not both. Put another way, the interference relies on the choice of paths taken by the photon being *indistinguishable*. If knowledge of the path (*welcher Weg information*) is complete the interference fringes are no longer detectable.

There can however be intermediate situations in which there is partial information about the path and correspondingly an interference pattern with reduced visibility. This can be demonstrated with the apparatus pictured in figure 12.12. Monochromatic light is incident through a lens onto the slits, which are each covered by a polarizer. The transmission axes of these polarizers are set at $+45°$ and $-45°$ to a line perpendicular

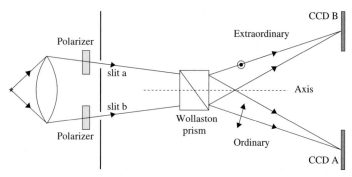

Fig. 12.12 An experiment to demonstrate partial information about the photon's path and the resulting reduced fringe visibility. The Wollaston prism and CCDs are mounted on a common frame which can be rotated about the axis indicated. The diagram is adapted from that of L. S. Bartell, *Physical Review* D21, 1698 (1980). Courtesy of Professor Bartell and the American Physical Society.

[2]See Section 10.5.2.

to the paper for slits a and b respectively. The Wollaston prism separates the ordinary and extraordinary components of the light so that they are focused by the lens onto the CCDs A and B respectively. The Wollaston prism[2] and the two CCDs are mounted in a frame that can be rotated about the axis indicated by the dotted line in the figure. If the amplitudes of the electric fields on slits a and b are both E, then their components with polarization perpendicular to the paper are both $E/\sqrt{2}$ and are extraordinary waves on entering the Wollaston prism. Similarly the components linearly polarized in the plane of the paper are both $E/\sqrt{2}$, and have ordinary polarization in the Wollaston prism. Consequently interfrence fringes are seen at both CCDs, and it is impossible to say through which slit any photon travelled.

Next suppose that the Wollaston prism and the detectors are rotated together through 45° around the dotted line axis. Light emerging from slit a now enters the Wollaston prism with ordinary polarization, while the light from slit b has extraordinary polarization. Therefore all the light from slit a(b) is directed to CCD A(B). The path is now known and the two slit fringes are gone from both CCDs. Finally the Wollaston-CCD array can be set at some intermediate orientation so that there is partial knowledge about the slit through which each photon passes. For example if the choice of orientation gives a 90% chance that any photon arriving at CCD A(B) comes from slit a(b), then the fringe pattern reappears but with a visibility reduced to around 60%.

12.10   Wavepackets and modes

In earlier chapters the emission of electromagnetic radiation from sources was pictured as taking the form of a stream of finite length wavepackets.

These wavepackets can now be recognized for what they are: the wave-trains describing individual photons emitted in an atomic transition. In the case of an excited state limited to decay to the ground state, the duration of wavepackets is determined by the lifetime, τ, of the excited state. Hence the corresponding quantum uncertainty in the photon energy is $\Delta E = h/\tau$. Any instrumental errors will lead to a larger overall uncertainty. Laser radiation is distinctive because a wavepacket contains large numbers of photons that have been emitted in phase. A detailed description of lasers is given in Chapter 14.

A further important connection can now be established between the uncertainty principle and the modes of an electromagnetic field introduced in Chapter 9. These modes are solutions of Maxwell's equations which satisfy the boundary conditions imposed by the particular optical system involved. They are the independent orthogonal states of the electromagnetic field in this system. One such mode is the Gaussian mode which propagates freely in a Fabry–Perot cavity. In the simple case of the electromagnetic field within a closed box discussed in Section 9.8 the modes have discrete wavelengths: the components of the wave vectors were shown to be integral multiples of π/L, where L is the box dimension considered. Thus the volume in wave vector space occupied by a single mode is

$$\Delta k_x \Delta k_y \Delta k_z = \pi^3/(L_x L_y L_z).$$

Rearranging this and noting that $\Delta k_i = \Delta p_i/\hbar$ gives

$$L_x L_y L_z \Delta p_x \Delta p_y \Delta p_z = \pi^3 \hbar^3.$$

This simple argument reproduces, apart from a numerical factor, eqn. 12.41 for the minimum volume in six-dimensional phase space within which a photon can be confined. Thus it can be inferred that, in general, and not just in the case of a rectangular box, the modes of the electromagnetic field for a particular set of boundary conditions represent the finest granularity into which the six-dimensional phase space can be divided consistent with the uncertainty principle.

12.10.1   Etendue

Etendue was defined in Chapter 4 as the product of the area and the solid angle of the light beam transmitted through an optical system. Multiplying the irradiance of the incident light beam by the etendue determines the radiant flux through the system. Here we examine the implications of the uncertainty principle on how finely the etendue can be resolved.

Beams in optical systems are usually paraxial so that the component of the photon momentum in the direction of the beam axis, the z-direction, is much greater than the transverse momentum components. Thus the uncertainty in this longitudinal component of momentum is almost the same as that in the total momentum,

$$\Delta p_z \approx \Delta p = \hbar \Delta k = \hbar \Delta \omega / c,$$

where $\Delta \omega$ is the spread in angular frequency of the light beam. Therefore the uncertainty in the longitudinal momentum is determined by the frequency spread.

Suppose the beam has a width in the lateral x-coordinate of Δx, and an angular spread of $\Delta \theta_x$. The uncertainty in the position of a photon is thus Δx, and that in the angle it makes with the x-axis is $\Delta \theta_x$. There are corresponding uncertainties in the y-direction. The beam etendue is then

$$\mathcal{T} = \Delta x \Delta y \Delta \theta_x \Delta \theta_y$$
$$= (\Delta p_x / p)(\Delta p_y / p) \Delta x \Delta y. \tag{12.44}$$

Now it was shown in Section 5.5.3 that the etendue from an aperture into its coherence area is

$$\mathcal{T}_c = \lambda^2. \tag{12.45}$$

Thus if the beam profile exactly matches the coherence area, it follows that

$$\Delta x \Delta y \Delta p_x \Delta p_y = \lambda^2 p^2 = h^2, \tag{12.46}$$

which reproduces the limit on this product imposed by the uncertainty principle. The result demonstrates that coherence area of the beam is the smallest meaningful division of the etendue.

12.11   Afterword

The interpretation of quantum phenomena introduced in this chapter and used hereafter is the generally accepted view originating with Bohr. In essence it accepts the probabilistic connection between waves and particles and does not consider any deeper explanation. Einstein, for one, was extremely uncomfortable with the idea that nature is probabilistic. Numerous experiments designed to seek flaws in the predictions of the standard interpretation have been carried out and have provided valuable insights into quantum behaviour. However, more than 80 years later, tests have yet to reveal any discrepancies between the predictions based on the standard interpretation and the data.

The remaining chapters of the book will build on the understanding developed in this present chapter of the dual nature of electromagnetic radiation and matter. In the 1920s and 1930s a new consistent *quantum mechanics* replaced the attempt by Bohr to describe atomic structure with an ad hoc mixture of classical mechanics and quantum conditions. The wholesale success obtained with this new quantum mechanics in predicting the details of atomic structure and spectra, including features not accounted for by the Bohr model, is described in Chapter 13 onwards. Chapter 13 also describes the associated discovery of the electron's intrinsic angular momentum, or *spin*, and relates how the circular polarization of electromagnetic radiation is connected to the photon spin. Photons and electrons obey contrasting quantum statistics, which also differ from classical statistics. In Chapter 14 the interaction of radiation with matter, the principles of lasers and their applications are outlined. Detectors of radiation are described in Chapter 15 while an account of modern communication systems based on optical fibres is presented in Chapter 16. Chapter 17 describes photonic crystals and how these possess band-gaps for light analoguous to the electronic band-gaps in semiconductors. Chapter 18 provides an account of the interactions of electromagnetic waves, particularly laser beams, with atoms. The standard semiclassical theory is used. A further step in developing the quantum theory of radiation is taken in Chapter 19, where creation and annihilation operators for photons are introduced, a step that is known as second quantization. Modern experimental techniques for manipulating photons are introduced in the same chapter, in particular the study of correlations between photons and the production of photons in entangled states. The final chapter considers a source enclosed within a cavity whose dimensions are comparable to the wavelength of radiation from the source. The source is a quantum dot and its emission can be manipulated in this environment, offering opportunities for quantum based secure cryptography.

12.12   Further reading

Crucial Experiments in Modern Physics by G. L. Trigg, published by Van Nostrand Reinhold Company (1971). This gives accounts of selected experiments with details drawn from the original papers.

Exercises

(12.1) Solar radiation falling on satellites causes the emission of electrons. In order to reduce the charging effect that this produces, the surface can be coated with platinum whch has a high work function, 6.33 eV. Calculate the cut-off frequency and wavelength for the photoelectric effect on platinum. Roughly how much of the solar spectrum's energy does this exclude?

(12.2) What is the wavelength shift of an X-ray when scattered through 17° by an electron initially at rest?

In figure 12.7 what is the origin of the left hand peak in each panel which appears undisplaced in wavelength from the incident X-rays?

(12.3) Calculate the wavelengths of a photon with energy 1 eV, and of an electron with kinetic energy 1 eV.

(12.4) Calculate the relative proportions of hydrogen atoms in the ground and first excited state in gas in thermal equilibrium at 100 000 K and at 1000 K. Would the Balmer and Lyman absorption lines be seen in the spectrum of radiation after passing through hydrogen gas at these two temperatures?

(12.5) In a certain atomic transition the parent and daughter states have lifetimes $3\,10^{-8}$ and $4\,10^{-8}\,$s. What is the natural width of the transition in terms of the photon energy?

(12.6) The Sun's surface temperature is around $6000\,$K. What is the peak wavelength of the spectrum and what is the surface irradiance (intensity)? What is the peak wavelength of radiation from a black body at room temperature, and of the cosmic background radiation at $2.75\,$K?

(12.7) Calculate the longest and shortest wavelengths in the Balmer, Paschen and Brackett series in the hydrogen spectrum.

(12.8) In positronium an electron is bound to a positron, the positron being the antiparticle of the electron which has the same mass as the electron, but opposite electric charge. What is the wavelength of the Lyman α-line of positronium?

(12.9) Calculate Planck's constant from the slopes of the lines drawn through the data points in figure 12.4.

(12.10) A camera is used to photograph a distant scene. calculate the uncertainty in the transverse momentum of a photon arriving at the image plane in terms of the $f/\#$, and the wavelength. Hence calculate the image resolution.

(12.11) A quasi-monochromatic light beam has a mean wavelength of $500\,$nm and a spread of $0.1\,$nm in wavelength. Its angular spread is $0.01\,$rad. What is the coherence volume?

(12.12) In the process of photoluminescence a material is irradiated with light and light of longer wavelengths is re-emitted by the material. Why is this light of longer rather than shorter wavelengths?

(12.13) In figure 12.12 the Wollaston prism and CCDs are rotated so that there is a 90% probability that the light reaching CCD A is from slit a. Calculate the visibility at the centre of the fringe pattern on either CCD. The illumination of the slits can be assumed equal.

(12.14) The cosmic microwave background last interacted with matter when the photons were still energetic enough to ionize hydrogen gas. After this decoupling era the wavelength of the photons stretched as the fabric of the universe expanded. If the photon energy at the spectral peak was $1.5\,$eV at the decoupling era what was the temperature in the universe at that time? By what factor has the universe since expanded?

Quantum mechanics and the atom

13.1 Introduction

Despite its successes the Bohr model of the atom is conceptually unsatisfactory in having quantum conditions grafted onto classical mechanics. Its successes lay in explaining the spectra of hydrogen-like atoms. Crucially it does not provide a way to calculate the relative intensities of spectral lines and fails to explain the spectra of other elements adequately. The more comprehensive and purely quantum mechanical analysis of atoms developed in the 1920s and 1930s is presented in this chapter. In the first part of the chapter the elements of quantum mechanics are introduced leading to Schroedinger's equation, a wave equation which describes the motion of material particles. Generally we shall only be concerned here with the motion of electrons.

Schroedinger's equation is applied first to determine the motion of an electron in a one-dimensional square electrostatic potential well. This simple case of a *quantum well* serves to demonstrate the basic properties of quantized bound states of electrons. Practical examples of quantum wells appear in quantum well lasers and optical modulators. Then the case of motion in an harmonic oscillator potential is treated, whose energy states turn out to resemble those of the electromagnetic field. After this the motion of an electron in a hydrogen-like atom is analysed. Quantization of the energy and angular momentum of the electron emerge in a natural way from the analysis.

Pauli's *exclusion principle* solved the riddle of why the electrons in an atom do not all enter the lowest energy level: there can never be more than one electron in any given quantum state. The exclusion principle and the related discovery of the intrinsic angular momentum, or *spin* of the electron are described in the next section. This leads into a discussion of multi-electron atoms, the splitting of energy levels and the selection rules governing transitions.

In the next section of the chapter measurements to determine the momentum and the angular momentum of photons are recounted. It emerges that the photon too has an intrinsic angular momentum (spin) and that in circularly polarized beams the photons have their spins aligned with their direction of travel. A comparison of the contrasting statistical behaviour of electrons and photons follows. Finally the interpretation of line widths and decay rates is considered in detail.

13.2   An outline of quantum mechanics

The new mathematical framework to deal with quantum phenomena was developed by Schroedinger, Heisenberg, Born and Jordan. This non-relativistic *quantum mechanics* is adequate for our purposes but is strictly applicable only if the material particles travel at velocities very much less than c. The key achievement was to arrive at a wave equation for the electron analogous to the electromagnetic wave equation for the photon. This wave equation replaces Newton's equation of motion for material particles. An underlying assumption is that the universal laws of conservation of energy, momentum and angular momentum should remain valid in quantum mechanics.

Dirac produced a relativistic theory which was developed into a comprehensive relativistic field theory of electromagnetic radiation and its interaction with charged particles by Feynmann, Tomonaga and Schwinger, and is known as quantum electrodynamics (QED). Dirac received the Nobel Prize in Physics in 1933; Tomonaga, Schwinger and Feynman shared this prize in 1965.

It is postulated that the behaviour of a single electron or a set of electrons is described by a complex wavefunction which contains all possible information that exists about the system. This wavefunction, $\Psi(q_n, t)$, is a function of time and all the independent variables, written as a set $\{q_n\}$. These variables could be the spatial coordinates for a single electron (plus its polarization – if it should have any). The interpretation of the wavefunction parallels the interpretation of electromagnetic waves when locating a photon: the probability for finding a system with variables in a range $dV = dq_1 dq_2...$ around q_1, q_2,... is defined to be

$$P(q_1, q_2, ...)\, dV = \Psi^* \Psi dV. \tag{13.1}$$

The wavefunction used is normalized, meaning that a numerical factor is inserted so that integrating PdV over the full range of the independent variables gives unity. In the case of a single electron PdV is simply the probability of finding the electron within the spatial volume dV. The formal development of quantum mechanics presented below has basic features that apply equally to photons and electromagnetic waves. The treatments of electrons and photons diverge because the wave equations are different and, as we shall see later, because any number of photons can share the same quantum state while only a single electron can ever occupy any given quantum state.

Quantities that are measurable for a particle or a system of particles are known as *observables*. Position, momentum, orbital angular momentum, polarization and energy are all observables. It goes without saying that the measurements of observables give real and not complex numbers. As a first example these ideas are illustrated for a free electron moving in one dimension. It is described by a wavefunction which is simply a plane wave travelling in the x-direction

$$\Phi_k = (1/\sqrt{L}) \exp\left[i(kx - \omega t)\right], \tag{13.2}$$

where L is a very large range in x to which the electron is restricted and which will be increased to infinity as required.[1] The operators for momentum and the total energy are defined as follows and indicated by placing hats over the respective symbols for the observables

$$\hat{p} = -i\hbar \frac{\partial}{\partial x}, \tag{13.3}$$

$$\hat{H} = +i\hbar \frac{\partial}{\partial t}. \tag{13.4}$$

When these operators act on a plane sinusoidal wavefunction they give

$$\hat{p}\Phi_k = -i\hbar \frac{\partial \Phi_k}{\partial x} = \hbar k \Phi_k = p\Phi_k, \tag{13.5}$$

$$\hat{H}\Phi_k = +i\hbar \frac{\partial \Phi_k}{\partial t} = \hbar \omega \Phi_k = E\Phi_k. \tag{13.6}$$

The quantities p and E appearing on the right hand side, without hats, are the values that can be obtained in measurements of the momentum and kinetic energy respectively.[2] It is argued that these operators for momentum and total energy should apply equally for wavefunctions in general because any wavefunction can be Fourier analysed into linear sums of sinusoidal waves.

The motion of any free electron would be described by a wavepacket made up of a superposition of plane sinusoidal waves, analogous to a photon wavepacket. The Fourier transform of the wavepacket is a frequency distribution from which an energy distribution can be calculated. Then the result of measuring the electron energy once would be some random value within this distribution. However making measurements on a large set of electrons with identical wavepackets would reproduce the distribution determined by Fourier analysis. The mean value of such a set of measurements is called the *expectation value*.

Because the operators are complex and the quantities measured are real it follows that the waves for electrons and other material particles must themselves be complex, unlike electromagnetic fields which are real. Here the wave equation for electrons will be introduced first and then solutions obtained for simple potentials.

13.3 Schroedinger's equation

Schroedinger constructed an operator equation with which to analyse the *non-relativistic* motion of electrons. The starting point is to write the law of conservation of energy for an electron moving in some electrostatic potential $V(\mathbf{r})$

$$E = V(\mathbf{r}) + p^2/2m, \tag{13.7}$$

where E is the total energy and $p^2/2m$ is the kinetic energy of the electron. Generalizing eqn. 13.3, the operator equivalent of

$$p^2 = p_x^2 + p_y^2 + p_z^2$$

[1] Plane waves provide a simple example for discussion and finite realistic wavepackets are all linear sums of plane waves. Plane sinusoidal waves extend to infinity and the range L is needed to give a normalizable wavefunction. The values of measurable quantities are correctly predicted when the limit $L \to \infty$ is taken. On occasion care is needed when taking the limit.

[2] This prediction of an exact value for the momentum appears to violate the uncertainty principle. However the likelihood of finding the electron in any interval, $\Phi^*\Phi \mathrm{d}x = \mathrm{d}x/L$, is the same everywhere. Then the position, which is the conjugate variable to the momentum, is indeterminate.

is simply

$$-\hbar^2\nabla^2 = -\hbar^2(\partial^2/\partial x^2 + \partial^2/\partial y^2 + \partial^2/\partial z^2).$$

This result can be used to convert eqn. 13.7 into a notional equation between operators

$$i\hbar\partial/\partial t = V(\mathbf{r}) - (\hbar^2/2m)\nabla^2. \tag{13.8}$$

A valid wave equation results if the operators act on the electron wavefunction $\Psi(\mathbf{r},t)$. The result is

$$i\hbar\partial\Psi(\mathbf{r},t)/\partial t = V(\mathbf{r})\Psi(\mathbf{r},t) - (\hbar^2/2m)\nabla^2\Psi(\mathbf{r},t), \tag{13.9}$$

which is called *Schroedinger's time dependent equation*. Its solution describes the motion of the electron in the chosen potential $V(\mathbf{r})$.

If, as here, the potential does not vary with time the solution factorizes to give

$$\Psi(\mathbf{r},t) = \psi(\mathbf{r})\exp\left(-iEt/\hbar\right) \tag{13.10}$$

which when substituted in Schroedinger's equation gives its *time independent* form

$$E\psi(\mathbf{r}) = V(\mathbf{r})\psi(\mathbf{r}) - (\hbar^2/2m)\nabla^2\psi(\mathbf{r}), \tag{13.11}$$

where E is the electron kinetic plus potential energy. In the case that the electron is free this reduces to

$$E\psi(\mathbf{r}) = -(\hbar^2/2m)\nabla^2\psi(\mathbf{r}). \tag{13.12}$$

A solution is

$$\psi_k(\mathbf{r}) = \exp\left(i\mathbf{k}\cdot\mathbf{r}\right),$$

and if this is substituted in eqn. 13.12 it gives, as expected, $E = \hbar^2k^2/2m$ or $E = p^2/2m$.

Schroedinger's equation is linear in ψ so that wavefunctions which satisfy the equation can be superposed to give another valid wavefunction just as solutions of the electromagnetic wave equation can be superposed. There are several crucial differences between, on the one hand Maxwell's equations and electromagnetic waves, and on the other Schroedinger's equation and electron waves. Schroedinger's equation is complex and the electron waves are complex and not directly measurable: Maxwell's equations are real and the electromagnetic fields are real and directly measurable. Note that it has often been useful in the preceding chapters to perform calculations using complex fields whose real parts are the actual electromagnetic fields. Another difference is that for Schroedinger's equation to apply the motion of the electron must be non-relativistic, whereas Maxwell's equations are fully relativistic. All the external influence on an electron is absorbed into a static potential V in Schroedinger's equation, which is only adequate to describe electrostatic fields. Despite

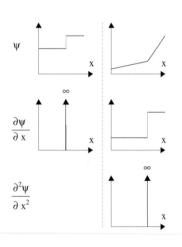

Fig. 13.1 The panels show unphysical discontinuities. In the left hand panel the wavefunction, has a discontinuity, it jumps up, which would require infinite momentum. In the right hand panel the derivative has a similar discontinuity, and this would require infinite energy.

these limitations a basic understanding of atomic states and their radiation is achieved by applying Schroedinger's equation.

Any solution of Schroedinger's equation must satisfy several simple requirements. Firstly the wavefunction must be finite everywhere in order that the probability of finding the electron is finite everywhere. The next requirement is that the wavefunction is continuous and single valued everywhere. If instead the wavefunction jumped discontinuously, as shown in the left hand panel of figure 13.1, the derivative would become infinite at that point. Thus a measurement of momentum made over a region including this point would yield an infinite momentum. Similarly the first derivative must be continuous everywhere. Were this to jump discontinuously, as shown in the right hand panel of figure 13.1, then the second derivative would be infinite. Referring back to eqns. 13.7 and 13.9 we see that this is impossible when both the energy and the potential are finite everywhere. These requirements on the continuity of the wavefunction and its derivative are essential tools when joining up solutions of Schroedinger's equation at boundaries where the potential changes. We shall see how boundary conditions directly affect the quantized values of energy and other measurable quantities.

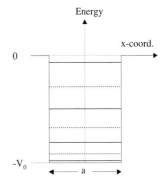

Fig. 13.2 The energy levels of eigenstates in the square potential well.

13.3.1 The square potential well

Before tackling the motion of an electron in the Coulomb potential of the nucleus the motion of an electron of mass m in a one-dimensional square potential well will be studied. This example provides an uncluttered first view of a quantum wavefunction for an electron in a potential well. The potential is drawn in figure 13.2, it has a value $-V_0$ over the region $-a/2 < x < a/2$ and is zero elsewhere. Within the attractive well Schroedinger's equation is

$$(-\hbar^2/2m)\mathrm{d}^2\psi/\mathrm{d}x^2 = (E + V_0)\psi \text{ (internal)}, \tag{13.13}$$

while outside the potential it becomes

$$(-\hbar^2/2m)\mathrm{d}^2\psi/\mathrm{d}x^2 = E\psi \text{ (external)}. \tag{13.14}$$

Bound states of the electron, for which E is negative and the kinetic energy, $(E + V_0)$, is positive are considered first. A solution inside the well which is symmetric about the origin is

$$\psi_i = A_i \cos(k_i x), \tag{13.15}$$

where $k_i = \sqrt{2m(E + V_0)}/\hbar$ and A_i is some constant. Externally

$$\psi_e = A_e \exp(\mp k_e x) \tag{13.16}$$

where $k_e = \sqrt{-2mE}/\hbar$ and A_e is another constant. The upper sign in the exponent is taken for $x > a/2$ and the lower sign for $x < -a/2$. The

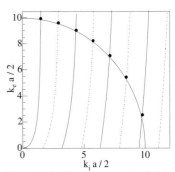

Fig. 13.3 Graphical method for obtaining solutions of Schroedinger's equation for the square well potential of depth V_0 and width a. k_i is the wavenumber within the well, and k_e that outside it. The curves are discussed in the text.

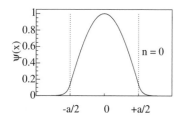

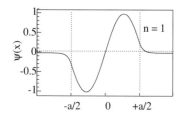

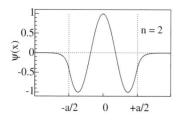

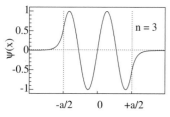

Fig. 13.4 The four wavefunctions of lowest energy satisfying the square well boundary conditions. They are labelled with the number of nodes within the well. Broken lines mark the well edges where classical motion would terminate.

opposite choices of sign for the exponentials would give wavefunctions growing exponentially with the distance from the well. These can be rejected because they grow infinitely.

Applying the requirements that the wavefunction and its first derivative are continuous at the wall at $x = a/2$ gives

$$A_i \cos (k_i a/2) = A_e \exp (-k_e a/2) \text{ and}$$
$$A_i k_i \sin (k_i a/2) = k_e A_e \exp (-k_e a/2).$$

Dividing one equation by the other gives

$$k_e = k_i \tan (k_i a/2). \tag{13.17}$$

From the definitions of k_i and k_e we also have

$$(k_i a/2)^2 + (k_e a/2)^2 = ma^2 V_0/(2\hbar^2). \tag{13.18}$$

The last two equations can be solved simultaneously either by computer or graphically as exhibited in figure 13.3 where $(k_e a/2)$ is plotted as a function of $(k_i a/2)$ for a given potential V_0. The relation found in eqn. 13.17 is represented by the full lines, while the quarter circle represents eqn. 13.18 with $ma^2 V_0/(2\hbar^2)$ taken to be 100. Simultaneous solutions to eqns. 13.17 and 13.18 lie at the points where these curves intersect.

A second set of wavefunctions which are antisymmetric about the origin also satisfy Schroedinger's equation for the square well. The waves inside the well have the form

$$\psi_i = B_i \sin (k_i x), \tag{13.19}$$

where B_i is some constant. Outside the well

$$\psi_e = B_e \exp (\mp k_e x) \tag{13.20}$$

where B_e is another constant. For these wavefunctions the continuity conditions lead to a different transcendental equation

$$k_e = -k_i \cot (k_i a/2). \tag{13.21}$$

This equation is plotted with broken lines in figure 13.3. On this plot the simultaneous solutions to eqns. 13.21 and 13.18 lie at the intersections of the broken lines and the quarter circle. Then on figure 13.2 the energy levels of all seven solutions are shown using full and broken lines for the states with even and odd wavefunctions respectively. Finally the wavefunctions of the four lowest energy (most tightly bound) states are plotted in figure 13.4.

The preceding analysis shows that bound states are restricted to discrete energies. Only then can the sinusoidal waves inside the well join smoothly onto a wave that decays exponentially outside the well. At other energies the requirement of continuity at the boundary makes it necessary to have a sum of a decaying and an *increasing* exponential outside the well. No matter how little the electron's energy differs from the discrete value picked out by the solution of Schroedinger's equation in figure 13.3 the exponentially increasing component of the wave outside the well will tend to infinity at an infinite distance and cannot describe electron states localized in the well. This restriction to states with *discrete* energies is a feature which distinguishes quantum mechanics from classical mechanics. Discrete energy states are met in atoms, molecules and in nuclei. Unbound electrons, that is to say electrons with positive energies, have wavefunctions that are oscillatory both inside and outside the potential well. The continuity conditions at the boundary can now be satisfied at any positive energy and so there is a continuum of allowed states extending from zero energy upwards.

Another departure from classical behaviour illustrated by the solutions of the square well potential is that the wavefunction of a bound electron does not vanish in the region outside the well where its kinetic energy has become *negative*. Classically the electron would be confined between the walls where its kinetic energy is positive and it would never penetrate beyond the well. The quantum wavefunction decays exponentially in this region and is analogous to the evanescent electromagnetic wave occurring in total internal reflection and described in Section 9.5. The parallel extends further to include frustrated total internal reflection. Figure 13.5 shows a potential barrier of finite width. Electrons incident from the left are reflected to give standing waves. In addition the electron's wavefunction penetrates the potential barrier, and at the far side this exponentially decaying wave joins smoothly onto an oscillatory wave that travels away from the boundary. Electrons can therefore travel through a region where their kinetic energy is negative and emerge on the far side, a possibility which is absolutely forbidden to them in classical mechanics. This purely quantum process is called *barrier penetration* or *tunnelling*.

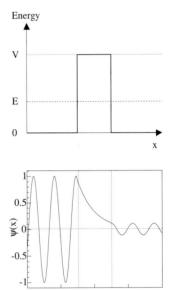

Fig. 13.5 Potential barrier and wavefunction penetration.

13.4 Eigenstates

The wavefunctions that are solutions of Schroedinger's equation for simple potentials like the square well, and including the case of the free electron with zero potential energy, are known as energy *eigenfunctions*. The corresponding energies are called energy *eigenvalues* and the electron is said to be in an *eigenstate* of energy. An eigenstate may be an eigenstate of several observables with each taking unique values for a given eigenstate. These are then known as *compatible* or *simultaneous* observables: examples are the energy, the angular momentum and a component of the angular momentum of an electron in a hydrogen atom. In an eigenstate the measurement of these compatible observables leaves the electron in the eigenstate. It is worth repeating that the eigenvalues of energy are discrete when the potential localizes the electron in a potential well, but continuous from zero up to any conceivable positive value when an electron is free.

The existence and the properties of eigenstates generalize to systems of electrons and other material particles. Such a system has a set of eigenstates $\{\phi_i\}$ of observables such as A with eigenvalues $\{a_i\}$ respectively. With the standard notation the operator corresponding to A is $\hat{A}$, and this acts on the wavefunction ϕ_i in the following way:

$$\hat{A}\phi_i = a_i\phi_i, \tag{13.22}$$

meaning that any measurement of the observable A on the eigenstate ϕ_i always gives the eigenvalue a_i.

13.4.1   Orthogonality of eigenstates

An intrinsic property of eigenstates is their *orthogonality* in the sense that the overlap integrals between the wavefunctions of any pair of them over all the free variables vanish. Suppose ϕ_i and ϕ_j are two such eigenfunctions of an electron; then writing Schroedinger's equation for ϕ_i and multiplying it by ϕ_j^*,

$$\phi_j^* \left(-\frac{\hbar^2}{2m}\nabla^2 + V \right) \phi_i = \phi_j^* E_i\phi_i,$$

then repeating the process with the wavefunctions the other way about

$$\phi_i \left(-\frac{\hbar^2}{2m}\nabla^2 + V \right) \phi_j^* = \phi_i E_j\phi_j^*.$$

Subtracting one equation from the other and integrating the result over a volume much larger than the potential well gives

$$(E_i - E_j) \int \phi_j^*\phi_i dV = -(\hbar^2/2m) \int (\phi_j^*\nabla^2\phi_i - \phi_i\nabla^2\phi_j^*)\, dV$$

$$= -(\hbar^2/2m) \int \nabla \cdot (\phi_j^*\nabla\phi_i - \phi_i\nabla\phi_j^*)\, dV.$$

Using Gauss' theorem the right hand side becomes a integral over a surface enclosing the volume considered,

$$(E_i - E_j) \int \phi_j^*\phi_i dV = -(\hbar^2/2m) \int (\phi_j^*\nabla\phi_i - \phi_i\nabla\phi_j^*) \cdot d\mathbf{S}.$$

This new surface integral is evaluated over a surface sufficiently remote from the potential well that the wavefunctions and their derivatives have vanished. Hence

$$\int \phi_j^*\phi_i\, dV = 0, \quad \text{if } E_j \neq E_i. \tag{13.23}$$

Eigenfunctions are usually normalized for convenience so that

$$\int \phi_i^*\phi_i\, dV = 1. \tag{13.24}$$

Thus

$$\int \phi_j^* \phi_i \mathrm{d}V = \delta_{ji}, \qquad (13.25)$$

where δ_{ji} is the Kronecker δ defined by

$$\delta_{ji} = 0 \text{ for } j \neq i, \qquad (13.26)$$
$$\delta_{ji} = 1 \text{ for } j = i. \qquad (13.27)$$

With a free electron

$$\phi_k = (1/\sqrt{L}) \exp\left[i(kx - \omega t)\right]$$

where the subscript k is a continuous variable rather than an integer label. Then using the Dirac δ function introduced in Chapter 7 and eqn. 7.21

$$\int \phi_k^* \phi_{k'} \mathrm{d}x = \delta(k - k'). \qquad (13.28)$$

Similar properties hold good for the modes of electromagnetic radiation in an optical setup, because these are the eigenstates of Maxwell's equations satisfying the boundary conditions imposed by the optical components. For example the standing electromagnetic waves possible in a Fabry–Perot cavity have discrete wavelengths and their waveforms are mutually orthogonal.

13.5   Expectation values

In the more general case that a system is not in an eigenstate of an observable, the value that is obtained by measuring the observable can only be predicted statistically. Quantum mechanics predicts the *expectation value* of an observable A, which is written $\langle \hat{A} \rangle$, and defined by the equation

$$\langle \hat{A} \rangle = \int \psi^* \hat{A} \psi \, \mathrm{d}V, \qquad (13.29)$$

where ψ is normalized. The equation is to be interpreted in this way. Suppose that the same measurement of A is made on each of a large number of systems which have been prepared in exactly the same way so that they have identical wavefunctions ψ – such a hypothetical collection of systems is called an *ensemble*. Then the average value of the observable measured over the ensemble equals the expectation value. In the case of an eigenstate of the observable A the expectation value is simply the eigenvalue of A for that eigenstate.

Any wavefunction ψ of a system which has an observable A can always be expanded as a linear superposition of the normalized eigenfunctions $\{\phi_i\}$ of A. Assuming for simplicity that the eigenvalues are discrete,

$$\psi = \sum_i c_i \phi_i. \qquad (13.30)$$

Then the expectation value of A in a state with wavefunction ψ is defined to be

$$\langle \hat{A} \rangle = \int \psi^* \hat{A} \psi \, dV$$

$$= \sum_{i,j} c_i^* c_j \int \phi_i^* \hat{A} \phi_j \, dV$$

$$= \sum_{i,j} c_i^* c_j a_j \int \phi_i^* \phi_j \, dV$$

$$= \sum_j c_j^* c_j a_j. \tag{13.31}$$

The interpretation of this result is that

$$P_j = c_j^* c_j$$

is just the probability that the measurement finds the value a_j, or equally the probability that the system is found to be in the eigenstate ϕ_j. When the eigenvalues are continuous as for a free electron described by eqn. 13.2

$$\psi = \int c(k) \phi_k \, dk. \tag{13.32}$$

A similar analysis to that given above then yields

$$\langle \hat{A} \rangle = \int c^*(k) c(k) a(k) \, dk, \tag{13.33}$$

where $a(k)$ is the value obtained when A is measured on the eigenstate with momentum k. Now

$$P(k) \, dk = c^*(k) c(k) \, dk \tag{13.34}$$

is the probability that the measurement finds a momentum eigenvalue lying between k and $k + dk$.

All actual measurements yield real values so that expectation values of observables have to be real. Thus for any observable A

$$\int \psi^* \hat{A} \psi \, dV = \left(\int \psi^* \hat{A} \psi \, dV \right)^*$$

$$= \int \psi \hat{A}^* \psi^* \, dV,$$

and then

$$\int \psi^* \hat{A} \psi \, dV = \int (\hat{A} \psi)^* \psi \, dV. \tag{13.35}$$

Operators which have this mathematical property for all wavefunctions of a system are called *hermitean*; thus observables are always represented by hermitean operators.

The expectation values of observables are unchanged if the wavefunction is multiplied by a phase factor $\exp(i\alpha)$ where α is real. Thus there is always a phase ambiguity in the wavefunctions.

13.5.1   Collapse of the wavefunction

If the measurement of the observable A on a system with wavefunction ψ gives the value a_i, the system must immediately thereafter be in the eigenstate with wavefunction ϕ_i, and no longer in the state with wavefunction ψ. A second measurement of the observable A will again give a_i, and so would further measurements. This result is profoundly different from anything met in classical mechanics. The change is discontinuous: up to the exact moment of the measurement the system is evolving according to the wavefunction ψ and immediately afterwards its wavefunction has become ϕ_i. This step is known as the *collapse of the wavefunction*.

Schroedinger highlighted the logical difficulty of an external observer causing the wavefunction to collapse by using a cat fable. The cat is locked in a box together with a mechanism which will release a lethal gas if and when a single radioactive nucleus decays. It is then argued that the wavefunction of the contents of the box should be a superposition of two wavefunctions: the first for the undisturbed mechanism and a live cat; the second for an activated mechanism and a dead cat. Later Schroedinger opens the box and observes the contents. At this instant the wavefunction of the contents collapses to either one that contains a live cat, or to another that contains a dead cat. The generally favoured resolution of this paradox of a cat simultaneously alive and dead is through what is called *decoherence*. Broadly speaking any interaction of a quantum system with its surroundings is equivalent to making a measurement. For example the air molecules striking the cat are sufficient to collapse its wavefunction.[3]

[3]See for example 'Decoherence and the the transition from quantum to classical' by W. J. Zurek, *Physics Today*, October 1991.

Carrying the idea of having wavefunctions for macroscopic objects to its logical conclusion requires the universe to possess a wavefunction. Then all observers are part of the wavefunction. Whether this view is valid, and how the wavefunction of the universe could collapse are questions that have led to considerable speculation.

13.5.2   Compatible or simultaneous observables

Eigenstates of a system are usually eigenstates of several observables, and it requires knowledge of all of these to completely specify an eigenstate. Here we consider the case where there are just two of these *compatible* observables, A and B. There is a set of eigenfunctions $\{\phi\}$ for which

$$\hat{A}\phi_j = a_j\phi_j; \;\; \hat{B}\phi_j = b_j\phi_j.$$

Suppose some eigenfunctions are *degenerate*, sharing the same eigenvalue for A, while each eigenvalue of B is different from all others. A measurement of A may then leave a system described by a superposition of the degenerate eigenfunctions. In this case a subsequent measurement of B

will result in the wavefunction collapsing into a single eigenfunction from this superposition, for example ϕ_k. Further measurements thereafter of A and B yield a_k and b_k respectively. The expectation value for the product of compatible observables

$$\int \psi^* \hat{A}\hat{B}\psi \, \mathrm{d}V = \sum_j c_j^* c_j a_j b_j = \int \psi^* \hat{B}\hat{A}\psi \, \mathrm{d}V,$$

holds true whatever arbitrary state ψ of the system is considered. Consequently the expectation value of $\hat{A}\hat{B} - \hat{B}\hat{A}$ always vanishes. This operator is called the *commutator* of A and B and is written with square brackets $[\hat{A}, \hat{B}]$. If, as in the case being considered

$$[\hat{A}, \hat{B}] = 0, \qquad (13.36)$$

$\hat{A}$ and $\hat{B}$ are said to commute.

The uncertainty principle revisited

The operators of conjugate variables like the position and momentum do not commute

$$[\hat{x}, \hat{p}]\psi = \left\{ x\left(-ih\frac{\partial}{\partial x}\right) - \left(-ih\frac{\partial}{\partial x} x\right) \right\}\psi = i\hbar\psi.$$

For any such pair of conjugate variables, which we denote by F and G

$$[\hat{F}, \hat{G}] = i\hbar. \qquad (13.37)$$

This relationship is now shown to be directly related to the uncertainty principle.

Writing δF for the operator $\hat{F} - \langle\hat{F}\rangle$, the *variance* is defined as

$$(\Delta F)^2 = \langle\hat{F}^2\rangle - \langle\hat{F}\rangle^2$$
$$= \langle[\hat{F}^2 - 2\langle\hat{F}\rangle\hat{F} + \langle\hat{F}\rangle^2]\rangle = \langle(\delta F)^2\rangle. \qquad (13.38)$$

With Hermitian operators real and imaginary parts separate thus[4]

$$\langle\delta F\delta G\rangle = \langle\delta F\delta G + \delta G\delta F\rangle/2 + \langle\delta F\delta G - \delta G\delta F\rangle/2.$$

Then using eqn. 13.37 again

$$[\delta F, \delta G] = i\hbar,$$

so that

$$|\langle\delta F\delta G\rangle|^2 > \hbar^2/4. \qquad (13.39)$$

Using the Schwarz inequality and eqn. 13.38

$$(\Delta F)^2(\Delta G)^2 = \langle(\delta F)^2\rangle\langle(\delta G)^2\rangle \geq |\langle\delta F\delta G\rangle|^2.$$

Finally using eqn. 13.39 to replace the right hand side in this equation gives

$$(\Delta F)^2(\Delta G)^2 \geq \hbar^2/4 \qquad (13.40)$$

which reproduces the uncertainty principle. In general we can say that if the operators of two observables do not commute, then simultaneous measurements of their observables will obey an uncertainty relation.

[4]We have

$$\langle FG\rangle^* = \langle(FG)^\dagger\rangle = \langle G^\dagger F^\dagger\rangle$$
$$= \langle GF\rangle,$$

where the last equality makes use of the Hermitean property. A similar result then follows $\langle\delta F\delta G\rangle^* = \langle\delta G\delta F\rangle$. Hence the expectation value of the sum $\langle\delta F\delta G + \delta G\delta F\rangle$ is real, and that of the difference is imaginary.

13.6 The harmonic oscillator potential

A frequently met dynamical system is that of a mass undergoing simple harmonic motion in the x-dimension under a restoring force linear in the displacement from the origin, kx. Its eigenstates are of particular interest because they are exact parallels to the eigenstates of black body radiation. This parallel will prove of value in Chapter 19. The potential energy is the integral of the form

$$V = \int_0^x kx \, dx = kx^2/2,$$

and this function is displayed in figure 13.6. Schroedinger's time independent equation for this potential is then

$$-(\hbar^2/2m) \, d^2\psi/dx^2 + (kx^2/2)\psi = E\psi.$$

Setting $\alpha^4 = km/\hbar^2$ and $s = \alpha x$, the equation can be rewritten

$$d^2\psi/ds^2 + (\lambda - s^2)\psi = 0, \tag{13.41}$$

where $\lambda = 2E/\hbar\omega_0$, with $\omega_0 = \sqrt{k/m}$ being the classical angular frequency of oscillation of the mass in this potential. This equation has analytic solutions of the form

$$\psi(s) = H(s) \exp(-s^2/2), \tag{13.42}$$

where $H(s)$ is a polynomial. Solutions with a term $\exp(s^2/2)$ are excluded because they diverge at infinity and are not confined to the potential well. The derivatives are

$$d\psi(s)/ds = [dH(s)/ds] \exp(-s^2/2) - s \, H(s) \exp(-s^2/2),$$
$$d^2\psi(s)/ds^2 = [d^2H(s)/ds^2] \exp(-s^2/2) - 2s \, [dH(s)/ds] \exp(-s^2/2)$$
$$+(s^2 - 1) \, H(s) \exp(-s^2/2),$$

and when these are substituted into eqn. 13.41 the result is

$$[d^2H(s)/ds^2] - 2s \, [dH(s)/ds] + (\lambda - 1) \, H(s) = 0. \tag{13.43}$$

Finite polynomial solutions to this equation can be obtained provided that λ takes the discrete values[5]

$$\lambda = 2n + 1, \tag{13.44}$$

where n is any non-negative integer. When λ is replaced by its defined value, $2E/(\hbar\omega_0)$, it immediately follows that the energy of the nth eigenstate is

$$E_n = \hbar\omega_0(n + 1/2). \tag{13.45}$$

The lowest energy solutions, labelled with the value of n as a subscript, are

$$H_0(s) = 1; \; H_1(s) = 2s; \; H_2(s) = 4s^2 - 2; \; H_3(s) = 8s^3 - 12s. \tag{13.46}$$

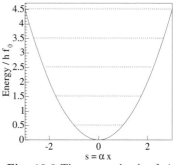

Fig. 13.6 The energy levels of eigenstates in the harmonic potential well.

[5]See Chapter 4 of the third edition of *Quantum Mechanics* by L. I. Schiff, published by McGraw-Hill Kogakusha Ltd., Tokyo (1968).

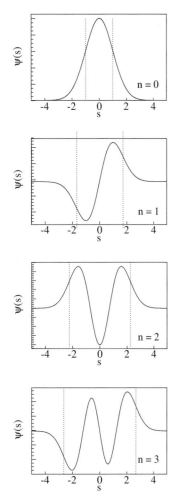

Fig. 13.7 The four wavefunctions of lowest energy in a harmonic well, labelled with the quantum number n. The vertical dotted lines indicate the points at which the kinetic energy is zero, and where the classical motion terminates.

These functions are called *Hermite* polynomials. There is a simple recurrence relation derived from eqn. 13.43 that can be used to generate further members of the sequence

$$H_{n+1}(s) = 2sH_n(s) - 2nH_{n-1}(s).$$

The resulting *Gauss–Hermite* solutions to the Schroedinger equation for the harmonic well, $H_n(s)\exp(-s^2/2)$, are shown in figure 13.7. Of these the lowest energy solution has the familiar Gaussian shape. Corresponding energy levels are indicated by the horizontal lines in figure 13.6.

The classical state of lowest energy would be that in which the mass is at rest at $x = 0$ and has zero kinetic energy. By contrast the lowest energy quantum state with the Gaussian wavefunction has a positive energy $\hbar\omega_0/2$. This is called the *zero point energy* and is a characteristic of quantum systems. In the square well potential the lowest energy state is displaced upward in energy from $-V_0$ showing that there also the particle is not at rest in its lowest energy state.

It is striking that the energy spectrum of the eigenstates in the harmonic well matches that postulated by Planck for the modes of the electromagnetic field of angular frequency ω_0 – apart from the displacement upward by the zero point energy. This feature will be considered further in Chapter 19, and it will emerge that the parallel is exact: when there are no photons in a mode of the electromagnetic field it too has a zero point energy $\hbar\omega_0/2$! In the intervening chapters this uniform dispacement upward by $\hbar\omega_0/2$ of all the energy levels of any mode of electromagnetic radiation can be safely neglected because only the transitions between states are discussed.

13.7   The hydrogen atom

Schroedinger's analysis of the electron motion within the Coulomb potential due to the nuclear charge provided a detailed and precise description of atomic structure, which incorporated all the successes of the Bohr model. In a hydrogen-like atom with a nucleus carrying a charge Ze and having mass M, the Coulomb potential felt by the single electron of mass m is $-Ze^2/(4\pi\varepsilon_0 r)$ at a distance r from the nucleus. This potential is drawn for the hydrogen atom in figure 13.8. Then the Schroedinger time independent equation is

$$-(\hbar^2/2\mu)\nabla^2\psi - Ze^2\psi/(4\pi\varepsilon_0 r) = E\psi, \qquad (13.47)$$

where $\mu = mM/(M+m)$ is the reduced mass of the electron and E is the total energy of the atom. The same approach is taken as that applied in seeking solutions to the square well potential. Although the analysis is more complicated the solutions in this case are analytic. Here only the results are discussed, but full details of the solution can be found in

Table 13.1 Table of radial eigenfunctions $R_{nl}(r)$.

n	l	$R_{nl}(r)$
1	0	$2(Z/a_0)^{3/2} \exp{(-Zr/a_0)}$
2	0	$(1/\sqrt{8})(Z/a_0)^{3/2} [2 - Zr/a_0] \exp{(-Zr/2a_0)}$
2	1	$(1/\sqrt{24})(Z/a_0)^{3/2} [Zr/a_0] \exp{(-Zr/2a_0)}$
3	0	$(\sqrt{4/3}/81)(Z/a_0)^{3/2} [27 - 18Zr/a_0 + 2(Zr/a_0)^2] \exp{(-Zr/3a_0)}$

many standard texts on atomic physics or quantum mechanics.[6]

The eigenfunctions separate into radial and angular components, which are best written in spherical polar coordinates

$$\psi_{nlm_l}(r, \theta, \phi) = R_{nl}(r)\, Y_{lm_l}(\theta, \phi). \qquad (13.48)$$

Each solution is identified by three integral *quantum numbers* n, l, m_l. The function R_{nl} contains an associated Laguerre polynomial $F(r)$, and Y_{lm} is a spherical harmonic function. In general these have the forms

$$R_{nl}(r) = \exp{(Cr/n)}\, r^l\, F(r), \qquad (13.49)$$

$$Y_{lm_l}(\theta, \phi) = \sin^{|m_l|}\theta\, \exp{(im_l\phi)}\, G(\theta), \qquad (13.50)$$

where C is a constant and $F(r)$ and $G(\theta)$ are polynomials in r and $\cos\theta$ respectively. The exact forms of the eigenfunctions are given in tables 13.1 and 13.2 for a few of the lowest values of n, l and m_l. These wavefunctions are *orthonormal* in the usual sense that the volume integrals over all space are

$$\int \psi_{nlm_l}^*\, \psi_{n'l'm_l'}\, \mathrm{d}V = \delta_{n,n'}\, \delta_{l,l'}\, \delta_{m_l,m_l'}. \qquad (13.51)$$

Valid combinations of the quantum numbers are restricted to the following values:

$$n = 1, 2, 3, \ldots\ldots;$$
$$l = 0, \ldots\ldots, n-2, n-1;$$
$$m_l = -l,\, -l+1,\, -l+2,\, \ldots\ldots,\, l-2,\, l-1,\, l. \qquad (13.52)$$

The energy of an electron, its orbital angular momentum and a component of its angular momentum, which can be chosen to be the z-component, are the three compatible observables and their eigenvalues are specified by the quantum numbers n, l and m_l respectively. The predicted energy eigenvalues duplicate those found with the Bohr model

$$E_n = -\mu Z^2 e^4/[(4\pi\varepsilon_0)^2 2\hbar^2 n^2], \qquad (13.53)$$

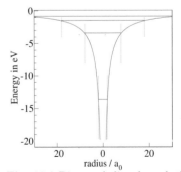

Fig. 13.8 Diametral slice through the Coulomb potential due to the hydrogen nucleus. The first few electron energy levels are shown. a_0 is the Bohr radius.

---

[6]For example the third edition of *Quantum Mechanics* by L. I. Schiff, published by McGraw-Hill Kogukusha, Tokyo (1968).

Table 13.2 Table of angular eigenfunctions $Y_{lm}(\theta, \phi)$.

l	m	$Y_{lm}(\theta, \phi)$
0	0	$1/\sqrt{4\pi}$
1	0	$\sqrt{(3/4\pi)}\cos\theta$
1	± 1	$\sqrt{(3/8\pi)}\sin\theta\exp(\pm i\phi)$
2	0	$\sqrt{(5/16\pi)}[3\cos^2\theta - 1]$
2	± 1	$\sqrt{(15/8\pi)}\sin\theta\cos\theta\exp(\pm i\phi)$
2	± 2	$\sqrt{(15/32\pi)}\sin^2\theta\exp(\pm 2i\phi)$

so that n is called the *principal* or energy quantum number.

Angular momentum operators can be constructed from the momentum operators as follows. In classical mechanics the z-component of the vector angular momentum, **L**, is

$$L_z = xp_y - yp_x,$$

from which the quantum mechanical operator can be obtained by replacing the position and momentum by their operator equivalents

$$\hat{L}_z = -i\hbar \left[x\frac{\partial}{\partial y} - y\frac{\partial}{\partial x} \right]. \tag{13.54}$$

The total orbital angular momentum operator, **L**, is given by

$$\hat{\mathbf{L}}^2 = \hat{L}_x^2 + \hat{L}_y^2 + \hat{L}_z^2. \tag{13.55}$$

$\hat{L}^2$ commutes with all the components L_x, L_y and L_z, but these components do not commute with each other. For example

$$[L_x, L_y] = i\hbar L_z. \tag{13.56}$$

This explains why the total orbital angular momentum and only one of its components can be compatible observables. For completeness the forms of these operators in spherical polar coordinates are

$$\hat{\mathbf{L}}^2 = -\hbar^2 \left[\frac{1}{\sin\theta}\frac{\partial}{\partial\theta}\left(\sin\theta\frac{\partial}{\partial\theta}\right) + \frac{1}{\sin^2\theta}\frac{\partial^2}{\partial\phi^2} \right] \tag{13.57}$$

and $\hat{L}_z = -i\hbar\frac{\partial}{\partial\phi}.$ (13.58)

When these operate on the wavefunctions they give

$$\hat{\mathbf{L}}^2\psi_{nlm_l} = l(l+1)\hbar^2\psi_{nlm_l} \tag{13.59}$$

and $\hat{L}_z\psi_{nlm_l} = m_l\hbar\psi_{nlm_l}.$ (13.60)

It follows that l specifies the magnitude, $\sqrt{l(l+1)}\hbar$, of the orbital angular momentum, while $m_l\hbar$ specifies its component in the z-direction.

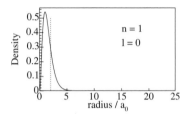

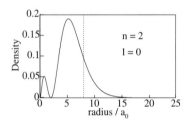

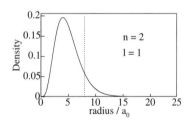

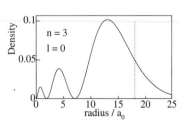

Fig. 13.9 Radial electron density distributions in the hydrogen atom for the lowest energy eigenstates. The dotted lines indicate where the kinetic energy changes sign.

The historical spectral notation is used in labelling the eigenstates of the electron: an electron with *orbital angular momentum* quantum number $l = 0, 1, 2, 3, 4, 5....$ is said to be in an s, p, d, f, g, h,.... state. A pair of electrons with quantum numbers $n = 2$, $l = 1$ are described as being in the *configuration* $2p^2$. The radial distribution of the electron probability density $r^2 R_{nl}^2$ is shown in figure 13.9 for a few values of n and l, where the factor r^2 is included to take account of the growth in the volume element as the radius increases, $\mathrm{d}V = r^2 \sin\theta \; \mathrm{d}\theta \, \mathrm{d}\phi \, \mathrm{d}r$. The dotted lines in this diagram indicate where the kinetic energy changes sign. Each wavefunction has $(n - l)$ radial nodes. Figure 13.10 shows diametral sections containing the z-axis through the $l = 1$ electron probability distributions. The three-dimensional distributions are obtained by rotating these planar distributions around the axis indicated by the broken line in figure 13.10. Note that the three 2p electronic wavefunctions give a combined electron density distribution that is spherically symmetric. Such spherically symmetric distributions always result when there is an electron in each of the $2l + 1$ substates for a given l value; that is to say when the electron *subshell* is full.

Figure 13.11 indicates, in a simplistic manner, the alignment of the orbital angular momentum vector about the z-axis for the eigenstates with $l = 2$. The orbital angular momentum vector makes a well defined angle with the z-axis $\cos^{-1}\{m_l / \sqrt{[l(l+1)]}\}$, but its azimuthal position is indeterminate, reflecting the fact that only one component of orbital angular momentum can be an observable compatible with the total angular momentum. The eigenstates have a definite *parity*, that is to say the result of reflecting the coordinates in the origin causes the wavefunction to change by a factor ± 1. A wavefunction with orbital angular momentum ℓ has a parity of $(-1)^\ell$.

The wavelengths of photons emitted in transitions between the energy levels are discrete. Therefore if the light from a source is passed through a diffraction grating the photon will strike a point in the image plane that is determined by its wavelength. Detecting this location yields the wavelength and hence identifies the eigenstates of the atom before and after the transition. In general if the lifetime of the state decaying is τ the uncertainty in the energy of the photon is at least $\hbar/\tau$. If other transitions exist with energies lying within this range it is no longer possible to identify the states involved from the photon wavelength alone. Eigenstates sharing the same value of n but different values of l and m_l have identical energies and such eigenstates are termed *degenerate*. As will be seen later the degeneracy is lifted by relativistic effects and by placing the atom in a constant magnetic or electric field.

The most intense lines in the hydrogen spectrum involve transitions in which l changes by unity

$$\Delta l = \pm 1, \tag{13.61}$$

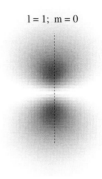

$l = 1; \; m = 0$

$l = 1; \; m = +1 \text{ or } -1$

Fig. 13.10 Diametral sections through the probability distributions in the hydrogen atom for electrons in the 2p shell. The lower distribution is toroidal in three dimensions.

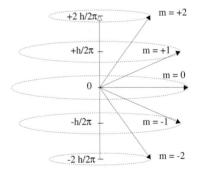

Fig. 13.11 Orientations of the angular momentum vector for the eigenstates with the orbital angular momentum quantum number equal to 2.

which is one example of a *selection rule* in optical spectra. The corresponding rule for the associated changes in the magnetic quatum number is

$$\Delta m = \pm 1, 0. \tag{13.62}$$

Transition rates for these *electric dipole* transitions are calculated in Chapter 18, where the selection rules are also discussed. Transitions between the states of a hydrogen atom in which the change in ℓ is different from those specified in eqn. 13.61 occur at far lower rates and are known collectively as *forbidden* transitions.

13.8   The Stern–Gerlach experiment

Spatial quantization in which a component of the electron's orbital angular momentum in some arbitrarily selected direction is quantized is another unexpected quantum feature. The details of spatial quantization are now developed and the experimental test of its consequences carried out by Stern and Gerlach will be described.

We focus first on a single electron in an atom. As it orbits the nucleus it forms a current loop which has correspondingly a magnetic moment

$$\mu = Ai, \tag{13.63}$$

where A is the area of the loop and i is the current. If the electron's orbit has radius r, the area $A = \pi r^2$, while the current is the product of the electron charge multiplied by the number of times it passes any given point in its orbit in one second. Suppose the electron velocity is v, then this current is

$$i = -e(v/2\pi r).$$

Hence the magnetic moment

$$\mu = -evr/2 = -\mu_B L/\hbar, \tag{13.64}$$

where L is the orbital angular momentum and μ_B is a natural unit of magnetic moment defined to be $e\hbar/2m$, and called the *Bohr magneton*. Vectorially

$$\boldsymbol{\mu} = -\mu_B \mathbf{L}/\hbar. \tag{13.65}$$

An external magnetic field applied in what we take to be the z-direction will break the spherical symmetry of the atom's environment and define a suitable *quantization* axis. The energy of the atom is altered by an amount

$$E_m = -\boldsymbol{\mu} \cdot \mathbf{B} = \mu_B m_l B, \tag{13.66}$$

where $m_l \hbar$ is the z-component of the orbital angular momentum. m_l is therefore called the *magnetic* quantum number. In a non-uniform magnetic field varying in the z-direction the atom will experience a force

$$F = -\mu_B m_l \frac{\partial B}{\partial z}. \tag{13.67}$$

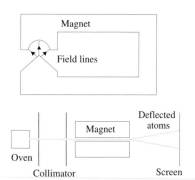

Fig. 13.12 The Stern–Gerlach experiment. The upper panel shows a section through the magnet perpendicular to the beam direction. The lower panel shows a vertical section containing the incident beam.

Stern and Gerlach carried out an experiment in 1922 which displayed this effect of spatial quantization directly. Their experiment is illustrated in figure 13.12. Atoms evaporated from liquid silver in an oven are collimated into a beam by a series of slits in metal plates. The atoms then travel between the poles of a magnet whose faces are shaped to give a magnetic field strength which varies strongly across the gap and hence has a large value of $\partial B/\partial z$. Beyond the magnet the atoms travel some distance to a screen where they are detected. When the magnet is off the atoms travel undeviated and form a line image of the collimator slit on the screen.

The force given by eqn. 13.67 lies in the z-direction, pointing from one pole piece to the other. In the classical view the magnetic moments of the atoms would be oriented in random directions and their angular deflections in the direction of the field gradient would simply broaden the collimator slit image. However when the magnet was turned on the image on the screen changed to a pair of line images of the collimator slit well separated from one another in the direction of the field gradient. This is precisely what is expected from eqn. 13.67, demonstrating the reality of spatial quantization. However a disturbing anomaly was noted. The number of components into which an atomic beam splits should be odd because $2l + 1$ is an odd number; but in practice the number of components seen with some atomic elements, including silver, is even. A simple explanation of this anomaly was revealed a few years later with the discovery of the spin of the electron.

13.9   Electron spin

In 1925 Uhlenbeck and Goudsmit proposed that the electron has an intrinsic angular momentum or *spin* whose magnetic quantum number is 1/2. Therefore the magnitude and the quantized component of this spin have values given by

$$\mathbf{s}^2 = (1/2)(1 + 1/2)\hbar^2 \ , \ m_s\hbar = \pm\hbar/2. \tag{13.68}$$

The total angular momentum, $\mathbf{j}$, of an electron must be the vector sum of its intrinsic and its orbital angular momenta

$$\mathbf{j} = \mathbf{s} + \mathbf{l}, \tag{13.69}$$

with magnetic quantum numbers m_j in the sequence $-j, -j + 1, \ ... \ , +j$. Overall an electron in an atom therefore requires four eigenvalues to specify its eigenstate fully (n, l, m_l, m_s) which doubles the number of available quantum states.

The spin of the electron is not associated with any mechanical motion of some internal structure within the electron. Modern experiments can probe for structure as small as 10^{-18} m and none has been detected in the electron: as far as we know the electron is point-like. The explanation

Table 13.3 Table of noble gas electron configurations. In the notation used here the initial number is the value of the principal quantum number. The letter signifies the orbital angular momentum in spectroscopic notation: s stands for $l = 1$, p for $l = 2$, and so on through d, f, g,... The superscript indicates the number of electrons sharing those two quantum numbers.

Element	Atomic number	Configuration
Helium	2	$1s^2$
Neon	10	$1s^2\ 2s^2\ 2p^6$
Argon	18	$1s^2\ 2s^2\ 2p^6\ 3s^2\ 3p^6$
Krypton	36	$1s^2\ 2s^2\ 2p^6\ 3s^2\ 3p^6\ 4s^2\ 3d^{10}\ 4p^6$

for why the splitting in the Stern–Gerlach experiment can produce an even number of lines is now clear. Any atom containing an odd number of electrons will automatically have a total angular momentumn quantum number, j, that is half integral making $(2j + 1)$ even. A further experimental observation is that the magnetic moment of the electron associated with its spin is almost exactly twice the value obtained by applying the relation, eqn. 13.65, deduced for the orbital angular momentum; its magnetic moment is thus μ_B rather than $\mu_B/2$.

A difficulty with both the Bohr and Schroedinger models as outlined so far is that in elements whose atoms contain many electrons these electrons would all drop into the lowest energy level. It would then be hard to explain why the chemical properties of the elements show a marked periodic behaviour as the atomic number increases through the periodic table. Contemporaneously with Uhlenbeck and Goudsmit, Pauli provided the essential idea which solved the problem of how electrons arrange themselves in atoms.

Pauli enunciated the *exclusion principle* which states that there can only ever be one electron in any given eigenstate.

This means that as the atomic number increases from one element in the periodic table to the next each additional electron enters the lowest energy, *empty* eigenstate. Each eigenstate with specific values of (n, l, m_l) can contain at most two electrons, one with $m_s = +1/2$, the other with $m_s = -1/2$.

The periodicity in the chemical behaviour of the elements is reflected in their ionization energies, shown here in figure 13.13 as a function of atomic number. This ionization energy is the energy required to detach the least well-bound electron from the atom so that it emerges with zero kinetic energy. Clear peaks are evident at the atomic numbers of the chemically inert noble gases: helium, neon, argon, etc. These elements have the electron configurations given in table 13.3. The inference is that

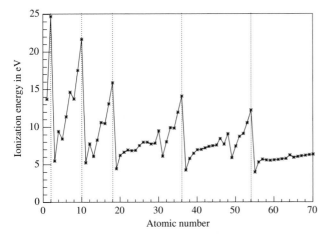

Fig. 13.13 The ionization energies required to remove the least well bound electron from the atomic elements are plotted against the atomic number. The inert noble gases with their completed electron shells are indicated by the dotted lines.

the electron configuration in which all the $2(2l+1)$ eigenstates for a given l-value are filled are extremely well bound. It was noted earlier that if all the magnetic sublevels for a given orbital angular momentum are filled then the product of the electron wavefunctions is spherically symmetric. For such a wavefunction the total orbital angular momentum operator has an expectation value zero. Further the spins of the two electrons paired in an eigenstate with the same (n, l, m_l) values are antiparallel and have a total spin of zero. Thus the *closed shells* in the noble gases have zero total orbital angular momentum, zero total spin and hence zero total angular momentum. The spherically symmetric electronic structure of atoms of noble gases is thus also exceptionally stable.

13.10   Multi-electron atoms

The alkali metals follow the noble gases in the periodic table and therefore have one additional electron which enters a previously empty $l = 0$ orbit. For example sodium, with 11 electrons, follows neon and sodium's electron configuration is $(1s^2\ 2s^2\ 2p^6)3s^1$. The lone 3s electron experiences an electric field due to the nuclear charge, Ze, surrounded by the $(Z - 1)$ electrons in the closed shells. If the electrostatic field of the $(Z - 1)$ electrons exactly cancelled that of $Z - 1$ protons, then the electric potential felt by the 3s electron would be $-e^2/4\pi\varepsilon_0 r$. Consequently the energy levels and the spectral lines produced when the 3s electron in a sodium atom is excited are similar to those of hydrogen; but the lines corresponding to transitions to any of the (full) 1s, 2s and 2p eigenstates are naturally absent. For similar reasons the other alkali metals show spectra with some resemblance to that of hydrogen.

Figure 13.9 illustrates that eigenfunctions with lower principal quantum number, n, penetrate closer to the nucleus than those of higher principal quantum number. Eigenstates with lower values of n are therefore less well shielded from the electric charge on the nucleus and hence more strongly bound in alkali metal eigenstates than the corresponding hydrogen eigenstates. There is also some dependence of the binding energy on the orbital angular momentum quantum number because the radial distribution given in eqn. 13.49 depends on l. In general the calculation of the eigenfunctions for individual electrons is more complicated than for hydrogen-like atoms. It is necessary to introduce an average electric potential which includes the potential due to the other electrons as well as that due to the nucleus. However the mutual Coulomb interaction of the electrons in the unfilled shells cannot alter their total orbital angular momentum, their total spin nor their total angular momentum. Thus the angular momentum observables for the light elements are the vector sums of the orbital angular momenta, of the spins and of the total angular momentum, and finally the magnetic component of the total angular momentum:

$$\mathbf{L} = \sum_i \mathbf{l}_i, \ \ \mathbf{S} = \sum_i \mathbf{s}_i, \ \ \mathbf{J} = \mathbf{L} + \mathbf{S}, \ \ M = \sum_i (m_{l_i} + m_{s_i}), \quad (13.70)$$

where the sums run over all the electrons. The atomic state of any light element atom can be characterized fully by the set of quantum numbers (L, S, J, M).[7] The possible values of J lie at integral steps within the range

$$|L - S| \leq J \leq L + S. \quad (13.71)$$

The labelling of the eigenstates of an atom is usually in the form of a *term symbol* $^{2S+1}X_J$, where X is the upper case spectroscopic label corresponding to the value of L; that is S, P, D, F, ... for $L = 0, 1, 2, 3$, etc. When there is a single active electron it can be useful to add the principal quantum number n thus, $n^{2S+1}X_J$. The superscript is called the *multiplicity*. It gives the number of eigenstates with different values of J for that combination of S and L provided $L \geq S$. For convenience, and where there is no ambiguity the principal quantum number or the multiplicity are often omitted.

There are more complex selection rules for the dominant transitions in multi-electron atoms

$$\Delta S = 0, \ \ \Delta L = 0, \pm 1, \ \ \Delta J = 0, \pm 1, \ \ \Delta M = 0, \pm 1; \quad (13.72)$$

but of these the transitions $J = 0 \rightarrow J = 0$, and $M = 0 \rightarrow M = 0$ for $\Delta J = 0$ are excluded. Underlying these rules is the requirement that in an allowed transition a single electron emits, or absorbs a single photon in an electric dipole transition.

Relativistic and quantum effects modify the energies of the eigenstates that have been calculated using a mean electrostatic potential for each

[7]This is known as *Russell–Saunders* coupling of angular momenta. In heavy elements the coupling between the spin and orbital angular momenta of individual electrons is stronger than the coupling between different electron spins or between different electron angular momenta. This produces a different coupling scheme, called j–j coupling. See for example *Quantum Physics of Atoms, Molecules, Solids, Nuclei and Particles* by R. Eisberg and R. Resnick, published by John Wiley and Sons, New York (1974).

electron. In addition the interaction of the magnetic moments of the orbital and intrinsic angular momenta leads to a displacement in energy proportional to both of these angular momenta

$$\Delta E \propto +\mathbf{L} \cdot \mathbf{S}.$$

This *spin–orbit splitting* of energy levels is illustrated for hydrogen in figure 13.14, where it is only 10^{-4} of the level spacing. In atoms with several electrons in an unfilled shell the spin–orbit coupling is larger; it also increases with increasing atomic number because the electric fields are stronger. Take for example the case of the 3P state of a sodium atom which splits into two states: in the $3P_{1/2}$ state the orbital and spin angular momenta are antiparallel, while in the $3P_{3/2}$ state they are parallel. The transitions giving the sodium yellow D-lines are $3P_{3/2} \rightarrow 3S_{1/2}$ and $3P_{1/2} \rightarrow 3S_{1/2}$ with wavelengths 588.995 nm and 589.592 nm respectively, which are easily resolved.

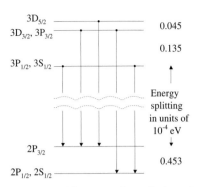

Fig. 13.14 The spin–orbit splitting of some hydrogen eigenstates, and the allowed transitions between the levels.

The inclusion of quantum and relativistic corrections in a systematic manner was achieved when Dirac replaced the Schroedinger equation with a relativistic wave equation of the electron. Later this was refined into quantum electrodynamics by Feynmann, Tomonaga and Schwinger and involved the quantization of the electromagnetic fields themselves. The simpler aspects of field quantization will be considered in Chapter 19.

13.10.1 Resonance fluorescence

Fluorescence was described in Chapter 11 as the prompt radiation emerging from a material when illuminated. The wavelength of the fluorescence can never be shorter than that of the incident radiation because the atoms excited can only re-emit photons of equal or lower energy than the incident photons. When a cell containing sodium vapour at around 250 K is illuminated by a sodium lamp the whole cell glows yellow in all directions. This is an example of *resonance fluorescence* involving the transition between the $3S_{1/2}$ to $3P_{1/2,3/2}$ levels of the sodium atom at a wavelength 589 nm. Photons in the incident light have exactly the right energy to excite the sodium atoms in the vapour cell from the $3S_{1/2}$ state into a 3P excited state and these atoms then promptly re-emit photons of the same energy so that they return to their ground state.[8]

[8]The time distribution of the decays follows $\exp\left(-t/\tau\right)$, where τ is the radiative lifetime. τ is 16.3 ns for both decays $3P_{1/2,3/2} \rightarrow 3S_{1/2}$.

13.10.2 Atoms in constant fields

An applied constant magnetic field $\mathbf{B}$ changes the energies of the eigenstates of an atom by an amount determined by the atom's magnetic moment μ. In the case of elements with low atomic numbers the energy displacement in a magnetic field $\mathbf{B}$ is

$$\Delta E = -\mu \cdot \mathbf{B} = +(\mu_{\mathrm{B}}/\hbar)(\mathbf{L} + 2\mathbf{S}) \cdot \mathbf{B}, \qquad (13.73)$$

where the factor 2 is required because the magnetic moment associated with spin is twice that which would be associated with an equal orbital angular momentum. If the magnetic field is sufficiently weak the magnetic interaction energy is much smaller than the spin-orbit interaction energy, so that the quantum states of atoms remain those having J and M quantized. The application of such an external magnetic field causes the splitting of spectral lines, in what is known as the *Zeeman effect*. With a field of a few teslas the individual Zeeman split lines can be resolved by a Fabry–Perot etalon. The relevant selection rules are

$$\Delta M = 0, \pm 1 \text{ (but not } M = 0 \rightarrow M = 0, \text{ if } \Delta J = 0 \text{)}. \qquad (13.74)$$

For example the transition in sodium $^2P_{3/2} \rightarrow ^2 S_{1/2}$ splits into six component lines. In the case that the total spin of the atom is zero the splitting is especially simple. Then the change in the energy of the atomic state produced by the magnetic field

$$\Delta E = (\mu_B/\hbar)\mathbf{L} \cdot \mathbf{B} = \mu_B BM. \qquad (13.75)$$

Hence the transition energy is different for each of the allowed transitions: $\Delta M = 0, \pm 1$. There is splitting of a line into three components, one of which is undisplaced while the other two are displaced equally from this, one up in frequency the other down in frequency. This is seen for example in the splitting of some neon lines and is known as *normal* Zeeman splitting. In other atoms whose total spin is non-zero the line splitting is more complicated and is known, for historic reasons, as *anomalous* Zeeman splitting. When the magnetic field is very large the magnetic interaction dominates over the spin-orbit interaction and the eigenstates have m_l and m_s quantized, rather than J and M. An applied electric field also produces splitting of spectral lines and this is known as the *Stark effect*.

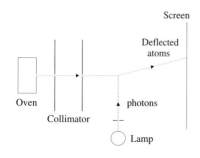

Fig. 13.15 Frisch's experiment to measure directly the linear momentum of photons.

13.11   Photon momentum and spin

In Chapter 9 the linear momentum of a light beam $\mathbf{p}$ was shown to be directly proportional to the energy of the beam E by considering the force exerted when a light beam is reflected from a conducting surface. It was found that in free space $E = pc$. Compton reinterpreted this as the relationship between the energy and momentum of each photon. Direct confirmation of this idea was obtained in 1933 by Frisch using the arrangement shown in figure 13.15. Atoms evaporating from a bath of liquid sodium stream through a collimator and head toward a cold screen half a metre away, where they adhere. The whole region is evacuated. Light from the sodium lamp shown in the diagram passes through a filter which transmits only the D-lines at 589 nm. Any sodium atom in the beam which is in the 3S state can absorb one of these photons and when this happens the atom absorbing the photon acquires a transverse momentum equal to the photon momentum of h/λ. The resulting transverse velocity of the atom is therefore

$$v = h/(\lambda M) = 0.0294 \mathrm{\ m\,s^{-1}},$$

where $3.82\,10^{-26}$ kg has been substituted for M, the mass of the sodium atom. At the temperature of the sodium bath, $700\,\mathrm{K}$, the mean velocity of the atoms is $900\,\mathrm{m\,s^{-1}}$ and hence the deflection would be $3.8\,10^{-5}$ rad, giving a displacement on the screen of about $16\,\mu$m. Some of the sodium atoms re-radiate a photon while in flight, but in any direction, so that $16\,\mu$m becomes the mean deflection.

More recently this same technique has been used extensively for isotope separation. Isotopes of an element have the same electric charge but different nuclear masses, hence different Rydberg constants, and therefore the spectral lines of isotopes will be separated slightly in wavelength. The apparatus used is conceptually similar to that illustrated in figure 13.15, with the oven now containing the mixture of isotopes, and with the sodium light source replaced by a laser of extremely narrow linewidth. This laser's wavelength is tuned precisely to the wavelength of an intense transition for the isotope of interest. Only atoms of that isotope are able to absorb the laser light and are then deflected. If a long path and a narrow collection slit are used the atoms of the isotope of interest can be filtered off and accumulated.

Poynting suggested in 1909 that circularly polarized electromagnetic waves should carry angular momentum. The structureless photon therefore, like the structureless electron, should have an intrinsic angular momentum. This idea was confirmed in 1936 by Beth and Harris who measured directly the torque produced when circularly polarized electromagnetic radiation has its sense of rotation reversed on passing through a half-wave plate.

The apparatus designed by Beth is drawn in the right hand panel of figure 13.16. A beam of plane polarized infrared light incident from below travels successively through a quarter-wave plate (QWP), a half-wave plate (HWP) and a second quarter-wave plate whose rear surface is silvered. After reflection the beam reverses its path and leaves where it entered. The HWP is freely suspended on a very fine quartz fibre through a hole in the upper QWP. In the simplest arrangement the slow axes of the two QWPs are set respectively parallel and perpendicular to the slow axis of the HWP. The resulting states of polarization are drawn in the left hand panel for each step in the light's path. Each pass through the HWP reverses the state of circular polarization of the beam and correspondingly its angular momentum. The reaction produces a torque on the HWP and this torque is doubled by having two passes through the HWP.

This static deflection is still too small to measure accurately. Beth therefore enhanced the deflection dynamically in the same way that chil-

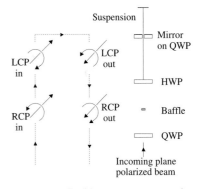

Fig. 13.16 Beth's experiment to observe the angular momentum of circularly polarized light. The right hand panel shows the optical equipment. The left hand panel indicates the state of polarization of the beam at each stage going in and out after reflection. Adapted from R. A. Beth, *Physical Review Letters* 50, 115 (1936), courtesy the American Physical Society.

dren pump a swing. He reversed the sense of circular polarization of the incoming radiation, and hence the torque, at the natural frequency of oscillation of the HWP on its suspension. The HWP then oscillated with an easily measurable angular displacement.

The observed deflection agreed with the assignment of unit intrinsic angular momentum to the photon in units of $\hbar$: the magnitude of the photon's spin and the quantized component are given by

$$\mathbf{s}^2 = (1)(1+1)\hbar^2 \text{ and } m_s\hbar = \pm\hbar.$$

It follows that a circularly polarized beam must be viewed as made up of photons with their spins either all parallel to the beam direction, or all anti-parallel, depending on the sense of the circular polarization. By contrast a plane polarized beam, being an equal superposition of left and right circularly polarized photons has no net angular momentum. Naturally Beth checked this conclusion too.

In any transition the usual mechanical conservation laws apply: not only linear momentum and total energy are conserved, but also the total angular momentum and its component along an applied magnetic field. The normal Zeeman effect provides a simple demonstration of the conservation of angular momentum. Observations can be made with a modification of the arrangement of a constant deviation spectrometer and a Fabry–Perot etalon described in Chapter 5, and shown here in figure 13.17. A neon source is located between the poles of an electromagnet and light emerges from a hole in one pole so that the source is viewed along the field direction. The choices of 15 mm for the Fabry–Perot etalon spacing and 1 T for the field give good results. After exiting the etalon the light passes through a quarter-wave plate and a Polaroid before entering the viewing telescope. Each spectral line is split into a triplet, the outer members corresponding to transitions in which $\Delta M = \pm 1$. The QWP plate converts states of circular polarization into states of plane polarization, while the Polaroid can be rotated to find whether the light reaching it is plane polarized. Settings are found at which the outer lines disappear: one line at one setting, the other at another setting 90° apart from the first. Evidently the light in the outer lines is circularly polarized: in one case left circularly polarized, and in the other right circularly polarized.

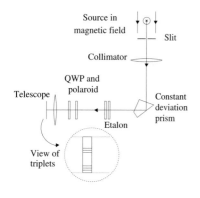

Fig. 13.17 Apparatus for viewing the polarization states of Zeeman split lines.

This result has a simple interpretation in terms of the conservation of angular momentum. The photons are emitted in transitions in which the change of the spin component of the atom, $\Delta M\hbar$, is $+\hbar$ and $-\hbar$. In order to conserve the quantized component of total angular momentum the photons have spin components $-\hbar$ and $+\hbar$ respectively, and give rise to the two outer lines.

Figure 13.18 summarizes in a schematic way the alignments met in parallel beams of right and left circularly polarized light in free space.

The velocity vector, the field vectors and their sense of rotation, as well as the spin orientation of the photons are all displayed. Notionally the direction of the spin vector could be pictured as lying on the cone sketched, but with indeterminate azimuth angle. Although the photon has unit intrinsic angular momentum only the two states with $m_s = \pm 1$ have been discussed because a photon travelling in free space is never in the $m_s = 0$ magnetic substate.   Such photons would contribute to components of the electric and magnetic fields along the direction of the wave's travel; components forbidden by Maxwell's equations. The ultimate reason for the absence of the $m_s = 0$ magnetic substate must be sought in quantum electrodynamics (QED), the quantum field theory of the electromagnetic field and its interaction with charges. In QED the absence of the longitudinal photons is directly related to the masslessness of the photon.

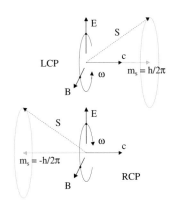

Fig. 13.18 The vector orientations in parallel beams of left and right circularly polarized light.

13.12   Quantum statistics

As noted previously, Schroedinger's equation is the equivalent for electrons of the electromagnetic wave equation. Correspondingly the modes of electromagnetic radiation discussed earlier are simply the *eigenstates for electromagnetic radiation*. However there is a striking contrast between the statistical properties of photons and electrons. On the one hand in Planck's successful calculation of the black body radiation spectrum any number of photons are permitted to share the same eigenstate (mode). Equally, lasers only function because astronomical numbers of photons may occupy the same mode. On the other hand Pauli's exclusion principle requires that any electron eigenstate never contains more than a single electron.

Surprisingly these disparate statistical behaviours have the following common origin. In quantum mechanics it is impossible to distinguish any particle of one species from another particle of the same species; neither one photon from another, nor one electron from another. If two particles of the same species are in a state with a wavefunction $\psi(1,2)$ there is no measurement that can distinguish between this state and one in which the particles are interchanged and whose wavefunction is $\psi(2,1)$. All the predictions for expectation values contain the product $\psi^*(1,2)\psi(1,2)$, from which it can be inferred that

$$\psi^*(1,2)\psi(1,2) = \psi^*(2,1)\psi(2,1). \qquad (13.76)$$

This has two possible solutions

$$\psi_s(2,1) = +\psi_s(1,2), \qquad (13.77)$$

$$\psi_a(2,1) = -\psi_a(1,2). \qquad (13.78)$$

The second, antisymmetric wavefunction ψ_a will vanish if the two particles are in the same eigenstate. Evidently this is the correct choice for pairs of electrons, because the exclusion principle comes out of it. There is no such restriction with the symmetric wavefunction, ψ_s. This then is the choice for photon wavefunctions. When there are many electrons present the wavefunction must be antisymmetric under the interchange of each pair of electrons separately. Correspondingly the wavefunction of multiple photons must be symmetric under the interchange of any pair of them. Photons are then said to obey *Bose–Einstein* statistics, while electrons obey *Fermi–Dirac* statistics, these statistics being named after their discoverers.

Planck's analysis showed that the mean number of photons in a state with energy E in equilibrium at temperature T K is

$$\langle n_{\text{photon}} \rangle = 1/[\exp(E/k_B T) - 1]. \tag{13.79}$$

A parallel analysis for electrons requires that account is taken of the exclusion principle, and would give a similar expression with -1 in the denominator replaced by +1. In addition, because the electrons carry a charge and have mass their number is conserved, unlike the number of photons. When this further constraint is taken into account we have

$$\langle n_{\text{electron}} \rangle = 1/\{ \exp[(E - \mu(T))/k_B T] + 1 \}, \tag{13.80}$$

where $\mu(T)$, the *chemical potential* depends on the material environment of the electrons. This occupancy is always less than unity and is exactly one half when $E = \mu(T)$. Note that classical particles obey Boltzmann statistics for which

$$\langle n_{\text{classical}} \rangle = 1/\{\exp[(E - \mu(T))/k_B T]\}, \tag{13.81}$$

where the chemical potential depends on the particle species and the environment. When the excitation energy E becomes large compared to $k_B T$ the populations are all sparse and the quantum distributions approach the classical distribution.

13.13   Line widths and decay rates

Suppose the probability per unit time for an atom to decay from an excited state, 2, to another state, 1, is γ. Each excited atom in a volume containing $N(t)$ such atoms at time t has an identical probability to decay, but we can never predict which will in fact decay in a further time dt. All that can be predicted is that the average rate of losing atoms from the excited state 2 is γ, thus

$$dN(t)/dt = -\gamma N(t). \tag{13.82}$$

Integrating this gives the well known exponential decay distribution

$$N(t) = N(0) \exp(-\gamma t). \tag{13.83}$$

Thus the intensity of light emitted by the volume of atoms must vary in a similar way with time

$$I(t) = I(0) \exp(-\gamma t). \tag{13.84}$$

and the electric field of the radiation will contain a time dependence $\exp(-\gamma t/2)$. The electric field will also be oscillating at an angular frequency, $\omega_0 = \Delta E/\hbar$, where ΔE is the energy difference between the two atomic states in the transition. Thus the electric field variation with time is

$$E(t) = E(0) \exp(-\gamma t/2) \cos(\omega_0 t). \tag{13.85}$$

The Fourier transform of this function has been given in eqn. 7.53. The spectral distribution of the electric field and of the intensity of light emitted by the atoms have Lorentzian shapes:

$$e(\omega) = (iE(0)/2)/[(\omega - \omega_0) + i\gamma/2], \tag{13.86}$$
$$P(\omega) = (\gamma/2\pi)/[(\omega - \omega_0)^2 + \gamma^2/4], \tag{13.87}$$

normalized to unit power over the whole angular momentum spectrum. This last equation can be recognized also as the spectral distribution for absorption obtained in Section 11.5.1 using wave theory. In that analysis atomic excitations were treated as mechanical oscillations of electrons within atoms.

The lifetime and the line width are linked together through the uncertainty principle. Some care is necessary to assist in the applying this connection. If the parent state decays into several lower energy states the overall decay rate has contributions from all the partial decay rates, γ_i. That is

$$\gamma_{\mathrm{p}} = \sum_i \gamma_i = 1/\tau_p, \tag{13.88}$$

where τ_{p} is the parent lifetime. The rate of decay into each daughter is

$$\mathrm{d}n_i/\mathrm{d}t = n_{\mathrm{p}}(\gamma_i/\gamma_{\mathrm{p}}) \exp(-\gamma_{\mathrm{p}}t) \tag{13.89}$$

where n_{p} is the initial number of parent atoms. Turning to the line width of a decay expressed as an energy, this is simply the uncertainty in the energy release. Both the parent and daughter state need to be taken into account because the daughter can also decay. The parent state has an uncertainty in energy $\hbar\gamma_{\mathrm{p}}$. That of the daughter is $\hbar\gamma_{\mathrm{d}}$, where γ_{d} is its decay rate.[9] Thus the total uncertainty in the energy release in the decay p→d is

$$\hbar\gamma = \hbar\gamma_{\mathrm{p}} + \hbar\gamma_{\mathrm{d}}. \tag{13.90}$$

In terms of their lifetimes

$$\gamma = 1/\tau_{\mathrm{p}} + 1/\tau_{\mathrm{d}}. \tag{13.91}$$

If the daughter is much shorter lived than the parent this line width will be much broader than the parent lifetime would lead us to expect.

[9] The daughter's decay rate γ_{d} must not be confused with the partial rate appearing in eqn. 13.88, which is its production rate in the parent's decay.

In Section 7.3.1 broadening of the line shape of an atomic transition in matter was discussed from the classical viewpoint. This needs some further comments here. The lifetimes and line shapes of atomic transitions are altered by the motion of the atom and by interactions with its neighbours. In a low pressure gas in a discharge tube the Doppler broadening dominates, while in a solid the effect of the electromagnetic fields of neighbouring atoms can dominate. When the effect, of whatever origin, is the same for all atoms in the source the broadening is called *homogeneous*: when the effect varies with the state of the atom the broadening is known as *inhomogeneous*. In a perfect crystal the electric fields acting on all atoms are identical so the broadening is homogeneous, while in an amorphous solid, like glass, the local fields vary and the broadening is inhomogeneous. Homogeneous broadening leads to a Lorentzian line profile, while inhomogeneous broadening leads to a Gaussian line profile, as in the case of Doppler broadening described by eqn. 7.56.

Atomic collisions produce broadening through two effects. Firstly the collision can knock an electron out of an excited state before it would have normally decayed, which is collisional de-excitation. Secondly collisions can cause abrupt phase changes in the wavepacket emitted, known as de-phasing. Apart from collisional de-excitation all these processes mentioned affect the line width but do not change the decay rate of the excited state. The coherence time of the radiation will be the inverse of the line width broadened by all contributing mechanisms.

This section has brought together results common to classical wave theory and quantum theory which show that despite the fundamental inadequacy of the classical view of atomic processes it does contain useful elements. We can see that the frequency of oscillation in the classical picture is in reality the transition frequency, while damping is the way the loss of atoms from the excited state is expressed classically. Often the classical picture is convenient and remains a useful interpretive tool.

13.14   Further reading

The third edition of *Quantum Mechanics* by L. I. Schiff, published by McGraw-Hill Kogakusha, Tokyo (1968).

Quantum Physics of Atoms, Molecules, Solids, Nuclei and Particles by R. Eisberg and R. Resnick, published by John Wiley and Sons (1974). This textbook follows the consequences of quantum theory into the realms given in its title.

Exercises

(13.1) What is the mean number of photons in an eigenstate/mode of energy 0.4 eV in a system in thermal equilibrium at a temperature 3000 K? Repeat the calculation for electrons and classical particles.

(13.2) What is the mean number of photons in an eigenstate/mode of energy 0.4 eV in a system in thermal equilibrium at a temperature 3 K. Repeat the calculation for electrons and classical particles.

(13.3) Show that an operator whose eigenvalues for a set of orthonormal eigenstates are all real is hermitian.

(13.4) Show that the momentum components p_x and p_y commute, that the coordinates x and y commute and that an unmatched momentum p_y and coordinate x commute.

(13.5) Check that the functions in eqn. 13.46 satisfy eqn. 13.43.

(13.6) An electron in a hydrogen atom drops from the $n = 3$ excited state to its ground state. What is the energy of the photon emitted? What is the momentum of the photon? What is the velocity with which the hydrogen atom recoils in order that momentum is conserved? What is the kinetic energy of the hydrogen atom? Discuss where this energy comes from.

(13.7) Calculate the angle between the orbital angular momentum vector and the quantization axis in an eigenstate with orbital angular momentum quantum number 4 and magnetic quantum number 3.

(13.8) Show that the wavefunctions of the three 2p electron states in a hydrogen-like atom give an overall isotropic electron distribution.

(13.9) Show that $[L_z, L_x] = i\hbar L_y$.

(13.10) Show that for a sinusoidal plane wave the expectation value of the momentum raised to any power m is $(\hbar k)^m$.

(13.11) Check that the operator $-i\hbar\partial/\partial\phi$ acting on the eigenfunctions in Table 13.2 returns $m_l\hbar$ in each case.

Lasers

14

14.1   Introduction

Einstein deduced that for radiation and matter to reach thermal equilibrium there must exist an additional process beyond the absorption and spontaneous emission of photons. This unexpected process is *stimulated* emission in which one photon incident on an atom causes the emission of a second photon coherent with the incident photon. This requires the atom to be initially in the appropriate excited state. Within lasers a cascade of stimulated emissions can produce an astronomical number of coherent photons, a process whose description – Light Amplification by the Stimulated Emission of Radiation – led to the acronym *laser*. The analogous microwave masers were invented first in 1953, and then in 1960 Maiman built the first ruby laser. Since that time lasers have become ubiquitous: in supermarket and library checkouts, in DVD readers and in optical fibre communications. Laser beams are more highly coherent and of greater radiance than any earlier sources; these are properties widely exploited in science and industry. Some lasers are pinhead sized, others fill huge laboratory halls. Their versatility is amazing: CO_2 lasers are used to drill holes in 2 cm thick steel plates with high precision, while other lasers weld detached retinas. These important and interesting devices are at the heart of modern optics and its applications.

In this chapter the major types of lasers will be discussed, concentrating on a basic example in each case. Some applications are included in this chapter, in particular a couple of representative spectroscopic techniques. The study of non-linear processes has developed rapidly through the use of laser beams, and these processes are discussed in the last section of the chapter.

In the section immediately below, the Einstein coefficients which quantify the emission and absorption processes are introduced and the relationships between them are deduced. Then the three prerequisites for most types of laser operation are discussed: these are stimulated emission, a population inversion and a resonant cavity. The He:Ne gas laser is used as the introductory example. This account includes discussions of laser gain, hole burning, laser speckles and optical beats. After this another more powerful gas laser, the CO_2 laser, and some of its applications are treated.

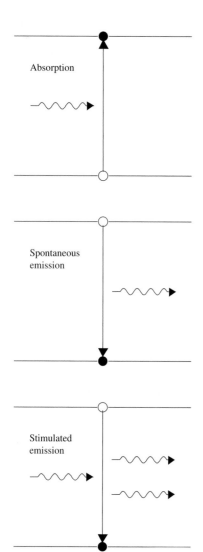

Fig. 14.1 The three photon–matter interactions: in the upper panel absorption; in the centre panel spontaneous emission; and in the lower panel stimulated emission.

There follows a section devoted to dye lasers which were the first tunable lasers. Two modern spectroscopic techniques are described, representative of techniques made feasible with the advent of tunable lasers. Then a modern laser interferometer is described which provides an example of the usefulness of the heterodyne technique.

Semiconductor lasers are small compared to gas or dye lasers and eminently suited to applications in consumer electronics and telecommunications. A brief account of semiconductor properties prefaces an account of the principal semiconductor laser types, including quantum well lasers. Discussion of their usage in telecoms is postponed to a later chapter.

Two *dielectric*, or *solid state*, lasers are described in the next section: these are the Nd:YAG and Ti:sapphire lasers. Working in tandem these lasers provide the extreme energy beams with which it is hoped to produce nuclear fusion in tritium/deuterium pellets by heating and implosion. They have also been used to provide the most precise optical frequency scales based on what are called optical combs.

In a final section non-linear processes are described. These include parametric processes such as those in which laser beams generate harmonics of their frequency in crystals. This discussion closes with an account of Raman and Brillouin scattering, whose applications required the intense beams that only lasers can provide.

14.2   The Einstein coefficients

The three processes by which matter emits or absorbs electromagnetic radiation are sketched in figure 14.1. Of these the upper pair are already familiar: the absorption and the *spontaneous* emission of a photon whose energy equals that gained or lost by the atom changing its quantum state. In 1916 Einstein realized that the process exhibited on the lower panel of the figure must also occur, and this became known as *stimulated* emission. Stimulated emission occurs when an atom is initially in an excited state and a photon is incident on it with an energy equal to the difference in energy between this state and some lower energy state. The incident photon stimulates the atom to drop into the lower energy state, and to emit a photon that is a copy of itself. The second photon has the same energy, direction, polarization and phase as the first: the pair are thus fully coherent and can be described by the same wavepacket. When a beam of coherent light passes through matter it excites some atoms: of these some decay spontaneously while others are stimulated to decay. The photons from stimulated decays rejoin the beam, but those emitted spontaneously generally leave it, appearing as scattered light. The simple relationships between the three processes in thermal equilibrium will now be deduced.

Suppose that there are $N_1(N_2)$ atoms of one atomic species per unit volume in the two states of energies $E_1(E_2)$ with $E_2 > E_1$, and that all other energy states can be ignored. The rate at which photons are absorbed is evidently proportional to N_1. It is also proportional to the energy density of photons of frequency $f = (E_2 - E_1)/h$, that is to the spectral energy density of the radiation $W(f)$ measured in $\mathrm{J\,m^{-3}Hz^{-1}}$. The absorption rate is

$$(\mathrm{d}N_2/\mathrm{d}t)_{\mathrm{ab}} = B_{12}N_1 W(f). \qquad (14.1)$$

Similarly the rate of stimulated emission is proportional to N_2 and $W(f)$,

$$(\mathrm{d}N_2/\mathrm{d}t)_{\mathrm{st}} = -B_{21}N_2 W(f). \qquad (14.2)$$

Finally the rate of spontaneous emission is proportional to N_2 but independent of $W(f)$,

$$(\mathrm{d}N_2/\mathrm{d}t)_{\mathrm{sp}} = -A_{21}N_2. \qquad (14.3)$$

A_{21}, B_{12} and B_{21} are constants known as the *Einstein coefficients*. Further $A_{21} = 1/\tau_{\mathrm{sp}}$, where τ_{sp} is the lifetime of the upper state against spontaneous decay.[1]

Evidently when the radiation is of sufficiently high intensity that effects non-linear in the electric field strength become important the simple analysis being followed here would be inadequate.

If the atoms and the radiation are in thermal equilibrium so that the numbers N_1 and N_2 do not change with time, then the three processes must balance giving

$$B_{12}N_1 W(f) - B_{21}N_2 W(f) - A_{21}N_2 = 0, \qquad (14.4)$$

whence

$$W(f) = A_{21}N_2/(B_{12}N_1 - B_{21}N_2). \qquad (14.5)$$

The Einstein coefficients will naturally be different for different atomic or molecular processes. However the coefficients are independent of the radiation density and Einstein used this fact to infer the relationship between the coefficients. He specialized to the case of thermal equilibrium with black body radiation. At a temperature T K in material of refractive index n the black body radiation spectral energy density is given by eqn. 12.7,

$$W(f) = (8\pi h f^3 n^3/c^3)/\left[\exp\left(hf/k_{\mathrm{B}}T\right) - 1\right]. \qquad (14.6)$$

Furthermore, in thermal equilibrium the ratio of the numbers of atoms in the two states is given by the Boltzmann relation

$$(N_2/g_2) = (N_1/g_1)\exp\left(-hf/k_{\mathrm{B}}T\right),$$

[1] As in Chapter 12 the spectral energy density is regarded as a function of the frequency. The spectral energy density can also be expressed in terms of the angular frequency, $W^\omega(\omega)$. Then for a given spectral interval:

$$W^\omega(\omega)\,\mathrm{d}\omega = W(f)\,\mathrm{d}f.$$

Now $\mathrm{d}\omega = 2\pi f$ and hence $W(f) = 2\pi W^\omega(\omega)$. It follows that in eqn. 14.1 $B_{12}W(f)$ must be replaced by $B_{12}^\omega W^\omega(\omega)$, where B_{12}^ω is the appropriate Einstein coefficient when spectra are given as a function of the angular frequency. Then $B_{12}^\omega = 2\pi B_{12}$, and similarly $B_{21}^\omega = 2\pi B_{21}$. The spontaneous emission coefficient A_{21} is unchanged.

where $g_{1,2}$ are the degeneracies of the two states. That is to say that $g_{1,2}$ are the numbers of equivalent quantum states having energies $E_{1,2}$. Thus

$$N_1 = (N_2 g_1/g_2) \exp(hf/k_B T). \tag{14.7}$$

Inserting this value of N_1 into eqn. 14.5 gives

$$W(f) = A_{21}/\{B_{12}(g_1/g_2)\exp(hf/k_B T) - B_{21}\}. \tag{14.8}$$

The equations 14.6 and 14.8 must be consistent, hence we arrive at two equalities

$$g_1 B_{12} = g_2 B_{21}, \tag{14.9}$$

$$A_{21} = (8\pi hf^3 n^3/c^3)B_{21}. \tag{14.10}$$

These relations between the Einstein coefficients hold true in general because the coefficients are determined by the atomic structure. In Chapter 18 they are calculated for a transition in hydrogen. The ratio of the rates of stimulated to spontaneous emission, again in thermal equilibrium, is thus

$$(\mathrm{d}N_2/\mathrm{d}t)_{\mathrm{st}}/(\mathrm{d}N_2/\mathrm{d}t)_{\mathrm{sp}} = B_{21}W(f)/A_{21}. \tag{14.11}$$

In equilibrium with black body radiation this ratio becomes

$$(\mathrm{d}N_2/\mathrm{d}t)_{\mathrm{st}}/(\mathrm{d}N_2/\mathrm{d}t)_{\mathrm{sp}} = 1/[\exp(hf/k_B T) - 1]. \tag{14.12}$$

It is illuminating to consider the mechanical analogy, of a pendulum being driven at its resonant frequency. The drive will do work on the pendulum for one choice of phase between the drive and the oscillations, but when this phase is changed by π energy is drawn from the pendulum. These alternatives are the classical counterparts of absorption and stimulated emission. Spontaneous emission is therefore the process without a classical analogue, rather than stimulated emission!

14.3   Prerequisites for lasing

Suppose a volume of neon gas is heated to 400 K, which is the temperature within a He:Ne laser. The relative probability given by eqn. 14.12 for stimulated emission of light at 632.8 nm, the He:Ne lasing wavelength, is only 10^{-25}. Raising the equilibrium temperature alone does not produce the cascade of stimulated emissions that define lasing. The other 'ingredients' needed for lasing action in addition to stimulated emission are generally these: a *population inversion* and a Fabry–Perot cavity. A population inversion is achieved when the number of atoms in some excited state of energy E_2 exceeds the number of those in a state of lower energy E_1. This provides a reservoir of atoms which can then be stimulated to emit photons coherently. A glance at eqn. 14.7

Table 14.1 Table of coherence times and lengths.

Source	τ_c	ℓ_c
Sunlight 0.4–0.7 μm	3 fs	900 nm
Hg lamp 435 nm line	100 fs	30 μm
Low pressure sodium lamp	5 ps	1.5 mm
Multimode He:Ne laser	0.6 ns	18 cm
Single mode He:Ne laser	1 μs	300 m

reveals that a population inversion cannot happen in thermal equilibrium: it requires instead some subtle *pumping* process which fills the upper state while keeping the lower energy state virtually empty. Once pumping has produced a population inversion any random photon of energy $(E_2 - E_1)$ can stimulate an excited atom to emit a second coherent photon; both photons can then stimulate two more excited atoms to emit two more coherent photons so there are now four fully coherent photons. The process of stimulated emission repeats again and again so that a stream of coherent photons builds up travelling through the *active* material. This single pass through the active material can be extended to multiple passes by enclosing the active material between two parallel mirrors, which define a Fabry–Perot cavity. If the reflectance of one mirror is 100%, and that of the other is slightly less than 100%, then photons which pass through the weakly transmitting mirror make up the external highly coherent laser beam.

The use of a cavity produces another significant effect: standing wave modes develop in the cavity with wavelengths determined by the cavity length. We can take over results proved in Section 5.9. A standing wave's free space wavelength, λ, and frequency, f, are given by

$$m\lambda = 2nL \text{ or } f = mc/(2nL), \tag{14.13}$$

where m is an integer, n is the refractive index and L the cavity length. Usually the natural emission lines of the atoms within the cavity have a much broader width than the separation of cavity modes. Lasing occurs at the frequencies of the cavity modes, the only frequencies at which radiation can build up within the cavity. If lasing occurs at only a single frequency the laser is called a single (longitudinal) mode laser, in a multi(-longitudinal) mode laser several standing waves at nearby frequencies lase. The coherence times and coherence lengths for various sources are shown in table 14.1; this makes clear the huge improvement in coherence that lasers bring.

The electric field distribution in standing waves inside a Fabry–Perot cavity has been discussed in Section 6.15. The simplest *transverse* mode, the TEM$_{00}$ mode is plane polarized, the *magnitude* of the electric field has azimuthal symmetry and falls off with a Gaussian dependence on the radial distance from the beam axis. This is the mode which lasers are generally designed to produce because of its compactness and symmetry around the beam axis. Higher order modes are less compact and are usually attenuated by some aperture, as for example the capillary tube of a helium–neon, He:Ne, laser. Stimulated emission is then inadequate to compensate this loss and the intensity of higher order modes is insignificant. Thus the ideal situation is for the laser to lase at one frequency (a single longitudinal mode), emitting a plane polarized beam with a TEM_{00} profile (a single transverse mode).

Lasers are sources in which stimulated emission accounts for almost all the emission and thus lasers emit long streams of identical photons. Beams from a single longitudinal mode laser can have a far narrower line width and consequently be more highly coherent than any beam from pre-existing optical sources. The lateral spread of many laser beams is due almost entirely to diffraction: that is to say they are *diffraction limited*. If the laser is lasing in a single longitudinal with TEM_{00} profile, then, as explained in Section 12.10.1, all the radiation is confined to the minimum possible etendue, λ^2. Thus the radiance (or equivalently the brightness) of laser beams can be orders of magnitude higher than those from sources previously available. The introductory example of a laser taken here is the helium–neon laser, a simple very widely used gas laser.

14.4   The He:Ne laser

A simplified diagram of the structure of a He:Ne laser is shown in figure 14.2, the glass envelope being filled with a gas mixture containing one neon atom for every five helium atoms, at a total pressure of typically 0.003 atmospheres (2.5 torr). A DC source at about 1 kV drives a steady discharge current of a few milliamps through the gas via a large series resistor. Electrons in this discharge excite the atoms by colliding with them. Those neon and helium energy levels of relevance here are dis-

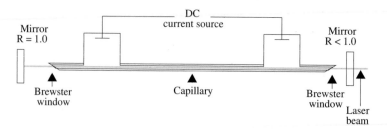

Fig. 14.2 Outline structure of a He:Ne laser.

played in figure 14.3. The pumping action starts with excitation of the more numerous helium atoms by electron collision into the 2^1S_0 level, which is displaced by only 0.05 eV from the neon 5s level. The label 5s indicates that one of the 2p electrons has been promoted to the 5s level in neon. There is a high probability that a helium atom in the 2^1S_0 level will transfer its whole energy to a neon atom when they collide, leaving the neon atom in the 5s level. Neon atoms excited into the 5s level decay first into the 3p level, then into the 3s level, and finally to their ground state. The transitions from the 5s state are partially forbidden and the neon atoms have therefore an extended lifetime in the 5s state. Once in the 3p state neon atoms return stepwise to the ground state. The first step is fast which implies that such radiation can be reabsorbed quickly as well. Having a narrow tube reduces trapping of this radiation within the gas, and so ensures the rapid emptying of the 3p state.

To summarize: the neon atoms are pumped into the 5s state and a population inversion is achieved between this and the almost empty 3p state. The 5s→3p transition is the lasing transition.

Any random photon whose wavelength corresponds to that of the 5s → 3p transition, 632.8 nm, can initiate a cascade of coherent photons through stimulated emission. When the direction of motion of this initial photon coincides with the tube axis the cascade of stimulated photons undergoes repeated reflections from the external mirrors shown in figure 14.2 and as a result standing wave modes develop. Then, if the *gain* due to stimulated emission for a round trip inside the cavity exceeds the sum of losses in the cavity and at the mirrors, that mode's intensity will build up and it will lase: an intense coherent beam at the mode frequency is emitted through the weakly transmitting mirror.

Referring again to figure 14.2 the gas is seen to be confined in a capillary of a few millimetres in diameter and typically 40 cm long. The end caps are glass plates inclined at the Brewster's angle to the tube axis so that the p-polarized radiation is fully transmitted, while the s-polarized light suffers a loss in intensity of over 14% at each surface. This loss greatly exceeds the gain per pass due to stimulated emision, with the result that the emerging laser beam is p-polarized. These windows also absorb infrared radiation which eliminates the lasing that would occur at 3.39 μm through another set of neon levels. The intensity of the beam from He:Ne lasers ranges from a fraction of a milliwat to a maximum of around 100 mW.

Pumping power is electrical and its agents are the electrons in the discharge which strike the atoms. Other pumping processes used to produce lasing are met below: charge injection is used in semiconductor lasers and optical pumping in some other solid state lasers.

A He:Ne laser is usually the first laser that students encounter and provides a simple, sturdy and stable source. The narrow line width and pure polarization state of its beam make it very valuable in interferometry. In applications such as bar-code reading, lecture room pointers and direction indicators for construction work the He:Ne laser has been

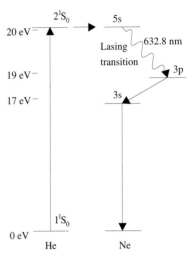

Fig. 14.3 He:Ne electron energy levels relevant to lasing at 632.8 nm.

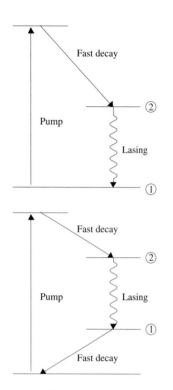

Fig. 14.4 Three and four level lasers. The upper state in the lasing transition is labelled 2, and the lower level is labelled 1.

largely replaced by semiconductor lasers. The latter are not only more compact, efficient and robust than He:Ne lasers, but also operate at the low voltages generally used in electronics.

14.4.1   Three and four level lasers

Lasers can be either three level, or four level, as indicated in figure 14.4, where in each case the slow transition from level 2 to level 1 is the lasing transition, while all other transitions are fast. The pumping of a three level system is more difficult because over half the atoms must be pumped from the ground state into level 2 in order to achieve a population inversion. In four level systems the energy of level 1 above the ground state should be much greater than $k_B T$ so that the equilibrium population of level 1 is negligible. With this condition fulfilled the fraction of atoms that need to be pumped into level 2 to produce a population inversion between levels 2 and 1 is quite modest. The He:Ne laser fits better into the class of four level lasers.

14.4.2   Gain

Even if a population inversion is established in an active medium enclosed in a cavity, lasing will still only occur if the gain in radiation at the lasing frequency per round trip due to stimulated emission exceeds all losses incurred within the cavity and at the mirrors. This requirement is investigated next.

The line shape of the lasing transition, $F(f)$, is taken to be normalized so that $\int F(f)\mathrm{d}f$ is unity; it might have either the Gaussian or Lorentzian profiles introduced in Section 7.3.1 or a combination of these known as a *Voigt profile*. Here we consider a beam of spectral intensity $I(f)$ $\mathrm{W\,m^{-2}Hz^{-1}}$ travelling into the active laser medium, and whose intensity, $I(f)$, does not vary across the line width of the transition. The spectral energy density of the radiation measured in $\mathrm{J\,m^{-3}Hz^{-1}}$ is

$$w(f) = I(f)(n/c),$$

where n is the refractive index of the medium. Then, taking into account the line shape of the transition, the effective spectral energy density of radiation, as far as the transition of interest is concerned, is[2]

$$W(f) = F(f)I(f)(n/c),$$

measured in $\mathrm{J\,m^{-3}\,Hz^{-1}}$. Now we calculate the net gain per round trip in the cavity taking account of stimulated emission and absorption.

---

[2]This result is strictly valid for homogeneously broadened lines where all the atoms can undergo transitions at any frequency across the line profile. However the conclusions drawn in this section are valid for both inhomogeneously and homogeneously broadened lines.

Spontaneous emission can be ignored because it is both incoherent and relatively isotropic. Losses through mirrors will be included later. If N_1 and N_2 are again the number densities of atoms in the lower and higher energy states respectively, a coherent beam travelling a distance Δz passes across $N_2 \Delta z$ excited atoms per unit area and so the change in its intensity in this distance is

$$\Delta I = -[(\mathrm{d}N_2/\mathrm{d}t)_{\mathrm{st}} + (\mathrm{d}N_2/\mathrm{d}t)_{\mathrm{ab}}]\, hf\, \Delta z.$$

Using eqns. 14.1 and 14.2 this becomes

$$\Delta I = (B_{21}N_2 - B_{12}N_1)F(f)I(f)(n/c)hf\Delta z. \qquad (14.14)$$

There is therefore an exponential dependence of the intensity on path length in the medium

$$I(z;f) = I(0;f)\exp[g(f)z],$$

where

$$g(f) = (B_{21}N_2 - B_{12}N_1)F(f)hfn/c$$

is called the *gain coefficient*. Using eqns. 14.9 and 14.10 this gain coefficient is

$$g(f) = A_{21}[N_2 - (g_2/g_1)N_1]\,F(f)(c^2/8\pi f^2 n^2). \qquad (14.15)$$

Positive gain requires $g_1 N_2 > g_2 N_1$, which makes explicit the need for a population inversion. We can simplify the notation by setting $\Delta N = N_2 - (g_2/g_1)N_1$ and $A_{21} = 1/\tau_{\mathrm{sp}}$ where τ_{sp} is the lifetime of the upper state against spontaneous decay. Using eqn. 7.54

$$F(f_0) = 2\pi P(\omega_0) = 4/\gamma,$$

where γ is the observed line width, and may be much broader than the natural line width A_{21}. With these substitutions the gain at the central frequency can be rewritten

$$g(f_0) = [\Delta N/(\tau_{\mathrm{sp}}\,\gamma)]\,(c^2/2\pi f_0^2 n^2) = \Delta N\,\sigma_0, \qquad (14.16)$$

where

$$\sigma_0 = (c^2/2\pi f_0^2 n^2)/(\tau_{\mathrm{sp}}\gamma). \qquad (14.17)$$

The magnitude of g depends on two factors: σ_0, which is specific to the particular transition; and the population inversion ΔN, which is determined by the pump power. The quantity σ_0 has the dimensions of area, and eqn. 14.16 shows it to be the stimulated emission cross-section for a single excess excited atom. Further, σ_0 is proportional to the spontaneous emission rate ($A_{21} = 1/\tau_{\mathrm{sp}}$), and inversely proportional to the observed line width (γ). It is clear that, thanks to the factor f_0^{-2} in σ_0, lasing becomes more difficult as the frequency increases.

The loss of intensity in the beam per unit length in the cavity due to scattering and other processes can be expressed as

$$I(z; f) = I(0; f) \exp(-\alpha z), \qquad (14.18)$$

where the absorption coefficient, α, is not expected to vary much across the line width. At the mirrors there are discrete losses which include the effects of scattering and absorption in the mirror, in addition to the transmission loss through the mirror. The overall gain in one pass along the whole length of the cavity and returning to the starting point is thus

$$G = \exp(2Lg) \exp(-2L\alpha) R^2, \qquad (14.19)$$

L being the length of the cavity and R the reflectance at the mirrors. There is evidently a threshold for lasing at which the gain, g_{th}, exactly balances the losses. Then

$$\exp(2Lg_{th}) \exp(-2L\alpha) R^2 = 1,$$

whence

$$g_{th} = \alpha - \ln R/L, \qquad (14.20)$$

and because R is usually close to unity we have to a good approximation that

$$g_{th} = \alpha + (1 - R)/L. \qquad (14.21)$$

When a laser is turned on the pumping process causes a population inversion and the gain rises. If the pumping rate is high enough the gain crosses its threshold value and lasing commences. A further increase in the pump rate causes a rise in the population inversion and a stronger laser beam. By contrast the gain stabilizes at its threshold value thanks to negative feedback: if the gain rises the laser intensity rises and this drives N_2 down, so that the gain falls; on the other hand if the gain falls the intensity falls, N_2 rises and so the gain rises. This is the standard condition for *continuous wave* (CW) lasing.

The laser intensity begins to *saturate* when the stimulated emission rate overtakes the spontaneous emission rate. Both processes then deplete the upper state population equally and further increases in intensity are hard to achieve. When these rates are equal the intensity of the internal laser beam, I_{sat}, is called the *saturation intensity*. The corresponding photon flux in the laser is

$$\phi_{sat} = I_{sat}/(hf_0),$$

where f_0 is the laser frequency. Now the spontaneous and stimulated emission rates become equal when

$$\sigma_0 \phi_{sat} = 1/\tau_{sp}. \qquad (14.22)$$

Substituting for ϕ_{sat} in terms of I_{sat} gives

$$I_{sat} = hf_0/(\sigma_0 \tau_{sp}). \qquad (14.23)$$

Line widths

The lasing transition may be one of several by which the upper state can decay. When these other decays have rates $1/\tau_i$ the lifetime of the upper state, τ, is given by

$$1/\tau = 1/\tau_{\mathrm{sp}} + \sum_i (1/\tau_i), \qquad (14.24)$$

and the natural line width in frequency is $1/(2\pi\tau)$. Environmental line broadening mechanisms were introduced in Section 7.3.1. Homogeneous line broadening is that due to processes acting identically on all the atoms, while inhomogeneous line broadening involves processes which vary from atom to atom. The total line width due to all effects is that denoted by γ in the analysis of the previous section.

The dominant line broadening process in the He:Ne laser is the Doppler shift resulting from the random thermal motion of the neon atoms. Like other inhomogeneous processes the Doppler broadening gives a line with the Gaussian shape described by eqn. 7.56. Of course the lifetime of the atomic states is unaffected by Doppler shifts.[3] Whenever homogeneous broadening is dominant the spectral lines retain the Lorentzian profile described by eqn. 7.54. In a Nd:YAG laser the lasing transition is a transition between energy states of the Nd^{3+} ions. These ions are well dispersed in a regular crystal in environments that are closely similar, hence the line broadening is homogeneous. However in an amorphous solid such as the glass matrix of a Nd:glass laser the environment is different for each Nd ion and the result is inhomogeneous broadening and a Gaussian line profile.

Another distinction needs to be appreciated, this time between the different effects that collisions can produce, although all give homogeneous broadening. On the one hand collisions with electrons and atoms can cause an excited atom to drop into a lower energy state. This reduces the lifetime of the atom as well as broadening the frequency spread of the spectral line. On the other hand, when an atom which is already radiating undergoes a collision, this alters the phase of the wavetrain randomly. The Fourier transform of such a wave has a broader frequency spread than that of an uninterrupted wave. However these phase-shifting collisions do not alter the decay rate or lifetime of the atomic state.

14.4.3 Cavity modes

Figure 14.5 shows the Doppler broadened line for the lasing transition in the He:Ne laser, and the positions of the cavity modes, which are equally spaced in frequency. The Doppler broadened profile has the shape given by eqn. 7.56 with FWHM of about 2 GHz in frequency, or $2\,10^{-3}$ nm in wavelength. With mirrors a distance L apart the cavity mode spacing in frequency is

$$\Delta f = c/(2nL), \qquad (14.25)$$

[3] The Doppler shift for velocities of $\sim\!3000\ \mathrm{ms^{-1}}$ in a hot gas is $v/c \sim\! 10^{-5}$, and the dilation of the lifetime is $1/\sqrt{[1-(v/c)^2]}$ or $\sim (1+10^{-10})$.

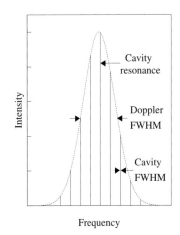

Fig. 14.5 The Doppler broadened line of the Ne transition at 632.8 nm. The cavity resonance positions are also shown; their individual widths are narrower than the lines drawn here.

where n the refractive index is close to unity for a gas. Taking a mirror separation of 40 cm gives a mode spacing of 0.375 GHz or $5\,10^{-4}$ nm. The FWHM of the modes can be taken from eqn. 5.58

$$\Delta f_c = f/\text{CRP} = c/(2nL\,\mathcal{F}), \qquad (14.26)$$

where $\mathcal{F}$ is the cavity finesse. Applying this equation to the case of a cavity of 40 cm length and finesse 400 gives a line width of about 1 MHz, or equivalently $1.4\,10^{-6}$ nm. The corresponding coherence time is $\tau_c = 1/(2\pi\Delta f_c)$, giving $1.7\,10^{-7}$ s.

Fabry–Perot cavities are the optical counterparts of resonant electric circuits and mechanical oscillators. The sharpness of a resonance is characterized by its Q *value*, defined as

$$Q = f/\Delta f_c, \qquad (14.27)$$

where f is the resonant frequency and Δf_c the FWHM of the resonance. If N is the number of stored photons we have a simple identity

$$Q = (hfN)Q/(hfN),$$

and substituting for Q on the right hand side gives

$$Q = f(hfN)/(hfN\Delta f_c) = \omega(hfN)/(hfN/\tau_c),$$

where as usual $\omega = 2\pi f$. Therefore

$$Q = \omega(\text{energy stored})/(\text{rate of energy loss}). \qquad (14.28)$$

The line width and coherence time evaluated one paragraph above are those of a *passive* cavity: the coherence time is determined by the rate at which the radiation in the TEM_{00} mode is lost through or around the mirrors at reflection. In a laser with an *active* cavity the coherent radiation in the mode is continually refreshed by stimulated emission so that a much longer coherence time and narrower line width can result. Working against this are effects which cause the optical path length of the cavity to fluctuate, and so cause the mode wavelength to fluctuate. Fluctuations in the gas temperature and density, mechanical vibrations and thermal expansion of the cavity all play a part. Even if all the physical parameters could be held constant an ultimate lower limit on the line width is imposed by phase fluctuations due to spontaneous emission. This limit will be evaluated below in Section 14.7.3 and comparisons made with the actual performance of specific lasers.

Lasing modes which differ according to the number of nodes they produce along the cavity are called *longitudinal modes*. This does not exhaust the choices of modes, because for each longitudinal mode there can be alternative transverse distributions of the mode intensity. The simplest mode is the TEM_{00} mode described in Section 6.15, a mode whose intensity falls off as a Gaussian in the radial distance from the beam axis. Profiles of the intensities of other transverse modes are complex in form and less compact around the beam axis than the TEM_{00} mode. With an appropriately small limiting aperture their losses are made larger than their round trip gain and they cannot lase. The radius w of the $1/e^2$ intensity point for the TEM_{00} mode can be obtained from eqn. 6.74. For simplicity we take the example of a marginally stable symmetric confocal cavity of length L. Then $z = L/2$ and $R = L$ in eqn. 6.74 so that the beam radius at the mirror

$$w = \sqrt{L\lambda/\pi},$$

with the waist, w_0, being a factor $\sqrt{2}$ smaller. Thus a 40 cm long confocal He:Ne laser operating at 632.8 nm would only make use of the gas discharge within a diameter of $\sim$0.5 mm around the optical axis. The semi-angle of the divergence of the TEM_{00} mode is given by eqn. 6.63

$$\theta = \lambda/w_0\pi,$$

which is 1.0 mrad for the case being considered.

The radiance of even a 1 mW He:Ne laser is astonishingly large: using the beam width and divergence just calculated this radiance is

$$\mathcal{R} = 10^{-3}/[(\pi w_0^2)(\pi\theta^2)] = 10^{-3}/\lambda^2$$

which gives $2.5\,10^9$ W m^{-2}sr^{-1}, four orders of magnitude greater than the radiance of the Sun.[4]

An alternative configuration preferred for the He:Ne laser is a *hemispherical* cavity with one plane mirror and the other having radius of curvature close to the separation (with $L < R$). This arrangement is fairly insensitive to misalignment, and lies within the region of stability in figure 6.34. With such a cavity the laser spot size can be tuned by altering $(R - L)$ and can be made a few times larger than that obtained with the confocal configuration. Mirrors are usually made of much greater diameter than the laser beam in order to make construction and alignment easier. The beam can then be clipped by an aperture, placed near to the spherical mirror of a hemispherical laser; while the glass capillary of a He:Ne laser can be the mode-clipping aperture.

Lasers are also made with unstable cavities with one small and one large mirror. The diverging beams can be turned to an advantage provided that the gain per pass compensates the loss of beam per pass. Having a large beam area means that much more active material is illuminated and laser power is high. Radiation escaping around the smaller mirror forms the external beam, albeit not having a Gaussian profile.

---

[4]An alternative route to this result would be to use the fact that the etendue deduced in Section 12.10 for a single mode is λ^2.

14.4.4   Hole burning

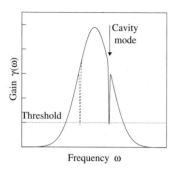

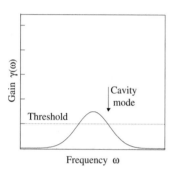

Fig. 14.6 Gain profiles for lasing where a single cavity mode lies within the gain profile. In the upper panel the line is inhomogeneously broadened, so that lasing causes hole burning. In the lower panel the line is homogeneously broadened and the whole profile is pulled down when lasing commences.

The discussion in Section 7.3.1 shows that the spectral lines of the neon atoms at 400 K in the low pressure gas of a He:Ne laser are Doppler broadened by about 2 GHz, while collision broadening is very much smaller. Thus the width of a spectral line emitted by a group of neon atoms all travelling with the same velocity is equal to the homogeneous line width, but is displaced by the Doppler shift. The upper panel of figure 14.6 shows the Doppler broadened line of the neon transition at 632.8 nm, which is also the same shape as the laser gain curve. For the purpose of illustration, only a single cavity mode is assumed to lie within the gain curve. Radiation in this mode can only stimulate emission from those neon atoms whose Doppler shifted frequency is close to the mode frequency. As explained in Section 14.4.2, at the onset of lasing the gain is reduced to the threshold value at the cavity frequency; elsewhere the gain profile is unaffected: the lowered gain mirrors a reduced population inversion. This effect, called *hole burning*, is shown in the upper panel of figure 14.6. A second hole also appears symmetrically placed with respect to the peak of the gain profile because the Doppler shift is reversed between alternate passes through the cavity. The width of the hole burnt in the Doppler gain profile is much narrower than the Doppler broadened line so that much potential gain goes to waste. This helps to explain why a He:Ne laser's efficiency in converting electric power at the plug to laser beam power is only 0.05%. Another reason contributing to the low efficiency is that a lot of energy is lost in ionizing helium atoms, which have a high ionization potential. As an example of an efficient laser we introduce the Nd:YAG laser here.

The lasing transition in a Nd:YAG laser is that of the Nd ions, which are diffused in a near perfect yttrium aluminium garnet crystal lattice. Line broadening is now due to environmental effects that are the same for all the Nd ions and hence homogeneous. Consequently a cavity mode located anywhere across the gain profile in a Nd:YAG laser can stimulate all the Nd ions to emit. When lasing commences it pulls down the whole of the gain profile until the gain at the frequency of the mode falls to the threshold value. This is the case shown in the lower panel of figure 14.6. A Nd:YAG laser is therefore much more efficient than a He:Ne gas laser.

Usually many longitudinal modes across the line profile compete for the pumped energy. In a He:Ne laser they act on separate sets of atoms with different velocities, so that many modes lase at the same time. Initially it was thought that with homogeneous line broadening only the mode nearest the gain peak would lase. However modes draw energy principally from atoms around their antinodes, and the antinodes of different modes are at different locations. As a result modes around the gain peak lase, but relatively fewer than with inhomogeneous broadening.

Hole burning has a positive value: it provides a signal by which a laser can be tuned to the centre of the natural line. Of course this implies a mechanism to control the laser frequency by, for example, altering the cavity length. If the two complementary holes burnt in the profile are brought to coincide at the centre of the profile, then the laser frequency coincides with the centre of the spectral line. In this condition the laser power drops appreciably because the components of the laser beam travelling in opposite directions now stimulate the same group of nearly stationary atoms. This minimum in the laser intensity, the *Lamb dip*, signals that the laser frequency is tuned accurately to the transition frequency.

14.4.5   Laser speckles

Laser speckles are the familiar bright spots seen when a laser beam falls on any surface that is not perfectly flat. The pattern changes and appears to scintillate as one's eye moves. This effect comes about because the laser beam is coherent across the area of each mode, which can mean across the whole beam. In this situation light reflected from all the surface imperfections illuminated will therefore interfere, and the eye perceives a pattern of bright and dark spots formed by this interference. The features of the interference pattern have a mean angular size determined by the resolution of the eye: this is λ/d, where λ is the laser wavelength and d the diameter of the observer's pupil. Even small involuntary eye movements change the relative path length from the observer's eye to different points on the surface by many wavelengths, and this causes the scintillations. Laser speckles have features in common with the speckles described for astronomical images in Section 8.4. In each case light encounters irregularities within its coherence area: irregularities in the surface viewed give rise to laser speckles, while astronomical speckles originate with variations in the refractive index of the atmosphere.

14.4.6   Optical beats

The possibility of observing beats between two coherent sources differing in frequency by Δf was discussed in Section 5.5.2. It was shown that the interference could be detected if the detector had a response time shorter than $1/\Delta f$. This interference was first observed optically by Javan, Ballick and Bond in 1962.[5] In their experiment two independent He:Ne lasers of slightly different length were operated so that each lased in a single longitudinal TEM_{00} mode at $1.153\,\mu\text{m}$ wavelength. It was arranged that the beams had parallel polarizations and then they were superposed using mirrors so that the combined beam struck a photomultiplier. The output of the photomultiplier was spectrum analysed

---

[5] A. Javan, E. A. Ballick and W. L. Bond: *Journal of the Optical Society of America* 52, 96 (1962)

and seen to contain the expected beat note at around 5 MHz. There is an apparent paradox because a photon originating from one laser at one frequency would only be expected to interfere with itself and not with a photon at another frequency from a different laser. The resolution of this quandary lies in asking whether it is possible to distinguish which laser any particular photon came from. If it is then there should be no interference. However, if as in Javan's observations it is not possible to identify which source the photons come from, then interference will be observed.

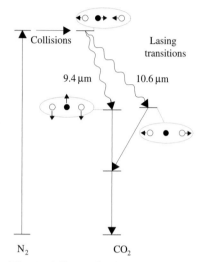

Fig. 14.7 Energy levels involved in the operation of a CO_2 laser. Sketches of the vibrational excitations are drawn near to the corresponding levels.

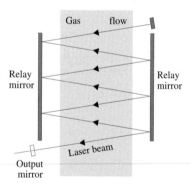

Fig. 14.8 Beam path and gas flow inside a CO_2 gas laser.

14.5 The CO_2 gas laser

CO_2 lasers contain a gas mixture at around 0.1 atmosphere pressure, consisting of $\sim 12\%$ CO_2, $\sim 8\%$ nitrogen, with the balance being helium. Lasing occurs between the vibrational energy levels of the CO_2 molecule shown in figure 14.7; the lasing wavelength can be either $10.6\,\mu$m or $9.4\,\mu$m. An electric discharge, which can be continuous or pulsed at a few kilohertz, provides energetic electrons which collide with and excite the nitrogen molecules. These in turn transfer their excitation energy in further collisions to CO_2 molecules. Some CO_2 molecules are then in the uppermost vibrational excited state shown in the diagram with decay rates of around $0.3\,\text{s}^{-1}$ for the two decays drawn there. After the lasing transition is completed the daughter states themselves are short lived, and so a population inversion is produced. Cartoons drawn close to the levels involved in lasing in figure 14.7 indicate the types of vibration occurring in these states. The helium atoms are useful both in helping to deplete the lower enegy states in collisions and in carrying off the heat generated to the walls of the laser tube. These lasers have modest efficiencies of around 10%, so there is a lot of heat to dissipate. In multi-kilowatt CO_2 lasers the gas is circulated through a heat exchanger in a closed loop, and the laser beam is folded as shown in figure 14.8. The relay mirrors are copper and the exit mirror is some material such as GaAs which has a low absorption coefficient for $10\,\mu$m radiation. Both the mirrors and the tube are water cooled.

Industry uses large numbers of CO_2 gas lasers for marking and cutting metals and ceramics, as well as for welding. As an example a commercial 5 kW pulsed CO_2 laser can cut 1 cm thick steel plates at the rate of $2\,\text{cm}\,\text{s}^{-1}$. The pulsed power density employed in drilling reaches $10^{12}\,\text{W}\,\text{cm}^{-2}$ so that the material is ablated, thus eliminating the swarf associated with mechanical tools. Much lower power CO_2 lasers are used in medicine, in particular for delicate vocal chord surgery.

14.6   Organic dye lasers

The active material of these lasers is a dye-loaded liquid or solid. A distinguishing feature is that the gain bandwidth is broad and within this range the wavelength is continuously tunable. In fact dye lasers were the first fully tunable lasers. Specializing to liquid lasers, the lasing material is a dye dissolved in water or in an organic solvent at a dilution of parts in 10^4, sufficient to isolate the dye molecules from one another and so ensure that they enjoy similar environments. These complex molecules possess a variety of molecular vibrational and rotational states which result in broad continuous bands of energy levels, as illustrated in figure 14.9. This gives dyes their characteristic property of being able to absorb and emit over a range of wavelengths. One such dye is rhodamine 6G which can lase at any wavelength in the range 570–610 nm; another is Na-fluorescein which can lase anywhere between 535 and 570 nm. The dye molecules must be first pumped to the S_1 band using radiation of wavelength shorter than the wavelength at which the laser is to operate. One electron within a molecule is then in an excited state. In the singlet (S) and triplet (T) levels its spin is aligned respectively anti-parallel and parallel to the spin of the rest of the molecule; both partners having spin $\hbar/2$. The interband transitions $S \leftrightarrow S$ and $T \leftrightarrow T$ are allowed and fast, while the $T \leftrightarrow S$ transitions have lifetimes of order 10^{-5} s. Excitation to the S_1 band is followed by a rapid de-excitation through several steps to the lowest level in the S_1 band in about 10^{-12} s. Thereafter the molecule makes an interband transition to some level in the S_0 band, the lifetime for this transition being a few times 10^{-9} s, and this is the lasing transition. Finally the molecule drops into the ground S_0 state in a further 10^{-12} s.

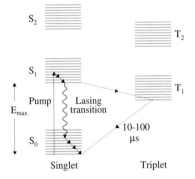

Fig. 14.9 Energy levels involved in the operation of a typical dye laser.

Dye lasers can be operated to give either pulsed or continuous wave lasing. Figure 14.10 shows the principal components of a CW dye laser. The dye is in the form of a stable continuous liquid jet crossing a resonant cavity. Energy is pumped into the dye by a flash lamp or another laser, often a frequency doubled Nd:YAG laser with output wavelength 532 nm; this pump light is focused onto the dye where it crosses the cavity. The cavity is defined by a mirror and an echelle grating, with the grating mounted in the Littrow configuration described in Section 6.9.4. In this arrangement the grating orientation selects the lasing wavelength and its resolving power determines the spectral width of the laser beam. Then in order to make full use of the resolving power of the grating the beam is expanded using a telescope so as to cover the entire grating width. The faces of the dye jet stream in the excitation region are inclined at the Brewster's angle to the beam in the cavity. Only the p-polarized light is then fully transmitted at these surfaces, with the result that only this polarization component achieves positive gain on a round trip in the laser cavity. This means that the laser beam electric field will lie in the plane of the diagram. Any misalignment of the electric fields of the pump and the dye laser reduces the efficiency of the pumping

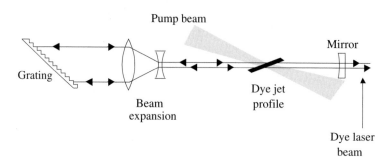

Fig. 14.10 Outline structure of a dye laser.

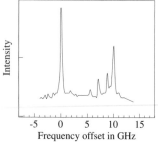

Fig. 14.11 The upper panel shows the hydrogen gas emission spectrum near the H_α line at room temperature. The Doppler broadening is ∼6 GHz. In the lower panel the saturation spectrum is shown, with a resolution improved by a factor of around 20 (courtesy Prof. Haensch).

process. Consequently the angle between the pump beam and the dye laser beam is kept as small as possible.

The line widths achieved with the layout shown in figure 14.10 are in the range 0.3–10 GHz. A layout which produces better resolution will be described later for the Ti:sapphire laser, another tunable laser. Then line widths in the range 1–10 MHz are obtained.

Unavoidably some small proportion of the dye molecules decay from the S_1 band into the T_1 band so that there is a steady accumulation of molecules in the T_1 band. These molecules then readily absorb light to make further $T \rightarrow T$ transitions, light that might otherwise contribute to stimulated emission. In order to sustain continuous operation the dye must be refreshed, and therefore it is pumped around a circuit. This circulation also avoids continuous exposure of the dye to the intense laser beam which can break down the dye molecules.

The existence of tunable lasers has made it possible to develop new spectroscopic techniques. Two examples of such techniques will now be examined: saturation spectroscopy in which the Doppler broadening of spectral lines is eliminated, and cavity ring-down spectroscopy which is used to study weak spectral lines.

14.6.1   Saturation spectroscopy

Saturation spectroscopy was first developed by Haensch, Levenson and Schawlow and by Bardé.[6] The upper panel of figure 14.11 shows part of

[6]T. W. Haensch, M. D. Levenson and A. L. Schawlow: *Physical Review Letters* 27, 707 (1971); C. Bardé: *Comptes Rendues Acad. Sciences*, Paris 271, 371 (1970).

the emission spectrum of hydrogen gas at room temperature around the H_α line at 656 nm. In the lower panel the same spectrum appears, this time taken using saturation spectroscopy. The Doppler broadening has now been eliminated and fine structure, not previously observable in the spectrum by optical methods, is clearly revealed. The Doppler width of a spectral line is given by eqn. 7.56, which predicts that in hydrogen gas at room temperature the width of these lines is around 6 GHz, compared to the natural line width ∼100 MHz for such allowed transitions.

Figure 14.12 contains a sketch of the equipment used in saturation spectroscopy. The input beam comes from a laser whose frequency can be tuned continuously over the range of interest; it could be either a pulsed dye or a Ti:sapphire laser. This beam is split in two beams and these daughter beams are directed in opposite directions through the cell containing the gas so that their paths overlap there. One of these beams,

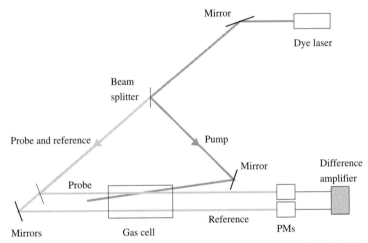

Fig. 14.12 Apparatus for saturation spectroscopy. Adapted from T. W. Haensch, I. S. Shain and A. L. Schawlow: *Physical Review Letters* 27, 707 (1971). Courtesy Professor Haensch and the American Physical Society.

the *pump beam*, is around ten to one hundred times more intense than the other, the *probe* beam. After emerging from the cell the probe beam falls on a detector whose output will provide a record of the absorption spectrum. The experiment involves scanning the dye laser frequency in small steps across the Doppler gain profile of the absorption line of the gas, and recording the detector current at each step.

The Doppler broadened profile of a transition is shown in the upper panel of figure 14.13, plotted as a function of the velocity of the atoms toward the probe beam. Suppose that the frequency of the laser, f, is lower than the transition frequency of the gas being investigated, f_0. Each beam will only be absorbed by atoms moving towards it with velocity v such that the beam appears blue-shifted as seen by the atoms

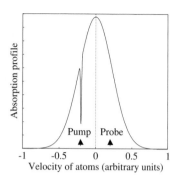

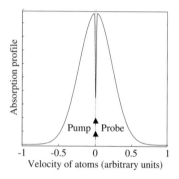

Fig. 14.13 The upper panel shows the spectral line plotted against the velocity of the atoms toward the probe beam. In the lower panel the pump/probe frequency coincides with the line centre.

to f_0. That is

$$f_0 = f(1 + v/c).$$

As a result the pump burns a hole in one side of the Doppler profile and the weaker probe samples the atoms on the opposite side of the profile, which is indicated in figure 14.13. During the laser frequency scan the attenuation of the probe beam emerging from the gas cell follows the Doppler profile *until* the laser frequency reaches the centre of the profile. At this stage the pump and the probe are both being absorbed by the population of atoms at rest, so that the population the probe samples is depleted by the action of the pump. Consequently the detector response follows the curve shown in the lower panel of figure 14.13 with a narrow dip centred on the transition frequency. The width of this dip is determined by the combination of the homogeneously broadened line width, Δf_h and the laser line width, Δf_l. Doppler broadening has been eliminated. A limit to the resolution attainable is set by the laser line width, which is around 1 MHz for modern dye lasers.

Thus far no mention has been made of the second probe beam in figure 14.12, which passes through a part of the gas cell well away from the pump beam. This probe falls on a second detector which therefore records the pure Doppler profile. Then the Doppler free spectrum can be isolated by subtracting the first detector signal from the second detector signal. Two further important components are not illustrated in the figure, which together suppress backgrounds. The first is an electrically driven mechanical windmill which is used to chop the pump beam; the second is a phase sensitive detector fed by the difference amplifier, which selects the component of this signal at the chopping frequency.

14.6.2　Cavity ring-down spectroscopy

The study of very weak transitions poses a new problem because the absorption that a beam experiences in passing through the sample becomes undetectable. For the same reason it is difficult to detect the presence of a gas through its absorption spectrum if this gas is at a low concentration in a gas mixture. An obvious way to enhance weak spectral features is to pass the the beam repeatedly through the gas sample. This method was taken to its limit in 1988 by O'Keefe and Deacon in what is called *cavity ring-down* (CRD) spectroscopy.[7] This technique is now widely used in environmental science studies. They enclosed the gas sample in a Fabry–Perot cavity with high reflectance mirrors so that the light traversed the gas very many times. The basic elements of the apparatus are shown in figure 14.14. In a single measurement step one laser pulse at the desired frequency is injected into a Fabry–Perot cavity containing the gas of interest. Unlike other applications of the Fabry–Perot cavity the coherence length of the pulse is kept much shorter than the cavity

---

[7]A. O'Keefe and D. A. G. Deacon: *Review of Scientific Instruments* **59**, 2544 (1988)

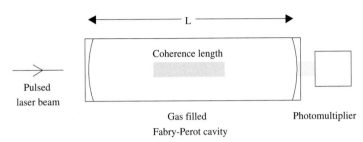

Fig. 14.14 Apparatus for cavity ring-down spectroscopy, with an optically stable Fabry–Perot cavity.

length so that multiple beam interference effects are eliminated. The pulse behaves like a shuttlecock bouncing to and fro between the cavity mirrors. On the far side of this cavity from the laser source a photomultiplier and its associated circuitry continuously record the instantaneous intensity of light passing through the mirror. Depending on the level of absorption of the laser pulse within the cavity, the signal at this detector will decay more or less slowly. The quantity of interest is the time it takes for the intensity of the light reaching the photomultiplier to fall by a factor e. From this *cavity ring-down* time the absorption coefficient of the gas at the wavelength of the laser can be determined. A series of such measurements is taken as the laser frequency is tuned over its spectral range, and from these measurements the absorption spectrum of the gas can be tabulated against frequency.

Obviously the ring-down time is also affected by the rate at which the light escapes from the cavity through the mirrors. If this rate is too high the decay due to absorption becomes undetectable. Hence the mirror reflectance needs to be close to unity, with $1-R$ typically less than 10^{-4}.

Suppose that the cavity is filled with a gas whose absorption coefficient at the laser frequency is α, and that the cavity has length L. Then the intensity change in the laser beam in one round trip is

$$r = \exp\left(-2\alpha L\right)R^2,$$

where we assume that the mirrors have equal reflectance. The time taken for one round trip is $2L/c$. Then the above equation can be rewritten as a function of t, the time elapsed since the pulse entered the cavity,

$$r = \exp\left(-c\alpha t\right)R^2.$$

Averaging over times long compared to one round trip, the intensity measured by the photomultiplier has this time dependence

$$I(t) = I(0)\exp\left(-c\alpha t\right)R^{2n}$$

where n is the number of round trips in time t. Now n is simply $ct/2L$ so we have

$$I(t) = I(0) \exp(-c\alpha t) R^{ct/L}.$$

Next taking logarithms, and recalling that when R is very close to unity the approximation $-\ln R = 1 - R$ is valid, we have

$$\ln[I(t)/I(0)] = -c\alpha t - (1 - R)(ct/L).$$

An exponential curve is fitted to the variation of the intensity with time and from this a decay constant, τ, can be extracted:

$$1/\tau = c\alpha + c(1 - R)/L, \tag{14.29}$$

which is called the cavity ring-down time. If the cavity is evacuated and the ring-down time remeasured at the same frequency the result is

$$1/\tau_0 = c(1 - R)/L. \tag{14.30}$$

Evidently the effective path length traversed by the pulse is

$$L_{\text{eff}} = L/(1 - R),$$

which for a cavity of length $0.5\,\text{m}$ and mirrors with reflectance 0.9999 is about $5000\,\text{m}$. This large amplification of the length traversed by the laser pulse means that very weak spectral features become detectable. The evacuated cavity ring-down time corresponding to these parameters is $16.7\,\mu\text{s}$. Subtracting eqn. 14.30 from eqn. 14.29 gives the absorption coefficient

$$\alpha = 1/c\tau - 1/c\tau_0. \tag{14.31}$$

Defining $\Delta\tau$ to be the precision of timing, then the minimum detectable absorption is

$$\alpha_{\text{min}} \approx (\Delta\tau/c\tau^2). \tag{14.32}$$

Taking $\Delta\tau$ to be a modest $1\,\text{ns}$, α_{min} is $10^{-8}\,\text{m}^{-1}$ for the case considered. Sensitivities a thousand times smaller are attained with ring-down spectrometers based on the design described here. The beauty of cavity ring-down spectroscopy is that, unlike standard absorption spectroscopy, it is insensitive to fluctuations in the laser intensity: only the ring-down time is being measured.

A novel application of CRD spectroscopy is in the determination of the reflectance of mirrors, like those used in gravitational wave interferometers, whose reflectances differ from unity by only one part in a million. The ring-down time of an empty cavity made from these mirrors is measured and inserted in eqn. 14.30, and this gives a precise value of the reflectance of the mirrors. No other methods of measuring reflectance appear to be competitive when the reflectance is that close to unity.[8]

---

[8]See C. J. Hood, H. J. Kimble and J. Ye: *Physical Review* A64, 033804 (2001).

14.6.3   A heterodyne laser interferometer

The design described here is of a modern precision Michelson interferometer used to measure distances with nanometre accuracy. The source used is a stabilized He:Ne laser with long coherence length. Heterodyning is a term taken from radio usage to describe the beating together of two electromagnetic beams of slightly different frequency: one is an unmodulated reference wave and the other wave is modulated. In this way the modulation is transfered to the beat note and can be recovered electronically. A key feature of the interferometer is to split the emission from a single laser into orthogonally polarized components with frequencies differing by a few megahertz. One way to produce the desired beams is to place the laser in a longitudinal magnetic field which Zeeman splits the emission line. The difference in their polarization is used to separate the beams so that one beam travels along one arm and the other beam along the other arm of the Michelson interferometer. After passing through the interferometer the beams interfere and give a beat wave at a few megahertz which carries the interference information. Electronics is used to count the integral number of fringes appearing at the centre of the field during the translation of the mobile mirror. How this measurement is used to extract the movement to a fraction of a fringe is now addressed.

Figure 14.15 shows the interferometer. The two beams are partially reflected at the beam splitter BS to provide a reference signal. The reflected beams travel through a polarizer P2 with its axis set at 45° to both polarizations. The two components emerge from P2 with identical polarization and, applying Malus' law, half their incident intensity. These components interfere at the photodetector D2 to give a *reference signal*

$$A_{\mathrm{ref}} = [\exp\left(-i\omega t\right) + \exp\left(-i\omega' t\right)]/\sqrt{2}.$$

The intensity averaged over times long compared to the optical period, but small compared to the beat period is

$$\overline{I_{\mathrm{ref}}} = AA^* = 1 + \cos\Delta\omega t,$$

where the beat angular frequency, $\Delta\omega = \omega' - \omega$. Capacitive coupling to the detector removes the DC component of the signal giving a *reference signal*

$$\overline{I_{\mathrm{ref}}} = \cos\Delta\omega t. \tag{14.33}$$

This result needs further comment to bring out some salient points. The first thing to note is that the interference between the two different wavelength beams produces beats with a beat period $2\pi/\Delta\omega \sim 1\,\mu\mathrm{s}$. A laser source with a very stable frequency is required in order to make its coherence time long compared to this beat period. Detectors are readily available with short enough response times, $\sim 1\,\mathrm{ns}$, so that they can record these fast beats. Detectors will be discussed in the following chapter.

A second beat signal is obtained from the radiation falling on the other detector D1 and this contains the information on the path difference between the interferometer arms. The two beam components that go straight through the first beam splitter are separated by the polarizing beam splitter PBS. One frequency component (with one polarization)

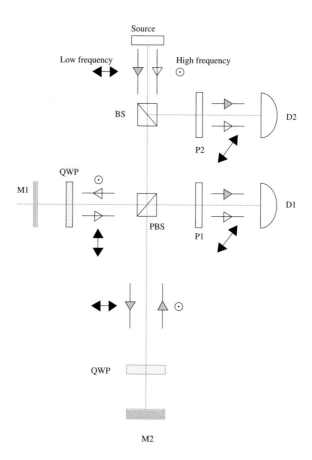

Fig. 14.15 Two-frequency Michelson interferometer. The cross-hatched boxes are quarter-wave plates, the solid boxes are mirrors and the open boxes are plane polarizers. Detectors are drawn as half moons. The two frequencies' paths are indicated by grey and open arrows, with their polarizations alongside.

is reflected from mirror M1 while the other frequency component (with orthogonal polarization) is reflected from mirror M2. Both beams make two passes through a quarter-wave plate which rotates their polarizations by 90°. The returning beams are therefore both directed by the PBS onto the 45° polarizer P1. They emerge from P1 with identical polarization and fall on the photodetector D1. The wave falling on D1 is thus

$$A_{\text{int}} = \{-\exp i\omega t + \exp\left[-i(\omega' t + \phi)\right]\}/\sqrt{2},$$

where ϕ is the phase change due to the difference in the paths via M1 and

M2. Repeating the analysis previously carried through for the reference signal we obtain the time averaged signal recorded by the detector D1

$$\overline{I_{\text{int}}} = \cos(\Delta\omega t + \phi). \tag{14.34}$$

This is called the *interferometer* signal. The reference and interferometer signals, given in eqns. 14.33 and 14.34, are multiplied electronically, yielding

$$
\begin{aligned}
I_{\text{h}} &= \cos(\Delta\omega t + \phi)\cos\Delta\omega t \\
&= \cos^2\Delta\omega t\cos\phi - \cos\Delta\omega t\sin\Delta\omega t\sin\phi \\
&= \{[\cos(2\Delta\omega t) + 1]\cos\phi - \sin(2\Delta\omega t)\sin\phi\}/2.
\end{aligned}
$$

The next processing step is to take the time average of I_{h} over an interval much longer than the beat period. This gives

$$I_{\text{c}} = (\cos\phi)/2. \tag{14.35}$$

A second similar electronic chain is used to multiply $\overline{I_{\text{int}}}$ by the quadrature of $\overline{I_{\text{ref}}}$, namely $\overline{I'_{\text{ref}}} = \sin\Delta\omega t$, and the time average of this product is taken. This yields

$$I_{\text{s}} = -(\sin\phi)/2. \tag{14.36}$$

Next ϕ is extracted from I_{c} and I_{s}, and finally if the integral fringe count is m, the total distance travelled by the mobile mirror, M1, is

$$t = (\lambda/2n)(m + \phi/2\pi), \tag{14.37}$$

where n is the refractive index of the air in the arms of the interferometer.

The precision is mainly limited by the degree of flatness of the interferometer mirrors. If these are flat to better than $\lambda/100$ then the path difference can be determined to $\lambda/100$.

A determination of the direction of travel requires some more circuitry and software analysis. The values of $\sin\phi$ and $\cos\phi$ are recorded at fixed time intervals and the changes between measurements calculated. Then for example a product is taken of $\cos\phi$ and the change in $\sin\phi$,

$$\Delta(\sin\phi)\cos\phi = \cos^2\phi\Delta\phi.$$

From eqn. 14.37 we can see that the sign of this product depends on the direction of travel of the mobile mirror.

Information about the mirror *velocity* can be extracted if the phase shift produced by the mirror movement is measured dynamically. While in motion the moving mirror induces a Doppler shift of the frequency of the reflected light ω by an amount $2v\omega/c$, where v is the mirror's instantaneous velocity. As a result the instantaneous difference between the angular frequency of the interferometer and reference signals is

$$\Omega = \Delta\omega + 2\omega v/c. \tag{14.38}$$

If this frequency difference is measured continuously it provides a continuous record of the mirror's velocity. In practice the mobile mirror is mounted onto the object whose motion is of interest.

Measurements of velocity using a single frequency laser beam are also feasible. However this *homodyne* interferometry has significant disadvantages relative to heterodyne operation. Although beating the returning beams still yields a Doppler shift in frequency it is now with respect to a baseline frequency shift, $\Delta\omega$, of zero hertz at zero velocity. Information as to the direction of motion is therefore lost with homodyne operation. A velocity of $1\,\mathrm{m\,s^{-1}}$ gives a Doppler shift of $3.2\,\mathrm{MHz}$ for a wavelength of $633\,\mathrm{nm}$, which means that in homodyne operation the electronics has to uniformly amplify signals over the range from DC to radio frequencies, which is costly. By contrast, in heterodyne operation the signal frequencies lie entirely in the radio band, so that the extraction of the velocity and its direction are straightforward.

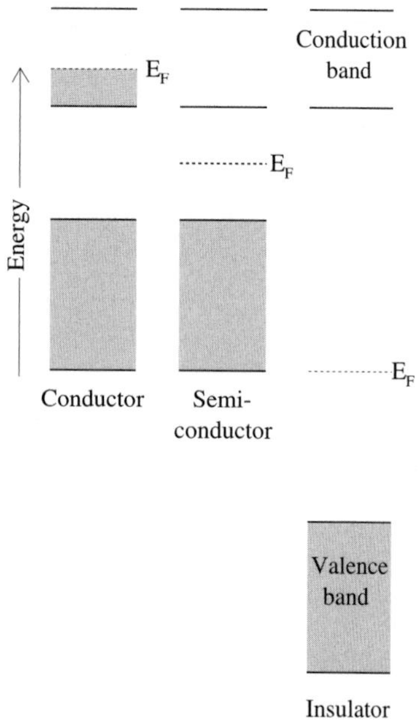

14.7 Introducing semiconductors

Semiconductor lasers were developed later than gas lasers, but are now the most widely used: of order a hundred million are manufactured each year. As a preliminary to discussing this important class of lasers a brief description of the relevant properties of semiconductors is given here.

In solids the neighbouring atoms are sufficiently close that their electromagnetic interactions alter the quantum states of the outermost electron shells. Instead of the single discrete energy state repeated for each atom in a gas there are states *shared across the crystal* which are all slightly different in energy and so cover a band of energies. Figure 14.16 shows typical examples of the energy bands occupied by the electrons from the outer unfilled shell in the cases of a conductor, a semiconductor and an insulator. At the absolute zero of temperature the electrons pack down to fill the energy levels below a limit E_F, called the *Fermi energy*, while all higher energy levels are empty. Those levels which are occupied by electrons at low temperature are shaded in figure 14.16. In insulators and semiconductors the lower, *valence* band is full while the upper, *conduction* band is empty. Thus if a voltage is applied across these materials no current flows because the Pauli principle prohibits another electron from entering an energy state already containing an electron. In metals the conduction band is partly filled and current can flow across the metal. More generally, at a temperature $T\,\mathrm{K}$ the probability of a level with energy E being occupied is given by the Fermi–Dirac formula, eqn. 13.80,

$$F(E) = \{\exp\left[(E - \mu(T))\right]/k_B T\,] + 1\}^{-1}, \qquad (14.39)$$

where $\mu(T)$ is the chemical potential. Note that $F(E)$ falls to exactly one half at an energy $\mu(T)$, and that the Fermi energy is simply $\mu(0)$. This distribution is illustrated in figure 14.17 for a semiconductor. The

Fig. 14.16 Conduction and valence bands for a conductor, a semiconductor and an insulator. In each case the broken line indicates the Fermi energy.

average thermal excitation at room temperature $k_{\mathrm{B}}T = 0.025\,\mathrm{eV}$. In the case of semiconductors the *band-gap* between the valence and conduction bands is sufficiently small, $\sim 1\,\mathrm{eV}$, so that the thermal excitation causes appreciable numbers of electrons to enter the bottom of the conduction band at room temperatures and leave empty energy levels (*holes*) in the valence band. If a voltage is applied across the semiconductor the electrons in the conduction band will move. Holes in the valence band behave as if they had an effective mass equivalent to that of an electron and positive charge. This behaviour is crucial to the analysis of semiconductor behaviour. The electrons and holes contribute equally to the current in a semiconductor. By contrast the band-gap in an insulator is so large that thermal excitation causes negligible promotion of electrons into the conduction band, and hence only tiny currents flow. Copper, a metal conductor, has an electrical conductivity of $\sim 6\,10^{7}\,\mathrm{S\,m^{-1}}$; semiconductors such as silicon and GaAs have electrical conductivities which are orders of magnitude lower, $4.3\,10^{-4}\,\mathrm{S\,m^{-1}}$ and $3\,10^{-7}\,\mathrm{S\,m^{-1}}$ respectively; insulators like glass have electrical conductivities that are less than $10^{-10}\,\mathrm{S\,m^{-1}}$.

When an electron drops from the conduction to the valence band in a semiconductor a photon may be emitted, and this transition is the lasing transition employed in semiconductor lasers. In semiconductors such as GaAs photon emission dominates over non-radiative de-excitation processes. It would be convenient if this were also true for silicon which is the semiconductor mostly widely used in electronics: unfortunately it is not. This difference in behaviour is connected with differences in the electron energy–wave number (E–k) distributions within the valence and conduction bands for GaAs and silicon. These distributions are compared in figure 14.18. A transition in which an electron drops from the conduction band to the valence band in silicon involves a change of $10^{10}\,\mathrm{m^{-1}}$ in the value of its wave vector k. However the value of the wave vector of the infrared photon emitted in the transition is far smaller, $10^{6}\,\mathrm{m^{-1}}$. Consequently the vector sum of the electron and photon wave vectors does not vanish, and as a result momentum would not be conserved. The only way that the overall momentum can be conserved is for the imbalance to be taken up by a *phonon*, that is to say, a quantum of mechanical oscillation of the silicon lattice. Such a complex radiative process has a characteristic time of around $1\,\mathrm{ms}$, making it far slower than competitive non-radiative processes by which the electron can reach the valence band.

In the corresponding radiative transition in GaAs the electron gets rid of energy without any change in momentum, and a photon alone is emitted. This radiative transition is therefore very fast, with characteristic time around $1\,\mathrm{ns}$. As a result the far slower non-radiative transitions are rare. GaAs is an example of what is called a *direct* band-gap semiconductor, while silicon is called an *indirect* band-gap semiconductor. Only the direct band-gap semiconductors with their efficient conversion of the

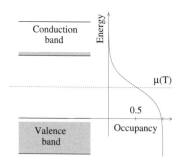

Fig. 14.17 Energy distribution in a semiconductor at a finite temperature.

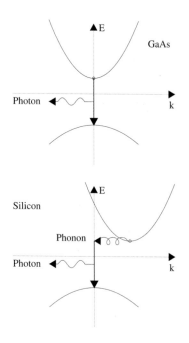

Fig. 14.18 The E-k distributions of electrons and holes in GaAs in the upper panel, and in silicon in the lower panel. E is the electron/hole kinetic energy and k the wave vector. Radiative transitions are shown in each case.

injected holes and electrons to photons offer any potential for constructing lasers.

The energy of the photons emitted is simply the band-gap energy. This is 1.43 eV in GaAs, corresponding to a wavelength of 865 nm, slightly outside the visible spectrum. Radiative transitions are suppressed if the semiconductor crystals have imperfections such as dislocations, because these provide sites for fast non-radiative processes. The *E*–*k* curve in figure 14.18 for electrons in the conduction band is described near the apex by an expression similar to that expected for free electrons,

$$E = k^2\hbar^2/(2m^*) + E_\mathrm{c}, \tag{14.40}$$

where E_c is the energy at the bottom of the conduction band. The mass is replaced by an *effective mass*, m^*, to account for the effect of the electron's interactions with the lattice ions. The effective mass of an electron in the conduction band of such semiconductors is $\sim 0.1\,m_\mathrm{e}$. In the valence band the corresponding relation is

$$E = -k^2\hbar^2/(2m^*) + E_\mathrm{v}, \tag{14.41}$$

where E_v is the energy at the top of the valence band. The electrons in the valence band behave as if their mass were negative. Consequently a hole in the valence band, which is the absence of an electron with negative mass, behaves like a particle with positive mass and charge.

The pure semiconductors which have been under discussion so far consist of one element or one uniform alloy. They are called *intrinsic* semiconductors and their conductivity is called intrinsic conductivity. Additional *extrinsic* conductivity can be obtained by *doping* intrinsic semiconductors with a sufficiently low concentration of atoms of other suitable elements such that the doping does not produce dislocations; the resulting extrinsic semiconductors have properties that can be tailored to make a wide variety of semiconductor lasers. GaAs is a semiconductor widely used in lasers and will be the example we specialize to here.

Gallium is an element in group III of the periodic table with three valence electrons per atom and arsenic is from group V with five valence electrons per atom. The Ga and As atoms in GaAs share eight valence electrons in a covalent bond. Replacing a small fraction of the gallium atoms with similar sized atoms from a group IV element, such as Sn or Si, leaves the crystal structure unchanged while each such *donor* atom provides one valence electron surplus to the requirement of forming the covalent bond. The extra electrons inhabit what are called *donor* quantum states: these electrons are only weakly bound to their parent ion in orbits which extend over many atoms. The donor states lie only of order 0.01 eV below the conduction band, so that at room temperature a high proportion of the donated electrons are thermally excited into the conduction band and can contribute to the current when an electric

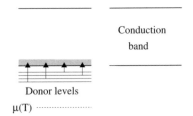

Donor levels

μ(T)

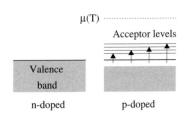

Fig. 14.19 Semiconductor energy levels produced by n- and p-doping. The arrows indicate the thermal excitation of electrons, as described in the text.

field is applied.

On the other hand if the gallium atoms are replaced by atoms such as zinc or cadmium with one fewer valence electrons, then vacant *acceptor* levels appear whose energies lie slightly above the valence band. At room temperature electrons from the valence band are thermally excited into these acceptor levels leaving holes in the valence band, and these holes are available to carry current. Both donor and acceptor levels are shown on figure 14.19. The correspondingly shifted Fermi levels are also indicated. Acceptor and donor doped GaAs are known as p-GaAs and n-GaAs respectively. The electrons in n-GaAs and the holes in p-GaAs are known as *majority carriers*, while the other carriers are called *minority carriers*. Doping improves the concentration of *carriers* dramatically, that is to say the concentration of electrons and holes that contribute to the current. Whereas in intrinsic GaAs the carrier density is around one in 10^{16} atoms, a typical doping level of one atom in 10^6 produces a carrier density of around one in 10^6 also. Heavy n/p-doping (above roughly one in 10^4) in which the conductivity approaches that of metals is labelled as n^+/p^+-doping; light n/p-doping (below roughly one in 10^9) in which the conductivity remains near the intrinsic conductivity is labelled as ν/π-doping.

Figure 14.20 shows the energy levels across a *diode junction* between p-GaAs and n-GaAs. When no external voltage is applied, which is the case illustrated in the upper panel, the electrons flow from the n-doped region until the Fermi levels are equalized on the two sides of the junction. The resulting relative negative voltage of the p-doped region increases the energy of the electrons in this region, and is called the *contact potential*, V_{J}. In the lower panel of this figure the same junction is *forward biased*, meaning that an external positive voltage, typically 1 or 2 volts, is applied to the p-doped region. This flattens the ramp in energy across the contact region. There is now a *population inversion* in the region of contact, which is maintained by the continuous *injection* of electrons from the n-doped side and holes from the p-doped side. In direct band-gap semiconductors like GaAs the injected electrons will drop into the injected holes when they meet at the junction, and photons having an energy equal to the band-gap energy, E_{g} will be emitted.

A device made in this way is a light emitting diode (LED), and is an incoherent source. Figure 14.21 shows a cross-section through a simple LED. Red, green and blue LEDs are built using semiconductors with appropriate band-gaps. Advanced designs have structures that can share features of the semiconductor lasers described below. LEDs are widely used in displays now seen on car dashboards, in hand torches and in traffic lights. LED lifetimes now exceed those of any competitive sources and high intensity arrays of LEDs are beginning to replace other lighting elements.

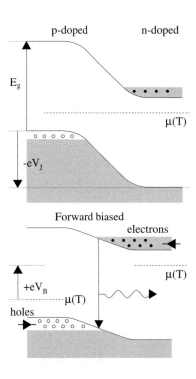

Fig. 14.20 Energy diagrams for a semiconductor junction diode: in the upper panel with no applied field; in the lower panel with forward bias. A radiative recombination is shown.

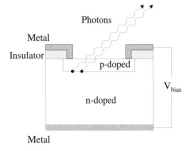

Fig. 14.21 A cross-section through a light emitting diode. The outer dimensions in this plane are a few microns.

The remaining ingredient required to turn a doped forward biased GaAs junction into a semiconductor laser is a Fabry–Perot cavity. A standard technique is to cleave the GaAs along crystal planes so that these form moderately high reflectance mirrors at the ends of the device. Such a simple *homojunction* diode made of bulk p- and n-doped material will not lase at room temperatures because the electrons and holes spread many microns from the junction where the inversion occurs.

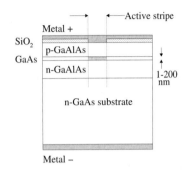

Fig. 14.22 A cross-section through a double heterostructure laser perpendicular to the laser beam's optical axis.

14.7.1   Double heterostructure lasers

The basic semiconductor lasers which operate efficiently in ambient conditions have the generic structure illustrated in figure 14.22. This shows a cross-section taken perpendicular to the direction of the laser beam. Lasing occurs in the GaAs region sandwiched between layers of n- and p-doped GaAlAs. In the GaAlAs alloy a proportion of the gallium atoms are replaced by aluminium atoms, aluminium being another group III element with the same number of valence electrons as gallium. There are two *heterojunctions* separating pairs of dissimilar semiconductors, n-GaAlAs/GaAs and GaAs/p-GaAlAs, thus making a *double heterostructure* (DH). The width of the positive metal strip in contact with the p-GaAlAs determines the region through which the electric current flows and hence the width of the region in the GaAs layer which lases. This active *stripe* has a rectangular cross-section, ~100 nm by a few microns. Its length along the beam direction, which lies perpendicular to the diagram, being of order 0.3 mm. The GaAs crystal is cleaved along crystal planes lying parallel to the diagram in order to form a reflective Fabry–Perot cavity. GaAs has a high refractive index, 3.6, which when substituted into eqn. 9.79 gives a reflectance of 0.32 at the air/cleaved GaAs interface.

Figure 14.23 shows the level structure when the DH is strongly forward biased, that is with the p-region made positive with respect to the n-region. The alloy GaAlAs has a larger band-gap than GaAs, therefore the GaAs layer forms a well in potential energy for both the electrons in the conduction band and also, paradoxically, for the holes in the valence band. Note that while the electron energies increase upward in the diagram, energies of holes (absence of electrons) increase in the opposite sense. Thus a hole in the valence band will have its lowest energy at the *top* of the valence band. Electrons and holes are injected, under the forward bias, into the thin GaAs layer, producing a dense population inversion there and so facilitate lasing. Another useful property of the double heterostructure is that the refractive index of GaAlAs is less than that of GaAs, and consequently the active region forms a waveguide to confine the radiation laterally. This is illustrated by the diagram in the lower part of figure 14.23.

Figure 14.24 shows a notional plot of the light intensity in the laser beam versus the injection current. At a low injection current spontaneous, and therefore incoherent, emission dominates; then at a current of order 10 mA the gain threshold for lasing is crossed and the light intensity rises steeply with a slope of order $0.5\,\mathrm{W\,A^{-1}}$. The gain curve for GaAs is centred on 860 nm and is about 10 nm wide so that there will be many possible lasing frequencies satisfying the condition for standing waves in the Fabry–Perot cavity

$$f = mc/2nL,$$

where m is an integer, n the refractive index and L the cavity length. Taking the laser length, which is also the cavity length, to be 0.3 mm and the GaAs refractive index of 3.6, then the mode spacing will be 140 GHz in frequency or 0.35 nm in wavelength. When the injection current is below threshold the thirty or more modes lying within the gain curve radiate equally, but when the injection current rises above the threshold only one or two modes near the peak of the gain curve will lase.

If ΔI is the excess current above the threshold value the number of electron/hole *annihilations* per second contributing to lasing is $\Delta I/e$. Then the laser beam power is

$$P = \eta(\Delta I/e)hf, \qquad (14.42)$$

where the overall efficiency of electron/hole annihilations to give photons in the laser beam, η, is around 0.3. An estimate of the threshold current density can be obtained by supposing that the threshold current of 10 mA is spread over the active stripe of area $300\,\mu\mathrm{m}$ by $3\,\mu\mathrm{m}$: this gives $1000\,\mathrm{A\,cm^{-2}}$. Diffraction at the exit side surface of the active stripe produces a rather wide angle beam of elliptical cross-section. Astigmatic lenses or an anamorphic pair of prisms, like those shown in Figure 2.19, are used to reshape and focus the beam. Although semiconductor lasers have active regions whose linear dimensions are roughly one thousandth those of gas lasers, nonetheless their beams have comparable powers, and this reflects the relative density of the population inversions in the two classes of laser.

In double heterostructures like that shown in figure 14.22, the alloys on either side of each junction must have crystalline structures which match to high precision. Otherwise there would be defects in the region of the junctions, and such defects would provide locations at which non-radiative electron/hole annihilations would proceed rapidly. This would rule out any prospect of lasing. Fortunately $\mathrm{Ga_{1-x}Al_xAs}$ crystals are all interpenetrating face centred cubic lattices whose lattice constant changes negligibly, from 0.564 to 0.566 nm, as x varies from zero to unity. This makes it possible for crystals to be grown with few defects at the boundary between different GaAlAs alloys. GaAlAs lasers are built whose lasing wavelength can be designed to lie anywhere in the interval 750–870 nm, corresponding to the range in x over which the band-gap remains direct.

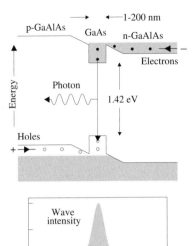

Fig. 14.23 The upper panel shows the energy levels in a double heterostructure GaAlAs laser. The lower panel shows wave intensity variation across the same section.

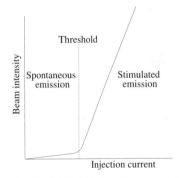

Fig. 14.24 Light intensity versus injection current for a typical DH laser.

Table 14.2 Table of ranges over which lasers can be constructed from various semiconductor alloys.

Alloy	Range in wavelength
GaAs	0.82–0.88 μm
$In_x Ga_{1-x} N$	0.36–0.60 μm
$In_x Ga_{1-x} As_y P_{1-y}$	1.0–1.7 μm
$Al_x Ga_{1-x} As$	0.68–0.92 μm
$Cd_x Pb_{1-x} S$	1.9–4.2 μm
$Cd_x Hg_{1-x} Te$	3.2–17 μm

Silica based optical fibres exhibit extremely low absorption near wavelengths of 1310 nm and 1550 nm. Semiconductor lasers operating at these wavelengths are made from alloys of another set of semiconductors, whose crystal lattices and lattice constants match equally well. These are the *quaternary* compounds $Ga_{1-x} In_x As_{1-y} P_y$ from which lasers are built whose wavelength may be set anywhere within the range from 1000 to 1700 nm. Table 14.2 gives the wavelengths available with lasers made from a selection of semiconductor alloys. The art of *band-gap engineering* has made it possible to design semiconductor lasers which operate at wavelengths anywhere in the near ultraviolet, visible and near infrared parts of the spectrum.

The application of semiconductor lasers as sources for optical fibre transmission in telecom applications will be developed in Chapter 16. For the present we can note here several advantages of semiconductor lasers for telecom applications:

- The etendue of a semiconductor laser beam is well matched to that required for efficient injection into optical fibres,
- Their compactness and low voltage requirements are compatible with standard modern fast electronics,
- The wavelengths of semiconductor lasers can be chosen to minimize simultaneously both the absorption losses and dispersion along optical fibres.

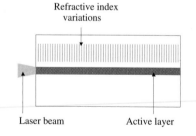

Refractive index variations

Laser beam   Active layer

Fig. 14.25 A cross-section through a DFB laser containing the beam's optical axis.

Any departure of a laser beam from the diffraction limited ideal is characterized by the M^2 parameter. We use the expressions introduced in Section 6.15. Suppose the actual beam waist and divergence are larger by a factor M than the ideal: $\Theta = M\theta$ and $W_0 = M w_0$. Then the expression for the beam width at a distance z from the waist becomes

$$W^2 = W_0^2 + z^2 \Theta^2, \qquad (14.43)$$

which is M^2 larger than the ideal.

14.7.2 DFB lasers

The lasers described so far use a cavity to define the lasing wavelength. An alternative is to build wavelength selection in by forming a diffracting structure called a *distributed feedback Bragg* (DFB) reflector within the laser. This is a region in one of the bounding layers whose refractive index varies cyclically with period Λ along the laser beam direction. It is therefore necessary to give the laser facets an antireflection coating so as to eliminate cavity resonances. A cross-section along the beam axis of such a DFB laser appears in figure 14.25. There will be strong Bragg reflection back along the beam axis at any free space wavelength λ for which

$$\lambda/n_{\text{eff}} = 2\Lambda,$$

where n_{eff} is the effective refractive index. At such wavelengths the superposition of the oppositely travelling waves produces standing waves along the beam axis. The strongest feedback is obtained by choosing this wavelength to lie at the peak of the gain profile. Lasing is then restricted to the mode at this wavelength. An important advantage of DFB lasers is that line widths are narrow, typically $10\,\text{MHz}$ or $2\,10^{-5}\,\text{nm}$ at $800\,\text{nm}$ wavelength.

14.7.3 Limiting line widths

The presence of spontaneous emission imposes a quantum mechanical limit on the line width of lasers. The laser frequency is taken to be f_{L} and its angular frequency is then $\omega_{\text{L}} = 2\pi f_{\text{L}}$. As we shall prove in Section 19.11 the ratio of the stimulated to the spontaneous emission in a mode equals the number of photons in the mode, n. Then the ratio of the respective electric field amplitudes is $\sqrt{n} : 1$. The vector sum of these fields is presented in figure 14.26. Each spontaneous photon has random phase, so that each of them changes the phase of the total field by $\Delta\phi \sim 1/\sqrt{n}$. Successive spontaneous photons emitted from the laser cavity cause the phase to perform a random walk with steps spaced at intervals equal to the cavity coherence time, τ_{c}. A random walk in phase space requires $1/\Delta\phi^2$ steps to give a total phase change of one radian, which in this case is just n steps. The time taken is called the *phase diffusion* time, τ_{D}, and equals $n\tau_{\text{c}}$. During the phase diffusion time the laser phase would alter by $\omega_{\text{L}}\tau_{\text{D}}$ in the absence of any spontaneous emission, hence the fractional uncertainty in the laser angular frequency due to the phase diffusion is $1/(\omega_{\text{L}}\tau_{\text{D}})$. The corresponding limiting laser line width is thus

$$\Delta\omega_{\text{L}} = 1/\tau_{\text{D}} = 1/(n\tau_{\text{c}}).$$

Now the laser power maintained is

$$P = n\hbar\omega/\tau_{\text{c}},$$

whence

$$n = P\tau_{\text{c}}/\hbar\omega.$$

Substituting this value of n into the expression for the limiting line width

$$\Delta\omega_{\text{L}} = \hbar\omega/(P\tau_{\text{c}}^2).$$

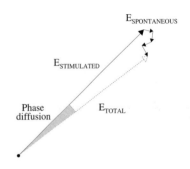

Fig. 14.26 Diffusion of the laser phase due to spontaneous emission.

This result must be multiplied by a factor $(N_2/\Delta N)$ because the number of atoms in the upper level, N_2, capable of spontaneous emission is greater than the population inversion, ΔN, responsible for stimulated emission:

$$\Delta f_{\mathrm{L}} = h f N_2 / (2\pi \Delta N P \tau_{\mathrm{c}}^2), \tag{14.44}$$

which is one form of the *Schawlow–Townes* relation.

This limiting line width is not attained for various reasons. Consider the case of a He:Ne laser, length $0.4\,\mathrm{m}$, a cavity finesse of 400, $1\,\mathrm{mW}$ power, and assume that $N_2/\Delta N$ is unity. Taking a result from Section 14.4.3 τ_{c} for this cavity is $1.7\,10^{-7}\,\mathrm{s}$, so that Δf_{L} has a value of about $\sim 10^{-3}\,\mathrm{Hz}$. This would be 10^{-17} times the laser frequency, and in order not to degrade this performance the *optical length* of the laser would need to be kept constant to a similar precision! Moderately well stabilized He:Ne lasers have line widths of order $1\,\mathrm{MHz}$, and with further refinements a width of $1\,\mathrm{Hz}$ is attainable. In the case of semiconductor lasers with shorter, lower reflectivity cavities τ_{c} is a factor 10^4 smaller so that the predicted Δf_{L} rises to $\sim 0.1\,\mathrm{MHz}$. Even this is not attainable because spontaneous emission causes the hole density to fluctuate and in turn these fluctuations cause refractive index fluctuations. Line widths of the DFB lasers are about a hundred times larger than the Schawlow–Townes limit.

Extremely narrow line widths are obtained using an empty, temperature controlled, Fabry–Perot cavity to provide a stable reference frequency. An active feedback system continuously drives the laser frequency onto this reference frequency. The Pound–Drever stabilization described in Section 8.12.1 is one implementation of this approach.

Laser intensities fluctuate as well as the frequency. In the case of a diode laser the effect is due primarily to fluctuations in bias current and temperature, as well as the statistical fluctuations of spontaneous emission and electron–hole recombinations. The overall effect is expressed as a *relative intensity noise* (RIN) measured in $\mathrm{dB\,Hz}^{-1}$. RIN makes an important contribution to the noise on long optical links, a topic discussed in Chapter 16.

14.8   Quantum well lasers

Quantum well lasers are even more compact and efficient than the semiconductor lasers described so far. The thickness of the active layer where lasing takes place is reduced from hundreds of nanometres to less than ten nanometres, so that it is less than the wavelength of a thermal electron.[9] As a consequence such a narrow active region behaves as a quantum well of the sort described in Section 13.3.1, and the motion of the electron, and equally that of a hole, is quantized in the z-direction *perpendicular* to the physical layers making up the laser.

Figure 14.27 shows the quantum well laser band structure and the discrete electron and hole energy levels within the well. The energy gap is seen to be increased by the sum of the electron and hole kinetic energies in their ground states.

The quantum well laser has advantages that stem from the very different density of electron states compared to a standard double heterostructure laser. Consider first the case of electrons in bulk semiconductor: they move freely in three dimensions with reduced effective masses, and they have two spin states. The dynamics is the same as that of photons in free space which was analysed in Section 9.8.1, and the photons also have two spin states. Therefore we can take over the expression for the density of states for photons and apply it to electrons in a standard DH laser. The density of states of electrons in the non-quantum well laser is thus

$$\rho(k)\, dk = k^2\, dk/\pi^2.$$

Re-expressing k in terms of kinetic energy E,

$$dk = \sqrt{m^*/(2E)}\, dE/\hbar,$$

and then

$$\rho(E)\, dE = \rho(k)\, dk = (2m^*)^{3/2}\sqrt{E}\, dE/(2\pi^2\hbar^3). \qquad (14.45)$$

This function is drawn as the broken line in the right hand panel of figure 14.28.

Turning to the quantum well case the density of electron states must be strictly zero for kinetic energies ranging from zero right up to the energy of the first quantum state, which is shown in the left hand panel of figure 14.28. At an electron energy equal to that of the first quantum state the density of states rises immediately to the value appropriate to electrons moving in three dimensions with the same energy; that is to say the density given by eqn. 14.45. When an electron has greater energy than that of the lowest energy quantum state the excess energy is due to motion in the transverse plane. The density of states resulting from this motion is

$$\rho_t(E_t)\, dE_t \propto k_t\, dk_t \propto dE_t, \qquad (14.46)$$

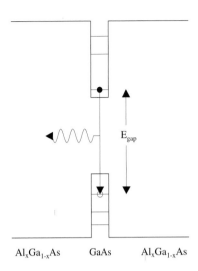

Fig. 14.27 The energy level structure of a quantum well laser, showing a radiative transition.

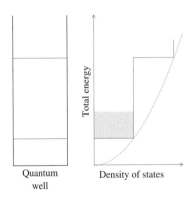

Fig. 14.28 Quantum well energy levels are shown in the left hand panel. In the right hand panel the density of states of the quantum well is shown as a full line and the density of states for an electron in bulk semiconductor by a broken line.

---

[9]In order to appreciate this consider an electron with thermal energy at room temperature, that is $0.024\,\mathrm{eV}$. Using an effective mass of $0.067\,m_e$, appropriate for an electron in the conduction band in GaAs, the de Broglie relation gives a wavelength of $31\,\mathrm{nm}$.

where k_t is the wave number and E_t the kinetic energy from this motion. Thus the number of states occupied increases linearly with the energy between each pair of quantum well levels. The shaded area in the right hand panel of figure 14.28 shows the occupied states when the maximum electron energy lies somewhere between the first and second quantum well energies.

The large density of states accessible exactly at the energy of the first quantum state means that the efficiency is much larger than that of DH semiconductor lasers with their smoothly rising density of states. In addition the quantization of the allowed states also gives a narrowing of the gain profile, while the energy gap, and hence the laser wavelength can be selected by making an appropriate choice of the quantum well thickness. This makes it possible to construct a sequence of lasers with small differences in wavelength, something which is useful in telecoms applications. Thanks to their smaller active volumes quantum well lasers can respond to modulation at frequencies higher than standard semiconductor lasers.

Another advantage of quantum well lasers is that because the volume of the active region is so much smaller, the threshold current density is drastically reduced to of order $30 \, \mathrm{A\,cm^{-2}}$. The restricted size also imposes limitations that need to be addressed.

Single quantum wells *saturate* at relatively low currents because the number of electrons and holes that can be stored in the smaller active region is correspondingly less. Therefore up to tens of quantum wells are laid down one above the other between the confinement layers, thus providing a larger active region without sacrificing the low threshold current. As a result the optical power remains linear up to correspondingly higher values of the injection current. These *multi-quantum well (MQW)* lasers are now widely used in telecom and other applications. The small cross-section of the active area perpendicular to the beam direction means that powers above about $100 \, \mathrm{mW}$ can damage the surface facets. This limitation is overcome by using arrays of quantum well lasers. Such arrays provide beams with incoherent powers of $50 \, \mathrm{W}$ with which to pump the solid state dielectric lasers described later in this chapter.

The advantages conferred by small size, such as low threshold current and high efficiency, can be enhanced still further. Quantum wires and quantum dots are structures in which two dimensions or all three dimensions are made less than the electron wavelength. Correspondingly quantum dots have threshold currents of microamps rather than milliamps. They are now coming into use, for example in tracking biological processes: biological samples in which quantum dots have been deposited are pumped with radiation of shorter wavelength than the lasing wavelength. The dots then lase and can be easily located.

14.8.1   Vertical cavity lasers

The semiconductor lasers described above are all examples of *edge emitting* lasers (EELs). Here attention is shifted to *vertical cavity surface emitting lasers* (VCSELs) whose beam is directed upward through the confining layers above it. Figure 14.29 shows a section through a VCSEL in the form of a mesa, a few microns tall, etched from the parent wafer. That the laser light now travels perpendicular to the MQW layers seems disadvantageous because it reduces the thickness of active material that the photons travel through per pass. However a large overall gain is recovered by inserting high reflectance *distributed Bragg reflector* (DBR) mirrors above and below the confining layers. The active region is therefore enclosed within a vertical cavity of very high finesse and the laser light makes correspondingly more passes through the active region. These distributed Bragg reflector mirrors are formed from alternate layers of high and low refractive index layers of GaAs and AlAs, and possess the properties described in Section 9.7.1 for such structures. Figure 9.20 illustrates that a reflectance very close to unity can be obtained over a narrow wavelength range with such a mirror. The gain is maximized by locating the MQWs at an antinode of the standing laser wave in the vertical cavity formed by the mirrors. A transparent upper electrode made of indium tin oxide (ITO) forms the exit window for the laser beam. Laterally, the guidance of the laser light is provided by total internal reflection at the air/VCSEL interface.

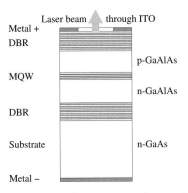

Fig. 14.29 The structure of a vertical cavity surface emitting laser. The structure is a few microns tall.

The short VCSEL cavity length, of order $1\,\mu$m, results in the mode spacing given by eqn 14.25 being much larger for VCSELs than for EELs. It then becomes easy to design the DBR mirror stop band, over which its reflectance is 100%, to include a single longitudinal lasing mode.

The novel VCSEL geometry brings other significant advantages. The lasing surface is potentially the whole surface area of the active region, which can be anything from 10 to $1000\,\mu$m^2. With a large surface area a larger laser power can be tolerated. The VCSEL aperture is naturally circular so that a circular TEM$_{00}$ beam is obtained automatically, while the large aperture means that the beam divergence is correspondingly reduced compared to EELs. Values of the beam quality parameter M^2 introduced in Section 14.7.1 are usually a little above unity. Thresholds, as for other MQW lasers, are as low as $1\,$mA and wallplug efficiencies are about 20%. VCSELs producing beams of a few milliwatts are suitable for telecom applications, where the orientation of their near diffraction limited beams makes for easy coupling into optical fibres.

Semiconductor lasers are grown on wafers of typically 4 inches diameter and each wafer yields many thousands of devices. The wafer must be cleaved to produce the individual EEL chips, and the functionality of the lasers is only then open to test. On the other hand the status of VCSELs can be monitored during the various stages of fabrication which makes for greater yields. Afterwards the wafer only needs dissecting. If an array of VCSELs is required, an area of a VCSEL wafer can be used directly because the VCSEL beams fire perpendicular to the wafer surface.

14.9   Nd:YAG and Nd:glass lasers

These lasers are examples of solid state dielectric lasers which have much larger volume than semiconductor lasers, and as a result are very much more powerful. They can be operated to give pulsed or CW beams. The lasing ions, Nd^{3+}, are dispersed at atomic concentrations of 0.5–1.5% in a matrix that can be glass or yttrium aluminium garnet ($Y_3Al_5O_{12}$ or briefly YAG). The energy levels involved are shown in figure 14.30, and the laser structure in figure 14.31. Pumping of the population inversion is now performed optically using a flashlamp, or another laser. When a semiconductor laser provides the pump beam the complete device is called a diode pumped solid state or *DPSS laser*. In the example shown a GaAs laser at 808 nm pumps the Nd ions into the $^4F_{5/2}$ level.

After reaching the $^4F_{5/2}$ level the Nd ion loses energy by collisional interactions with other lattice ions and drops into the $^4F_{3/2}$ level in about 10^{-12} s. A Nd ion in the $^4F_{3/2}$ level has a long effective radiative lifetime (550 μs in YAG) because the lasing transition, $^4F_{3/2} \rightarrow ^4I_{11/2}$, with a change of four units in angular momentum is highly forbidden. The lasing wavelengths in the different matrices are very similar: 1064 nm with a YAG matrix, and 1054 nm with glass. Further collisional interactions take the ions from the $^4I_{11/2}$ state into the ground state. The $^4I_{11/2}$ state is 0.26 eV above the ground state in energy so that the fraction of Nd ions in this state in thermal equilibrium at room temperature is negligible. A very long lifetime of the upper lasing level is a feature which Nd lasers share with other solid state lasers, and these lasers are, as a result, able to accumulate very large population inversions. Thus a high photon density can be achieved and very intense beams.

A Nd:YAG crystal used to produce a 10 mW CW beam is only a few cubic millimetres in volume; at the other extreme metre diameter blocks of Nd:glass are used in high power lasers. YAG has the advantage over glass as the host matrix that it has a regular crystal lattice. The ions' environments are thus very similar and the lasing line is homogeneously broadened to only 0.45 nm. This compares with the inhomogeneous broadening of 28 nm in an amorphous glass matrix. Cooling is also easier with YAG because its thermal conductivity is much larger, again a consequence of its crystal structure. However glass can be cast in large, cheap, uniform blocks which are ideal for producing pulses of extreme power, provided that an adequate interval is left between pulses for cooling.

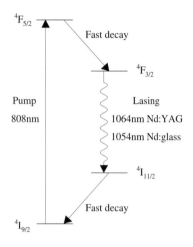

Fig. 14.30 Energy levels involved in the operation of Nd:YAG or Nd:glass laser.

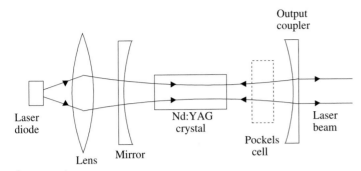

Fig. 14.31 Outline structure of a laser diode pumped Nd:YAG laser.

In figure 14.31 the left hand mirror must transmit the diode laser input beam at 808 nm and reflect the Nd laser beam at 1064 nm, while the right hand mirror has reflectance slightly less than unity at 1064 nm to provide the output beam. Both faces of the Nd:YAG crystal are antireflection coated in this laser. The efficiency of converting the diode laser power to laser power at 1064 nm is typically 0.40, which, taken with a similar diode efficiency, gives an efficiency for converting wallplug power to power in the final laser beam of about 0.16. The technique of frequency doubling is discussed later in the chapter: for the present we note that a β-barium borate crystal inserted in the Nd:YAG laser cavity at the position indicated for the Pockels cell will convert a high fraction of the laser power at 1064 nm to a coherent beam at 532 nm. The use of the Pockels cell itself shown in the figure is described in the next section.

14.9.1 Q switching

Q switching is the method used to cause lasers to emit a train of short pulses, each pulse having very large instantaneous power. In particular Q switching is used to exploit the long lifetimes of the upper state in the lasing transitions of solid state lasers like Nd:YAG lasers. The technique involves pumping the upper level for a time comparable to its lifetime while holding off the cavity feedback and then turning on the feedback to produce a sudden burst of photons. During the pumping phase there will be the usual gain in beam intensity from a single pass through the active material but this gain is not handed on to the next pass.

Either an active or, more interestingly, a passive *shutter* within the laser cavity is used to cyclically alter the *Q*-value of the cavity. The shutter blocks or diverts the laser beam and then briefly opens. During the intervals when the shutter is closed the pumping continues and the population of the long lived upper state steadily builds up. When the shutter opens the gain swiftly rises to a large value and the laser beam intensity follows suit. The intense stimulated emission depletes the population of the upper lasing level rapidly, and the result is a short high intensity pulse.

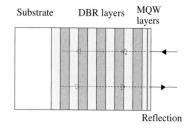

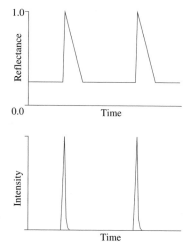

Fig. 14.32 Passive Q switching. The upper panel shows the SESAM structure. The central panel shows the variation of its reflectivity with time and the lower panel shows the consequent pulsing of the laser beam intensity.

Active shutters can take several forms: a mirror on a mount which can be rotated to block the beam; or an acousto-optical modulator of the sort described in Section 7.6 can be used to deflect the beam. Alternatively a Pockels cell and polarizer can be used. When a voltage is applied to the Pockels cell it alters the polarization of the beam so that this is no longer fully transmitted by the polarizer and the gain is held below the threshold for lasing.

Q switching can also be effected passively. One method is to replace one of the cavity mirrors by a *semiconductor saturable absorber mirror* (SESAM) of which one example is shown in figure 14.32. A SESAM is a distributed Bragg reflector having a set of quantum wells within the surface layer of the mirror; the energy gap of the quantum wells is chosen to match exactly the energy of the lasing transition. The quantum wells initially absorb the incident radiation arriving from within the cavity and this action holds the gain below unity. During this stage the population inversion builds up steadily within the solid state laser cavity. Eventually the quantum wells in the SESAM become *saturated* with their populations equalized. At this point the reflectance of the SESAM climbs to unity because no more photons can be absorbed by the quantum wells. There is a corresponding sharp increase in the round trip gain and a high power pulse is produced by the laser. This quickly depletes the population inversion in the active material and the quantum wells empty rapidly. The sequence of events is portrayed in figure 14.32. These plots are appropriate for a case in which the peak gain is well above threshold and the cavity ring-down time is short, of order 10^{-8} s. The rapid rise in intensity is followed by an exponential decay with a time constant set by the ring-down time.

Typical values for the energy and duration of the pulses produced by a Q switched Nd:YAG laser are 1 kJ and 1 ns, giving an instantaneous power of 10^{12} W. Pulses as short as 100 fs are achieved and the magnitude of the electric field in such pulses is huge. The power, P, and the peak electric field, E, are related through

$$P/A = \varepsilon_0 c E^2/2 \tag{14.47}$$

where A is the beam's area of cross-section. Thus the electric field amplitude is

$$E = 27.4\sqrt{P/A}. \tag{14.48}$$

A beam of 10^{12} W focused onto a spot of radius $50\,\mu$m produces an electric field there of $3\,10^{11}$ V m^{-1}. This stupendous value can be put in context by comparing it with the electric field acting on the electron in an isolated hydrogen atom

$$E_{\text{Hy}} = e/[4\pi\varepsilon_0 a_0^2] = 5.1\,10^{11}\,\text{V m}^{-1},$$

where a_0 is the Bohr radius. Interatomic fields in solids are considerably smaller, of order 10^5–10^8 V m^{-1}. Q switched laser beams are evidently intense enough to affect atomic energy levels and distort atomic and molecular structure. In such circumstances the response of matter to radiation will no longer remain proportional to the applied field. The non-linear behaviour of materials under small constant applied electromagnetic fields was introduced in Chapter 10, and effects produced by the intense electromagnetic fields in laser beams will be described in the sections below.

14.10   Ti:sapphire lasers

Titanium doped sapphire (Al_2O_3) solid state lasers are extremely versatile: they can be tuned over a wide frequency range, and are capable of producing pulses of extremely short duration, lasting only a few optical cycles. In this pulsed mode of operation they can also be used, paradoxically, as the basis for advanced frequency standards.

Ti:sapphire lasers are pumped by another laser, generally a Nd laser frequency doubled to 532 nm. The Nd laser is itself in turn pumped

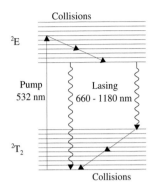

Fig. 14.33 Energy levels involved in the lasing action of a Ti:sapphire laser.

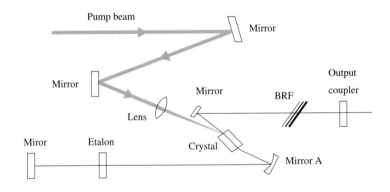

Fig. 14.34 Outline of the structure of a Ti:sapphire laser. (Courtesy CDP Systems, Moscow, Russia and Mark Goossens, Edinburgh Instruments, 2 Bain Square, Livingstone, EH54 7DQ, UK.)

by a semiconductor laser array. Figure 14.33 shows the broad spread of the electronic and vibrational levels of the titanium ion, Ti^{3+} in a sapphire matrix. Once the Ti ion has been pumped into the ^2E band it undergoes collisional de-excitation that quickly brings it to the bottom of that band. Next the lasing transition between bands occurs, for which the ion has a long lifetime of 3.2 μs. Finally further collisions carry the ion rapidly to the ground state. The interband transition can have any wavelength from 660–1180 nm, so that useful gain is available across the whole of this range. The crystals used are generally cylindrical and can range from a few millimetres in diameter and a millimetre in thickness to metre sized crystals, depending on the power required. The titanium

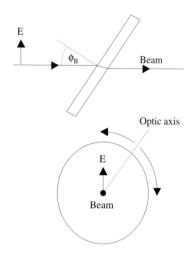

Fig. 14.35 Birefringent filter used in tuning a Ti:sapphire laser. The upper panel is a side view; the lower is a view along the beam. Rotating the disk in its own plane, as indicated, changes the refractive index for the extraordinary component of the beam and hence the difference between indices appearing in eqn. 14.49.

doping used is at an atomic concentration of about 0.1%. Figure 14.34 illustrates the components of a tunable Ti:sapphire laser. The mirror labelled A at the lower right has high reflectance for the range 660 to 1180 nm, but transmits the pump light at 532 nm.

Commonly Ti:sapphire lasers are tuned over portions of the full range using a sequence of components. A first wavelength selection is made with *birefringent filters* (BRF): these are wave plates of uniaxial material inclined at the Brewster's angle, ϕ_B, to the laser beam. Only the p-polarized waves are transmitted without loss so that the laser beam has p-polarization. The fact that the BRF plates and the crystal surfaces are inclined at their Brewster's angles to the beam leads to astigmatism, and this is corrected by the combination of curved mirrors and a folded beam path shown in the figure. Figure 14.35 shows side and end views of one BRF plate whose optic axis lies in the plane of the plate. After crossing this plate of thickness t the difference between the optical paths travelled by the ordinary and extraordinary waves is

$$\Delta s = \Delta n\, t / \sin \phi_B, \tag{14.49}$$

where Δn is the difference between the refractive indices of the ordinary and extraordinary waves. If Δs equals an integral number of waves at some wavelength λ, then a beam at this wavelength passing through the plate would emerge with the same polarization as on entry. After successive round trips radiation of this wavelength becomes p-polarized, and suffers no further losses from reflection. However light of a nearby wavelength which enters the BRF plate with p-polarization emerges elliptically polarized and suffers partial reflection at the exit surface of the plate. This loss inhibits lasing except at the loss-free wavelengths.

In a multi-plate filter the thicknesses of the second and third BRF plates are integral multiples of the thickness of the thinnest. The transmittance for plates of thickness t, $2t$ and $9t$ is plotted in figure 14.36 as a function of the frequency. In the lowest panel the overall transmittance of the three such plates in series is shown. It can be seen that the transmittance remains at unity only at frequencies picked out by the thinnest plate, while the other plates reduce the range around these frequencies over which the transmittance remains close to unity. The combination gives the free spectral range of the thinnest plate and the bandwidth of the thickest. If $\Delta\lambda_{fsr}$ is the free spectral range then

$$m\lambda = \Delta s,$$
$$(m-1)(\lambda + \Delta\lambda_{fsr}) = \Delta s,$$

where m is an integer. Subtracting the last equation from the previous equation gives

$$\lambda - (m-1)\Delta\lambda_{fsr} = 0.$$

Then using eqn. 14.49

$$\Delta\lambda_{fsr} \approx \lambda/m = \lambda^2/\Delta s \approx \lambda^2/\Delta n t. \tag{14.50}$$

Finally

$$\Delta f_{\text{fsr}} = c/\Delta nt. \qquad (14.51)$$

If the thinnest plate is a 0.1 mm thick sheet of calcite, for which $\Delta n = 0.172$, the free spectral range given by eqn. 14.51 is roughly 17 THz. When tuning the laser wavelength the filters are rotated in unison so that the transmittance curves translate together right or left in the figure. Fine tuning uses an etalon consisting of a solid plate with reflective coatings on each face: tuning involves tilting this etalon with respect to the beam axis. A silica etalon of thickness 0.5 mm would provide a free spectral range, $c/(2nt)$, of 200 GHz.

The intracavity tuning system just described is more complex than the grating tuning introduced when describing the dye laser. However the grating is more lossy than the birefringent filter plus etalon combination. Commercial systems mostly use this intracavity tuning for both dye and Ti:sapphire lasers.

14.11   Optical Kerr effect and mode locking

When the electromagnetic fields of a laser beam become comparable in amplitude to interatomic fields in solids the analysis carried through in Section 11.5.1 needs modification. It was assumed there that the restoring force on the electron was linear, which implies that the potential has a quadratic dependence on electron displacement, however strong the force. This potential is indicated by the broken line in figure 14.37. In general this can only be correct for small displacements, and a more realistic potential has a shape like that drawn with a full line in the same figure. Therefore the electronic polarization must be expressed as an expansion in powers of the applied electric field

$$P_i = \varepsilon_0 \left[\sum_j \chi_{ij} E_j + \sum_{j,k} \chi_{ijk}^{(2)} E_j E_k + \sum_{j,k,l} \chi_{ijkl}^{(3)} E_j E_k E_l + ... \right],$$

where χ_{ij} is the susceptibility at low field values, which if we assume the material is isotropic at low field values we can take to be a scalar χ. Then

$$\mathbf{P} = \varepsilon_0 \chi \mathbf{E} + \mathbf{P}_{\text{NL}}. \qquad (14.52)$$

The successive non-linear coefficients get progressively smaller: the components of $\chi^{(2)}$ are $\sim 10^{-12}$ m V^{-1} and the components of $\chi^{(3)}$ are $\sim 10^{-21}$ m^2 V^{-2} for solids when the applied field is oscillating at optical frequencies. Therefore the higher order terms in P_{NL} become appreciable when the electric field approaches the electric field felt by an electron in an atom. Evidently the refractive index of a material can also have a nonlinear dependence on the intensity of a laser beam traversing it. In turn

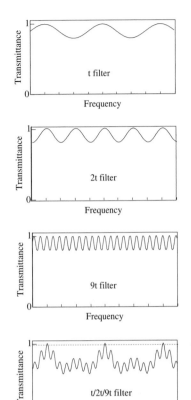

Fig. 14.36 Sketches of the transmittance through the layers of a birefringent filter. From the top downward in sequence: transmittance through plates of thickness t, 2t, 9t and through all three plates in series. The dotted line indicates notionally the transmittance at which the gain would be at its threshold value.

this means that a laser beam can affect its own phase, an effect known as *self-phase modulation*.

When the material is centrosymmetric at the crystal level, that is to say it is symmetric under the parity transformation $\mathbf{r} \to -\mathbf{r}$, then all the even coefficients in eqn. 14.52 must vanish.[10] Such materials include liquids, amorphous solids and cubic crystals. Now the relative permittivity is given by

$$\varepsilon_r = 1 + P/\varepsilon_0 E, \tag{14.53}$$

and the refractive index by $n = \sqrt{\varepsilon_r}$. Then to a good approximation we have, for these materials,

$$n = n_0 + 3\chi^{(3)} E^2/(8n_0), \tag{14.54}$$

where n_0 is the refractive index at low field values and $\chi^{(3)}$ is the appropriate element of the tensor. It follows that the refractive index depends linearly on the laser beam intensity, I,

$$\Delta n = n - n_0 = n_2 I. \tag{14.55}$$

This result was quoted earlier in Section 10.7.2 for constant applied electric fields. The electric dipoles in matter possess mechanical inertia and will therefore respond differently to constant and optical fields. At optical frequencies the response is to the mean square field value.

The phase change induced by the optical Kerr effect in a laser beam of angular frequency ω_0 over a distance L in these materials is

$$\phi = \omega_0(nL/c - t)$$
$$= (2\pi L/\lambda)(n_0 + n_2 I) - \omega_0 t. \tag{14.56}$$

Consequently the variation of the beam intensity with time in a laser pulse leads to a variation of frequency with time. The instantaneous frequency is then

$$\omega = -\mathrm{d}\phi/\mathrm{d}t = \omega_0 - (2\pi L n_2/\lambda)\mathrm{d}I/\mathrm{d}t. \tag{14.57}$$

This self-phase modulation is most marked in the very short pulses produced by mode locking.

14.11.1 Mode locking

Mode locking is the form of Q switching introduced by Spence, Keane and Sibbett [11] which is used to produce laser pulses as short as $5\,10^{-15}\,\text{s}$ or $5\,\text{fs}$. How short such a pulse is can be appreciated by noting that at

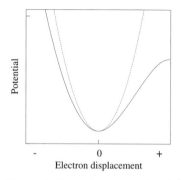

Fig. 14.37 Electron potentials in a solid as a function of displacement from equilibrium. The quadratic potential shown by the broken line gives a linear restoring force. The more realistic potential shown as the full line gives a non-linear restoring force.

---

[10]See Section 10.7.1.
[11]D. E. Spence, P. N. Keane and W. Sibbett, *Optics Letters* **16**, 42 (1991).

a wavelength of 700 nm it contains only two complete wave cycles. In order to achieve mode locking a shutter is required in the cavity which opens at intervals exactly equal to the time taken for the photons to return to it. The layout of a mode-locked Ti:sapphire laser is pictured in figure 14.38, in which the Ti:sapphire crystal itself becomes the passive switching element. Ti:sapphire has such a large Kerr effect coefficient

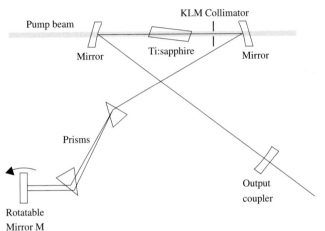

Fig. 14.38 Kerr lens mode locked Ti:sapphire laser.

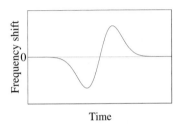

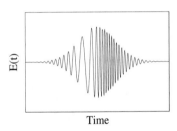

Fig. 14.39 A positively chirped pulse. The upper panel shows the frequency offset variation with time; the lower shows the electric field variation with time. For clarity a very short pulse only thirty or so cycles long is shown. At a wavelength of 800 nm such a pulse would have a duration of 100 fs.

(n_2 is $3.2\ 10^{-20}\,\mathrm{m^2W^{-1}}$) that the strong electric field within an intense laser beam alters the refractive index of the crystal appreciably. A laser beam with the simple $\mathrm{TEM_{00}}$ mode structure has a Gaussian distribution of intensity as a function of the distance off axis. In this mode the refractive index of the Ti:sapphire will develop a similar radial variation so that the laser crystal becomes in effect a graded index lens like those described in Section 4.6. If the beam is a short pulse the focusing produced is sufficient to squeeze the beam through the collimator located between the crystal and mirror. By contrast a CW mode has a weaker electric field. It will not be focused and will be partially blocked by the collimator: hence it will have a reduced gain relative to that of a short pulse. The net result is that the beam circulating in the cavity quickly evolves into a single short pulse. This passive mechanism is known as *Kerr lens mode locking* (KLM).

The self-focusing of the laser beam is accompanied by self-phase modulation. In the case of Ti:sapphire which has a positive Kerr coefficient, the wave angular frequency, ω, falls as the pulse intensity rises and then rises as the intensity falls. A Gaussian pulse is used for illustration in figure 14.39. The waveform is shown in the lower panel and the frequency shift along the pulse evaluated with eqn. 14.57 is illustrated in the upper panel. Pulses like this, in which the frequency varies with time, are called *chirped* pulses. In this case the chirp is defined to be positive because the frequency rises with time. If the frequency falls with time the pulse is called negatively chirped. Both eqn. 14.57 and figure 14.39 show that chirping through self-phase modulation introduces a wider range of frequency components into the pulse than existed before.

This positive chirping is cancelled by group velocity dispersion at the prism pair and mirror shown in figure 14.38. The light emerging from the second prism is dispersed and so spread across the final mirror M. It is then possible to alter the relative path lengths in the cavity for light of different wavelength by rotating this mirror about an axis perpendicular to the diagram. If the initial pulse before chirping had a pure Gaussian distribution in time and angular frequency then the widths in time and angular frequency would satisfy eqn. 7.61, $\sigma_\omega = 1/2\sigma_t$, and it would be what is called a *bandwidth limited* pulse. After chirping the range of frequencies is increased and the pulse is no longer bandwidth limited. Compressing such a pulse until it is once again bandwidth limited produces a pulse shorter than the initial pulse with a correspondingly broader spread in frequency than that of the initial pulse. This sequence of chirping followed by compression produces shorter, and hence more intense, bandwidth limited pulses.

14.12   Frequency combs

While femtosecond pulses are of immense direct value in studying fast physical, chemical and biological processes, they also offer improvements in the precision in the measurement of frequency, and hence of time. The pioneering work of T. W. Haensch and J. L. Hall in this field and in precision spectroscopy led to them sharing the 2005 Nobel Prize for Physics with R. Glauber.[12] We start by analysing the frequency content of the mode locked laser output shown in the upper panel of figure 14.40. Pulses are spaced at intervals, t_{rep}, the photon round trip time in the cavity, and the FWHM of each pulse is called t_{p}. The carrier waves within the pulse envelopes travel at the phase velocity, ω/k, whilst the pulses themselves travel at the group velocity, $\mathrm{d}\omega/\mathrm{d}k$. The phase of the carrier wave at the peak of a pulse is called the *carrier-envelope offset*, ϕ_{ceo}. The change in this offset per round trip in the laser cavity is

$$\Delta\phi_{\mathrm{ceo}} = \int_0^L \omega(\mathrm{d}x/v_{\mathrm{g}} - \mathrm{d}x/v_{\mathrm{p}}), \qquad (14.58)$$

where L is the round trip length and $v_{\mathrm{g}}(v_{\mathrm{p}})$ the group (phase) velocity. Using eqn. 11.42 this can be rewritten

$$\Delta\phi_{\mathrm{ceo}} = \int_0^L (\omega^2/c)\,(\mathrm{d}n/\mathrm{d}\omega)\,\mathrm{d}x. \qquad (14.59)$$

It is useful to define an angular frequency

$$\omega_{\mathrm{ceo}} = \Delta\phi_{\mathrm{ceo}}/t_{\mathrm{rep}}. \qquad (14.60)$$

---

[12]See for example D. J. Jones, S. A. Diddams, J. K. Ranka, A. Stentz, R. S. Windeler, J. L. Hall and S. T. Cundiff: *Science* 288, 635 (28th April 2000); Th. Udem, R. Holzwarth and T. W. Haensch: *Nature* 416, 233 (2002).

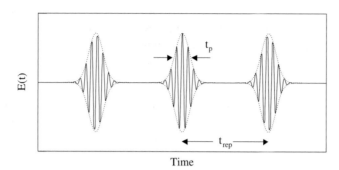

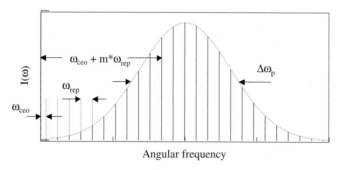

Fig. 14.40 The upper panel shows the pulse train from a mode locked laser. The lower panel shows the angular frequency components of this pulse train drawn with solid lines. The latter is called an optical comb.

Then an individual pulse has an electric field of the form

$$A(t) \exp\left[-i(\omega_c + \omega_{ceo})t\right].$$

where ω_c is the mean angular frequency of the carrier waves in the pulse, and $A(t)$ is the pulse envelope. The electric field of the whole pulse train is the convolution of the above individual pulse shape with δ-functions spaced at intervals t_{rep}

$$E(t) = A(t) \exp\left(-i\omega_c t\right) \otimes \sum_m \delta(t - mt_{rep}) \exp\left(-i\omega_{ceo}t\right). \qquad (14.61)$$

It is the equivalent frequency distribution which is of interest. The Fourier transform of the first component in this convolution is

$$\int A(t) \exp\left[i(\omega - \omega_c)t\right] \mathrm{d}t = a(\omega - \omega_c), \qquad (14.62)$$

where $a(\omega)$ is the Fourier transform of $A(t)$. For simplicity the sequence of pulses is assumed to continue indefinitely. We can then use the Poisson summation theorem to re-express part of the second component of the convolution

$$\sum_m \delta(t - mt_{rep}) = (1/t_{rep}) \sum_n \exp\left(-i\omega_{rep}nt\right), \qquad (14.63)$$

where $\omega_{\text{rep}} = 2\pi/t_{\text{rep}}$. Although this theorem requires that m and n should both be integers running from minus to plus infinity, the actual pulse trains are sufficiently long for the summation formula to be quite reliable here. Thus the whole of the second component in the convolution is

$$\sum_m \delta(t - mt_{\text{rep}}) \exp(-i\omega_{\text{ceo}}t) = (1/t_{\text{rep}}) \sum_n \exp[-i(n\omega_{\text{rep}} + \omega_{\text{ceo}})t],$$

(14.64)

whose Fourier transform is $2\pi \sum_n \delta(\omega - n\omega_{\text{rep}} - \omega_{\text{ceo}})/t_{\text{rep}}$. In Chapter 7 it was shown that the Fourier transform of a convolution is the product of transforms, hence the Fourier transform of the electric field given in eqn. 14.61 is

$$e(\omega) = 2\pi[a(\omega - \omega_{\text{c}})/t_{\text{rep}}] \sum_n \delta(\omega - n\omega_{\text{rep}} - \omega_{\text{ceo}}).$$

(14.65)

This function is displayed in the lower panel of figure 14.40, and forms what is called an *optical comb* with very narrow spectral lines at the frequencies

$$\omega = n\omega_{\text{rep}} + \omega_{\text{ceo}}.$$

The envelope is centred at the angular frequency ω_{c}, the mean angular frequency of the carrier waves. It would generally have a more complex shape than the smooth Gaussian chosen here. In terms of frequency the previous equation becomes

$$f = nf_{\text{rep}} + f_{\text{ceo}}.$$

(14.66)

14.12.1   Optical frequency measurement

The precision achieved in the measurement of wavelengths using traditional interferometry is parts in 10^{11}, and is unlikely to improve. Measurements of frequency are in principle more precise because they only involve counting pulses in one second. The wavelength is then easy to deduce using $\lambda = c/f$. The optical comb provides a simple and elegant method for measuring the frequency of any stabilized optical source.

The measurement involves comparison with the internationally recognized frequency reference provided by the ^{133}Cs transition at 9.19 GHz. There are three steps in the procedure, all involving counts made at microwave frequencies with standard high speed electronics calibrated against the ^{133}Cs microwave transition.

In the first step light from the stabilized laser source at frequency f_{opt} is beaten against light from the Ti:sapphire laser. A crude frequency selection only is needed to pick out the *tooth* in the comb nearest in frequency to the stabilized source. This mth tooth is at a frequency $f_{\text{ceo}} + mf_{\text{rep}}$, so the beat frequency is

$$f_{\text{beat}} = f_{\text{ceo}} + mf_{\text{rep}} - f_{\text{opt}}.$$

(14.67)

The repetition frequency f_{rep} depends on the cavity length, and for a 1.5 m long folded cavity the round trip time would be 10 ns, making f_{rep} equal to 100 MHz. The difference frequency f_{beat} will also be a microwave frequency so that both f_{beat} and f_{rep} can be measured electronically. Evidently f_{ceo} is also required before f_{opt} can be determined. It can be measured by producing a second harmonic of one of the low frequency teeth and beating this against a high frequency tooth.[13] Suppose these teeth are the nth and the $2n$th, then the beat frequency is

$$2(nf_{rep} + f_{ceo}) - (2nf_{rep} + f_{ceo}) = f_{ceo}, \qquad (14.68)$$

which is also in the microwave range and can be measured electronically. Inserting the measured microwave frequencies, f_{beat}, f_{rep} and f_{ceo}, into eqn. 14.67 gives f_{opt}.

In Chapter 18 we shall examine how the transitions of an isolated ion are used to provide a reference frequency for a stabilized laser. An optical comb is then used to transfer the measurement of laser frequency to the microwave region. This total system forms an optical clock. In such a clock the analogue of the pendulum swing is one cycle of the electromagnetic radiation from the ion. All the other components are the equivalent of the escapement and gear train.

14.13   Extreme energies

The amplification achievable within any laser is limited by the heating and the mechanical stresses produced by the high electric fields. In 1985 Strickland and Mourou[14] devised a way to circumvent this difficulty: the technique is known as *chirped pulse amplification* (CPA). Three steps are required and these are illustrated in figure 14.41, which is based on the design of the ASTRA facility at the Rutherford Appleton Laboratory. A short pulse from a laser is first chirped and stretched by a large factor. At ASTRA an input pulse 20 fs long is stretched to 530 ps so that the intensity is reduced more than ten thousandfold. The pulse stretcher shown uses a pair of diffraction gratings and steering mirrors. The pulse is diffracted and reflected back along its original path in such a way that the highest frequency (*blue*) component of the pulse has travelled 20 cm further than the lowest frequency (*red*) component. Next this positively chirped pulse is passed through several Ti:sapphire rods pumped by a Nd:YAG laser, from which it emerges with high intensity. In the third step the pulse is compressed as shown using parallel metal gratings arranged so that the red component now travels the longest path. During compression the intensity of the pulse is amplified by the compression factor.

---

[13]We will be meeting techniques for generating second harmonics later in this chapter.

[14]D. Strickland and G. Mourou: *Optical Communications* **56**, 219 (1985).

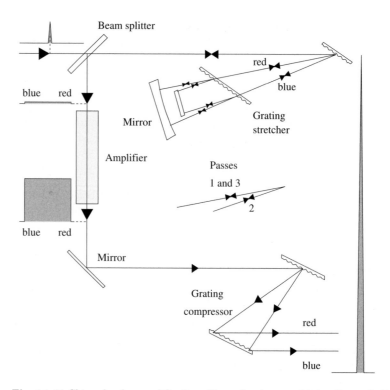

Fig. 14.41 Chirped pulse amplification. The pulse shape and intensity are indicated at each stage. The inset indicates the sequence of the passes which the beam follows in the stretcher. Courtesy Professor P. A. Norreys and CCLRC Rutherford Appleton Laboratory.

The ASTRA facility at the Rutherford Appleton Laboratory produces some of the most powerful laser pulses using CPA. Typically a nanojoule 20 fs pulse from a mode locked Ti:sapphire laser is stretched to 530 ps. Pulses are amplified from 1 nJ to 1.5 J by Ti:sapphire crystals pumped by Nd:YAG lasers. After compression to 40 fs the pulses have peak powers of typically 40 TW. If this degree of amplification were applied directly to an unstretched pulse the pulse would destroy the amplifier crystals. This output pulse is indeed so intense that the path of the final beam has to be evacuated to prevent self-focusing in the air. Focused intensities of $10^{18}\,\mathrm{W\,cm^{-2}}$ are easily attained. Programs are underway in several laboratories to attempt to ignite nuclear fusion by imploding and heating tritium/deuterium pellets with similar intense laser beams.

14.14   Second order non-linear effects

We now return to analyse the second order non-linear effects that are important in non-centrosymmetric crystals. For simplicity any higher order effects are ignored in this analysis. Then taking the laser electric field to be $E\cos(\omega t)$ the non-linear polarization is

$$P_{NL} = \varepsilon_0\chi^{(2)}E^2\cos^2(\omega t) = \varepsilon_0\chi^{(2)}E^2\left[1+\cos(2\omega t)\right]/2. \quad (14.69)$$

This result contains an *optically rectified* term $\chi^{(2)}E^2/2$ as well as a second harmonic component. The quantum interpretation of the second harmonic component is that two photons of angular frequency ω are absorbed and a single photon of angular frequency 2ω is emitted. This process is used to convert the electromagnetic radiation at 1064 nm wavelength present in a Nd:YAG laser to radiation at 532 nm wavelength by placing a suitable crystal with a high value of $\chi^{(2)}$ within the laser cavity. Its location would be that indicated for the Pockels cell in figure 14.31. The output is proportional to E^2 so that a pulsed laser will give a higher yield of the second harmonic radiation than a CW laser of equal mean power. A useful second harmonic component can also be obtained from some semiconductor lasers by operating them at high power.

Dispersion in the crystal convertor causes the harmonic waves generated at angular frequency 2ω to travel at a different phase velocity from the parent waves at angular frequency ω. Consequently the waves at angular frequency 2ω which originate at different locations along the beam's path will be out of phase with one another and in long crystals there will be destructive interference. This difficulty can be finessed by using a convertor crystal which is birefringent: wave velocities at the two wavelengths can then be made equal provided that it can be arranged that one has ordinary, and the other has extraordinary polarization. We specialize to the case of a negative uniaxial crystal, β-barium borate (BBO), for which $n_e < n_o$. Figure 14.42 shows its refractive indices as a function of wavelength. At each wavelength the extraordinary refractive index takes a value somewhere between the two extremes depending on the direction of the wave vector with respect to the crystal axis. The star on the diagram indicates the choice of the refractive index for extraordinary waves at 532 nm wavelength which is equal to the refractive index of ordinary waves at 1064 nm wavelength. How this choice is implemented is demonstrated with the index ellipsoids at the two wavelengths shown in figure 14.43. The wave vector direction drawn there makes the refractive indices of extraordinary waves at 532 nm wavelength and ordinary waves at 1064 nm wavelength equal to each other. Referring back to Section 10.4.2 and figure 10.5 may be helpful here. This is called *type I* phase matching, and schematically the polarization choices can be expressed in equivalent ways

$$o(2\lambda) + o(2\lambda) \rightarrow e(\lambda),$$
$$o(\omega) + o(\omega) \rightarrow e(2\omega),$$

where for example $o(\omega)$ indicates a wave with ordinary polarization at angular frequency ω. The alternative *type II* phase matching is obtained with the combination

$$o(\omega) + e(\omega) \rightarrow e(2\omega).$$

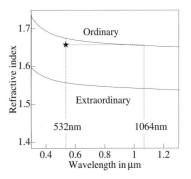

Fig. 14.42 Ordinary and extraordinary refractive indices for BBO. Type I phase matching is obtained at the position of the star for frequency doubling of a Nd:YAG laser beam.

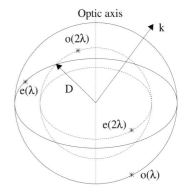

Fig. 14.43 Index ellipsoids for BBO at wavelengths 532 nm and 1064 nm. The orientation of the wave vector is that required to give type I phase matching when frequency doubling a Nd:YAG laser beam.

Second harmonic generation in BBO is efficient: in practice 50% of the input energy can be converted to second harmonic energy. The direction the wave vectors make with the optic axis can be obtained using eqn. 10.23. The refractive index of the extraordinary wave at angular frequency 2ω, $n(2\omega)$ is given by

$$1/n^2(2\omega) = \cos^2\theta/n_{\mathrm{o}}^2(2\omega) + \sin^2\theta/n_{\mathrm{e}}^2(2\omega).$$

For type I phase matching θ is selected such that $n(2\omega)$ is the same as the ordinary refractive index at frequency ω, $n_{\mathrm{o}}(\omega)$. Then the previous equation can be rewritten

$$1/n_{\mathrm{o}}^2(\omega) = \cos^2\theta/n_{\mathrm{o}}^2(2\omega) + \sin^2\theta/n_{\mathrm{e}}^2(2\omega).$$

Rearranging this gives the angle between the wave vector and the crystal axis needed to achieve phase matching

$$\sin^2\theta = [\, n_{\mathrm{o}}^{-2}(\omega) - n_{\mathrm{o}}^{-2}(2\omega)\,] \,/\, [\, n_{\mathrm{e}}^{-2}(2\omega) - n_{\mathrm{o}}^{-2}(2\omega)\,], \qquad (14.70)$$

which was used to create figure 14.42.

Another useful technique for generating a coherent beam of a new frequency is to direct two intense laser beams of two different frequencies at a convertor crystal with a high value of $\chi^{(2)}$, such as β-barium borate. Suppose that these beams have angular frequencies ω_1 and ω_2, and wave numbers $\mathbf{k}_1$ and $\mathbf{k}_2$ respectively. Then the resulting second order non-linear component of polarization within the crystal is

$$P_{\mathrm{NL}} = \varepsilon_0 \chi^{(2)} \,[\, E\cos(\mathbf{k}_1\cdot\mathbf{r} - \omega_1 t) + E\cos(\mathbf{k}_2\cdot\mathbf{r} - \omega_2 t)\,]^2. \qquad (14.71)$$

In this product there are components of angular frequencies $2\omega_1$, $2\omega_2$, $\omega_1 + \omega_2$ and $\omega_1 - \omega_2$. The term of interest here is

$$P_{\mathrm{NL}}(\omega_1 + \omega_2) = \varepsilon_0 \chi^{(2)} E^2 \,[\cos[\,(\mathbf{k}_1 + \mathbf{k}_2)\cdot\mathbf{r} - (\omega_1 + \omega_2)t\,], \qquad (14.72)$$

which shows that a new beam at angular frequency ω_3 and wave number $\mathbf{k}_3$ can be produced for which

$$\omega_3 = \omega_1 + \omega_2, \qquad (14.73)$$
$$\mathbf{k}_3 = \mathbf{k}_1 + \mathbf{k}_2, \qquad (14.74)$$

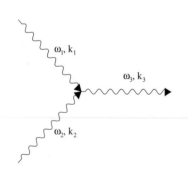

Fig. 14.44 Photon energy–momentum balance in a convertor crystal.

provided that ω_3/k_3 is equal to the phase velocity of electromagnetic waves at angular frequency ω_3 in the crystal. Equations 14.73 and 14.74 are known as the *phase matching* conditions. In quantum terms they specify that energy and momentum are conserved when photons of energy–momentum $\hbar(\omega_1, \mathbf{k}_1)$ and $\hbar(\omega_2, \mathbf{k}_2)$ are absorbed and one with energy–momentum $\hbar(\omega_3, \mathbf{k}_3)$ is emitted. The process is displayed in figure 14.44. If there is perfect phase matching then the phase of a wave created at a depth x and at time t in a crystal of thickness L will, on emerging at time T from the end of the crystal, be

$$[\,(k_1 + k_2)x - (\omega_1 + \omega_2)t\,] + [\,k_3(L - x) - \omega_3(T - t)\,] = k_3 L - \omega_3 T.$$

As a consequence the waves generated at all depths in the crystal are all in phase as they exit the crystal.

Some conversion will still take place even if phase matching is not perfect. In this situation the wave at frequency ω_3 produced at a vector distance $\mathbf{x}$ measured from the mid-point along the crystal will be out of phase with a wave produced at the mid-point by $(\mathbf{k}_1 + \mathbf{k}_2 - \mathbf{k}_3) \cdot \mathbf{x}$. The magnitude of this wave produced in a layer of thickness $\mathrm{d}x$ at a distance x from the mid-point is

$$\mathrm{d}A = \cos\left(\Delta k x\right)\mathrm{d}x,$$

where $\Delta k x = (\mathbf{k}_1 + \mathbf{k}_2 - \mathbf{k}_3) \cdot \mathbf{x}$. Integrating over the crystal thickness

$$
\begin{aligned}
A &= \int_{-L/2}^{+L/2} \cos\left(\Delta k x\right)\mathrm{d}x \\
&= (2/\Delta k L)\sin\left(\Delta k L/2\right)L \\
&= L\operatorname{sinc}(\Delta k L/2).
\end{aligned}
\tag{14.75}
$$

With this degree of mismatch the intensity is

$$I(\Delta k) = I(0)\operatorname{sinc}^2(\Delta k L/2), \tag{14.76}$$

and the ratio $I(\Delta k)/I(0)$ is plotted in figure 14.45. The net conversion rate can be maximized by choosing a crystal length such that

$$L_{\mathrm{c}} = \pi/\Delta k, \tag{14.77}$$

which is sometimes called the coherence length.

The second order non-linear effect can equally well be used to generate radiation at angular frequencies ω_1 and ω_2 from a beam of angular frequency ω_3. The phase matching conditions deduced above also apply to this process. Devices based on the reverse process are sketched in figure 14.46: the upper panel shows a *parametric* amplifier, and the lower panel a *parametric* oscillator. In both cases a crystal such as BBO is the conversion material. The *pump* beam at angular frequency ω_3 provides the input power. In the parametric amplifier a separate weak input *signal* beam at angular frequency ω_1 is amplified. There is no input signal beam to the parametric oscillator; its signal beam develops as a standing wave in the cavity shown. An accompanying *idler* beam at angular frequency ω_2 is produced in both the parametric amplifier and in the oscillator. Its direction and energy are determined by the phase matching conditions.

Optical parametric devices are employed to generate coherent, high intensity beams at frequencies for which there are no suitable lasers. The only limitation on their frequency coverage is that the crystal employed should be transparent to all three beams: pump, signal and idler.

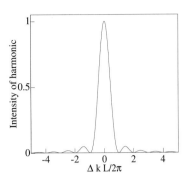

Fig. 14.45 Intensity of harmonic generation as a function of the phase mismatch in a crystal such as BBO.

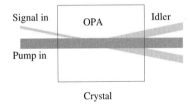

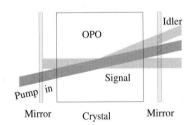

Fig. 14.46 Outline diagrams for the function of an optical parametric amplifier and an optical parametric oscillator.

Lithium triborate and BBO are two crystals widely used in parametric amplifiers, being transparent over the ranges 0.16–2.60 μm and 0.19–3.5 μm respectively. Tuning of the signal and idler frequencies over a wide range is rather easy: the crystal is rotated so that the angle between its optic axis and the beam axis is altered and this alters the phase matching conditions. Heating in parametric devices is much less than in a laser, hence signal beams of very high power can be produced by pumping the crystal with laser beams. In applications that require a pair of entangled photons the idler and signal beams assume equal importance. This important topic will be developed in Chapter 19.

Third order non-linear processes give rise to analogous four wave mixing in which the incident waves of angular frequency ω_1 and ω_2 produce waves of angular frequencies ω_3 and ω_4, satisfying the condition

$$\omega_1 + \omega_2 = \omega_3 + \omega_4. \tag{14.78}$$

Alternatively three incident waves can mix to produce a fourth. Four wave mixing has made itself evident in telecom applications when typically 40 optical beams equally spaced in frequency are transmitted long distances down the same optical fibre. Four wave mixing then causes cross-talk between the information carried on adjacent sets of beams. The way this cross-talk is avoided will be discussed in the following chapter.

14.14.1   Raman scattering

Rayleigh scattering from matter is accompanied by related weaker processes called Raman and Brillouin scattering which, unlike Rayleigh scattering, involve a change in the internal energy of the scatterer. Raman scattering, which we treat first, can involve electronic excitation of atoms, or vibrational and rotational excitation of molecules, or the excitation of larger structures. The transitions involved in Raman scattering from a molecule are exhibited in figure 14.47. There is a two step process: a photon is absorbed raising the molecule to a higher energy state and a second photon is immediately emitted leaving the molecule in a quantum state differing from its initial state only in its vibrational/rotational energy. If the emerging photon has more (less) energy than the incident photon the transition is known as an *anti-Stokes* (*Stokes*) transition. The energy levels drawn with full lines in the upper panel are molecular rotational/vibrational levels. Broken lines are used to denote the intermediate states. As shown in the lower panel of figure 14.47, the Stokes and anti-Stokes lines are equally displaced from the Rayleigh line. Of the pair the Stokes line has the higher intensity because the initial state for this transition has the lower energy and hence the larger population. In the quantum picture all three scattering types are two step processes involving a virtual intermediate state.[15] In transparent materials from

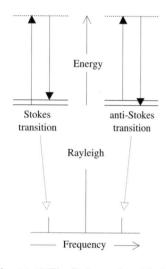

Fig. 14.47 The Stokes and anti-Stokes transitions are shown in the upper panel; the lower plot is of the frequencies of these lines and the Rayleigh line.

---

[15]See for example Chapter 8 of the second edition of *The Quantum Theory of Light* by R. Loudon, published by Oxford University Press (2000).

which the scattering occurs ultraviolet radiation is required to produce excited states of the constituent atoms. It is only the extreme tails of these states that are accessed when light or infrared radiation is incident on such materials. Being so far off resonance the resulting intermediate states have very short lifetimes and are described as *virtual* states.

The energy change produced in a nitrogen gas molecule by Raman scattering is ~0.3 eV or ~1.0 meV depending on whether its vibrational or rotational quantum state changes. The corresponding frequency shifts between the incident and scattered radiation are ~70 THz and ~230 GHz. Again specializing to nitrogen gas, the intensities of the vibrational and rotational Raman lines are respectively 0.07% and 2.3% of the Rayleigh intensity. Raman scattering has become relatively simple to study with laser beams: they have the narrow bandwidth that permits easy discrimination between Raman and Rayleigh scattering; and they have sufficient intensity to excite copious Raman scattering.

Raman scattering provides information on molecular states that is complementary to information obtained by standard spectroscopic techniques. The reason is that the overall change in quantum numbers in Raman scattering, being a two step process, can circumvent the selection rules for a single direct transition between the initial and final state. Properties of such otherwise inaccessible states can be studied following their excitation by Raman scattering.

The vibrational excitations within condensed matter can also be treated quantum mechanically, and the quanta of this mechanical energy are known as *phonons*.[16] Within the fused silica of optical fibre a very large number of possible vibrational and rotational excitations are available thanks to the variability in the chemical bonds between the atoms in this amorphous material. Raman scattering involves the exchange of a phonon between the photon and *non-propagating* or *optical* modes of excitation of the amorphous structure. In these modes adjacent ions vibrate in anti-phase which gives the pairs a large electric dipole moment. Interaction with light and infrared radiation is thus enhanced. The intensity of the light scattered from silica is plotted against the energy change in figure 14.48. The prominent Stokes' peak is displaced by about 0.055 eV or 13.2 THz in frequency from the Rayleigh line. Raman scattered light builds up in intensity over the long distances traversed in telecom fibres to such an extent that *stimulated* Raman scattering can become important.

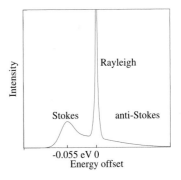

Fig. 14.48 The spectrum of light after traversing a long span of optical fibre.

---

[16] These entities have already been mentioned in connection with indirect band-gap transitions.

14.14.2   Brillouin scattering

In contrast to Raman scattering Brillouin scattering involves the transfer of energy to *acoustic* modes of vibration in the material as distinct from the optical modes excited in Raman scattering. In these modes adjacent ions oscillate with similar phase, so the pair's induced electric dipole moment is small.[17] The quanta of acoustic modes have thermal energies $\sim 0.024\,\text{eV}$ at room temperature. Absorption and emission of phonons in Brillouin scattering produces Stokes and anti-Stokes lines having displacements from the Rayleigh scattering line which are orders of magnitude smaller than in Raman scattering. Brillouin scattering can be pictured in terms of electromagnetic waves scattering from acoustic waves. This was the framework used in Section 7.6 to analyse scattering from the uniform acoustic waves produced by a piezo-electric crystal. In that analysis the uniform acoustic wave patterns were shown to Bragg scatter the incident light. More usually a laser beam passing through matter is scattered from a multiplicity of acoustic waves that carry the thermal energy of the material.

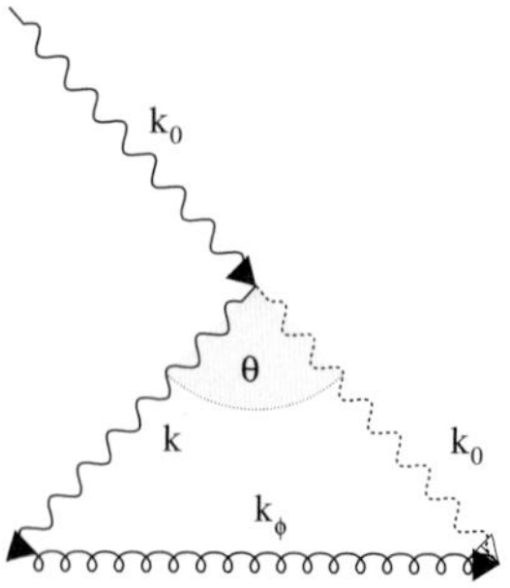

Figure 14.49 shows an incoming photon of frequency f and wave number $\mathbf{k}_0$ being scattered by an acoustic wave with acoustic velocity v_a in a material of refractive index n. The phonon exchanged has energy E_ϕ, momentum $\mathbf{p}_\phi$, wave vector $\mathbf{k}_\phi$. After the exchange the photon emerges with wave vector $\mathbf{k}$, travelling at an angle θ to its original direction. The exchange of an acoustic phonon hardly changes the photon energy so that $k \approx k_0$ and then applying conservation of momentum to the exchange gives

$$k_\phi = 2k_0 \sin\left(\theta/2\right).$$

Being massless the phonons have energy given by

$$
\begin{aligned}
E_\phi = v_\text{a} p_\phi &= v_\text{a} \hbar k_\phi \\
&= 2 v_\text{a} \hbar k_0 \sin\left(\theta/2\right) \\
&= 2 v_\text{a} \sin\left(\theta/2\right) \left[\,hn/\lambda\,\right],
\end{aligned}
\tag{14.79}
$$

where λ is the free space wavelength of the electromagnetic radiation. Here this result will be applied to backward scattering in an optical fibre. We take the acoustic wave velocity to be $6000\,\text{m s}^{-1}$, the refractive index to be 1.5, and the wavelength in free space to be $1.5\,\mu\text{m}$. Then the frequency shift

$$\Delta f = E_\phi / h = 2 v_\text{a} n / \lambda,$$

is $12\,\text{GHz}$. Brillouin scattering in an optical fibre contributes to the central scattering peak in figure 14.48, which includes the Rayleigh scattering and near elastic scattering from the density and chemical fluctuations

Fig. 14.49 Brillouin scattering.

---

[17]See Chapter 2 of the second edition of *Solid State Physics* by J. S. Blakemore, published by Cambridge University Press (1985), for details on the acoustic and optical modes of crystals.

in the fused silica. In a gas at room temperature and pressure the Brillouin scattering involves frequency shifts that are around 1 GHz from the Rayleigh scattering peak.

14.14.3   Stimulated Raman and Brillouin scattering

Despite their low rates both Brillouin and Raman scattering impose maximum limits on the beam intensities used in optical fibre communications. The scattered radiation builds up continuously along the multi-kilometre spans of single mode fibre so that eventually stimulated emission initiated by this scattered radiation commences. Any further increase in the injected power into the fibre simply fuels the stimulated Brillouin or Raman scattering (SBS or SRS). A widely used arrangement described in Chapter 16 is to transmit multiple beams of closely spaced wavelengths, each carrying separate information, along a single fibre. Stimulated Brillouin scattering can then cause interference between channels and degrade the information carried.

The acoustic waves from which Brillouin scattering occurs continue for several nanoseconds so that the Brillouin scattering of a laser beam is strong when its coherence time is as long or longer. That is to say, the bandwidth should be less than $\sim 50\,\text{MHz}$. Only radiation scattered close to the forward and backward directions is trapped in a single mode fibre core. That in the exact forward direction has zero frequency shift and can be safely neglected. The remaining Brillouin scattering in single mode optical fibre is dominantly backward, which reduces some of its impact on communications. Over multi-kilometre paths the threshold for SBS is around 10 mW: excess beam intensity is lost to SBS and the useful beam intensity is clamped near this threshold value. Fortunately this threshold can be raised by using a broad bandwidth source, so that with $\sim 1\,\text{GHz}$ bandwidth the threshold rises to $\sim 1\,\text{W}$. Stimulated Raman scattering has a corresponding threshold of $\sim 2\,\text{W}$ which is independent of the laser bandwidth because this scattering is from optical phonons. Raman scattering is however predominantly forward.

On the positive side high optical power can be pumped into fibres in order to amplify or excite lasing through stimulated Brillouin or stimulated Raman scattering. Such lasers are relatively easy to integrate into optical fibre systems. It will be shown in Chapter 16 how stimulated Raman scattering can also be used to compensate the attenuation experienced by laser beams travelling very long distances in optical fibre.

14.15   Further reading

Laser Fundamentals, second edition by W. T. Silfvast, published by Cambridge University Press (2004). This text provides a comprehensive and up to date account of many types of modern lasers.

Detailed analyses of the properties of cavities used for lasers and many other topics in laser physics appear in the fifth edition of *Principles of Lasers* by O. Svelto, published by Plenum Press, New York (1998).

The Physics of Semiconductor Devices, fourth edition, by D. A. Fraser, published by the Clarendon Press, Oxford (1986). This provides a clear and concise account of the semiconductor physics underlying semiconductor lasers.

Exercises

(14.1) The FWHM of the Doppler line of a He:Ne laser is 1.5 GHz. The cavity length is 50 cm and the reflectance of the mirrors is 0.99. What is the cavity finesse? How many longitudinal modes lie within the FWHM of the Doppler line?

(14.2) A laser rod is to be pumped by a linear flash lamp mounted parallel to it and of similar length and aspect ratio. Flexible metal sheeting is used to reflect the light from the lamp onto the rod. How should the tube and lamp be positioned and into what shape should the sheeting be formed to maximize the pumping efficiency?

(14.3) The mirrors of a one metre long Fabry–Perot cavity have reflectance $(1-10^{-6})$. Calculate the cavity ring-down time. What is the alteration in the ring-down time if the cavity is filled with a gas whose absorption coefficient is $10^{-6}\,\mathrm{m}^{-1}$?

(14.4) The values of the refractive indices of BBO relevant to frequency doubling the output of a Nd:YAG laser are: $n_o(532) = 1.675$, $n_e(532) = 1.555$, $n_o(1064) = 1.655$, $n_e(1064) = 1.542$. Calculate the walk-off angle between the frequency doubled beam and the pump beam in the BBO crystal.

(14.5) Obtain an expression relating the Fabry–Perot cavity ring-down time and the threshold gain for the same cavity used as a laser.

(14.6) A He:Ne laser with a 40 cm long cavity has mirrors of reflectance 0.99. Calculate the threshold gain at the central wavelength of the gain profile 632.8 nm. Hence estimate the population inversion in atoms per m^3. Assume that the absorption coefficient is zero, that the upper state lifetime in the lasing transition is $10^{-7}\,\mathrm{s}$ and that the gain profile FWHM is 2 GHz.

(14.7) A Nd:YAG laser has a ring=down time, τ_c, of 20 ns, a gain profile of width 200 GHz and the lifetime of the excited state against spontaneous decay is $230\,\mu\mathrm{s}$. Estimate the population inversion in ions per m^3 required to produce lasing. The refractive index is 1.5.

(14.8) Show for a He:Ne laser whose thermal energy corresponds to a temperature of 400 K that the Doppler broadened profile has a FWHM of 1.5 GHz.

(14.9) Why are the lasing transitions in Nd:YAG and in Nd:glass respectively homogeneously and inhomogeneously broadened?

(14.10) Is it possible for lasing to occur when the population of the lower level is greater than that in the upper state?

(14.11) Calculate the frequency shift in light of wavelength $1\,\mu\mathrm{m}$ which undergoes backward Brillouin scattering in a crystal. Take the velocity of acoustic waves to be $3000\,\mathrm{m\,s}^{-1}$ and the refractive index to be 1.5.

(14.12) A laser system used to generate an optical comb emits a stream of 800 nm pulses 20 ps long with a 100 MHz repetition rate. What is the frequency separation of the teeth of the comb? How broad in frequency is the envelope of the comb?

(14.13) Suppose the quantum well of a quantum well laser is sufficiently deep that the electron wavefunction has nodes very close to the walls of the well. The well thickness, taken to lie in the z-direction, is L_z. Show that the the kinetic energy of an electron in the well in the z-direction is $n^2h^2/(8m^*L_z^2)$ where m^* is the electron's effective mass and n is an integer.

(14.14) Consider a semiconductor in which the conduction band electrons have an effective mass $0.1m_e$ and the holes an effective mass $0.5\,m_e$ in the valence band. Calculate the change in lasing transition energy and frequency between a DH laser and a QW laser for which the well is 14 nm thick.

Detectors

15

15.1 Introduction

The two main processes used to detect electromagnetic radiation at visible and near visible wavelengths both involve single photon absorption on a single electron. One process is the photoemission of an electron from the surface of an alkali metal or semiconductor; the other is the excitation of an electron from the valence to the conduction band of a semiconductor. Both processes produce free charge carriers so that an applied electric potential causes a current to flow under continuous illumination. This *photocurrent* is in principle linearly proportional to the intensity of the incident light and in practice linearity is maintained over a wide range of intensity.

Several physical properties determine the usefulness of *photodetectors*. These properties include the sensitivity in responding to light, measured in amps of current per watt of optical power incident, and how this sensitivity varies with wavelength. There is a small *dark current* which flows even if there is no incident radiation, and this constitutes a background to any measurement. The speed of response of the phodetector to changes in illumination determines what rate of data transfer is achievable with the detector. Finally there is noise on any measurement with a photodetector due both to the random arrival of photons, and to thermal excitation of electrons in the detector and its associated circuitry. Both dark current and noise impose limits on the usefulness of detectors when the radiation levels are weak. All these properties are discussed as needed for each detector type. In several types of detector the applied potential is made sufficiently large so that a free charge created by the initial photon absorption acquires enough energy to produce a cascade of charges in collisions with atoms. As a result a measurable electrical pulse can even be produced following the absorption of a single photon.

Semiconductor detectors are described immediately below, commencing with the simpler photoconductors. This case is used to introduce the parameters used to quantify light detection efficiency: these are the quantum efficiency and responsivity. The next section is used to discuss photodiodes, which are widely used as detectors in receiving signals carried on electromagnetic radiation transmitted along optical fibres. Possible sources of noise will be enumerated and analysed, and the various parameters used for quantifying noise will also be described. A section on solar cells follows. Avalanche photodiodes are discussed next: in these devices high electric fields are used to accelerate the photoproduced free charges to energies large enough that they ionize atoms inside the photodiode, thus creating an avalanche of charges. Avalanche photodiodes therefore have better sensitivity at low light levels than simple photodiodes. Another type of photodiode, the Schottky photodiode, forms the next topic. Schottky photodiodes have very fast responses and are therefore used for very high speed communications. The final topic on semiconductor detectors concerns the two-dimensional arrays of photodiodes used as sensors in digital cameras and telescopes.

The remainder of this chapter is used to describe detectors which rely on photoemission; the first step in detection being the ejection of electrons from the surface of a *photocathode*, which is an alkali metal or semiconductor surface with a low work function. In a photomultiplier the *photoelectrons* are accelerated toward another electrode, called a *dynode*, from which they eject several secondary electrons. This step of acceleration and multiplication is repeated at a sequence of up to 14 dynodes, so that as many a 10^8 electrons may be produced from a single incident photon. Photomultipliers are fast and very sensitive, well suited for use in measuring very low light levels and for counting individual photons. Microchannel plates are compact versions of the amplifying dynode stages in photomultipliers and in addition have imaging capabilities. They and their application in image intensifiers and night vision devices will be treated at the end of the chapter.

15.2   Photoconductors

These simple detectors consist of a semiconductor across which a voltage is applied. In the dark they have a high resistivity. When an incident photon whose energy exceeds the band-gap is absorbed it can raise an electron from the valence to the conduction band, leaving a hole behind. This process provides charge carriers that increase the conductivity of the semiconductor. If the band-gap energy is E_g then only light of wavelength shorter than λ_c is effective, where

$$\lambda_c = hc/E_g = 1.24 \ \mu\text{m}/[\,E_g \text{ in eV}\,]. \tag{15.1}$$

Direct and indirect band-gap semiconductors are equally useful. The band-gaps are 1.11 eV in silicon and 1.43 eV in gallium arsenide. A useful measure of the conversion probability is the *quantum efficiency*, which is simply the number of electrons which pass into the external circuit from the detector divided by the number of photons incident. Clearly if the photon energy is less than the band-gap energy this is identically zero. If each photon absorbed produces one electron then the quantum efficiency, η, is

$$\eta = (1 - R)\,[\,1 - \exp(-\alpha t)\,]. \tag{15.2}$$

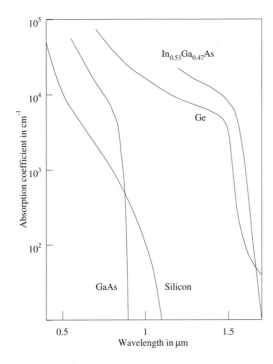

Fig. 15.1 Absorption coefficients of several semiconductors: adapted from J. M. Senior, *Optical Fibre Communications: Principles and Practice*, second edition 1992, published by Pearson/Prentice-Hall International, Englewood Cliffs, New Jersey. Courtesy Dr Senior and Pearson Publishing.

In this equation the first factor is the fraction of light transmitted through the semiconductor surface of reflectance R. The second factor is the fraction of the light entering the semiconductor which is absorbed in its thickness t, α being the absorption coefficient for the radiation incident. Figure 15.1 shows the wavelength dependence of the absorption coefficients of representative semiconductors. On crossing the threshold wavelength the absorption coefficient, α, rises sharply for a direct band-gap semiconductor, such as gallium arsenide, and then rises less rapidly as the wavelength decreases further. In the case of silicon, an indirect band-gap semiconductor, the discussion in Section 14.7 shows that momentum conservation in the absorption process cannot be achieved exactly at threshold without involving a phonon. Hence the absorption coefficient in silicon rises much more slowly with decreasing wavelength. This has the consequence that the absorption length in silicon changes a lot across the visible spectrum, being $\sim 5\,\mu$m at 700 nm and $\sim 0.2\,\mu$m at 400 nm. Recall that 90% of the radiation is absorbed in 2.3 absorption lengths. The quantum efficiency given by eqn. 15.2 is an upper limit. It does not take into account losses of electrons and holes which are trapped by crystal defects and then recombine. Such defects are especially dense near the surface, making this loss severe when the absorption length is

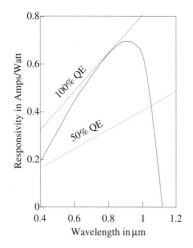

Fig. 15.2 Responsivity of a silicon photoconductive cell. The straight lines indicate the responsivity with 50% and 100% quantum efficiency.

Photoconductor separating electrodes

Positive Electrode

Negative Electrode

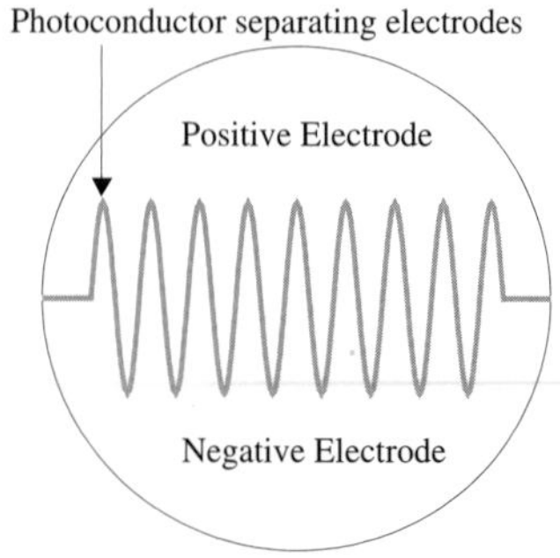

Fig. 15.3 Photoconductive cell viewed face on.

a fraction of a micron or less.

The current produced by a flux of ϕ photons per second is

$$I = \phi e \eta G, \tag{15.3}$$

which includes a factor G called the *photoconductive gain*. This gain can be greater or less than unity and depends on two response times: the first is the time taken for a free charge to recombine in the bulk semiconductor, t_r, and the second is the time taken by a charge to travel between the electrodes supplying the applied voltage, t_x. Suppose that t_r exceeds t_x, then the photoproduced charges leave the detector before recombining. As each departs through the electrical contact a similar charge is drawn in from the other electrical contact in order to keep the semiconductor neutral. This repeats until such time as the daughter or further descendant recombines. Thus the current is enhanced by a factor

$$G = t_r/t_x, \tag{15.4}$$

and this gain can in practice reach several hundreds. On the other hand if t_r is less than t_x only a fraction of the charges escape recombination before they reach the electrodes.

The incident optical power in a flux of ϕ photons per second at a wavelength λ is

$$P = \phi(hf) = \phi hc/\lambda. \tag{15.5}$$

The sensitivity of a detector is quantified by the *responsivity*, defined as the current per unit incident optical power

$$S = I/P = Ge\eta\lambda/(hc). \tag{15.6}$$

In figure 15.2 the responsivity is plotted for a silicon photoconductor, taking $R = 0$ and $G = 1$. Two straight lines are drawn which correspond to constant quantum efficiencies of 100% and 50%. The linear rise of responsivity with wavelength obtained at constant quantum efficiency comes about because, as the wavelength rises there are more photons per watt of beam power. Near threshold at $1.11\,\mu$m the responsivity of a silicon detector rises rapidly and peaks at approximately 850 nm. As the wavelength falls further the absorption becomes concentrated in the surface layer and then trapping of free charges at surface crystal defects and their subsequent recombination drastically reduces the number available to contribute to conduction. Thus the responsivity falls off rapidly at shorter wavelengths.

Cadmium sulphide has a responsivity which closely matches the shape of the response of the human eye as a function of wavelength and for this reason has been widely used for controlling street light switches.[1]

---

[1]Its days may be numbered because the EU has issued a directive aimed at reducing the usage of heavy metals including cadmium. This directive bans the deployment of such photosensors.

The dark resistance of a typical photoconductive cell is $\sim 1\,\text{M}\Omega$ while in daylight this drops to $\sim 1\,\text{k}\Omega$. Alloys of HgCdTe have band-gaps which are narrower. By varying the relative proportions of cadmium and tellurium photoconductive cells can be designed using these alloys to detect radiation at any wavelength over the range 1–$25\,\mu\text{m}$.

The photoconductor cell shown in figure 15.3 is suitable for use at wavelengths for which the absorption length is short. A layer of photoconductor is deposited on a ceramic base and two metal electrodes are evaporated above this with an *interdigitated* gap left between them. This shape maximizes the surface area while minimizing the electrode separation. A compact device results whose resistance under illumination is low. The surface is covered with a transparent anti-reflecting coating which also protects the semiconductor. A light meter circuit based on such a cell is pictured in figure 15.4.

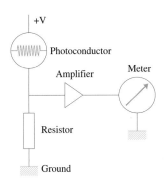

Fig. 15.4 Photoconductive cell light meter.

15.3 Photodiodes

The commonest detectors for visible and near visible wavelength radiation are photodiodes of various types. They operate like semiconduc-

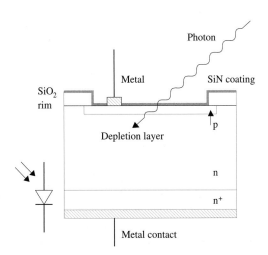

Fig. 15.5 A section through a photodiode. At the lower left is the symbol used for a photodiode in electronics.

tor LEDs or lasers in reverse: light incident on the *depletion layer* at a diode junction produces current carriers in the form of electron–hole pairs. Direct and indirect band-gap semiconductors are equally useful in photodiodes as in photoconductors. Figure 15.5 shows a section through a silicon photodiode: the symbol at the bottom left is that used for photodiodes in electronics. With no light incident there is a small dark current through the detector due to the high energy tail of thermally excited electrons crossing the band-gap. Figure 15.6 shows the junction region of such a photodiode under a reverse bias V. The upper panel shows the electron energy levels, the centre panel shows the electric field distribution and the lower panel the charge distribution. For this example the density of acceptor atoms in the p-doped region is chosen to be twice that of the donor atoms in the n-doped region. Typically the junction voltage in the absence of any bias voltage, V_j, is around 0.8 volts.

The reverse bias widens the depletion layer by sweeping free charge from more donor and acceptor atoms, and leaving them ionized. The depletion layer is highly resistive because the free charge has been substantially eliminated, in contrast to the bulk of the doped regions. Pretty much all the applied voltage is developed across the depletion layer, which is typically 1–3 μm thick. Because the doping density in the p-silicon is chosen for illustration to be twice that in the n-silicon, it follows that the depletion layer reaches twice as far into the n-doped region as into the p-doped region.

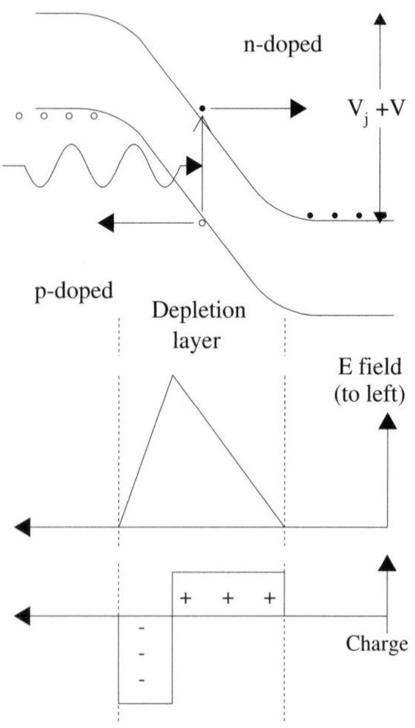

Returning to the upper panel a photon with energy greater than the band-gap is absorbed and produces an electron-hole pair; the electric field drives these carriers across the depletion layer and this constitutes the current. However when photons are absorbed outside the depletion layer, for example in the p-doped region, any electrons (which do not recombine on the way) will diffuse slowly until they reach the depletion layer, and only then do they feel the electric field and move quickly across this layer. Evidently this slows down the response of the photodiode and every effort must be made to minimize the fraction of absorptions taking place outside the depletion layer. Electrons diffuse about three times faster than holes in silicon therefore the light is incident through the p-doped layer as shown in figure 15.5. The p-doped region is also made thin in order to minimize the number of photons absorbed outside the depletion layer.

Fig. 15.6 The upper panel shows the energy levels of a photodiode with a junction potential V_j under reverse bias V. The centre panel shows the electric field distribution, and the lower panel shows the unbalanced charge.

The upper surface of the photodiode is *passivated*, that is to say coated to protect the p-layer from the atmosphere. Silica or silicon nitride is used, with the thickness selected to minimize the reflection from the surface. The heavily doped n$^+$ layer shown in figure 15.5 is there to give a low resistance contact, that is an *ohmic* contact, to the metal electrode. The depletion layer should ideally match the absorption length in thickness, but this can be especially difficult with silicon whose absorption length changes rapidly across the visible spectrum.

Example 15.1

Here we show how to relate the applied potential, the doping levels of the semiconductor and the thickness of the depletion layer. Applying Maxwell's equation 9.13 to the n-doped part of the depletion layer

$$\mathrm{d}E_\mathrm{n}/\mathrm{d}x = eN_\mathrm{d}/(\varepsilon_0\varepsilon_\mathrm{r}),$$

where E_n is the electric field in the x-direction normal to the junction surface, N_d is the number density of donor ions and ε_r is the relative permittivity of the semiconductor. Integrating from the edge of the depletion layer to the junction gives

$$E_\mathrm{n} = eN_\mathrm{d}x/(\varepsilon_0\varepsilon_\mathrm{r}),$$

so that the electric field increases as shown in the centre panel in figure 15.6 until the junction is reached. Then it falls off linearly through the negatively ionized acceptor ions in the p-doped part of the depletion layer. Integrating once more gives the potential drop across the n-doped region

$$V_\mathrm{n} = eN_\mathrm{d}x_\mathrm{n}^2/(2\varepsilon_0\varepsilon_\mathrm{r}),$$

where x_n is the thickness of the depletion layer on the n-doped side. Repeating this calculation for the p-doped region of the depletion layer gives

$$V_\mathrm{p} = eN_\mathrm{a}x_\mathrm{p}^2/(2\varepsilon_0\varepsilon_\mathrm{r}),$$

where N_a is the number density of acceptor ions. Now in addition the depletion layer is overall neutral so that

$$N_\mathrm{a}x_\mathrm{p} = N_\mathrm{d}x_\mathrm{n}. \tag{15.7}$$

Taking the ratio $V_\mathrm{p}/V_\mathrm{n}$ and using eqn. 15.7 to eliminate $x_\mathrm{n}/x_\mathrm{p}$ gives

$$V_\mathrm{p} = (N_\mathrm{d}/N_\mathrm{a})V_\mathrm{n}.$$

Then the total potential across the depletion layer is

$$V_\mathrm{n} + V_\mathrm{p} = [eN_\mathrm{d}x_\mathrm{n}^2/(2\varepsilon_0\varepsilon_\mathrm{r})][1 + N_\mathrm{d}/N_\mathrm{a}]. \tag{15.8}$$

This is of course equal to $V_\mathrm{j} + V$. With equal doping, $N_\mathrm{d} = N_\mathrm{a} = N$,

$$V_\mathrm{n} + V_\mathrm{p} = eNx^2/(4\varepsilon_0\varepsilon_\mathrm{r}),$$

where x is now the full thickness of the depletion layer.

15.3.1   Dark current

In addition to the photocurrent through a pn junction there is current carried by *thermally excited* free charges, and this current will now be evaluated. This current has two opposing components. In the p-doped region electrons can be thermally excited into the conduction band and then cross the junction into the lower energy conduction band in the n-doped region. This is called the *thermal current* which, because it is the result of thermal excitation across the semiconductor band-gap, is independent of the bias applied. The other current is in the opposite direction. Electrons from the conduction band of the n-doped region can be thermally excited sufficiently to be able to enter the conduction band of the p-doped region. There they recombine with holes. This *recombination current* is determined by the difference in the conduction band energies and is therefore strongly dependent on the bias applied to the junction. At zero bias the two currents are in balance, while under forward bias the recombination current rises rapidly.

The number density of the charge carriers around the junction is shown in figure 15.7 when a forward bias is applied. n_p and p_p are the densities of electrons and holes in the p-doped region respectively; n_n and p_n are the corresponding densities in the n-doped region. The vertical scale is logarithmic. In equilibrium the densities are constrained by the semiconductor equation[2]

$$n_\mathrm{p}p_\mathrm{p} = n_\mathrm{n}p_\mathrm{n} = n_\mathrm{i}^2, \tag{15.9}$$

where n_i is the density of carriers in the intrinsic semiconductor, which in silicon is $2\,10^{16}\,\mathrm{m}^{-3}$. If the p- and n-doping are at the level $10^{21}\,\mathrm{m}^{-3}$ the majority carrier densities, n_n and p_p, are $10^{21}\,\mathrm{m}^{-3}$. As a result the minority carrier densities are heavily suppressed to $4\,10^{11}\,\mathrm{m}^{-3}$. The density of minority carriers is highest at the edge of the depletion layer and dies away with distance from the junction. At large distances from the depletion layer the densities of both carrier types are constant.

The electron densities at the two edges of the depletion layer of an unbiased junction are related by the Boltzmann equation

$$n_\mathrm{p} = n_\mathrm{n} \exp\left(-eV_\mathrm{j}/k_\mathrm{B}T\right), \tag{15.10}$$

where V_j is the junction potential and $T\,\mathrm{K}$ the absolute temperature. When a *forward bias V* is applied the minority carrier density increases due to charge being injected across the junction and

$$n_\mathrm{p} + \Delta n_\mathrm{p} = n_\mathrm{n} \exp\left[-e(V_\mathrm{j} - V)/k_\mathrm{B}T\right].$$

Using the previous equation to eliminate n_n gives

$$n_\mathrm{p} + \Delta n_\mathrm{p} = n_\mathrm{p} \exp\left(eV/k_\mathrm{B}T\right),$$

so that

$$\Delta n_\mathrm{p} = n_\mathrm{p}[\exp\left(eV/k_\mathrm{B}T\right) - 1], \tag{15.11}$$

[2]See the fourth edition of *The Physics of Semiconductor Devices* by D. A. Fraser, published by the Clarendon Press, Oxford (1968).

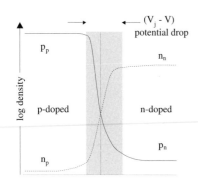

Fig. 15.7 Carrier distributions across a p–n junction with the depletion layer shaded. The junction is forward biased so that there is a substantial injection of carriers across the junction.

with a similar expression for the increase of holes in the n-doped region at the junction. The excess charge at the boundary of the depletion layer diffuses steadily into the bulk semiconductor producing a current proportional to the excess charge densities

$$I_0 = I_r \left[\exp\left(eV/k_B T\right) - 1 \right]. \tag{15.12}$$

From the viewpoint of detecting radiation in a photodiode under reverse bias this current is an unwanted *dark current*. Its limiting value with large reverse bias, I_r, is called the *saturation dark current*. The magnitude of I_r is proportional to n_p and p_n, which eqn. 15.10 shows are exponentially dependent on temperature. At room temperature this dark current is $\sim 1\,\mathrm{nA}$ for a silicon detector of area one square centimetre. A fall of $10\,\mathrm{K}$ causes the dark curent to fall by a factor of two. Devices can be fitted with thermo-electric cooling that can easily cool a detector by $40\,\mathrm{K}$. Heavy doping is also helpful: an increased majority carrier density leads to a higher recombination rate with the result that there are fewer minority carriers to support the dark current. It is straightforward to compensate the dark current in a measurement. Most simply the current can be measured with the detector exposed to the source and then shielded; the difference of these measurements will have the dark current removed, apart from its fluctuations. Evidently when the light incident on the detector carries data in the form of modulation at high frequency the background including dark current can be removed by using capacitive coupling in the readout electronics.

15.4   Photodiode response

Figure 15.8 illustrates how the total current varies with the applied voltage across the photodiode for several levels of illumination. A flux of ϕ photons at a wavelength shorter than the threshold wavelength produces a photocurrent

$$I = e\eta\phi, \tag{15.13}$$

where, as before, η is the quantum efficiency. There is no gain of the type seen in photoconductors because all the carriers recombine on leaving the depletion layer.[3]

When measuring light intensities a photodiode is operated with reverse bias, which is the bottom left quadrant on figure 15.8. A linear variation of current with light intensity is obtained. At a large reverse bias the free electrons can be accelerated sufficiently so that they become energetic enough to initiate ionization: there is a corresponding rapid rise in current, known as *avalanche breakdown*. The broken straight line drawn in the lower left quadrant of figure 15.8 is a *load line* showing the variation of current with voltage when the photodiode drives a fixed resistive load. If the resistance is R_L the load line slope is $1/R_L$. Consequently the smaller that R_L is, the steeper the load line will be, and

[3] Although both an electron and a hole are produced this does not mean that the current is twice that given by eqns. 15.3 or 15.13. You can imagine that the electron and hole travel round the external circuit in opposite senses and annihilate at some point. Then it becomes clear that together their paths pass once around the circuit, not twice.

hence the greater the range of linear response. The range of linear response obtained in practice typically extends over input optical powers from a fraction of a nanowatt up to tens of milliwatts. The saturation dark current is also indicated on the figure.

Notice that in the lower right hand quadrant of the figure the current is opposite to the applied bias voltage. In order to appreciate how this comes about we need to refer to figure 15.6. First note that a small *forward bias* V only reduces the height of the step in energy between the conduction bands in the p- and n-doped regions to $V_j - V$. It does not reverse the step. Thus photoelectrons continue to flow toward the n-doped region provided that the bias remains less than the junction potential and hence they flow against the bias voltage. In the lower right hand quadrant of figure 15.8 energy is *drawn* from the photocell, and this is the mechanism by which solar cells produce electrical power from sunlight.

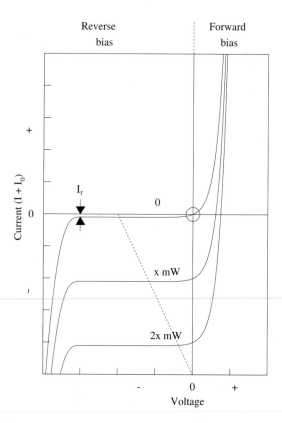

Fig. 15.8 Current versus voltage through a photodiode for three intensities of the incident light. The highest curve is for no illumination and the other two curves are for intensities x mW and $2x$ mW. I_r is the saturation dark current. The broken line is the load line discussed in the text.

15.4.1   Speed of response

How quickly a photodiode can respond to changes in the intensity of incident radiation is determined by two factors. The first is the time taken to sweep charge across the depletion layer; the second is the electrical response time of the detector and amplifier. We continue to consider the case of a silicon photodiode. Usually the electric field in the depletion layer is high enough, above $1\,\mathrm{MV\,m^{-1}}$, so that the electrons reach the limiting *saturation* velocity of around $10^5\,\mathrm{m\,s^{-1}}$, while the holes are travelling roughly half as fast. With a $3\,\mu\mathrm{m}$ depletion layer the collection time is $\sim$30 ps. In practice collection takes longer because electron–hole pairs are also produced by absorptions outside the depletion layer. One of the two charges then has to diffuse to the depletion layer before being swept across it. This diffusion is a comparatively slow process. The electrical parameters of the circuit are the capacitance of the diode junction and the resistance of the load, which is taken to be $50\,\Omega$ here. The capacitance is

$$C = \varepsilon_0 \varepsilon_\mathrm{r} A / d,$$

where ε_r, the relative permittivity of silicon, is 11.8, the area, A, is taken to be $1\,\mathrm{mm^2}$ and d, the thickness of the depletion layer, is $3\,\mu\mathrm{m}$. Inserting these values gives a capacitance of 35 pF. Then the RC time constant is 1.7 ns, which is longer than the collection time. Such a device could detect modulation of radiation at frequencies approaching one gigahertz. In operation with monomode optical fibres the sensitive area of the detector is matched to the cross-section of the light mode in the fibre, or about $1000\,\mu\mathrm{m^2}$; a shallower depletion layer is used too. Then both the collection time and the RC time constant are a few picoseconds.

Photodiodes, known as *pin photodiodes*, are constructed with a very lightly doped nearly intrinsic layer between the p- and n-layers. With large enough bias the depletion layer extends across the whole intrinsic layer. One immediate advantage is the increase in quantum efficiency from having a wide depletion layer. If the doped regions are made relatively thin the proportion of absorptions that take place outside the depletion layer is reduced. The full collection time is then only very little longer than the drift time across the depletion layer.

Those photodiodes for use at the telecoms wavelengths (around 900 nm, and from 1200 to 1700 nm) are either AlGaAs/GaAs diodes sensitive over the range 700–900 nm, or InGaAsP/InP diodes covering the range 900–1700 nm. The radiation incident will have a narrow wavelength range so that it becomes practical to select the chemical content of the semiconductor through which the radiation enters such that its band-gap is greater than the photon's energy. This layer is transparent to the incident radiation and is therefore called a *window layer*; any delay associated with charges diffusing from this layer into the depletion layer is eliminated. The efficiency and speed can be further enhanced using structures of the sort already met in the context of lasers. Distributed

Bragg reflectors are built on both surfaces of the photodiode providing a Fabry–Perot cavity at the wavelength of interest, with the junction lying at an antinode of the cavity. The remainder of the cavity is mostly taken up by window layers. With this structure the depletion layer can be made thin enough to give a very fast response while having a high quantum efficiency because the radiation traverses it many times.

15.4.2 Noise

The output from the photodiode plus amplifier is only useful to the extent that the noise they generate does not impair the signal carried by the radiation incident on the photodiode. This signal may simply be the intensity of the radiation which is to be measured. However in communication applications information is carried in the form of high frequency modulation of the radiation. A filter is then used to pass frequency components in the detector output restricted to some *bandwidth* around the signal frequency. This frequency filter will naturally also pass the noise in this limited frequency range. The device noise as specified by manufacturers is expressed per unit $\sqrt{\text{Hz}}$. Then if the photodiode and amplifier have bandwidth B the noise is $\sqrt{B}$ times this value.

A first contribution to the noise comes from the randomness in the arrival of photons. This randomness is quantum mechanical in origin and has been discussed in Chapter 12. A second contribution to noise comes from the thermal motion of charges within any resistance whether in the detector or the amplifier. A final contribution is less well understood and is called *1/f noise*, a term which describes its frequency distribution. This third component of the noise will be ignored here because it is only of significance at low frequencies below a few hundred hertz.

The number of photons arriving at a detector per unit time in a beam of constant intensity is random and therefore has a Poisson distribution. This noise is called the *shot noise*. If the resolution time of the detector is τ then the mean number of photoelectrons in this interval is

$$\overline{n} = \phi \eta \tau \tag{15.14}$$

where, as before, ϕ is the photon flux and η is the quantum efficiency of the detector. The probability of there being n photoelectrons in a given time interval when the average is $\overline{n}$ will have a Poissonian distribution

$$\mathcal{P}(n) = \overline{n}^n \exp\left(-\overline{n}\right)/n! \tag{15.15}$$

with a mean square deviation equal to $\overline{n}$. Then the mean current is

$$\overline{I} = e\overline{n}/\tau, \tag{15.16}$$

with the variance on the current due to the shot noise being

$$\sigma_{\text{s}}^2 = (e/\tau)^2 \overline{n} = e\overline{I}/\tau. \tag{15.17}$$

This result can be recast in terms of the bandwidth, B, of the detector's response. Using eqn. 7.65

$$\sigma_{\mathrm{s}}^2 = 2e\overline{I}B. \tag{15.18}$$

A similar expression applies to the dark current because the thermal excitation giving rise to it is also a random process

$$\sigma_{\mathrm{d}}^2 = 2eI_{\mathrm{d}}B, \tag{15.19}$$

where I_{d} is the *dark current* flowing in the absence of any illumination.

Thermal noise, also called *Johnson noise*, is caused by the thermal motion of mobile charges in any electrical device, in this case the load resistor, R_{L}. At a temperature T K the mean thermal energy associated with movement along the direction of the current is $k_{\mathrm{B}}T/2$, where k_{B} is the Boltzmann constant. Equating this to the energy stored in the diode capacitance gives

$$C\overline{V^2}/2 = k_{\mathrm{B}}T/2, \tag{15.20}$$

where $\overline{V^2}$ is the mean square noise voltage. The corresponding mean square noise current through the load resistance R_{L} is then

$$\sigma_{\mathrm{j}}^2 = \overline{V^2}/R_{\mathrm{L}}^2 = k_{\mathrm{B}}T/R_{\mathrm{L}}^2C, \tag{15.21}$$

which is the Johnson noise current. If a transimpedance amplifier is used the feedback resistance divided by the open loop gain should be inserted in place of R_{L} when calculating the Johnson noise. CR_{L} is the time constant of the circuit comprising the photodiode and its load, which can be related to the electrical bandwidth using eqn. 7.66. This gives $B = 1/(4R_{\mathrm{L}}C)$ so that the previous equation can be rewritten

$$\sigma_{\mathrm{j}}^2 = 4k_{\mathrm{B}}TB/R_{\mathrm{L}}. \tag{15.22}$$

The Johnson and the shot noise are uncorrelated which means that the total mean square noise current is

$$\sigma^2 = \sigma_{\mathrm{s}}^2 + \sigma_{\mathrm{j}}^2. \tag{15.23}$$

A useful measure for assessing the significance of the noise is the *signal to noise power ratio* (SNR)

$$\mathrm{SNR} = \overline{I^2}/\sigma^2. \tag{15.24}$$

There is a corresponding signal to noise ratio between currents. It is worth noting here that power is the relevant quantity in recording information electronically or otherwise.

A parameter which enables the user to calculate the noise expected from a detector in a particular application is the *noise equivalent power*

(NEP). This is the input optical power that would exactly equal the noise per unit bandwidth in the absence of any signal:

$$\text{NEP} = \sqrt{(\sigma_0^2/B)/S}, \tag{15.25}$$

in $\text{W Hz}^{-1/2}$. S is the detector responsivity and σ_0 is given by

$$\sigma_0^2 = \sigma_j^2 + \sigma_d^2. \tag{15.26}$$

A final useful noise parameter, the *detectivity* D^* measured in $\text{m Hz}^{1/2}\,\text{W}^{-1}$, is the inverse of the NEP per unit area of the detector,

$$D^* = \sqrt{\text{Area}}/\text{NEP}. \tag{15.27}$$

Detectivity is useful when comparing the noise of detectors whose sensitive areas are of different size. All these noise parameters vary rapidly with the wavelength of the radiation detected, and are therefore specified at particular wavelengths by manufacturers. A silicon photodiode of area $1\,\text{mm}^2$ might typically have an NEP of 10^{-14} or $10^{-15}\,\text{W}/\sqrt{\text{Hz}}$ at 950 nm wavelength.

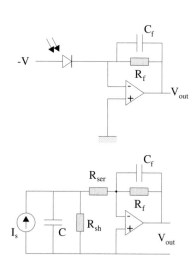

15.4.3   Amplifiers

The currents from photodiodes are usually amplified before further analysis. A commonly used amplifier is the *transimpedance amplifier*, shown in figure 15.9, which uses an operational amplifier with feedback to produce a voltage output from a current input. The upper panel shows the basic circuit, while the lower panel shows the equivalent circuit treating the photodiode as a pure current source with discrete impedance elements. R_{sh} is the shunt resistance of the photodiode, for example $100\,\text{M}\Omega$, C is its capacitance, and R_{ser} is the small series resistance which we may neglect here. R_{f} is the feedback resistor and C_{f} the feedback capacitor. An operational amplifier has several important properties that are relevant here. First the gain of the operational amplifier with no feedback, the *open loop gain*, is very high at low frequencies and rolls off with increasing frequency. The gain bandwidth product is usually specified being, for example 1 MHz, meaning unit gain at 1 MHz and 10^6 gain at 1 Hz. In practice the device is used at frequencies where the gain remains very large. Secondly the input impedance of the operational amplifier is very high, $\sim 10^{12}\,\Omega$ so that very little current enters the amplifier, and almost all the current from the photodiode flows through the feedback path. Finally the negative and positive input terminals are at almost identical potentials, and here where the positive terminal is at ground potential, the voltage on the negative terminal, V_{in}, is very small. Paradoxically, the input impedance presented to the photodiode by the transimpedance configuration is small, which means that the share of the photodiode current flowing in the shunt circuit is much less than that flowing through the feedback circuit. For simplicity the whole of

Fig.   15.9 Photodiode and transimpedance amplifier. The upper panel shows the basic circuit and the lower panel the equivalent circuit.

the photodiode current will be assumed to flow through the feedback circuit. Therefore the voltage developed between the output terminal and ground is

$$V_{\text{out}} = IZ_{\text{f}} + V_{\text{in}}, \tag{15.28}$$

where Z_{f} is the feedback path impedance. This result holds independent of the precise gain, just so long as this gain remains large. Now

$$1/Z_{\text{f}} = 1/R_{\text{f}} + j\omega C_{\text{f}},$$

where ω is the angular frequency of the modulation of the radiation falling on the photodiode. Then

$$Z_{\text{f}} = R_{\text{f}}/(1 + j\omega\tau_{\text{f}}),$$

where $\tau_{\text{f}} = C_{\text{f}}R_{\text{f}}$. Consequently the amplitude of the output voltage developed is

$$V_{\text{out}} = -IR_{\text{f}}/\sqrt{(1 + \omega^2\tau_{\text{f}}^2)}. \tag{15.29}$$

It is the frequency response of the feedback circuit that is critical here, rather than that determined by the photodiode capacitance and shunt resistance. The cut-off frequency can be made high by keeping C_{f}, and hence τ_{f}, small.

If A is the open-loop gain of the operational amplifier, then

$$V_{\text{out}} = -AV_{\text{in}}$$

where V_{in} is the voltage between the input terminals of the operational amplifier. Substituting for V_{out} using eqn. 15.28 this becomes

$$V_{\text{in}} = IZ_{\text{f}}/(A+1).$$

Hence the load seen by the photodiode is small. Consequently the load line in figure 15.8 will be steep. In turn, this means that full use can be made of the intrinsic linear response of the photodiode over a wide range of the radiation intensity. This is the desirable property that the use of a transimpedance amplifier has achieved. Limits to this range of linear response are discussed in the next section.

There are contributions to the output voltage noise from shot noise in the detector, Johnson noise in the feedback resistor and amplifier noise, σ_{a}. The mean square deviation of the output voltage, the noise, is then

$$\sigma_v^2 = (\sigma_{\text{s}}^2 + \sigma_{\text{j}}^2)R_{\text{f}}^2/(1 + \omega^2\tau_{\text{f}}^2) + \sigma_{\text{a}}^2. \tag{15.30}$$

As before $\sigma_{\text{s}}^2 = 2eIB$, while the Johnson noise is $\sigma_{\text{j}}^2 = 4k_{\text{B}}TB/R_{\text{f}}$. At low frequencies these contributions exceed the amplifier noise substantially.

Linear range

Photodiode current is linearly proportional to the intensity of the incident radiation over a range in intensity of typically 10^6. The lower limit to this useful range is reached when the photocurrent is comparable to the noise floor made up of the shot noise on the dark current and the Johnson noise. This limit can therefore be determined from the NEP and the bandwidth of the detector. An upper limit to the range over which response is linear is generally reached when the current is so large that the voltage across the body of the photodiode outside the depletion layer becomes appreciable with respect to the bias applied. This saturation current is given by

$$I_s = (V_j + V)/R_L,$$

where V_j is the junction potential, V the reverse bias, and R_L is the load resistance. The corresponding input power at this point is called the *saturation power*: $P_s = I_s/S$, where S is the detector's responsivity.

15.4.4　Solar cells

Solar cells are photodiodes used to generate electrical power from solar radiation. The intensity of solar radiation reaches about $1000\,\mathrm{W\,m^{-2}}$ at midday in summer at European latitudes, so that solar radiation is seen as a promising source of renewable energy. Crystalline silicon and GaAs have threshold wavelengths of $1.12\,\mu\mathrm{m}$ and $870\,\mathrm{nm}$ respectively, hence photodiodes made from these semiconductors are sensitive to a large fraction of the Sun's radiation. Silicon is preferred because the existing infrastructure in the electronics industry reduces fabrication costs. The overall efficiency for producing electrical power from solar power using commercially available cells is at best 25%. Amorphous silicon is cheaper to make but has a threshold at $730\,\mathrm{nm}$ wavelength, so that the efficiency is only 12% at best. The highest efficiency is achieved with cells that consist of several layers each tailored for a different range of wavelengths, with experimental devices achieving 40% conversion efficiency.

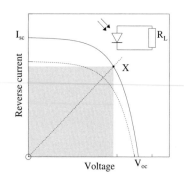

Fig. 15.10 Solar cell current and voltage variation for two light intensities. The broken diagonal line is the load line for the circuit drawn.

Referring to figure 15.8, the working region of solar cells is in the lower right hand quadrant with energy being drawn from the photodiode. This quadrant is reproduced in figure 15.10 with current–voltage characteristics drawn in for two levels of illumination. The inset shows a simple circuit with a load resistor, R_L. I_{sc} is the current with a short circuit ($R_L = 0$), and V_{oc} is the open circuit voltage developed ($R_L = \infty$). The point X is the working point where the load line and the characteristic cross. In this condition the shaded area, which is the current times voltage, gives the output power. An optimal choice of the load gives an output power of around $0.8 I_{sc} V_{oc}$.

Here we make an estimate of the efficiency of a silicon solar cell for converting solar power to electrical power. The equation for the current in the photodiode is

$$I = I_r[\exp\left(eV/k_BT\right) - 1] - I_\ell,$$

where I_ℓ is the photocurrent caused by the incident radiation and I_r is the saturation dark current. Rearranging this gives

$$I + I_\ell + I_r = I_r \exp\left(eV/k_BT\right).$$

Then on open circuit

$$I_\ell + I_r = I_r \exp\left(eV_{oc}/k_BT\right),$$

so that

$$V_{oc} = (k_BT/e) \ln\left[\,(I_\ell + I_r)/I_r\,\right] \approx 0.0259 \ln\left(I_\ell/I_r\right) \qquad (15.31)$$

at room temperature. We consider crystalline silicon as the photodiode material with a dark current which can be as low as $1\,\mathrm{pA\,cm^{-2}}$. Its responsivity is $0.4\,\mathrm{A\,W^{-1}}$ so that a daylight intensity of $100\,\mathrm{mW\,cm^{-2}}$ produces a short circuit current density of $40\,\mathrm{mA\,cm^{-2}}$. Using eqn. 15.31 the corresponding open circuit voltage at $300\,\mathrm{K}$ is $0.63\,\mathrm{V}$. Thus the overall efficiency for converting solar power to electric power is

$$\eta_{solar} = 0.8 V_{oc} I_{sc}/P_{solar} = 0.20.$$

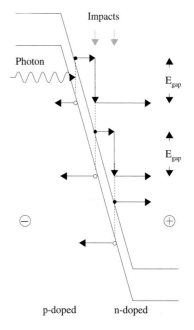

Fig. 15.11 Photon conversion, shown together with the first two electron impact ionizations at the start of an avalanche in a photodiode.

15.5   Avalanche photodiodes

When a large reverse bias is applied to a photodiode the charge carriers released by photons are accelerated sufficiently so that they produce further electron–hole pairs by *impact ionization* on the lattice atoms. Two impact ionizations are shown in figure 15.11. Each ionization creates a new electron–hole pair and the incident electron loses an energy equal to the band-gap energy. If the electric field is large enough an avalanche of charge is produced, so that both the current and the responsivity are increased: of course the quantum efficiency is unchanged. When the incident radiation is so weak that the shot noise is very much less than the Johnson noise then operation in the avalanche mode can be used to increase the output current substantially, while not affecting the noise appreciably. Using eqns. 15.18 and 15.22 this condition will hold provided that

$$2eIB \ll 4k_BTB/R_L \quad \text{that is } I \ll 0.052/R_L,$$

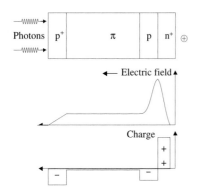

Fig. 15.12 Avalanche photodiode: the top panel shows the structure; the centre panel the electric field distribution; the bottom panel the charge distribution due to the depletion of the majority carriers.

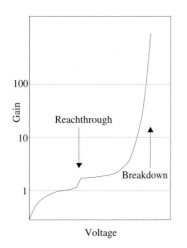

Fig. 15.13 Gain variation against the voltage applied to an avalanche photodiode. At reachthrough the depletion layer extends all across the π-layer.

at room temperature. Avalanches take time to develop and to decay away, hence the depletion layer should be made as thin as possible so that the avalanche terminates quickly. On the other hand a high quantum efficiency requires a thick depletion layer. This conflict is resolved in *avalanche photodiodes* (APDs) by having separate absorption and multiplication layers, as illustrated in figure 15.12. The uppermost panel shows the doping pattern along the photodiode; the middle panel shows the electric field distribution; and the lower panel shows the charge distribution.

Absorption takes place across the wide weakly doped π-doped layer. Therefore the reverse bias is increased to the point that the electric field spills across the p-layer and reaches right through the weakly doped π-layer. Electrons generated in the π-layer are swept into the thin region of high electric field around the p–n$^+$ junction, and there the avalanche develops. The gain in silicon for which the current remains linearly proportional to the radiation intensity can be as large as 1000. Figure 15.13 shows the gain as a function of the applied voltage. At high enough voltages the avalanche becomes uncontrolled and a large pulse of current is produced, independent of whether it is only a single photon or many photons that are being absorbed. This breakdown voltage is ∼150 V for silicon APDs.

Both holes and electrons can cause impact ionization, and the yield from holes relative to that from electrons is called k. The value of k is small, 0.02, for silicon, which means that an avalanche in silicon initiated by electrons propagates once in the direction of travel of the electrons, and then terminates quickly. This gives the fastest possible APD response. If k is close to unity then components of the avalanche can propagate in both directions and the avalanche only dies away slowly. The current after multiplication is

$$J = GI = eG\eta\phi, \tag{15.32}$$

where G is the gain achieved in the multiplication step. The shot noise is

$$\sigma_{\text{s}}^2 = \overline{J^2} - (\overline{J})^2 = G^2\,[\,\overline{I^2} - (\overline{I})^2\,].$$

Then using eqn. 15.17 to replace the variance in I,

$$\sigma_{\text{s}}^2 = G^2(2e\overline{I}B) = 2e\overline{J}GB, \tag{15.33}$$

An additional factor F is needed to take account of *excess shot noise* occuring in the multiplication process,

$$\sigma_{\text{s}}^2 = 2e\overline{J}BGF. \tag{15.34}$$

F can be held down to 2–5 with appropriately designed APDs. Thus the signal to noise ratio is now

$$\text{SNR} = \overline{J^2}/\,[\,2e\overline{J}BGF + \sigma_{\text{j}}^2\,].$$

The InGaAs/InP APDs used as detectors at telecom wavelengths around $1.31\,\mu$m and $1.55\,\mu$m have gains of 10–40 and breakdown voltages of $\sim$50 V. Such APDs with sensitive areas, $\sim$1000 μm^2, matched to optical fibres have dark currents of $\sim$10 nA, capacitances $\sim$1 pF, and bandwidths of a few gigahertz.

APDs operated above the breakdown voltage make simple and robust detectors of single photons. Absorbed photons give rise to uniform large pulses and these pulses can be counted to determine the number of photons. A suitable circuit is needed to protect the APD from runaway avalanches and to quench the avalanches. When operated in this *Geiger mode* silicon APDs have gains of 10^{5-8}. The rise time of the pulse is short, $\sim$10 ps. However there is a dead period of $\sim$1 μs after each count during which time the electric field within the APD is insufficiently large for the absorption of a new photon to produce a pulse.[4] There is also a stream of *dark counts* initiated by electrons that have been thermally excited into the conduction band. These dark counts can be suppressed by cooling the detector, either through tens of degrees using a thermoelectric cooler or down to 77 K with liquid nitrogen.

APDs are widely used wherever intensities are weak, for example in laser ranging and fluorescence spectroscopy. Laser ranging consists of directing a laser pulse at a distant mirror, and timing the interval between the outgoing and returning pulse. This return pulse is necessarily weak. In fluorescence spectroscopy a material is first irradiated by a laser and then the spectrum of its fluorescence is measured. From this spectrum the chemical composition of the material may be determined. The technique is valuable when the sample of material is delicate or minute, as for example scrapings from an historical or artistic artefact. In these two applications APDs of very large area, $\sim$1 cm^2, are usually employed.

Individual photons can also be detected using photomultiplers, and these will be discussed later in this chapter. For the present we can note that the silicon APD has a much better quantum efficiency across the visible spectrum, $\sim$70 % compared to $\sim$15 %. APDs are in addition more compact, easier to cool, require a relatively low voltage, and are insensitive to magnetic fields. Photomultipliers, on the other hand, are simpler to use, give gains of 10^8 without any external amplifier, have short dead times, and can be made with sensitive areas of up to $\sim$600 cm^2.

[4]When high speed operation is necessary the avalanche is quenched early by reducing the voltage early and then the voltage is rapidly reapplied. In this way the dead time can be reduced to $\sim$ 10ns.

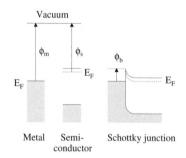

Fig. 15.14 The left hand panel shows the energy levels of separated metal and n-doped semiconductor; the right hand panel shows the energy levels for the Schottky junction. The broken line in the semiconductor region indicates the energy of the donor state electrons.

15.6 Schottky photodiodes

Schottky photodiodes are high speed photodiodes in which a doped semiconductor is in contact with a thin metal layer. We consider the case shown in figure 15.14 of an n-doped semiconductor. In this example the junction will be a *Schottky junction* if the work function in the metal, ϕ_m is larger than that of the semiconductor, ϕ_s. The left hand panel of

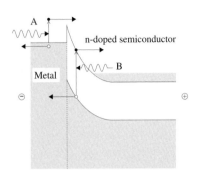

Fig. 15.15 Schottky photodiode with reverse bias. The alternative absorption processes are both shown.

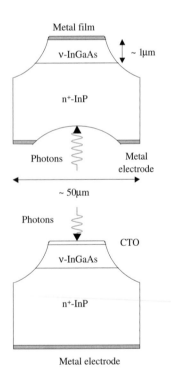

Fig. 15.16 Schottky n-doped InGaAs-metal photodiodes: showing both back illumination and illumination through the conductive transparent metal oxide.

figure 15.14 shows the electron energy levels in the isolated metal and isolated semiconductor, while the right hand panel shows the energy levels at an unbiased junction. A potential barrier is established at the junction, a *Schottky barrier*, which prevents electrons flowing across the junction. The barrier height is

$$\phi_{\mathrm{b}} = \phi_{\mathrm{m}} - \phi_{\mathrm{s}} + \Delta E. \qquad (15.35)$$

ΔE is the small difference between the energy at the base of the conduction band in the semiconductor and the Fermi level, the latter lying close to the donor energy level. A depletion layer develops in the semiconductor near the metal, in which the electrons have been swept out, leaving positively charged donor ions. Electrons are attracted from inside the bulk of the metal and form a sheet of charge in the metal at the junction. Figure 15.15 shows the energy levels at an n-doped semiconductor–metal junction under reverse bias. Note that the Schottky barrier height is unaffected by the applied voltage.

There are two distinct methods used to detect radiation using Schottky diodes. In one approach photons of energy greater than ϕ_{b} are absorbed on electrons in the metal, and this process is illustrated by the absorption labelled A in figure 15.15. Electrons gain enough energy to surmount the Schottky barrier and their flow across the junction constitutes a photocurrent. This Schottky barrier height can be small, much less than the band-gap energy in the semiconductor. Consequently it is possible to detect radiation well into the infrared. For example the Schottky junction between PtSi and p-doped silicon has a barrier height of only 0.22 eV, which corresponds to a threshold wavelength of 5.6 μm. (In this type of Schottky photodiode the semiconductor is p-doped, and holes cross the Schottky barrier.)

The second method used to detect radiation with a Schottky photodiode is illustrated by the absorption labelled B in figure 15.15. Now the photon is absorbed in the depletion layer, in the same way as in a pn photodiode. For example the active region could be lightly doped ν-GaAlAs covered with a layer of gold only 10 nm thick. A cross-section through such a detector is shown in the upper panel of figure 15.16, where the illumination enters through the heavily doped n^{+}-InP ohmic substrate. InP is transparent at wavelengths longer than 920 nm and InGa$_{0.5}$As$_{0.5}$ has a threshold wavelength for absorption at 1650 nm so the detector illustrated would be sensitive between 920 nm and 1650 nm, which includes the telecom wavebands. Front illuminated Schottky photodiodes, like that shown in the lower panel of figure 15.16, are constructed with a conductive transparent oxide layer in place of metal. Cadmium tin oxide is one such material having a transmittance of 90% at the telecom wavelengths, while indium tin oxide is transparent across the visible spectrum.

An important advantage of Schottky photodiodes over other photodiodes is their higher speed of response. Front illumination gives absorptions close to the metal so that only the electrons, which have the greater mobility, travel across the depletion layer. In addition the slow diffusion of carriers generated outside the depletion layer is eliminated because the material adjacent to the depletion layer is a conductor. As a result Schottky photodiodes are capable of operating at ~100 GHz.

15.7 Imaging arrays

Two-dimensional rectangular arrays of semiconductor photodiodes built on a common substrate are the dominant devices used in imaging applications today. The arrays found in digital cameras, camcorders and astronomical telescopes are almost all silicon based *charge coupled devices, CCDs*, with 1–100 million individual identical square sensitive elements or *pixels*. The lens system focuses the scene of interest onto the surface of the array. Pixel side length depends on the application and generally lies somewhere in the range 5–25 μm. Digital SLR cameras have CCDs with sensitive regions of area 24 mm×36 mm or 16 mm×24 mm. The re-

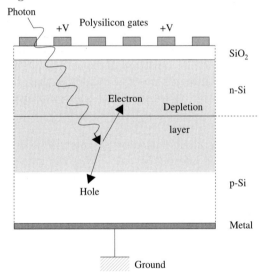

Fig. 15.17 Part of a section taken through a column of pixels in a CCD array. The depletion layer is shaded.

lated CMOS arrays are also silicon photodiode arrays in which each pixel also contains a few transistors that provide processing power. CMOS arrays share the *complementary metal oxide semiconductor* structure of mainstream silicon electronics. Figure 15.17 shows a section through a CCD taken along a column of pixels. The substrate is p-doped silicon with a surface layer of n-doped silicon, and at their junction a depletion layer forms. A 0.1 μm thick insulating layer of silicon dioxide (silica) is grown on top of the n-doped layer. The pixel photodiodes are de-

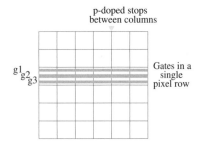

Fig. 15.18 Face-on view of a CCD showing the gates for a single row of pixels and the channel stops that isolate the columns.

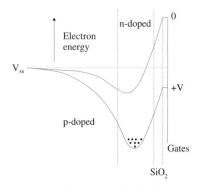

Fig. 15.19 Potential wells for electrons in a CCD with a buried channel.

fined and controlled by the voltages applied to *gates*. These gates are semi-transparent indium tin oxide (ITO) or polysilicon electrodes which extend in the row direction across the whole width of the array. There are three such gates per pixel, shared by all the pixels in a row. Each column of pixels is isolated from its neighbours by p-doped implants in the surface of the silicon which run the length of the detector between the columns, known as *channel stops*.

Figure 15.18 shows a face-on view of such a detector with a notional number of rows and columns. During the recording phase a positive voltage is applied to the central gate of each pixel and zero voltage to its two outer gates. Electrons produced in a pixel will therefore end up in the deeper potential well below the central gate. Capture of electrons in the central well continues throughout the exposure so that the charge stored under a pixel is directly proportional to the integrated intensity of the radiation falling on the area of the pixel.

The two potential wells are illustrated in figure 15.19 showing the standard *buried channel* obtained by suitable choices of the doping levels and gate voltages. The figure shows how this arrangement keeps the captured electrons away from the silicon/silicon dioxide interface. If the electrons were allowed to accumulate against the silicon/silicon dioxide interface the many defects at this interface would trap them so that they could not be read out. The substrate potential V_{ss} can be set to zero or, as indicated in the figure, set positive but lower than the positive gate voltage. With the latter choice the CCD operates in what is called the *inverted mode*.

An estimate of the combination of parameters needed to produce a buried channel can be inferred by calculating the gate voltage with respect to the substrate which depletes the whole n-doped channel of width x_n. Writing eqn. 15.8 again we have

$$V = [eN_d x_n^2/(2\varepsilon_0\varepsilon_r)][1 + N_d/N_a].$$

Taking the density of the donors N_d to be $3\,10^{21}\,\mathrm{m}^{-3}$ and x_n to be $1\,\mu\mathrm{m}$ gives

$$V = 2.3(1 + r),$$

where r is the dopant ratio N_d/N_a. The p-doping is made lighter so that the depletion layer extends well into the substrate. A suitable value for the dopant ratio is 8, which requires an applied voltage of 21 V. Before each exposure the electrons are flushed from the pixel leaving the n-doped layer fully depleted. Then setting the gate voltage lower to ~ 10 V the potential acquires the shape shown in the figure.

An estimate of the maximum charge stored per pixel can be made as follows. The capacitor storing the electrons is effectively a $0.5\,\mu\mathrm{m}$ layer of silicon having relative permittivity 11.68 in series with a $0.1\,\mu\mathrm{m}$ layer of silicon dioxide having a relative permittivity 3.9. Each layer has a capacitance

$$C = \varepsilon_0 \varepsilon_r A/d,$$

where A is its area, d its thickness and ε_r its relative permittivity. Taking the pixel area to be $10\,\mu\text{m} \times 10\,\mu\text{m}$ gives a series capacitance equal to $12.8\,\text{fF}$. Then, assuming the potential is $4\,\text{V}$, the maximum charge that can be stored is around $50\,\text{fC}$ or $300\,000$ electrons. When the charge is greater than the storage capacity the excess charge overflows into the neighbouring pixels, a process known as *blooming* with consequent image deterioration. In order to eliminate blooming an additional well is provided alongside each column of pixels and any excess charge flows into this *drain*.

During the exposure to record the scene projected by the camera onto the CCD, electrons are also being thermally excited, first into traps at the insulator/silicon interface and thence into the silicon conduction band. The contribution of this dark current must be estimated and then subtracted from the total charge stored; in addition the gates have to be pulsed to remove any dark charge before each exposure. The dark current with the substrate voltage set to zero is typically $1\,\text{nA}\,\text{cm}^{-2}$ at room temperature and has a strong dependence on the absolute temperature. Thermo-electric cooling substantially reduces the dark charge. Cooling to temperatures around $150\,\text{K}$ cuts the dark charge accumulation rate by a factor 10^7, so that a $25\,\mu\text{m} \times 25\,\mu\text{m}$ pixel picks up only a few electrons per hour. This is the degree of suppression necessary in making long astronomical exposures.

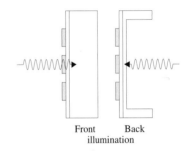

Front   Back
illumination

Fig. 15.20 Comparison of front illumination of a standard CCD and back illumination of a thinned CCD.

With inverted mode operation the dark current is drastically reduced in comparison with normal mode operation. Holes now travel to the silicon/silicon dioxide interface where their potential energy is low, and in so doing they neutralize the surface traps there. In this mode the remaining dark current is due to the few defects in the bulk of the silicon: typically it falls to $10\,\text{pA}\,\text{cm}^{-2}$ at room temperature, and is correspondingly less if the CCD is cooled.

15.7.1 Quantum efficiency and colour

A proportion of the incident light is absorbed in the polysilicon gates and silicon dioxide insulating layer. In addition there are losses because charge can be trapped at the interfaces on surface crystal defects. These effects are worst at the blue end of the spectrum because the absorption length in silicon drops to $0.2\,\mu\text{m}$ at $400\,\text{nm}$ wavelength. In order to recover quantum efficiency in the blue the substrate can be thinned by careful etching to leave a final thickness of around $15\,\mu\text{m}$, and this is then *back illuminated* through the bare surface of the thinned substrate. Figure 15.20 compares the structure of front and back illuminated CCDs. Back-illuminated CCDs are often anti-reflection coated to

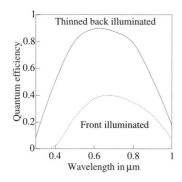

Fig. 15.21 Quantum efficiency for silicon CCDs as a function of wavelength. The quantum efficiency of thinned back illuminated CCDs with anti-reflection coating is about twice that of the front illuminated standard CCD.

increase the surface transmission. The quantum efficiency of a thinned, back illuminated, anti-reflection coated CCD is compared to that of a front illuminated CCD in figure 15.21. In order to support the fragile thinned CCD it is mounted with the front face against a rigid frame. There is an improvement in quantum efficiency over front illumination of about a factor two. For reference, the quantum efficiency of the human eye in daylight is much poorer, peaking at 0.03 at $555\,\mu$m. Standard panchromatic film maintains a quantum efficiency of about 0.02 across the whole visible spectrum. The exposed substrate is heavily p-doped to maintain good electrical contact which must now be made through the edges of the CCD. Crucially the resulting high field in the substrate keeps electrons from the surface where they would become trapped.

There are several ways by which colour images can be recorded with CCDs. The standard method adopted in cameras is to place filters

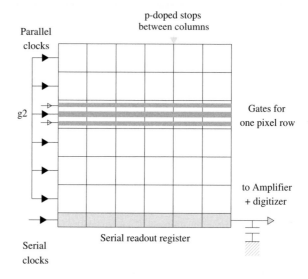

Fig. 15.22 Components of CCD readout. The parallel clock lines for the g2 gates are shown. For clarity the g1 and g3 clock lines are only indicated for one row.

over each pixel but this unfortunately reduces quantum efficiency. Usually there are two green-sensitive pixels for each pair of blue- and red-sensitive pixels in order to mimic the eye response. Camcorders however have three separate CCDs, one for each colour.

15.7.2   CCD readout

The components of a CCD used to read out the charge stored in each pixel when reconstructing the image are indicated in figure 15.22. After exposure the charge stored in each row of pixels is passed one step down the column to the row beneath it in the diagram, with the charge in the lowest row being transferred to a serial output register consisting of pixels called *transfer elements* which are optically shielded by an opaque surface film. After this the charges in the output register are moved one element at a time to the right so that each charge passes in turn from the bottom right of the figure into an amplifier and digitizer. This sequence

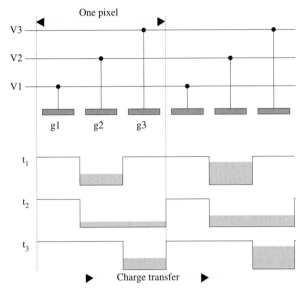

Fig. 15.23 Charge transfer from one gate to the next in three phase transfer. Positive voltage is applied to the lines V1, V1 and V3 in the sequence: V2 at t_1; V2 and V3 at t_2; V3 at t_3; V3 and V1 at t_4; V1 at t_5; V1 and V2 at t_6.

is repeated until the charges from every row have all been amplified and digitized. The time consuming step is the digitization. Analogue to digital convertors giving 12 bit precision (that is 1 part in 2^{12}) run at typically 50 million samplings per second (MSPS), and thus the full frame from a one megapixel camcorder can be read out in $\sim$20 ms.

The upper part of figure 15.23 shows the voltage drive connections to the gates along a column of pixels used for *three phase* readout.[5] Positive voltage is applied sequentially to the three lines (V1, V2, V3, V1, V2,...) so that the potential wells containing the stored electrons move along the column. The movement of these potential wells is illustrated in the lower part of the same figure at consecutive phases: t_1, t_2 and t_3. Three such cycles of voltage changes carry the charges forward by one pixel. Charge transfer is extremely efficient, with a loss per transfer of 10^{-6} achieved in scientific devices.

[5] Examples of these gates are shown shaded in figure 15.22.

A difficulty with this *full frame architecture* is that during the readout the exposure continues and this can produce smears at locations of high image intensity. Some form of external shuttering is therefore necessary. An alternative approach to eliminate the unwanted sensitivity during the read out time is to modify the CCD structure so that part of each pixel's area is used as a storage capacitor that is shielded from radiation by an opaque surface film. These areas form columns between the active pixel columns. With this *inline transfer architecture* the first step in the readout is to transfer, simultaneously, the contents of each pixel into its storage capacitor. Thereafter the readout proceeds as described previously. Exposure times are then electronically controllable and can be easily made as short as $1\,\mu$s.

It is evident that the surface area of a pixel remaining clear after taking into account inline transfer, drains and blocking between columns has become rather small. A lot of the quantum efficiency lost in this way with a front illuminated CCD can be recovered by using a microlens array above the CCD. Each pixel's lens collects as much of the light incident on the pixel area as possible and focuses it onto the residual active area. The resulting quantum efficiency is comparable to that of back illuminated CCDs.

15.7.3   Noise and dynamic range

When the charge on a pixel is N electrons the corresponding shot noise in the charge collected is $\sqrt{N}e$. Taking the previous estimate made for the charge required to fill the well completely, namely 300 000 electrons, the shot noise on this charge is $\sqrt{300\,000}$ or $\sim$550 electrons. The corresponding dark charge accumulated in an exposure lasting Δt seconds with a dark current density of $I\,\mathrm{A\,m^{-2}}$ and a pixel area of $A\,\mathrm{m^2}$ is

$$Q = \Delta t I A.$$

Therefore during a 1/500th second exposure of a $100\,\mu\mathrm{m}^2$ area pixel whose dark current of $1\,\mathrm{nA\,cm^{-2}}$ the dark charge accumulated is about 12 electrons. The range of charge per pixel which is both detectable and linearly proportional to the input light intensity therefore runs from about 10 to 300 000 electrons, giving a *dynamic range* of 30 000. This range can be improved by cooling the CCD.

15.7.4   EM-CCDs

A modified CCD, the *electron multiplying CCD (EM-CCD)*, can be read out at megapixels per second while keeping the readout noise less than one electron per pixel. At similar readout rates the readout noise from a standard CCD would be tens of electrons per pixel. The other source of noise, the dark current, can be rendered negligible by cooling EM-CCDs to liquid nitrogen temperatures. Biological and astronomical research benefit from this new capability: for example the *lucky imaging* procedure described in Section 8.5.1. Surveillance imaging with EM-CCDs also gives improved image quality in difficult lighting conditions.

A sketch of the EM-CCD architecture developed by e2v Technologies is shown in figure 15.24. Taking a particular example from the e2v range,

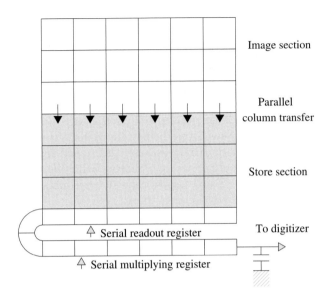

Fig. 15.24 Architecture of an EM-CCD: a Low Light Level CCD (L3CCD) developed by e2v Technologies, Chelmsford, UK.

the EM-CCD is divided into an image area of 536×528 pixels and a second identically pixelated, but optically shielded, storage area. There is a serial readout register, like that for a standard CCD, containing 536 *transfer elements*. Its output is fed into a new serial *multiplying register* containing an equal number of transfer elements. The electric potential applied between adjacent elements in the multiplying register when transfering charge is increased to around 40 V, so that the accelerated electrons can cause impact ionization. This is still a relatively modest voltage compared to that applied in an avalanche photodiodes. As a result each transfer between elements produces a small average multiplication factor of typically 1.015. However along a register containing 536 such elements the overall multiplication will have grown to 2800. The readout noise is therefore scaled down relative to the signal charge on an image pixel by this multiplication factor. Depending on the application the transfer voltage can be set to give any multiplication factor from unity up to several thousand.

Suppose that n electrons were stored in a pixel, then taking a total multiplication factor of M, the output is Mn electrons. Because the multiplication is a stochastic process the mean square deviation on the count is FMn, where F is a factor that approaches 2 when Mn is large. The signal to noise ratio (including the statistical fluctuations of the signal) is thus

$$\mathrm{SNR} = n/\sqrt{(Fn + FD + R^2/M^2)} \qquad (15.36)$$

where D is the dark current induced charge per pixel, which is also multiplied, and R is the root mean square readout noise per pixel. Evidently the effect of the readout noise per pixel is suppressed by a factor around M, so that when M is large an EM-CCD becomes virtually free of readout noise. If the EM-CCD is also cooled to liquid nitrogen temperatures the dark current induced noise becomes negligible too. The price paid for the reduction in readout noise is an increased statistical error, indicated by the factor F. It is equivalent to throwing away half the photons when large numbers of photons are incident on the detector. This loss can be recovered when the EM-CCD is used in the *single photon counting* mode. Exposures are then made sufficiently short that each pixel receives either a single photon or no photons, and the factor F is exactly unity. Studies show that at low counts per pixel, of order 5, the factor F remains close to unity.

In figure 15.24 the frame consisting of the pixels exposed to light is the upper unshaded area. After an exposure the content of the full frame is transferred to the storage area, shown gray, shielded from any light. The contents of the image section are transferred in parallel, down the columns, into the corresponding locations in the store. In the example chosen, with 528 pixels per column, and a transfer rate of $1\,\text{Mpixel}\,\text{s}^{-1}$, this parallel transfer is completed in under $1\,\text{ms}$. Either during or before the next exposure the store is read out. Taking a transfer rate of $10\,\text{Mpixel}\,\text{s}^{-1}$ the serial readout from the store, containing 536×528 pixels, will be completed in $28\,\text{ms}$. This readout rate can be increased substantially by reading out sections of the EM-CCD in parallel. Each such section is read out through its own short serial readout register and multiplying register. For example the 240×240 pixels of the e2v Technologies CCD220 are read out through eight independent paths. In this way *frame readout* at kHz rates is achieved.

Impact amplification is very sensitive to small changes in temperature and in the voltage causing the impact ionization. The temperature needs to be maintained constant to $0.1\,\text{K}$ and the voltage must be stable to millivolts in order that the overall amplification remains constant to 1%.

15.7.5   CMOS arrays

These arrays, also known as *active pixel arrays*, differ from CCD arrays in having two or three transistors fabricated into the structure of each pixel. The transistors are used to convert charge to voltage at the pixel itself and to perform other operations. CMOS stands for complementary metal oxide semiconductor, the standard structure used in electronic chips. Existing infrastructure then makes it cheap to produce CMOS arrays at the system level. In addition row and column gates are fabricated across the whole array, so that any pixel can be addressed independently by applying voltage to just the row gate and column gate for this pixel. This makes it possible to read out pixels much more rapidly than with a CCD. More of each pixel is occupied by non-sensitive components than is the case for CCDs so that microlens arrays are essential to achieve a high quantum efficiency.

Disadvantages of the on-pixel processing are that the signals from individual pixels under uniform illumination are more variable in CMOS arrays than in CCD arrays, and in addition these signals are noisier. However both blooming and smearing are automatically eliminated with CMOS arrays. It is also possible to make a partial read out from selected areas of interest from a CMOS array rather than always reading out everything, which is unavoidable with a CCD. CCDs are still found in most digital cameras, with the penetration of CMOS arrays being higher for DSLR cameras. One final point is that all the electronics, such as the clock drives, which are external to a CCD can be incorporated on the CMOS array chip, giving a camera on a chip.

This completes the description of semiconductor photodetectors. For the remainder of the chapter attention is centred on that other broad class of photodetectors, the vacuum tube detectors. Photoemission from alkali metal or semiconductor surfaces provides the first step in photon detection.

15.8 Photomultipliers

Photomultipliers are vacuum tube detectors of radiation in which photons incident on a thin *photocathode* eject electrons via the photoeffect, as in Millikan's experiment.The photoelectrons are then multiplied within the device by a factor up to 10^8. Figure 15.25 pictures an axial section through a photomultiplier contained within a highly evacuated cylindrical glass envelope. Light is shown incident on the photocathode material deposited on the inside of the end window, making this an *end window tube*. Photons of energy greater than the work function, W, of the photocathode can eject electrons into the evacuated space.

Photoelectrons from the photocathode are accelerated by an applied potential toward the first in a series of dynodes, labelled D1. These dynodes are electrodes coated with a good *secondary emitter* from which an energetic electron will eject several secondary electrons on impact. Along the dynode chain the potential increases by equal steps so that multiplication is repeated at each dynode. Having a vacuum eliminates collisions with gas molecules which would otherwise halt this multiplication. The output pulse of charge arriving at the anode produces a voltage pulse across the load. Copper coated with CuBeO(Cs) is an excellent secondary emitter whose yield peaks at $\sim$5 electrons for incident electrons with energy around 100 to 200 eV. The voltage steps are therefore made 100 to 200 V using a chain of equal resistors as illustrated in the diagram.

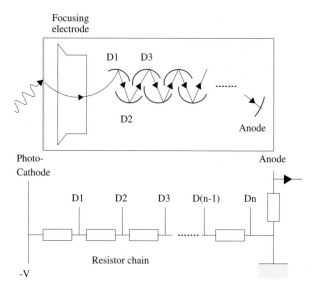

Fig. 15.25 End window photomultiplier.

Typically there are eight to fourteen dynodes and the amplification is from 10^6 to 10^8. This overall gain, G, is extremely sensitive to the voltage, V, because the gain at each dynode is roughly proportional to the voltage for that step, thus

$$G \propto (V/n)^n,$$

where there are n dynodes. Then

$$\mathrm{d}G/G = n\mathrm{d}V/V,$$

so that with ten dynodes a 1% change in voltage leads to a 10% change in gain.

The wavelength below which photoemission can occur is

$$\lambda_{\mathrm{max}} = hc/W = 1.24\,\mu\mathrm{m}/W(\text{in eV}), \qquad (15.37)$$

where W is the work function of the surface. Alkali metals have photoemission thresholds in the near infrared and visible part of the spectrum. The quantum efficiency peaks at wavelengths just below the threshold wavelength and declines rapidly as the wavelength falls further. It requires alloys consisting of several such alkali metals, each with a different threshold, to obtain a surface whose quantum efficiency is consistently high over a wide range of wavelengths. One material commonly used for detection across the visible spectrum is Na_2KCsSb, whose threshold is at $850\,\mu$m, and whose spectral response is given the appellation *S20*.

The responsivity of a photomultiplier using an S20 photocathode on a borosilicate glass window is shown by the full line in figure 15.26, where the straight lines indicate responsivities corresponding to quantum efficiencies of 10% and 20%. This sensitivity range can be extended into the ultraviolet by the use of a quartz window on the photomultiplier. Photomultipliers generally have quantum efficiencies which are in the range 10–25%, much lower than those of photodiodes. Photocathodes made from compounds of group III and V elements are useful for detecting radiation in the near infrared covering the telecom bands. The shot noise and Johnson noise are given by the same formulae as for the photodiodes, namely eqns. 15.18 and 15.22. There is excess noise due to the statistical fluctuations in the multiplication along the dynode chain, which usually increases the shot noise by a factor of up to 2.

Photomultipliers have a wide dynamic range over which the anode current remains linearly proportional to the intensity of the radiation incident on the photocathode. The lower useful limit is set by the noise on the dark current coming from thermionic emission, mainly at the photocathode and first dynode. Anode dark currents vary widely at room temperature: with an S20 photocathode and multiplication by 10^6 the anode dark current is between $\sim 1\,\mathrm{pA}$ and $\sim 1\,\mathrm{nA}$. The dependence of dark current on the absolute temperature follows the Richardson–Dushman formula

$$I = DT^2 \exp\left(-W/k_\mathrm{B}T\right), \tag{15.38}$$

where W is the work function of the photocathode, and D is $1.2\,10^6$ $\mathrm{A\,m^{-2}K^{-2}}$. Cooling the photomultiplier is therefore very effective in reducing this background. The dark current increases by a large factor if the photocathode is exposed to daylight, even in the absence of applied voltage, and requires hours to return to its unexposed value.

An upper limit to the region of linear response is reached when the current through the last steps in the dynode chain becomes appreciable in comparison to the current in the parallel resistor chain. At this limit the voltage distribution along the dynode chain is altered. This in turn affects the amplification. Nonetheless the response can be linear over ranges of 10^6 in incident intensity, very comparable to what is obtained with photodiodes.

A fundamental weakness of photomultipliers is that an applied magnetic field can seriously affect the multiplication. In magnetic fields the electron paths spiral along the magnetic field lines and hence can be deflected away from the following dynode. This is an effect from which the compact solid state devices are relatively immune.

15.8.1   Counting and timing

At extremely low light intensities the pulses due to individual photon absorptions are apparent: there is no longer a continuous current. The pulse of anode current initiated by a single photon absorption has a duration that is determined by the spread in transit time between electrons

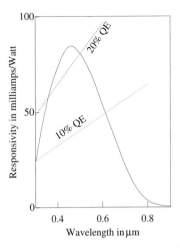

Fig. 15.26 The responsivity of an S20 photocathode inside a borosilicate window. Lines are drawn for quantum efficiencies of 10% and 20%.

which have followed longer and shorter paths down the dynode chain. The layout of the dynodes shown in figure 15.25 is designed to minimize this variation. Typically the FWHM of fast photomultiplier pulses is about 2 ns. If the photomultiplier has a gain of G and the pulse of current produced by a single photon lasts for Δt, then the anode current is $Ge/\Delta t$ and the voltage developed across a resistive load R_{L} is

$$V = GeR_{\mathrm{L}}/\Delta t.$$

Taking a gain of 10^7, a pulse lasting 5 ns and a load of $50\,\Omega$ then the pulse height is 16 mV. The inherent fast response, short dead time and high amplification of photomultipliers therefore make it relatively straightforward to detect individual photons and to time their arrivals when the photon rate is even as high as 100 MHz.

By measuring the arrival time to high precision (for example to 1 ns) it becomes possible to detect coincidences between photons arriving at different detectors, and alternatively to measure the distribution of the time interval between successive photons. Experimental work to be described in Chapter 18 which depended for its success on the ability to discriminate single photon pulses and to time photon pulses accurately was carried out with fast photomultipliers. Figure 15.27 indicates the distribution in the peak pulse voltage from a photomultiplier when single photons are recorded with the intervals between successive photons being much longer than the response time of the photomultiplier. The three curves are distributions obtained with equal numbers of photons in each case, but with different anode voltages. The higher the applied voltage, the broader and flatter the pulse height distribution becomes. It is the randomness inherent in the multiplication along the dynodes that is responsible for the width of the distributions. Pulses contributing to the small sharp rise at the lowest pulse heights in figure 15.27 are spurious noise pulses from such effects as thermal emission at dynodes. In order to count the photons the pulses are fed to a *pulse height discriminator* which only produces an output pulse when the input pulse exceeds a *threshold* as indicated by the arrow in figure 15.27. These output pulses from the discriminator are then counted electronically. Looking at figure 15.27 it is evident that the count rate increases as the anode voltage increases. However the applied voltage must be not be set so high that the noise pulses exceed the threshold for counting.

Figure 15.28 shows how the counting rate varies with the voltage applied to the photomultiplier for three beam intensities. When the applied voltage is very low the gain is low and the pulse height is rarely large enough to exceed the threshold. The count rate reaches about 50% of the true rate when the applied voltage approaches a value at which the height of the average pulse is equal to the threshold voltage. At that stage the count rate is increasing steeply with voltage because the amplification increases like a high power of the voltage. After this steep rise there is a slowly rising plateau where the increase is due to noise

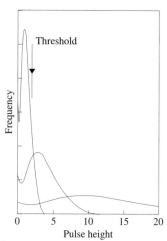

Fig. 15.27 Photomultiplier pulse height distributions for three different anode voltages. The number of pulses is the same for all three plots.

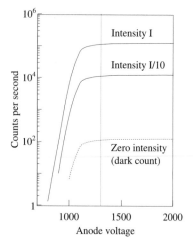

Fig. 15.28 Photomultiplier count rate versus applied voltage for beams of relative intensity I, $0.1I$ and 0. The vertical line indicates the 'plateau' condition for valid count rates.

pulses crossing the threshold. In order to maintain a valid count rate the photomultiplier voltage should be set about 10% above the knee. The counts recorded will include some spurious counts from other origins such as cosmic rays, radioactive material in the photomultiplier and noise so that it is necessary to make a pulse count with no light incident and to subtract this background count from that taken with the light on. After subtraction there is only a small residual noise, which comes from fluctuations in the background counts. Similar procedures to those described in this section would also be used in photon counting and timing with APDs.

15.9 Microchannel plates and image intensifiers

A single *microchannel* is an analogue version of the dynode chain in a photomultiplier. It consists of a narrow diameter glass tube whose bore is coated with a good secondary emitter. The voltage to produce the acceleration is applied between the ends of the tube. Then an electron entering the negative end of the microchannel will be accelerated along the microchannel and when it strikes the tube wall it can eject secondary electrons. This process repeats many times along the microchannel so that a pulse of electrons emerges from the positive end of the tube. Figure 15.29 illustrates the development of the electron cascade along a microchannel. A *microchannel plate* (MCP) is a glass plate typically penetrated by several million parallel identical microchannels: looking at either face the openings form a close packed regular array of identical circular apertures. The ratio of the channel diameter to the plate thickness, is about 1:50, with the choice of channel diameter ranging from $5\,\mu$m upward. Each face of the plate is coated with a metal film so that there is good electrical contact to the throat of each channel.

Fig. 15.29 Development of an electron cascade along an individual microchannel.

The raw materials for the construction of a plate are uniform cylindrical glass rods with lead glass cladding over a borosilicate crown core. These rods are stacked together, and the stack is heated and stretched. The resultant long multiple rod is sliced into lengths which are again stacked together, heated and stretched. After enough repetitions of this sequence the resulting solid block now contains millions of individual rods fused together, and this block is sliced to produce individual plates. Finally the softer borosilicate crown glass cores are etched away in a hydrogen atmosphere. This process also chemically reduces the inside surface of the lead glass in each channel to lead oxide, a material which is a good secondary emitter. Generally the block is sliced so that the channels are tilted around 10 degrees away from the normal to the plate's surface. Without this tilt the electrons, which are incident almost perpendicular to the surface, could travel through a channel without once striking its wall. The area of the channel openings makes

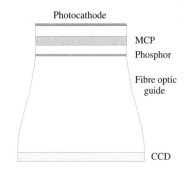

Fig. 15.30 Image intensifier based on a microchannel plate.

up a fraction of generally between 0.4 and 0.7 of the whole channel plate surface. The gain obtained when a potential of 1000 V is applied across a single plate is about 10^4, and with a matched pair of MCPs in series the gain rises to about 10^7.

Microchannel plates are widely used in image intensifiers. One such device is illustrated in figure 15.30. It shows a section through the intensifier containing the optic axis. The incident light from a camera lens falls on the photocathode surface which is located parallel to, and about 1 mm distant from the MCP. A MCP used in this way could typically have diameter 25 mm, a thickness $250\,\mu$m and have pores of diameter $10\,\mu$m. A potential of 200 V accelerates the photoelectrons ejected from the photocathode onto the MCP front face, and a voltage of 1000 V applied across the MCP itself produces the multiplication in the channels. After this the electrons emerging from the output face of the MCP are accelerated for the last time through a final potential of 2–3 kV to impinge on a phosphor coated glass surface a few millimetres distant. In this way an intensified monochrome replica of the image projected on the photocathode is produced at the phosphor. The components so far described form a simple image intensifier as used in night sights and can produce a useful image in starlight when the illuminance is only 10^{-3} lx.

Figure 15.30 includes further components that have not yet been mentioned. A solid block of optical fibres is used to transfer the light emerging from the phosphor onto a CCD array. This final stage provides the capability to store and subsequently process the enhanced image produced at the phosphor. The image resolution is defined by the granularity of the MCP, and the granularity of the fibre block is made fine enough so that it does not degrade this resolution. An image intensifier is enclosed in a shell that is evacuated to a very high vacuum, $< 10^{-5}$ torr. This is necessary in order to suppress the number of ions produced in electron–gas molecule collisions: these ions get accelerated by the electric fields and on impact damage the photocathode and MCP.

An assembly consisting of a photocathode and an MCP with an anode faceplate has the same overall function as a photomultiplier, and is called an MCP-PMT. These devices have very similar quantum efficiency and responsivity to photomultipliers because they use similar phosphors. Their timing resolution is better because of the shorter electron path in a MCP-PMT. When the pores have diameters less than $10\,\mu$m the transit time spread, is only 50 ps wide, so that MCP-PMTs have a time resolution of ~ 10 ps. Furthermore they can be gated on and off electronically in times of order 10 ns simply by reversing the relatively low cathode-MCP potential. This gives MCP-PMTs a rate capability approaching that of PMTs. However MCP-PMTs have areas under 1 cm^2, which is roughly one thousandth that of large PMTs. Finally the MCP-PMT is more robust and can operate in magnetic fields that would render a photomultiplier useless.

15.10   Further reading

Detection of Light; from the Ultraviolet to the Submillimeter, second edition,by G. H. Rieke, published by Cambridge University Press (2003), provides a comprehensive and detailed account of detectors extending from the ultraviolet well into the infrared.

Semiconductor Detector Systems by H. Spieler, published by Oxford University Press (2005), gives a thorough account of the electronic systems into which detectors are integrated.

Exercises

(15.1) A silicon photoconductor is cuboidal in shape with terminals in the form of metal layers over two opposite faces. Radiation of $1\,\mu$m wavelength is incident on a side face which is 2 mm wide and 1 mm long in the direction extending between the contacts. How thick should the silicon be in order to absorb 90% of the radiation transmitted through the air/silicon interface? The resistivity of the silicon is $3000\,\Omega$ m and its relative permittivity is 11.8. Calculate the electrical time constant of the device.

(15.2) A voltage of 200 mV is applied to the photoconductor described in the last question. Using a mobility, μ, of 0.135 m^{2}V^{-1}s^{-1} calculate the transit time of electrons through the photoconductor. If the recombination time is 0.1 ms what is the photoconductive gain?

(15.3) Estimate the fraction of light reflected from a silicon/air interface. Then determine the quantum efficiency and responsivity of the photoconductor described in the last two questions.

(15.4) If the photoconductor described in the last three questions is exposed to light modulated at a frequency 0.1 MHz would the photoconductor output show this modulation?

(15.5) A photodetector has a bandwidth, B, of 100 MHz, a dark current, i_d, of 4 nA at 300 K. The optical power falling on the detector's sensitive area is $2\,\mu$W, for which the responsivity is 0.6 A W^{-1}. The detector current flows through a load of resistance,

R, equal to $50\,\Omega$. Calculate the shot noise, the Johnson noise and the signal to noise ratio. What is the NEP of this detector?

(15.6) If the light intensity to which the detector described in the last question is exposed is increased to 2 mW recalculate the shot noise, Johnson noise and SNR.

(15.7) A photomultiplier has a photocathode area of $200\,$mm^2, and a quantum efficiency of 20% at 500 nm wavelength. The anode voltage is 2 kV giving an average multiplication of 5 at each of the ten dynodes. The anode dark current is 3 nA and the load R is $50\,\Omega$. Calculate the gain, responsivity, NEP and detectivity.

(15.8) With which, if any, photodetector is it possible to discriminate between the arrival of one or two photons simultaneously?

(15.9) Why is it that photoconductors produce gain whereas photodiodes do not? In the case of photoconductors why does high gain imply a slow response?

(15.10) Show that the product of the bandwidth and the responsivity for a photoconductor equals $e\eta\lambda/(hc\tau_x)$, where τ_x is the spread in the transit time.

(15.11) The quantum efficiency of a silicon APD is 0.8 at 800 nm. The incident radiation power on the APD is $0.1\,\mu$W, and the current from the APD is $50\,\mu$A. Determine the APD gain.

Optical fibres

16.1   Introduction

Optical fibres were introduced in Section 2.3. Standard glass fibre is cylindrically symmetric in cross-section, with a central core surrounded by cladding with a slightly lower refractive index. Light is guided along the fibre within the core by total internal reflection at the interface between core and cladding. In what is called *single mode* fibre the core is so fine ($\sim 8\,\mu$m diameter) that at near infrared wavelengths only a single optical mode propagates within the core along the fibre. Modern single mode fibre has been developed to have very low attenuation and also low dispersion for single mode propagation at wavelengths in the near infrared. Thanks to these properties it became possible to transmit information at very high rates over exceptionally long distances on optical fibre. Information is transfered to a laser beam using for example a modulator like that shown in figure 10.17 and the beam is injected into the core of an optical fibre. At the far end of the fibre connection a photodetector recovers the initial electrical signal from the beam. Correspondingly high frequency electrical signals carried on conductors are attenuated much more rapidly because the skin effect confines the current to a thin surface layer and hence leads to high resistance.

In *wavelength division multiplexing* (WDM) typically tens of laser beams equally spaced in wavelength, each carrying a separate information stream, are transmitted on a single fibre. The data rate per beam is many gigabits per second ($\mathrm{Gb\,s^{-1}}$), where a gigabit is 10^9 bits. Each beam therefore carries as much information as several satellite links. Fibre links also have a longer life than satellites and avoid the awkward time delay inherent in satellite communication. Millions of kilometres of single mode optical fibre now carry the major share of the data, video and audio traffic around the world. For example optical fibre links across the Atlantic and Pacific can support transfer of many terabits of information per second ($\mathrm{Tb\,s^{-1}}$), one terabit being 10^{12} bits.

Over short distances information transmission on optical fibre is less competitive for a number of reasons. Coaxial cable and twisted pairs then offer comparable data rates; the cost and complexity of the electrical/optical conversion at the fibre ends becomes more significant; and fibre connections require more sophisticated tools and testing as well as a cleaner environment. Consequently the distribution of information over the shorter multiple paths to offices and homes, the so-called *last mile*, is as yet limited.

In comparison with electrical connections optical fibre has other advantages that are of importance in many applications. Optical fibre is insensitive to electromagnetic interference, and does not generate any, nor are there the problems with ground loops that affect long electrical links. It is also lighter and occupies far less space, and is made from some of the commonest elements in the Earth's crust.

In the first section below attenuation in optical fibre is discussed. Then the theory of propagation along optical fibre will be outlined, showing in what circumstances transmission can be limited to a single optical mode. Single and multimode fibre properties are discussed including their attenuation and dispersion. A short section used to introduce signalling concepts is followed by a discussion of suitable sources and detectors. Then follows an account of devices used to connect fibres and route radiation between fibres. After this a simple link with a laser source and photodiode detector is considered: the noise sources are described and hence the launch power needed to give a low bit error rate is deduced.

In the next section long haul links are discussed. Methods of signal amplification on fibre, dispersion compensation and wavelength division multiplexing are all covered.

Solitons are waves which travel without change of shape: they may be electromagnetic waves on optical fibre, water waves or other waves. Electromagnetic soliton waves offer advantages over rectangular pulses in carrying information over optical fibre, and currently their exploitation in information transfer is beginning. Soliton propagation is analysed in Appendix D, and a section in this chapter is devoted to describing their potential in communication over optical fibre.

A growing number of sensors either incorporate optical fibres as the sensor or use optical fibre to transmit or collect radiation. Examples are treated in the last section of this chapter, and they include fibre optic gyroscopes and optical current transformers. The ancillary devices needed to manipulate radiation in these applications are similar to those used in communications.

16.2   Attenuation in optical fibre

A principal reason why long haul communication traffic is carried predominantly on optical fibre lies in the exceptionally low absorption loss which near infrared radiation suffers in pure silica. Standard single mode optical fibre is made from pure silica with a core doped with a few percent of GeO_2. The core has diameter of around $8\,\mu$m and the surrounding cladding has an outer diameter of $125\,\mu$m. Typically the refractive index of the core is a fraction of a percent larger than that of the cladding. Attenuation in this specialized optical fibre is sufficiently small that it is generally expressed in $dB\,km^{-1}$, a dB being defined by eqn. 1.33. Figure 16.1 shows the attenuation along single mode optical fibre plot-

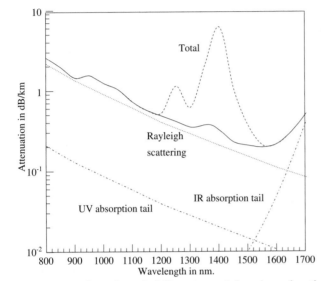

Fig. 16.1 Losses in single mode optical fibre at near infrared wavelengths. The plot is adapted from T. Miya, Y. Terunuma, T. Hosaka and T. Miyashita, *Electronics Letters* 15, 106 (1979) and published by the Institute of Engineering and Technology; by courtesy of the publishers.

ted against wavelength for near infrared radiation. The upper broken curve indicates the attenuation achieved through the use of ultra-pure materials, for which the loss approaches $0.2\,dB\,km^{-1}$ at 1550 nm wavelength, corresponding to about a 4.5% loss per kilometre. By contrast passage through tens of metres of ordinary window glass attenuates the intensity of incident light to an unusable level. The absorption minimum found near 1550 nm lies in a gap between atomic absorption peaks in the ultraviolet and molecular absorption peaks in the infrared. There is an additional unavoidable loss imposed by Rayleigh scattering from small scale inhomogeneities in density or chemical composition that are frozen in when the glass cools during manufacture. Contamination from metals has to be held below parts per billion. For example one part per billion of copper would produce further attenuation peaking around 850 nm with an amplitude of $1\,dB\,km^{-1}$. The remaining absorption peaks, following the upper broken line in the figure, are due to water trapped in the glass. Photons in radiation at these wavelengths have the correct energy to excite the vibrations of the OH ions trapped within the glass. The solid curve in figure 16.1 shows the attenuation currently achieved

Table 16.1 ITU bands.

Band	Wavelength in nm
O	1260–1360
E	1360–1460
S	1460–1530
C	1530–1565
L	1565–1625
U	1625–1675

by almost totally excluding water from the glass. Whereas transmission had been restricted to wavelengths close to 1310 nm or 1550 nm, now there is a 400 nm wide band of wavelengths over which the attenuation is low enough for long haul useage. Table 16.1 lists the wavelength bands defined by the *International Telecommunication Union* (ITU).

As an example of modern trans-oceanic optical fibre connections the Apollo link installed in 2003 consists of four fibre pairs across the Atlantic, each pair with a capacity of 800 gigabits of data per second.

High frequency electrical signals carried on coaxial metal cables are, by comparison, severely attenuated by the *skin effect*. As the frequency rises the current in a conductor is confined to an ever thinner surface layer. This causes the resistance of the cables to increase and hence the attenuation of the signal to increase as its frequency rises. The loss in modern video coaxial cables is 10 dB over 100 m at 100 MHz, and at 1 GHz the loss is 33 dB. For short links cheap plastic fibre is also available which can be used with visible or near infrared radiation: the attenuation of such fibres ranges from $\sim$20–200 dB km^{-1} depending on the type of plastic used.

16.3   Guided waves

In Chapter 2 guidance of light along optical fibres was explained in terms of rays undergoing total internal reflection at the core/cladding boundary. The analysis carried out next is based on wave theory and will be confined to the case of *step index* fibre, that is fibre in which there is a step in refractive index between a core of uniform refractive index and cladding of a different uniform refractive index. This analysis reveals the lateral distribution of the light propagating along the fibre and shows that with a small enough ratio between the core diameter and the wavelength only a single optical mode is trapped and travels along the fibre. The simpler case of light trapped in a plane sheet of glass sandwiched between two sheets of lower refractive index was treated in Section 9.9, and the reader may like to refer to that section before reading on.

The electromagnetic wave travelling along the fibre is taken to have angular frequency ω so that the electric field is

$$E = U \exp\left(-i\omega t\right). \tag{16.1}$$

Inserting this into the wave equation, eqn. 9.36, and using cylindrical polar coordinates (r, ϕ, z) with the z-axis pointing along the fibre axis gives

$$\partial^2 U/\partial r^2 + (1/r)\partial U/\partial r + (1/r^2)\partial^2 U/\partial \phi^2 + \partial^2 U/\partial z^2 + k^2 U = 0, \tag{16.2}$$

where the magnitude of the wave vector k is $n\omega/c$, n being the refractive index. In single mode fibre the refractive indices of the core, n_1, and the cladding, n_2, differ by a fraction of one percent. It is then justified to make the simplifying approximation that the waves travel in the z-direction with transverse electric and magnetic fields. This is known as the *weak guiding regime*, in which

$$U = u(r) \exp\left(i\ell\phi\right) \exp\left(i\beta z\right). \tag{16.3}$$

In the weak guiding regime the modes can be linearly polarized. Two orthogonal choices are possible, for example with $\mathbf{E}$ in the x- or y-directions. This justifies the way that the electric field was presented in eqn. 16.1 as a scalar, from which the vector field $\mathbf{E}$ can be recovered by multiplying by a unit vector in the transverse $r\phi$ plane. The integer ℓ specifies how many times the waveform repeats around the fibre axis, and β, the *propagation constant*, is the wave vector component along the direction of travel. Substituting the waveform 16.3 into the wave equation, eqn. 16.2, gives

$$\mathrm{d}^2 u/\mathrm{d}r^2 + (1/r)\mathrm{d}u/\mathrm{d}r + (k^2 - \beta^2 - \ell^2/r^2)u = 0. \tag{16.4}$$

Equation 16.3 is a form of Bessel's equation, with solutions in which the radial component of the wave vector is $\sqrt{(k^2 - \beta^2)}$. Within the core and cladding the wave numbers are $k_1 = n_1\omega/c$ and $k_2 = n_2\omega/c$ respectively. Now if the wave is to be guided along the fibre its transverse waveform inside the core has to be a standing wave, which means that $\sqrt{(k_1^2 - \beta^2)}$ is real. At the same time the transverse waveform in the cladding must be an evanescent wave, with $\sqrt{(k_2^2 - \beta^2)}$ imaginary. Thus the propagation constant should lie in the range given by

$$k_2 < \beta < k_1. \tag{16.5}$$

In the core the solution to eqn. 16.4, that is to say the mode, takes the form of a Bessel function of the first kind,

$$u(r) = E_{01} J_\ell(k_\mathrm{r} r), \tag{16.6}$$

where $k_\mathrm{r} = \sqrt{(k_1^2 - \beta^2)}$. This solution oscillates approximately like a damped sinusoid. In the cladding the solution is a modified Bessel function of the second kind

$$u(r) = E_{02} K_\ell(\kappa_\mathrm{r} r), \tag{16.7}$$

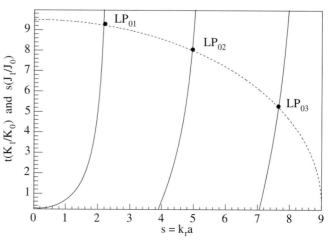

Fig. 16.2 Solutions to the characteristic equation for the propagation of LP_{0m} modes on optical fibre. The parameters are defined in the text. V is set to 9.0.

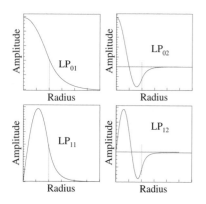

Fig. 16.3 Radial waveforms of the LP_{01}, LP_{02}, LP_{11} and LP_{12} modes. The dotted lines indicate the core radius in each case.

where $\kappa_r = \sqrt{(\beta^2 - k_2^2)}$. This solution resembles a decaying exponential. How the core and cladding solutions fit together is determined by the requirements, given in eqns 9.24 and 9.25, that the tangential components of the electric and magnetic field are continuous at the surface between the core and the cladding; that is at radius a. These conditions can be satisfied simultaneously provided that[1]

$$k_r a \frac{J_{\ell\pm1}(k_r a)}{J_\ell(k_r a)} = \pm\kappa_r a \frac{K_{\ell\pm1}(\kappa_r a)}{K_\ell(\kappa_r a)}, \tag{16.8}$$

which is the *characteristic equation* for a given choice of ℓ in eqn. 16.3. At this point it is useful to introduce a quantity, called the *normalized frequency*, V, which is defined by

$$V^2 = k_r^2 a^2 + \kappa_r^2 a^2 = (\omega^2/c^2)(n_1^2 - n_2^2)a^2. \tag{16.9}$$

For the lowest order modes with $\ell = 0$, the characteristic equation reduces to

$$sJ_1(s)/J_0(s) = tK_1(t)/K_0(t), \tag{16.10}$$

where $s = k_r a$ and $t = \kappa_r a$. This equation is solved graphically in figure 16.2 taking V to have the value 9.0. The solid curves show the variation of the left hand side of eqn. 16.10 with s: note that they cross the s-axis wherever $J_1(s)$ is zero. The broken line curve shows the corresponding behaviour of the right hand side of the equation. Solutions to eqn. 16.10 occur where the curves intersect, and these locations are marked by solid dots and labelled with the modes' names. Figure 16.3 shows the radial dependence of the electric field amplitudes for several modes. The name

---

[1]The proof of this statement can be found in the fifth edition of *Optical Electronics in Modern Communications* by Amnon Yariv, published by Oxford Universty Press (1997).

LP$_{\ell m}$ signifies a linearly polarized mode, with the first subscript indicating the number of nodes around the z-axis and the second subscript indicating the number of radial maxima in intensity. The cross-sections showing one choice for the polarizations of the LP$_{01}$ and LP$_{11}$ modes are sketched in figure 16.4. There are two degenerate modes in each case which have the same spatial distributions, but orthogonal polarizations.

Solutions of the characteristic equation are eigenstates, or modes, that would travel without changing their form along the fibre in the absence of absorption in the glass. In the case of any other waveforms the boundary conditions require that the electric field in the cladding increases with the radial distance. These waves therefore spread into the cladding and are eventually lost.

Looking at figure 16.2 it is evident that the normalized frequency V determines the number of the freely propagating modes. The LP$_{0m}$ mode can only be a solution to the characteristic equation provided that V is larger than the value of $k_r a$ at which mth zero of $J_1(k_r a)$ occurs. These zeroes are located at values of $k_r a$ equal to 0, 3.83, 7.02, etc. For the LP$_{1m}$ modes the cut-offs are at the zeroes of $J_0(k_r a)$, namely when $k_r a$ equals 2.405, 5.552, etc., while other modes have even larger cut-off values in $k_r a$. If a mode is launched along a fibre and the radiation has a frequency such that the value of V is just smaller than the cut-off for the mode, then radiation leaks gradually into the cladding without returning. Such modes are called leaky or lossy modes.

When V is less than 2.405 only the LP$_{01}$ mode can propagate freely, and the fibre is called *single mode* or *monomode*. Single mode operation is the standard for telecom usage because of its simplicity: the waveform has a cross-section not very different from the Gaussian cross-section of the mode preferred for laser operation; interference between modes and dispersion between modes travelling with different group velocities are eliminated. Using eqn. 16.9 it follows that V will be sufficiently small to exclude all but the LP$_{01}$ mode when

$$\sqrt{(n_1^2 - n_2^2)}(\omega/c)a < 2.405. \qquad (16.11)$$

Thus at 1310 nm, which has been a preferred telecom wavelength,

$$\sqrt{(n_1^2 - n_2^2)}\, a < 0.5 \ \mu\text{m}.$$

If the core radius were only a micron then it would be difficult to manufacture and hard to inject much light into such a fibre. Consequently the refractive indices of the core and cladding have to be almost equal in order to have a reasonably large core radius. There will also be a shortest wavelength for which an optical fibre operates as a single mode fibre. This cut-off wavelength is

$$\lambda_{\min} = 2\pi a \sqrt{(n_1^2 - n_2^2)}/2.405 = 2.61a\sqrt{(n_1^2 - n_2^2)}.$$

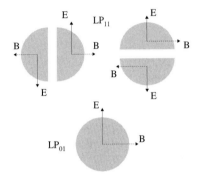

Fig. 16.4 Sketches of the cross-sections and field orientations in the LP$_{01}$ and LP$_{11}$ modes. In each case drawn there are orthogonal modes which have the electric and magnetic field directions exchanged.

The numerical aperture of a step index fibre is given by eqn. 2.41

$$NA = \sqrt{(n_1^2 - n_2^2)}, \tag{16.12}$$

which will necessarily be small for a single mode fibre because n_1 and n_2 differ so little.

An estimate of the number of modes propagating freely in a multimode fibre can be made as follows. Using eqn. 4.13 the etendue of an optical fibre is

$$\mathcal{T} = \pi^2 a^2 (n_1^2 - n_2^2). \tag{16.13}$$

Now the etendue per mode is given by eqn. 12.45, so that an estimate of the number of modes that freely propagate is $\mathcal{T}/\lambda^2$ per polarization state. Hence the total number of freely propagating modes is

$$N = 2\mathcal{T}/\lambda^2 = 2\pi^2 a^2 (n_1^2 - n_2^2)/\lambda^2 = V^2/2, \tag{16.14}$$

This estimate is a good approximation when V is large.

For future reference an *effective area* of the cross-section of an optical mode travelling along an optical fibre is defined here as

$$A_{\text{eff}} = \left[\int I(\mathbf{r}) \, \mathrm{d}A \right]^2 / \int I^2(\mathbf{r}) \, \mathrm{d}A \tag{16.15}$$

where $I(\mathbf{r})$ is the intensity of the optical mode at a point $\mathbf{r}$ on the cross-section. Launching light into the core of a single mode fibre requires a very tight and well aligned beam, which lasers are well suited to provide.

16.4   Fibre types and dispersion properties

In a widely used standard single mode fibre, Corning$^\circledR$ SMF-28^{TM}, the core diameter is $8.2\,\mu$m with refractive index 0.36% higher than that of the cladding, while the effective group velocity is c/1.4677 at 1310 nm. The cladding has an outer diameter of $125\,\mu$m, which is adequate to fully contain the evanescent tail of the LP_{01} mode. At 1310 nm the NA of this monomode fibre is 0.14, and the effective area of the LP_{01} mode is $8\,\mu$m^2. The loss at 1550 nm is less than $0.22\,\text{dB}\,\text{km}^{-1}$.

Multimode fibre has a much larger core diameter than single mode fibre. The difference between the refractive indices of the core and cladding is made larger too. $50\,\mu$m is a common choice for the core diameter, with the result that the numerical aperture is larger and generally lies in the range 0.2–0.5. At visible and near infrared wavelengths there are up to several thousand modes freely propagating along multimode fibre. Most modes are far from the cut-off wavelength and therefore have weak evanescent tails. The cladding outer diameter is standardized at

$125\,\mu$m in common with single mode fibre, and is adequate to contain these weak evanescent tails of the modes in multimode fibre.

A protective buffer layer and a tough cover of outer diameter $250\,\mu$m are standard for both single mode and multimode fibre. The buffer has a higher refractive index than the cladding in order to prevent modes in the cladding from being trapped there by total internal reflection at the cladding/buffer interface.

The dispersion in multimode fibre is simpler to analyse in terms of ray paths, rather than by calculating the dispersion between the modes. In the upper panel of figure 16.5 the paths of rays are shown within the core of a *step index* multimode fibre. In this class of fibre the refractive index changes sharply between core and cladding, just as in standard single mode fibre: this is shown in projection to the right of the panel. An axial ray has the shortest path, while those striking the interface at the critical angle of incidence $\theta_c = \sin^{-1}(n_2/n_1)$ have the longest path. The difference between the times taken by the axial and critical rays to travel an axial distance L along the fibre is

$$\Delta t = (L/\sin\theta_c - L)/v = L(n_1 - n_2)/(n_2v),$$

where $v = c/n_1$. Then

$$\Delta t = n_1\Delta n L/(n_2 c),$$

where $\Delta n = n_1 - n_2$. Dispersion is far less in the multimode *graded index* fibre shown in the lower panel of figure 16.5. Its refractive index has the quadratic dependence on radius shown on the right of the panel. Wide angle rays travel through material of lower refractive index off axis and follow curved paths similar to those of rays in the GRIN lenses described in Section 4.6. In combination these effects help to reduce the difference between the optical path lengths of the wide angle rays and the axial ray. The corresponding time difference now drops to

$$\Delta t = n_1(\Delta n)^2 L/(8n_2^2 c),$$

where n_1 is the axial refractive index.

The spread in transit times due to intermodal dispersion on step index and graded index multimode fibre can now be estimated for typical fibres. We take the axial refractive index in the core to be 1.50 and the refractive index of the cladding to be 1.47. Then over 1 km of step index fibre this spread is 102 ns, and over graded index fibre it is 0.26 ns.

The single allowed mode travelling along a single mode fibre is of course free of intermodal dispersion. There remains the chromatic dispersion caused by the variation of the refractive index of both core and cladding with wavelength. This dispersion has two principal components. First the *material dispersion* due to the change of the refractive

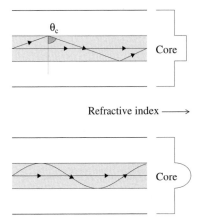

Fig. 16.5 Sketches of the paths of rays travelling in the core of step index (above) and graded index (below) multimode fibre. On the right the refractive index profiles are drawn.

indices with wavelength, and second the *waveguide dispersion* due to the change of the distribution of the LP_{01} mode between core and cladding with wavelength. The group velocity of radiation is given by eqn. 11.41

$$v_\mathrm{g} = \mathrm{d}\omega/\mathrm{d}k$$

where ω is the angular frequency and k is the wave number. Then the difference in transit times between light of free space wavelength λ and $\lambda + \Delta\lambda$ over a path length L is

$$\Delta t = [\,\mathrm{d}(L/v_\mathrm{g})/\mathrm{d}\lambda\,]\,\Delta\lambda = L\,D_\mathrm{m}\,\Delta\lambda, \qquad (16.16)$$

where

$$D_\mathrm{m} = \mathrm{d}(1/v_\mathrm{g})/\mathrm{d}\lambda$$

is known as the *group velocity dispersion*. Using eqn. 11.43 to substitute for $1/v_\mathrm{g}$ gives

$$D_\mathrm{m} = -(\lambda/c)\,(\mathrm{d}^2 n/\mathrm{d}\lambda^2). \qquad (16.17)$$

In addition a change in wavelength affects the solution of eqn. 16.10 and hence the distribution of the mode energy between the core and cladding. This in turn affects the group velocity. The resulting waveguide dispersion will depend on the degree to which the mode penetrates the cladding and hence on the ratio a/λ. Figure 16.6 indicates the two contributions to the chromatic dispersion in single mode fibre expressed as the spread in transit time in picoseconds over one kilometre for a change in wavelength of one nanometre. The material contribution is

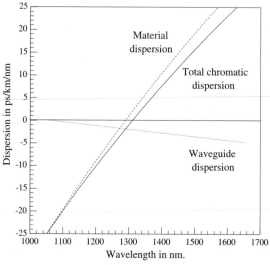

Fig. 16.6 Chromatic dispersion in standard single mode fibre.

drawn with a broken line, the waveguide contribution with a dotted line, and the total dispersion with a full line. These plots are appropriate for a standard silica fibre with a GeO_2 doped core of diameter $\sim 8\,\mu m$. Very conveniently the zero dispersion wavelength coincides with the wavelength, 1310 nm, at which the absorption in optical fibre (between two OH absorption peaks) has a minimum. Hence wavelengths around 1310 nm were initially preferred for long haul transmission. The wavelength at which the chromatic dispersion vanishes can be altered by changing the doping level or by manipulating the radial distribution of the doping. For example reducing the core diameter shifts the zero dispersion point to higher wavelengths. However the energy density over the core increases, which reduces the overall power level at which non-linear effects become important.

A comparison between the spread in transit time due to dispersion in multimode and single mode fibres requires a choice of bandwidth for the source. As an average case take a bandwidth of 0.4 nm (5 GHz). This yields a spread in transit times of picoseconds per kilometre, which is around a thousand times less than that on multimode fibres. Hence information can be transmitted at around a thousand times higher rate on single mode fibre. Its low attenuation and low dispersion at near infrared wavelengths accounts for the dominance of the use of single mode optical fibre in long haul communications.

The near coincidence between the wavelength where the absorption is least and that at which the dispersion vanishes is not just due to chance. Referring back to Chapter 11 and in particular to figures 11.3 and 11.4 it is seen that the behaviour of the refractive index and absorption are linked. Replotting the refractive index as a function of wavelength between a pair of absorption resonances yields the diagram drawn in the upper panel of figure 16.7. In this region the absorption always has a minimum. The first and second differentials of the refractive index are shown in the lower panels of figure 16.7. Evidently the material dispersion contribution to the group velocity dispersion, given by eqn. 16.17, will vanish close to the absorption minimum.

Asymmetries in the fibre cross-section, composition and stress lead to differences between the group velocities of the orthogonal LP_{01} modes resulting in *polarization mode dispersion* (PMD). If the refractive indices of the two polarizations differ by Δn then the phase difference over a path length L is $2\pi L \Delta n / \lambda$, where λ is the free space wavelength. Hence the initial phase relation between the two polarization states repeats after a distance $\lambda / \Delta n$, which is known as the beat length. Over other path lengths a beam that was initially plane polarized emerges with elliptical polarization. Temperature and pressure affect this birefringence and the result is that the polarization state of the emerging light can fluctuate rapidly. Photodetectors are polarization insensitive so that this fluctuation does not affect the actual detection process. However the dispersion between the polarization components becomes important

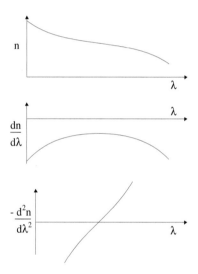

Fig. 16.7 The refractive index and its first two derivatives in the wavelength region between two absorptive resonances.

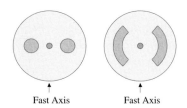

Fast Axis   Fast Axis

Fig. 16.8 Panda and bow tie polarization maintaining fibre cross-section. The small circle at the centre of each is the core.

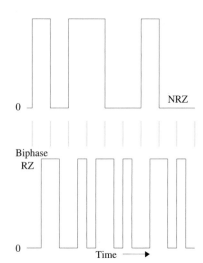

Fig. 16.9 Binary codes: non-return to zero and biphase return to zero. The markers indicate the end of each bit period.

at extreme data rates. Single mode fibres are now manufactured with polarization mode dispersion below $0.1\,\mathrm{ps}/\sqrt{\mathrm{km}}$ which leads to a differential delay of less than 1 ps on a 100 km long link.

In applications which require that the state of polarization is well defined *polarization maintaining* fibres are used. Surprisingly these fibres are designed to have large built-in stresses and hence large linear birefringence. Sections through polarization maintaining fibre types are shown in figure 16.8, for the so-called *bow tie* and *panda* fibres. The regions indicated by the heavier shading contain glass with a different thermal expansion coefficient from the rest of the cladding and this produces stresses when the fibre cools during manufacture. The difference between the travel times over a few metres of these fibres for the LP_{01} modes polarized along the fast and slow axes is around a few picoseconds. As a result these modes move in and out of phase very rapidly as they travel along the fibre. Only light linearly polarized exactly along the slow or fast axis of the fibre retains its linear polarization as it travels along the fibre. Poor alignment will mean that the light emerging from the other end of the fibre will have unstable polarization. When the alignment is off by an angle θ Malus's law predicts that the intensities of the polarization components along the two axes are proportional to $\cos^2\theta$ and $\sin^2\theta$. Thus an alignment better than $1.8°$ will keep the intensity ratio above 1000:1 and avoid such instability.

16.5   Signalling

Computer data, video and audio information are all presented for transmission down a link as electrical pulses. These are in the form of a string of binary *bits*: one voltage level represents a 0-bit and another level a 1-bit. The information is transferred to a laser beam that is injected into the fibre core. This transfer is made either by using the electrical pulses to modulate the laser voltage or to operate a modulator like that shown in figure 10.17 which acts on the laser beam. One common code used is the *non-return to zero* (NRZ) code, in which a binary string 101100100 appears as shown in the upper panel of figure 16.9. The upper level for a 1-bit is 'light on' and the lower level for a 0-bit is 'light off'. The highest data rates are tens of $\mathrm{Gb\,s^{-1}}$ (gigabits per second). This is still relatively small compared to laser frequencies of 300 THz so that each bit contains huge numbers of optical waves. A biphase *return to zero* (RZ) code is shown in the lower panel for the same bit string, and requires double the switching speed. In this coding scheme a 0-bit is signalled by a transition down during the time slot for that bit, while an upward transition signals a 1-bit. At the far end of the fibre link the light falls on a photodetector. In order to recover the bit stream from the output of the detector a discriminator is required. This compares the signal voltage to a voltage level half way between those expected for the 0- and

1-bits. It is clear that the discrimination must be made in the case of the NRZ code at a time close to the mid-point of each pulse period, which requires synchronization between the incoming pulses and the discriminator. Circuits for this purpose locate the pulse edges and adjust the timing continuously. They have therefore to be capable of interrogating the incoming pulses at frequencies much higher than the pulse frequency.

After travelling along many kilometres of optical fibre the pulses are distorted by the combined effects of noise, attenuation, dispersion and time jitter. One simple method used to assess the quality of the signal emerging after detection is to display the voltage level repetitively on a fast oscilloscope. The resulting display has the appearance shown in figure 16.10 and is known as an *eye diagram*. The number of pulses sampled is very large. In the upper panel the *eyes* are open and a voltage level can be set which will discriminate reliably between a 1-bit and a 0-bit. Many causes will lead to departures from this ideal: noise makes the pulse height vary, dispersion rounds the sharp edges of the pulses and timing jitter moves pulses forward and backward in time. Taken together they produce the partial closure of the eyes seen in the lower panel of figure 16.10. Discrimnation is then less reliable. Repeaters can be incorporated along links to limit the signal degradation. At these repeaters the incoming optical signals are converted back to electrical signals, which are then reshaped as square pulses and retimed, before being converted back to optical signals to travel the next section of the link.

Information to be transmitted is first packaged into blocks containing a fixed number of bits, some of the time slots being reserved for bits which are not part of the raw data. Of these additional bits some are used to identify the data destination and its priority. Other bits are used to assist in the detection and correction of errors in transmission. These error recovery bits are assigned values so that they form specific patterns with the data bits. Whenever the optical signal is converted to an electrical signal the data is checked by looking for these patterns. Only limited types of error can be corrected, in other cases a retransmission is requested for the block of data involved.

Amplitude modulation of a laser beam of frequency f with a sinusoidal wave at a much lower frequency Δf_m would give a waveform

$$[1 + \delta \cos (\Delta f_m t)] \cos (ft) =$$
$$\cos (ft) + (\delta/2) \cos [(f + \Delta f_m) t] + (\delta/2) \cos [(f - \Delta f_m) t],$$

containing components at frequencies $f \pm \Delta f_m$. A modulation deep enough to turn off the carrier wave at frequency f would have $\delta \sim 1$, so that these components would have large amplitudes comparable to that of the carrier. The pulses used in modulating laser beams transmitted along optical fibres are rectangular, which contain Fourier components of much higher frequency than the bit rate Δf_m. With this caveat in mind, Δf_m will be taken as an estimate of the bandwidth of the modulated

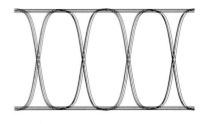

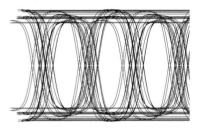

Fig. 16.10 Oscilloscope eye diagrams. In the upper diagram the eyes are open, but in the lower diagram the noise, dispersion and time jitter have partially closed the eyes.

beam. The corresponding wavelength spread is $\Delta\lambda_m = \Delta f_m \lambda^2/c$, where the wavelength in free space is $\lambda = c/f$. Using eqn. 16.16 the dispersion over a link of length L and group velocity dispersion D_m is therefore

$$\Delta t_m = \Delta f_m L D_m \lambda^2/c. \qquad (16.18)$$

This is to be compared with the duration of the signal pulses at frequency Δf_m, namely $\Delta t_p = 1/\Delta f_m$. If the signal is to be recognizable we must have

$$\Delta t_m \ll \Delta t_p, \qquad (16.19)$$

which can be re-expressed as

$$\Delta f_m^2 \ll c/(L D_m \lambda^2). \qquad (16.20)$$

As an example consider transmission over a 100 km link of standard single mode fibre with group velocity dispersion 17 ps/km/nm at 1550 nm wavelength.[2] This gives

$$\Delta f_m \ll 8.6\,10^9 \text{ Hz}. \qquad (16.21)$$

Bearing in mind the earlier caveat we can see that operation over such a link will be restricted to data rates well below $1\,\mathrm{Gb\,s}^{-1}$.

There is an additional contribution to the bandwidth from the laser line width, Δf_ℓ, so that the overall bandwidth is $\sqrt{[(\Delta f_m)^2 + (\Delta f_\ell^2)]}$. In any case the laser linewidth should be kept small compared to the modulation frequency, irrespective of the requirement to limit dispersion.

16.6   Sources and detectors

Semiconductor lasers have properties that make them ideal sources for communications over single mode optical fibre: firstly they have high brightness and an etendue matched to this optical fibre. The transverse mode of these lasers is the TEM$_{00}$ mode, whose Gaussian profile matches well with the LP$_{01}$ mode of optical fibre. A semiconductor laser typically injects a beam of many milliwatts power into the core of a single mode optical fibre. Alignment is obviously critical because the cross-sectional areas of both a fibre core and a semiconductor laser beam are small. Manufacturers supply laser packages in which the laser, a microlens and a fibre are already aligned and sealed. A length of the fibre called a *pigtail* extends from package and can be used to couple the source into a fibre link. If the laser is edge emitting the lens will need to be astigmatic to convert the laser beam profile to a circular shape. Another approach is to use anamorphic prism pairs as pictured in figure 2.19. A further key advantage of semiconductor lasers is that their line widths are very narrow compared to any other source and this, we have seen, is important in limiting the effects of dispersion. A typical choice of source is an InGaAsP/InP DFB laser of the type described in Section 14.7.2. By choosing the chemical composition appropriately the operating wavelength can be located between 1000 and 1700 nm. A bandwidth of 10 MHz is usual for a DFB laser.

At high frequencies, it is often preferable to use an external modulator rather than modulate the laser directly. The reason for this preference is that when a laser is turned on it takes some time for its the gain to reach equilibrium and during this interval the laser frequency chirps. A similar effect occurs when the laser is turned off. Travelling wave versions of the Mach–Zehnder modulator described in Section 10.7.1, can be used to modulate light at very high frequencies. These modulators have metal waveguides on which the electrical pulses propagate along the modulator arms in phase with the optical waves. In this way modulation rates of $40\,\mathrm{Gb\,s^{-1}}$ are achieved.

LEDs are well suited to use as sources for injecting light into multimode fibres. They are incoherent sources emitting in many modes with line widths of order 50 nm. They have comparable etendue, or put another way the many modes of the LED match with the many modes which can propagate freely on the multimode fibre. With so many modes the illumination across the fibre becomes more uniform and stable than it would with just a few modes filled. Using multimode fibre and LEDs is far less expensive than using single mode fibre and lasers. Besides this cost advantage the alignment of components is far less critical, and the components are themselves less delicate. Therefore for transmission over short distances the multimode fibre and LED combination is preferred. However the relatively large dispersion of multimode fibre and the broad linewidth of LEDs render them both unsuitable for long haul, high rate information transfer.

The photodetectors used to convert the optical signals to electrical signals are usually InGaAs/InP photodiodes. These may be pin photodiodes, or avalanche photodiodes when the highest sensitivity is required. The properties of these detectors have been discussed at length in Chapter 15. The impact of the noise produced by sources and detectors, and of detector sensitivity will be discussed later.

16.7   Connectors and routing devices

The devices described here are those commonly used in communication over single mode fibre, but they are often used in other applications and with other fibre types. Components such as lasers, photodiodes and routing devices are often *connectorized* by manufacturers so that the device package has one or more pigtails of monomode fibre, perhaps a metre in length. The fibre end could be fitted with a standard connector, or it could be a bare end, depending on the user's requirements.

Permanent fibre to fibre connections are made with mechanical or fusion splices. In either case the two fibre ends are prepared by stripping back the buffer layer and cleaving the core and cladding. Generally the cleaver consists of a steel anvil with a groove into which the fibre end is introduced. There is a hinged diamond cutter which is brought down to touch the fibre and the action cleaves the fibre at right angles to its length. This surface is checked with a microscope so as to detect and reject cleaved surfaces which show crazing or chipped edges.

A fusion splicer has several components: saddles into which the cleaved ends of the two fibres are clamped; precision controls to align and bring the fibre ends together; an arc welder; and a microscope used to monitor the processes. When the alignment is satisfactory the electric arc is fired across the fibre joint in order to produce a weld. The loss at a well made fusion splice is only $\sim$0.1 dB. Mechanical splices are made by inserting the fibres into opposite ends of a precision capillary. Connections between two sets of fibres are made using a silicon substrate into which precision grooves have been etched parallel to a crystal axis. Each pair of fibre ends to be connected are placed touching one another in one of the grooves. All the fibre ends are clamped in place by a second silicon block. Mechanical splices can be assisted by using a gel between the fibre ends with a refractive index matching that of the fibre. The loss at a well made mechanical splice is $\sim$0.25 dB.

Non-permanent junctions are made using connectors, and a typical example used with single mode fibre is shown in figure 16.11. Alignment depends on a precision ceramic ferrule through which a capillary tube runs. The stripped end of the fibre is fed through this capillary so that it protrudes slightly. Then the protruding end is ground down and polished flush with the ferrule surface. Two such connectors fit into either side of a precision ceramic sleeve and just touch one another. The connectors are secured to this sleeve by screw or bayonet latches similar to those used in electrical connectors. Losses at good connectors are only $\sim$0.6 dB, and only vary by $\sim$0.2 dB over a thousand insertions and withdrawals.

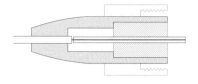

Fig. 16.11 Fibre optic connector. The connector body (cross-hatched) supports the precision ceramic ferrule through which the stripped fibre is threaded. The boot (shaded) grips the fibre cover and relieves the strain. Outside there is a threaded (or bayonet) latch.

Losses are also incurred if the fibre is bent appreciably, because then the angle of incidence at the core/cladding interface is changed and some light can escape into the cladding. Along a single turn of 30 mm diameter on a typical monomode fibre the loss is less than 0.5 dB. However the loss mounts rapidly as the turn diameter is reduced below this value. Bends can also be useful. When light is injected into an optical fibre some energy can enter leaky modes which may later re-enter the core at a connector or other device and disturb the operation of the link. These lossy modes can be shed easily by bending the fibre loosely over a snake-like former at a point along its length near the source.

16.7.1 Directional couplers

Directional couplers are devices used to couple light from one fibre to another. Two examples are shown in figure 16.12. The directional coupler in the upper panel is constructed from two fibres stripped of their protective coating. These are wound around one another, heated to near the melting point, stretched and then cooled. Within the fused region modes extend over the whole fibre so that radiation arriving in one core may exit through the other.

The second directional coupler shown in figure 16.12 consists of a waveguide structure built on a substrate of lower refractive index. Evanescent waves couple light between the light guides. When input A is illuminated from one end the variation of power in the two guides in the overlap region is

$$I_A(z) = I_A(0)\left[\cos^2(Cz) + b^2 \sin^2(Cz)\right],$$
$$I_B(z) = I_A(0)\, a^2 \sin^2(Cz),$$

where z is measured from the start of the overlap region, $a^2 + b^2 = 1$ and C is the strength of coupling. Perfect coupling has $b = 0$, and in this case an overlap of length $\pi/2C$ transfers all the power from input A to output B. The waveguides have rectangular cross-sections of a few microns side length so that radiation couples efficiently to or from single mode optical fibre butted against an entry or exit port.

Couplers can be designed to combine light of different wavelengths onto a single fibre. In an arrangement used in fibre amplifiers, light from a pump laser of wavelength 980 nm entering through input B is fully transferred to output A, whilst infrared light at 1550 μm which enters through input A emerges from output A. Another coupling device, the *star coupler*, is made by fusing several fibres together rather than two; in this case the input from one fibre is shared equally among all the outgoing fibres.

More sophisticated waveguide devices will be met as we proceed. The trend in manufacturing is to concentrate as many manipulations of light as possible within devices built on a single substrate, thus avoiding interconnections by fibre. These units are called *planar waveguide circuits*.

16.7.2 Circulators

The symbol used to indicate a three port circulator is shown on the left in figure 16.13. Light entering port 1 in the left hand panel exits only at port 2; light entering at port 2 exits only at port 3; light entering port 3 is lost. On the right a four port circulator is shown. Figure 16.14 shows the internal structure of a three port circulator. It incorporates two birefringent plates, BP1 and BP2, a Faraday rotator FR, which

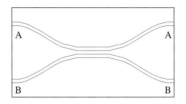

Fig. 16.12 Directional couplers. The upper panel shows two stripped fibres that have been welded together. The lower panel shows a planar waveguide coupler.

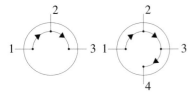

Fig. 16.13 Symbolic representations of three and four port circulators. The arrows indicate the permitted light paths within each circulator.

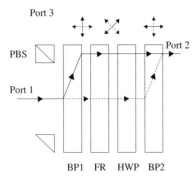

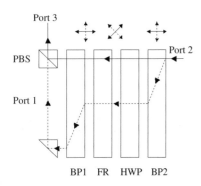

Fig. 16.14 Paths of light through a circulator. In the upper panel the paths from port 1 to port 2 are shown, and in the lower panel those from port 2 to port 3. The polarizations are indicated for each path. The labels are explained in the text.

gives a 45° rotation, a half-wave plate, HWP, and a polarizing beam splitter, PBS. In the upper panel light travels from port 1 to port 2, and in the lower panel from port 2 to port 3. Sketches are included of the polarization states as the light enters and leaves the rotator and as it leaves the half-wave plate; all as seen seen from port 1. The solid line polarization vectors correspond to the solid line path, and the broken line polarization vectors correspond to the broken line path.

Light entering port 1 with ordinary polarization travels straight through BP1 along the broken line, while light with extraordinary polarization walks off along the solid line. Each of these beams has its polarization rotated by 45° in the rotator, and then in travelling through the half-wave plate their polarizations get reflected about its optical axis. These two steps, taken together, interchange the polarizations of the two beams, and consequently there is a compensating walk-off in BP2 which reunites the beams at port 2. Travelling in the reverse direction, as shown in the lower panel of figure 16.14, the beams recover their initial polarization before entering BP1. The beam entering BP2 with extraordinary polarization (broken line) therefore walks off twice while the beam with ordinary polarization is undeflected. Both beams exit through port 3. When light enters port 3 each polarization component is deflected away from port 2 and is lost. Consequently the only connected paths are $1 \rightarrow 2$ and $2 \rightarrow 3$.

16.7.3 MMI devices

An example of these compact waveguide devices is shown in figure 16.15. The entry and exit waveguides have rectangular cross-section of a few microns side length in order to match the area of the modes in single mode optical fibres. Linking these waveguides is a broad slab in which multiple modes propagate between its entry and output ports. The length of this slab can be chosen such that the *multiple mode interference* (MMI) produces an image of the input ports coincident with the output ports. This makes a 2×2 coupler. MMI devices can also be designed to combine and split the incident beams in other more complicated ways: 1×n, 4×4 and so on. In the remainder of this section the principles of MMI operation will be demonstrated by analysing the simple 2×2 coupler.

Within the central slab each mode has a waveform

$$E = E_0 \exp\left[i(k_\mathrm{t}x + \beta z - \omega t)\right], \tag{16.22}$$

using the coordinates shown in figure 16.15. Each mode that propagates without loss has nodes at the side walls, so that

$$k_\mathrm{t}W = m\pi, \tag{16.23}$$

where k_t is the transverse x-component of the wave vector $\mathbf{k}$, W is the width of the wide section and m is an integer describing the mode order.

Also

$$k^2 = k_t^2 + \beta^2, \qquad (16.24)$$

where β is the propagation constant. With the slab geometry shown the waves propagate through it in a direction close to its long axis, and $k_t^2 \ll k^2$. Using eqn. 16.23 to replace k_t in eqn. 16.24 we then have to a good approximation

$$\begin{aligned}
\beta_m &= \sqrt{(k^2 - m^2\pi^2/W^2)} \\
&= k\,[\,1 - (m\pi/kW)^2/2\,] \\
&= 2\pi n/\lambda - m^2\pi\lambda/(4nW^2) \qquad (16.25)
\end{aligned}$$

for the mth order mode. Here λ is the wavelength in free space and n is the refractive index of the waveguides and slab. A *beat length* is defined over which the lowest two modes recover their initial phase difference,

$$L_{12} = 2\pi/(\beta_1 - \beta_2) = 8nW^2/(3\lambda). \qquad (16.26)$$

The relative phase between the first and the jth modes also returns to its initial value after a distance

$$L_{1j} = 2\pi/(\beta_1 - \beta_j) = 3L_{12}/(j^2 - 1).$$

Consequently all the modes recover their initial phase relation to the first mode after a distance $3L_{12}$. As a result the intensity pattern across the slab at this plane is the same as across its entry surface. Light entering the MMI through the lower left (right) hand port would all exit through the upper left (right) hand port. At a distance $3L_{12}/2$ from the entry surface the modes with m odd are all in phase again while the m even modes are exactly π out of phase: the intensity pattern is then the mirror image of that at entry. In an MMI of this length the light entering the left hand input would exit at the right hand port, and vice versa.

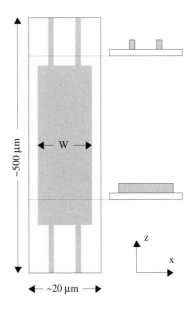

Fig. 16.15 MMI waveguide structure. On the left is a plan view and on the right cross-sections are shown through the structure. The waveguide sections shown are several microns in width and height.

The function of the central slab is to make the interference maxima fall on the output ports. These maxima are not displaced laterally when small errors are made in the length of the slab. Construction is therefore relatively less critical than for directional couplers. Further, the ports of MMI devices can be placed well apart which means that there is negligible coupling from one input port back into another input port.

16.8 Link noise and power budget

A key requirement for any optical fibre link is that the information transmitted is received with negligible errors. The accuracy of the transfer is quantified by the *bit error rate* (BER), that is the probability that a 1-bit is incorrectly assigned as a 0-bit, or vice-versa. It is measured by transmitting and checking the reception of a long sequence of random bits. For video transfer a BER of below 10^{-9}, meaning less than one bit being incorrect in 10^9, is adequate, while for data transfers a BER of below 10^{-12} is required. A limit on the bit error rate translates into a corresponding limit on the signal to noise ratio at the receiver, and in turn this imposes requirements on the total noise and the attenuation over the link. In this section the relationship between SNR and BER will be deduced and an example of the *power budget* calculated for a link. The link considered consists of monomode fibre with 0.3 dB/km attenuation, and the laser is a 1310 nm wavelength DFB laser.

Let the average electrical current for 1-bits be i_s, while that for the 0-bits is zero, above some constant background. The noise current is taken to have a Gaussian distribution with rms value σ. Then if the discriminator level is set at $i_s/2$, the chance of a 1-bit being recorded as a 0-bit is

$$\mathrm{BER} = [1/(\sqrt{2\pi}\sigma)] \int_{-\infty}^{i_s/2} \exp\left[-(i - i_s)^2/2\sigma^2\right] \mathrm{d}i.$$

Now make the substitution $i_s - i = j$, so that

$$\mathrm{BER} = [1/(\sqrt{2\pi}\sigma)] \int_{i_s/2}^{\infty} \exp\left(-j^2/2\sigma^2\right) \mathrm{d}j,$$

which is evidently also the probability that a 0-bit is recorded as a 1-bit. Next put $\xi = j/(\sqrt{2}\sigma)$, and the above equation becomes

$$\mathrm{BER} = (1/\sqrt{\pi}) \int_{i_s/(2\sqrt{2}\sigma)}^{\infty} \exp\left(-\xi^2\right) \mathrm{d}\xi$$

$$= (1/2)\left[1 - \mathrm{erf}(i_s/(2\sqrt{2}\sigma))\right], \quad (16.27)$$

where $\mathrm{erf}(z)$ is the *error function*.[3] This BER is plotted against the ratio i_s/σ in figure 16.16. In order to obtain a BER suitable for video transfer of below 10^{-9} the ratio i_s/σ must be above 12. Thus the SNR of the electrical power from the detector must satisfy

$$\mathrm{SNR} = (i_s^2/\sigma^2) > 144 \ (21.6\,\mathrm{dB}).$$

The noise is made up of contributions from the detector and the laser. Using eqns. 15.18 and 15.22 the sum of the shot noise and the Johnson noise in the detector current is

$$\sigma_{(\mathrm{s+j})}^2 = (2\eta e^2/hf)P_\mathrm{d}\Delta f + (4k_\mathrm{B}T/R_\mathrm{L})\Delta f, \quad (16.28)$$

where η is the detector quantum efficiency, P_d is the optical power falling on the detector, Δf is the bandwidth and R_L the load resistance.

Laser intensity noise, known as the *relative intensity noise* (RIN), arises from the unavoidable spontaneous emission accompanying stimulated emission which was discussed in Section 14.7.3. If the mean optical power of the laser is P_0 then the mean square deviation of this power is

$$\sigma_\mathrm{L}^2 = (\mathrm{RIN})\,P_0^2\,\Delta f. \quad (16.29)$$

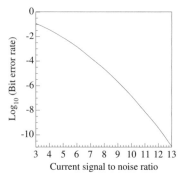

Fig. 16.16 Logarithm of the bit error rate versus the signal to noise ratio for the detector current.

---

[3]See the *Handbook of Mathematical Functions* by M. Abramowitz and I. A. Stegun, published by Dover Publications Inc., New York (1972).

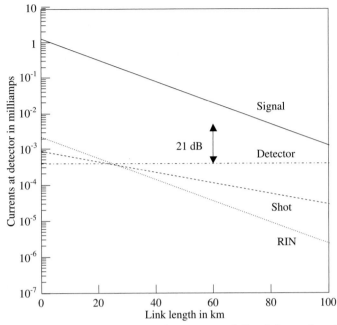

Fig. 16.17 Signal and noise currents along an optical fibre link as a function of link length. A margin of 21 dB in the power SNR is indicated.

Relative intensity noise is around $10^{-13}\,\text{Hz}^{-1}$ ($-130\,\text{dB}\,\text{Hz}^{-1}$) for DH lasers, around $3\,10^{-15}\,\text{Hz}^{-1}$ for DFB lasers, and around $10^{-16}\,\text{Hz}^{-1}$ for neodymium YAG lasers. Taking the attenuation over a link of length L as $\exp{(-\alpha L)}$, and using eqns. 16.28 and 16.29, the total mean square noise in the detector current is

$$\sigma^2 = (\eta e/hf)^2(\text{RIN})P_0^2 \exp{(-2\alpha L)}\Delta f$$
$$+(2\eta e^2/hf)P_0 \exp{(-\alpha L)}\Delta f + (4k_B T/R_L)\Delta f, \quad (16.30)$$

where we have used $P_d = P_0 \exp{(-\alpha L)}$. The three contributions are plotted separately in figure 16.17 as functions of the distance along the fibre path, together with the detector current for a 1-bit. Typical values have been used for all parameters: 2.5 mW of optical power is launched by the 1310 nm DFB laser into the link; the RIN is $3\,10^{-15}\,\text{Hz}^{-1}$; the fibre attenuation is 0.3 dB/km; the quantum efficiency of the detector is 0.5; the bandwidth is 1 GHz and the load is 1 kΩ at 300 K. Up to a distance of 80 km the current SNR remains above 12, and equivalently the power SNR remains above the 21.6 dB required to maintain a BER below 10^{-9}. Some further allowance must be made for the losses in connectors and splices. The noise generated by stimulated Raman and Brillouin scattering has been discussed in Chapter 14. This noise would be unimportant in the case considered above, but its growth as the launched power increases limits the maximum usable power to of order one watt.

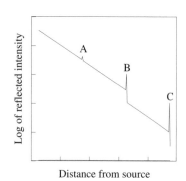

Fig. 16.18 Optical time domain reflectometry. The logarithm of the reflected intensity is plotted versus distance from the source. At A there is a good splice/connector, at B a poor splice/connector and at C a bare fibre end.

Actual power losses occurring along an optical fibre link are measured by using *optical time domain reflectometry* (OTDR). One component of the reflectometer is a laser which emits a train of short pulses down the link. A second component is a receiver which records the intensity and timing of light returning after each pulse is emitted. The delay between signal launch and return is converted to a distance along the fibre at which the reflection occured. A typical display seen on a reflectometer screen appears in figure 16.18. This shows the logarithm of the accumulated reflected intensity from many pulses versus the distance down the link at which reflection took place. Over the greater part of the plot the reflected intensity is due to Rayleigh back scattering. Rayleigh scattering is directly proportional to the beam intensity, so that it falls off exponentially with distance in the same way as the beam intensity. The attenuation per kilometre of fibre is determined from the gradient of these sections of the plot. At the point labelled A the tiny blip and negligible drop indicate a well made splice or a good connector, while the large spike and loss thereafter at point B reveal a poor connection. Finally at point C there is a reflection from a bare fibre end.

Poor connections can give rise to considerable noise in optical systems with lasers. Each reflective surface can form a Fabry–Perot cavity with a laser mirror facet, and standing waves will build up in these cavities. Radiation trapped in this way can easily disturb the laser operation and increase its relative intensity noise.

16.9   Long haul links

The discussions in Sections 16.4, 16.5 and 16.8 have shown that transmission over optical links of 50–100 km range on single mode fibre can achieve a very low BER at gigahertz data rates. Beyond this range the effects of attenuation and dispersion become severe. These difficulties have been successfully overcome so that transoceanic optical links now carry most of the communications traffic between continents. The ingredients contributing to this achievement are now outlined and then described in detail.

Fortunately amplification can be carried out optically while the signal is still on its way. This avoids the complications that would be involved in converting the signal to an electrical signal, amplifying it electronically and finally converting back to an optical signal. However electronic *regeneration* has the advantage that the signal can be re-shaped and re-timed, processes which may be necessary after a sequence of optical amplifier stages. Dispersion can be eradicated by using lengths of fibre with opposing and so compensating dispersions.

Alongside increasing the reach of optical links a parallel advance has been made in improving the data rate by transmitting data separately on many laser beams of different wavelengths along a single fibre. At first this *wavelength division multiplexing* (WDM) was limited to two wavelengths, at 1310 nm and 1550 nm. Nowadays a typical system would use 48 lasers whose wavelengths are spaced apart by 0.8 nm (100 GHz) around 1550 nm, spanning the C-band (1530–1565 nm), which is termed *dense WDM* (DWDM). A limit to the total bandwidth obtainable is reached when the data rate in Gb s^{-1} approaches the channel spacing in gigahertz, because any further increase in the data rate on each channel or decrease in the spacing in laser frequency will cause adjacent channels to overlap and interfere.

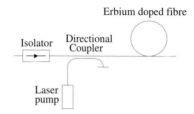

Fig. 16.19 Erbium doped fibre amplifier station.

In this limit the O- to U-bands, extending from 1260–1675 nm, could together provide a total bandwidth of approaching 60 THz on one single mode optical fibre.

16.9.1   Fibre amplifiers

The most mature amplifier technology, that of the *erbium doped fibre amplifier* (EDFA), uses stimulated emission to amplify the attenuated signal beam. Figure 16.19 shows the components of such an amplifier. Its key component is a length of fibre doped with 100–500 parts per million (ppm) of erbium which is spliced into the link at the point where amplification is required. An independent local laser source feeds into the doped fibre via a directional coupler, the laser wavelength being chosen so that the Er^{3+} ions are pumped into a long lived excited state. The incoming signal beam is then amplified by stimulated emission from Er^{3+} ions in these excited states. Figure 16.20 shows the energy levels of the Er^{3+}, which are broadened by the inhomogeneous electric fields within the glass. The upper states involved in stimulated emission lie within the E_2 band and have long lifetimes of about 12 ms. These states can be pumped directly by a laser of wavelength 1480 nm or indirectly via the E_3 level using a laser of wavelength 980 nm. If the attenuated signal beam has a wavelength within the range emitted in the $E_2 \rightarrow E_1$ transitions, namely 1450–1650 nm, it will stimulate emission from the excited ions and hence be amplified along the length of the erbium doped fibre. The gain produced is largest at about 1535 nm and is useful over the range 1470–1630 nm, thus covering the whole of the C-band. A virtue of the EDFA is that it can simultaneously amplify all the optical beams travelling along the same fibre whose wavelengths lie within this broad gain spectrum. This property has been central in the development of wavelength division multiplexing. Typically 100 mW of pump power injected into 10 m of erbium doped fibre will produce 20 dB gain in a microwatt signal anywhere across the C-band (1530–1565 nm).

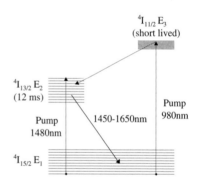

Fig. 16.20 Er^{3+} energy levels involved in erbium doped fibre amplification. Alternative pumping schemes are indicated.

At the same time that the signal is amplified in a fibre amplifier photons spontaneously emitted by the erbium ions in the $E_2 \rightarrow E_1$ transition are also multiplied if their paths are directed along the core of the fibre amplifier. Any of this amplified spontaneous emission (ASE) within the bandwidth of the signal beats with the signal to give noise. The isolator, shown in figure 16.19, prevents the backward propagation of ASE light to amplifiers earlier in the link.

A competitive technique for providing amplification makes use of the Raman effect described in Section 14.14.1. This method of amplification requires no special doping of the fibre and operates along the whole length of the link. It is the Stokes transition that is utilized. Referring to figure 14.47 a laser pumps the intermediate state in the Stokes transition and the photons in the signal beam stimulate the downward transition from this state. This stimulated emission amplifies the signal beam. Figure 14.48 shows the distribution in frequency of the radiation emitted in the Stokes transition in optical fibre. The gain in the signal beam produced by Raman amplification will have a similar dependence on frequency. Therefore the gain curve has a broad peak which is centred 13.2 THz below the frequency of the laser pump, that is at 100 nm higher in wavelength, and its width is around 80 nm. In the Stokes transition the energy transferred induces mechanical oscillations in the glass lattice and hence it is independent of the laser wavelength. Therefore the central wavelength of the Raman gain curve is always 100 nm above the laser pump wavelength. This makes it possible to amplify beams of any wavelength across the O- to U-bands simply by using a pump laser whose wavelength is 100 nm shorter. A set of pump lasers spaced ∼100 nm apart in wavelength can provide a relatively flat Raman gain curve with any desired wavelength coverage.

Raman amplification is weak relative to amplification in erbium doped fibre over a similar length of fibre, but now the whole length of the link provides amplification rather than the tens of metres used in EDFA. The high power pump lasers needed for Raman amplification, producing around 1 W, are now widely available; this has removed a crucial advantage of EDFA. In comparison with the long lived excited state involved in EDFA the Raman excited states are short lived. The fluctuations in laser pump power are therefore mirrored in Raman amplification, whereas they are smoothed out in erbium doped amplifiers. If the Raman pump beam and the signal beam travel together along a link the fluctuations would get imprinted on the signal beam with a corresponding huge increase in noise. This catastrophe is avoided by the simple trick of injecting the pump beam from the opposite end of the fibre to the signal beam. Then the effect of fluctuations is averaged over the whole time the data beam spends travelling along the link, which is order 0.25 ms for a 50 km span.

16.9.2 Dispersion compensation

Across the C-band, for which fibre amplifiers were first introduced, there is considerable group velocity dispersion in standard single mode fibre. We have seen in Section 16.5 how this puts an upper limit on the modulation frequency, well below what is attainable at 1310 nm wavelength where the group velocity dispersion vanishes. Solutions to this problem required the development of new fibre types differing in their radial doping profile from standard single mode fibre. Figure 16.21 shows the refractive index profile for *non-zero* (NZ) dispersion shifted fibre. This has a core of smaller cross-section and hence the effective area of the LP_{01} mode is smaller too. The group velocity dispersion of NZ dispersion shifted fibre is plotted as a function of wavelength in figure 16.22 and is reduced to about 4 ps/km/nm at 1550 nm wavelength.[4]

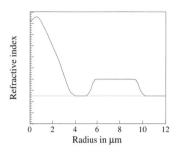

Fig. 16.21 Refractive index profile of non-zero dispersion shifted fibre. Courtesy Scott R. Bickham and Michael M. Sauer, Corning Inc.

It might seem preferable to have aimed for zero dispersion across the C-band. Nonetheless this must be avoided because exact zero dispersion is a condition which gives rise to excessive crosstalk between channels in wavelength division multiplexing. The origin of this crosstalk lies in the non-linear interaction causing *four wave mixing* between the beams at different wavelengths travelling along the same fibre. These are similar to the interactions discussed in Section 14.14. Four wave mixing is a third order process: waves of frequencies f_n and f_{n+m} interact to give satellite waves of frequencies $f_n - (f_{n+m} - f_n)$ and $f_{n+m} + (f_{n+m} - f_n)$. When the laser frequencies are equally spaced, which is the norm in dense wavelength division multiplexing, the frequencies of the generated channels coincide with original channel frequencies. Then if the group dispersion is zero the generated channels will remain in phase with the original channels as they travel along the fibre and so interfere and garble the data on all channels. Having some small residual dispersion causes the generated channels to drift in and out of phase with the original channels so that they are incoherent and do not interfere. The necessary small residual dispersion, ∼4 ps/km/nm, has then to be corrected in order to maintain the signal bandwidth. A simple approach is to terminate each section between amplification stages by a relatively short length of fibre having a very large dispersion of the opposite sign to the main length. Such *dispersion compensating* fibre would have a group velocity dispersion of −100 ps/km/nm. Alternatively a link can contain equal lengths of fibre with equal and opposite small dispersions.

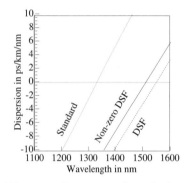

Fig. 16.22 Dispersions of standard single mode fibre, dispersion shifted fibre and non-zero dispersion shifted fibre.

Purely electronic signal processing methods have been developed to provide dispersion compensation on shorter links at high data rates. These are very valuable for upgrading the many links installed with non-dispersion shifted fibre. They offer dispersion compensation ∼1000 ps/nm.[5]

---

[4]The ITU recommends specifications to help achieve compatibility between optical fibres and devices from different suppliers. Thus ITU-T G.652 covers standard single mode fibre optimized at 1310 nm, G.652.C covers low water peak fibre.

[5]My thanks to Professor Ian Bennion for pointing out this progress to me.

16.10   Multiplexing

The number of laser beams carried at the same time on one single mode optical fibre on long haul optical fibre links is typically 48 with their frequencies equally spaced at intervals of 100 GHz, that is at intervals of 0.8 nm in wavelength. This is known as *dense wavelength division multiplexing* (DWDM), and the layout for such a link is illustrated in figure 16.23. The details relating to dispersion compensation and amplifica-

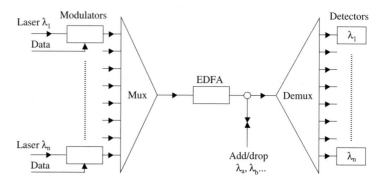

Fig. 16.23 Optical link with wavelength division multiplexing. The erbium doped amplifier and the add/drop module are indicated.

tion are left out; instead the emphasis in the figure and in the following sections is on the multiplexing/demultiplexing of laser beams of several wavelengths onto/from a single optical fibre, and the routing of individual wavelengths. Compact, sophisticated devices have been developed to perform these functions. Several passive and active devices are described below. Passive devices are considered which use thin film interference filters or fibre Bragg gratings to select wavelengths, as well as array waveguide gratings which can simultaneously combine or separate beams of multiple wavelengths. Moving now to *active* devices, arrays of micron sized mirrors provide examples of a *micro electro-mechanical system* (MEMS) which route multiple beams and can simultaneously switch their routing. All the devices described below have important applications outside communications: for example in optical fibre sensors, adaptive optics and projection TV. As mentioned earlier one objective being pursued by manufacturers is to integrate many components into monolithic waveguide devices and so eliminate the interconnections between components, which are delicate and expensive to make.

16.10.1   Thin film filters and Bragg gratings

Both these types of device can be used to select an individual beam from among the many beams multiplexed onto a single fibre, even when the separations in wavelength are as little as 0.4 nm. The operation of thin film interference filters has been analysed in Section 9.7 and the filters required in the present context must be apodized in order to suppress all the beams nearby in wavelength to the one selected. A simple demultiplexer using several thin film transmission filters is shown in figure 16.24. Each filter on the glass block transmits one wavelength channel.

A related device is the *fibre Bragg grating* (FBG), in which a cyclic variation in the refractive index along the fibre core (the Bragg grating) forms the interference filter. If the grating pitch is Λ and the mean refractive index of the core is n, then light reflected from successive grating planes is in phase if its free space wavelength λ_B is such that

$$\lambda_B = 2\Lambda n. \tag{16.31}$$

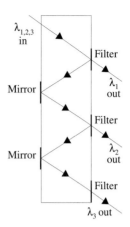

Fig. 16.24 Thin interference films on a glass block, used to demultiplex three wavelengths.

The grating pattern is imprinted on the fibre core using an ultraviolet pulsed laser and a diffraction grating. Although the imprinting is enhanced by the GeO_2 doping of the core, the modulation obtained in the refractive index, $\Delta n/n$, is still only about 10^{-4}, with the consequence that the relative amplitude of the wave reflected from each grating cycle is correspondingly small. It requires a grating extending for millimetres along the optical fibre in order to fully reflect the incident radiation at the Bragg wavelength. An estimate of the bandwidth can be made as follows. Let the grating be of length L, and contain N grating cycles, then multiplying eqn. 16.31 by N gives

$$N\lambda_B = 2nL. \tag{16.32}$$

Suppose that changing the wavelength by $\Delta\lambda$ produces destructive interference between the reflections from the two ends of the grating. Then

$$(\lambda_B + \Delta\lambda)(N - 1/2) = 2nL. \tag{16.33}$$

Subtracting eqn. 16.32 from eqn. 16.33 gives

$$\lambda_B/2 = (N - 1/2)\Delta\lambda,$$

so that to a good approximation

$$\Delta\lambda = \lambda_B/2N = \lambda_B^2/(4nL). \tag{16.34}$$

Thus a 1 mm long grating of refractive index 1.5 at a wavelength of 1.55 μm has a bandwidth 0.4 nm. A more detailed analysis shows that the reflectance distribution as a function of wavelength has side lobes like those shown for a thin film grating in the lower panel of figure 9.21. Apodization, that is the elimination of these side lobes, is achieved by a technique identical to that used in the case of thin film gratings. The refractive index is modulated by a Gaussian envelope so that its variation is like that shown in the upper panel of figure 9.21.

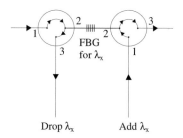

Drop λ_x Add λ_x

Fig. 16.25 Optical add/drop module for a wavelength λ_x.

A fibre Bragg grating is shown in figure 16.25 being used with two circulators to drop a beam of one wavelength, λ_x, from multiplexed beams and to add a similar beam. This add/drop module is also pictured schematically in figure 16.23 at an intermediate station along a fibre link. Beams of many wavelengths enter port 1 of the first circulator in figure 16.25 and exit at port 2. While travelling along the fibre connecting the two circulators the beam of wavelength λ_x is reflected back by the Bragg fibre grating, it re-enters port 2, exits at port 3 and is dropped for local reception. In the reverse process a locally generated beam of wavelength λ_x enters port 1 of the second circulator, is reflected by the same fibre Bragg grating and then exits at port 3 together with the beams of other wavelengths coming from the first circulator. Evidently by imprinting several gratings at different points along the fibre linking the circulators it is possible to add/drop several wavelengths simultaneously.

Chirped fibre Bragg gratings are gratings in which the grating pitch changes along the fibre. In this way the path length of the reflected radiation varies with wavelength. Such chirped fibre gratings provide a way by which dispersion can be compensated on existing links made from non-dispersion shifted fibre. If the Bragg wavelength changes by $\Delta\lambda_B$ over a grating of total length L, then the dispersion compensation achieved is

$$D = 2nL/(c\Delta\lambda_B), \tag{16.35}$$

where n is the refractive index of the fibre. Taking a length of $0.1\,\mathrm{m}$, a refractive index 1.5 and a wavelength change of $0.4\,\mathrm{nm}$, the compensation is $2500\,\mathrm{ps/nm}$.

16.10.2   Array waveguide gratings

Array waveguide gratings (AWG) are compact devices that use diffraction to multiplex beams of different wavelengths arriving on individual fibres onto a single fibre, or alternatively to demultipex them from a single fibre onto individual fibres. A demultiplexer, typically a few centimetres across, is illustrated in figure 16.26. The upper panel shows the waveguide structure built onto a substrate. An input fibre carrying many wavelengths feeds the single input waveguide which leads to the first coupling slab. From there multiple waveguides lead to a second coupling slab. On the far side of this second slab several waveguides emerge, each carrying a single wavelength from among those present in the input. Details of the second coupling slab appear in the lower panel of the same figure, in which the ports marked a and b are entry ports and those marked 0 and 1 are exit ports.

Both the entry and exit faces of the coupling slabs have circular outlines, which are of equal radius and are centred on the centre of the opposite face of the same slab. In the first slab the light from the input waveguide spreads out due to single slit diffraction at its junction with

the slab. The illumination at the exit face is uniform and, thanks to
the geometry, the phase is constant over this exit face. Connecting the

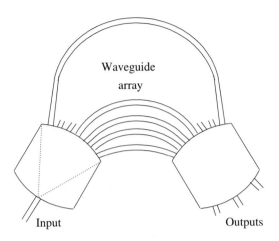

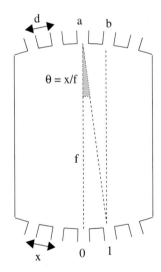

Fig. 16.26 Array waveguide grating. The upper panel shows both slabs and the
waveguide array. The lower panel shows details of the second slab. The ports labelled
a and b are the input ports, those labelled 0 and 1 are output ports.

coupling slabs is the *waveguide array*, the lengths of the waveguides increasing by a fixed amount from one waveguide to the next. This has the
consequence that there is a fixed phase difference between light entering
the second slab from adjacent waveguides, exactly as if the light were
emerging from the slits of a diffraction grating. The number of output
ports from the second slab equals the number of wavelengths to be demultiplexed. It is arranged that the increment in path length between
successive waveguides is such that the principal diffraction maxima for

each of these wavelengths falls on its own unique exit port.

Suppose that the wavelengths multiplexed in the input are spaced at intervals $\Delta\lambda$, that the central wavelength is λ_0 in free space, and that the fixed path increment between successive waveguides in the array is ΔL. ΔL will typically be of order $10\,\mu$m. Now the entry ports on the second slab (...a,b,....) are equidistant from the central exit port, labelled 0. Thus waves of wavelength λ_0 will arrive in phase at port 0 provided that the optical path length increment between successive waveguides is an integral number of wavelengths: that is provided that

$$n\Delta L = m\lambda_0, \tag{16.36}$$

where n is the refractive index of the waveguides and m is the diffraction order, a positive integer. With this arrangement the mth order principal maximum for wavelength λ_0 falls on the central exit port.

The design must also be such that the mth principal maximum for the wavelength next to the central wavelength, $\lambda_0 + \Delta\lambda$, falls on the adjacent exit port, labelled 1. This requires that the difference in optical path to port 1 via any pair of adjacent array waveguides such as those labelled a and b must be exactly m times the wavelength $\lambda_0 + \Delta\lambda$. Now the difference between these paths across the second slab alone is

$$\Delta s = nd \sin\theta, \tag{16.37}$$

where d is the spacing between adjacent waveguides as they enter the second slab, and θ is the angle shown in figure 16.26. This angle is small, so that $\sin\theta$ can be approximated by x/f where f is the radius of curvature of the slab's input and exit faces and x is the spacing between successive exit ports. Then

$$\Delta s = nxd/f. \tag{16.38}$$

To this optical path difference we must add the difference in optical length between the array waveguides feeding ports a and b, namely $n\Delta L$. If this total path difference satisfies the equality

$$n\Delta L + nxd/f = m(\lambda_0 + \Delta\lambda), \tag{16.39}$$

then indeed the mth order principal maximum for wavelength $\lambda_0 + \Delta\lambda$ falls on port 1. Subtracting eqn. 16.36 from eqn. 16.39 gives

$$nxd/f = m\Delta\lambda. \tag{16.40}$$

Using eqn. 16.36 to replace m gives

$$\Delta\lambda = x\lambda_0 d/(f\Delta L). \tag{16.41}$$

We must not overlook the possibility that secondary maxima will cause crosstalk between the channels. Array waveguide gratings are constructed with a large number of waveguides, in this case 100 waveguides.

Each principal maximum is then sufficiently narrow that it and a few nearby secondary maxima fall on the designated exit port. Referring to Section 6.5 the intensities of the remoter secondary maxima falling on adjacent exit ports are negligible compared to the intensity of the principal maxima. As a result crosstalk between the wavelength channels is avoided.

16.10.3   MEMS

The micro electro-mechanical devices of interest here are arrays of up to several thousand identical mirrors, typically $10\,\mu m \times 10\,\mu m$ in area, and all independently controlled. An example was described in detail at the end of Chapter 2. Their movements are implemented by either piezoelectric or electrostatic actuators. Such devices are constructed using methods similar to those perfected for microchip manufacture. A simple example is the 4×4 optical cross-connect shown in figure 16.27. It consists of a square array containing 16 mirrors. When a voltage is applied across the terminals of a given mirror, electrostatic attraction raises the mirror into an upright position, while if unenergized it lies flat. Any combination of one-to-one connections between the input and output ports can be obtained by raising the four appropriate mirrors. Switching from one configuration to another takes only $15\,\mu s$. This MEMS router has a number of advantages: the device is compact and there is negligible crosstalk between channels. Switching speeds are low compared to electronic switching, so this method can only be used where switching is infrequent.

A second application of mirror arrays is that of selective switching in DWDM. In this example the mirrors switch demultiplexed wavelength channels between the output fibres at will. The complete demultiplexer plus wavelength switch is drawn in figure 16.28: in the upper panel a plan view is shown, and in the lower panel a side view. For clarity only a pair of wavelengths are used in the illustration. Divergent light from the fibre end terminating some link is collimated so as to fall as a parallel beam on the diffraction grating. The diffracted beams of the different wavelengths diverge from one another and are focused by the second lens onto separate mirrors in a linear array of mirrors. The lens is placed a distance equal to its focal length from the mirrors so that the light in each demultiplexed beam arrives focussed at a mirror. Each beam returns to the lens, after which it is once again parallel. The grating returns the reflected demultiplexed beams along their common input path in the upper, plan view. Beam separation is achieved by tilting the mirrors as in the lower, side view. The response of the mirrors is *analogue*, meaning that their tilt angles can be varied continuously over some range by varying the applied voltage. One mirror is shown tilted up, the other tilted down so that the two wavelengths (λ_1 and λ_2) are separated in the vertical plane. In the side view the two wavelengths are seen to exit through additional lenses that focus them onto two different fibres

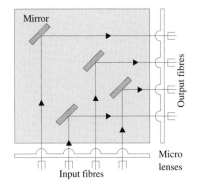

Fig. 16.27 MEMS mirror array used as a cross-connect.

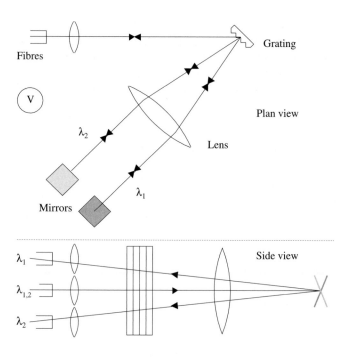

Fig. 16.28 Wavelength selective switching illustrated with two wavelengths. The side view is that seen from the point indicated in the plan view by the letter V.

stacked vertically around the input fibre. In an actual demultiplexer there could be, for example, 48 channels. The time required to reset the routing is $\sim$1 ms. Optical cross-connects operate more quickly because they are *binary*: they have just two mirror tilt settings, and they can therefore be driven hard against stops.

16.11   Solitons

Solitons are isolated waves that travel without changing their shape in media. The first example taken note of was a water wave on a canal, observed by Scott Russell in 1834. His attention was drawn to the bow wave of a canal barge which parted from the barge's prow when the barge was halted. To his surprise this single well-defined wave carried on down the canal for one or two miles without changing shape. The preservation of the waveform requires that the spreading of that waveform due to dispersion should be exactly cancelled by some non-linear process. An analogous type of cancellation was described in Section 14.11.1. In that example self-phase modulation caused by the non-linear Kerr effect in a crystal is used to cancel the effect of dispersion in a laser and hence produce mode-locking.

In optical fibre also the self-modulation due to the Kerr effect produces chirping of the opposite sense to that produced by positive group velocity dispersion. This is indicated in the upper pair of panels in figure 16.29. Positive group velocity dispersion ($D_m > 0$) causes high frequencies to travel faster than low frequencies, while eqn. 14.57 shows that the Kerr effect delays the high frequencies in the pulse when $n_2 > 0$. Self-modulation grows with the intensity so that at some intensity there can be perfect cancellation of the dispersion. When this is the case the waveform travels without chirping or lengthening of its envelope, and appears as shown in the lower panel of the figure.

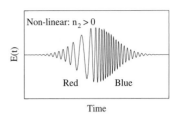

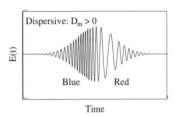

In Appendix D the wave equation is deduced for electromagnetic waves travelling along optical fibre, taking account of both dispersion and the Kerr effect. This equation, known as the non-linear Schroedinger equation, has a *fundamental soliton* solution of the form given in eqn. D.12. This is

$$E = \mathrm{sech}[(t - k_1 z)/t_0] \exp{(i\pi z/4z_0)}, \qquad (16.42)$$

where t is the time, z the position, $1/k_1$ is the group velocity of the pulse, and z_0 is called the *soliton period*. The shape of the envelope is quite similar to a Gaussian. There are also solutions whose initial shape at $z = 0$ is

$$E = N\mathrm{sech}(t/t_0), \qquad (16.43)$$

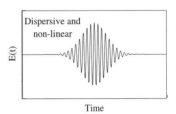

where N is an integer greater than or equal to two. Only the fundamental soliton travels without changing shape. Those having larger values of N do change shape as they travel, but return to their original shapes after a path length equal to the soliton period, z_0. As shown in Appendix D, at a wavelength of 1550 nm z_0 can be expressed in terms of the pulse length t_0 in picoseconds, and the group velocity dispersion, D_m, expressed in ps/km/nm:

$$z_o \approx t_0^2/D_m \ \ \mathrm{km}. \qquad (16.44)$$

Fig. 16.29 The upper panel shows the effect of fibre non-linearity on a pulse. The centre panel shows the effect of dispersion along the fibre. In the lower panel the combined effects compensate, leaving the waveform unchanged at any point along the fibre.

With a pulse of 20 ps duration and a dispersion of 1 ps/km/nm this yields a soliton period of 400 km, which is much longer than a single span between amplifiers. A final result taken from the appendix is that the power in a fundamental soliton is given by

$$P_0 \approx 1000/z_0 \ \ \mathrm{mW}. \qquad (16.45)$$

With the parameters chosen this comes to 2.5 mW, quite similar to the power launched in non-soliton transmission.

Mollenauer, Stolen and Gordon[6] were the first to investigate soliton propagation on optical fibre, launching 1550 nm laser pulses of 7 ps duration over 700 km of single mode fibre, which corresponded to about half the soliton period. The autocorrelations of the output pulses launched

[6]L. F. Mollenauer, R. H. Stolen and J. P. Gordon, *Physical Review Letters* 45, 1095 (1980)

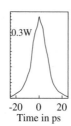 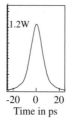 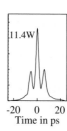 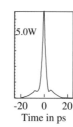 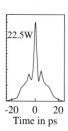

Fig. 16.30 The autocorrelations of optical pulses after travelling over 700 km of optical fibre. Each is labelled with the input power. The details are explained in the text. Adapted from L. F. Mollenauer, R. H. Stolen and J. P. Gordon: *Physical Review Letters* 45, 1095 (1980). Courtesy Professor Mollenauer and the American Physical Society.

with different power levels are shown in figure 16.30. Of these the second pulse has the power predicted for an $N = 1$ soliton and indeed is identical in shape to the input pulse. The next three output pulse shapes are those for initial pulses whose energies were approximately 4, 9 and 16 times larger. These output pulses have the shapes expected for $N = 2$, 3 and 4 solitons respectively. Finally the leftmost output pulse was obtained with an input pulse of a much lower energy. Its self-modulation is therefore negligible and it has simply broadened due to chromatic dispersion.

When a pulse has an energy and a shape similar to that of a fundamental soliton its form converges toward that of the fundamental soliton as it propagates. The residual components that distinguished the original pulse from the soliton disperse and form a smooth weak background.

16.11.1   Communication using solitons

Fundamental solitons are immune to dispersion so it is attractive to use them to replace the rectangular pulses of standard data transmission; this is illustrated in figure 16.31. Although optical solitons have been considered for this role for decades, commercial soliton based transmission systems only appeared in 2005.

The preferred source of optical solitons for telecoms is a CW semiconductor laser whose output is modulated by a lithium niobate modulator. If the tails of solitons overlap this causes cross-phase modulation so that the pulses drift toward one another as they travel over a link. It is therefore necessary to have a pulse width less than half the bit period. Naturally solitons lose energy in the same way as any other pulses and when this happens the dispersion will no longer be fully compensated by self-phase modulation. The Raman amplification described in Section 16.9.1 can maintain the pulses continuously near their optimum energy because it provides amplification over the whole fibre length.

The simple approach for soliton based communications outlined above requires the use of low dispersion fibres. However four wave mixing would then lead to interference between the channels multiplexed onto the same fibre. The solution adopted to eliminate this problem is one already described for non-soliton transmission: the link is made up of alternating lengths of fibre with cancelling group velocity dispersions. The resulting *dispersion managed* solitons therefore do change in form as they propagate, but it is arranged that after each cycle containing a pair of fibres with opposite dispersions the pulses return to their initial shape. There is an accompanying relaxation on the requirement on the pulse height. Locally dispersion is the dominating effect and pulses are nearly Gaussian in shape.[7]

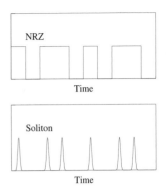

Fig. 16.31 The upper panel shows a non-return to zero pulse train. In the lower panel the corresponding soliton pulse train is shown.

16.12 Fibre optic sensors

Optical fibres are used in many sensors simply to carry light to and from the sensor. It is the returning radiation that is monitored and any variation in this can be used to infer how the sensor's environment is changing. Here the discussion is limited to examples in which the optical properties of the fibre are altered by the environment so that the fibre is also the sensor. Sensors to detect temperature, pressure and strain based on fibre Bragg gratings are considered first. Then two sensors that exploit unique properties of optical fibre are described: the *fibre optic gyroscope* (FOG) and the *optical current transformer* (OCT) used in measuring current flow along very high voltage power lines.

16.12.1 Fibre Bragg sensors

Changes in external pressure, temperature and strain all alter the length of an optical fibre. Hence if a fibre is imprinted with a Bragg grating the wavelength at which Bragg reflection occurs can provide an indication of such changes. A long fibre can be built into a structure, such as a dam, with the fibre having several gratings spaced at intervals along its length. If the pitch of each grating is different from that of every other grating it then becomes possible to monitor the environment along the whole fibre because each reflection can be recognized by its unique wavelength. Figure 16.32 shows such a distributed sensor. The grating spectrometer images the reflections from the Bragg gratings, which all have different wavelengths, onto separate areas of a CMOS array. When, for example, the temperature of one of the Bragg gratings changes, its pitch changes and this alters the wavelength reflected. In turn the location where this reflected light is imaged by the grating spectrometer on the CMOS array changes, and from the displacement the temperature can be determined.

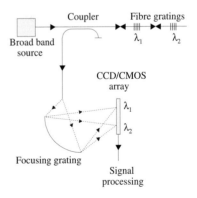

Fig. 16.32 Fibre Bragg grating sensor.

---

[7]An informed expert account of the advances that have led to practical soliton transmission is given by L. F. Mollenauer and J. P. Gordon in *Solitons in Optical Fibers* published by Academic Press, New York (2006).

An alternative method used in monitoring oilwell temperatures is to detect the Raman back-scattering from a plain optical fibre. The Stokes and anti-Stokes intensities are proportional to their parent state populations

$$I_s = A \exp\left(-E_1/k_B T\right); \quad I_{as} = A \exp\left(-E_2/k_B T\right)$$

where E_1 and E_2 are the parent state energies. Evidently the intensity ratio depends only on the temperature T K and the known energy difference $E_2 - E_1$:

$$I_s/I_{as} = \exp\left[(E_2 - E_1)/k_B T\right].$$

Hence the temperature can be extracted from this intensity ratio. A high powered pulsed laser is used to illuminate the fibre. Then, because the delay between the pulse leaving and returning is determined by the distance to the scattering point, the Raman scattering from different sections of the fibre can be identified by time slicing the returning light. Thin film interference filters are needed to isolate the Raman scattering from the stronger Rayleigh back scattering.

16.12.2   The fibre optic gyroscope

The principles of the Sagnac interferometer described in Section 5.7.4 find a modern application in fibre optic gyroscopes. Figure 16.33 shows the common form, an interferometer fibre optic gyroscope (I-FOG). Any rotation of the fibre loop in its own plane relative to the inertial frame of the fixed stars produces a phase difference between coherent light beams travelling around the fibre loop in the clockwise and counterclockwise senses. This phase difference was given in eqn. 5.39. In an I-FOG the loop is a multiturn coil of optical fibre which gives a large phase difference while keeping the device compact. The phase difference becomes

$$\phi_s = 8\pi\Omega N A/c\lambda, \tag{16.46}$$

where the rotation rate is Ω, the free space wavelength is λ and the coil has N turns each of area A.

Long distance navigation requires headings (directions) with a precision of $\sim 0.01\,^\circ$/hr, equivalent to $\sim 10^{-7}\,\mathrm{rad\,s^{-1}}$ in Ω.[8] Fibre optic gyros with this sensitivity have been available for some years. Less precision, $\sim 10^\circ$/hr, is required when monitoring vehicle motion in crash simulation. Fibre optical gyros have major advantages over mechanical gyros, being more rugged and having no moving parts. They have largely replaced mechanical gyros, with many hundred thousands in current use.

The interferometric fibre optic gyroscope shown in figure 16.33 uses an *open loop* design in which the phase difference, ϕ_s, is measured directly.

---

[8]The most widely used modern navigational aid is GPS, but this can only give positions, and is not accessible to submarines.

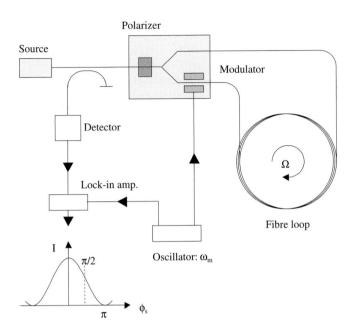

Fig. 16.33 Interferometric fibre optic gyroscope. At the bottom left is the output of the lock-in amplifier plotted against the phase difference between the counter-rotating beams emerging from the fibre loop.

The coil would typically have a diameter of 0.1 m and be 1 km in length. A laser is the source and its beam is divided equally between the counter-rotating beams travelling round the loop. After completing this circuit the beams are recombined and pass through a directional coupler to a photodetector. The intensity at the photodetector is

$$I = 2I_0(1 + \cos \phi_s), \qquad (16.47)$$

where I_0 is the intensity of either beam. This dependence is sketched in the bottom left hand corner of figure 16.33. Such a design for an I-FOG would have poor sensitivity because the rate of change of the intensity

$$\mathrm{d}I/\mathrm{d}\phi_s = -2I_0 \sin \phi_s \qquad (16.48)$$

is close to zero at low rotation rates. Ideally the working point, that is to say the point at which the rotation is zero, should be moved to $\phi_s = \pi/2$. Then two advantages accrue: the sensitivity is at its maximum and the sign of the rotation can be determined by whether ϕ_s increases or decreases from $\pi/2$. A controlled, constant phase shift of this sort is not a practical proposition and instead frequency modulation is used to achieve a similar result. The lithium niobate modulator indicated in figure 16.33 modulates the light beams at an angular frequency ω_m: the counter-clockwise beam is modulated before, and the clockwise beam is modulated after passing round the loop. Therefore the phase difference becomes

$$\Delta\phi = \phi_s + \phi_m \cos\left[\omega_m(t - \tau/2)\right] - \phi_m \cos\left[\omega_m(t + \tau/2)\right], \quad (16.49)$$

where ϕ_m is the amplitude of this modulation and τ is the transit time around the loop. The above expression reduces to

$$\Delta\phi = \phi_s + 2\phi_m \sin\left(\omega_m\tau/2\right)\sin\left(\omega_m t\right).$$

This is simplified by choosing the modulation frequency such that $\omega_m = \pi/\tau$. Then

$$\Delta\phi = \phi_s + 2\phi_m \sin\left(\omega_m t\right), \quad (16.50)$$

and the intensity at the photodiode becomes

$$I = 2I_0(1 + \cos\Delta\phi). \quad (16.51)$$

The end product of introducing modulation is that the working point is oscillating along the cosine curve in figure 16.33 at an angular frequency ω_m. This means that in order to determine ϕ_s the component of I varying at angular frequency ω_m must be extracted. The lock-in amplifier shown in figure 16.33 is the component used to do this. It receives two inputs: the first input is the output current from the photodetector; the second, reference input, is a voltage from the oscillator driving the modulator. Lock-in amplifiers filter off any signal from the first input which is at the frequency of the reference input, and output a signal proportional to this component. We can expand the current, I, into its Fourier components, retaining only the constant term and the term with angular frequency ω_m, giving

$$\begin{aligned} I &= 2I_0\left[1 + J_0(2\phi_m)\cos\phi_s - 2J_1(2\phi_m)\sin\left(\omega_m t\right)\sin\phi_s\right] \\ &\approx 2I_0\left[1 + J_0(2\phi_m) - 2J_1(2\phi_m)\sin\left(\omega_m t\right)\phi_s\right], \quad (16.52) \end{aligned}$$

where J_0 and J_1 are Bessel functions of the first kind.[9] In this case the lock-in amplifier output is proportional to

$$V_L = 4I_0 J_1(2\phi_m)\sin\phi_s \approx 4I_0 J_1(2\phi_m)\phi_s, \quad (16.53)$$

and from this the desired phase ϕ_s can be calculated. $J_1(2\phi_m)$ can be maximized by choosing $2\phi_m$ to be $1.8\,\mathrm{rad}$, and this in turn maximizes the sensitivity.

From eqn. 16.52 it is apparent that the noise in the detector current sets the limit to the precision with which ϕ_s can be determined. Usually the limiting contribution is from shot noise. Then using eqn. 15.18 the noise to signal ratio on the current is $\sqrt{(2eB/I_0)}$. Taking a current, I_0, of $100\,\mu\mathrm{A}$ and a bandwidth, B, of $1\,\mathrm{Hz}$ this ratio is $S = 6\,10^{-8}$. To

---

[9]See eqns 9.1.42 and 9.1.43 from the *Handbook of Mathematical Functions* by M. Abramowitz and I. A. Stegun, published by Dover Publications, New York (1964).

a good approximation the minimum detectable phase angle is equal to S rad. Inserting this value in eqn. 16.46 gives the minimum detectable rotation rate

$$\Omega_{\min} = S\left[\lambda c/(8\pi NA)\right].$$

Choosing $NA = 100\,\mathrm{m}^2$ and $\lambda = 1550\,\mathrm{nm}$, gives $\Omega_{\min} \approx 10^{-8}\,\mathrm{rad\,s}^{-1}$ or $0.002\,^\circ/\mathrm{hr}$, which is adequate for long distance navigation.

A number of other sources of noise and bias have had to be understood and their effects eliminated in order to achieve this sensitivity in practice. Firstly there is the effect of the variation of the fibre properties along its length which cause the beam polarizations to alter as they travel. Therefore a polarizer is placed after the laser and polarization maintaining fibre is used in the loop. Secondly light which is Rayleigh scattered from either beam can induce self- and cross-modulation of the beams with attendant random variations in the phases. To avoid this a source is used whose coherence length is relatively short, only several microns, and this inhibits the interaction between the beams as they pass each other within the loop. Thirdly, any small temperature gradients will affect the beams differently and it is necessary to wind the fibre coil so that turns symmetrically placed with respect to one another about the mid-point of the coil are in good thermal contact. Finally temperature variations of the laser source can affect its output and hence distort the phase measured. A suitable laser source is formed by a length of erbium doped fibre having reflective facets at each end, which itself is pumped by another laser. This fibre laser has the required broad bandwidth of ~50 nm and is far less temperature sensitive (only 5 parts per million change in wavelength per degree kelvin) than a a multimode laser. Even then its temperature must be finely controlled. When these safeguards are implemented the fibre optic gyroscope becomes a compact and robust navigational tool.

16.13   Optical current transformer

Current sensors are needed in power distribution systems to both detect failures and to meter the current flow accurately for revenue assessment. Electrical transformers on 100 kV lines have to be extremely well insulated: they are therefore heavy, difficult to install and correspondingly expensive. The equivalent optical current transformer is simply an optical fibre end wound around the current carrying cable. Detection of the current is made possible by the Faraday effect discussed in Section 10.7.3. The magnetic field produced in the optical fibre by the current flowing in the cable causes the plane of polarization of the light travelling in the fibre to rotate. What is measured is the overall rotation of the plane of polarization of light going round the loop. Unlike an electromagnetic transformer there is no hysteresis, and no requirement for insulation.

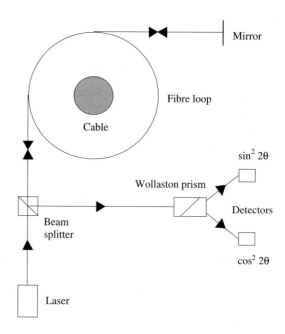

Fig. 16.34 Optical current transformer. The relative intensities at the two detectors are proportional to $\cos^2 2\theta$ and $\sin^2 2\theta$, where θ is the rotation of the plane of polarization of the laser beam in one pass through the fibre loop.

Plane polarized light from a laser is transmitted along the fibre, and its plane of polarization rotates under the influence of the magnetic field through an angle given by eqn. 10.53

$$\theta = \mathcal{V} \int \mathbf{B} \cdot \mathrm{d}\mathbf{L} \tag{16.54}$$

where $\mathcal{V}$ is the Verdet constant, $\mathbf{B}$ is the magnetic field and $\mathrm{d}\mathbf{L}$ is an element of length along the fibre axis. The integral is taken over the whole length of the fibre. Assuming a closed fibre loop of N turns and a current of I amps, then using Ampere's law, eqn. 9.10, to replace the integral in the previous equation we get[10]

$$\theta = \mathcal{V}(\mu_0 N I), \tag{16.55}$$

a result which is independent of the actual shape of the loop. For example at 820 nm Verdet's constant for optical fibre is $\sim 4\,\mathrm{rad}\,(\mathrm{T\,m})^{-1}$, so that with ten turns of fibre wound around a cable carrying 1 kA the plane of polarization rotates through 2.8°. An optical current transformer is drawn in figure 16.34. Light is reflected back along the optical fibre from a mirror deposited on the fibre's end facet. As seen from the laser, the rotation of the plane of polarization due to the Faraday effect is in the same sense on the return path as on the outward path. This doubles the rotation.

[10]It is possible to neglect the displacement current because the frequency is low.

Figure 16.34 shows that the returning beam is incident on a Wollaston prism, which separates the beam into components whose polarizations are parallel and perpendicular to the intial laser beam's polarization. These components have intensities proportional to $\sin^2 2\theta$ and $\cos^2 2\theta$, where θ is the rotation angle given by eqn. 16.55. Single mode fibres are used, but even these have some residual linear and circular birefringence. Various refinements in the design of the optical current transformer are therefore required in order to eliminate the unwanted rotation of the plane of polarization caused by birefringence.

The right and left circularly polarized components of the beam both reverse their senses of rotation on reflection from the mirror at the end of the fibre. Therefore the effects of circular birefringence on the outward and return paths simply cancel. Unfortunately the effect of linear birefringence would be doubled. In order to avoid this doubling a 45° Faraday rotator is placed in front of the mirror: the mirror/rotator combination is a *Faraday mirror* which has the property that it rotates the polarization plane by 90°. As a result the components of the electric field along orthogonal axes, E_x and E_y, are exchanged between the outward and return journeys at each point along the fibre path, apart from the small rotations induced by circular birefringence and the Faraday effect. There is therefore partial cancellation of the effects of linear birefringence.

Near total cancellation is achieved by the further step of replacing the fibre by *spun fibre*. During manufacture the fibre is spun around its optic axis while it is being drawn and still molten. This process leaves a *large* residual circular birefringence so that the plane of polarization of radiation travelling along the fibre rotates rapidly. The spin period along the fibre is made sufficiently short that the linear birefringence changes negligibly over this distance. It follows that in one complete rotation of the polarization of the radiation travelling along the fibre the linear birefringence is essentially constant, and so the net effect of the linear birefringence per full rotation is nil. With these improvements a precision of one part in a thousand is obtained in measuring currents up to several hundred kiloamps.

16.14　Further reading

Understanding Fiber Optics by J. Hecht, published by Pearson/Prentice Hall (2006). This text contains a comprehensive discussion of all aspects of fibre optics in communications; not overburdened with mathematics.

　Solitons in Optical Fibers: Fundamentals and Applications by L. F. Mollenauer and J. P. Gordon, published by Elsevier Academic Press (2006). A lively and interesting account of this burgeoning field by two pioneers.

　Guided Wave Optical Components and Devices, edited by B. P. Pal and published by Elsevier Academic Press (2005). This book contains a compilation of articles by experts.

Exercises

(16.1) Light of wavelength 1550 nm suffers a loss of 20 dB/km along a plastic optical fibre. A beam of 1 W power is injected into a 10 km length of this fibre. How many photons arrive per minute at the far end?

(16.2) A step index multimode glass fibre has core and cladding with respective refractive indices 1.48 and 1.46 at 1310 nm wavelength. Calculate the intermodal dispersion over 5 km of this fibre. Estimate the highest data rates that such step index fibres can support and express this in $\text{Mb}\,\text{s}^{-1}\text{km}$. A (quadratic) graded index fibre has the same axial and cladding refractive indices as the step index fibre. The chromatic group velocity dispersion on the graded index fibre is 200 ps/km/nm at 1310 nm, and the source used has a bandwidth of 40 nm. Calculate the intermodal and chromatic contributions to the dispersion over 5 km of this graded index fibre. Hence estimate the maximum achievable data rate.

(16.3) An LED emits radiation of mean wavelength 900 nm and a spread of 40 nm down a 10 km length of fibre for which the material dispersion $\text{d}^2 n/\text{d}\lambda^2$ is $5\,10^{10}\,\text{m}^{-2}$. What is the corresponding broadening of pulses transmitted over this fibre length?

(16.4) Consider the case of a 50 km link using a 1550 nm laser source of RIN equal to $-145\,\text{dB}\,\text{Hz}^{-1}$ which injects 1.5 mW into the fibre whose attenuation is $0.2\,\text{dB}\,\text{km}^{-1}$. The detector has a bandwidth 1.5 GHz, a quantum efficiency of 0.4 and drives a $1\,\text{k}\Omega$ load. Calculate the noise contributions on the signal at the load due to shot noise, detector noise and RIN. What is the margin in dB between the signal power and the noise? If there is a connection close to the detector which loses 10 dB of optical power how does this affect the margin?

(16.5) Explain how a fibre Bragg grating with varying pitch can chirp incident light beams.

(16.6) Two identical single mode fibres are aligned, their ends separated a small air gap. What is the loss of optical power through this connection in dB? You may ignore multiple reflections.

(16.7) What is the free spectral range of an array waveguide grating? Calculate this for an AWG used at 1550 nm in which the difference in the length of adjacent waveguides is 10 μm and their refractive index is 1.5.

(16.8) Following on from the previous question, an array waveguide grating has slabs and guides with the same refractive index 1.5 at the operating wavelength 1.55 μm. The second slab has length 1.1 cm and the difference in length between successive guides is 63 μm. Both the entry and exit ports to the second slab are spaced apart, centre-to-centre, by 20 μm. What is the channel spacing, the diffraction order and the free spectral range? How many channels can be used?

(16.9) Soliton pulses of 10 ps duration and wavelength 1550 nm are injected into a single mode optical fibre with a chromatic group velocity dispersion 3 ps/km/nm. What is the soliton period? What is the soliton power required? What is the pulse energy and how many photons does it contain?

(16.10) An optical fibre has a core of diameter 10 μm and refractive index 1.480. The cladding has refractive index 1.475. What is the numerical aperture? What is the cut-off wavelength for single mode operation?

(16.11) Show explicitly that if the slab in the MMI waveguide structure pictured in figure 16.15 has a length $3L_{12}/2$ rather than $3L_{12}$ the outputs are exchanged left with right.

(16.12) Estimate the number of modes that can freely propagate over multimode fibre of numerical aperture 0.5, core diameter 50 μm at 1300 nm wavelength.

Photonic crystals

17.1 Introduction

In 1987 Yablonovitch and John independently[1] pointed out that there can be *photonic band-gaps* for electromagnetic waves in dielectrics, analogous to the electronic band-gaps present in semiconductors. Electronic band-gaps come about because the electron waves scattered from a regular array of ions will interfere, and when the reflected waves from parallel layers of ions are in phase there is coherent reflection from all the layers of ions, that is to say Bragg reflection. These waves cannot then propagate through the dielectric. The corresponding lattice from which electromagnetic radiation can be reflected is a periodic spatial variation in the relative permittivity and hence in the refractive index of a dielectric structure. Examples of simple periodic arrays formed from two different dielectrics are shown in figure 17.1 for one, two and three dimensions: these are known as *photonic crystals*. There will be coherent Bragg reflection when the optical path over one period of the photonic crystal is an integral multiple of half-wavelengths.

The analysis of electron waves travelling in a crystal lattice is complicated by the mutual interactions of the electrons. Photons, by contrast, do not interact with one another so the analysis for electromagnetic waves is easier. The length scales are widely different in the two cases: the spacings between atoms in solids are of order 0.5 nanometres, while dielectric layers of a few hundred nanometres' thickness are chosen to match the wavelengths of visible light, or the near infrared radiation used in telecoms. Electron waves in crystals obey Schroedinger's equation in which the electric potential is due to the array of ions making up the crystal. Electromagnetic waves in a photonic crystal obey Maxwell's equations with the spatially varying relative permittivity playing, as we shall see, the part of a potential.

One-dimensional photonic crystals are simply the multilayer dielectric filters met in Section 9.7.1, now renamed in order to emphasize the common features they share with two- and three-dimensional photonic crystals. The band-gap in such a filter is the range of wavelengths over which electromagnetic radiation is evanescent in the layered dielectric,

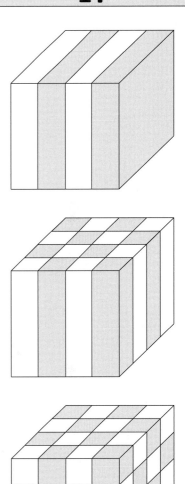

Fig. 17.1 Sections of one-, two- and three-dimensional photonic crystals.

---

[1] E. Yablonovitch, *Physical Review Letters* **58**, 2059 (1987); S. John, *Physical Review Letters* **58**, 2486 (1987).

and so fully reflected from it. Put another way, photons with energies in the corresponding band-gap in energy are all reflected. Some three-dimensional photonic crystals have omni-directional band-gaps, meaning that they reflect light incident from any direction, provided its wavelength lies within the band-gap.

The existence of local defects in the crystal structure of a semiconductor can result in local energy levels appearing within an electronic band-gap. For instance the presence of dopant atoms produces donor or acceptor levels in the band-gap. Similarly, defects in a photonic crystal can give rise to modes that lie within a photonic band-gap and whose electromagnetic waves are localized around the defect. A filter mentioned in Section 9.7.1 contains a defect in a one-dimensional optical lattice. Its structure is $(HL)^m(HH)(LH)^m$, where H(L) indicate layers of high(low) refractive index dielectric, all with the same optical thickness, $\Lambda/2$. Light of wavelength 2Λ lies midway across the range of wavelengths undergoing Bragg reflection from the $(HL)^m$ stacks. However this is also a wavelength at which standing waves will develop in the Fabry-Perot etalon formed by the (HH) cell. Thus there is an allowed local electromagnetic wave whose photons have energy $hc/2\Lambda$ located inside the (HH) cell. This wave is evanescent in the $(HL)^m$ stacks, decaying exponentially with the distance from the outer surface of the (HH) cell.

A few artificial three-dimensional crystals with band-gaps in the near infrared and visible region have been constructed. However the principal practical applications to date involve one- and two-dimensional photonic crystals which are easier to construct. Photonic crystals are observed quite often in nature: some butterfly wings and insect shells have vivid coloration thanks to near 100% reflection of light with wavelengths lying within narrow band-gaps. Opals owe their *iridescence* to narrow band-gaps, and one form of artificial three-dimensional photonic crystal which will be described later is the *inverse opal* structure.

In the first section below *Bloch waves* are introduced. These are modes of the electromagnetic field in photonic crystals, equivalent to the sinusoidal plane waves within a homogeneous dielectric. Some necessary, but limited background material on crystal structure is inserted as appropriate. There follows a thorough analysis of the propagation of electromagnetic waves in one-dimensional photonic crystals including off-axis beams. This yields the essential concepts that will be used when we turn to examine multi-dimensional photonic crystals. However the parallel analysis of the higher-dimensional photonic crystals is sufficiently complex that it will not be pursued here.

The concluding part of the chapter mainly concerns two-dimensional photonic crystals, and their applications. Photonic crystal slabs, which are simply two-dimensional photonic crystals generally under one wavelength thick, are described first. Their applications include the improvement of LED efficacy, silicon-on-insulator waveguides and compact optical modulators. The refraction near a band-gap edge produces a super prism effect and a negative refractive index in two-dimensional photonic crystal slabs. Both these suprising effects are explained. After that photonic crystal fibres are described: they find applications as sources of intense white light and are components of the optical clocks to be met in Chapter 18. In a final section a short account of three-dimensional photonic crystal structures is given.

17.2   Bloch waves

Because the relative permittivity varies in a periodic manner in photonic crystals, the plane sinusoidal electromagnetic wave modes

$$\mathbf{E} = \mathbf{E}_0 \exp\left[i(\mathbf{k} \cdot \mathbf{r} - \omega t)\right], \tag{17.1}$$

that are used as basis states for waves propagating in a homogeneous and isotropic dielectric have to be replaced by waves which are modulated in the same periodic manner as the crystal. Within an extended photonic crystal reflections at the regularly spaced internal interfaces impose the periodicity of the crystal lattice on the waves. Let the variation of the relative permittivity be such that

$$\varepsilon_{\mathrm{r}}(\mathbf{r} + \mathbf{R}_n) = \varepsilon_{\mathrm{r}}(\mathbf{r}). \tag{17.2}$$

where $\mathbf{R}_n$ is any one of the possible displacements from one point in the crystal lattice to any other identical point. Then Bloch's theorem states that all solutions to Maxwell's equations in this photonic crystal are linear superpositions of the orthogonal modes[2]

$$\mathbf{E}_{\mathbf{k}}(\mathbf{r}) = \mathbf{u}_{\mathbf{k}}(\mathbf{r}) \exp\left[i(\mathbf{k} \cdot \mathbf{r} - \omega t)\right], \tag{17.3}$$

where

$$\mathbf{u}_{\mathbf{k}}(\mathbf{r} + \mathbf{R}_n) = \mathbf{u}_{\mathbf{k}}(\mathbf{r}). \tag{17.4}$$

The modes presented in eqn. 17.3 are called *Bloch states* or *Bloch waves*.

In order to proceed we need first to examine the basic properties of crystal lattices. In any crystal lattice the vectors $\mathbf{R}_n$ can always be specified in terms of three *primitive lattice vectors* which we can call $\mathbf{a}_1$, $\mathbf{a}_2$ and $\mathbf{a}_3$.[3] Thus

$$\mathbf{R}_n = n_1 \mathbf{a}_1 + n_2 \mathbf{a}_2 + n_3 \mathbf{a}_3, \tag{17.5}$$

where n_1, n_2 and n_3 are integers.

---

[2] A proof of Bloch's theorem is given in *Optical Waves in Crystals* by A. Yariv and P. Yeh, published by John Wiley and Sons, New York (1984). In fact Bloch proved the analogous result for electron waves in a periodic potential.

[3] See for example the eighth edition of *Introduction to Solid State Physics* by C. Kittel and published by John Wiley and Sons, New York (2005).

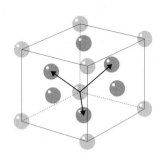

Fig. 17.2 Primitive basis vectors of the face-centred cubic crystal lattice: $\mathbf{a}_1$, $\mathbf{a}_2$ and $\mathbf{a}_3$. The face-centred lattice locations are made darker for clarity.

A simple crystal lattice, the face-centred cubic lattice, shown in figure 17.2 will be used as our example. In an opal structure the dielectric spheres would be large enough to touch and form a continuous rigid structure. Alternatively in an inverse opal structure the spheres would be hollows in a solid matrix. The inverse opal is of particular interest because it is one of the few three-dimensional lattices with an omnidirectional band-gap that can readily be assembled artificially on the optical scale. In figure 17.2 the primitive lattice vectors for the face-centred cubic lattice, $\mathbf{a}_1$, $\mathbf{a}_2$ and $\mathbf{a}_3$, are indicated by the three arrowed lines.

We define a set of primitive reciprocal lattice vectors in a *reciprocal lattice* as follows:

$$\mathbf{b}_i = 2\pi(\mathbf{a}_j \wedge \mathbf{a}_k)/(\mathbf{a}_i \cdot \mathbf{a}_j \wedge \mathbf{a}_k), \qquad (17.6)$$

with i, j and k running cyclically through 1, 2, 3. It is worth emphasizing that the vectors $\mathbf{b}_i$ lie in $\mathbf{k}$-space and have dimensions reciprocal to spatial dimensions. The reciprocal lattice for a face-centred cubic lattice is the body-centred cubic lattice shown in figure 17.3. The vectors drawn are the primitive reciprocal lattice vectors: $\mathbf{b}_1$, $\mathbf{b}_2$ and $\mathbf{b}_3$. The scalar products of these vectors with the primitive lattice vectors are

$$\mathbf{a}_i \cdot \mathbf{b}_j = 2\pi\delta_{ij}. \qquad (17.7)$$

General reciprocal lattice vectors are multiples of these,

$$\mathbf{G}_m = m_1\mathbf{b}_1 + m_2\mathbf{b}_2 + m_3\mathbf{b}_3, \qquad (17.8)$$

with m_1, m_2 and m_3 also being integers. Then for all lattice vectors and all reciprocal lattice vectors we have, using eqn. 17.7,

$$\mathbf{G}_m \cdot \mathbf{R}_n = 2N_{mn}\pi, \qquad (17.9)$$

where N_{mn} is an integer.

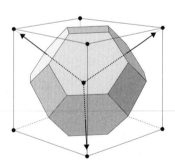

Fig. 17.3 First Brillouin zone and the primitive reciprocal lattice vectors of the face-centred cubic lattice: $\mathbf{b}_1$, $\mathbf{b}_2$ and $\mathbf{b}_3$.

We now assert that the general form of $\mathbf{u}_\mathbf{k}(\mathbf{r})$ which satisfies Bloch's theorem is

$$\mathbf{u}_\mathbf{k}(\mathbf{r}) = \sum_i \mathbf{c}_i(\mathbf{k}) \exp\left(i\mathbf{G}_i \cdot \mathbf{r}\right). \qquad (17.10)$$

Under a displacement by the lattice vector $\mathbf{R}_j$ this becomes

$$\mathbf{u}_\mathbf{k}(\mathbf{r} + \mathbf{R}_j) = \sum_i \mathbf{c}_i(\mathbf{k}) \exp\left[i\mathbf{G}_i \cdot (\mathbf{r} + \mathbf{R}_j)\right]$$

$$= \sum_i \mathbf{c}_i(\mathbf{k}) \exp\left(i\mathbf{G}_i \cdot \mathbf{r}\right), \qquad (17.11)$$

where eqn. 17.9 has been used. As required $\mathbf{u}_\mathbf{k}(\mathbf{r})$ reproduces itself under any lattice vector displacement.

Key features, that we now prove, are that there is in effect considerable duplication among these general Bloch waves, and that we only need to use a very restricted, well-defined set from among them. A Bloch wave can be rewritten as follows

$$\mathbf{E_k}(\mathbf{r}) = \sum_i \mathbf{c}_i(\mathbf{k}) \exp\left(i\mathbf{G}_i \cdot \mathbf{r}\right) \exp\left[i(\mathbf{k} \cdot \mathbf{r} - \omega t)\right]$$

$$= \sum_i \mathbf{c}_i(\mathbf{k}) \exp\left[i(\mathbf{G}_i - \mathbf{G}) \cdot \mathbf{r}\right] \exp\left\{i[(\mathbf{k} + \mathbf{G}) \cdot \mathbf{r} - \omega t]\right\}, \quad (17.12)$$

where $\mathbf{G}$ is any reciprocal lattice vector. Every sum or difference of reciprocal lattice vectors is another reciprocal lattice vector. Here we put $\mathbf{G}_i - \mathbf{G} = \mathbf{G}'_i$ so that

$$\mathbf{E_k}(\mathbf{r}) = \sum_i \mathbf{c}_i(\mathbf{k}) \exp\left(i\mathbf{G}'_i \cdot \mathbf{r}\right) \exp\left\{i[(\mathbf{k} + \mathbf{G}) \cdot \mathbf{r} - \omega t]\right\}$$

$$= \mathbf{u_{k+G}}(\mathbf{r}) \exp\left\{i[(\mathbf{k} + \mathbf{G}) \cdot \mathbf{r} - \omega t]\right\}, \quad (17.13)$$

where $\mathbf{u_{k+G}}(\mathbf{r})$ also satisfies eqn. 17.4. Finally putting $\mathbf{k}' = \mathbf{k} + \mathbf{G}$, we have

$$\mathbf{E_k}(\mathbf{r}) = \mathbf{u_{k'}}(\mathbf{r}) \exp\left[i(\mathbf{k}' \cdot \mathbf{r} - \omega t)\right] \equiv \mathbf{E'_{k'}}(\mathbf{r}), \quad (17.14)$$

where $\mathbf{E'_{k'}}(\mathbf{r})$ and $\mathbf{E_k}(\mathbf{r})$ are called *equivalent Bloch waves*. We see that the wave vector of the Bloch wave can equally be assigned to be any one of the discrete set made up of $\mathbf{k}$ and $\mathbf{k} + \mathbf{G}$ using any reciprocal lattice vector $\mathbf{G}$.[4] The values of the coefficients $\mathbf{c}_i(\mathbf{k})$ can always be determined by solving Maxwell's equations for the particular crystal structure, subject to any boundary conditions. The result will depend on both the form of the crystal lattice and the distribution of relative permittivity in the basic cell. Equivalent Bloch waves with wave vectors $\mathbf{k}$ and $\mathbf{k}' = \mathbf{k} + \mathbf{G}$ have identical group velocities because $d\mathbf{k}'/d\omega = d\mathbf{k}/d\omega$. Therefore, as expected, photons, and the energy that they carry, propagate identically in equivalent Bloch waves.

An important simplification in handling Bloch waves will be illustrated using figure 17.4. This shows a single reciprocal lattice vector $\mathbf{G}$, where the dotted lines indicate the planes

$$(\mathbf{G} \pm \mathbf{k})^2 = \mathbf{k}^2, \quad \text{or}$$
$$2\mathbf{k} \cdot \mathbf{G} = \pm G^2.$$

Any Bloch wave whose wave vector $\mathbf{k}$ ends outside the shaded region can be reproduced by an equivalent Bloch wave whose wave vector $\mathbf{k}'$ differs from $\mathbf{k}$ by some multiple of $\mathbf{G}$ and ends within the shaded region. When the planes corresponding to all the reciprocal lattice vectors are drawn in two or three dimensions they enclose an area or volume, respectively, called the first *Brillouin zone*. The Bloch waves whose wave vectors lie within the first Brillouin zone then form a complete set of modes from which all physical waveforms in the photonic crystal can be constructed.

[4]Note that plane sinusoidal waves with wave vectors $\mathbf{k}' = \mathbf{k} + \mathbf{G}$ are simply those produced when a plane sinusoidal wave with wave vector $\mathbf{k}$ is diffracted or Bragg scattered from a lattice with reciprocal lattice vectors $\mathbf{G}$. This point is developed in Exercise 17.6. Within an infinite lattice the effect of mutual scattering among these modes maintains the Bloch wave.

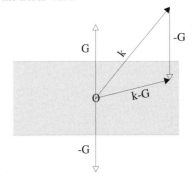

Fig. 17.4 First Brillouin zone in one dimension. O is the origin from which the wave vectors are measured.

The outer surface of the first Brillouin zone of a face-centred cubic lattice is shown in figure 17.3.

In the following two sections an analysis of the propagation of electromagnetic waves in one-dimensional photonic crystals will be carried through. Extending this analysis to two and three dimensions is straightforward but requires computer codes to handle the increased complexity. Happily the interpretation of many features met in the one-dimensional case apply to the higher-dimensional photonic crystals also.

Rough estimates of the central wavelengths and spread of the photonic band-gaps in two and three dimensions can be made quite simply. The Bragg condition provides an estimate of the central wavelength, λ, in air. For the first band-gap, due to scattering at a lattice plane we have

$$\lambda = [2d\overline{n}] \cos \theta, \tag{17.15}$$

where d is the lattice spacing, $\overline{n}$ is the mean refractive index, and θ is the angle the wave vector makes with the normal to the lattice plane. Expressed in terms of the angle of incidence in air, θ_{air}, this becomes

$$\lambda = 2d\sqrt{(\overline{n}^2 - \sin^2 \theta_{\text{air}})}. \tag{17.16}$$

The width of a band-gap can be inferred crudely by setting the angles to zero in eqn. 9.114. This gives

$$\Delta\lambda/\lambda \sim \Delta n/\pi\overline{n}, \tag{17.17}$$

where Δn is the difference between the refractive indices of the two dielectrics. A rough estimate of the refractive index contrast required to give a clear band-gap can be made by setting $\Delta\lambda/\lambda$ to 0.20 in this equation. The estimate for this ratio, $(\overline{n} + \Delta n/2)/(\overline{n} - \Delta n/2)$, is 1.9, which is substantial. At near infrared wavelengths silicon is transparent and has a refractive index of 3.5, so that silicon in air structures are suitable for photonic crystals. This choice benefits from the available experience in manufacturing micron-sized silicon structures. The band-gaps for different viewing directions may or may not overlap depending on how similar the crystal plane separations are for different crystal axes. Thus the more closely the shape of the Brillouin zone resembles a sphere (circle) the more likely overlap is. Clearly if the ratio of the relative permittivities, and hence of the refractive indices is large the chances of overlap go up, and, equally, if there is an overlap the width of the omni-directional band-gap increases.

17.3   Dispersion relations

The photonic crystal considered here is one-dimensional, consisting of a stack of plane dielectric layers of infinite lateral area with refractive indices alternately n_1 and n_2, and with equal physical widths $\Lambda/2$. In addition the dielectrics are non-absorbing. For simplicity it is assumed that n_1 and n_2 do not vary with wavelength, so that any dispersion arises from the periodicity of the refractive index alone. The following analysis builds on that developed for lossless interference filters in Chapter 9, but with changed emphasis. We are now interested in propagation through a periodic medium with such a large number of layers that end effects can safely be ignored.

Suppose that the electric field of the forward travelling waves in the nth layer is U_n and that of the backward travelling wave is V_n. In a periodic medium these fields are made up of an indefinite number of reflected and multiply reflected waves in equilibrium. Thus the fields at corresponding points in the nth layer pair and $(n+1)$th layer pair can only differ by one overall phase factor:

$$\begin{bmatrix} U_{n+1} \\ V_{n+1} \end{bmatrix} = \begin{bmatrix} U_n \exp(ik\Lambda) \\ V_n \exp(ik\Lambda) \end{bmatrix} \tag{17.18}$$

where k is the wave number and $k\Lambda$ is called the *Bloch phase*. Using eqn. 9.88 we also have that

$$\begin{bmatrix} U_{n+1} \\ V_{n+1} \end{bmatrix} = \mathbf{M} \begin{bmatrix} U_n \\ V_n \end{bmatrix}, \tag{17.19}$$

where $\mathbf{M}$ is the wave-transfer matrix. Combining the last two equations gives

$$\mathbf{M} \begin{bmatrix} U_n \\ V_n \end{bmatrix} = \begin{bmatrix} \exp(ik\Lambda) & 0 \\ 0 & \exp(ik\Lambda) \end{bmatrix} \begin{bmatrix} U_n \\ V_n \end{bmatrix}, \tag{17.20}$$

or

$$[\mathbf{M} - \mathbf{I}\exp(ik\Lambda)] \begin{bmatrix} U_n \\ V_n \end{bmatrix} = 0, \tag{17.21}$$

where $\mathbf{I}$ is the diagonal *unit matrix* $\begin{bmatrix} 1 & 0 \\ 0 & 1 \end{bmatrix}$. Equation 17.21 is an *eigenvalue equation* (as is Schroedinger's equation) that determines which modes propagate in the photonic crystal. It has valid solutions when the determinant of the total matrix vanishes, that is when

$$[M_{11} - \exp(ik\Lambda)][M_{22} - \exp(ik\Lambda)] - M_{12}M_{21} = 0.$$

Noting that the general form of $\mathbf{M}$ for a lossless stack in eqn. 9.100 has unit determinant, the above equation reduces to

$$1 - [M_{11} + M_{22}]\exp(ik\Lambda) + \exp(2ik\Lambda) = 0.$$

Taking the roots gives

$$\exp(ik\Lambda) = \text{Trace}[\mathbf{M}]/2 \pm \sqrt{\{(\text{Trace}[\mathbf{M}]/2)^2 - 1\}}. \tag{17.22}$$

On examining eqn. 17.22 we can see that whenever $|\text{Trace}[\mathbf{M}]| > 2$ the phase factor is real, which means that the wave decays with distance and is evanescent in the photonic crystal. Otherwise, when there is unattenuated transmission we can write

$$\cos(k\Lambda) = \text{Trace}[\mathbf{M}]/2. \tag{17.23}$$

The value of the trace for lossless dielectric layers can be taken from eqn. 9.111. For the present, consideration is limited to light at normal incidence. Then in eqn. 9.111 $u_{1,2}$ reduce to the refractive indices $n_{1,2}$, and the last equation becomes

$$\cos(k\Lambda) = \cos\phi_1\cos\phi_2 - [n_1/n_2 + n_2/n_1][\sin\phi_1\sin\phi_2]/2. \tag{17.24}$$

Here $\phi_i = \pi n_i \Lambda/\lambda = n_i\Lambda\omega/(2c)$, with λ being the wavelength in air, and ω the angular frequency. Equation 17.24 is the *dispersion relation* connecting the wave vector and frequency of electromagnetic radiation at normal incidence in this non-absorbing one-dimensional photonic crystal. For convenience a Bragg wavelength, λ_B, is defined to be the wavelength for which the reflections from consecutive *double layers* would be in phase at normal incidence:

$$\lambda_B = (n_1 + n_2)\Lambda = 2\overline{n}\Lambda, \tag{17.25}$$

where $\overline{n}$ is the mean refractive index. The corresponding Bragg angular frequency

$$\omega_B = \pi c/(\overline{n}\Lambda), \tag{17.26}$$

so that

$$\Lambda = \pi c/(\overline{n}\omega_B). \tag{17.27}$$

Substituting this value for Λ in the expression for ϕ_i gives

$$\phi_i = (\pi/2)(n_i/\overline{n})(\omega/\omega_B). \tag{17.28}$$

Finally replacing ϕ_i in eqn. 17.24 gives

$$\cos(k\Lambda) = \cos[\pi n_1\omega/(2\overline{n}\omega_B)]\cos[\pi n_2\omega/(2\overline{n}\omega_B)]$$
$$-(n_1/n_2 + n_2/n_1)\sin[\pi n_1\omega/(2\overline{n}\omega_B)]\sin[\pi n_2\omega/(2\overline{n}\omega_B)]/2 \tag{17.29}$$

which, to repeat, is the dispersion relation for electromagnetic waves propagating perpendicular to the layers in a one-dimensional photonic crystal. This relationship is shown by the full lines in figure 17.5 which chart the Bloch waves across the first Brillouin zone. There are regions of angular frequency centred at ω_B, $2\omega_B$... within which $\cos(k\Lambda) > 1$, making the electromagnetic waves evanescent within the photonic crystal. These are the *photonic band-gaps* analogous to the electronic band-gaps in solids seen in figure 14.16. If the photonic crystal were replaced by a uniform dielectric with the same mean refractive index the dispersion relation, $\omega = \pm kc/\overline{n}$, would follow the broken straight lines. These lines are reflected at the Brillouin zone boundary for the comparison:

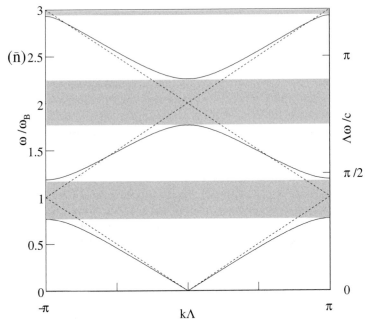

Fig. 17.5 Dispersion curves and band-gaps for a stack of alternate layers of dielectric of refractive index 1.5 and 3.5. Light is incident normal to the surface.

Brillouin zones are undefined in a homogeneous dielectric.

Notice that if the diagram in figure 17.5 were continued to the right, beyond the edge of the Brillouin zone at $k\Lambda = +\pi$, each dispersion curve would repeat the segment starting at $k\Lambda = -\pi$, where the wave vectors are exactly one primitive reciprocal lattice vector less. The dispersion curves must therefore have zero slope at the edge of the Brillouin zone, otherwise there would be a discontinuity in the group velocity $d\omega/dk$ at that point.

A useful insight into the behaviour around the Brillouin zone boundary comes from recalling that the total wave is made up of rightward and leftward moving waves which are partially reflected at each discontinuity. At the exact zone boundary these two wave trains have identical phase velocities and their superposition is a standing wave with zero group velocity. Moving a little away from the zone boundary, the left- and right-going wavetrains have slightly different velocities. The total wave has an overall slow drift, that is to say, a small group velocity. There is a clear fall in the frequency of the radiation in the photonic crystal relative to that in the homogeneous dielectric below the band-gap, and an equally clear increase above the band-gap. These changes are due to the difference in the distributions of the modes in the crystal below and above the band-gap. The form of these modes can be obtained using a variational principle:[5] the functional

$$\frac{\int |\nabla \wedge \mathbf{E}|^2 \, d\mathbf{V}}{\int \varepsilon_0 \varepsilon_r |\mathbf{E}|^2 \, d\mathbf{V}} \tag{17.30}$$

is minimized while keeping the mode orthogonal to all lower energy modes. This requires that the denominator is maximized by concentrating the electric field within the regions of higher relative permittivity, while minimizing the numerator by limiting the spatial oscillations. Evidently regions of dielectric with a large relative permittivity play the part of electrostatic potential wells for charged particles. A second important point is that a mode whose angular frequency lies close to ω_B has a wavelength in the dielectric close to 2Λ and hence the same periodicity as the relative permittivity. Therefore the antinodes of a mode just below the band-gap will coincide with the centres of the layers of higher relative permittivity. This is illustrated in the lower panel of figure 17.6. The next least energetic mode, the one immediately above the gap, must be orthogonal to this mode. Hence its antinodes have to coincide with the centres of the layers with the smaller relative permittivity, as shown in the upper panel of figure 17.6. The energies of these adjacent modes, given by $\int \varepsilon_0 \varepsilon_r |\mathbf{E}|^2 \, d\mathbf{V}$, are therefore different. If the relative permittivities of the two dielectrics are almost equal the gap is small and it opens up as the ratio between the relative permittivities, the *dielectric contrast*, increases.

In three-dimensional photonic crystals the crystal layer spacings can be different along different crystal axes. As a result the central wavelength of the band-gap will depend on the direction from which the radiation is incident. The more nearly equal the different spacings are, the more nearly the first Brillouin zone approaches a spherical shape, and the closer the band-gaps for different crystal planes will coincide in wavelength. Crystals with a face-centred cubic lattice have Brillouin zones looking like that shown in figure 17.3, and approximate quite well to a spherical shape. Band-gaps for all angles of incidence have been obtained at wavelengths in the near infrared with silicon/air photonic crystals having face-centred cubic symmetry. These will be discussed later.

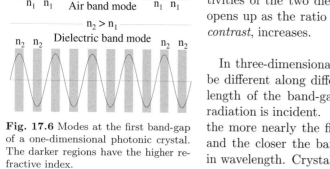

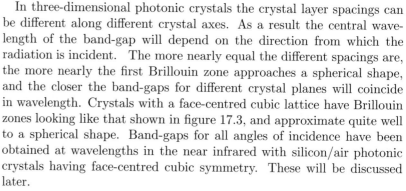

Fig. 17.6 Modes at the first band-gap of a one-dimensional photonic crystal. The darker regions have the higher refractive index.

17.3.1   Off-axis beams

When light is incident on a one-dimensional photonic crystal at angles away from the normal the optical path lengths through the layers change and the frequency of the band-gap changes. We must now use eqn. 9.112 for the trace of the transfer matrix. The dispersion relation, eqn. 17.24,

---

[5]See Chapter 2 of of the second edition of *Photonic Crystals: Molding the Flow of Light* by J. D. Joannopoulos, S. G. Johnson, J. N. Winn and R. D. Meade, published by Princeton University Press (2008).

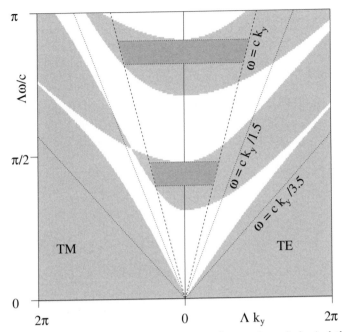

Fig. 17.7 Dispersion plot for a stack of dielectric layers of equal physical thickness, alternating between refractive indices 1.5 and 3.5. Λ is the total physical thickness of two consecutive layers, ω is the angular frequency of the radiation incident and k_y the component of its wave vector in a plane parallel to the interfaces. Regions of free propagation are left clear.

is then replaced by

$$\cos(k\Lambda) = \cos\phi_1 \cos\phi_2 - [u_1/u_2 + u_2/u_1][\sin\phi_1 \sin\phi_2]/2. \quad (17.31)$$

Here $\phi_i = \pi n_i (\Lambda/\lambda) \cos\theta_i$, and $u_i = n_i \cos\theta_i$ for TE (s-)polarized waves, while $u_i = n_i/\cos\theta_i$ for TM (p-)polarized waves, with θ_i being the angle the wave vector within dielectric i makes with the normal. A mode with this dispersion relation propagates freely provided that $|\mathrm{Trace}[\mathbf{M}]| \leq 2$. When the modulus of the trace is greater than 2 the wave is evanescent and reflected fully from the photonic crystal. Figure 17.7 shows a dispersion plot for light incident on a stack of layers with refractive indices 1.5 and 3.5, all of equal physical thickness $\Lambda/2$. k_y is the component of the wave vector *parallel* to the layer surfaces. Regions over which free propagation occurs are left clear while the band-gaps are shaded. The left and right halves show the dispersion plots for TM(p) and TE(s) polarizations respectively. In such plots all the modes which share the same angle of incidence populate straight lines through the origin. The y-axis corresponds to light at normal incidence, and is simply figure 17.5 compressed sideways. Three other lines through the origin are drawn in each half of the figure. They correspond to light travelling parallel to the interface in air, and in the two dielectrics respectively: each line is

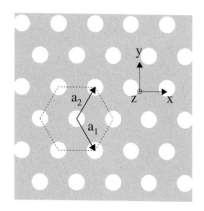

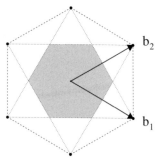

Fig. 17.8 Triangular patterned photonic crystal showing the lattice above, and the reciprocal lattice below. The first Brillouin zone is shaded.

labelled $\omega = ck_y/n$, where n is the relevant refractive index. These are known as *light lines*. In a homogeneous dielectric all freely propagating modes would lie above the light line for that dielectric. In the case of TM(p-)polarization there is one angle of incidence, the Brewster angle, at which waves invariably propagate. The modes concerned would lie in the left half of the figure on a straight line threading through the gap between forbidden regions.

In the usual case that light is incident from a source in air onto the stack, the accessible modes must lie above the light line for air. Modes lying above the light line in air but which are evanescent in the photonic crystal are indicated by the heavily shaded regions in figure 17.7. Consequently light arriving in air whose angular frequency lies within either of these ranges is totally reflected, whatever its angle of incidence.

17.4   Two-dimensional photonic crystals

Figure 17.8 shows a cross-section through a typical two-dimensional photonic crystal having a triangular array of air holes in an otherwise uniform dielectric. The holes run through the full depth of the crystal which is in the direction perpendicular to the diagram. There is interference between electromagnetic waves reflected from the array, and consequently there can be photonic band-gaps. In practice two-dimensional photonic crystals are made in one of two distinct forms. There are *photonic crystal fibres* in which a pattern of holes extends along a glass fibre whose length may be metres, and there are *photonic crystal slabs* which are of order one wavelength in thickness.

Silicon is relatively transparent at telecom wavelengths, which has led to current interest in developing optical interconnects via waveguides in two-dimensional photonic silicon crystal slabs. A very useful property of these waveguides is that the group velocity along them can be tuned to a very low value by designing the crystal pattern appropriately. Electro-optic effects within the dielectric are therefore considerably enhanced because the light spends a longer time in the dielectric. A modulator, described later, which makes use of silicon photonic crystal slabs is extremely compact, being only $100\,\mu$m long.

In the following section the fundamental properties of two-dimensional photonic crystals will be discussed. The dispersion relations of these photonic crystals are naturally more complex than those for one-dimensional crystals, and their solutions usually require specialized techniques. What will be presented here is therefore qualitative and relies on making inferences based on the results of the analysis of the one-dimensional case. After this, the uses of silicon-on-silica and free-standing silicon photonic crystal slabs as waveguides and modulators are discusssed. There follows an explanation of negative refraction and the super prism effect.

The improvement in the efficacy of LEDs brought about by patterning the emissive surface as a photonic crystal is then detailed. Last of all, photonic crystal fibres and their applications are described.

17.4.1 Two-dimensional band-gaps

The photonic crystals considered here have a regular pattern either of air holes in the dielectric, or of dielectric posts in air. The structure is uniform in the direction perpendicular to the plane of the pattern, which we define to be the z-direction. As an example the photonic crystal in figure 17.8 has a triangular lattice of air holes in a dielectric that is otherwise isotropic and homogeneous. Its primitive lattice vectors, both of length a, are indicated. The pattern can be expressed as the variation of the relative permittivity of the form

$$\varepsilon_{\mathrm{r}}(\boldsymbol{\rho}) = \varepsilon_{\mathrm{r}}(\boldsymbol{\rho} + \mathbf{R}), \tag{17.32}$$

where $\boldsymbol{\rho}$ is the vector location of any point in the cross-sectional plane, which is here the xOy plane, and $\mathbf{R}$ is any linear combination of the primitive lattice vectors. Then Bloch's theorem states that any electromagnetic wave within the crystal has an electric field of the form

$$\mathbf{E} = \mathbf{E}_{\mathbf{k}}(\boldsymbol{\rho}) \exp\left[i(\beta z + \mathbf{k}_{\mathrm{p}} \cdot \boldsymbol{\rho} - \omega t)\right], \tag{17.33}$$

where $\mathbf{E}_{\mathbf{k}}(\boldsymbol{\rho})$ has the same periodicity as $\varepsilon_{\mathrm{r}}(\boldsymbol{\rho})$ and where the wave vector $\mathbf{k} = \boldsymbol{\beta} + \mathbf{k}_{\mathrm{p}}$, with $\boldsymbol{\beta}$ lying along the z-direction and $\mathbf{k}_{\mathrm{p}}$ lying in the xOy plane. A similar expression holds for the magnetic field. TE and TM modes are defined to have $\mathbf{H}$ and $\mathbf{E}$ pointing respectively in the z-direction. The way the fields are distributed so as to concentrate the $\mathbf{D}$ field in the dielectric with the larger relative permittivity can be very different for the TE and TM modes. Thus their band structures and gaps differ: a connected lattice of high-ε_{r} regions favours band-gaps among the TE modes, while TM band-gaps are more likely if the lattice consists of isolated high-ε_{r} regions.[6] The case in which electromagnetic radiation travels in or close to the xOy plane will be considered next, being relevant for radiation within photonic slabs. At the other end of the angular range, when radiation travels close to the z-direction in a photonic crystal fibre, the distinction between the TE and TM modes disappears.

17.5 Photonic crystal slabs

Silicon-on-silica photonic crystal slabs have considerable potential for use in monolithic photonic devices. Silicon is transparent and has a refractive index of 3.5 at telecom wavelengths while a silica substrate has

---

[6]For further details see Chapter 5 of the second edition of *Photonic Crystals: Molding the Flow of Light* by J. D. Joannopoulos, S. G. Johnson, J. N. Winn and R. D. Meade, published by Princeton University Press (2008).

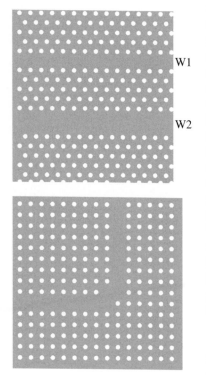

W1

W2

Fig. 17.9 Photonic crystal slabs with linear defects.

a refractive index of 1.5. Total internal reflection at the silicon interfaces with air and silica ensures that modes propagating at small angles to the slab plane remain within the slab. This is known as *index guidance*. Index guidance is improved if the substrate is partly etched away to leave a silicon slab that is largely free-standing in air. Figure 17.8 also shows, in addition to a photonic crystal slab, the corresponding reciprocal lattice and first Brillouin zone. This Brillouin zone is approximately circular, which makes it likely that a slab of this geometry can have a band-gap for all directions of propagation lying in the plane of the slab. A silicon slab in air with the geometry shown in figure 17.8 can have a band-gap for TE modes, but not for TM modes. Evidently this TE band-gap can only be exploited if the incident radiation is plane polarized. Suitable ratios between the hole radius, r, the hole spacing centre-to-centre, a, and the slab thickness, t, are required if there is to be a wide band-gap in the TE modes. These are $r/a \approx 0.3$ and $t/a \approx 0.6$. The corresponding TE band-gap is centred at angular frequency $\omega_c \approx 0.3(2\pi c/a)$ and has relative width $\Delta\omega/\omega_c \approx 0.2$. With this slab thickness, of around one half-wavelength, the mode structure transverse to the plane of the slab is restricted to a standing half-wave. In common with one-dimensional photonic crystals the band-gap widens as the dielectric contrast between the dielectrics increases.

Omitting one or more adjacent rows of holes across the slab can create a waveguide at some wavelength within the band-gap. Figure 17.9 displays examples of such structures. Those with linear defects formed by missing out n consecutive rows of holes or pillars are labelled Wn. The dispersion plot for such a slab waveguide is shown in Figure 17.10, where a single TE waveguide mode associated with the defect has entered the band-gap. This mode is indicated by the full line curve. If this mode is excited by pumping radiation at the appropriate frequency into one end of the waveguide channel from an optical fibre, then the radiation propagates along the waveguide and exits from the far end of this channel. The mode is evanescent in the body of the slab outside the waveguide channel. A power loss along such a waveguide as small as $\sim 10\text{dB/cm}$ has been attained. Although large compared to losses along a single mode optical fibre, this is nonetheless sufficiently small to make devices practical in which the paths are measured in millimetres. By contrast the loss from such a mode lying on the part of the dispersion curve just above the light line in air in figure 17.10 is $\sim 1000\,\text{dB/cm}$ due to loss of radiation through the slab surface. Coupling electromagnetic radiation from a silicon ridge waveguide directly into a waveguide in a silicon photonic crystal slab would be very lossy: in the ridge waveguide a mode has diameter $\sim 3\,\mu\text{m}$ while in the slab waveguide a mode diameter is only $\sim 0.5\,\mu\text{m}$. Instead the ridge waveguide area is tapered down to match the slab waveguide, the tapering being gentle enough that little power is lost, that is to say it is *adiabatically* coupled.

The lower panel in figure 17.9 shows a waveguide turning through $90°$

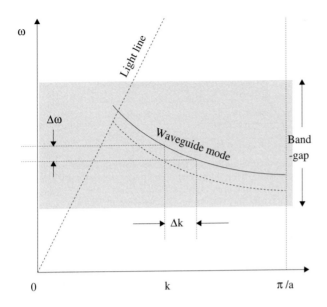

Fig. 17.10 Dispersion curve for a waveguide mode in a photonic crystal slab. Changing the refractive index displaces this from the solid to the broken curve

in about one wavelength. If radiation whose wavelength matches that of the waveguide mode is injected into one end of the waveguide it propagates around this sharp bend with low loss. By comparison the minimum turn radius in single mode fibre for which the power loss is acceptable is much larger, of order a few centimetres. Evidently this ability to steer near infrared radiation in silicon photonic crystal slab waveguides, with relatively low losses, makes them potential interconnects for monolithic optoelectronic devices.

17.5.1   Slow light

The dispersion curve in figure 17.10 shows how the group velocity, $v_g = \mathrm{d}\omega/\mathrm{d}k$, of the mode propagating along a linear waveguide vanishes at the edge of the Brillouin zone ($k = \pi/a$). Group velocities as small as $c/100$ have been obtained by operating near the zone edge.

It will emerge that the phase delay along such a waveguide is correspondingly enhanced by a factor c/v_g. This feature was exploited by Ray T. Chen and colleagues in constructing a compact Mach–Zehnder photonic crystal modulator for infrared radiation fabricated from a silicon-on-insulator wafer. The active region of the arms of their device is sketched in figure 17.11. Both arms are W1 photonic crystal waveguides having a triangular pattern of holes. A modulating voltage is applied across one of the arms. Outside the active region the optical circuit is completed by silicon rib waveguides. The authors engineered the waveg-

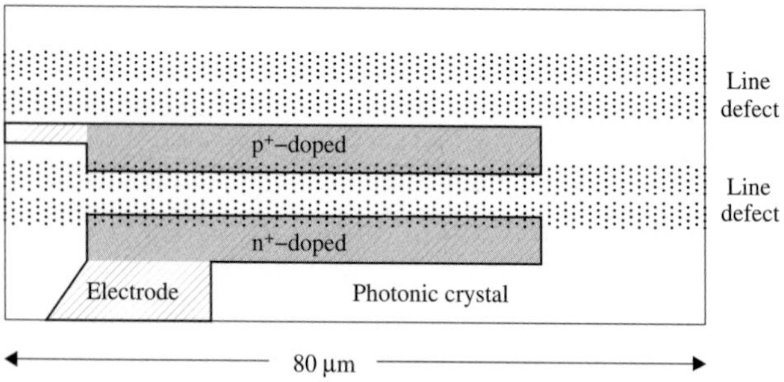

Fig. 17.11 Photonic crystal modulator. Drive electrodes are cross-hatched. Between the doped regions the silicon is intrinsic. Diagram adapted from L. Gu, W. Jiang, X. Chen, L. Wang and R.T. Chen, *Applied Physics Letters* 90, 071105 (2007). Courtesy Professor Ray T. Chen and the American Institute of Physics.

uides so that the dispersion curve of the propagating mode had a section of small constant slope, making the group velocity low and the dispersion negligible in that region. This they achieved by suitable choices of the width of the waveguide and the diameter of the holes adjacent to it. The pattern period is 400 nm, the hole diameter 220 nm and the silicon thickness 260 nm. Those areas shaded in the figure are n^+- and p^+-doped regions on either side of the lower waveguide, which is itself intrinsic silicon. Electrodes are indicated by cross-hatching within a bold outline and make the lower waveguide into a *pin* diode.

In its passive state the diode is held off by reverse bias of -1 V. When instead a forward bias of $+2$ V is applied across the diode charge flows into the intrinsic region that forms one arm of the Mach–Zehnder. There is a consequent rapid change in the arm's refractive index, typically $\Delta n \sim 0.001$ in a time ~ 0.5 ns. This change displaces the dispersion curve of the waveguide mode in figure 17.10 from the full line to the broken line. If n is the refractive index for this mode, $\omega = ck/n$. Then the separation of the two curves in ω at a fixed value of the wave vector k is simply

$$\Delta\omega = ck\Delta n/n^2.$$

This displacement is labelled in figure 17.10. Now the quantity of interest is the change in wave vector produced by the applied voltage because it determines the phase shift between radiation travelling along the two arms of the Mach–Zehnder interferometer. This change, at constant frequency , is also labelled on the figure. Its value is

$$\Delta k = \Delta\omega/v_{\mathrm{g}} = ck\Delta n/(n^2 v_{\mathrm{g}}),$$

v_{g} being the group velocity. The resulting interference between light that travelled through the two arms is destructive when the arm length over which the voltage is applied, L, is such that $\Delta kL = \pi$. Taking Δk from the previous equation, there will be destructive interference when

$$L = \pi/\Delta k = n^2 v_{\mathrm{g}}\pi/(ck\Delta n)$$

$$= (n/\Delta n)(v_{\mathrm{g}}/c)\lambda/2, \qquad (17.34)$$

where λ is the wavelength of the mode in free space. The enhancement factor, c/v_{g} reported was 10, so that $L \approx 100\,\mu$m. Such a modulator is only 1% of the length of the lithium niobate modulator described earlier in Chapter 10. This new compact modulator lends itself to integration in monolithic photonic devices. Modulation rates of GHz have been achieved, comparable to that attained using lithium niobate modulators.

Non-linear effects are also enhanced when light travels slowly in photonic crystal waveguides. In the case of the Kerr effect the enhancement is quadratic in the slow-down factor: a first factor of c/v_{g} comes from the increased time the light spends in the waveguide, and the second factor from the increase in intensity of the light. The functionality of photonic crystal devices is obtained solely through structuring, so that the choice of the operational wavelength and the bandwidth are at the designer's discretion. Devices relying on intrinsic material properties, such as an atomic transition, lack this flexibility.

By comparison the reduction in group velocity in electromagnetically induced transparency (EIT) described in Chapter 18 is far greater than that so far observed in photonic crystal waveguides. However the complexity of the experimental arrangements for observing EIT limits its practical application. An inherent feature of EIT is that it is restricted to wavelengths across the width of the atomic transition involved, whereas the wavelength and bandwidth of the mode of a photonic crystal waveguide can be engineered via the structure. EIT's restricted bandwidth implies a correspondingly slow response: very short pulses are not delayed at all. Another relevant difference is that in EIT the energy of the incident radiation is mostly transferred to the atomic cloud on entry, and is re-emitted as it exits. By contrast the energy of the radiation injected into a photonic crystal waveguide remains in the electromagnetic field. Consequently the radiation intensity in the photonic crystal waveguide is increased by the slowing-down factor.

17.5.2 Anomalous refraction effects

In addition to the existence of band-gaps arising from the periodic modulation of photonic crystals, there are related anomalous refraction effects. Two such effects are now described, first negative refraction and then the super prism effect.

The method of ray reconstruction when light enters a homogeneous dielectric from air will be presented in a slightly novel way in preparation for tackling these unexpected effects. Let the wave vector of the incident radiation be $\mathbf{k}_0$, then if the homogeneous dielectric has refractive index n the wave vector after refraction will be $\mathbf{k}$, where $k = nk_0$. Circles are drawn in the upper panel of figure 17.12 with radii k_0 for air and nk_0 for the dielectric. These are called isofrequency surfaces because a wave

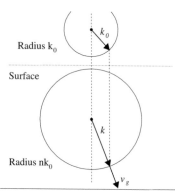

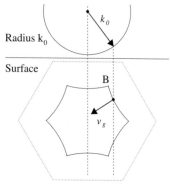

Fig. 17.12 Construction of the refraction of rays entering from air into (upper panel) a homogeneous dielectric, and into (lower panel) a photonic crystal.

vector of radiation at the given frequency will extend from the origin to touch the corresponding circle in air or in the dielectric. The key point to recall is that the component of the wave vector parallel to the surface is unchanged after refraction. A line is therefore drawn from the tip of $\mathbf{k}_0$ perpendicular to the interface, and where it cuts the isofrequency surface for the dielectric is the end point of $\mathbf{k}$. This procedure is simply presenting Snell's law geometrically. In the homogeneous dielectric the group velocity lies in the same direction as the phase velocity, that is along $\mathbf{k}$. Note that the upper intersection of the projection from $\mathbf{k}_0$ onto the dielectric's isofrequency surface can be safely ignored because it would correspond to light travelling *from* the dielectric into the air.

Now we turn to the case of light incident from air onto a photonic crystal slab with a triangular pattern of air holes. We consider light travelling parallel to the plane of the slab. At low frequencies such that the wavelength is much larger than the repeat distance in the crystal pattern the isofrequency surface would also be circular in the crystal. However at a frequency approaching the band-gap frequency the isofrequency surface takes on the hexagonal symmetry shown in the lower panel of figure 17.12. For guidance, the outline of the first Brillouin zone is superposed as a broken line. The intersection of the projection from $\mathbf{k}_0$ onto the isofrequency surface for the photonic crystal is indicated by a dot. The direction of the group velocity, given by $d\omega/dk$, lies perpendicular to the isofrequency surface at this point. Now these isofrequency surfaces close up as the frequency increases so that the direction of the group velocity is inward, as drawn. It follows that the refracted beam points leftward rather than pointing rightward, as would always be the case in a homogeneous dielectric. It is as if the photonic crystal slab had a *negative refractive index*.

An equally strange effect is apparent when the incident beam is directed so that the projection from $\mathbf{k}_0$ intersects the photonic crystal's isofrequency surface near the cusp at B. A small change in the wavelength can then cause the point of intersection to move right across the cusp with an accompanying very large change in the angle of refraction. This is called the *super prism effect*.

17.5.3 LED emission

The efficacy of LEDs has been severely limited because much of the radiation produced is trapped by total internal reflection at the emissive surface and subsequently re-absorbed within the LED. Surface emitting, rather than edge emitting, LEDs suffer less from this problem, and lend themselves to improvement, so these will be the type considered here. The materials used in LEDs have high refractive indices, for example the GaN used to make blue LEDs has a refractive index of 3.5. The critical angle at an interface with air is $16.6°$ or $0.29\,\mathrm{rad}$. Assuming isotropic emission within the LED the fraction of light escaping is

$$(1/2) \int_0^{0.29} \sin\theta \, d\theta = 0.02,$$

or double this if the base of the LED is a perfect reflector. Palliatives such as corrugating the emissive surface give some improvement. Better still the emissive surface of the LED can be etched or otherwise treated to apply a pattern of holes and turn this face into a photonic crystal slab about half a wavelength thick. Radiation from the active material readily penetrates this layer at all angles of incidence. As shown in figure 17.13 part of this radiation will be diffracted inside the photonic crystal. This changes the wave vector so that the angle of incidence on the air interface becomes less than the critical angle. The wave vectors labelled $\mathbf{k}_1$ and $\mathbf{k}_2$ are those before and after diffraction respectively, while $\mathbf{G}$ is a reciprocal lattice vector. If the bottom surface of the LED is a good reflector then radiation internally reflected once or more has repeated opportunities to be diffracted inside the crystal layer and so emerge from the surface.

Surface emitting LEDs with the surface patterned as a photonic crystal dissipate less of the applied electrical power as heat so that cooling is less of a problem. They can also be made with large surface areas, bringing corresponding advantages as light sources. Large area red, green and blue LEDs emitting $\sim 80 \, \mathrm{lm} \, \mathrm{W}^{-1}$ are manufactured, for example by Luminus Devices[7] and have lifetimes of over 100,000 hours, exceeding that of linear fluorescent tubes. Such LEDs are increasingly used for illuminating LCD screens, in automotive lighting and in the home.

17.6 Photonic crystal fibres

Photonic crystal fibres, *PCF*s, otherwise known as *microstructured fibres*, are optical glass fibres of a uniform chemical composition in which light is guided by an array of holes running the length of the fibre. The cross-sections through two such fibres, shown in figure 17.14, reveal an ordered triangular array of holes. Construction of PCFs starts with the fusing together of a set of pure silica capillary tubes and solid rods. The resulting block is then used as the feed material in a drawing tower like that employed in making standard optical fibre. Both types shown here are produced with an outer cladding diameter of $125 \, \mu\mathrm{m}$ in order to be compatible with installation procedures used for standard fibre. In the upper panel of figure 17.14 is an example of those fibres whose core is left solid by omitting one or more holes at the centre of the pattern. Around this core the region of perforated glass constitutes a cladding of lower mean refractive index, and the slight difference in refractive index

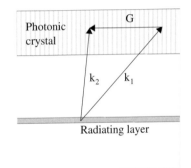

Fig. 17.13 Section of a surface emitting LED with a photonic crystal slab etched into the emissive face.

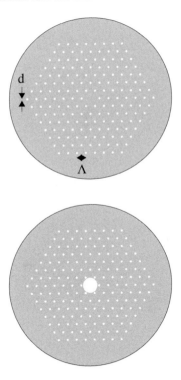

Fig. 17.14 Cross-sections through photonic crystal fibres. The uper panel shows an example in which light is index guided along the core. In the fibre shown in the lower panel the light is band-gap guided.

[7]Luminus Devices, 1100 Technology Park Drive, Billerica, MA 01821, USA.

between core and cladding guides the light along the core, just as in a standard single mode fibre. The spacing of the holes and their diameter can be varied so as to tune the dispersion and the cross-sectional area of the modes propagating on the fibre over a much wider range than is possible with single mode fibre. One application opened up by this flexibility is the production of fibre which carries a single mode however short the wavelength of the radiation. The criterion that only the LP_{01} mode propagates becomes, using the definition in eqn. 16.9,

$$V = (2\pi a/\lambda)\sqrt{[n_1^2 - n_2^2]} < 2.405. \qquad (17.35)$$

Here n_1 should be interpreted as the mean refractive index over the area of the fundamental mode, a its radius and n_2 the refractive index of the cladding mode. As the wavelength decreases the difference between the refractive indices falls, so that if V is less than 2.405 at some wavelength it remains less than 2.405 for all shorter wavelength. Specializing to the situation that the holes have the triangular pattern seen in figure 17.14 with a single missing central hole, this fibre becomes *endlessly single mode* provided $d/\Lambda < 0.43$ where Λ is the hole spacing and d the hole diameter.

In the lower panel of figure 17.14 the core is by contrast hollow which rules out index guiding of light within the core. On the other hand the two-dimensional optical crystal surrounding the core can have band-gaps that will prevent radiation over a range of wavelengths from penetrating through the cladding. The radiation is *band-gap guided* along the core. In principle, because the radiation travels principally in the air filled core with only an evanescent tail in the cladding, the attenuation of band-gap guided light could be very low. This would be ideal for communications. However the lowest attenuation achieved, around 2 dB/km, is as yet uncompetitive when compared to around 0.2 dB/km achieved with standard single mode fibre.

The effect of exposing microstructured fibre to intense short laser pulses is to generate a broad continuum of spatially coherent light. Pulses from a Nd:YAG laser at 1060 nm can produce a *supercontinuum* spectrum whose intensity is uniform from ~ 400 nm to ~ 1600 nm. The process by which this broad spectrum is generated is complex and involves self-phase modulation, Raman scattering and four wave mixing. A supercontinuum spectrum is easiest to obtain if the fibre has low dispersion because the phase matching required for four wave mixing is then guaranteed over a wide range of wavelengths. In addition the fibre is designed so that the wavelength at which the dispersion vanishes is close to the pump wavelength.[8] Commercially available supercontinuum sources are almost 1000 times brighter than an incandescent source, and in addition are pulsed and possess lateral spatial coherence. Such sources have made feasible new studies in biology and other disciplines. Supercontinuum fibre has also become an ingredient of the optical clocks described in Chapter 18.

[8]This is just the condition to be avoided in dense wavelength division multiplexing. See Section 16.9.2.

17.7   Three-dimensional photonic crystals

Three-dimensional photonic crystals can have band-gaps which are truly omni-directional, so that light within a limited range of wavelengths and incident from any direction will be totally reflected. Such omnidirectional band-gaps are only possible if the spacings of the differently oriented crystal planes are as far as possible equal, which makes the first Brillouin zone close to spherical. A crystalline form which is close to fulfilling this requirement is the face-centred cubic crystal shown in figure 17.2; its Brillouin zone is also shown in figure 17.3. A diamond lattice is somewhat superior, its crystal structure consisting of two interpenetrating face-centred cubic lattices. There are two further requirements that secure omni-directional band-gaps. Firstly the band-gaps arising from different crystal planes have to be broad enough so as to overlap one another. From the discussion of Section 17.2 we can see that this will require that the dielectric contrast is large. Secondly the structure should consist of veins of the dielectric with higher relative permittivity which favours wide TE band-gaps, and these should connect compact regions of this dielectric, so as to favour wide TM band-gaps.[9] Then overlap of the TE and TM band-gaps is achievable. Artificial three-dimensional photonic crystals with lattice cells around one micron in size are however relatively difficult to manufacture. Consequently almost all practical applications make use of photonic crystal slabs.

Three-dimensional photonic crystals for use at near infrared wavelengths have been constructed in silicon up to tens of layers deep. A variety of methods are used: electron beam lithography with reactive etching as used in microelectronics, holographic lithography and self-assembly. Two three-dimensional crystal structures will now be described, both having the face-centred cubic symmetry.

The construction of an inverse opal photonic crystal involves a first step of crystal self-assembly. Uniformly sized spheres of silica are deposited from a solution onto a substrate. The spheres assemble spontaneously on the substrate into the face-centred cubic lattice and afterwards this material is dried. At this stage the crystalline form in silica is very close to that of natural opal. The material is then heated so as to sinter the spheres. Necks form linking the adjacent spheres and make the structure rigid. Next silicon is introduced by chemical vapour deposition to fill the interstices between the silica spheres. Finally the silica spheres themselves are etched away, leaving a very open *inverse opal* structure in silicon with an omni-directional band-gap whose fractional bandwidth $(\Delta\lambda/\lambda)$ is around 0.05. For a mean pore size of 330 nm the peak reflectivity occurs at around 550 nm. A section through an in-

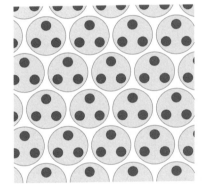

Fig. 17.15 Section through an inverse opal. The white connected regions are the walls of the silicon shells. The three spots in each of the spherical cups are holes linking adjacent pores. These holes mark the regions over which the silica spheres became sintered together.

---

[9]See Chapter 5 in *Photonic Crystals: Molding the Flow of Light*, second edition, by J. D. Joannopoulos, S. G. Johnson, J. N. Winn and R. D. Meade, published by Princeton University Press (2008).

verse opal photonic crystal is shown in figure 17.15, sliced through the centre of one layer of spherical holes. Any small variation in the size of the pores across the crystal alters the band-gap and hence the colour seen in reflection. Although the band-gaps for different viewing angles overlap this overlap is partial and a change in viewing direction elicits a corresponding change in the colour seen. It is this feature which characterizes iridescence. In a natural opal crystal the silica spheres vary in size sufficiently so that there can be additional attractive variations in colour across the surface. Inverse opals are also made from dielectrics other than silicon. However if the dielectric contrast is less than about 8 the omni-directionality of the band-gap is lost.

Figure 17.16 shows a section through the second example of a three-dimensional photonic crystal having a silicon framework as constructed by Joannopoulos and colleagues.[10] This is made in layers which are alternately a solid silicon sheet punctured by a triangular pattern of holes and then a layer of free-standing silicon posts. The latter are indicated by the broken circles in the diagram. Each pair of layers requires a separate sequence of electron beam lithography and reactive etching of the silicon. Between successive pairs of layers the patterns of holes and posts are displaced laterally in a sequence that is indicated by the letters A–D on the diagram. The letters A and D are about $1\,\mu$m apart and each double layer is about $500\,$nm thick. After three displacements the pattern duplicates the original pattern, giving the structure face-centred cubic crystal symmetry. The large dielectric contrast between silicon and air combined with an open structure ensure an omni-directional band-gap centred at a wavelength around $1.3\,\mu$m. The fractional band-gap width of the hole-and-post structure ($\Delta\omega/\omega$) is around 0.21, much larger than that of the inverse opal structure using the same material. With appropriate choices of cell size the band-gaps can be positioned to cover other ranges of telecom wavelengths throughout which silicon remains transparent.

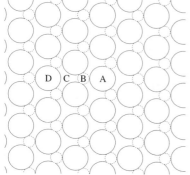

Fig. 17.16 Section through a *hole-and-post* photonic crystal. The broken circles indicate the posts in the second layer. The letters indicate the corresponding positions of a hole centre over a cycle of four hole layers.

17.8　Further reading

Photonic Crystals: Molding the Flow of Light, second edition, by J. D. Joannopoulos, S. G. Johnson, J. N. Winn and R. D. Meade, published by Princeton University Press (2008) is a very readable general text on photonic crystals.

'Photonic Structures in Biology' by P. Vukusic and J. R. Sambles, in *Nature* volume 424, pages 852-855 (2003) reviews natural photonic structures.

'Slow light in photonic crystal waveguides' by T. F. Krauss, *Journal of Physics D: Applied Physics* **40**, 2666 (2007), published by the Institute of Physics. This article gives a clear and compact review of the physical

---

[10]M. Qi, E. Lidorikis, P. T. Rakich, S. T. Johnson, J. D. Joannopoulos, E. P. Ippen and H. I. Smith, *Nature* **429**, 538 (2004).

principles and of the potential for optical signal processing.
'Slow light in photonic crystals' by T. Baba in *Nature Photonics* 2, 465
(2008). This is another modern review.
Nanophotonic Materials : Photonic Crystals, Plasmonics, and Metamaterials, edited by R. B. Wehrspohn, H.-S. Kitzerow and K. Busch and
published by Wiley-VCH (2008). This is a more advanced text.

Exercises

(17.1) Dielectrics of refractive indices 1.5 and 3.5 are used
to construct a multilayer filter with alternate layers
of the two dielectrics, each pair having altogether a
physical thickness of 100 nm. Estimate the central
wavelength and the width of the band-gap at an
angle of incidence equal to $20°$. What is the Bragg
wavelength for this filter?

(17.2) The primitive vectors of a body-centred cubic lattice connect a body centre to the nearest corners.
For a cube of unit side length they are

$$\mathbf{a}_1 = (\mathbf{i} + \mathbf{j} - \mathbf{k})/2,$$
$$\mathbf{a}_2 = (-\mathbf{i} + \mathbf{j} + \mathbf{k})/2,$$
$$\mathbf{a}_3 = (\mathbf{i} - \mathbf{j} + \mathbf{k})/2.$$

What are the reciprocal lattice vectors, and what
points do they connect on this lattice

(17.3) Using the notation of Section 17.3 the ingoing
waves on a section of a simple plane filter are U_1
and V_2, while the outgoing waves are U_2 and V_1.
An S-matrix is defined by

$$\begin{bmatrix} U_2 \\ V_1 \end{bmatrix} = S \begin{bmatrix} U_1 \\ V_2 \end{bmatrix} = \begin{bmatrix} t_{12} & r_{21} \\ r_{12} & t_{21} \end{bmatrix} \begin{bmatrix} U_1 \\ V_2 \end{bmatrix}.$$

Express the components of the S-matrix in terms
of the components of the corresponding M-matrix:

$$\begin{bmatrix} U_2 \\ V_2 \end{bmatrix} = M \begin{bmatrix} U_1 \\ V_1 \end{bmatrix} = \begin{bmatrix} A & B \\ C & D \end{bmatrix} \begin{bmatrix} U_1 \\ V_1 \end{bmatrix}.$$

(17.4) Suggest ways in which an L3 cavity in a photonic
crystal can be loaded with photons.

(17.5) Compare the length of the Mach–Zehnder interferometer described in Section 17.5.1 with that of
a similar interferometer which lacks the photonic
crystal patterning.

(17.6) An incident plane wave $\exp[i(\mathbf{k}_0 \cdot \mathbf{r} - \omega t)]$ is scattered by a crystal producing plane waves with wave
vector $\mathbf{k} = \mathbf{k}_0 + \Delta\mathbf{k}$. Suppose that the modulus of
the amplitude of the scattered wave produced by
an element $d\mathbf{r}$ of the crystal with coordinate $\mathbf{r}$, relative to the centre of the basic cell it is located in,
is $h(\mathbf{r})d\mathbf{r}$. Also suppose that the basic cells have
centres at locations $\boldsymbol{\rho}_i$. Show that the amplitude
of the scattered wave is

$$\sum_i \exp[-i\Delta\mathbf{k} \cdot \boldsymbol{\rho}_i] \, H(\Delta\mathbf{k}),$$

where $H(\Delta\mathbf{k})$ is the Fourier transform of $h(\mathbf{r})$. Also
show that allowed values of $\Delta\mathbf{k}$ for which the scattering from the whole lattice adds constructively
are just the reciprocal lattice vectors.

Quantum interactions

18

18.1   Introduction

In this chapter the interaction of electromagnetic radiation with atoms is analysed semiclassically: the atoms are treated as quantum mechanical objects and the electromagnetic radiation as classical waves. This approach is successful in explaining a wide range of phenomena. A principal defect of the semiclassical analysis is that spontaneous emission must be put in by hand; it does not emerge naturally. Beyond this, inferences based on knowledge of the quantum aspect of radiation are used here to give guidance on the interpretion of the semiclassical analysis. Applications are described which exploit the interaction of coherent light in the form of laser beams with low density gases of atoms or ions cooled by laser beams. A representative sample of the phenomena so far observed are discussed.

In the first section below a calculation of the rates for electric dipole transitions in atoms is made subject to two simplifying assumptions. First that the radiation has low intensity, such that non-linear effects are negligible, and second that the spectral distribution is flat across the linewidth of the transition considered. The susceptibility of dilute gases is also calculated.

Then follows a study of the interaction of laser beams with atoms when the laser frequency is close or equal to one of the atoms' dipole transition frequencies. In this case, by contrast, there are very many photons per mode and the spectrum of the radiation can be much narrower than the transition line width. The consequences are striking. Atoms illuminated in this way initially absorb laser radiation and enter the excited state; then the radiation stimulates them to emit and return to the ground state. This sequence repeats as long as the atoms are illuminated by the laser. The oscillations are known as *Rabi oscillations*, which celebrates Rabi's earlier discovery of the equivalent effect produced with microwaves.

In quantum electrodynamics (QED), which is the quantum theory of electromagnetic interactions, both the electromagnetic field and the atom are quantized and treated as a total system. One idea borrowed from this approach is to view an atom and the photons within a laser beam whose frequency is tuned close to a transition frequency as a *dressed state* of the atom. Several optical effects are described in the next two sections of the chapter, namely Mollow fluorescence and the Autler–Townes effect, whose explanations are straightforward in terms of dressed states.

Normally a *probe* laser beam at a frequency matching an allowed transition from the ground state to an excited state E, will be absorbed in a short distance by an assembly of atoms. However in certain circumstances when a second laser is used to simultaneously pump a different allowed transition to the same excited state E, the atoms become transparent to the probe beam. This property is known as *electromagnetically induced transparency* (EIT). It is equally surprising that the probe laser beam can be slowed to a walking pace in a cloud of such atoms, and can even be stalled for an interval as long as several milliseconds. These effects are customarily observed in gases, particularly at very low temperatures, but have been produced in a few cases in particular crystals. A section of the chapter is devoted to the discussion and explanation of EIT.

The following dilemma was faced in attempts to obtain the ultimate in precision measurements of optical frequencies. In principle a suitably narrow spectral line could be measured by observing the resonance fluorescence from a single stationary ion illuminated by a laser. As the laser frequency is scanned in steps across the absorption line the resonance fluorescence could be measured and hence the line shape determined. Unfortunately a state whose fluorescence has narrow line width has necessarily a low decay rate so that the signal is hard to detect against experimental background. The paradoxical way of avoiding the dilemma was suggested by Dehmelt in 1982 and involves a quantum effect known as *shelving*.

Shelving is also one of the techniques that has made it possible to build optical clocks whose 'pendulum beats' are the oscillations of radiation emitted in a very narrow width, optical transition of a single ion. A second essential ingredient in the construction of such an optical clock is the ability to cool ions to very low temperatures so that their transition frequencies are only affected by the second order Doppler shifts. The final element required for an optical clock is an *optical comb* of the sort described in Section 14.12.1. Optical combs are used to convert optical frequencies at $\sim 10^{15}$ Hz to known multiples of microwave frequencies, for which electronic counting techniques already exist.

The subsequent two sections of the chapter cover in turn the laser cooling of ions, and then shelving. In a final section the example of an optical clock based on a single, near stationary, trapped ion is used to illustrate how the devices and techniques mentioned above are integrated togetther.

A fully quantum mechanical treatment in which the radiation is also quantized yields essentially identical results for the phenomena discussed in this chapter. In the chapter after this one the process of quantizing the electromagnetic field will be presented with applications.

18.2   Transition rates

The Einstein coefficients introduced in Section 14.2 quantify the transition rates for the processes of spontaneous and stimulated emission and of absorption. Here explicit expressions are derived for the Einstein coefficients where the transitions occur between atomic states in the presence of radiation whose spectral distribution is flat across the line width of the transition. The analysis is semiclassical: the atomic states are quantized while the radiation is treated as classical waves.

A simple example is considered in which a single electron undergoes transitions in an atom between a pair of levels with energies W_1 and $W_2(> W_1)$, and with wavefunctions $\Psi_1(\mathbf{r}, t)$ and $\Psi_2(\mathbf{r}, t)$ respectively. All other states of the atom are ignored so that the atom is an idealized *two state atom*. The angular frequency of radiation emitted or absorbed in transitions between these states is

$$\omega_0 = (W_2 - W_1)/\hbar. \tag{18.1}$$

The Schroedinger equation of this atom in isolation is

$$\widehat{H}\Psi(\mathbf{r}, t) = i\hbar\, \partial\Psi(\mathbf{r}, t)/\partial t. \tag{18.2}$$

Here H is the total energy, or Hamiltonian, and $\widehat{H}$ the corresponding operator. Thus the eigenstates of the isolated atom satisfy

$$\Psi_i(\mathbf{r}, t) = \psi_i(\mathbf{r}) \exp\left(-iW_i t/\hbar\right), \tag{18.3}$$

where $i = 1, 2$. The spatial wavefunctions $\psi_i(\mathbf{r})$ are independent of time, and are orthogonal and normalized: that is to say the integrals over all space satisfy the equation

$$\int \psi_i^*(\mathbf{r})\psi_j(\mathbf{r})\, dV = \delta_{ij}, \tag{18.4}$$

where δ_{ij} is the Kronecker delta function defined by eqns. 13.26 and 13.27. Any coherent state of this atom can be described by a linear superposition of the two wavefunctions

$$\Psi(\mathbf{r}, t) = \sum_i c_i\psi_i(\mathbf{r}) \exp\left(-iW_i t/\hbar\right), \tag{18.5}$$

where $\sum_i |c_i|^2 = 1$ so that the overall probability of finding the electron in one or other of the two energy levels is unity. The atom is exposed to an oscillating electromagnetic field whose electric field is

$$\mathbf{E} = \mathbf{E}_0 \cos\left(\mathbf{k} \cdot \mathbf{r} - \omega t\right). \tag{18.6}$$

The principal effect is a contribution to the potential energy of the atom given by

$$\Delta W = e\mathbf{E} \cdot \mathbf{r}, \tag{18.7}$$

where $\mathbf{r}$ is the vector distance of the electron from the nucleus, taken here to be the origin of coordinates. Because the electromagnetic wave and its interaction are being treated classically the operator corresponding to this potential energy is simply a scalar quantity ΔW. Now the wavelength of visible and near visible radiation is much larger than the atomic size, which is of order $0.1\,\mathrm{nm}$, and so the factor $\mathbf{k} \cdot \mathbf{r}$, which is less than or equal to $2\pi r/\lambda$, is very small. Therefore, to an excellent approximation, the electric field can be written as $\mathbf{E}_0 \cos \omega t$, and the interaction energy of eqn. 18.7 can be rewritten as

$$\Delta W = e\mathbf{E}_0 \cdot \mathbf{r} \cos \omega t. \tag{18.8}$$

It will not be possible to use a complex extension of the electric field because the atomic quantum wavefunctions are themselves inherently complex. If the electric field were made complex any equations would mix atomic wavefunctions, whose real and imaginary components have physical meaning, with electric fields whose imaginary components are only there for computational convenience. The real and imaginary parts of any predictions would lack a clear interpretation.

Taking this energy contribution into account, eqn. 18.2 is replaced by

$$[\,\widehat{H} + \Delta W\,]\Psi(\mathbf{r}, t) = i\hbar\, \partial\Psi(\mathbf{r}, t)/\partial t. \tag{18.9}$$

Substituting the general wavefunction from eqn. 18.5 into this equation gives

$$\widehat{H} \sum_i c_i \psi_i \exp\left(-iW_i t/\hbar\right) + \Delta W \sum_i c_i \psi_i \exp\left(-iW_i t/\hbar\right)$$
$$= \sum_i W_i c_i \psi_i \exp\left(-iW_i t/\hbar\right) + i\hbar \sum_i \dot{c}_i \psi_i \exp\left(-iW_i t/\hbar\right), \tag{18.10}$$

where $\dot{c}_i = \partial c_i/\partial t$. Using eqn. 18.2 this reduces to

$$\Delta W \sum_i c_i \psi_i \exp\left(-iW_i t/\hbar\right) = i\hbar \sum_i \dot{c}_i \psi_i \exp\left(-iW_i t/\hbar\right). \tag{18.11}$$

Multiplying this equation by ψ_1^* and integrating the result over all space gives

$$\Delta W_{11}\, c_1 \exp\left(-iW_1 t/\hbar\right) + \Delta W_{12}\, c_2 \exp\left(-iW_2 t/\hbar\right) = i\hbar\, \dot{c}_1 \exp\left(-iW_1 t/\hbar\right), \tag{18.12}$$

where

$$\Delta W_{ij} = \mathbf{E}_0 \cdot \mathbf{D}_{ij} \cos \omega t, \tag{18.13}$$

$$\text{and } \mathbf{D}_{ij} = e \int \psi_i^* \mathbf{r} \psi_j \, dV \tag{18.14}$$

[1] Strictly speaking the electric dipole moment matrix element.

is the electric dipole moment *between* the two atomic states.[1] Then

$$\mathbf{D}_{ji} = \mathbf{D}_{ij}^*. \tag{18.15}$$

Multiplying eqn. 18.12 by $\exp\left(iW_1 t/\hbar\right)$, and using eqn. 18.1 gives

$$\Delta W_{11} c_1 + \exp\left(-i\omega_0 t\right) \Delta W_{12}\, c_2 = i\hbar\dot{c}_1. \tag{18.16}$$

It is straightforward to show that ΔW_{11} and ΔW_{22} are identically zero. The electron density in any atomic state should be the same if the coordinates are simply reversed through the origin: $\mathbf{r} \rightleftharpoons -\mathbf{r}$, which is called the *parity* transformation. Consequently

$$\psi_i^*(-\mathbf{r})\psi_i(-\mathbf{r}) = \psi_i^*(\mathbf{r})\psi_i(\mathbf{r}),$$

and hence

$$\psi_i^*(-\mathbf{r})\,[-\mathbf{r}]\,\psi_i(-\mathbf{r}) = -\psi_i^*(\mathbf{r})\,[\,\mathbf{r}\,]\,\psi_i(\mathbf{r}).$$

This means that the integral of eqn. 18.14 would reverse sign under the parity transformation when $i = j$. Therefore the integral must vanish when $i = j$, and so must ΔW_{11} and ΔW_{22}. Then eqn. 18.16 reduces to

$$\exp\left(-i\omega_0 t\right) \Delta W_{12}\, c_2 = i\hbar \dot{c}_1. \tag{18.17}$$

The remaining term ΔW_{12} is an interaction energy due to the *electric dipole coupling* between the two states of the atom. How large this interaction energy is determines how fast the transitions between the two states take place. An angular frequency Ω_0, called the *Rabi frequency*, is now defined in terms of this interaction energy by putting

$$\hbar\Omega_0 \exp\left(i\nu\right) = \mathbf{E}_0 \cdot \mathbf{D}_{12}, \tag{18.18}$$

where Ω_0 and ν are both real. For convenience the overall complex quantity is given a symbol

$$\mathcal{V} = \Omega_0 \exp\left(i\nu\right) = \mathbf{E}_0 \cdot \mathbf{D}_{12}/\hbar. \tag{18.19}$$

Then eqn. 18.13 can be rewritten as

$$\Delta W_{12} = \hbar\mathcal{V} \cos\left(\omega t\right), \tag{18.20}$$

and finally substituting for ΔW_{12} in eqn. 18.17 gives

$$\mathcal{V} \cos\omega t \exp\left(-i\omega_0 t\right) c_2 = i\dot{c}_1. \tag{18.21}$$

A similar derivation gives a parallel expression

$$\mathcal{V}^* \cos\omega t \exp\left(+i\omega_0 t\right) c_1 = i\dot{c}_2. \tag{18.22}$$

For atoms initially in the lower state $c_1 = 1$ and $c_2 = 0$. Then the transition probability over a time interval t is $|c_2|^2/t$. When the transition rate is low the above pair of simultaneous equations can be solved iteratively. First $c_1(t)$ is set to unity and $c_2(t)$ calculated; then this value of $c_2(t)$ is used in the equations to give an improved value of $c_1(t)$, and so on. Only the first step is followed here, which is adequate for low enough intensity fields such that effects non-linear in the electric field strength can be neglected. This restriction is of course identical to that made when defining the Einstein coefficients in Section 14.2. Replacing c_1 by unity in eqn. 18.22

$$i\dot{c}_2 = \mathcal{V}^* \cos{(\omega t)} \exp{(i\omega_0 t)}$$
$$= (\mathcal{V}^*/2)\{\exp{[i(\omega_0 + \omega)t]} + \exp{[i(\omega_0 - \omega)t]}\}.$$

Integrating this result with respect to time gives

$$c_2 = (-\mathcal{V}^*/2)\{\,[\exp{[i(\omega_0 + \omega)t]} - 1]/(\omega_0 + \omega)$$
$$+ [\exp{[i(\omega_0 - \omega)t]} - 1]/(\omega_0 - \omega)]\,\}. \qquad (18.23)$$

Of the two terms in the above equation, the one divided by the sum of two optical frequencies $(\omega_0 + \omega)$ is very much smaller than the other divided by the difference of two nearby optical frequencies $(\omega_0 - \omega)$. Therefore the former term can be neglected with no great loss in accuracy: this is what is known as the *rotating wave* approximation, and is used systematically throughout the remainder of the chapter. In this approximation the relation given in eqn. 18.13 becomes

$$\Delta W_{12} = \mathbf{E}_0 \cdot \mathbf{D}_{12} \exp{(i\omega t)}/2 = [\hbar\mathcal{V}/2] \exp{(i\omega t)}, \qquad (18.24)$$

where eqn. 18.18 has been used in making the second equality. Now continuing the calculation of c_2,

$$c_2(t) = -i\mathcal{V}^* \exp{[i(\omega_0 - \omega)t/2]} \sin{[(\omega_0 - \omega)t/2]}/(\omega_0 - \omega). \qquad (18.25)$$

Therefore when a group of atoms is irradiated by the laser the fraction of atoms transferred to the upper state in time t is

$$|c_2(t)|^2 = \Omega_0^2 \sin^2{[(\omega - \omega_0)t/2]}/(\omega - \omega_0)^2. \qquad (18.26)$$

In the case that the atomic dipoles have random orientations with respect to the applied electric field the average value of the square of the electric dipole moment along the direction of the electric field is

$$\mu^2 = |D_{12}|^2 \overline{\cos^2{\theta}} = |D_{12}|^2/3, \qquad (18.27)$$

where θ is the angle between the electric field and the dipole directions. For perfect alignment of the atomic dipole and the electric field vector

$$\mu^2 = |D_{12}|^2, \qquad (18.28)$$

Using eqn. 18.18 we get

$$\Omega_0^2 = E_0^2 \, \mu^2/\hbar^2. \qquad (18.29)$$

Then inserting this value for Ω_0 in eqn. 18.26, the average fraction of atoms transferred to the upper state in time t is

$$\overline{|c_2(t)|^2} = [(E_0^2 \, \mu^2/\hbar^2)] \sin^2{[(\omega - \omega_0)t/2]}/(\omega - \omega_0)^2. \qquad (18.30)$$

The spectral energy density of the radiation, $W_{em}(f)$ in $Jm^{-3}Hz^{-1}$, is related to the electric field strength, E_0, in this way:

$$\varepsilon_0 \varepsilon_r E_0^2/2 = \int W_{em}(f)\,df, \qquad (18.31)$$

where $f = \omega/2\pi$. Thus the previous equation can be rewritten

$$\overline{|c_2(t)|^2} = [\,2\,\mu^2/(\varepsilon_0\varepsilon_r\hbar^2)\,]$$
$$\times \int W_{em}(f)\{\sin^2[(\omega - \omega_0)t/2]/(\omega - \omega_0)^2\}df. \quad (18.32)$$

Next we make use of the second basic assumption, namely that the spectral energy density W_{em} is constant across the atomic line width. The integral of interest is thus

$$I = \int \{\sin^2[(\omega - \omega_0)t/2]/(\omega - \omega_0)^2\}\,d\omega/(2\pi). \qquad (18.33)$$

The main contribution to this integral comes from the interval $|\omega - \omega_0|t < 2\pi$ or so. Thus if t is made sufficiently long the limits of the integral can be extended to $\pm\infty$ without affecting the integral's value appreciably. In this case the integral[2] equals $t/4$. Finally the average fraction of atoms transferred in time t is

$$\overline{|c_2(t)|^2} = [\,\mu^2/(2\varepsilon_0\varepsilon_r\hbar^2)\,]\,W_{em}(f)t. \qquad (18.34)$$

Consequently the Einstein coefficient B_{12} given by eqn. 14.1 reduces to

$$B_{12} = \overline{|c_2(t)|^2}/[\,W_{em}(f)t\,]$$
$$= \mu^2/(2\varepsilon_0\varepsilon_r\hbar^2). \qquad (18.35)$$

Now the spectral energy density W_{em} used in the definition of B_{12} is the sum over all orientations of the polarization. Hence μ^2 takes the value $D_{12}^2/3$. Thus

$$B_{12} = D_{12}^2/(6\varepsilon_0\varepsilon_r\hbar^2). \qquad (18.36)$$

The spontaneous decay rate given by the Einstein coefficient A_{21} can be deduced despite the absence of spontaneous decays from the semiclassical analysis: using eqn. 14.10 gives

$$A_{21} = 2\omega^3 D_{12}^2 n/(3\varepsilon_0 hc^3), \qquad (18.37)$$

where n is the refractive index. Thus if the wavefunctions of the initial and final states are known then D_{12} and μ can be calculated. It is now straightforward to make a number of predictions about processes in which weak broad sources interact with two-state atoms. First the electric dipole selection rules already outlined in Section 13.7 are investigated. Then follows an exercise to calculate the rate for the transition 2p→1s in hydrogen. Lastly the susceptibility of a gas is derived.

---

[2]Integral 3.821/9 in the 5th edition of *Table of Integrals, Series and Products* by I. S. Gradshteyn and I. M. Ryzhik, edited by A. Jeffrey, and published by Academic Press, New York (1994). $\int_{-\infty}^{\infty}\{\sin^2(ax)/x^2\}\,dx = a\pi$, $a > 0$.

18.2.1   Selection rules

It can be seen from eqn. 18.36 that the size of the Einstein coefficients is determined by the overlap integral defining D_{12}. This overlap integral given in eqn. 18.14 is responsible for the electric dipole selection rules outlined in Chapter 13.

We consider the simplest case only, that of a hydrogen atom and use the expressions for the wavefunctions given in eqn. 13.48. Taking the electric field direction to define the polar z-axis, and with the electron having polar coordinates (r, θ, ϕ) with respect to the nucleus, the *angular* part of the integral in eqn 18.14 is

$$D_{\mathrm{ang}} = \int Y^*_{\ell',m'}(\theta, \phi) \, \cos\theta \, Y_{\ell,m}(\theta, \phi) \, \sin\theta \, \mathrm{d}\theta \, \mathrm{d}\phi, \qquad (18.38)$$

where (ℓ,m) and (ℓ',m') are the angular momentum quantum numbers of the two states. All such integrals vanish identically unless both $\Delta\ell = \ell' - \ell = \pm 1$ and $\Delta m = m' - m = 0, \pm 1$. The reader may like to check a simple case.

What might be called a geometric interpretation of these selection rules can be obtained in the following way. First note that the vector $\mathbf{r}$ has Cartesian components $x = r\sin\theta\cos\phi$, $y = r\sin\theta\sin\phi$ and $z = r\cos\theta$. Thus $x+iy$, $x-iy$ and z have the same angular dependence as the spherical spherical harmonic functions for unit angular momentum which appear in table 13.2: $Y_{1,+1}$, $Y_{1,-1}$ and $Y_{1,0}$ respectively. Therefore we can legitimately treat $\mathbf{r}$ as a spherical harmonic with unit angular momentum in these angular integrations. The quantity being integrated in eqn. 18.38 is thus the overlap of a spherical harmonic of orbital angular momentum ℓ' with the product of two spherical harmonics: one with angular momentum ℓ and and the other with unit angular momentum. Now we have seen that the angular momenta of atoms are quantized and that they add vectorially. Putting these facts together we can infer that the overlap integral simply vanishes whenever the vector addition is impossible. In this way a selection rule $\Delta\ell = \pm 1$ emerges. The selection rules for the magnetic quantum number arise in a similar way.

Other weaker transitions are still possible in which the change in orbital angular momentum is different from that occurring in an electric dipole transition. These transitions are possible thanks to the slight departure of the actual electric field from the constant value assumed when writing eqn. 18.8. In that case only the first term in the expansion

$$\exp{(i\mathbf{k}\cdot\mathbf{r})} = 1 + i\mathbf{k}\cdot\mathbf{r} - (\mathbf{k}\cdot\mathbf{r})^2/2 + \ldots$$

was considered. The contribution of the $(n+1)$th term is smaller by a factor $(\mathbf{k}\cdot\mathbf{r})^n/n!$ in amplitude, and by a factor $(\mathbf{k}\cdot\mathbf{r})^{2n}/(n!)^2$ in intensity compared to the electric dipole. These are known as *forbidden transitions*.

Example 18.1

Here we calculate the rate for the transition of an electron in a hydrogen atom from the $n = 2$ to the $n = 1$ state. As a result of the spin-orbit interaction the electron occupies eigenstates of the total angular momentum and its quantized component. The states with the electron excited to the $n = 2$ states are, in spectroscopic notation, $2P_{3/2}$, $2P_{1/2}$, and $2S_{1/2}$. There are electric dipole transitions $2P_{3/2} \rightarrow 1S_{1/2}$ and $2P_{1/2} \rightarrow 1S_{1/2}$, but the decay of the $2S_{1/2}$ state is not an allowed dipole transition because the orbital angular momentum of the electron does not change. The calculation of the decay rate is simplest for the $P_{3/2}$ state with a quantized angular momentum component $3/2$. This has a stretched configuration with the orbital angular momentum and spin aligned, in the sense that their quantized components are both at their maximum. Its wavefunction is therefore a simple product of orbital and spin terms: $\psi(n = 2; l = 1; m_\ell = +1) \chi(m_s = +1/2)$. The wavefunction of the final state, $1S_{1/2}$, is also simple: $\psi(n = 1; l = 0; m_\ell = 0) \chi(m_s = +1/2)$. Spatial wavefunctions of the hydrogen atom are given in tables 13.1 and 13.2. Those required for the calculation are

$$\psi_p = \psi(n = 2; l = 1; m_\ell = +1)$$
$$= r \exp(-r/2a_0) \sin\theta \exp(i\phi)/(8\sqrt{\pi}a_0^{5/2}), \tag{18.39}$$
$$\psi_s = \psi(n = 1; l = 0; m_\ell = 0)$$
$$= \exp(-r/a_0)/(\sqrt{\pi}a_0^{3/2}). \tag{18.40}$$

The Einstein coefficient for the transition between these states is given by eqns. 18.14 and 18.37 using the spatial wavefunctions just presented

$$A = [2\omega^3/(3\varepsilon_0 hc^3)] \left| \int \psi_p^* e\mathbf{r}\psi_s \, dV \right|^2$$
$$= [2\omega^3/(3\varepsilon_0 hc^3)] e^2 (|X|^2 + |Y|^2 + |Z|^2), \tag{18.41}$$

where $dV = r^2 \sin\theta \, d\theta \, d\phi \, dr$ and

$$X = \int \psi_p^* x\psi_s \, dV, \quad Y = \int \psi_p^* y\psi_s \, dV, \quad Z = \int \psi_p^* z\psi_s \, dV.$$

Evaluating the first of these integrals

$$X = [1/(8\pi a_0^4)] \int r^2 \exp(-3r/2a_0) \sin^2\theta \cos\phi \exp(-i\phi) \, dV$$
$$= [1/(8\pi a_0^4)] \left[\int_0^\infty \exp(-3r/2a_0) r^4 dr \right] \left[\int_0^\pi (\cos^2\theta - 1) d(\cos\theta) \right]$$
$$\left[\int_0^{2\pi} \{[1 + \exp(-2i\phi)]/2\} \, d\phi \right]$$
$$= [1/(8\pi a_0^4)][(2a_0/3)^5(24)][4/3][\pi] = 4a_0(2/3)^5. \tag{18.42}$$

The Y and Z integrals have the same radial and polar angle components as the X integral so that

$$Y = ... \int_0^{2\pi} \{[1 - \exp(-2i\phi)]/2i\} \, d\phi = X/i, \qquad (18.43)$$

while

$$Z = ... \int_0^{2\pi} \exp(-i\phi) \, d\phi = 0. \qquad (18.44)$$

Collecting terms in eqn. 18.41 gives

$$A = [2\omega^3/(3\varepsilon_0 hc^3)] \, 32(2/3)^{10} \, e^2 a_0^2 = 6.26 \, 10^8 \, \text{s}^{-1}. \qquad (18.45)$$

The decay rates of the six 2p-states of a hydrogen atom, that is the four $2P_{3/2}$ and the two $2P_{1/2}$ states, all have this same value.

18.2.2   Electric susceptibility

A classical analysis of the response of a gaseous dielectric to a time varying electric field was given in Section 11.5.1. The electrons were regarded as oscillating at some (undetermined) natural frequencies under the restoring force provided by the Coulomb attraction of the nucleus. It was made clear in Chapters 12 and 13 that those resonance frequencies are in fact the transition frequencies between atomic states, while the classical damping reflected the decay rate of the atomic state. In this section the semiclassical model of the atom–radiation interaction is used to obtain an expression for the susceptibility of a dilute gas.

As before a two level atom is considered. Using the rotating wave approximation eqn. 18.22 becomes

$$(\mathcal{V}^*/2) \exp(-i\Delta\omega t) c_1 = i\dot{c}_2, \qquad (18.46)$$

where we have put

$$\Delta\omega = \omega - \omega_0. \qquad (18.47)$$

The decay of the upper level by spontaneous emission has to be put in by hand

$$(\mathcal{V}^*/2) \exp(-i\Delta\omega t) c_1 - i\gamma c_2/2 = i\dot{c}_2, \qquad (18.48)$$

where $\gamma = 1/A_{21}$. Without the applied field this collapses to $\dot{c}_2 = -\gamma c_2/2$, with the intended decay rate γ. Again assuming c_1 is close to unity, which is appropriate with low intensity radiation, the above equation becomes

$$\dot{c}_2 + \gamma c_2/2 + (i\mathcal{V}^*/2) \exp(-i\Delta\omega t) = 0. \qquad (18.49)$$

Substituting the long term solution at the driving frequency,

$$c_2 = c_0 \exp(-i\Delta\omega t),$$

into this equation gives

$$c_0 = (\mathcal{V}^*/2)/(\Delta\omega + i\gamma/2). \tag{18.50}$$

Thus

$$\begin{aligned} c_2 &= (\mathcal{V}^*/2)\exp\left(-i\Delta\omega t\right)/(\Delta\omega + i\gamma/2) \\ &= (\mathbf{E}_0 \cdot \mathbf{D}_{12}^*/2\hbar)\exp\left(-i\Delta\omega t\right)/(\Delta\omega + i\gamma/2), \tag{18.51} \end{aligned}$$

where eqn. 18.19 was used in making the second equality. The wavefunction in eqn. 18.5 becomes

$$\begin{aligned} \Psi(\mathbf{r}, t) &= \psi_1(\mathbf{r})\exp\left(-iW_1 t/\hbar\right) + c_2\psi_2(\mathbf{r})\exp\left(-iW_2 t/\hbar\right) \\ &= \{\,\psi_1(\mathbf{r}) + c_2\psi_2(\mathbf{r})\exp\left(-i\omega_0 t\right)\,\}\exp\left(-iW_1 t/\hbar\right). \tag{18.52} \end{aligned}$$

Hence the electric dipole moment of the atom is

$$\begin{aligned} \mathbf{p}(t) &= -e\int \Psi^* \mathbf{r}\Psi\,\mathrm{d}V \\ &= -\mathbf{D}_{12}^* \exp\left(i\omega_0 t\right)c_2^*(t) - \mathbf{D}_{12}\exp\left(-i\omega_0 t\right)c_2(t). \end{aligned}$$

Replacing c_2 with the help of eqn. 18.51 gives

$$\begin{aligned} \mathbf{p}(t) = &-\mathbf{D}_{12}(\mathbf{E}_0 \cdot \mathbf{D}_{12}^*/2\hbar) \\ &\times [\exp\left(+i\omega t\right)/(\Delta\omega - i\gamma/2) + \exp\left(-i\omega t\right)/(\Delta\omega + i\gamma/2)], \end{aligned}$$

so the component along the direction of the electric field is

$$\begin{aligned} p(t) = &-[|(\mathbf{E}_0 \cdot \mathbf{D}_{12})|^2/(2\hbar E_0)] \\ &\times [\exp\left(+i\omega t\right)/(\Delta\omega - i\gamma/2) + \exp\left(-i\omega t\right)/(\Delta\omega + i\gamma/2)]. \end{aligned}$$

Averaging over the angular distribution of the dipoles gives

$$\begin{aligned} \overline{p(t)} = &\,[E_0\mu^2/(2\hbar)]\,[\exp\left(+i\omega t\right)/(\omega_0 - \omega + i\gamma/2) \\ &+ \exp\left(-i\omega t\right)/(\omega_0 - \omega - i\gamma/2)]. \tag{18.53} \end{aligned}$$

This result can be compared with the polarization expressed in terms of the susceptibility $\chi(\omega)$ for the electric field

$$E_0\cos(\omega t) = (E_0/2)\,[\exp\left(i\omega t\right) + \exp\left(-i\omega t\right)].$$

Rewriting eqn. 9.3 for this purpose

$$N\overline{p(t)} = (\varepsilon_0 E_0/2)\,[\chi(+\omega)\exp\left(-i\omega t\right) + \chi(-\omega)\exp\left(i\omega t\right)], \tag{18.54}$$

where N is the number of atoms per unit volume. The comparison with eqn. 18.53 yields

$$\chi(\omega) = (N\mu^2/\varepsilon_0\hbar)/(\omega_0 - \omega - i\gamma/2). \tag{18.55}$$

A corresponding result of the classical analysis is given in eqn. 11.20. The superiority of the analysis presented here is clear: $\chi(\omega)$ is directly calculable for a given transition from a knowledge of the state wavefunctions, and such predictions are then found to agree with experimental measurements.

18.3   Rabi oscillations

These are the high frequency oscillations of the population of atomic states when an intense laser beam illuminating atoms has a frequency near to the frequency for a transition between the lower energy state and a higher energy state. Initially the radiation is absorbed so that the atoms move into the higher energy state. At this point the laser beam stimulates the atoms to emit and return to the ground state. This sequence repeats indefinitely during the time the laser stays on. It is the dominant effect provided that the laser beam is sufficiently intense so that few spontaneous emissions occur during each cycle; that is provided $\Omega_0 \gg \gamma$, the decay width. In addition to the higher beam intensity, another difference from the situation analysed in the preceding sections is that the laser line width is much narrower than that of the atomic transition and is narrow enough that the radiation incident can often be considered as monochromatic.

The analysis of these Rabi oscillations will be made for a two state atom, so we can re-use the expressions 18.21 and 18.22 derived above for the relations between the amplitudes of the two states, c_1 and c_2. These can be rewritten

$$(\mathcal{V}/2)\left[\exp\left(i\omega t\right) + \exp\left(-i\omega t\right)\right]\exp\left(-i\omega_0 t\right)c_2 = i\dot{c}_1, \qquad (18.56)$$

$$(\mathcal{V}^*/2)\left[\exp\left(i\omega t\right) + \exp\left(-i\omega t\right)\right]\exp\left(i\omega_0 t\right)c_1 = i\dot{c}_2. \qquad (18.57)$$

In the rotating wave approximation only the exponents in $\Delta\omega = \omega - \omega_0$ survive, so we have

$$(\mathcal{V}/2)\exp\left(i\Delta\omega t\right)c_2 = i\dot{c}_1, \qquad (18.58)$$

$$(\mathcal{V}^*/2)\exp\left(-i\Delta\omega t\right)c_1 = i\dot{c}_2. \qquad (18.59)$$

The analysis will be very different from that followed in the case of low intensity radiation because now $|c_2|$ ranges from zero to unity. It is therefore necessary to extract a general solution of these simultaneous equations. Eliminating c_1 gives

$$\ddot{c}_2 + i\Delta\omega\dot{c}_2 + (\Omega_0^2/4)c_2 = 0. \qquad (18.60)$$

A solution of the form $c_2 = \exp\left(i\lambda t\right)$ can be attempted, which when substituted into this equation gives

$$-\lambda^2 - \Delta\omega\lambda + \Omega_0^2/4 = 0. \qquad (18.61)$$

This equation has two roots

$$\lambda_\pm = (1/2)\{-\Delta\omega \pm \Omega\}, \qquad (18.62)$$

where

$$\Omega = \sqrt{(\Delta\omega^2 + \Omega_0^2)} \qquad (18.63)$$

is called the *generalized Rabi frequency*. Thus the complete solution has the form

$$c_2(t) = c_+ \exp(i\lambda_+ t) + c_- \exp(i\lambda_- t), \qquad (18.64)$$

where the coefficients, c_+ and c_-, are determined from the initial conditions. The first condition is that at $t = 0$, $c_2 = 0$. Hence $c_- = -c_+$. Secondly, after a very short time Δt

$$c_2(\Delta t) = c_+ \left[i\lambda_+ \Delta t - i\lambda_- \Delta t \right] = ic_+ \Omega \Delta t. \qquad (18.65)$$

In the same approximation eqn. 18.59 becomes

$$\dot{c}_2(\Delta t) = -i(\mathcal{V}^*/2)c_1(\Delta t) = -i(\mathcal{V}^*/2), \qquad (18.66)$$

whence

$$c_2(\Delta t) = -i(\mathcal{V}^*/2)\Delta t. \qquad (18.67)$$

Comparing eqns. 18.65 and 18.67 gives

$$c_+ = -(\mathcal{V}^*/2\Omega). \qquad (18.68)$$

Thus the full solution for c_2 is

$$\begin{aligned}
c_2(t) &= -(\mathcal{V}^*/2\Omega) \exp(-i\Delta\omega t/2) \left[\exp(i\Omega t/2) - \exp(-i\Omega t/2) \right] \\
&= -i(\mathcal{V}^*/\Omega) \exp(-i\Delta\omega t/2) \sin(\Omega t/2). \qquad (18.69)
\end{aligned}$$

In turn

$$c_1(t) = \exp(i\Delta\omega t/2) \left[\cos(\Omega t/2) - i(\Delta\omega/\Omega) \sin(\Omega t/2) \right]. \qquad (18.70)$$

Then the fraction of atoms in the excited state is

$$|c_2(t)|^2 = (\Omega_0/\Omega)^2 \sin^2(\Omega t/2). \qquad (18.71)$$

This fraction oscillates at the generalized Rabi frequency Ω. The greater the offset of the laser frequency from the transition frequency becomes the smaller will be the ratio (Ω_0/Ω) in eqn. 18.71 and hence the weaker the oscillations in c_2. Figure 18.1 displays how the population of the upper state develops with time for three choices of the laser detuning from the transition frequency: $\Delta\omega = 0$, $\Omega_0/2$ and Ω_0. In these respective cases $|c_2(t)|^2$ is $\sin^2(\Omega_0 t/2)$, $0.8 \sin^2(\sqrt{5/16}\Omega_0 t)$ and $0.5 \sin^2(\Omega_0 t/\sqrt{2})$. Reference to eqn. 18.18 shows that increasing the laser power increases Ω_0 and hence Ω also. This has two effects: first the oscillations speed up; secondly if the laser frequency is off resonance, so that $\Omega \neq \Omega_0$, the amplitude of the Rabi oscillations increases.

The Rabi oscillations can be readily observed with a detector placed out of the laser beam. This detector will capture the fluorescence, that is the spontaneous emission from the upper state, whose intensity will be proportional to the population of the upper state. A filter transmitting a narrow band of wavelengths placed in front of the detector would be adequate to exclude extraneous light. Interesting effects are produced

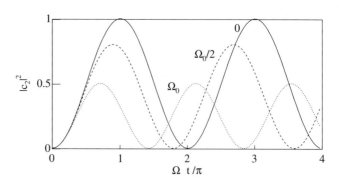

Fig. 18.1 Time dependence of the upper state population for different offsets of the pump angular frequency from the transition angular frequency. The offsets are $\Delta\omega = 0$, $\Omega_0/2$ and Ω_0.

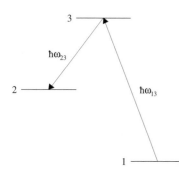

Fig. 18.2 Excitation of level 2 by π pulses in sequence for the allowed transitions $1 \rightarrow 3$ and $3 \rightarrow 2$, where the transition $1 \rightarrow 2$ is forbidden.

when the laser beam is pulsed using an optical modulator like those described in Section 10.7.1. If the pulse duration is precisely π/Ω, eqn. 18.71 shows that if all the atoms are initially in the lower state the pulse transfers the maximum number to the upper state; and if the laser is also on resonance all the atoms are transferred. This type of pulse is called a π-*pulse*. Such pulses can be used to transfer atoms into states not linked to the ground state by any fast electric dipole transition. Referring to figure 18.2 a first π-pulse from a laser emitting at angular frequency ω_{13} transfers the atoms from the state labelled 1 to the state labelled 3. Then a second π-pulse from another laser at angular frequency ω_{23} transfers these atoms to the final state labelled 2. These pulses do not necessarily need to be square pulses of duration π/Ω, all that is required is that the integral over the pulse duration $\int \Omega \, dt = \pi$. This way of exciting states that are not directly accessible is more efficient than the Raman excitation described in Chapter 14.

A pulse with double the duration of a π-pulse is called a 2π-pulse and should simply return the atoms to the initial state. In principle all of the radiation absorbed is therefore re-emitted as stimulated emission, and so there is no net absorption. This process is therefore known as *self-induced transparency*. In order to approach this ideal the pulse now does have to be a flat-topped pulse – which is hard to achieve. There is also inevitably some reduction in the intensity of the emerging pulse due to spontaneous emission.

Laser pumping of atomic transitions has an additional effect: it produces *power broadening* of the spectral line width from γ to $\sqrt{(\gamma^2 + 2\Omega_0^2)}$. A *saturation* intensity is defined as that intensity at which the power broadening increases the line width by a factor $\sqrt{2}$: $\Omega_{0s} = \gamma/\sqrt{2}$. The saturating electric field amplitude can be obtained by substituting for Ω_0 in eqn. 18.18

$$E_{0s} = \hbar\gamma/(\sqrt{2}\,|D_{12}|), \tag{18.72}$$

which is also the electric field of the incident beam in air. The saturation energy flux is

$$I_s = \varepsilon_0 c E_{0s}^2/2 = \varepsilon_0 c \hbar^2 \gamma^2/(4\,D_{12}^2). \tag{18.73}$$

We can replace one γ with A_{21} in air, and then use eqn. 18.37 to replace A_{21}. This gives

$$I_s = 2\pi^2 c \hbar \gamma / 3\lambda^3 \tag{18.74}$$

for the saturation intensity.

18.4 Dressed states

The preceding analysis shows that an atom is strongly influenced by an intense laser beam tuned to a frequency close to one of its transition frequencies. Any mode of the electromagnetic field of a laser is populated by large numbers of photons, whereas the same mode would, by contrast, generally be empty in thermal radiation. It will emerge that the energies of the atomic states involved in the transition are then altered by the presence of the laser beam to an extent that depends on the electric field strength. In these conditions a more appropriate description of the system as a whole is to use *dressed* states, rather than *bare* non-interacting states. These dressed states incorporate the energy shifts produced by a laser mode coupled to the atomic transition. This fruitful approach will be presented here in a simplified manner only.

We start from the expressions given above in eqns. 18.69 and 18.70 for the coefficients of the two *bare* atomic states when the laser is on resonance:

$$c_1(t) = (1/2)\left[\exp\left(-i\Omega_0 t/2\right) + \exp\left(i\Omega_0 t/2\right)\right], \tag{18.75}$$
$$c_2(t) = (1/2)\left[\exp\left(-i\Omega_0 t/2\right) - \exp\left(i\Omega_0 t/2\right)\right]\exp\left(-i\nu\right). \tag{18.76}$$

Then the wavefunction of the illuminated atom is

$$\begin{aligned}
\Psi = (1/2)\{\ &\psi_1 \exp\left(-i\omega_1 t + i\Omega_0 t/2\right) \\
&+ \psi_1 \exp\left(-i\omega_1 t - i\Omega_0 t/2\right) \\
&- \psi_2 \exp\left(-i\omega_2 t + i\Omega_0 t/2\right) \\
&+ \psi_2 \exp\left(-i\omega_2 t - i\Omega_0 t/2\right)\},
\end{aligned} \tag{18.77}$$

where as before $\Psi_j(\mathbf{r}, t) = \psi_j(\mathbf{r}) \exp\left(-i\omega_j t\right)$, and the factor $\exp\left(-i\nu\right)$ has been absorbed in ψ_2. This can be re-expressed as

$$\Psi = [\Psi_+ + \Psi_-]/\sqrt{2}, \tag{18.78}$$

where the new wavefunctions are

$$\begin{aligned}
\Psi_+ = \frac{1}{\sqrt{2}}&\left[\psi_1 \exp\left(-i\omega_1 t\right) + \psi_2 \exp\left(-i\omega_2 t\right)\right] \\
&\times \exp\left(-i\Omega_0 t/2\right),
\end{aligned} \tag{18.79}$$

$$\begin{aligned}
\Psi_- = \frac{1}{\sqrt{2}}&\left[\psi_1 \exp\left(-i\omega_1 t\right) - \psi_2 \exp\left(-i\omega_2 t\right)\right] \\
&\times \exp\left(+i\Omega_0 t/2\right).
\end{aligned} \tag{18.80}$$

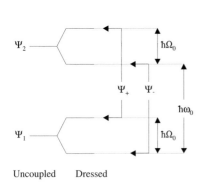

Uncoupled   Dressed

Fig. 18.3 Uncoupled (Ψ_1, Ψ_2) and dressed atomic energy levels (Ψ_+, Ψ_-), for the case that the pump angular frequency equals the transition angular frequency ω_0.

These new states form an orthonormal pair:

$$\int \Psi_\pm^* \Psi_\pm \, dV = 1; \quad \text{and} \quad \int \Psi_\pm^* \Psi_\mp \, dV = 0.$$

The energies of the two components in each of these wavefunctions can be extracted from their time variation. The wavefunction Ψ_+ has components of energy $\hbar(\omega_1 + \Omega_0/2)$ and $\hbar(\omega_2 + \Omega_0/2)$, while Ψ_- has components of energy $\hbar(\omega_1 - \Omega_0/2)$ and $\hbar(\omega_2 - \Omega_0/2)$. Figure 18.3 shows the resulting energy levels corresponding to the components of Ψ_+ and of Ψ_-.

The two components of Ψ_+ differ in energy by exactly the photon energy, $\hbar\omega_0$. An inference can be made that these are components of a dressed atom: the atom and the mode of the electromagnetic field enveloping it together constitute the dressed atom. The higher energy atomic level is accompanied by one photon less in the mode than the lower energy atomic level, giving the two components of the dressed atom the same total energy. A similar inference can be made for the components of Ψ_-.

In the intense laser beams used the number of photons in the mode will be very large. Then a new interpretation emerges: Ψ_+ and Ψ_- are adjacent levels in a *tower* made up of pairs of dressed states, in which the number of photons increases by one from one pair to the next pair. An equally important feature is that the members of each pair are separated in energy by $\hbar\Omega_0$, where Ω_0 is the Rabi frequency. These inferences are found to be justified when the full quantum mechanical analysis is carried through. The splitting of levels seen in figure 18.3 is known as the *AC Stark effect* or *dynamic Stark splitting*.

The pairs of dressed states can be expressed in terms of the bare states in this symbolic manner:[3]

$$|\Psi_+; n\rangle = \left[\,|\Psi_1\rangle|(n+1)\rangle + |\Psi_2\rangle|n\rangle\,\right]/\sqrt{2}, \qquad (18.81)$$

$$|\Psi_-; n\rangle = \left[\,|\Psi_1\rangle|(n+1)\rangle - |\Psi_2\rangle|n\rangle\,\right]/\sqrt{2}, \qquad (18.82)$$

where the symbol $|\Psi_i\rangle$ represents a bare state with wavefunction Ψ_i; the symbol $|n\rangle$ represents a state of the laser mode containing n photons. The dressed state described by the symbol $|\Psi_+; n\rangle$ has an energy $\hbar\Omega_0$ greater than that of the dressed state described by the symbol $|\Psi_-; n\rangle$.

These levels are all available but it is the number of photons in the mode that determines which levels are populated.[4]

[3]This notation will be extensively used in the final chapters. State vectors are described in detail for use in conjunction with that chapter in Appendix E.

[4]The case of zero photons appears problematic because this level would not be split. However if there are only a few photons in the mode the laser intensity must be low and the splitting correspondingly small: in the limit that there are no photons in the mode, the laser is off and there is no question of splitting.

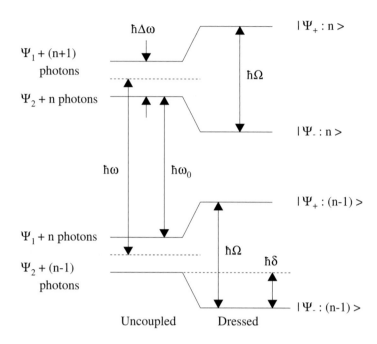

Fig. 18.4 Uncoupled and dressed states in the case that the pump laser angular frequency ω exceeds the transition angular frequency ω_0 by $\Delta\omega$. The dressed states form a tower of manifolds.

When the laser angular frequency differs from the transition frequency the expressions for the displacements of the energy levels become more complicated. Suppose now that the angular frequency of the laser, ω is no longer equal to ω_0, but is larger. Then an analysis parallel to that just carried through produces a level structure like that pictured in figure 18.4. The dressed states shown are related symbolically to the bare states in this way

$$|\Psi_+; n\rangle = \cos\theta|\Psi_1\rangle|(n+1)\rangle + \sin\theta|\Psi_2\rangle|n\rangle, \qquad (18.83)$$
$$|\Psi_-; n\rangle = \sin\theta|\Psi_1\rangle|(n+1)\rangle - \cos\theta|\Psi_2\rangle|n\rangle, \qquad (18.84)$$

$$\cos\theta = \sqrt{(\Omega + \Delta\omega)/2\Omega}, \qquad (18.85)$$
$$\sin\theta = \sqrt{(\Omega - \Delta\omega)/2\Omega}. \qquad (18.86)$$

The change in the separation in energy between the states Ψ_1 and Ψ_2 due to the pumping is $2\hbar\delta = \hbar(\Omega - \Delta\omega)$. As the laser detuning is increased this displacement falls and for very large detuning such that $\Delta\omega \gg \Omega_0$, $\hbar\delta \approx \hbar\Omega_0^2/4\Delta\omega$. This displacement therefore contributes a small *light shift* to levels that are not being pumped in an atom.

18.4.1   Mollow fluorescence

A portion of a tower of dressed states is shown on the left hand side of figure 18.5. The number of atoms in each manifold depends on the

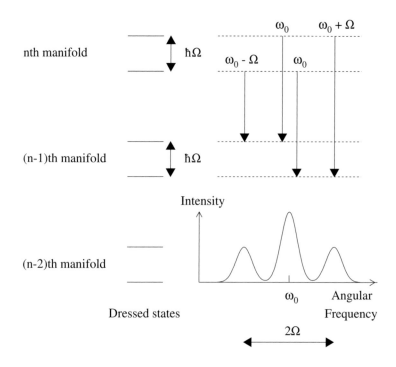

Fig. 18.5 Mollow fluorescence: on the left is a tower of dressed states; at the top right the typical transitions are shown; below these the resulting spectrum is shown.

laser intensity, and of course the laser intensity also determines the separation $\hbar\Omega$ between pairs of states. When an atomic gas is illuminated by a laser on resonance with some transition from the ground state the spectrum of its fluorescence shows a three-fold peak, an effect which is known as *Mollow fluorescence*. This spectrum is illustrated on the lower right of figure 18.5 while above it are shown examples of the transitions that contribute to it. One peak is centred at the angular frequency of the resonance, the other peaks are equally displaced up and down in frequency from this by Ω. Each populated level that emits fluorescence contributes one transition to the central peak and one transition to one of the outer peaks: therefore the central peak has an area equal to the sum of the areas of the outer peaks. The lines are also power broadened to widths $\sqrt{(\gamma^2 + 2\Omega_0^2)}$. When the laser intensity is gradually reduced the peaks merge into the central peak, while at the same time the peaks shrink in width. Finally when the laser intensity is very low, only the resonant fluorescence peak remains with natural width γ.

18.4.2 The Autler–Townes effect

Observation of the related Autler–Townes effect requires the use of two lasers. One intense laser beam pumps transitions between a pair of atomic states of a gas, so that these can be considered to be dressed states. Transitions from these dressed states to a third state are produced by weaker radiation from a second laser, a *probe* laser whose frequency is tunable. Transitions to the third state are at a frequency very different from the pump frequency so that this state is unaffected by the pumping, apart from a tiny light shift.

The probe laser's frequency is scanned in small steps from well below to well above the transition frequency and the absorption of its beam measured at each step. In the upper panel of figure 18.6 the dressed levels and the isolated third state are shown, together with transitions in which photons are absorbed from the probe beam. The lower panel shows the absorption spectrum found during the scan of the probe laser. There are two peaks separated by the Rabi frequency of the pump; this splitting is known as the *Autler–Townes effect*. When the intensity of the pump laser is reduced the splitting is reduced and the two peaks merge together.

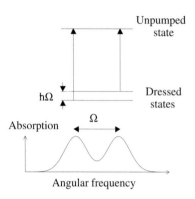

Fig. 18.6 Autler–Townes splitting of an absorption line: the transitions are shown above and the absorption spectrum below.

The response of the pumped atoms to the probe can be estimated by calculating the susceptibility due to the transitions between the pair of dressed states and the isolated state shown in figure 18.6. Let the probe laser angular frequency be ω, and the angular frequencies of the two transitions from the dressed states be $\omega_0 \pm \Omega/2$. Then using eqn. 18.55 we obtain for the susceptibility

$$\chi = \frac{g}{\Omega/2 - \Delta\omega - i\gamma/2} + \frac{g}{-\Omega/2 - \Delta\omega - i\gamma/2}, \tag{18.87}$$

where

$$g = N\mu^2/(\varepsilon_0\hbar), \tag{18.88}$$

assuming equal dipole moments and spontaneous decay rates for the two transitions. As usual $\Delta\omega = \omega - \omega_0$. Then

$$\chi = \frac{g(2\Delta\omega + i\gamma)}{(\Omega^2/4 - \Delta\omega^2 - i\gamma\Delta\omega + \gamma^2/4)}. \tag{18.89}$$

This result does indeed show two separated absorption peaks. Their angular frequencies are given by

$$\omega = \omega_0 \pm [\Omega/2 - \gamma^4/(16\Omega^3) + ...],$$

when γ is small compared to Ω. For future reference note that here the absorptive part of the susceptibility at resonance (where $\Delta\omega = 0$) is $4g\gamma/(\gamma^2 + \Omega_0^2)$, and that this remains sizeable.

18.5 Electromagnetically induced transparenc

In *electromagnetically induced transparency* (EIT) laser pumping at one wavelength renders a material transparent at a second wavelength to which it is usually opaque.[5] A significant proportion of the studies of EIT have been made with atomic gases of the alkali metals sodium and rubidium, which have a particularly suitable level structure. The effect has been observed with a few materials with special crystal structures but with no other condensed matter materials. Patterns of atomic levels which have been used to demonstrate EIT are illustrated in figure 18.7: Λ, V and cascade. For simplicity the following discussion will be restricted to the case of the Λ configuration which is depicted below the other two. The coupling transition, $2 \rightarrow 1$, and probe transition, $3 \rightarrow 1$,

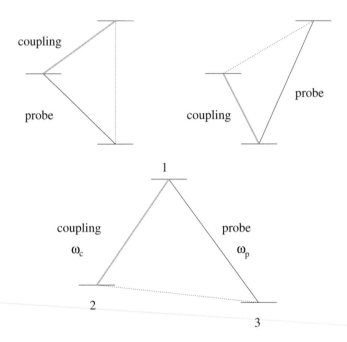

Fig. 18.7 Atomic level configurations for which electromagnetically induced transparency can be produced. The transition indicated with a dotted line is forbidden.

are both allowed electric dipole transitions, while the transition $2 \rightarrow 3$ is forbidden. Normally the beam from a *probe* laser at angular frequency

---

[5]See, for example, K.-J. Boller, A. Imamoglu and S. E. Harris, 'Observations of electromagnetically induced transparency'; *Physical Review Letters* 66, 2593 (1991); L. V. Hau, S. E. Harris, Z. Dutton and C. H. Behroozi,'Light speed reduction to $17\,\mathrm{m\,s^{-1}}$ in an ultracold atomic gas', *Nature* 397, 594 (1999); L. V. Hau 'Taming light with cold atoms', *Physics World* 14(9), 35 (2001); S. E. Harris, 'Electromagnetically induced transparency', *Physics Today* 36 (July 1997).

ω_{p} would be totally absorbed within centimetres along a container filled with the gas. However if a second *coupling* laser simultaneously pumps the coupling transition at angular frequency ω_{c} the gas mysteriously becomes transparent at the probe frequency. It was seen in the preceding section that a reduction of opacity at the resonance frequency was produced by the Autler–Townes effect as a result of the splitting of the absorption peak. However in electromagnetically induced transparency, the transparency is complete and is achieved in a *qualitatively* different manner involving quantum interference. This will be explained in the following paragraphs.

Not only is the gas completely transparent to the probe beam but the probe pulse undergoes a dramatic reduction in group velocity, slowing down to a walking pace in the gas. A probe pulse of length of order a kilometre in air can be compressed to a length of micrometres in the gas. It is even possible to, in a sense, halt the pulse and release it after a delay of order a few milliseconds.

In sodium or rubidium the two lower levels, 2 and 3, are usually chosen to be either the hyperfine or the Zeeman split levels of the ground state $(nS_{1/2})$; these are only separated by of order $[1\,\mathrm{GHz}]h$ in energy. The transitions to the upper state are the well known D-lines at 589 nm and 795 nm respectively for sodium and rubidium. Figure 18.8 shows a sketch of the experiment performed by Hau and colleagues.[6] A cloud of $\sim 10^6$ sodium atoms is trapped and cooled to $0.5\,10^{-6}\,\mathrm{K}$, using lasers and magnetic fields. The probe and coupling beams are polarized, in the senses indicated by arrows in the diagram, in order to excite transitions from two hyperfine split levels of the ground state into a common upper magnetic substate of the $3\mathrm{P}_{3/2}$ level. A third laser beam perpendicular to both the others traverses the cloud and falls on a CCD. The absorption in the gas cloud produces an image of the cloud on the CCD, which is cigar shaped, $200\,\mu\mathrm{m}$ long and $50\,\mu\mathrm{m}$ across. Images of the cloud are only taken in intervals during which the probe and coupling lasers are off. When these two lasers are on the intensities of their beams emerging from the gas region are measured continuously with photomultiplier tubes, giving a time resolution of around one nanosecond.[7]

When only the probe laser is on and is tuned to resonance with the probe transition the probe beam is fully absorbed within the cloud of atoms. As noted earlier, when the coupling laser is turned on the cloud becomes transparent to the probe laser. This change comes about because the atoms reach a particular superposition of the states 2 and 3, called the *dark state*. The dark state and the orthogonal *bright state* are

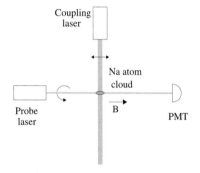

Fig. 18.8 Sketch of apparatus used by Hau and colleagues to observe electromagnetically induced transparency in a cloud of ultracold sodium atoms.

[6] A review is given in 'Taming light with cold atoms', by L. V. Hau, *Physics World* 14(9), 35 (2001).
[7] See Section 15.8 for details of photomultipler operation.

symbolically

$$|\text{dark}\rangle = (\Omega_p/\Omega)|2\rangle - (\Omega_c/\Omega)|3\rangle, \qquad (18.90)$$

$$|\text{bright}\rangle = (\Omega_p/\Omega)|3\rangle + (\Omega_c/\Omega)|2\rangle, \qquad (18.91)$$

where Ω_c and Ω_p are the respective Rabi frequencies of the coupling and probe beams, while $\Omega = \sqrt{(\Omega_p^2 + \Omega_c^2)}$. From eqn. 18.71 we see that the pumping rate between a pair of atomic energy levels is proportional to the Rabi frequency squared. Hence the *amplitude* for pumping from level 2 within the dark state into level 1 is $\Omega_c(\Omega_p/\Omega)$, and that for pumping from level 3 within the dark state into level 1 is $-\Omega_p(\Omega_c/\Omega)$, and these amplitudes cancel exactly. It follows that any atom reaching the dark state will remain in that state and be perfectly transparent to both the laser beams. On the other hand the bright state together with state 1 will form a tower of dressed states, and the absorption of the probe will show Autler–Townes splitting.

What is unexpected is that all the atoms end up in the dark state irrespective of their initial state. In order to understand this result we need to examine the progress of atoms initially in the bright state. The lasers will certainly pump these atoms betweeen the lower levels and level 1. However on each occasion that the atoms reach level 1, some will decay spontaneously and of these some will enter the dark state, where they stop. There is a one-way street into the dark state: transparency is inevitable. This process takes a few radiative lifetimes of the upper state.

The transition to the dark state is more rapid if only the coupling laser is on initially, and the probe laser is turned on afterwards. The probe laser beam is considered to be pulsed, which is the case of most interest here. In this sequence, provided the gas is sufficiently cold, the atoms are initially all in the lowest energy state, 3, which eqns. 18.90 and 18.91 show to be precisely the dark state when the probe laser is off. When the probe pulse arrives the atoms undergo a shift from the initial dark state 3 to a dark state that contains a superposition of states 2 and 3, and then when the probe pulse has passed they revert back to state 3. This evolution of the dark state requires only a small increase in the amplitude of state 2, so that equilibrium is reached rapidly in a few cycles of the coupling laser. The electromagnetic wave making up the probe pulse and the atoms in the evolving dark state form a single coherent system known as a *dark state polariton*.

The susceptibility of the cloud of atoms can be calculated in a way similar to that used for the Autler–Townes effect. There is a crucial difference compared to that case: the lower energy states are now coupled in the dark state. This makes the calculation more complex and it will not be attempted here.[8] The result, neglecting small terms, is that

$$\chi = g\Delta\omega \frac{-\Delta\omega^2 + \Omega_c^2/4 + i\Delta\omega\gamma'/2}{(\Delta\omega^2 - \Omega_c^2/4)^2 + \Delta\omega^2(\gamma')^2/4}. \tag{18.92}$$

As before the contraction used for the probe transition is

$$g = N\mu_p^2/(\varepsilon_0\hbar), \tag{18.93}$$

where μ_p^2 is the mean squared dipole moment for the probe transition.

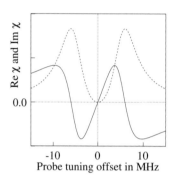

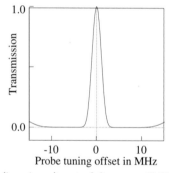

Fig. 18.9 The real (imaginary) part of the susceptibility is shown in the upper panel by a solid (broken) line. In the lower panel the transmission through the gas cloud is shown. All versus the offset of the probe frequency from the transition frequency.

A parameter γ' has been introduced which expresses the rate at which the coherence of the dark state is lost through spontaneous decays and atomic collisions. The real and imaginary parts of the susceptibility are shown in the upper panel of figure 18.9 as a function of the detuning of the probe laser from the transition frequency. Typical values are used in making this plot: 6 MHz for $\gamma'/2\pi$ and 12 MHz for $\Omega/2\pi$. The behaviour is very striking when the probe is tuned to resonance: instead of the absorption being at its strongest, the absorption vanishes so that the gas is predicted to be transparent. In order to emphasize this feature the transmittance for a layer of gas is shown in the lower panel of

---

[8]A clear presentation can be found in *Quantum Optics* by M. O. Scully and M. S. Zubairy, published by Cambridge University Press (1997).

figure 18.9. In a pioneer experiment by Harris and colleagues[9] in which a 10 cm long cell of warm strontium gas was used, the absorption of the probe laser at 337.1 nm changed from $\exp(-20)$ to $\exp(-1)$ when the coupling laser was applied.

This cancellation of the absorption at zero detuning is a purely quantum interference effect. There are two indistinguishable paths which are producing interference. These are not paths of light in space but instead alternative sequences of transitions by which the atoms can reach the higher energy state 1: in terms of the bare states these sequences are: $3 \rightarrow 1$ and $3 \rightarrow 1 \rightarrow 2 \rightarrow 1$. Thanks to the strong pumping by the coupling laser, the amplitude for the second path is as large as that of the first path and the interference between these paths is destructive. It is this key property which distinguishes electromagnetically induced transparency from the Autler–Townes effect, in which a drop and not a cancellation of absorption can be observed at resonance.

If the atoms are in motion then the Doppler frequency shifts vary from atom to atom and this can smear out the oscillations in the susceptibility seen in figure 18.9. At 1000 K the spread of Doppler shifts is about 1 GHz, while at 10^{-6} K it is only 32 kHz for sodium atoms emitting the D-line. EIT is therefore most readily produced in low density, ultra-low temperature gases. However, in warm gases these Doppler shifts can be made to cancel. The trick is to use probe and coupling transitions with almost equal frequency and to make the probe and coupling beams copropagating. The Doppler shifts of the beams then cancel in the sense that they induce transitions to the same energy level. Turning now to the possibilities for observing EIT in condensed matter, the first thing to note is that the disturbances of the phases of the atomic states due primarily to electron–phonon interactions are far worse. Coherence times for dark states would generally be less than 1 ns. It has therefore only been possible to produce EIT in a few special crystals.

18.5.1   Slow light

In addition to the transparency achieved in EIT, it is equally surprising that when the laser probe frequency is on resonance the phase velocity of the probe pulse is equal to c while its group velocity becomes very small. We can show this using eqn. 18.92. Suppose the probe laser is tuned close enough to the resonance that $\Omega_c^2 \gg \Delta\omega^2$. Then eqn. 18.92 predicts that the real part of the susceptibility is

$$\chi_r = 4g\Delta\omega/\Omega_c^2. \tag{18.94}$$

[9]K.-J. Boller, A. Imamoglu and S. E. Harris, *Physical Review Letters*, 66 2593 (1991).

Now the refractive index is $n = \sqrt{\varepsilon_r} = \sqrt{1 + \chi}$, so that its real part

$$n_r = 1 + \chi_r/2 = 1 + 2g\Delta\omega/\Omega_c^2, \qquad (18.95)$$

and the phase velocity exactly at resonance is c. Using eqn 11.42 the group velocity is given by

$$1/v_g = n_r/c + (\omega/c)(\mathrm{d}n_r/\mathrm{d}\omega),$$

so that close to resonance we have, to a good approximation,

$$1/v_g = 1/c + 2g\omega/(c\Omega_c^2) = 2g\omega/(c\Omega_c^2). \qquad (18.96)$$

Then using eqn. 18.93 to replace g in this equation gives

$$
\begin{aligned}
v_g &= \Omega_c^2 c/(2g\omega) \\
&= \Omega_c^2 \hbar\varepsilon_0 c/(2N\omega\mu_p^2), \qquad (18.97)
\end{aligned}
$$

where N is the density of atoms per unit volume and μ_p^2 is the mean squared dipole moment for the transition. The value of the group velocity can be made extremely small by reducing the coupling laser power which in turn reduces the Rabi frequency Ω_c. Looking back at eqn. 18.90 it is seen that this in turn alters the proportions of the atoms in levels 2 and 3 that make up the dark state. Therefore the intensity of the coupling laser must be changed slowly enough for the atoms to be able to follow the evolving dark state; such a change is called adiabatic.

The group velocity of the probe laser was reduced to $17 \, \mathrm{m\,s^{-1}}$ in the experiments of Hau and colleagues.[10] Thus a $2.5 \, \mu s$ pulse of the probe beam which is $750 \, \mathrm{m}$ long in air is reduced to a length of around $42 \, \mu m$ in the cloud of atoms. In those experimental conditions the relative permittivity of the gas, ε_r, is close to unity, and it follows that the maximum electric field amplitude, E_0, changes little on entering the cloud of atoms. In turn this means that the energy *density* in the electromagnetic field, $\varepsilon_0\varepsilon_r E_0^2/2$, does not change on entering the cloud of atoms. Therefore the effect of the compression of the probe pulse on entering the cloud of atoms is to reduce its total energy by a factor c/v_g; which for the example quoted is a factor 17 million. The energy of the probe pulse is transferred via the atoms to the coupling laser beam. On leaving the cloud the reverse process restores energy to the probe pulse.

In the extreme case that the coupling laser beam is turned off completely while the probe pulse is within the cloud, then eqn. 18.97 shows that the velocity of the probe is reduced to zero. The energy of the probe pulse is then almost totally transferred to the coupling beam and the tiny residual energy fraction is held by the dark state atoms located where the probe pulse expires. These same atoms have encoded within

[10]L. V. Hau, S. E. Harris, Z. Dutton and C. H. Behroozi, 'Light speed reduction to $17 \, \mathrm{m\,s^{-1}}$ in an ultracold atomic gas', *Nature* 397, 594 (1999).

them the information content of the probe pulse. What this means is that the oscillations of the probe pulse's waveform along its direction of travel are preserved in the proportions of atoms in level 2 and level 3 and in their phases, while the alignment of the angular momenta of the atoms preserves the sense of polarization of the probe pulse. The dark state polariton has then collapsed so that it consists solely of atoms in a coherent state without any electromagnetic wave component. Because the states 2 and 3 are only connected by a forbidden transition this purely atomic dark state polariton can persist after the coupling laser is turned off for a period of order the lifetime of the forbidden decay linking them. In several experiments the coupling laser has been again switched on after an interval of a few milliseconds; at which moment the probe pulse revived, then completed the remainder of its journey through the cloud of atoms and emerged to be detected.

In the following two sections techniques are described which are key elements in the operation of optical clocks. A final section is used to describe an optical clock using these techniques together with the optical comb described in Section 14.12.1.

18.6   Trapping and cooling ions

Laser beams are widely used to manipulate and cool atoms and ions, and to manipulate macroscopic particles. Here the example selected for discussion is the trapping and cooling of ions. This is of particular interest because the optical transition of an isolated, cooled ion provides the high precision reference frequency in an optical clock. First the ions need to be trapped.

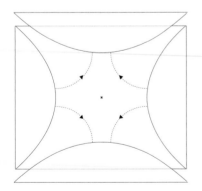

Fig. 18.10 Paul trap showing the direction of the electric force on an ion in one half cycle of the applied radio frequency field. The star marks where the micromotion vanishes.

Ions cannot be trapped electrostatically because it is impossible to form a local minimum in a purely electrostatic potential: this is a statement of Earnshaw's theorem. However electric fields oscillating at radio frequencies can produce the equivalent of a stable electrostatic potential well in which an ion can be trapped. Figure 18.10 shows an axial section through one type of radio frequency trap, the *Paul trap*. There are metal electrodes forming a pair of endcaps and a central ring, all being hyperboloids of revolution. The trap is contained in an evacuated container cooled to 4 K by liquid helium. A radio frequency voltage is applied between the endcaps and the ring: in one half cycle the electric force on an ion in the trap points in the direction indicated in the diagram, while in the next half-cycle this force is reversed. The net effect is an electric potential well with a minimum at the trap centre. Traps are typically of millimetre dimensions with an electric field of $\sim 100\,\mathrm{Vm}^{-1}$ oscillating at $\sim 10\,\mathrm{MHz}$. With hyperboloid electrodes the potential produced by an RF field is quadratic in both the radial and axial directions, so that the motion of a trapped and cooled ion will be the same as that analysed in Section 13.6. We take the axis of symmetry to be the z-axis. Once an ion is cooled its remaining *secular motion* in such a well is quantized and has a total kinetic energy given by

$$E = \sum_{x,y,z} \hbar\omega_i(n_i + 1/2), \qquad (18.98)$$

where ω_x, ω_y and ω_z are the angular frequencies for the motion along orthogonal axes, while n_x, n_y and n_z are integers. In the present case where there is azimuthal symmetry, $\omega_y = \omega_x$.

In addition to the secular motion in the well there is residual micromotion at the radio frequency, but this motion vanishes at the well centre. It is difficult to illuminate the ions when the trap has the structure shown in the figure. More practical traps dispense with either the endcaps or the ring: however a higher voltage is then required to hold the ions and the potential is no longer strictly quadratic in shape. Atoms are injected into the trap from a heated solid, and ionized by a low energy electron beam. These hot ions are cooled to thermal velocities by collisions with helium at very low pressure, and then this helium gas is pumped off; after which, one or more stages of laser cooling bring the ions close to rest near the trap centre. Only a single ion is required to provide a reference frequency, and this will reside at the trap centre and will have no micromotion. Such ions are kept continuously in the cooled trapped state for many days at a time.

Laser cooling of the trapped ions was first proposed by Wineland and Dehmelt in 1975. Two steps of cooling are considered here. In the first, *Doppler* step, a laser is used to illuminate the ions, and its frequency is tuned to just below that of a strong allowed transition. As seen by the ions travelling towards the laser beam the laser frequency is blue-shifted, and for some ions the shift makes the laser frequency equal to the transition frequency. These particular ions each absorb a laser photon in a head-on collision and their momenta are thereby reduced. Subsequently these ions spontaneously emit a photon, but the direction of emission is random so that the net momentum acquired by the cloud of ions through recoils is zero. By using two counter-propagating beams the ions travelling in both senses can be cooled. Furthermore, by arranging the laser beam axis to make equal angles with the three axes of the trap the cooling can be extended to motion along all three axes.

The minimum temperature achievable in this way is determined by the line width, γ, of the transition being used for cooling. In order to show this we consider an atom with a transition angular frequency ω_0 illuminated by a laser of angular frequency $\omega = \omega_0 + \Delta\omega$. Then the transition rate given by eqn. 18.51 is

$$\sigma = |c_2|^2 = \sigma_0/(\Delta\omega^2 + \gamma^2/4), \qquad (18.99)$$

where σ_0 has a fixed value for a given transition. Cooling is maximized when the laser is tuned so that

$$\Delta\omega = -\gamma/2. \qquad (18.100)$$

Before an ion absorbs a photon its kinetic energy is $mv^2/2$, where m is its mass and $\mathbf{v}$ its velocity. After absorbing the photon the ion's kinetic energy becomes

$$[\,m\mathbf{v} \pm \hbar\mathbf{k}\,]^2/(2m) = mv^2/2 \pm \hbar\mathbf{v}\cdot\mathbf{k} + R, \qquad (18.101)$$

where the alternative signs apply to absorption of a photon from the two oppositely directed laser beams. $\mathbf{k}$ is the photon wave vector for one beam, and $R = (\hbar k)^2/2m$. R is also the *recoil energy* that an ion receives when it spontaneously emits a photon with wave vector $\mathbf{k}$. The Doppler shifted angular frequencies of the beams, as seen by the ions, are respectively

$$\omega = \omega_0 - \gamma/2 \mp \mathbf{v}\cdot\mathbf{k}. \qquad (18.102)$$

Thus the rates for the transitions given by eqn. 18.99 are

$$\begin{aligned}
\sigma_{h/t} &= \sigma_0/[\,(-\gamma/2 \mp \mathbf{v}\cdot\mathbf{k})^2 + \gamma^2/4\,] \\
&\approx \sigma_0/(\gamma^2/2 \pm \mathbf{v}\cdot\mathbf{k}\gamma) \\
&= (2\sigma_0/\gamma^2)/(1 \pm 2\mathbf{v}\cdot\mathbf{k}/\gamma) \\
&\approx (2\sigma_0/\gamma^2)(1 \mp 2\mathbf{v}\cdot\mathbf{k}/\gamma),
\end{aligned} \qquad (18.103)$$

where the subscripts h and t refer to collisions with the $\mathbf{k}$ and $-\mathbf{k}$ beams. In order to obtain the net cooling rate these transition rates must be multiplied by the energy changes $\pm\hbar\mathbf{v}\cdot\mathbf{k} + 2R$, where the additional energy R relative to eqn. 18.101 is acquired when the ion recoils in the subsequent spontaneous emission. Then the net cooling rate is

$$\begin{aligned}
\mathrm{d}E/\mathrm{d}t &= \sigma_h(+\hbar\mathbf{v}\cdot\mathbf{k} + 2R) + \sigma_t(-\hbar\mathbf{v}\cdot\mathbf{k} + 2R) \\
&= (8\sigma_0/\gamma^2)\,[\,R - \hbar(\mathbf{v}\cdot\mathbf{k})^2/\gamma\,] \\
&= (8\sigma_0/\gamma^2)\,[\,R - (\hbar v^2 k^2/\gamma)\cos^2\theta\,],
\end{aligned} \qquad (18.104)$$

where θ is the angle the ion's velocity makes with the laser beam's direction. The cooling evidently ceases when the term in square brackets is zero. Taking the average over all inclinations of the ion's momentum with respect to the laser beam's direction, the limit at which cooling ceases is given by

$$R = \hbar k^2 v^2 \overline{\cos^2\theta}/\gamma = \hbar k^2 v^2/3\gamma. \qquad (18.105)$$

Now the recoil energy R is $(\hbar k)^2/2m$ so that in this limit

$$\hbar/(2m) = v^2/3\gamma. \qquad (18.106)$$

An estimate of the minimum kinetic energy and temperature reached in the Doppler phase of cooling is given by

$$3k_{\mathrm{B}}T/2 = mv^2/2, \qquad (18.107)$$

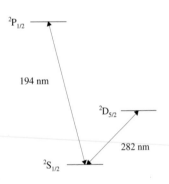

Fig. 18.11 Energy levels of the mercury ion. The transition at 194 nm is an allowed dipole transition, the transition at 282 nm is forbidden.

so that

$$k_BT = \hbar\gamma/2. \tag{18.108}$$

In the case of Hg ions cooled using the $^2P_{1/2} \to {}^2S_{1/2}$ transition with wavelength 194 nm and width 70 MHz the minimum temperature attainable would be around 1 mK. The recoil energy of the ion, R, expressed in terms of a frequency is written hf_r, where $f_r = 25$ kHz in the case of an Hg ion absorbing radiation at 194 nm. The quantum of vibrational energy of the secular motion of the ion in the trap is hf_s where the frequency of vibration of the ion, f_s is of order 3 MHz. Thus if the line width is around 70 MHz, the ion's energy after Doppler cooling amounts to an excitation of around 20 quanta of its secular vibrational motion.

A lower final temperature can be achieved by using a transition that has a narrow line width. In the case of a mercury ion there is a convenient electric quadrupole transition $^2D_{5/2} \to {}^2S_{1/2}$ at 282 nm with a natural line width of 1.7 Hz and a lifetime of 90 ms. This is shown in figure 18.11. The cooling process is modified because this line width is so narrow.

Figure 18.12 shows the absorption spectrum when the wavelength of a laser beam illuminating a Doppler cooled mercury ion cloud is scanned through wavelengths around 282 nm. Absorption is observed at resonance and at *sidebands* offset from the resonance by multiples of $\omega_s = 2\pi f_s$. These sidebands are due to transitions in which the ion also gains or loses one or more quanta of vibrational energy of its secular motion. The explanation why discrete peaks have replaced the usual continuous, Doppler broadened absorption peak proceeds as follows. The waveform of the radiation at the ion is, apart from a constant,

$$E = \cos(\omega_0 t - kz). \tag{18.109}$$

where the displacement of the ion due to its secular motion, z, is

$$z = z_0 \cos(\omega_s t). \tag{18.110}$$

Now the amplitude of the secular motion of the ion is only of order 10 nm, much smaller than the wavelength of the radiation. Therefore $kz \ll 1$, which defines what is called the *Lamb–Dicke regime*. In this Lamb–Dicke regime we can take $\sin kz = kz$ and $\cos kz = 1$, so that expanding the right hand side of eqn. 18.109 at the ion

$$
\begin{aligned}
E &= \cos(\omega_0 t) + kz\sin(\omega_0 t) \\
&= \cos(\omega_0 t) + kz_0\cos(\omega_s t)\sin(\omega_0 t) \\
&= \cos(\omega_0 t) + (kz_0/2)\{\sin[(\omega_0 + \omega_s)t] + \sin[(\omega_0 - \omega_s)t]\}.
\end{aligned}
$$

Therefore the radiation absorbed (or emitted) by the atom is dominantly at the unshifted frequency, with sidebands having intensities lower by a factor $(\pi z_0/\lambda)^2$. This is precisely what would be observed if the ions were point-like oscillators of frequency f_s. In addition, thanks to the

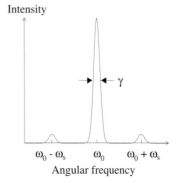

Intensity

Angular frequency

$\omega_0 - \omega_s \qquad \omega_0 \qquad \omega_0 + \omega_s$

Fig. 18.12 Absorption spectrum around the 282 nm transition for Doppler cooled mercury ions. Only the stronger, first sidebands are shown.

narrow line width of the transition, the sidebands are well resolved from the central peak.

These features can be taken advantage of in the second stage of cooling. The laser frequency is tuned to coincide with that of the lower sideband of the 282 nm transition. Each absorption of a photon of energy $\hbar(\omega_0 - \omega_s)$ is nearly always followed by the spontaneous emission of a photon of energy $\hbar\omega_0$, which guarantees cooling. After cooling the ions are predominantly in the lowest energy state for the secular motion, with only the *zero point* energy $\hbar\omega_s/2$. With a secular frequency of 3 MHz the corresponding temperature would be 74 μK.

Suppose that after Doppler cooling alone the ions are exposed to a laser tuned close to the $^2D_{5/2} \rightarrow {}^2S_{1/2}$ resonance. Then the central peak is addressed, and the secular motion is eliminated. Only the second order Doppler broadening due to time dilation[11] remains. The fractional change in wavelength is thus $(v/c)^2$ for radiation emitted or absorbed by the ion. Recalling the relation which relates the ion kinetic energy to its equivalent temperature, T K,

$$mv^2/2 = 3k_\mathrm{B}T/2, \tag{18.111}$$

we find that the equivalent temperature is $\sim$1 mK, so that the velocity is only $\sim$0.2 m s^{-1}. Hence the second order Doppler shift is of order one part in 10^{18}. This sets an ultimate limit on the precision that could be attained using the 282 nm transition of a Doppler-cooled, trapped mercury ion as a frequency or time standard.

Two or more ions can be held in a linear version of the Paul trap in which the ions settle along a line with well defined separations. A pair of Hg ions trapped in this way have been used to demonstrate a type of Young's two slit experiment.[12] Two ions a few microns apart scatter radiation from a laser beam of wavelength at 194 nm onto an imaging detector. Fringes are observed across the area of this detector due to interference between radiation scattered by the two ions.

18.7   Shelving

The 282 nm transition of mercury ions is an example of an electric-dipole forbidden transition possessing an exceptionally narrow natural line width, 1.7 Hz in 1064 GHz, which can be used as an ultra-precise frequency reference. Put another way, it can be used as the pendulum in an optical clock. A *single* measurement of an optical transition with a narrow line width of say, 1 Hz, offers a potential precision of one part

---

[11]See Section 1.9

[12]U. Eichmann, J. C. Bergquist, J. J. Bollinger, J. M. Gilligan, W. M. Itano, D. J. Wineland and M. G. Raizer: *Physical Review Letters* 70 2359 (1993).

in 10^{15} compared to one part in 10^{10} for a single measurement of a microwave transition of width $1\,\mathrm{Hz}$. However the study and use of such narrow optical transitions presents difficulties that the use of *shelving* has overcome.

Suppose a cloud of mercury ions is illuminated in a Paul trap by a laser at $282\,\mathrm{nm}$ ($1064\,\mathrm{GHz}$) so as to excite the transition to the $^2D_{5/2}$ level. In principle the central frequency and the width of the transition could be determined by scanning the laser frequency across the resonance in small steps and observing the fluorescence at each step. However each mercury ion can only emit a few photons per second; of these rare photons a single detector intercepts only a few percent; and of this subset the detector registers a fraction determined by its quantum efficiency. Hence the background counts from noise in the detector overwhelm the signal counts from the photons of interest.

The resolution of the dilemma was discerned by Dehmelt in 1982. We refer back to figure 18.11 which shows a simplified diagram of the energy levels of mercury, an ion to which Dehmelt's method has been successfully applied. What is done is to pump the ions on the fast dipole transition to the $^2P_{1/2}$ level using a laser tuned to $194\,\mathrm{nm}$, while also illuminating them with a probe laser of wavelength $282\,\mathrm{nm}$. Then paradoxically a detector is used to detect the fluorescence at $194\,\mathrm{nm}$, rather than fluorescence at $282\,\mathrm{nm}$, using appropriate filters.

Consider what happens when the intensity of the laser at $194\,\mathrm{nm}$ is sufficient to saturate the transition to the $^2P_{1/2}$ level whose width is $70\,\mathrm{MHz}$. The populations of the $^2S_{1/2}$ and $^2P_{1/2}$ levels are then equal, and around $\sim 10^8$ fluorescent photons will be emitted per second at $194\,\mathrm{nm}$ by a single ion. A detector could in principle intercept all the photons emitted within a cone covering some 5% of the total solid angle, and in practice still records of order 10^4 photons per second despite losses in the optics and the low quantum efficiency at the wavelengths involved. This stream of fluorescent photons at $194\,\mathrm{nm}$ from a single ion will come to an abrupt halt on each occasion that a photon from the laser at $282\,\mathrm{nm}$ is absorbed. At that instant the ion is transferred to the $^2D_{5/2}$ state where it remains for around $90\,\mathrm{ms}$ before decaying back to the $^2S_{1/2}$ ground state. Once the ion reaches the $^2S_{1/2}$ state again the stream of fluorescence at $194\,\mathrm{nm}$ resumes. The way the fluorescence at $194\,\mathrm{nm}$ varies with time is shown in figure 18.13 for a single trapped mercury ion illuminated in this way. The intervals during which the ion is *shelved* in the $^2D_{5/2}$ state are those with no fluorescence at $194\,\mathrm{nm}$. The distribution of the duration of the shelving intervals, t_d, is proportional to $\exp\left(-t_d/\tau\right)$, where τ is the lifetime of the $^2D_{5/2}$ state. This allows the lifetime of the $^2D_{5/2}$ state to be measured in an elegant manner. Of equal importance for our purpose, shelving can be used to help determine the exact frequency of the envisaged clock transition, $^2D_{5/2} \rightarrow {}^2S_{1/2}$.

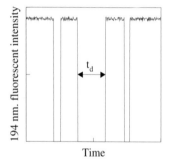

Fig. 18.13 Intensity of the fluorescence at $194\,\mathrm{nm}$ from a single cooled trapped Hg ion, showing typical shelving periods. The distribution of the duration of shelving times, t_d, depends on the lifetime of the $^2D_{5/2}$ level.

The method employed is to scan the frequency of the *probe* laser across this transition in small steps. At each step the number of shelvings will be proportional to the intensity of the transition to the $^5D_{5/2}$ level. After a complete scan the distribution of the number of shelvings per unit time is plotted against the frequency of the probe laser. The shape of this distribution will follow the line shape of the transition, with the line centre lying at the wavelength for which the shelving rate is maximum. Instead of having to detect the transition through rare fluorescence at 282 nm, the transition is now signalled by the cliff edges in the count of fluorescence photons at 194 nm. Thanks to this approach the forbidden transition at 282 nm can be probed with the statistics of the allowed transition at 194 nm: a statistical leverage called *quantum amplification*.

18.8   Optical clocks

The basic features of an optical clock are these:

- A cooled, trapped ion having an optical transition to the ground state with very narrow natural line width. This provides the frequency standard.
- The interrogation of this reference transition by means of a stabilized laser using the shelving technique. After each scan the central frequency of the laser is reset to the transition frequency of the ion.
- An optical comb of the sort described in Section 14.12.1. The optical comb is used to express the stabilized laser frequency as a known multiple of a microwave frequency, for which electronic counting techniques already exist.

The example of a clock based on a ^{199}Hg ion will be described. There are tiny but significant energy shifts between the atomic energy levels of different isotopes arising from the difference in the nuclear mass. These are avoided by using a single isotope. Such devices were developed by a team at the National Institute for Standards and Technology at Boulder, Colorado.[13] A much simplified sketch of such an ion clock is presented in figure 18.14. The sequence followed in interrogating the ion is to probe it with a tunable stabilized laser advancing in small frequency steps through the clock frequency transition at 282 nm. At each step the ion is first brought to the ground state, then the probe laser is turned on for a short time ~50 ms. Immediately afterwards the pump laser at 194 nm is turned on and depending on whether the flood of fluorescence at 194 nm is detected or not, it is possible to say whether shelving has occurred. This interrogation cycle is repeated tens of times at each frequency step along the scan. By having the lasers on at separate times the light shift

---

[13]See for example 'An optical clock based on a single trapped ^{199}Hg$^+$ ion' by S. A. Diddams, Th. Udam, J. C. Bergquist, E. A. Curtis, R. E. Drullinger, L. Holberg, M. W. Itano, W. D. Lee, C. W. Oates, K. R. Vogel and D. J. Wineland in *Science* 293, 825 (August 2001).

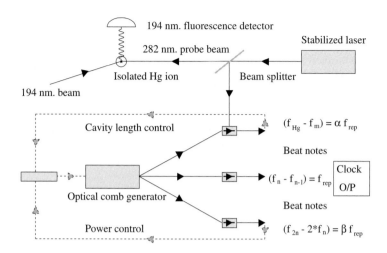

Fig. 18.14 Components of an optical clock based on a single cooled, trapped ^{199}Hg ion. The electronic control paths are indicated with broken lines. Courtesy Professor Wineland.

produced by one laser will not affect the energy levels seen by the other laser. The plot of the shelving rate against frequency is stored and used to reset the reference probe frequency so that the centre of its scan coincides with the exact centre of the $^{2}D_{5/2} \rightarrow {}^{2}S_{1/2}$ transition. In this way the central frequency of the stabilized laser is being continuously updated to agree with the transition frequency, f_{Hg}, thus making this latter accessible to external measurement or comparison.

An optical comb, of the type described in Section 14.12.1, is used to transfer frequencies from the optical to the microwave part of the spectrum for which there are standard electronic counting techniques. The comb is generated when femtosecond pulses from a mode-locked laser pass through a microstructured fibre. A comb is used which covers more than an octave in frequency, and has a tooth spacing f_{rep} of $\sim$1 GHz. The frequency of the mth tooth is

$$f_m = m f_{rep} + f_{ceo},$$

where f_{ceo} is not immediately known. Two beat frequencies are used to lock f_{rep} so that there is a simple, precisely determined numerical relationship between it and the stabilized laser frequency, f_{Hg}. Then when f_{rep} is measured by microwave electronics the ion transition frequency f_{Hg} can be obtained by a simple multiplication. One of the beat frequencies is the difference with the mth tooth, $(f_{Hg} - f_m)$, which is forced to be an integral multiple of $f_{rep}/100$ by adjusting the length of the cavity of the mode-locked laser:

$$f_{Hg} = f_m + n_1 f_{rep}/100$$

At the same time the beat frequency $(2f_n - f_{2n})$, which equals f_{ceo}, is forced to be an integral multiple of $f_{rep}/100$ by altering the power of the mode-locked laser:

$$f_{\text{ceo}} = n_2 f_{\text{rep}}/100$$

Finally the beat frequency between adjacent teeth $(f_n - f_{n-1})$ provides f_{rep} the reference frequency, around 1 GHz, which is passed to the microwave electronics for measurement. The outcome of these steps is

$$\begin{aligned}
f_{\text{Hg}} &= f_m + n_1 f_{\text{rep}}/100 \\
&= m f_{\text{rep}} + f_{\text{ceo}} + n_1 f_{\text{rep}}/100 \\
&= m f_{\text{rep}} + n_2 f_{\text{rep}}/100 + n_1 f_{\text{rep}}/100,
\end{aligned} \qquad (18.112)$$

in which all the integers m, n_1 and n_2 are known.

At the present time when the international standard clock is based on a microwave transition of ^{133}Cs, the frequency of the optical clock is determined by the microwave counters whose counting rate is based on a ^{133}Cs clock. If the optical clock were to become the primary standard, then it would be necessary to define its frequency to be equal to the best current measurement of this frequency in calibrations made with existing standard microwave clocks. Then the frequency f_{Hg} appearing in the previous equation would have an internationally defined value from which f_{rep} would be determined. In turn this would fix the length of the second to be exactly

$$1/f_{\text{rep}} = (m + n_1/100 + n_2/100)/f_{\text{Hg}}. \qquad (18.113)$$

Optical clocks have been built and compared with one another and with atomic clocks so that the precision of optical clocks is well studied, and currently lies in the range between one part in 10^{14} and one part in 10^{15}. The ultimate limit is imposed by the variation of second order Doppler shift due to the ion's residual secular motion: this was discussed in Section 18.6 and found to be one part in 10^{18}.

18.9   Further reading

Laser Cooling and Trapping by H. J. Metcalf and P. van der Straten, published by Springer (1999). This text covers the cooling and trapping of neutral atoms rather than ions, and physical applications.

Exercises

(18.1) Estimate the amplitude of the secular motion of Hg ions in a Paul trap if their vibration frequency is 1 MHz.

(18.2) Write down the expression for the susceptibility of a cloud of atoms when the incident radiation is tuned to an atomic transition. Evaluate this quantity for an atomic density of $10^{16}\,\text{m}^{-3}$, a resonance at wavelength 590 nm and line width 10 MHz. Assume that the dipole moment for the transition is twice the Bohr radius times the electron charge. What is the absorption length in the cloud?

(18.3) Show that the expression given for the saturation intensity in eqn. 18.74 reduces to $I_s = 208/(\tau\lambda^3)\,\mathrm{Wm^{-2}}$ where the lifetime against spontaneous decay is measured in nanoseconds and the wavelength in micrometres. What is the saturation intensity for pumping the transition of the previous question?

(18.4) Suppose that a DC electric field is applied to a two-state atom. What happens and how is the analysis of Section 18.2 affected?

(18.5) A laser beam of power density $P\,\mathrm{Wm^{-2}}$ is tuned to an atomic resonance of two-state atoms whose electric dipole moment is D_{12} for that transition. Show that the Rabi frequency is $(D_{12}/\hbar)\sqrt{(2P/\varepsilon_0 c)}$. If D_{12} is $2ea_0$ where a_0 is the Bohr radius, then deduce the power density required to produce a Rabi frequency of 5 MHz. How powerful a laser is required if the target area of the sample of gas is $10^{-8}\,\mathrm{m^2}$?

(18.6) Two-state atoms are pumped with a laser whose electric field has amplitude $10\,\mathrm{kV\,m^{-1}}$ to produce a transition at 589 nm, the laser being tuned to 589.0001 nm. If the dipole moment for the transition is $2.5ea_0$, where a_0 is the Bohr radius, calculate the mixing angle θ of the bare states in the dressed states.

(18.7) In an experiment to produce EIT the target cloud of ultra-cold sodium atoms has a density $3.3\,10^{18}\,\mathrm{m^{-3}}$. The cloud is exposed to coupling and probe lasers tuned to excite transitions from two different hyperfine split levels of the ground state to the same upper state, both wavelengths being around 589 nm. The coupling pump intensity is $500\,\mathrm{Wm^{-2}}$. The dipole moment for the transitions is $2.5ea_0$ where a_0 is the Bohr radius. Calculate the group velocity of the probe beam close to resonance.

(18.8) The Autler–Townes splitting of a probe transition from one atomic state to a ground state at $4.0\,\mu m$ wavelength is 50 MHz. A pump laser of wavelength $3.5\,\mu m$ provides a power density $0.40\,\mathrm{W\,cm^{-2}}$ and is tuned to the transition from another excited state to the same ground state. Estimate the dipole moment for the pump transition at $3.5\,\mu m$ and the decay rate of the associated excited state.

(18.9) The Autler–Townes splitting of a probe transition from one atomic state to a ground state at $0.55\,\mu m$ wavelength is 200 MHz. The pump laser provides a power density $2.5\,\mathrm{MW\,m^{-2}}$ and is tuned to the transition from another excited state to the same ground state at wavelength $0.6\,\mu m$. Estimate the dipole moment for the pump transition at $0.6\,\mu m$ and the decay rate of the associated excited state.

(18.10) Show that in EIT with the probe tuned very close to resonance the group velocity is

$$v_\mathrm{g} = (\mu_\mathrm{c}/\mu_\mathrm{p})^2 P_\mathrm{c}/(N\hbar\omega)$$

where P_c is the coupling laser power per unit area, μ_c and μ_p are the respective dipole moments for the coupling and probe transitions, and ω is the probe transition angular frequency. Show that the probe velocity is determined by how quickly the coupling laser beam can supply energy to pass all the atoms in the beam path through the common upper state.

The quantized
electromagnetic field

19.1 Introduction

In the preceding chapters the interaction of radiation with matter has been treated semi-classically: in any calculations radiation was regarded as waves with photons appearing simply at absorption or emission. However there are many quantum mechanical effects in optics that can only be understood using a theory in which not only atomic states are quantized but also the electric and magnetic fields. This more complete theory in which the electromagnetic field is quantized will be outlined first. Then in the remainder of the chapter this theory is used to interpret experiments involving photon correlations and the entanglement of photons.

It was noted earlier that Maxwell's equations hold the same place in electromagnetism that Schroedinger's equation holds in mechanics, with the fields $\mathbf{E}$ and $\mathbf{H}$ being the analogues of the wavefunction ψ. The step considered here in which the fields, $\mathbf{E}$ and $\mathbf{H}$, and also ψ, can be quantized is known as *second quantization* to distinguish it from the familiar first step in which the kinematic quantities, energy and momentum, were quantized – that is to say they became operators.

Second quantization will be carried through in the two sections immediately following this one. Then first order coherence will be re-examined in terms of quantized fields, using Young's two slit experiment as an example. Second order coherence and correlations between two different photons are introduced in the next section. After this the two quantum states are introduced whose properties closely match respectively those of beams from lasers and from thermal sources. This leads to a discussion and interpretation of the pioneering experiments of Hanbury Brown and Twiss to observe photon bunching and to measure stellar radii using photon correlations.

It emerges that pairs of photons can exist in correlated states that lack classical parallels. The classical state of two photons is simply expressed as a quantum state: a photon X in mode i and a photon Y in mode j, are in an overall state that can be written $|X\rangle_i|Y\rangle_j$. One example of a quantum state which has no classical equivalent is $|X\rangle_i|Y\rangle_j + |X\rangle_j|Y\rangle_i$, in which the two photons are said to be *entangled*. A section is used to describe how entangled photon pairs are produced by spontaneous parametric down conversion in non-linear crystals. The Hong-Ou–Mandel (HOM) interferometer uses a beam splitter to manipulate entangled photons: its application to demonstrate the indistinguishability of photons and its use to measure short coherence times are described in the following section. Then the Franson–Chiao experiment with entangled photons is described. This experiment confirms that path indistinguishability is essential for the observation of multi-photon as well as single photon interference effects. The concept of *complementarity*, originated by Bohr, extends the uncertainty principle to situations in which the complementary pieces of information are no longer limited to the values of conjugate kinematic quantities such as a particle's momentum and position. An experiment is used to illustrate this idea: one member of an entangled pair is used to select or alternatively erase an interference pattern formed by the other photon.

In a final section the calculation made in Section 18.2 of the transition rate of a two-level atom in a weak electromagnetic field is carried through again, with the field now quantized. This approach reveals that spontaneous emission is emission stimulated by the intrinsic fluctuations of the electromagnetic field in the vacuum. Thus spontaneous and stimulated emission are intimately related, and not disparate effects.

19.2   Second quantization

Modes of the electromagnetic field are discussed in Chapter 9.

Quantization of the fields integrates the particle–photon and the wave–electromagnetic field aspects of electromagnetic radiation. The classical electromagnetic field in free space, or dielectric, which we intend to quantize is first resolved into plane polarized sinusoidal plane waves. The real electric and magnetic fields of such a wave are presented here in a form that is designed to ease the introduction of quantization:

$$E_x = 2\zeta_\omega\, a \cos(kz - \omega t)$$
$$= \zeta_\omega\,[\,a(t)\,\exp(ikz) + a^*(t)\,\exp(-ikz)\,], \tag{19.1}$$
$$B_y = E_x/v, \tag{19.2}$$

where $\omega = kv$, $v = c/\sqrt{\varepsilon_r}$ and ε_r is the relative permittivity. $a(t) = a\exp(-i\omega t)$. The absorption of the dependence on time into $a(t)$ will prove its usefulness later on. The constant $\zeta_\omega = \sqrt{\hbar\omega/2\varepsilon_0\varepsilon_r V}$ has the dimensions of volts/metre. V is a reference volume[1] within which the processes of interest take place. Its exact size and shape are not of importance: in all cases it cancels from formulae for measurable quantities. The last element of the classical picture needed here is the total energy in the electromagnetic field in the volume V:

$$H = \int_V (\varepsilon_0\varepsilon_r E^2/2 + B^2/2\mu_0)\,\mathrm{d}V$$

[1]Actual waves are finite in extent whereas the plane waves into which Fourier analysis resolves them are infinite in extent. Thus it is necessary when handling plane waves to introduce a finite reference volume V covering the region of the actual waves. Physical quantities like energy density are evaluated per unit volume so the actual size of V need not be specified.

$$= \int_V \varepsilon_0 \varepsilon_r E^2 \, \mathrm{d}V. \qquad (19.3)$$

After substituting for E some terms contain integrals of the form

$$\int_{-Z}^{+Z} \exp\left(\pm 2ikz\right) \mathrm{d}z = \mp \sin\left(2kZ\right)/k \qquad (19.4)$$

where $\pm Z$ are the boundaries in z of the volume of integration. The other terms are linearly proportional to Z. Then because the reference volume V is much larger than the wavelength the oscillatory terms become negligible compared to those increasing linearly with Z. The integral therefore reduces to

$$\begin{aligned} H &= 2\zeta_\omega^2 \left\{ \varepsilon_0 \, \varepsilon_r a(t) a^*(t) \right\} V \\ &= \hbar\omega \left\{ a^*(t)\, a(t) \right\}. \end{aligned} \qquad (19.5)$$

Thanks to the judicious choice of the waveform in eqn. 19.1 this result hints at how the wave and quantum aspects come together. If in this energy equation $a^*(t)a(t)$ were the number of photons the equation would match Planck's formula.

The excitation of a mode of the electromagnetic field has a mathematical parallel in the excitation of a simple harmonic oscillator (SHO), whose quantization was presented in Section 13.6. Quantization of the SHO gives equally spaced energy levels which mimic the equally spaced energy levels in a mode of the electromagnetic field when 1, 2, 3,..... photons are present. A comparison of the energies of the two systems will now be used to bring out their formal mathematical resemblance; after which we follow the standard steps used to quantize the SHO, but now applied to the electromagnetic field.

The particular SHO considered consists of a unit mass oscillating under a restoring force $\omega^2 Q(t)$ proportional to the displacement $Q(t)$. Thus its angular frequency of oscillation is ω. Its total potential plus kinetic energy is:

$$H = (1/2)\{\, \omega^2 Q(t)^2 + P(t)^2 \,\}, \qquad (19.6)$$

where $P(t) = \dot{Q}(t)$ is the momentum of the unit mass. Returning now to the electromagnetic field we rewrite the complex electric field in terms of real quantities, $q(t)$ and $p(t)$

$$a(t) = (\omega q(t) + ip(t))/\sqrt{2\hbar\omega}, \qquad (19.7)$$
$$a^*(t) = (\omega q(t) - ip(t))/\sqrt{2\hbar\omega}, \qquad (19.8)$$

and then the energy equation, eqn 19.5, becomes

$$H = (1/2)\{\, \omega^2 q(t)^2 + p(t)^2 \,\}, \qquad (19.9)$$

which makes explicit the parallel with the SHO energy appearing in eqn. 19.6. Therefore $q(t)$ and $p(t)$ can be regarded at least mathematically as a *position* coordinate and the corresponding *momentum*. They are referred to in the present context as *quadratures* because they are components of the field which are $\pi/2$ out of phase: $q = \sqrt{2\hbar/\omega}\, a \cos(\omega t)$ and $p = -\sqrt{2\hbar\omega}\, a \sin(\omega t)$. Quantizing the electromagnetic field can therefore follow the same procedure as used for the SHO in Section 13.6.

The essential step is to associate quantum mechanical *operators* $\hat{q}(t)$ and $\hat{p}(t)$ with the variables q and p. These operators act on the *state vectors* describing the modes[2] of the electromagnetic field. Then, for consistency, these operators must obey the usual commutator relation connecting position and momentum operators which was presented in eqn. 13.37

$$[\hat{q}(t), \hat{p}(t)] = i\hbar. \tag{19.10}$$

This tells us that the electromagnetic field variables p and q are not compatible observables: a state which is an eigenstate of one is not an eigenstate of the other.

A suitable simple state vector for a mode of the electromagnetic field will be written $|n\rangle$, in which the argument is the number of photons in the mode. When the complete electromagnetic field is considered the overall state vector will be the product of state vectors, one for each mode of the electromagnetic field; each state vector being labelled to indicate the wave vector, frequency and the polarization of the mode.

The state vector must not be pictured as a wavefunction describing the mode shape in space; instead it describes the photon content of a mode. Choices of the basis modes other than sinusoidal plane waves are possible, and for these choices the labelling would be different. For each state vector $|n\rangle$ there is a dual state vector $\langle n|$ whose properties are described in Appendix E. Referring back to Section 13.6 the distribution of q is that shown for s in figure 13.7 with n equal to 0, 1, 2 and 3. The wavefunction of the state described by the state vector $|n\rangle$ is

$$\psi(s) = \langle s|n\rangle, \tag{19.11}$$

where $|s\rangle$ describes a state in which the distribution in q is a delta function at $q = s$. The states of a mode containing different numbers of photons are orthogonal, and for convenience each state vector can be defined as normalized: thus

$$\langle m|n\rangle = \delta_{mn}. \tag{19.12}$$

The operators of interest in the case of the electromagnetic field are defined by rewriting eqns. 19.7 and 19.8 in terms of operators

$$\hat{a} = (\omega\hat{q} + i\hat{p})/\sqrt{2\hbar\omega}, \tag{19.13}$$

$$\hat{a}^{\dagger} = (\omega\hat{q} - i\hat{p})/\sqrt{2\hbar\omega}. \tag{19.14}$$

---

[2] Appendix E contains some introductory material about state vectors.

The operator $\hat{a}$ will turn out to be an *annihilation* operator which removes a photon from a mode, whilst $\hat{a}^\dagger$ is a *creation* operator which creates an additional photon in the mode. Using eqn. 19.10 these definitions produce the commutation relation

$$[\hat{a}, \hat{a}^\dagger] = 1. \tag{19.15}$$

Different modes of the electromagnetic field are independent, and using i and j to label any two modes, we must have

$$[\hat{a}_i, \hat{a}_j^\dagger] = \delta_{ij}, \quad [\hat{a}_i^\dagger, \hat{a}_j^\dagger] = [\hat{a}_i, \hat{a}_j] = 0. \tag{19.16}$$

The electric field operator for a given mode is constructed from the classical electric field given in eqn. 19.1, replacing $a(t)$ with the operator $\hat{a} \exp(-i\omega t)$ and $a^*(t)$ with the operator $\hat{a}^\dagger \exp(+i\omega t)$

$$\hat{E} = \zeta_\omega [\hat{a} \exp(i\xi_\omega) + \hat{a}^\dagger \exp(-i\xi_\omega)], \tag{19.17}$$

where $\xi_\omega = kz - \omega t$, $\zeta_\omega = \sqrt{\hbar\omega/2\varepsilon_0 \varepsilon_r V}$ and $\omega = kv$.[3] The magnetic field is obtained from eqn. 19.2 with operators replacing simple fields. In the experiments described below the radiation is a parallel beam, and we take the beam to travel in the z-direction. Only ω is then needed to label a mode rather than ω and $\mathbf{k}$. Plane polarization along the x-direction can be assumed, allowing the component label on the electric field to be omitted. Generalizing the previous equation to give the electric field operator of a beam containing a number of frequencies,

$$\hat{E} = \sum_\omega \zeta_\omega [\hat{a}_\omega \exp(i\xi_\omega) + \hat{a}_\omega^\dagger \exp(-i\xi_\omega)], \tag{19.18}$$

where $\hat{a}_\omega$ and $\hat{a}_\omega^\dagger$ are the respective annihilation and creation operators for photons of frequency ω. This can be separated as follows:

$$\hat{E} = \hat{E}^+ + \hat{E}^-, \tag{19.19}$$

where

$$\hat{E}^+ = \sum_\omega \hat{E}_\omega^+ = \sum_\omega \zeta_\omega \hat{a}_\omega \exp(+i\xi_\omega), \tag{19.20}$$

$$\hat{E}^- = \sum_\omega \hat{E}_\omega^- = \sum_\omega \zeta_\omega \hat{a}_\omega^\dagger \exp(-i\xi_\omega). \tag{19.21}$$

$\hat{E}^+$ only contains *annihilation* operators $\hat{a}_\omega$, and is by convention called the *positive frequency* component of the field operator $\hat{E}$. $\hat{E}^-$ contains only *creation* operators $\hat{a}_\omega^\dagger$ and is called the *negative frequency* component. Using eqn. 19.15 we get the commutation relation between the angular frequency components of $\hat{E}^+$ and $\hat{E}^-$,

$$[\hat{E}_\omega^+, \hat{E}_\omega^-] = \zeta_\omega^2. \tag{19.22}$$

Some comments are also needed here about the way the time dependence is being handled in this chapter which contrasts with the familiar

[3]The space-time arguments on $\hat{E}(z,t)$ are omitted here in order to keep the notation simple.

Schroedinger representation that was used in the earlier chapters on quantum optics. The time dependence has been moved to the operators $\hat{E}^-$ and $\hat{E}^+$, while the modes of the electromagnetic field on which they operate are unchanging. This is called the *Heisenberg* representation, in contrast to the *Schroedinger* representation, in which the time dependence is all in the wavefunctions, and the operators are time independent. Comparisons between the two representations are made in Appendix F. In introducing the time dependent operators it was useful to show the time dependent factors $\exp(\pm i\xi_\omega)$ explicitly, with the operators $\hat{a}^\dagger_\omega$ and $\hat{a}_\omega$ being time independent. In what follows the time dependent factors will be separated whenever this is necessary or helpful.

Replacing the classical variables by operators in eqn. 19.9 gives the energy operator for a mode of the electromagnetic field:

$$\hat{H} = (1/2)(\omega^2\hat{q}\hat{q} + \hat{p}\hat{p}) = \hbar\omega\,(\hat{a}^\dagger\hat{a} + 1/2). \tag{19.23}$$

Then

$$\hat{H}|n\rangle = \hbar\omega\,(\,\hat{a}^\dagger\hat{a} + 1/2\,)|n\rangle = (n+1/2)\,\hbar\omega\,|n\rangle. \tag{19.24}$$

Pursuing the analogy with the harmonic oscillator, $\hat{a}^\dagger\hat{a}$ is interpreted as the operator that yields the number of photons in the mode, so that the energy of a mode containing n photons is $E_n = (n+1/2)\hbar\omega$. This looks reasonable apart from an unexpected additional energy of $\hbar\omega/2$ per mode. Consequently the vacuum – the state with zero photons in any mode – has energy:

$$H_0 = \sum_\omega \hbar\omega/2, \tag{19.25}$$

where the sum runs over all modes of the electromagnetic field. In quantum theory the vacuum is therefore not a static passive state, but has what is called a *zero point* energy. The number of modes is infinite so that the total energy in the vacuum is infinite. However this energy is not accessible to measurement because all the energies that are measurable are always differences in energy, and from these the zero point energy cancels.

From the existence of this *zero point* energy Casimir inferred that two closely-spaced, neutral, conducting surfaces in vacuum should feel a mutual attractive force. Casimir realized that within the gap the modes of the electromagnetic field are restricted by the boundary conditions on **E** and **B** at the conducting surfaces. Within the gap the allowed modes have discrete, closely spaced values of **k**; but outside the gap any value of **k** is possible. This means that, with fewer modes the total zero point energy density of the electromagnetic field within the cavity is less than outside. Hence there is a net inward force on the conductors. Though small, this *Casimir force* has been measured and found to be of the magnitude calculated by Casimir. This and effects noted later make it clear that the energy of a mode containing n photons is indeed $(n + 1/2)\hbar\omega$ and not $n\,\hbar\omega$.

It is now appropriate to check the consistency of the interpretation of $\hat{a}$ and $\hat{a}^\dagger$ as annihilation and creation operators respectively. Consider the effect of the operator $\hat{a}$ on a mode containing n photons. The energy of the new state $\hat{a}|n\rangle$ is given by:

$$\hat{H}\,\hat{a}|n\rangle = \hbar\omega\,(\hat{a}^\dagger\hat{a} + 1/2)\hat{a}|n\rangle = \hbar\omega\,\hat{a}\,(\hat{a}^\dagger\hat{a} - 1/2)|n\rangle,$$

where we have used the commutation relation, eqn. 19.15, to move one $\hat{a}$ operator past the $\hat{a}^\dagger$ operator. Continuing,

$$\hat{H}\,\hat{a}|n\rangle = \hbar\omega\,\hat{a}\,(n - 1/2)|n\rangle = (n - 1/2)\hbar\omega\,\hat{a}|n\rangle. \qquad (19.26)$$

Evidently the new state, $\hat{a}|n\rangle$, contains one less photon, which justifies the identification of $\hat{a}$ as an annihilation operator. The new state $|n-1\rangle$ including a normalization factor is given by

$$\hat{a}|n\rangle = \sqrt{n}\,|n - 1\rangle. \qquad (19.27)$$

Similarly we can show that the operator $\hat{a}^\dagger$ creates an additional photon,

$$\hat{a}^\dagger|n\rangle = \sqrt{n + 1}|n + 1\rangle. \qquad (19.28)$$

Finally the assignment of $\hat{a}^\dagger\hat{a}$ as the *number* operator $\hat{n}$ can be checked

$$\hat{n}|n\rangle = \hat{a}^\dagger\hat{a}|n\rangle = \sqrt{n}\,\hat{a}^\dagger|n - 1\rangle = n|n\rangle. \qquad (19.29)$$

Although the *occupation number* of photons in the mode is an observable, neither a nor a^* represent observables for the state $|n\rangle$.

Collecting the results for any empty mode

$$\hat{a}|0\rangle = 0, \quad \hat{a}^\dagger|0\rangle = |1\rangle, \quad \hat{H}|0\rangle = (\hbar\omega/2)|0\rangle. \qquad (19.30)$$

The electric field operator, eqn 19.18, is a sum of terms each containing a single annihilation or a single creation operator. Therefore its expectation value is

$$\langle n|\hat{E}|n\rangle = 0. \qquad (19.31)$$

This simple result reproduces the classical conclusion that the average electric field of a monochromatic plane wave is zero. On the other hand the expectation value of the square of the electric field does not vanish. If a single mode is considered the subscripts appearing in eqn 19.18 can be dropped:

$$
\begin{aligned}
\langle \mathbf{E}^2 \rangle &= \langle n|\hat{E}^2|n\rangle \\
&= \zeta_\omega^2\,\langle n|\,[\exp(2i\xi_\omega)\,\hat{a}\hat{a} + \hat{a}\hat{a}^\dagger + \hat{a}^\dagger\hat{a} + \exp(-2i\xi_\omega)\,\hat{a}^\dagger\hat{a}^\dagger\,]\,|n\rangle \\
&= (\hbar\omega/\varepsilon_0\varepsilon_{\mathrm{r}}V)\,(n + 1/2),
\end{aligned}
\qquad (19.32)
$$

where we have used $\langle n|\hat{a}\hat{a}|n\rangle = \langle n|\hat{a}^\dagger\hat{a}^\dagger|n\rangle = 0$, $[\hat{a}, \hat{a}^\dagger] = 1$, and $\hat{a}^\dagger\hat{a} = \hat{n}$. Rearranging this gives the total energy in a single mode,

$$\varepsilon_o \varepsilon_{\mathrm{r}} V \langle \mathbf{E}^2 \rangle = (n + 1/2)\hbar\omega, \qquad (19.33)$$

An empty mode contributes an energy $\hbar\omega/2$ so that even in a vacuum the mean square values of the fields do not vanish. In fact when all modes are included $\langle \mathbf{E}^2 \rangle$ is infinite. However, as was pointed out earlier, only energy differences are accessible to direct measurement.

The *vacuum fluctuations* of the electromagnetic field produce small changes in the energy levels of atomic states, and the change is slightly different for each different atomic configuration. Most famous of all is the Lamb shift: the resulting relative displacement of $0.11\,\mu$eV (27 MHz) between the $2s_{1/2}$ and $2p_{1/2}$ states in the hydrogen atom. The displacement was evaluated theoretically and then measured by Lamb in 1947. The quantitative agreement he found was crucial early evidence of the precision and reliability of the predictions of quantized field theory. The quantitative confirmation from similar effects and from the Casimir effect make it certain that the zero point energy of the electromagnetic field in the vacuum really exists. Second quantization underpins a full and consistent understanding of the electromagnetic field.

Beams from lasers and thermal sources all show fluctuations of amplitude and frequency due to variations in the number and energy of the photons emitted per unit time. If the internal physical state of the source – chemical content, temperature and pressure – are stable then the character of these fluctuations will not change with time. As pointed out in Chapter 7, light sources of interest are usually *stationary*. These give beams whose statistical properties observed over a time long compared to the wave period are independent of when the sampling is done. The average value of any measurable quantity taken in this way for a given source is thus equal to the quantum mechanical expectation value of the observable. Thus for a stationary source, where averaging is made over a time long compared to the wave period, and then repeated after an interval τ,

$$\langle \hat{E}^-(t)\hat{E}^+(t)\rangle = \langle \hat{E}^-(t+\tau)\hat{E}^+(t+\tau)\rangle, \qquad (19.34)$$

independent of τ.

19.2.1   Continuous variables

In beams from actual sources there is a spread of frequencies rather than the discrete set of frequencies tacitly assumed above. Wave vectors will point in one direction, along the z-direction, the waves being linearly polarized. Then the natural normalization volume V is a tube along the beam direction of area of cross-section $A\,\mathrm{m}^2$ and of length $L\,\mathrm{m}$.

The expressions for the electric field operators will be obtained by converting the expressions obtained above into ones using continuous variables. With the chosen normalization volume the wave vector spacing between travelling wave modes, Δk, is $2\pi/L$. It follows that the mode spacing in angular frequency, $\Delta\omega$, is $2\pi v/L$. We introduce operators $\hat{a}(\omega)$ and $\hat{a}^\dagger(\omega)$ appropriate for a continuous distribution. Then in order to have consistency, the expression for the number of photons in an angular frequency interval $\Delta\omega$ undergoes this conversion:

$$\hat{a}_\omega^\dagger \hat{a}_\omega \rightarrow \hat{a}^\dagger(\omega)\hat{a}(\omega)\Delta\omega. \qquad (19.35)$$

Thus

$$\hat{a}_\omega \rightarrow \sqrt{\Delta\omega}\,\hat{a}(\omega), \quad \text{and} \quad \hat{a}_\omega^\dagger \rightarrow \sqrt{\Delta\omega}\,\hat{a}^\dagger(\omega). \qquad (19.36)$$

Complementing this the summation over modes is replaced by an equivalent integration

$$\Delta\omega \sum_\omega \rightarrow \int d\omega, \qquad (19.37)$$

and

$$\delta_{\omega\omega'} \rightarrow \Delta\omega \delta(\omega - \omega'). \qquad (19.38)$$

Then it follows using eqn. 19.16 and this last result that

$$[\hat{a}(\omega), \hat{a}^\dagger(\omega')] = \delta(\omega - \omega'). \qquad (19.39)$$

For simplicity we consider the electric field expressed in eqn. 19.18 at the fixed point $z = 0$. When the photon distribution is continuous in frequency it is replaced by

$$\hat{E}^+(t) = [1/2\pi] \int_{-\infty}^{\infty} \chi_\omega \hat{a}(\omega) \exp(-i\omega t)d\omega, \qquad (19.40)$$

$$\hat{E}^-(t) = [1/2\pi] \int_{-\infty}^{\infty} \chi_\omega \hat{a}^\dagger(\omega) \exp(i\omega t)d\omega, \qquad (19.41)$$

where $\chi_\omega = 2\pi\zeta_\omega/\sqrt{\Delta\omega} = \sqrt{\pi\hbar\omega/[\varepsilon_0\varepsilon_r Av]}$.[4] When consideration is restricted to beams with a range of angular frequencies small compared to the central value ω_0 this is called the *narrow bandwidth* approximation. Then the term χ_ω can be regarded as a constant and taken outside the above integrals. With a beam directed along the z-direction the dependence on position and time is always through the combination $kz - \omega t = -\omega s$, so that the electric field operators away from $z = 0$ are readily evaluated by replacing t with s. The Fourier transform of $\hat{a}(\omega)$ is

$$\hat{a}(t) = [1/2\pi] \int_{-\infty}^{\infty} \hat{a}(\omega) \exp(-i\omega t)\,d\omega, \qquad (19.42)$$

which when substituted in the previous expressions gives the simpler, useful forms for narrow bandwidth beams

$$\hat{E}^+(t) = \chi_\omega \hat{a}(t), \qquad (19.43)$$

$$\hat{E}^-(t) = \chi_\omega \hat{a}^\dagger(t). \qquad (19.44)$$

[4]In vacuum $\chi_\omega = \sqrt{\pi\hbar\omega/[\varepsilon_0 Ac]}$.

The commutator involving these time dependent operators is

$$[\hat{a}(t), \hat{a}^\dagger(t')] = \delta(t - t')/2\pi. \tag{19.45}$$

For an *individual* component mode of angular frequency ω

$$\hat{a}(t + \tau) = \hat{a}(t)\exp(-i\omega\tau); \quad \hat{a}^\dagger(t + \tau) = \hat{a}^\dagger(t)\exp(i\omega\tau). \tag{19.46}$$

The expectation value of $\hat{a}^\dagger(\omega)\hat{a}(\omega)\mathrm{d}\omega$ is the number of photons with angular frequencies between ω and $\omega + \mathrm{d}\omega$ integrated over time. Therefore the integrated number of photons within this angular frequency range in a pulse having a state vector $|\psi\rangle$ is

$$f(\omega)\mathrm{d}\omega = \langle\psi|\hat{a}^\dagger(\omega)\hat{a}(\omega)|\psi\rangle\,\mathrm{d}\omega, \tag{19.47}$$

while the mean flux of photons in the pulse is

$$F = \int_0^\infty \mathrm{d}\omega\, f(\omega)/T, \tag{19.48}$$

where T is the pulse duration.

19.3   First order coherence

The rate at which photons are recorded by a photodetector can be obtained using Fermi's golden rule, eqn. G.6, derived in Appendix G. The matrix element given there splits into an atomic term involving the emission of an electron and an electromagnetic field term involving the annihilation of a photon. Thus

$$M_{if} = \langle\,[\text{electron} + \text{ion}]|\hat{O}|[\text{atom}]\,\rangle\,\langle f|\hat{E}^+|i\rangle \tag{19.49}$$

where $|i\rangle$ and $|f\rangle$ are respectively the initial and final states of the electromagnetic field alone, and $\hat{O}$ is the electron operator. Then the rate at which photons are absorbed by a detector is

$$\gamma = G \sum_f |\langle f|\hat{E}^+|i\rangle|^2, \tag{19.50}$$

where the contribution of the detector is absorbed into a constant G. G contains the product of the efficiency of the detector to detect radiation of the frequency incident and a geometric factor determined by the sensitive area of the detector and its location. The detector will need to be chosen so that the rate is useful, but thereafter G cancels from the quantities of interest used in measuring degrees of coherence. The above equation can be rewritten:

$$\gamma = G \sum_f \langle i|\hat{E}^-|f\rangle\,\langle f|\hat{E}^+|i\rangle. \tag{19.51}$$

The sum over accessible final states can be replaced by the sum over all states without affecting the equality because the additional states all have $\langle i|\hat{E}^-|f\rangle$ equal to zero. Then the closure relation eqn. E.10 gives

$$\gamma = G\,\langle i|\hat{E}^-\hat{E}^+|i\rangle. \qquad (19.52)$$

This is the key, simply interpreted result. It shows that the rate of detecting photons is proportional to the expectation value of the field operator product $\hat{E}^-\hat{E}^+$, and so to the number of photons in the initial state. For compactness, expectation values like $\langle i|\hat{E}^-\hat{E}^+|i\rangle$ will generally be truncated to $\langle\hat{E}^-\hat{E}^+\rangle$.

These results will now be applied to the familiar situation of Young's two slit experiment. The operator $\hat{E}^+$ can destroy a photon arriving from either slit, so we write

$$\hat{E}^+ = \hat{E}_1^+ + \hat{E}_2^+, \qquad (19.53)$$

where the subscripts label the slits. The intensity (the number of photons detected) at the screen is

$$\begin{aligned} \langle I\rangle &= G\langle\hat{E}^-\hat{E}^+\rangle = G\,[\,\langle\hat{E}_1^-\hat{E}_1^+\rangle + \langle\hat{E}_2^-\hat{E}_2^+\rangle + \langle\hat{E}_1^-\hat{E}_2^+ + \hat{E}_2^-\hat{E}_1^+\rangle\,] \\ &= \langle I_1\rangle + \langle I_2\rangle + 2G\,\Re e\langle\hat{E}_1^-\hat{E}_2^+\rangle, \end{aligned} \qquad (19.54)$$

where $\langle I_1\rangle$ and $\langle I_2\rangle$ are the intensities when one slit is illuminated and the other blocked. The quantum *degree of first order coherence* is defined in a way similar to the classical form so that the experiment dependent factor G is cancelled out:

$$g^{(1)}(1,2) = \frac{\langle\hat{E}_1^-\hat{E}_2^+\rangle}{\sqrt{\langle\hat{E}_1^-\hat{E}_1^+\rangle\langle\hat{E}_2^-\hat{E}_2^+\rangle}}. \qquad (19.55)$$

The value of the degree of first order coherence between a single stationary beam at times t and $t+\tau$ is

$$g^{(1)}(\tau) = \frac{\langle\hat{E}^-(t)\hat{E}^+(t+\tau)\rangle}{\langle\hat{E}^-(t)\hat{E}^+(t)\rangle}. \qquad (19.56)$$

A generally used definition of the coherence time of radiation takes the form

$$\tau_c = \int_{-\infty}^{+\infty} |g^{(1)}(\tau)|^2\,\mathrm{d}\tau. \qquad (19.57)$$

The coherence time is infinite for a pure sine wave since this is coherent everywhere.

Equations 19.55 and 19.56 have classical equivalents in eqns. 7.31 and 7.33 respectively. Classical time averages evolve to quantum expectation values.

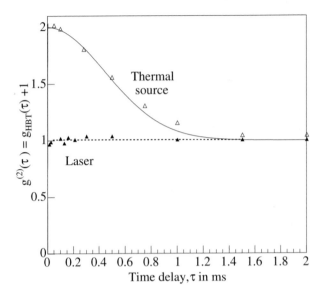

Fig. 19.1 The correlation between the arrival times of successive photons. Adapted from Arecchi, Gatti and Sona, *Physics Letters* 20, 27. Copyright 1966, with permission of Elsevier and Professor Arecchi.

19.4   Second order coherence

In 1954 Hanbury Brown and Twiss (HBT) made the quite surprising prediction that the arrival times at a detector of photons from a thermal source would show strong correlations. Explicitly the probability of successive photons arriving within a time short compared to the coherence time would be double that expected if the distribution of arrival times were random. Figure 19.1 shows results of a modern measurement of the time intervals between successive photons, in one case from a laser, and in another from a standard thermal source. A rise in probability is seen as the time interval between successive photons tends to zero, but only for the thermal source.

It seems paradoxical that in the case of the laser, whose first order coherence extends over a much longer period of time, the distribution of the arrival times of the photons is entirely random. Evidently second order coherence between pairs of photons is very different from first order coherence in which a photon interferes with itself. These correlations will be explained in quantum mechanical terms. Two preliminary steps toward the explanation are required: the correlations measured have to be expressed quantum mechanically, and the properties of thermal and laser light also need to be expressed in the quantum framework.

The matrix element for detecting photons at locations labelled 1 and 2 is $\langle f|\hat{E}_2^+\,\hat{E}_1^+|i\rangle$. Therefore the rate at which *coincidences* occur between photons arriving at the two separate detectors is given by

$$
\begin{aligned}
\gamma &= G\sum_f |\langle f|\hat{E}_2^+\,\hat{E}_1^+|i\rangle|^2 \\
&= G\sum_f \langle i|\hat{E}_1^-\,\hat{E}_2^-|f\rangle\langle f|\hat{E}_2^+\,\hat{E}_1^+|i\rangle \tag{19.58}
\end{aligned}
$$

where G includes the effect of the efficiency and the geometric area illuminated of each detector. Using the closure relation eqn. E.10 gives

$$
\gamma = G\,\langle i|\hat{E}_1^-\,\hat{E}_2^-\,\hat{E}_2^+\,\hat{E}_1^+|i\rangle. \tag{19.59}
$$

This ordering of the operators with the annihilation operators preceding (to the right of) the creation operators has emerged naturally and is known as *normal ordering*. In addition the annihilation operators are in chronological order and the creation operators in the reverse chronological order. This is known as being *time ordered*. In order to determine whether coincidences are occurring more or less frequently than the random rate, their rate is divided by the random rate. The random rate is obtained by taking the product of the individual count rates

$$
\gamma_0 = \langle i|\hat{E}_1^-\,\hat{E}_1^+|i\rangle\langle i|\hat{E}_2^-\,\hat{E}_2^+|i\rangle. \tag{19.60}
$$

The *degree of second order coherence* is then defined (leaving out the initial state label i) as:

$$
g^{(2)}(1,2) = \gamma/\gamma_0 = \frac{\langle\hat{E}_1^-\,\hat{E}_2^-\,\hat{E}_2^+\,\hat{E}_1^+\rangle}{\langle\hat{E}_1^-\,\hat{E}_1^+\rangle\langle\hat{E}_2^-\,\hat{E}_2^+\rangle}. \tag{19.61}
$$

Experimental factors such as the efficiency and size of each detector cancel out when this ratio is taken. When a single detector is used to detect correlations between photons separated by a fixed time interval, τ, the corresponding result is

$$
g^{(2)}(\tau) = \frac{\langle\hat{E}^-(t)\hat{E}^-(t+\tau)\hat{E}^+(t+\tau)\hat{E}^+(t)\rangle}{\langle\hat{E}^-(t)\hat{E}^+(t)\rangle\langle\hat{E}^-(t+\tau)\hat{E}^+(t+\tau)\rangle}. \tag{19.62}
$$

If the electromagnetic field is stationary so that the mean intensity remains constant

$$
g^{(2)}(\tau) = \frac{\langle\hat{E}^-(t)\hat{E}^-(t+\tau)\hat{E}^+(t+\tau)\hat{E}^+(t)\rangle}{\langle\hat{E}^-(t)\hat{E}^+(t)\rangle^2}, \tag{19.63}
$$

which is independent of t. The classical equivalents of these results have field values in place of the expectation values of operators.

In the narrow bandwidth approximation the electric field operators can be replaced using eqns. 19.43 and 19.44 to simplify the expressions for the degrees of first and second order coherence:

$$g^{(1)}(\tau) = \frac{\langle \hat{a}^\dagger(t)\hat{a}(t+\tau)\rangle}{\langle \hat{a}^\dagger(t)\hat{a}(t)\rangle}, \tag{19.64}$$

$$g^{(2)}(\tau) = \frac{\langle \hat{a}^\dagger(t)\hat{a}^\dagger(t+\tau)\hat{a}(t+\tau)\hat{a}(t)\rangle}{\langle \hat{a}^\dagger(t)\hat{a}(t)\rangle^2}. \tag{19.65}$$

19.5   Laser light and thermal light

Laser light and black body radiation are two extreme states of the electromagnetic field. A laser beam has very long wavepackets that approximate to pure sinusoidal waves; such beams are highly coherent. In contrast black body radiation consists of radiation of all wavelengths having short wavepackets with random phases. The radiation from most sources has a limited spectral range, but possesses statistical behaviour similar to black body radiation due to the random phases of the wavepackets. This radiation is therefore known as *thermal* or *chaotic* radiation. Those states of the electromagnetic field introduced in Section 19.2, $|n_i\rangle$, which contain a specific number, n_i, of photons in a mode i are known as *Fock states* and are mathematically simple. However special experimental techniques are needed to produce for example single photons on demand.[5] In the following two subsections the mathematical construction of coherent laser-like, and thermal states from simple Fock states will be described and their detailed coherence properties deduced.

[5]See Section 20.7.1.

Thermal radiation differs fundamentally from coherent radiation in not being in a *pure* quantum state. Pure states are linear sums of Fock states in the form $\sum_n p_n |n\rangle$ and the expectation of an observable A

$$\langle \hat{A}\rangle = \sum_{m,n} p_m^* p_n \langle m|\hat{A}|n\rangle \tag{19.66}$$

shows interference between the Fock states. On the other hand thermal radiation is an incoherent mixture of number states, whose probabilities are P_n. In this case the expectation value of the same observable is

$$\langle \hat{A}\rangle = \sum_n P_n \langle n|\hat{A}|n\rangle, \tag{19.67}$$

where there is no interference between the number states.

19.5.1   Coherent (laser-like) states

The coherence time of a very well stabilized laser can last for one second or longer, so that the beam emitted is nearly a pure sinusoidal wave extending over a distance that would reach to the Moon. Its electric field over this interval is of the form $a \exp(i\xi_\omega)$. It is tempting to infer that coherent quantum state are eigenstates of the electric field operator. In fact they are eigenstates of its positive frequency component, $E^+(t)$; that is to say they are eigenstates of the annihilation operator. The exact form of the state vector for a coherent state is

$$|\alpha\rangle = \exp\left(-|\alpha|^2/2\right) \sum_{n=0}^{\infty} (\alpha^n/\sqrt{n!})|n\rangle, \tag{19.68}$$

where $\alpha = |\alpha| \exp(i\theta)$ can be any complex number. The sum over Fock states appearing in eqn. 19.68 shows that the number of photons in a coherent state is variable. This does not however mean that in a given laser beam the number of photons is varying: rather it means that the ensemble of laser beams described by one coherent state vector contain different numbers of photons. The indeterminacy is like that affecting the impact point of a photon in Young's double slit experiment. The distribution of the number of photons around the mean is random, with a Poissonian distribution. The properties of these coherent states will now be examined in more detail.

The coherent states are normalized

$$\langle\alpha|\alpha\rangle = \exp\left(-|\alpha|^2\right) \sum_n \left[(\alpha^*)^n \alpha^n/n!\right] = 1, \tag{19.69}$$

where the second equality uses the expansion $\exp x = 1 + x + x^2/2! + \dots$ Then the effect of the annihilation operator is

$$\hat{a}_\omega |\alpha\rangle = \exp\left(-|\alpha|^2/2\right) \sum_{n=1}^{\infty} [\alpha^n/\sqrt{(n-1)!}]|(n-1)\rangle$$

$$= \alpha \exp\left(-|\alpha|^2/2\right) \sum_{n=0}^{\infty} [\alpha^n/\sqrt{n!}]|n\rangle$$

$$= \alpha |\alpha\rangle. \tag{19.70}$$

Similarly

$$\langle\alpha| \hat{a}_\omega^\dagger = \alpha^* \langle\alpha|. \tag{19.71}$$

Using eqn 19.18, the effect of the electric field operator $\hat{E}^+$ on a coherent state is

$$\hat{E}^+ |\alpha\rangle = \zeta_\omega \exp(i\xi_\omega) \hat{a}_\omega |\alpha\rangle = \alpha \zeta_\omega \exp(i\xi_\omega) |\alpha\rangle, \tag{19.72}$$

which confirms the earlier presumption that a coherent state is an eigenstate of the (positive frequency component of the) electric field operator. The hermitian conjugate relation is

$$\langle\alpha| \hat{E}^- = \langle\alpha|\zeta_\omega \exp(-i\xi_\omega) \alpha^*. \tag{19.73}$$

Hence the expectation value for the electric field is

$$\langle\alpha|\hat{E}^+ + \hat{E}^-|\alpha\rangle = 2 \zeta_\omega|\alpha| \cos(\xi_\omega + \theta). \tag{19.74}$$

As was intended the choice of coefficients in eqn. 19.68 has produced a state for which the expectation value of the electric field is a stable sinusoidal wave. The probability that there are exactly n photons in this state is

$$P(n) = |\langle n|\alpha\rangle|^2 = \exp\left(-|\alpha|^2\right)|\alpha|^{2n}/n!, \tag{19.75}$$

which is a Poissonian distribution for a mean of $|\alpha|^2$. Although the coherent states are normalized they are not orthogonal

$$\langle\alpha|\beta\rangle = \exp\left[(-|\alpha|^2 - |\beta|^2 + 2\alpha^*\beta)/2\right]. \tag{19.76}$$

This result reflects the fact that coherent states drawn from the ensembles described by $|\alpha\rangle$ and $|\beta\rangle$ may contain the same number of photons. The expectation value of the photon number squared

$$\langle n^2\rangle = \langle\alpha|\,\hat{a}^\dagger\hat{a}\hat{a}^\dagger\hat{a}\,|\alpha\rangle = |\alpha|^2\langle\alpha|\,\hat{a}\hat{a}^\dagger\,|\alpha\rangle, \tag{19.77}$$

where eqns. 19.70 and 19.71 have both been used. Then

$$\begin{aligned}\langle n^2\rangle &= |\alpha|^2\langle\alpha|\,[\hat{a}^\dagger\hat{a} + 1]\,|\alpha\rangle\\ &= |\alpha|^4 + |\alpha|^2\\ &= \langle n\rangle^2 + \langle n\rangle,\end{aligned} \tag{19.78}$$

and the variance

$$\Delta n^2 = \langle n^2\rangle - \langle n\rangle^2 = \langle n\rangle, \tag{19.79}$$

as expected for a Poissonian distribution. The variance quantifies the particle noise produced, for example the noise at a detector due to the variation in the number of photons arriving in a given time interval. It is identically the shot noise already met in Chapter 15. The squared electric field expectation is evaluated similarly, giving

$$\langle E^2\rangle = \langle E\rangle^2 + \zeta_\omega^2. \tag{19.80}$$

with variance $\Delta E^2 = \zeta_\omega^2$. This result is independent of the number of photons so that as the number of photons ($|\alpha|^2$) increases the uncertainty becomes less significant and the approximation to a pure sinusoidal wave improves. Figure 19.2(a) shows the time dependence of the electric field with its uncertainty indicated by upper and lower limits. The same electric field vector and its uncertainty are also shown on an Argand diagram in figure 19.3. The uncertainty in the phase angle $\phi = (\xi_\omega - \theta)$ when this angle is close to $0°$ can be written[6]

$$\Delta\phi = 1/(2\sqrt{n}), \tag{19.81}$$

hence the product $\Delta\phi\Delta n = 1/2$. ϕ and n are conjugate variables in the same sense that momentum and position are conjugate variables. For states of the electromagnetic field in general

$$\Delta\phi\Delta n \geq 1/2, \tag{19.82}$$

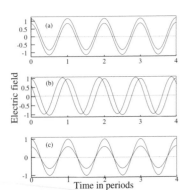

Fig. 19.2 The uncertainty spread in (a) a coherent wave; (b) the same wave after squeezing the amplitude uncertainty; (c) the same wave after squeezing the phase uncertainty. In each case the waves lie in the space between the two extremes shown.

---

[6]See Chapter 4 of the third edition of *The Quantum Theory of Light* by R. Loudon, published by Oxford University Press (2000).

which is the form the Heisenberg uncertainty principle takes for this pair of conjugate variables. The definition of this quantum optical phase is full of pitfalls and the reader should refer to advanced texts for details.[7] In the case of the simple Fock states the number of photons is exact but the phase of the field is completely indeterminate. It is possible by manipulating coherent beams to make the phase uncertainty smaller while keeping the product equal to $1/2$. Such states are known as *squeezed* states. Figure 19.2(b) shows the limits of a coherent wave after the amplitude uncertainty has been squeezed, while (c) shows the corresponding plot for a coherent wave after the phase uncertainty has been squeezed.

When a mode contains a large number of coherent photons, m, the equivalent classical electric field $E_{\text{cl}} \cos(kx - \omega t)$ can be deduced using eqn. 19.74,

$$E_{\text{cl}} = 2\zeta_\omega \sqrt{m} = \sqrt{(2m\hbar\omega/\varepsilon_0\varepsilon_\text{r}V)}. \tag{19.83}$$

Then $\sqrt{2\hbar\omega/\varepsilon_0\varepsilon_\text{r}V}$ is in some sense the *electric field per photon*.

Input for determining the degree of first order coherence of a coherent state is obtained using eqns. 19.72 and 19.73. Then

$$\langle \hat{E}^-(t)\hat{E}^+(t+\tau)\rangle = |\alpha|^2\zeta_\omega^2 \exp(-i\omega\tau). \tag{19.84}$$

Thus the degree of first order coherence is

$$g^{(1)}(\tau) = \frac{\langle \hat{E}^-(t)\hat{E}^+(t+\tau)\rangle}{\langle \hat{E}^-(t)\hat{E}^+(t)\rangle} = \exp(-i\omega\tau), \tag{19.85}$$

and the degree of second order coherence is

$$g^{(2)}(\tau) = \frac{\langle \hat{E}^-(t)\hat{E}^-(t+\tau)\hat{E}^+(t+\tau)\hat{E}^+(t)\rangle}{\langle \hat{E}^-(t)\hat{E}^+(t)\rangle^2} = 1, \tag{19.86}$$

which shows that light in the state described by eqn. 19.68 is indeed fully coherent in first and second order.

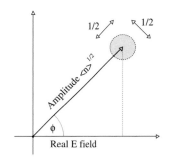

Fig. 19.3 Argand diagram showing the uncertainty in the electric field amplitude for a coherent state.

19.5.2   Thermal light

Thermal light includes black body radiation as well as radiation from sources with a restricted frequency range in which the photons at any frequency have random phase relative to each other. In Chapter 12 it was shown that in black body radiation the probability of having n photons in a given mode at temperature T K is

$$P(n) = \exp(-n\hbar\omega/k_\text{B}T)[1 - \exp(-\hbar\omega/k_\text{B}T)], \tag{19.87}$$

and the mean is

$$\langle n \rangle = 1/[\exp(\hbar\omega/k_\text{B}T) - 1]. \tag{19.88}$$

---

[7]See *Measuring the Quantum State of Light* by Ulf Leonhardt, Cambridge University Press (1997).

It follows that the distribution can be expressed alternatively as

$$P(n) = \langle n \rangle^n / [\langle n \rangle + 1]^{n+1}. \tag{19.89}$$

This form is therefore valid for the number of photons within a single mode from a thermal source. Putting $z = \langle n \rangle/(1 + \langle n \rangle)$ we have

$$\langle n^2 \rangle = \sum_n P(n) n^2$$
$$= (1 + \langle n \rangle))^{-1} \sum_n z^n n^2. \tag{19.90}$$

Then using $\sum_n z^n = 1/(1-z)$ and

$$\sum z^n n^2 = z \frac{d}{dz} \left[z \frac{d}{dz} \sum z^n \right],$$

we obtain

$$\langle n^2 \rangle = 2\langle n \rangle^2 + \langle n \rangle, \tag{19.91}$$

and the variance

$$\Delta n^2 = \langle n \rangle^2 + \langle n \rangle. \tag{19.92}$$

The second term in the variance is the particle noise, which also appeared in the variance of the photon number count of coherent states, while the first term is called the *wave noise*. Figure 19.4 shows the distribution in the number of photons from a laser and a thermal source in a single mode when they both have an average photon count of 5 in that mode.

A simple relation can now be derived between the degrees of first and second order coherence for thermal radiation. The second order correlation contains contributions of the form:

$$M = \langle \hat{a}^\dagger(t)\hat{a}^\dagger(t+\tau)\hat{a}(t+\tau)\hat{a}(t) \rangle. \tag{19.93}$$

Non-zero contributions are only possible when the pair of photons created enter exactly the same two modes from which photons were annihilated. Otherwise the initial state is not recovered after the annihilations and creations, and thus the amplitude is zero. The operator pairings that can satisfy this requirement are: (a) first/third and second/fourth; or (b) first/fourth and second/third. In pairing (a) $\hat{a}^\dagger(t)\hat{a}(t+\tau)$ and $\hat{a}^\dagger(t+\tau)\hat{a}(t)$ involve different[8] modes which are incoherent so the amplitude factorizes to $\langle \hat{a}^\dagger(t)\hat{a}(t+\tau) \rangle \langle \hat{a}^\dagger(t+\tau)\hat{a}(t) \rangle$. Pairing (b) factorizes for the same reason. Thus

$$M = \langle \hat{a}^\dagger(t)\hat{a}(t) \rangle \langle \hat{a}^\dagger(t+\tau)\hat{a}(t+\tau) \rangle + \langle \hat{a}^\dagger(t)\hat{a}(t+\tau) \rangle \langle \hat{a}^\dagger(t+\tau)\hat{a}(t) \rangle.$$

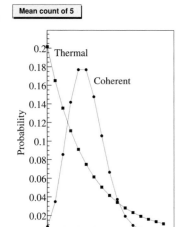

Mean count of 5

Fig. 19.4 The distribution in counts for a thermal and a coherent source when the average count is 5 in each case.

---

[8]The likelihood of the modes being the same is negligibly small because the choice is a random one among the many modes in thermal radiation

Then because the radiation is stationary $\langle \hat{a}^\dagger(t+\tau)\hat{a}(t+\tau)\rangle = \langle \hat{a}^\dagger(t)\hat{a}(t)\rangle$ and hence

$$M = \langle \hat{a}^\dagger(t)\hat{a}(t)\rangle^2 + |\langle \hat{a}^\dagger(t)\hat{a}(t+\tau)\rangle|^2. \tag{19.94}$$

Then applying eqns. 19.64 and 19.65 the degree of second order coherence

$$g^{(2)}(\tau) = 1 + |g^{(1)}(\tau)|^2, \tag{19.95}$$

which is the general result for radiation from a thermal source. The typical Lorentzian line shape in intensity of a source containing isolated stationary atoms was introduced in eqn. 7.54:

$$P(\omega) = \frac{\gamma/2\pi}{(\omega - \omega_0)^2 + \gamma^2/4} \tag{19.96}$$

where ω_0 is the central angular frequency, γ is the full width at half maximum. The time dependent correlation is the Fourier transform of this

$$\langle \hat{a}^\dagger(t)\hat{a}(t+\tau)\rangle = \int d\omega P(\omega)\exp(-i\omega\tau), \tag{19.97}$$

apart from a constant. Using tables of Fourier transforms[9] gives

$$\langle \hat{a}^\dagger(t)\hat{a}(t+\tau)\rangle = \exp(-i\omega_0\tau - \gamma|\tau|/2). \tag{19.98}$$

Thus the degree of first order coherence is

$$g^{(1)}(\tau) = \frac{\langle \hat{a}^\dagger(t)\hat{a}(t+\tau)\rangle}{\langle \hat{a}^\dagger(t)\hat{a}(t)\rangle} = \exp(-i\omega_0\tau - \gamma|\tau|/2). \tag{19.99}$$

Then making use of the general relationship between the degrees of second and first order coherence 19.95 gives

$$g^{(2)}(\tau) = 1 + |g^{(1)}(\tau)|^2 = 1 + \exp(-\gamma|\tau|). \tag{19.100}$$

At zero time difference this correlation coefficient is exactly 2, which comes about because there are exactly two possible pairings of the photons.

In most sources the atoms are close enough to interact with each other. Equally the atoms are in motion and their radiation is as a result Doppler shifted. Both these effects lead to a broader emission line which generally has a Gaussian or near Gaussian shape. The power spectrum is then given by eqn. 7.56

$$f(\omega) = \exp[-(\omega - \omega_0)^2/(2\sigma^2)]/(\sqrt{(2\pi)}\sigma), \tag{19.101}$$

---

[9]See page 359 of the fifth edition of *Table of Integrals, Series, and Products* by I. S. Gradshteyn and I. M. Ryzhik, edited by A. Jeffrey, and published by Academic Press, New York (1994).

where σ is the rms width of the line. Then repeating the above analysis for the Gaussian line shape gives the corresponding time dependent correlation

$$\langle \hat{a}^\dagger(t)\hat{a}(t+\tau)\rangle = \exp\left(-i\omega_0\tau - \sigma^2\tau^2/2\right), \tag{19.102}$$

apart from a constant, and

$$g^{(2)}(\tau) = 1 + \exp\left(-\sigma^2\tau^2\right). \tag{19.103}$$

At zero time difference this too gives a correlation of exactly 2. The correlation will have a similar dependence on τ when the radiation is filtered out from a source having a broad spectrum, that is to say *chaotic thermal radiation.*

The simple factor of 2 difference between the values of $g^{(2)}(0)$ for laser and thermal light can be explained directly from the Bose–Einstein statistics of photons. This requires that the wavefunction for two photons is overall symmetric under their interchange. For thermal light we have

$$\phi(x_1, x_2) = [\psi_1(x_1)\psi_2(x_2) + \psi_1(x_2)\psi_2(x_1)]/\sqrt{2},$$

where $\psi_{1/2}$ are orthonormal wavefunctions. Then the probability of detecting both photons at the same point, x, is

$$P(x, x) = |\phi(x, x)|^2 = 2|\psi_1(x)\psi_2(x)|^2 = 2P_1(x)P_2(x).$$

In the case of laser light the photons have the same wavefunction so that

$$\phi(x_1, x_2) = \psi(x_1)\psi(x_2),$$
$$\text{and } P(x, x) = |\psi(x)\psi(x)|^2 = [P(x)]^2,$$

hence the factor 2.

19.6   Observations of photon correlations

An experiment of the sort originally carried out by Hanbury Brown and Twiss to observe correlations between photons from a thermal source is shown in figure 19.5. Light from a source was filtered to isolate a spectral line and was focused on a pinhole. The emerging thermal light, having a narrow spread in wavelength, was divided by a beam splitter into two equal beams that illuminated separate photomultipliers. The electronic signals from the detectors were passed through separate bandwidth limited amplifiers which removed the steady time-average component of the signals but unfortunately also removed most of the wanted high frequency signal. The size of the aperture, the sensitive area of the detectors and their spacing was such that the detectors covered the transverse coherence area of the aperture and so accepted exactly one transverse mode as explained in Section 5.5.3. The electrical outputs from the detectors were multiplied electronically and this product integrated over a fixed time giving a final output proportional to the time-average. In their experiment Hanbury Brown and Twiss moved PM2 laterally and showed that when the two coherence areas overlapped there were correlations between the intensities detected by PM1 and PM2. As the overlap diminished so the correlations died away.

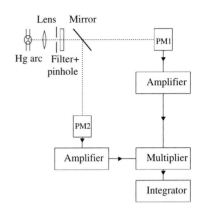

More recent versions of the experiment have a similar configuration, but use modern electronics, with which a variable time delay can be imposed between the signals from PM1 and PM2 before multiplication. In this case the coherence areas coincide throughout the measurements. The signal from one detector is delayed by a time τ so that instantaneously the multiplier output is $[I(t) - \langle I \rangle][I(t+\tau) - \langle I \rangle]$ and its time average is $\langle I(t)I(t+\tau) \rangle - \langle I \rangle^2$. This is compared to the mean squared intensity at one detector $\langle I \rangle^2$, giving

$$g_{\text{HBT}}(\tau) = \frac{\langle I(t)I(t+\tau) \rangle}{\langle I \rangle^2} - 1. \tag{19.104}$$

Fig. 19.5 The experimental apparatus used by Hanbury Brown and Twiss to observe the photon correlations from a thermal source.

This is seen to be $g^{(2)}(\tau) - 1$. Then for a chaotic source with a Gaussian spectrum we can use the prediction of eqn. 19.103

$$g_{\text{HBT}-\text{thermal}}(\tau) = \exp\left(-\sigma^2 \tau^2\right), \tag{19.105}$$

while for a coherent (laser) source eqn. 19.86 gives

$$g_{\text{HBT}-\text{coherent}}(\tau) = 0. \tag{19.106}$$

These predictions are shown in figure 19.1 together with experimental data taken in an experiment using laser and thermal sources; the agreement is excellent. There are no correlations in the case of the coherent beam so there is no bunching. However with incoherent beams there is more randomness in the arrival times and hence *photon bunching* will occur. When the time delay is reduced to zero the degree of second order coherence, $g^{(2)}(0)$, for a thermal source is thus exactly twice that for a laser. At the other extreme with a time delay larger than the coherence time the expectation value $\langle \hat{a}^\dagger(t+\tau)\hat{a}(t) \rangle$ vanishes and as a result the correlations disappear.

19.6.1   Stellar correlation interferometer

After the successful detection of correlations with thermal beams Hanbury Brown and Twiss built a stellar interferometer to make measurements of the angular sizes of stars by means of photon correlations. In their measurements in the 1950s they achieved a resolution far exceeding that possible with Michelson stellar interferometers. As shown in figure 19.6 two large searchlight mirrors were used to focus starlight onto separate photomultipliers PM_1 and PM_2 located at $\mathbf{r_1}$ and $\mathbf{r_2}$ respectively. The time average product of the signals was divided by the product of the time averages of the individual signals, which is essentially the ratio of the number of coincidences between photons arriving at the detectors

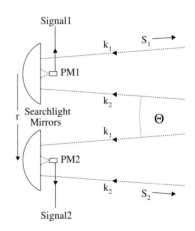

Fig. 19.6 Stellar correlation interferometer.

to the product of the numbers arriving at the two detectors.

For the moment suppose the interferometer views a pair of point sources, S_1 and S_2. A coincidence is observed if one photon with wave vector $\mathbf{k_1}$ from S_1 has arrived at PM_1 and one with wave vector $\mathbf{k_2}$ from S_2 at PM_2 or vice versa. These two possibilities are indistinguishable, and hence the amplitude for coincidences is the sum of these two amplitudes

$$\exp\left(i\mathbf{k}_1 \cdot \mathbf{r}_1 + i\mathbf{k}_2 \cdot \mathbf{r}_2\right) + \exp\left(i\mathbf{k}_2 \cdot \mathbf{r}_1 + i\mathbf{k}_1 \cdot \mathbf{r}_2\right)$$
$$= \exp\left(i\mathbf{k}_1 \cdot \mathbf{r}_1 + i\mathbf{k}_2 \cdot \mathbf{r}_2\right)\left[1 + \exp\left(i(\mathbf{k}_2 - \mathbf{k}_1) \cdot (\mathbf{r}_1 - \mathbf{r}_2)\right)\right],$$

ignoring overall constants. The rate of coincidences is proportional to the modulus squared

$$R = |1 + \exp\left(i(\mathbf{k}_2 - \mathbf{k}_1) \cdot (\mathbf{r}_1 - \mathbf{r}_2)\right)|^2/2 = 1 + \cos\left(k\Delta r\theta\right)$$

where the separation laterally of the detectors is Δr and the sources subtend an angle θ at the detectors. This treatment has ignored coincidences between two photons from the same source, one arriving at each detector. For a thermal source the inclusion of this contribution changes the rate to

$$R = 3 + \cos\left(k\Delta r\theta\right) = 1 + \cos^2\left(k\Delta r\theta/2\right). \tag{19.107}$$

The correlation pattern observed when viewing a star requires this result to be integrated over the area of the star.

Michelson's stellar interferometer described in Chapter 7 gives similar information from the first order interference pattern, but those measurements are very sensitive to changes in Δr. If Δr changes by even half a wavelength during the measurement it is enough to erase the phase information on which the basic Michelson inteferometer measurement relies. The HBT interferometer only requires that the path changes should be much smaller than the coherence length for the light passing the filters used. However astronomers later devised ways described in Section 8.6 which removed the difficulties with the basic Michelson stellar interferometer. The stellar correlation interferometer only enjoyed a temporary vogue, during which detector separations of up to 200 m gave an angular resolution of $2\,10^{-9}$ rad.

19.7   Entangled states

These are states of radiation – or material particles – which are only conceivable within quantum theory; they have no classical counterpart. A Young's slit experiment provides a simple example. If the flux is kept low enough that only a single photon is present in the apparatus at any given moment the electromagnetic field at the slits is in the entangled state

$$|\psi\rangle = [\,|1\rangle_1\,|0\rangle_2 + |0\rangle_1\,|1\rangle_2\,]/\sqrt{2}, \qquad (19.108)$$

where the subscripts indicate the slit (1 or 2) and the arguments indicate the number of photons at each slit (0 or 1). The entities entangled are the electromagnetic fields at the two slits and the overall state vector exhibits the purely quantum feature that the photon is in one or other state with probability 0.5. It was shown in Chapter 7 that if measurements are made to determine which slit each photon passes through, then the interference pattern disappears. If the photon is located at slit 1 then the overall state vector becomes $|\psi'\rangle = |1\rangle_1|0\rangle_2$; whilst if the photon is located at slit 2 the state vector is $|\psi''\rangle = |0\rangle_1|1\rangle_2$. Once the photon is located at a slit there is no entanglement and no *two slit* interference. When two or more photons are entangled the correlations between them must be measured to reveal their entanglement. Examples are given in the following sections.

19.7.1 Beam splitters

The symmetric, 50:50, non-absorbing beam splitter described in Section 9.7.2 is the standard *passive* device for producing entangled states. Figure 19.7 shows the input and output *ports* labelled with field operators. The operator $\hat{a}_i$ annihilates a photon at port i, and $\hat{a}_i^\dagger$ is the corresponding creation operator. Then using eqns. 9.115, 9.116 and 9.118 from Section 9.7.2 and replacing electric fields by operators

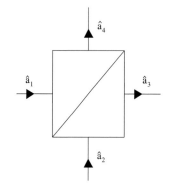

Fig. 19.7 Beam splitter with annihilation operators for input and output ports.

$$\hat{a}_3 = (\hat{a}_1 + i\hat{a}_2)/\sqrt{2}, \qquad (19.109)$$
$$\hat{a}_4 = (\hat{a}_2 + i\hat{a}_1)/\sqrt{2}, \qquad (19.110)$$

using the amplitudes given in eqn. 9.118. It is important to include the field at *both* input ports, even when only one port is illuminated because the vacuum cannot be ignored as it was in a classical analysis. As usual

$$[\hat{a}_i, \hat{a}_i^\dagger] = 1. \qquad (19.111)$$

The field operators at the input ports are certainly independent, hence

$$[\hat{a}_1, \hat{a}_2^\dagger] = 0. \qquad (19.112)$$

Equally those at output ports should be independent. Using eqns. 19.109 and 19.110 we can check that this is so:

$$[\hat{a}_3, \hat{a}_4^\dagger] = 0. \qquad (19.113)$$

Note that this equality only holds because the field operators at *both* input ports are included in eqns. 19.109 and 19.110. This confirms what was expressed above about the necessity of including the vacuum field in quantum analyses. Suppose port 1 only is illuminated. The state vector of the incident radiation is

$$|1\rangle_1|0\rangle_2 = \hat{a}_1^\dagger|0\rangle_1|0\rangle_2. \qquad (19.114)$$

From eqns 19.109 and 19.110 $\hat{a}_1 = (\hat{a}_3 - i\hat{a}_4)/\sqrt{2}$, and hence $\hat{a}_1^\dagger = (\hat{a}_3^\dagger + i\hat{a}_4^\dagger)/\sqrt{2}$. The corresponding emerging state is thus

$$[(\hat{a}_3^\dagger + i\hat{a}_4^\dagger)/\sqrt{2}]\,|0\rangle_3|0\rangle_4 = (|1\rangle_3|0\rangle_4 + i|0\rangle_3|1\rangle_4)/\sqrt{2} \qquad (19.115)$$

which is evidently entangled.

19.7.2   Spontaneous parametric down conversion

Experiments to illuminate fundamental concepts in quantum mechanics have made much use of states of the electromagnetic field in which two photons have been entangled. These are also states which offer one starting point for the development of practical quantum cryptography. A well tried *active* means for generating entangled pairs of photons at an adequate rate for experiments is spontaneous parametric down conversion (SPDC). This process is the reverse of the process for frequency summing introduced in Section 14.14. Intense coherent radiation from a laser, the *pump*, impinges on a non-linear crystal whose susceptibility increases with the electric field strength. The process of interest involves the absorption of a *pump* photon (p) and the simultaneous emission of a pair of photons. The latter are called the *signal* (s) and *idler* (i) with by convention $\omega_s \geq \omega_i$. Conservation of energy requires that the angular frequencies satisfy the relation

$$\omega_p = \omega_s + \omega_i, \qquad (19.116)$$

and conservation of momentum requires that in addition

$$\mathbf{k}_p = \mathbf{k}_s + \mathbf{k}_i. \qquad (19.117)$$

This *phase matching* of the pump and outgoing waves means that at any point $(\mathbf{r},\, t)$

$$\exp\left[i(\mathbf{k}_s \cdot \mathbf{r} - \omega_s t)\right]\exp\left[i(\mathbf{k}_i \cdot \mathbf{r} - \omega_i t)\right] = \exp\left[i(\mathbf{k}_p \cdot \mathbf{r} - \omega_p t)\right],$$
$$(19.118)$$

so that in classical terms the product of the outgoing waves matches the input wave. The severity of the momentum matching requirement becomes clear when eqn. 19.117 is rewritten in terms of the refractive indices at the different frequencies

$$\hat{\mathbf{k}}_p n_p \omega_p = \hat{\mathbf{k}}_s n_s \omega_s + \hat{\mathbf{k}}_i n_i \omega_i, \qquad (19.119)$$

where $\hat{\mathbf{k}}_p$ is the unit vector along the pump direction and n_p is the refractive index at the pump frequency. However, as we saw in Section 14.14, the refractive index of birefringent materials varies with direction and polarization so that phase matching is achievable. The use of β-barium borate (BBO), a negative uniaxial crystalline material, was already discussed there. BBO has many other practical advantages. Phase matching with high yields per pump photon is possible over the range 210–3000 nm and it has low absorption ($< 5\%$/cm) over most of

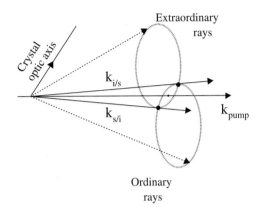

Fig. 19.8 Type II degenerate SPDC. The circles are where the ordinary and extraordinary rays of a given, in this case equal, wavelength cross the screen. The entangled pair selected by pinholes are labelled with wave vectors $\mathbf{k}_{i/s}$ and $\mathbf{k}_{s/i}$. The dotted lines indicate the plane containing the optic axis and beam.

this range. BBO crystals withstand high pump photon fluxes and can be made very large (5 cm long) and uniform, and are stable in air when anti-reflection coated. Another comparably useful material is potassium di-hydrogen phosphate (KDP).

The pump beam is usually polarized in the plane containing the optic axis of the crystal. Then alternative choices for the polarization of the signal and idler can be selected. In type I SPDC the signal and idler are both plane polarized perpendicular to the pump. In type II SPDC one photon has the same polarization as the pump (extraordinary ray), the other is polarized perpendicularly (ordinary ray). It is possible in addition, with an appropriately cut crystal, to choose the beam directions so that the signal and idler have the same frequency. This is called *degenerate* SPDC.

Figure 19.8 shows the layout for type II SPDC, for example in a BBO crystal pumped by 351.1 nm UV from an argon–ion laser. Momentum matching requires that the signal photon direction must lie somewhere on a cone whose apex is the conversion point, and the idler follows a complementary path on another such cone. The cone angles depend on the signal and idler wavelengths. In the figure a particularly important configuration is shown in which the cones for degenerate type II SPDC cross one another. Pinholes in an opaque screen placed at the cross-over points select signal and idler photon pairs. Interference filters placed behind the pinholes then limit the selection to photons of nearly the same frequency (degenerate).

One pinhole will pass an idler photon and the other a signal photon; but which goes through which hole is undetermined. The photons are therefore entangled and because they have different polarization (ordinary and extraordinary) they are said to be *polarization* entangled. Choosing a thin crystal, so that no phase lag develops between the ordinary and extraordinary rays, their overall state vector is

$$[\,|E\rangle_1|O\rangle_2 + |O\rangle_1|E\rangle_2\,]/\sqrt{2}. \tag{19.120}$$

Any phase lag at the exit surface of the crystal due to the difference between the phase velocities of the ordinary and extraordinary rays can be compensated by passing the signal and idler beams first through half-wave plates so that they exchange polarizations and then through a second BBO crystal identical to that used for SPDC. By using appropriately aligned quarter and half-wave plates other entangled states are readily obtained.

In order for there to be second order coherence between the two beams the coherence time should be longer than the time interval between the detection of the signal and idler photons. A typical filter bandwidth employed is $10\,\mathrm{nm}$, giving a coherence time of $\sim\!100\,\mathrm{fs}$.

The beauty of SPDC is that it provides entangled pairs of photons at a high controllable rate at wavelengths which the experimenter can select in the visible, near UV or near IR region. In the following sections several experiments using SPDC and beamsplitters will be used to illustrate the quantum principles underlying first and second order interference effects.

19.8   The HOM interferometer

The interferometer shown in figure 19.9 was used by Hong, Ou and Mandel to show directly the *indistinguishability* of photons. They injected degenerate signal and idler photons from type I SPDC into the opposite entry faces of a 50:50 beam splitter BS. The photons have the same polarization. From eqns. 19.109 and 19.110 it follows that

$$\hat{a}_1 = (\hat{a}_3 - i\hat{a}_4)/\sqrt{(2)}, \tag{19.121}$$
$$\hat{a}_2 = (\hat{a}_4 - i\hat{a}_3)/\sqrt{(2)}. \tag{19.122}$$

Consequently

$$\hat{a}_1^\dagger \hat{a}_2^\dagger = (\hat{a}_3^\dagger + i\hat{a}_4^\dagger)(\hat{a}_4^\dagger + i\hat{a}_3^\dagger)/2$$
$$= i(\hat{a}_3^\dagger \hat{a}_3^\dagger + \hat{a}_4^\dagger \hat{a}_4^\dagger)/2. \tag{19.123}$$

Thus the state prepared when the photons exit from the beam splitter will be

$$|1\rangle_1|1\rangle_2 = \hat{a}_1^\dagger \hat{a}_2^\dagger |0\rangle = i[\,|2\rangle_3|0\rangle_4 + |0\rangle_3|2\rangle_4\,]/2. \tag{19.124}$$

As a result both photons should exit through the same face of the beam splitter, and Hong, Ou and Mandel sought to verify this prediction. A complementary way to approach this result is to recall that the phase difference between the reflected and transmitted light is $\pi/2$ so that the amplitudes for two reflections and for two transmissions differ in phase by π and so cancel precisely. In either view the underlying idea being tested is that the amplitudes may be added because the idler and signal photons are *indistinguishable*.

351.1 nm radiation from an argon–ion laser induces type I SPDC in a KDP crystal labelled NL. In type I SPDC the signal and idler have the same (ordinary) polarization. M1 and M2 are mirrors and BS is a beam splitter which can be displaced as indicated by the arrows. D3 and D4 are fast photomultipliers. In front of each of these are pinholes and interference filters, labelled IF, which define the degenerate signal and idler beams. The electronic pulses associated with photons arriving at D3 and D4 were counted separately, and also the coincidences between pulses at D3 and D4.

When the beam splitter is exactly centred so that the apparatus is left–right symmetric both photons are expected to arrive at D4 or both at D3, never one at D3 and one at D4. D3/D4 coincidences become possible once the beam splitter is displaced sideways, as indicated by the two short arrows in the diagram. Figure 19.10 shows the D3/D4 coincidence rate as a function of the beam splitter position. When its position is symmetric with respect to the detectors the only coincidences remaining come from a slight lack of overlap between the idler and signal fields at the pinholes.

When the beam splitter is moved sideways the reflection paths change in relative length by some distance $s = c\tau$. In this case the coincidence rate becomes

$$P(\tau) \propto 1 - \exp\left[-(\tau\Delta\omega)^2\right], \tag{19.125}$$

where $\Delta\omega$ is the bandwidth of the interference filters. It confirms that $P(\tau)$ vanishes when the path difference, and τ, go to zero. In figure 19.10 the coincidence rate rises as the path difference increases from zero until eventually when the time difference is greater than the coherence time the rate becomes constant. The fit with eqn. 19.125 shown in figure 19.10 gives the bandwidth of the signal and idler radiation. At half-height the width of the dip in the figure is 16 μm, corresponding to a time delay of 100 fs, which is the coherence time, or equivalently the time duration of the photon wavepackets. Unlike measurements of coherence time using first order coherence, this technique does not require path differences to be held constant to a fraction of a wavelength during the measurement.

In modern versions of the HOM interferometer the light may be steered by monomode optical fibres rather than mirrors, while the beam splitter

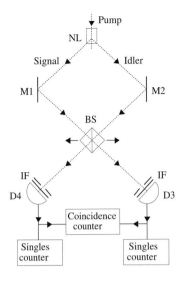

Fig. 19.9 HOM interferometer used to study photon correlations. NL is the non-linear crystal. BS is a 50:50 beam splitter with arrows showing the motion used. The interference filter IF and pinholes select signal and idler.

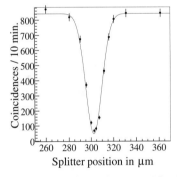

Fig. 19.10 Coincidences with the HOM interferometer as a function of the beam splitter position. The experimental results are compared to the predictions described in the text. Diagram adapted from C. K. Hong, Z. Y. Ou and L. Mandel, *Physical Review Letters* 59 2044 (1987). Courtesy Professor Hong and the American Physical Society.

is replaced by a 50:50 fibre coupler. Signal and idler photons are selected by pinholes and focused by lenses onto the fibre ends. Interference filters behind the pinholes give the usual tight restriction on the idler and signal frequency range.

19.9   Franson–Chiao interferometry

An experiment proposed by Franson demonstrates in a striking manner that the observation of two photon correlations depends simply on the alternative paths being indistinguishable. The results have a bearing on the validity of quantum mechanics. Figure 19.11 shows the layout of the experiment carried out by Kwiat, Steinberg and Chiao, in which a 351.1 nm beam produces SPDC in the KDP crystal labelled NL. The signal and idler photons of wavelength 702.2 nm travel through separate Mach–Zehnder interferometers and are detected at the photomultipliers, DS and DI respectively. The rate of coincidences between photons detected at DS and DI are recorded as well as the singles rates at DS and DI. Pinholes and interference filters, IF, are used to select degenerate entangled photon pairs. These Mach–Zehnder interferometers are constructed from 50:50 beam splitters, and totally internally reflecting prisms whose reflecting faces are labelled I and S. The long arms of each Mach–Zehnder form *optical trombones* whose path lengths can be varied by displacing the prisms labelled S and I laterally.

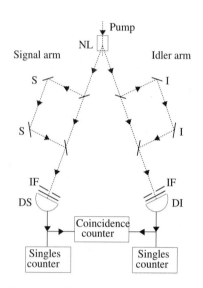

Fig. 19.11 Franson–Chiao interferometer used to observe two photon interference in which the idler and signal are not mixed.

In this experiment the conditions for interference at the detectors and for correlations between the two detectors are quite different. As regards first order interference the long and short paths in each Mach–Zehnder differ by 63 cm giving a delay, T_{d}, of 2.1 ns. This is much longer than the coherence time, T_{c}, of 120 fs determined by the bandwidth ($\sim$10 nm) of the filters in front of the detectors. Thus first order interference between the two arms of a single Mach–Zehnder will not be observed.

Turning now to correlations the situation is more delicate. Firstly the coherence time of the pump is 200 μs so that a photon travelling along the long path in one Mach–Zehnder remains coherent with its partner taking the short path in the other Mach–Zehnder. Secondly the time resolution, T_{r}, of the detectors is around 1 ns. Thus an electronic gate a little longer than T_{r}, but much less than T_{d}, was imposed on coincidences between DS and DI in order to exclude those in which one photon travelled along a long arm and the other along a short arm. The only remaining coincidences possible are just those between photons travelling along the two short arms or between photons travelling along the two long arms.

Figure 19.12 shows the variation of count rates as the length of the longer arm in the signal Mach–Zehnder is varied, all the other three arms being held fixed. The upper broken line gives the singles rate on one arm; the lower full line shows the coincidence rate between the signal and idler arms as a function of the position of the signal prism SS. As expected there is no sign of interference in the singles rate. Despite the absence of first order interference the coincidence rate also shown in figure 19.12 *does* oscillate as the length of the long arm in the signal Mach–Zehnder is changed. This result demonstrates that the observation of first order coherence is not a prerequisite for observing second order correlations.

The explanation of how such correlations can occur is as follows. Firstly, there are only two combinations of signal and idler paths through the Mach–Zehnders that fall within the electronic gate: short–short and long–long. There is an obvious difference between these combinations which at first sight appears to distinguish between them and would rule out the observed correlations: the time interval between SPDC taking place and photon detection is longer by many nanoseconds for the long–long coincidences than it is for short–short coincidences. However the conversion time in the crystal is *not* measured and so the coincidences resulting from the short–short and long–long combinations are *not* distinguishable.

The photons detected and giving coincidences are therefore in an entangled state described by the state vector

$$|\psi\rangle = (1/2)\left[\,|\psi_{\text{short}}\rangle_{\text{s}}\,|\psi_{\text{short}}\rangle_{\text{i}} + |\psi_{\text{long}}\rangle_{\text{s}}\,|\psi_{\text{long}}\rangle_{\text{i}}\,\right] \qquad (19.126)$$

where the subscripts s and i indicate the signal and idler respectively. The phase difference between the two entangled components is

$$\phi = \omega_{\text{s}}\Delta L_{\text{s}}/c + \omega_{\text{i}}\Delta L_{\text{i}}/c \qquad (19.127)$$

where ΔL_{s} and ΔL_{i} are the path differences within the signal Mach–Zehnder and within the idler Mach–Zehnder respectively. The phase in the entangled state can be expanded as

$$\phi = (\omega_{\text{s}} + \omega_{\text{i}})(\Delta L_{\text{s}} + \Delta L_{\text{i}})/(2c) + (\omega_{\text{s}} - \omega_{\text{i}})T_{\text{dd}}/2, \qquad (19.128)$$

where $T_{\text{dd}} = \Delta L_{\text{s}}/c - \Delta L_{\text{i}}/c$. The second term on the right hand side varies with the difference between the frequencies of the signal and idler, and can render the phase incoherent. T_{dd} is therefore made much shorter than the coherence time imposed by the filters, T_{c}. Then the second term becomes negligible and we have

$$\phi \approx \omega_{\text{p}}\Delta L/(2c) \qquad (19.129)$$

where $\Delta L = \Delta L_{\text{s}} + \Delta L_{\text{i}}$ and $\omega_{\text{p}} = \omega_{\text{s}} + \omega_{\text{i}}$ is the pump angular frequency. Thus the coincidence rate is, apart from constants,

$$P = (1/2)|1 + \exp{(i\phi)}|^2 = 1 + \cos{(\omega_{\text{p}}\Delta L/2c)}. \qquad (19.130)$$

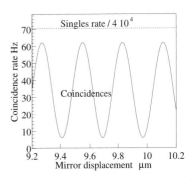

Fig. 19.12 Coincidence rate (solid line) and singles rate (broken line) in the measurement by P. G. Kwiat, A. M. Steinberg and R. Y. Chiao. The plots show smooth fits through the data appearing in *Physical Review* A47 2472 (1993). Courtesy Professor Chiao and the American Physical Society.

The interplay of T_{dd} and T_{c} is quite subtle. Note that

$$T_{\text{dd}} = \Delta L_{\text{long}}/c - \Delta L_{\text{short}}/c,$$

where ΔL_{short} (ΔL_{long}) is the difference in length between the two short (two long) arms of the Mach–Zehnders. Then supposing the idler photon arrives first at its detector, the difference in arrival times of the signal photon at its detector for long–long and short–short coincidences differ by T_{dd}. If this exceeds the coherence time T_{c} imposed by the filter then there can be no interference.

624 *The quantized electromagnetic field*

This accurately predicts the repeat distance observed in the data appearing in figure 19.12.

The experiment just described was suggested by Franson in order to make a test of *hidden variable* theories. Hidden variables would be properties of photons and other particles which are not observable (cannot be measured) but whose values would determine exactly the result of any measurement for which quantum mechanics can only ascribe probabilities for different possible results. Hidden variable theories are thus deterministic in contrast to quantum mechanics. It is impossible to distinguish between quantum mechanics and hidden variable theories in measurements of first order interference just because the hidden variables are hidden. The theories do differ in their predictions for certain correlations. In the Franson–Chiao experiment the maximum visibility in the coincidence interference pattern possible with hidden variable theories is only 0.5, far below the observed value of 0.80 ± 0.04. However a visibility of precisely 1.0 is what is predicted by quantum mechanics, the photons being anti-correlated for one path difference and fully correlated for another. The failure to reach a visibility of exactly unity seems to lie rather in experimental imperfections than in some defect of quantum mechanics: at the minima in the data the cancellation is not exact because of imperfections in alignment and the finite size of the pinholes. Many other experimental searches for evidence of hidden variables have been carried out, all with negative results.

19.10   Complementarity

It was stated in Chapter 13 that for any dynamical degree of freedom the uncertainty principle restricts the precision with which the corresponding pair of conjugate variables (e.g. x and the x-component of momentum) can be measured. An extension of the uncertainty principle called the principle of *complementarity* is required in explaining some interference effects. This states that precise knowledge of one of a conjugate pair of variables implies that all possible outcomes of measuring the partner variable are equally probable. In Young's two slit experiment the uncertainty principle and the principle of complementarity are equivalent. Any method used to detect the slit through which the photon passes will disturb the photon momentum and change its phase randomly so that the interference pattern is destroyed. Thus information on which path is travelled, the so-called *welcher Weg* information, destroys the interference pattern.

The principle of complementarity goes beyond the uncertainty principle when the definition is extended to include complementary pieces of information which are not linked dynamically. Information on correlations is the determining factor in whether second order interference effects are seen in the experiment performed by Kim, Kulik, Shih and Scully which is described next.

19.10.1   Delayed choice and quantum erasure

The apparatus is pictured in figure 19.13. A 351.1 nm beam from an argon–ion laser passes through a pair of slits in a screen placed directly in front of a BBO crystal. Type II SPDC is then produced in two regions, A and B, 0.7 mm apart within the crystal, each region being 0.3 mm wide. The signal and idler photons, having orthogonal polarization, are separated by a Glan–Thompson prism (not shown) placed directly behind the BBO crystal. Slits and interference filters are positioned in front of each detector which ensure that only degenerate photons of wavelength around 702.2 nm are detected. The signal beams from both slits are detected by the single detector D0 located in the focal plane of a lens. A slit is rigidly attached in front of D0 with its length perpendicular to the diagram. D0 and its slit could be moved together perpendicular to the lens axis in the plane of the diagram, that is to say in the y-direction. A prism is used to separate the two idler beams which then travel to separate beam splitters, BSA or BSB. The beams *reflected* from BSA and BSB fall on detectors DA or DB respectively. Detection of a photon at DA (DB) identifies the slit at which the conversion occurred as A (B). The beams *transmitted* by BSA and BSB are reflected by mirrors MA and MB respectively onto a final beam splitter BS and the entangled output beams from this beam splitter fall on two detectors D1 and D2. In contrast to the hits on DA and DB which directly identify at which slit the conversion occurred, hits on D1 and D2 give no information about where the conversion occurred.

In the experiment the coincidences between photons arriving at the signal and idler detectors were recorded; these are D0/DA, D0/DB, D0/D1 and D0/D2. The path differences are compensated by electronic time delays applied to the pulses from D0 on their way to the coincidence circuits. By design the optical path from the BBO crystal to the signal photon detector D0 is made much shorter than the optical path to any of the other detectors. As a result the signal photon is detected well before the information on which path the idler photon has selected is known.

The coincidences D0/DA and D0/DB identify that the conversion occurred at slit A and slit B respectively, and so they are termed WS *(which slit) coincidences*. The other coincidences detected, D0/D1 and D0/D2, can be due to photons from a conversion at either slit, so these are labelled ES *(either slit) coincidences*.

Figure 19.14 shows the rates for ES coincidences plotted against the displacement of detector D0 in the y-coordinate. D0/D1 coincidences

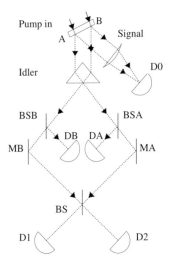

Fig. 19.13 Delayed coincidence and quantum erasure experiment. A and B are locations in a BBO crystal at which SPDC is produced. The signal photons fall on the photomultiplier D0 which can move laterally.

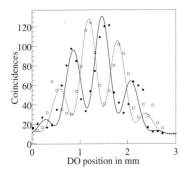

Fig. 19.14 The coincidence rates for D0/D1 (solid circles) and D0/D2 (open circles) as a function of the position of D0. The theoretical predictions are discussed in the text. The diagram is adapted from that of Y. H. Kim *et al, Physical Review Letters* 84, 1 (2000). Courtesy Professor Scully and the American Physical Society.

appear as solid circles, D0/D2 coincidences as open circles. Both these ES coincidence rates (D0/D1 and D0/D2) display a variation that is evidently a two slit interference pattern, with exactly one half-wave offset between the patterns for D0/D1 and D0/D2.

By contrast the WS coincidence rate (D0/DA and D0/DB) distributions do not show a two slit interference pattern when D0 is displaced laterally. The explanation is that these coincidences immediately tell from which slit the signal photon originated. However in recording the WS coincidence data the decision about which slit, A or B, the signal photon came from is delayed until the later detection of the idler photon. Such a measurement is therefore known as a *delayed choice* experiment.

The elimination of the interference pattern through making a delayed choice cannot be explained using the dynamical form of the uncertainty principle. The information being applied is simply the knowledge that the idler photon was detected by DA or DB. However the effect is interpretable using the principle of complementarity. In this more global view the complementary pieces of information are the identity of the detector (DA or DB) recording the idler photon, and the location of the signal photon in y. Then knowing one piece of information exactly requires that all outcomes for the other are equally probable.

A further step in refinement is made when taking the ES coincidence data. In this data set the information identifying the slit where conversion occurs, which was available at the time the idler photon was reflected from either BSA or BSB, is now destroyed. The information is destroyed by entangling the amplitudes with a beam splitter BS. This or any similar experimental procedure in which potential path information is deliberately destroyed is called *quantum erasure*. As has been seen, quantum erasure restores the interference pattern.

A general principle that the results of the experiments described here help to illustrate is that the quantum state of the system in an experiment includes not only what is known about the system, but also what is knowable from auxiliary experiments provided these would not disturb the experiment in question.

Analysis

A detailed analysis of the experiment proceeds as follows. The entangled initial state vector is

$$|i\rangle = |iA\rangle|sA\rangle + |iB\rangle|sB\rangle, \qquad (19.131)$$

where for example $|iA\rangle$ is the state vector of an idler photon from slit A. The operator whose expectation value is proportional to the D0/D1 coincidence rate is the normal ordered product of the operators for annihilating the photons at D0 and D1 and the operators for creating them there:

$$[\,\hat{a}^\dagger_{iA}(t_i)\hat{a}^\dagger_{sA}(t_{sA}) - i\hat{a}^\dagger_{iB}(t_i)\hat{a}^\dagger_{sB}(t_{sB})\,]\,[\,\hat{a}_{iA}(t_i)\hat{a}_{sA}(t_{sA}) + i\hat{a}_{iB}(t_i)\hat{a}_{sB}(t_{sB})\,],$$

in which the factors $\pm i$ account for the relative phase shift between the idler photon from slit B which is reflected, and the idler photon from slit A which is transmitted at the beam splitter. Among these terms

$$\hat{a}_{iA}(t_i) = \hat{a}_{iA}\exp(-i\omega_i t_i), \; \hat{a}_{iB}(t_i) = \hat{a}_{iB}\exp(-i\omega_i t_i)$$
$$\hat{a}_{sA}(t_{sA}) = \hat{a}_{sA}\exp(-i\omega_s t_{sA}), \; \hat{a}_{sB}(t_{sB}) = \hat{a}_{sB}\exp(-i\omega_s t_{sB}).$$

The phase shift due to the difference in path lengths from the two slits A and B to D0 is $\phi = \omega_s(t_{sA} - t_{sB})$. Then the rate of D0/D1 coincidences is, apart from constants,

$$\begin{aligned} R(\mathrm{D0/D1}) &= \langle i|\,[\,\hat{a}^\dagger_{iA}(t_i)\hat{a}^\dagger_{sA}(t_{sA}) - i\hat{a}^\dagger_{iB}(t_i)\hat{a}^\dagger_{sB}(t_{sB})\,] \\ &\quad [\,\hat{a}_{iA}(t_i)\hat{a}_{sA}(t_{sA}) + i\hat{a}_{iB}(t_i)\hat{a}_{sB}(t_{sB})\,]\,|i\rangle \\ &= 2 + i\exp(i\phi) - i\exp(-i\phi) \\ &= 2(1 - \cos(\psi)) = 4\sin^2(\psi/2), \end{aligned} \tag{19.132}$$

where $\psi = \pi/2 - \phi$. The calculation of the D0/D2 coincidence rate proceeds exactly as for the D0/D1 coincidences with the labels A and B now interchanged, giving

$$\begin{aligned} R(\mathrm{D0/D2}) &= 2 - i\exp(i\phi) + i\exp(-i\phi) \\ &= 4\cos^2(\psi/2). \end{aligned} \tag{19.133}$$

When the detector D0 is moved laterally a distance y the corresponding value of the phase factor is

$$\psi = 2\pi y d/f\lambda,$$

where d is the separation between the centres of slits A and B, and f is the focal length of the lens. The slits have a finite width a so that the coincidence rate is further modulated by the standard single slit diffraction pattern. Finally eqns. 19.132 and 19.133 become

$$R(\mathrm{D0/D1}) = 4\sin^2\frac{\pi dy}{f\lambda}\,\mathrm{sinc}^2\frac{\pi ya}{f\lambda}, \tag{19.134}$$

$$R(\mathrm{D0/D2}) = 4\cos^2\frac{\pi dy}{f\lambda}\,\mathrm{sinc}^2\frac{\pi ya}{f\lambda}. \tag{19.135}$$

These distributions are the basis of the superposed full and broken lines compared to the data in figure 19.14.

19.11   Transition rates

Transition rates will be calculated here for the simple two-level atom considered in Section 18.2, this time treating the electromagnetic field quantum mechanically as well as the atom. The results of this fully quantum mechanical calculation expose the fact that spontaneous and stimulated emission are two aspects of the same process. Spontaneous emission is emission driven by the field fluctuations in the vacuum, in the same way that stimulated emission is driven by the ambient electromagnetic field. For simplicity the conditions are assumed to be such that Fermi's golden rule, eqn. G.8, can be used to calculate the rates. Firstly the coupling between the electromagnetic field and the atom must be sufficiently weak, meaning that the Rabi frequency is very much smaller than the transition rates. This is often true, with exceptions such as an atom illuminated by a laser beam on or near resonance. Secondly the frequency distribution of the radiation must be continuous and flat across the transition line width. The atom is taken to lie in an unbounded space. Then, in order to make use in Chapter 20 of the results obtained here, the atom is supposed to be immersed in a material of refractive index n and relative permittivity $\varepsilon_r = n^2$, rather than in vacuum.

Fermi's golden rule, eqn. G.8, gives the transition rate for emission:

$$\gamma = (2\pi/\hbar^2)|M_{if}|_\omega^2 \, [\mathrm{d}N/\mathrm{d}\omega]_\omega, \qquad (19.136)$$

where $|M_{if}|_\omega$ is the matrix element for the transition considered. $[\mathrm{d}N/\mathrm{d}\omega]_\omega$ is the density of final states of the electromagnetic field. Both the matrix element and density of states are calculated for radiation of angular frequency ω determined by energy conservation

$$\hbar\omega = (W_2 - W_1), \qquad (19.137)$$

where $W_{1,2}$ are the energies of the two atomic states with wavefunctions $\psi_{1,2}$. The matrix element for an electric dipole transition is

$$M_{if} = \langle m; 2| \, e\mathbf{E} \cdot \mathbf{r} \, |m+1; 1\rangle, \qquad (19.138)$$

where the state vector $|\ell; k\rangle$ represents a state with ℓ photons in a mode of angular frequency ω, and an atom in a state k. Re-using the notation of Section 18.2 we have

$$M_{if} = \langle m| \, \mathbf{E} \, |m+1\rangle \cdot \mathbf{D}_{21}, \qquad (19.139)$$

where the electric dipole moment

$$\mathbf{D}_{21} = e \int \psi_2^* \mathbf{r} \psi_1 \, dV.$$

Now using eqn. 19.18 to re-express the electric field as an operator

$$\langle m| \, \hat{\mathbf{E}} \, |m+1\rangle = \zeta_\omega \langle m|[\hat{a}_\omega \exp{(+i\xi_\omega)} + \hat{a}_\omega^\dagger \exp{(-i\xi_\omega)}]|m+1\rangle)$$
$$= \zeta_\omega \sqrt{(m+1)} \exp{(+i\xi_\omega)} \, \varepsilon,$$

where ε is a unit vector in the direction of the electric field. The modulus of the matrix element has the same value for either direction of the reaction,

$$|M_{if}|^2 = |M_{fi}|^2 = \zeta_\omega^2 (m+1)\mu^2, \qquad (19.140)$$

where μ^2 is the mean squared value of the electric dipole moment. Thus the rate for emission is

$$\gamma_{\rm em} = (2\pi/\hbar^2)(\hbar\omega/2\varepsilon_0\varepsilon_{\rm r}V)\,(m+1)\,\mu^2 [{\rm d}N/{\rm d}\omega]_\omega. \qquad (19.141)$$

Expanding the density of states gives

$$[{\rm d}N/{\rm d}\omega]_\omega = \rho(\omega)V \qquad (19.142)$$

where $\rho(\omega)$ is the density of states per unit volume. We will use eqn. 9.128 to obtain the density of the final states in unbounded space. All orientations of the atomic dipole with respect to the electric field vector are equally probable so that μ^2 is equal to $D_{12}^2/3$. Then the rate of emission becomes

$$\gamma_{\rm em} = [\pi D_{12}^2\omega/3\hbar\varepsilon_0\varepsilon_{\rm r}]\,(m+1)\,\rho(\omega). \qquad (19.143)$$

A similar calculation, starting for consistency with a state that contains m photons again, gives the absorption rate

$$\gamma_{\rm ab} = [\pi D_{12}^2\omega/3\hbar\varepsilon_0\varepsilon_{\rm r}]\,m\,\rho(\omega). \qquad (19.144)$$

When there are no photons in the initial state eqn. 19.143 collapses to

$$\gamma_{\rm sp} = [\pi D_{12}^2\omega/3\hbar\varepsilon_0\varepsilon_{\rm r}]\rho(\omega), \qquad (19.145)$$

which must therefore be identified as the spontaneous emission rate. Then using eqn. 9.128 for the density of states in unbounded space this becomes

$$\gamma_{\rm sp} = [\pi D_{12}^2\omega/3\hbar\varepsilon_0\varepsilon_{\rm r}][n^3\omega^2/\pi^2c^3] = [n\omega^3/3\pi\hbar\varepsilon_0c^3]\,D_{12}^2, \qquad (19.146)$$

where $\varepsilon_{\rm r}$ was replaced by n^2. This result is precisely the Einstein coefficient for spontaneous emission, A_{21}, which was defined in eqn. 14.3 and presented after the semi-classical calculation in eqn. 18.37. Comparing eqn. 19.145 with eqn. 19.143 we see that the difference is

$$\gamma_{\rm st} = [\pi D_{12}^2\omega/3\hbar\varepsilon_0\varepsilon_{\rm r}]\,m\,\rho(\omega). \qquad (19.147)$$

This we assert is the rate of stimulated emission. Stimulated emission is thus driven by the number of photons present through the factor m, while spontaneous emission is driven by the vacuum, or to be more explicit by the electromagnetic field fluctuations in the vacuum. These are the same fluctuations mentioned first in Section 13.6 and then discussed in more detail in this chapter. They are responsible for the non-vanishing zero point energy and the non-vanishing mean square value of the electric field in the vacuum.

For completeness we calculate the Einstein B coefficients from γ_{st} and γ_{ab} and then show their consistency with the relationships between the Einstein coefficients deduced in Section 14.2. Recall that the energy in the mode of the electromagnetic field of angular frequency ω when m photons present is $m\hbar\omega$, and that the spectral energy density is $W(\omega)$. Then the density of states per unit volume

$$\rho(\omega) = W(\omega)/(m\hbar\omega). \tag{19.148}$$

Substituting this expression for $\rho(\omega)$ in eqn. 19.147 gives

$$\gamma_{\mathrm{st}} = [\pi D_{12}^2/(3\hbar^2\varepsilon_0\varepsilon_{\mathrm{r}})]W(\omega). \tag{19.149}$$

Now using the definition given in eqn. 14.2

$$\gamma_{\mathrm{st}} = B_{21}^\omega W(\omega), \tag{19.150}$$

with the superscript ω in B_{21}^ω indicating that this is the appropriate form of the Einstein coefficient when the spectral energy density is expressed as a function of the angular frequency rather than of frequency. Then using eqn. 19.149

$$B_{21}^\omega = \pi D_{12}^2/(3\hbar^2\varepsilon_0\varepsilon_{\mathrm{r}}). \tag{19.151}$$

In turn

$$B_{21} = B_{21}^\omega/(2\pi) = D_{12}^2/(6\hbar^2\varepsilon_0\varepsilon_{\mathrm{r}}). \tag{19.152}$$

Next combining this result with that of eqn. 19.146 gives

$$A_{21} = \gamma_{\mathrm{sp}} = (8\pi h f^3 n^3/c^3)B_{21}, \tag{19.153}$$

which reproduces the connection between these Einstein factors inferred earlier in Chapter 14 when matter is in thermal equilibrium with black body radiation. Finally comparing eqns. 19.144 and 19.147 we deduce the other relationship between the Einstein coefficients, for the case assumed here that the states are non-degenerate: namely that $B_{21} = B_{12}$.

19.12   Further reading

The Quantum Theory of Light, third edition by R. Loudon, published by Oxford University Press (2000). This standard text develops topics introduced here and in the previous chapter more fully and rigorously, including, for example, a detailed analysis on squeezed states.

Introductory Quantum Optics, by C. Gerry and P. Knight, published by Cambridge University Press (2005). This recent book covers quantum optics including, at greater depth, the fully quantum (Jaynes–Cummings) model of photon–atom interactions and covers interesting microwave experiments in cavity QED.

Optical Coherence and Quantum Optics, by L. Mandel and E. Wolf, published by Cambridge University Press (1995). A rigorous source book on the theory and techniques of optical coherence.

Exercises

(19.1) When the counting time in the HBT experiment for a thermal source is extended beyond the coherence time several modes, m in number, will be involved. Show that the variance in the photon count is $\langle N \rangle + \langle N \rangle^2/m$ where N is the total photon count.

(19.2) If the quantum efficiency of the photon detector is η the average count changes from $\langle M \rangle$ to $\langle m \rangle = \eta \langle M \rangle$. Show that the variance in the photon count for a thermal source becomes $\eta^2 \langle \Delta M^2 \rangle + \eta(1 - \eta)\langle M \rangle$. The first term is the variance expected for perfect efficiency $\times \eta^2$ and the second is the *partition noise*.

(19.3) Prove eqn. 19.80.

(19.4) Prove that $(\hat{a})^n|n\rangle = \sqrt{n!}\,|0\rangle$ and that $(\hat{a})^n|n-1\rangle = 0$.

(19.5) Prove eqn. 19.15 using eqns. 19.13 and 19.14.

(19.6) Deduce that the magnetic field operator $\hat{\mathbf{B}}(t) = \sum_\omega \mathbf{k} \wedge \mathbf{e}_\omega (\zeta_\omega/\omega)[\hat{a}_\omega \exp(-i\xi_\omega) + \hat{a}_\omega^\dagger \exp(i\xi_\omega)]$ where $\mathbf{e}_\omega$ is the unit polarization vector of the electric field component with frequency ω and wave vector $\mathbf{k}$.

(19.7) $|\{\alpha\}\rangle$ is a state made up of many coherent states in different modes of the electromagnetic field. In the case that there are two of these modes with frequencies $\omega(1)$ and $\omega(2)$, $|\{\alpha\}\rangle = |\alpha_{\omega(1)}\rangle|\alpha_{\omega(2)}\rangle$. Show that $g^{(2)}(\tau) = 1$ for such states.

(19.8) Show that the total energy operator for states of the electromagnetic field can be written: $\sum_n E_n|n\rangle\langle n|$, where E_n is the energy with n photons in the mode.

(19.9) In Young's double slit experiment the initial state can be expressed as $|in\rangle = (1/\sqrt{2})(\hat{a}_1^\dagger + \hat{a}_2^\dagger)|0\rangle$ where $\hat{a}_i^\dagger$ creates a photon at slit i. The intensity on the screen is proportional to $\langle in|\hat{E}^-\hat{E}^+|in\rangle$. Show that the intensity distribution across the screen is $[1 + \cos(\mathbf{k}\cdot\Delta\mathbf{r})]$ where $\Delta\mathbf{r}$ is the vector difference between the paths from the slits to the point on the screen being considered.

(19.10) Photons are separated with equal probability between two paths, 1 and 2, which might be the two ouput ports of a beam splitter. If the total number of photons incident is n what is the expectation for the product $n_1 n_2$ of the numbers entering each arm?

(19.11) Sketch Argand diagrams like that of figure 19.3 corresponding to the squeezed states represented in figure 19.2.

(19.12) Show that the phase mismatch in the generation of second harmonic waves in the forward direction in a distance $\mathrm{d}x$ is $\Delta k\mathrm{d}x = (2\omega/c)(n_{2\omega} - n_\omega)\mathrm{d}x$ where $n_{2\omega}$ is the refractive index of the crystal for the pump and n_ω that for both idler and signal. In a crystal of length L show that the intensity falls of with the phase mismatch like $\mathrm{sinc}^2(\Delta k L/2)$.

Quantum dots, optical cavities and cryptography

20

20.1 Introduction

Quantum dots are three-dimensional structures of order 10 nm in size, with unusual optical properties. Here the discussion is restricted to semiconductor quantum dots which are embedded in another semiconductor which has a larger electron band-gap. In this case a quantum dot forms a three-dimensional electric potential well for electrons and holes inside the surrounding semiconductor. The effect of its dimensions being smaller than the wavelength of thermal electrons is that the motion of electrons within the quantum dot is quantized in all three directions. This extends to three dimensions the one-dimensional quantization of electron motion discussed earlier in connection with quantum well lasers. Thus electrons in quantum dots occupy discrete energy states, just as they would in an atom. What would have been the valence and conduction bands in bulk semiconductor are now resolved into a relatively few discrete energy levels, as shown in figure 20.1. In the unexcited state of the quantum dot all the valence levels are full while all the conduction levels are empty. If an electron is excited into a conduction level it leaves a hole in a valence level, and the pair become bound by the Coulomb attraction between them. This bound state is called an *exciton*, for which the symbol X is used. Exciton binding energies are small compared to the spacing of the valence and conduction levels. The semiconductors used for the quantum dots, and in which they are embedded, are usually those used in the lasers and detectors met in earlier chapters, so that transitions like that shown in figure 20.1 involve visible or near infrared radiation.

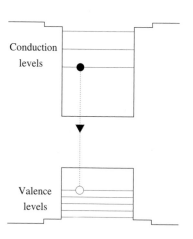

Fig. 20.1 Quantum dot energy levels with ground state exciton decay.

Exciton states can be excited either by incident radiation or by applying a voltage to inject charge carriers. The exciton transition of most interest, shown in figure 20.1, is the decay of the ground state (lowest energy) exciton, in which the electron returns to fill the vacant hole. It is the final step in exciton decay chains and it provides a prominent spectral line. This decay will be referred to as the *ground state exciton decay*. Exciton wavefunctions are similar in size to the quantum dot which means that their dipole moments are correspondingly large. Quantum dots can therefore be regarded, for many purposes, as large atoms whose excited states are the excitons.

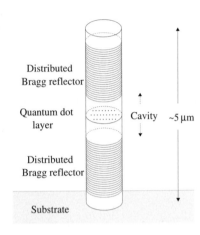

Distributed
Bragg reflector

Quantum dot
layer

Cavity   ~5 μm

Distributed
Bragg reflector

Substrate

Fig. 20.2 Pillar microcavity with embedded quantum dot.

[1]Studies at microwave frequencies involve atoms in highly excited states with electrons having principal quantum numbers of order 100 and consequently very large orbits. In these *Rydberg atoms* when an electron undergoes a microwave transition between two such orbits the electric dipole moment for the transition is correspondingly large.

Figure 20.2 shows an example of a device in which quantum dots are contained in an *optical cavity* whose dimensions are typically a few hundred microns. This ensures that the cavities have only a few modes at wavelengths in the visible and near infrared. The semiconductor pillar is a few microns in height. Distributed Bragg reflectors define the cavity in the growth direction. Lateral containment in the cavity is provided by total internal reflection at the curved semiconductor/air interface. A layer of quantum dots is embedded inside the section of the pillar forming the optical cavity. When radiation from a quantum dot matches a cavity mode in wavelength, the coupling between source and cavity mode can dominate the system's behaviour. In the case that the coupling is strong, there can be Rabi oscillations in which energy oscillates between the source and the cavity mode. With weak coupling the rate and directionality of the photon emission can be enhanced: but where there is no match the rate can be suppressed.

The use of the quantum theory of radiation is the key to explaining these coupling effects: when applied in this context it is called *cavity quantum electrodynamics (CQED)*. Experimental studies of CQED were initiated at microwave frequencies using millimetre sized cavities with metal walls. In this chapter only the more recently accessible optical effects will be discussed. However the conceptual framework introduced applies equally well at microwave frequencies. The electric dipole moments of excitons are larger than those of atoms in comparable states of excitation because the excitons have dimensions of order the size of the quantum dot. Consequently the electric dipole coupling of the exciton to cavity modes is enhanced compared to that of an atom by a factor of several hundred. This property has made it feasible to study the effects of cavity/source coupling with quantum dots at optical frequencies.[1]

Quantum dots and their properties are described in more detail in the section immediately following this one. One device is described in which Rabi oscillations are produced by laser excitation, not relying on coupling to an enclosing cavity. Then there follows an account of the different geometries of the optical microcavities in which quantum dots have been confined.

In the succeeding part of the chapter the conditions for strong and weak coupling between a cavity mode and an enclosed source are determined. Rabi oscillations in such systems are discussed. After this the case of weak coupling is examined, showing how the decay rate of an excited state of a quantum dot can be altered appreciably. This, the *Purcell effect*, will be analysed and an example of its experimental observation described.

The final part of this chapter begins with the description of a method for transmitting *quantum keys* for use in cryptography. Cryptographic keys are long trains of bits used to scramble and unscramble data. In the method of transmision considered here the bits are encoded as the polarization states of single photons using the protocol devised by Bennett and Brossard in 1984. The quantum nature of radiation guarantees that any attempt to intercept and re-send the key during its transmission between the parties exchanging information can be detected by them. Key distribution is then said to be unconditionally secure. Implementation of this protocol requires a practical source of single photons *available on demand*. One quantum dot source, under development, which goes far toward this ideal is discussed. More speculative proposals for quantum computing would impose the further requirement that the photons are indistinguishable. How far the source in question fulfils this added requirement is also examined.

At room temperatures the thermal broadening of exciton transitions is $\sim 5\,\text{meV}$. To complicate spectra further, the non-uniformity within a layer of quantum dots like that shown in figure 20.2 is such that the energies of the ground state exciton transition are spread over $\sim 50\,\text{meV}$. All studies described below were carried out with the cavities cooled cryogenically to 4-10 K. The individual exciton transitions then have line widths of $\sim 1\,\mu\text{eV}$. Only at such temperatures can single transitions from individual quantum dots be identified in the luminescence spectrum, and the effects of cavity confinement be examined systematically. The lifetimes, $\sim 1\,\text{ns}$, at cryogenic temperatures are sufficiently long that exciton states can be manipulated electronically. This control is important for some of the studies discussed below.

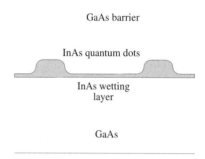

Fig. 20.3 Self-assembled quantum dots formed by the Stranski–Krastanow process.

20.2 Quantum dots

The standard method for growing semiconductor quantum dots is by self-assembly. This process will be described for InAs quantum dots grown on a GaAs substrate. InAs is deposited on a single crystal GaAs substrate by molecular beam epitaxy, *MBE*: in this process beams from furnaces containing the respective elements are directed at the substrate under very high vacuum. InAs and GaAs share the same crystalline form (zincblende) with lattice sizes that differ appreciably, by 7.2%. Even so the InAs laid down follows the crystal structure of the substrate, layer by layer. This is what constitutes epitaxial growth. However the strain and surface energy are such that when the InAs is a little under two monomolecular layers thick the InAs spontaneously clumps into thicker islands, each of small area, leaving a thin *wetting layer* of InAs connecting them. These islands are the quantum dots, and the formation step is called the *Stranski–Krastanow* process after its discoverers. InAs deposition is then terminated and a further layer of GaAs is laid down, burying the quantum dots within a GaAs slab. Figure 20.3 shows a section through such a structure. The size, shape and surface density of the quantum dots can be controlled through the rate of deposition, the temperature of the substrate, and reprocessing steps. Typical densities employed range from a few to hundreds of quantum dots per square micron of the substrate surface. In their final form the dots are usually lens-like or pyramid-like; typically 10 to 50 nm across and a few nanometres tall in the growth direction. They are non-uniform but the variation in size can be held as low as 5%. InAs and GaAs have band-gaps of 0.36 and 1.43 eV respectively with the InAs quantum dots forming potential wells for electrons and holes, as shown in figure 20.4.[2] Quantum dot di-

[2]These are type I quantum dots. In type II quantum dots, such as GaInAs/InP, there is a potential well for electrons inside the quantum dot, while at the same time there is a potential hill for holes. Then the electron in the exciton stays within the dot while the hole's wavefunction lies outside, but concentrated around the dot.

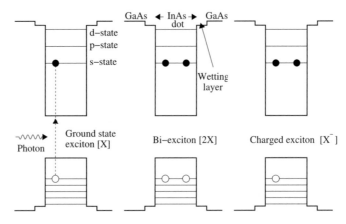

Fig. 20.4 Excitation of the ground state exciton, a bi-exciton and a charged exciton. The levels are those for an InAs quantum dot in a GaAs matrix, for which the photon indicated would have an energy ~1.1 eV.

[3]In bulk InAs the binding energy of the ground state s-wave exciton is given by the hydrogen atom formula, eqn. 12.30. This must be modified to take account of the relative permittivity of InAs, $\varepsilon_r = 15.2$. Also the reduced mass of the electron–hole pair, $m = 0.023 m_e$, replaces the electron–proton reduced mass. The resulting binding energy is $13.6 \, (m/\varepsilon_r^2)$ or 1.35 meV. Correspondingly the exciton Bohr radius in bulk InAs is 35 nm. In a quantum dot which restricts the exciton's size the exciton binding energy is larger.

mensions are all smaller than the electron wavelength, which is 160 nm at 10 K. It follows that electron and hole states are quantized with discrete energies.[3] These states are also illustrated in figure 20.4. The left hand diagram shows photo-excitation with an electron raised from the highest valence level to the lowest conduction level. The electron and the hole it left behind become bound as an exciton: in the case shown this is a *ground state exciton*. A transition from the ground state exciton to the unexcited quantum dot, what is called the decay of the ground state exciton, produces in the case of InAs/GaAs quantum dots a photon of energy ~1.1 eV (wavelength ~1130 nm). The binding energy of the exciton due to the Coulomb attraction between the electron and hole is relatively small, ~20 meV for an InAs dot of diameter 20 nm. The conduction s-, p- and d-states are spaced at intervals of 50–100 meV, depending on the dot size, shape and strain, while the valence levels are more densely packed. These single particle levels merge into the continuum of levels in the wetting layer indicated in figure 20.4. Higher energy exciton states are made up from an electron and hole in their higher energy levels. Examples of further excited states are illustrated by the central and right hand panels in figure 20.4. A bi-exciton (*XX* or *2X*) is shown in the central panel in which electrons have been excited into both s-level spin states. The right hand panel shows a charged exciton with a single attached electron (X^-).

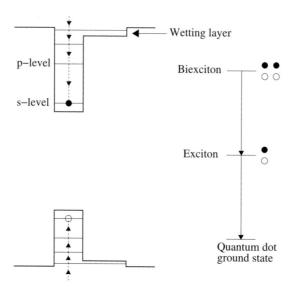

Fig. 20.5 On the left is a sketch of the discrete energy levels replacing the valence and conduction bands in a quantum dot. The transitions indicated by broken lines show how carriers produced in the semiconductor around the dot enter and cascade down the energy states of the dot. On the right selected excitonic transitions are shown.

Transitions from the various exciton levels to the unexcited InAs/GaAs quantum dot typically extend in energy from 1.1 eV to 1.3 eV, or in wavelength from 1100 nm down to 850 nm.[4] At room temperature the energies of the exciton transitions are smeared by thermal interactions between the exciton and its environment so that the line widths broaden to ~ 5 meV. Therefore the semiconductor devices are cooled to ~ 10 K, at which temperature the individual line widths are a few μeV. It is then possible to readily disentangle lines in the spectra. For example, the pair of transitions drawn in figure 20.5 can be resolved: a bi-exciton to ground state exciton transition and the subsequent decay of the ground state exciton.

Excitons can be produced either by photons or by charge injection. The resulting luminescence emerging from the quantum dot is then called photo- or electro-luminescence respectively. When InAs quantum dots in a GaAs matrix are irradiated by a laser beam which can excite electrons into the GaAs conduction band, the carriers produced near the dots will be attracted into the potential well as indicated in figure 20.5. This process is completed in ~10 ps. By contrast the lifetimes of the exciton states at cryogenic temperatures are of order a nanosecond, corresponding to a line width, already noted, of a few μeV.

Quantum dots are not uniform in size, shape, composition or strain; all factors which affect the energies of the quantized electron levels. Hence spectral lines emitted by an assembly of quantum dots are inhomoge-

[4]Replacing InAs by InGaAs moves the spectrum further into the near infrared to 1300 nm.

neously broadened to $\sim 50\,\mathrm{meV}$ ($\sim 50\,\mathrm{nm}$ in wavelength). It is therefore necessary to restrict the excitation to at most a few dots in order to simplify the assignment of spectral lines to specific transitions. This level of localization can be obtained by having few dots per unit area and then restricting the incident light or current by baffles in the cavity structure. In any commercial application charge injection may be preferred because of its compatibility with standard electronics. However most studies of quantum dots with or without cavities have been made using laser excitation. This approach has the advantages that the photon energy can be selected to excite particular states and that the beam can be readily pointed at any area of interest.

20.2.1 Rabi oscillations involving quantum dots

Rabi oscillations have been observed by Zrenner and colleagues[5] from individual quantum dots exposed to a laser beam whose frequency was tuned to the frequency of the exciton decay transition. This experiment does not involve specifically cavity effects. The quantum dots were built

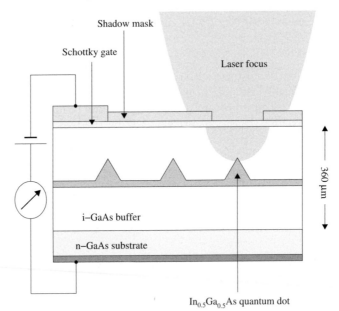

Fig. 20.6 Single quantum dot photodiode. Adapted from figure 1(a) in A. Zrenner, P. Ester, S. M. deVasconcellos, M. C. Huebner, L. Lackmann, S. Stufler and M. Bichler, *Journal of Physics: Condensed Matter* **20**, 454210 (2008). Courtesy Professor Zrenner and the Institute of Physics Publishing Ltd.

[6] See Section 15.6 for details on Schottky photodiodes.

into the n-i-Schottky photodiode[6] shown in figure 20.6 and held at 4.2 K. As will be explained below, the Rabi oscillations were detected through

---

[5] A. Zrenner, P. Ester, S. M. deVasconcellos, M. C. Huebner, L. Lackmann, S. Stufler and M. Bichler, *Journal of Physics: Condensed Matter* **20**, 454210 (2008).

the variations they induce in the photocurrent.

The InGaAs quantum dots were embedded in a 360 nm thick intrinsic GaAs layer, itself grown on an n-doped GaAs contact. Above the GaAs was a 5 nm thick semi-transparent titanium Schottky contact. In turn this was covered by an 80 nm thick opaque aluminium mask having apertures of width 100–500 μm. A microscope objective of NA 0.7 focused the beam from a tunable Ti:sapphire laser through one such aperture onto a single quantum dot. The applied bias voltage produced an electric field of 35 kV m^{-1}, large enough to ionize any excitons created by the laser beam. Thus the photocurrent produced as a result of this ionization was proportional to the rate at which excitons were created.

Zrenner and colleagues used the Stark effect produced by the applied bias voltage to tune the exciton energy. First they measured the dependence of the frequency of radiation from exciton decay on the bias voltage. Then they set the bias voltage so that the frequency of this radiation coincided with the laser frequency.

The strong coupling between the intense laser electromagnetic field and the quantum dot caused Rabi oscillations between the ground state exciton and the unexcited state of the quantum dot. Ionization of the excitons by the applied electric field shortens their lives: at a typical bias of 0.8 V the resulting *dephasing time* is 50 ps. In order that the photocurrent should faithfully follow the exciton population the laser was pulsed on for periods of 2.3 ps at a repetition frequency of 80 MHz. Firstly the effect of having such short pulses is that an exciton produced in a pulse would very rarely decay during that pulse. Secondly an exciton produced in one pulse would almost always decay during the 12.5 ns interval before the next pulse. Under these conditions the mean photocurrent is exactly proportional to the exciton population induced by each pulse. According to eqn. 18.71 the population of the exciton state following an individual pulse is proportional to $\sin^2(\Omega_0\Delta t/2)$, where Δt is the pulse duration, and Ω_0 is the Rabi angular frequency, given by the product of the electric field strength and the dot's electric dipole moment, $\mathbf{E}\cdot\mathbf{D}_{12}/\hbar$. Figure 20.7 shows the variation in photocurrent as the intensity, and hence electric field of the laser pulse is increased. The abscissa of figure 20.7 is marked in units of $\Omega_0\Delta t$, which is the area of the pulse. We see that following pulses of area π, 3π, 5π,... the exciton population is at its maximum and after pulses of area 2π, 4π, 6π,... the exciton level is again empty. The current expected from a sequence of π pulses is readily calculated. Each π pulse should excite one exciton in the quantum dot illuminated, and when this exciton is ionized a single charge should circulate round the external circuit. Hence, with 80 million pulses per second the current predicted would be 80 million electrons per second, which is 12.8 pA. This agrees very well with the peak photocurrent seen in figure 20.7.

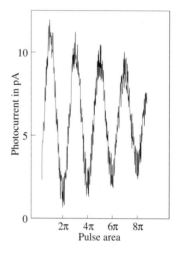

Fig. 20.7 Rabi photocurrent oscillations. Adapted from figure 3(a) in A. Zrenner *et al.*, *Journal of Physics: Condensed Matter* **20**, 454210 (2008). Courtesy Professor Zrenner and the Institute of Physics Publishing Ltd.

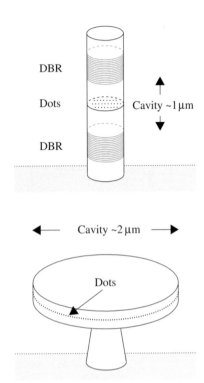

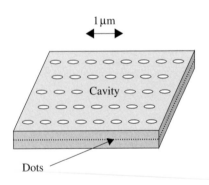

Fig. 20.8 Pillar, whispering gallery and photonic crystal microcavities.

20.3   Optical microcavities

Three types of semiconductor microcavity commonly used to host quantum dots are displayed in figure 20.8. That shown in the uppermost panel is the micropillar mentioned in the introduction to this chapter. Such micropillars are a few microns tall and typically between 100 nm and 5 μm in diameter. The cavity could be a GaAs slab between two distributed Bragg reflectors made up of many alternate layers of GaAs and AlAs, each layer being one quarter wavelength in optical thickness. One or more layers of self-assembled quantum dots are located in antinodal planes of the electric field of the cavity mode. One arrangement is to have a slab of optical thickness equal to one wavelength and place the dots at the antinode of electric field in its midplane. Unexpectedly this places antinodes of the electric field at the surfaces of the distributed Bragg reflectors, rather than nodes. The reason is that the electromagnetic field penetrates the reflectors, so that the nodes of the electric field lie effectively about one quarter wavelength inside each reflector.[7] Confinement at the pillar walls is provided by total internal reflection. This mechanism is very effective because the refractive index of GaAs is 3.5 at near infrared wavelengths. A mode confined inside a micropillar ideally has a Gaussian profile across the cross-section of the pillar, like the mode within a Fabry–Perot etalon described in Chapter 6. This minimizes losses at the side walls.

A disk microcavity is shown in the central panel of figure 20.8. Radiation is confined within the disk by total internal reflection at the upper and lower faces as well as at the edge. Again a layer of quantum dots is placed at an antinodal plane within the cavity. Radiation can be reflected repeatedly at the edge face so that it travels around the disk. When the wavelength of the radiation is such that the accumulated phase change in one complete circuit around the disk is an integral multiple of 2π the result is a stable cavity mode. The radiant energy in these modes is concentrated near the periphery of the disk. Such cavity modes are called *whispering gallery* modes because they are analoguous to the acoustic modes of such structures as the whispering gallery beneath the dome of St Paul's cathedral in London.

The final example of a microcavity shown in the bottom panel of figure 20.8 is formed by a defect in a two-dimensional photonic crystal. Quantum dots are laid down in the midplane of the crystal, whose optical thickness is one half wavelength. Some of the dots lie within the area of the cavity formed by the defect. Confinement of radiation within the plane of the crystal is effected by the band-gap of the crystal pattern surrounding the defect. In the growth direction confinement is obtained

---

[7] See 'Basics of dipole emission from a planar cavity' by R. Baets, P. Bienstman and R. Bockstaete in the *Cargese Lectures on Confined Photon Systems* 1998, edited by H. Benisty *et al.*, published by Springer-Verlag, Berlin(1999).

through total internal reflection at the upper and lower faces of the crystal.

Figure 20.9 exhibits two examples of defects in photonic crystals for which the associated cavity modes have high Q values. In the upper panel the linear defect with three consecutive holes missing is called an *L3 defect*. The cavity mode is confined within the area of this defect. By slightly displacing the holes adjacent to the defect along its axis, and by making these holes slightly smaller than the rest, the cavity mode acquires a smooth Gaussian profile along this axis, with minimal penetration into the crystal surrounding the defect. This property confers a long lifetime on photons trapped in this cavity mode. Another structural strategy for obtaining a high Q value is shown in the lower panel of figure 20.9. In this case the cavity is formed locally within an extended linear defect by varying the separation of the holes parallel to the defect axis. The change from a uniform spacing need only be a few percent to give a Gaussian shaped mode confined to a few cycles of the pattern along the linear defect.

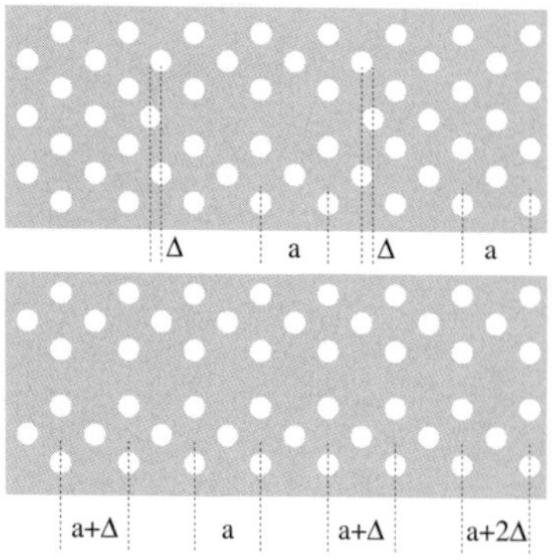

Fig. 20.9 Photonic crystal structures, with defects providing high Q cavities. a is the regular spacing between adjacent hole centres; Δa is the offset introduced locally to form the optical cavity.

20.4 Strong and weak coupling

In Section 20.2.1 Rabi oscillations of a quantum dot driven by a laser were described. Rabi oscillations, and Rabi splitting of energy levels, can also occur when a quantum dot, or an atom, is enclosed in an optical cavity and a mode of the cavity is resonant with a transition of this source. Then the energy of excitation cycles between the source and the cavity mode. This regime requires strong coupling between the source and the cavity mode. If the coupling is weak its principal effect is that the decay rate of the source from its excited state is altered. The physical conditions corresponding to the weak and strong coupling regimes are now examined for the case of the quantum dot in an optical cavity.

The rate of decay of the photon population from the cavity, γ_c, is the first of three parameters relevant to assessing whether the coupling between cavity mode and quantum dot is strong or weak. The second parameter is the rate of decay of the exciton ground state, γ. This includes all decays whether or not a photon enters the cavity mode, as for example a non-radiative decay. The third parameter is the Rabi angular frequency, given by eqn. 18.18,

$$\Omega_0 = \mathbf{E} \cdot \mathbf{D}_{12}/\hbar,$$

where $\mathbf{E}$ is the electric field and $\mathbf{D}_{12}$ the electric dipole moment of the source. The Rabi frequency quantifies the strength of the coupling between cavity mode and quantum dot.

If $\Omega_0 \gg \gamma$ and $\Omega_0 \gg \gamma_c$ there can be many Rabi oscillations during the lifetimes both of a photon in the cavity and of the exciton. In this case photons can be emitted by the quantum dot into the cavity mode, reabsorbed by the quantum dot and then re-emitted in a sequence that continues over many cycles. This defines the *strong coupling* regime. At the other extremes if $\gamma_c \gg \Omega_0$ or if $\gamma \gg \Omega_0$ oscillations will be frustrated. In the first case through the more rapid loss of the photons from the cavity and in the second case through the more rapid spontaneous decays of the excitons. These are the conditions for *weak coupling*. A case of interest is that in which the exciton decay is much faster than the Rabi oscillations and at the same time much slower than the cavity decay rate. Then the exciton decays into an effectively empty cavity, so that its decay is spontaneous. Now because the spectrum of the cavity differs from that of free space the rate of spontaneous decay can be altered. This effect was first described, for microwaves, by Purcell.

In the following two sections analyses of both weak and strong coupling are presented for completeness. This provides the necessary framework for subsequently interpreting experimental observations of weak coupling, and its applications.

20.5   Rabi oscillations in cavities

A simple system is analysed here consisting of a two-level atom in a cavity which has only a single mode at a frequency equal to that of the atomic transition. The formalism and results apply equally to the case of a quantum dot in an optical microcavity. We shall make use here of the concept of dressed states introduced in Section 18.4. Then by taking results from Chapter 19 we make a calculation of the energies of such states in which the electromagnetic field is treated quantum mechanically.

Dressed states of an atom can be written, following the notation of Section 18.4

$$|\psi_\pm; m\rangle = (1/\sqrt{2})[|\psi_1\rangle|m\rangle \pm |\psi_2\rangle|(m-1)\rangle]. \tag{20.1}$$

where $|\psi_j\rangle$ is the state vector for the atom in level j, and $|m\rangle$ is the state vector of the cavity mode containing m photons. The energy operator is

$$\hat{H} = \hat{H}_0 + \hat{\mathbf{E}} \cdot (e\mathbf{r}),$$

where H_0 is the energy in the absence of any atom/cavity mode interaction, and $\mathbf{E} \cdot (e\mathbf{r})$ is the energy contributed by the electric dipole interaction. Then the energies of the dressed states of the atom are

$$\langle\psi_\pm|\hat{H}_0|\psi_\pm\rangle + e\langle\psi_\pm|\hat{\mathbf{E}} \cdot \mathbf{r}|\psi_\pm\rangle. \tag{20.2}$$

The electric field operator is given by eqn. 19.18,

$$\hat{\mathbf{E}}(t) = \varepsilon\zeta_\omega[\hat{a}_\omega \exp(+i\xi) + \hat{a}_\omega^\dagger \exp(-i\xi)] \tag{20.3}$$

where ε is the unit polarization vector pointing in the direction of the electric field. Also $\zeta_\omega = \sqrt{\hbar\omega/2\varepsilon_0\varepsilon_r V}$, where V is the cavity volume, ω is the angular frequency of the mode and ε_r the relative permittivity of the dielectric forming the cavity. If the interaction between the atom and the electromagnetic field were turned off their total energy would be simply

$$H_0 = W_1 + (m + 1/2)\hbar\omega,$$

where W_1 is the energy of the lower state. The additional interaction energy is

$$\Delta E = \langle \psi_\pm; m|\hat{\mathbf{E}} \cdot (e\mathbf{r})|\psi_\pm; m\rangle. \tag{20.4}$$

We can immediately drop the exponential factors from $\mathbf{E}$.[8] Then

$$\Delta E = \zeta_\omega \langle \psi_\pm; m| [\hat{a}_\omega + \hat{a}_\omega^\dagger] \, \varepsilon \cdot (e\mathbf{r})|\psi_\pm; m\rangle$$
$$= (1/2) \zeta_\omega \{ [\langle \psi_1|\langle m| \pm \langle \psi_2|\langle(m-1)|] [\hat{a}_\omega + \hat{a}_\omega^\dagger]$$
$$\varepsilon \cdot (e\mathbf{r}) [|\psi_1\rangle|m\rangle \pm |\psi_2\rangle|(m-1)\rangle] \}.$$

Then applying eqns. 19.27 and 19.28

$$\Delta E = (1/2) \zeta_\omega \{ \pm\sqrt{m} \langle \psi_2|\varepsilon \cdot (e\mathbf{r})|\psi_1\rangle \pm \sqrt{m} \langle \psi_1|\varepsilon \cdot (e\mathbf{r})|\psi_2\rangle \}.$$
$$= \pm(1/2) \zeta_\omega \sqrt{m} \, \varepsilon \cdot [\mathbf{D}_{21} + \mathbf{D}_{12}]. \tag{20.5}$$

For simplicity we take the electric dipole moment to be real, so that $\mathbf{D}_{21} = \mathbf{D}_{12}^* = \mathbf{D}_{12}$. Thus

$$\Delta E = \pm \zeta_\omega \sqrt{m} \, D_{12}, \tag{20.6}$$

where we take the electric dipole moment to be *aligned* with the direction of the electric field in the cavity mode. Then the Rabi splitting between the two states of the dressed atom is

$$\hbar\Omega_0 = 2\Delta E = 2\zeta_\omega \sqrt{m} \, D_{12}. \tag{20.7}$$

When the number of photons in the cavity mode is very large the Rabi splitting can be re-expressed in terms of the classical electric field, $E_{\rm cl}$, using eqn. 19.83. This gives $\hbar\Omega_0 = E_{\rm cl} D_{12}$, which reproduces the result from the semi-classical calculation in Section 18.4. At the other extreme no photons accompany the atom in its excited state. In that case the level splitting becomes

$$\hbar\Omega_{\rm v} = 2\zeta_\omega D_{12}, \tag{20.8}$$

which is called the *vacuum Rabi splitting*.

The energy levels for the the uncoupled atom and cavity mode are shown on the left hand side of figure 20.10; on the right are the energy levels for the dressed atom. At the extreme left of the figure are written the number of photons in the cavity mode when the atom is in its ground state: the atom in its excited state is accompanied by one fewer photons. When a very large number of photons are present the energy

[8]Firstly the electric field varies so little across the atom that the spatial variation in $\exp(\pm i\xi)$ can be ignored. Secondly, the energy difference between the bare states is the photon energy so that the overall time dependent factor is unity. Writing only the time dependent components

$$\langle \psi_1|\mathbf{r}|\psi_2\rangle \exp(-i\xi)$$
$$= \exp[i(E_1/\hbar - E_2/\hbar + \omega)t].$$

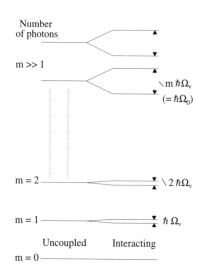

Fig. 20.10 On the left are the energy levels of an uncoupled atom and cavity mode; on the right are the energy levels when the interaction between the atom and cavity mode is taken into account. The level structure on the right is called the Jaynes–Cummings ladder.

states form the towers drawn in figure 18.3.

Both the strength of the quantum dot/cavity mode coupling and the Rabi splitting depend on the cavity volume, V, through the factor ζ_ω in the last few equations. The electric field in the cavity mode is largest, and hence the coupling is strongest, when the photons are confined to as small a volume as possible. However if each dimension is less than one half wavelength there are no modes at all. Consequently the volume must exceed $(\lambda/2n)^3$, where λ is the free space wavelength and n is the refractive index of the material forming the cavity. In the case that the dielectric is GaAs with refractive index 3.5 this limiting volume, considering a wavelength in air of $1\,\mu$m, is $0.023\,\mu$m^3.

We can now calculate the vacuum Rabi splitting when the cavity enclosing the quantum dot is empty of radiation. The parameters chosen are these: an exciton transition wavelength of $1\,\mu$m; a matrix with a relative permittivity equal to 12.25; a cavity of volume $1\,\mu$m^3; and an exciton dipole moment d equal to the electron charge times 1.0 nm or $1.6\,10^{-28}$ C-m. Then the electric field in the semiconductor is

$$E = \sqrt{(\hbar\omega/2\varepsilon_r\varepsilon_0 V)} = 0.303\,10^5\,\text{Vm}^{-1},$$

and the vacuum Rabi splitting is

$$\hbar\Omega_v = Ed = 48\,\mu\text{eV}.$$

Rabi oscillations are damped by losses of the population from the upper state or through the upper state going out of phase with the lower state so that coherence is lost. These causes of damping, ignored in the preceding analysis, are now enumerated for the case of quantum dots in semiconductor cavities. Population losses occur when photons fail to enter the cavity mode or are lost subsequently from the cavity. These losses can happen because a photonic band-gap confining the mode is not omni-directional, or because the photon strikes a cavity surface at an angle less than the critical angle. Alternatively there can be a population loss through decays to a third state or through non-radiative decays. Together these *population loss* mechanisms cause the upper state population to fall off with time like $\exp(-t/T_1)$. The rate, $1/T_1$, will evidently approach the sum of the exciton decay rate and the cavity loss rate, $\gamma + \gamma_c$. A second distinct set of processes which damp the Rabi oscillations are elastic interactions that an exciton has with its surroundings. Such interactions alter its phase without changing its state. These are called *pure dephasing* processes with time constant T_2'. Overall the total damping of the Rabi oscillations due to population loss and pure dephasing is called *dephasing*. The time constant, T_2 for the overall damping is given by

$$1/T_2 = 1/T_1 + 1/T_2'. \tag{20.9}$$

The observed dephasing times of excitons at $10\,$K are in the range $100\,$ps to a few nanoseconds.

20.6   Weak coupling

The system being considered is again a two-level atom, or equally a quantum dot, interacting with a single cavity mode. The lifetime of the excited state is now much shorter than the Rabi period, $\gamma \gg \Omega_0$, so that Rabi oscillations can be ignored. It becomes legitimate to treat the atom/dot-cavity mode interaction perturbatively. If in addition the cavity line width is much greater than transition line width, $\gamma_c \gg \gamma$, two consequences follow. Firstly the atom/dot decays into an effectively empty cavity so that its decay is spontaneous. Secondly the spectral distribution of radiation in the cavity is continuous across the transition line width. In such a case Fermi's golden rule can be used to calculate the spontaneous decay rate. The calculation would resemble that already made in Section 19.11 for the decay of an atom in an unbounded medium.

The crucial change required compared to Section 19.11 is that the density of modes in unbounded space is replaced by the density of states in the cavity. Use of the term *density of states*, when there is just one mode, is surprising. However the mode has a finite lifetime and hence it has a spread in energy and angular frequency. We use the Lorentzian form, eqn. 7.54, for the distribution in ω, the angular frequency of the mode:

$$\mathrm{d}N/\mathrm{d}\omega = \frac{(\gamma_c/2\pi)}{[(\omega - \omega_c)^2 + \gamma_c^2/4]}. \tag{20.10}$$

where ω_c is the cavity mode's central angular frequency and γ_c the full width at half maximum of the distribution. Its integral, $\int [\mathrm{d}N/\mathrm{d}\omega]\,\mathrm{d}\omega$, is unity so, indeed, there is only one mode. At resonance between the atomic transition and the cavity mode

$$[\mathrm{d}N/\mathrm{d}\omega]_{\omega=\omega_c} = 2/(\pi\gamma_c) = 2Q/(\pi\omega_c), \tag{20.11}$$

where, as usual, $Q = \omega_c/\gamma_c$ quantifies the sharpness of the cavity resonance.

The general expression for the spontaneous decay rate has already been calculated and is given in eqn. 19.145. In the present case the electric dipole will be assumed to be fully aligned with the cavity electric field so that $D_{12}^2/3$ is replaced by D_{12}^2, where D_{12} is the atom's electric dipole moment between the two states. Equation 20.10 divided by the cavity volume V gives the density of states per unit volume. Inserting this we have the spontaneous decay rate of the atom or quantum dot:

$$\gamma_{\mathrm{cav}} = [(\pi D_{12}^2\omega/(\hbar\varepsilon_0\varepsilon_r V)]\,\frac{(\gamma_c/2\pi)}{[(\omega - \omega_c)^2 + \gamma_c^2/4]}. \tag{20.12}$$

Close to resonance $\omega \approx \omega_c$, so that we can replace ω by $\gamma_c Q$, giving

$$\gamma_{\mathrm{cav}} = [D_{12}^2 Q/(2\hbar\varepsilon_0\varepsilon_r V)]\,\frac{\gamma_c^2}{[(\omega - \omega_c)^2 + \gamma_c^2/4]}. \tag{20.13}$$

This result can be compared to the decay rate in an unbounded medium of refractive index n, given by eqn. 19.146

$$\gamma_{\text{open}} = n\omega^3 D_{12}^2/(\varepsilon_0 \pi \hbar c^3). \tag{20.14}$$

Then the ratio between these two spontaneous decay rates

$$\gamma_{\text{cav}}/\gamma_{\text{open}} = [3Q(\lambda/n)^3/(16\pi^2 V)]\frac{\gamma_c^2}{[(\omega - \omega_c)^2 + \gamma_c^2/4]}. \tag{20.15}$$

Note that the factor 3 appears because in an unbounded medium the polarization vectors of the modes are isotropically distributed with respect to any atomic dipole direction, whilst in a cavity the dipole is taken to be fully aligned with the mode's field. When the atomic transition and cavity mode are in exact resonance, this ratio is called the *Purcell factor*,

$$F_{\text{P}} = \gamma_{\text{on}}/\gamma_{\text{open}} = [6Q/\pi^2]\{(\lambda/2n)^3/V\}. \tag{20.16}$$

The Purcell factor quantifies the enhancement in the decay rate of the atom/quantum dot produced by the cavity when the two are in resonance.

The critical components of the Purcell factor are the cavity Q value, and the size of the cavity expressed in terms of the minimum cavity size, $V/(\lambda/2n)^3$. Having a large Q value ensures that the density of states at resonance is large. Having a small cavity ensures that the energy density of each mode is large; which means that the vacuum fluctuations are also large. Taken together these choices will maximize the Purcell factor.

Far off resonance the density of final states in the tail of the Lorentzian distribution becomes very small: eqn. 20.12 reduces to

$$\gamma_{\text{off}} = [D_{12}^2\omega/(2\hbar\varepsilon_0\varepsilon_r\gamma_c V)][\gamma_c/(\omega - \omega_c)]^2. \tag{20.17}$$

If we make a crude approximation with $\omega - \omega_c$ replaced by ω, then

$$\gamma_{\text{off}} \approx D_{12}^2/(2\hbar\varepsilon_0\varepsilon_r QV), \tag{20.18}$$

and the ratio

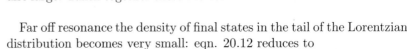

$$\gamma_{\text{off}}/\gamma_{\text{open}} \approx [3/(2\pi^2 Q)]\{(\lambda/2n)^3/V\}. \tag{20.19}$$

This crude estimate shows that off resonance there can be considerable suppression of the decay rate when the cavity Q value is large enough to overwhelm any enhancement produced by having a small cavity.

Figure 20.11 sketches the spectra involved for the case just considered, in which the cavity line width, γ_c, is much greater than the atom's transition line width, γ. The cavity profile shown follows the variation of the density of states per unit volume in the cavity. Over this narrow wavelength range the corresponding density of states per unit volume in unconfined space would be almost constant.

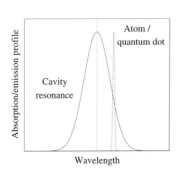

Fig. 20.11 The spectra of the cavity mode and of the atomic (quantum dot) transition.

Table 20.1 Table of representative microcavity properties.

Property	Photonic crystal	Micropillar	Microdisk
Volume (V)	$(\lambda/n)^3$	$10(\lambda/n)^3$	$10(\lambda/n)^3$
Q value	20 000	8000	12 000
Purcell factor	1000	40	100

20.6.1   Cavity Purcell factors

We can review the conditions under which the Purcell factor would be significant in the case of a quantum dot in an optical cavity. Crucially the cavity resonance would need to be sharp, and the cavity dimensions would be made as small as the wavelength, so as to maximize Q/V. At a temperature $\sim 10\,\mathrm{K}$ an exciton line width of a few μeV at a wavelength $\sim 1\,\mu$m makes $\lambda/\Delta\lambda > 10^5$. Some representative values of the volume, Q value and calculated Purcell factor are given in table 20.1 for the three types of microcavity shown in figure 20.8. These cavity Q values are much less than 10^5 so that the cavity spectrum is continuous across the exciton line width, as assumed when making the analysis in the previous section. Observed Purcell factors are generally smaller than calculated values. Of the three types of cavity described earlier photonic crystal cavities are the most easily integrated into existing optoelectronic systems: they can be formed using standard techniques for fabricating electronic chips.

A clean demonstration of the Purcell effect in a photonic crystal was given by Hsu and colleagues.[9] They used a 180 nm thick GaAs layer on a substrate with InGaAs quantum dots located in the layer's midplane. An array of holes was etched through the GaAs, following the pattern shown in the upper panel in figure 20.9. Several L3 linear defects, with three missing holes, were included. The pattern period had $a = 300$ nm, with holes of radius $0.31a$, giving a band-gap extending from 0.9 to 1.2 μm in wavelength. Holes adjacent to the defect were offset by $0.10a$. Then the substrate was etched away to leave the GaAs layer as a suspended membrane. The L3 defect cavity had only a modest Q value of about 300. There were only one or two quantum dots per cavity.

A sketch of the apparatus used in the photoluminescent study of the quantum dots is shown in figure 20.12. The beam from a 1.55 eV (800 nm) Ti:Sapphire laser was focused onto the crystal through a microscope objective. Luminescence captured by this same lens was analysed using a three-stage monochromator which restricted the spectral range at output to tens of nanometres around the ground state exciton decay wavelength. The authors studied the time dependence of the

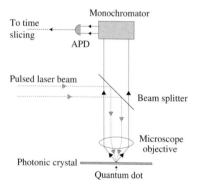

Fig. 20.12 Observation of photoluminescence from a a single quantum dot embedded in a photonic crystal. Courtesy Professor T. M. Hsu.

[9]W.-H. Chang, W.-Y. Chen, H.-S. Chang, T.-P. Hsieh, J.-I. Chyi and T.-M. Hsu, *Physical Review Letters* 96, 117401 (2006).

photoluminescence from individual quantum dots excited by pulses of 0.2 ps duration from the Ti:sapphire laser. Their detector was a silicon avalanche photodiode. The results for ground state exciton decays in three different quantum dots are shown in figure 20.13. One quantum dot is located in the bulk material outside the area of the photonic crystal pattern; a second, QD1, lies inside a cavity within the photonic crystal and its exciton transition is on resonance with the cavity mode; while for a third quantum dot, QD2, the exciton transition is off resonance from the cavity mode. The corresponding exciton lifetimes are 0.65 ns,

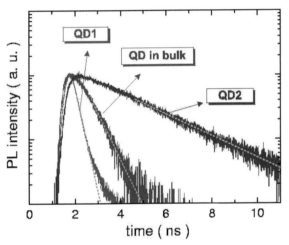

Fig. 20.13 Time resolved photoluminescence from quantum dots illustrating the Purcell effect. One dot is in bulk material, the others are inside cavities in a photonic crystal. The data labelled QD1 is from an exciton transition nearly resonant with the cavity mode, while that labelled QD2 is from a non-resonant exciton transition. Courtesy Professor T. M. Hsu and the American Physical Society.

0.21 ns and 2.52 ns respectively. These results show first that the decay of an exciton on resonance with a cavity mode becomes more rapid when compared to an exciton in an unbounded medium. Secondly an exciton decay far off resonance with the cavity mode is strongly inhibited.

One important consequence of the Purcell effect is that most of the spontaneous emission from an exciton on resonance with a cavity mode is funnelled into the cavity mode. In the example being discussed the fraction of the spontaneous emission on resonance and captured by the cavity mode is

$$\beta = (\gamma - \gamma_{\mathrm{PC}})/\gamma. \tag{20.20}$$

Here γ is the total decay rate as measured in bulk material $(1/0.21)\,\mathrm{ns}^{-1}$ and γ_{PC} is the rate at which emitted radiation escapes through the photonic crystal via *leaky* modes. This last rate, γ_{PC}, is similar to the decay rate measured off resonance, when very little radiation can enter the cavity mode. Hence γ_{PC} can be taken to be $(1/2.52)\,\mathrm{ns}^{-1}$. With these values β comes out to be approximately 0.92. Quite generally, for all microcavity types, a value of β close to unity is expected when the Purcell factor is large.

In striking contrast the values of β for typical lasers are orders of magnitude smaller. For example only a tiny fraction of the spontaneous emission from a He:Ne laser is directed at its end mirrors. Instead most spontaneous emission misses the mirrors. This is one reason why most of the input power to a He:Ne laser is wasted.[10] By inference room temperature lasers based on quantum dots within microcavities, in which an exciton transition is resonant with a cavity mode, ought to be highly efficient. In addition the constraint on the electron motion in quantum dots means that quantum dot lasers should be relatively insensitive to temperature variation. This last property is a significant advantage over other semiconductor lasers which usually require precise temperature control. The quantum dot lasers so far constructed bear out these predictions.[11]

[10]Another reason is discussed in Section 14.4.4.

20.7   Quantum cryptography

Quantum-based protocols for the distribution of keys for cryptography have the potential for being unconditionally secure against any eavesdropper. The standard BB84 protocol requires a source of *single photons on demand* for its straightforward implementation. Quantum dot devices are seen as sources potentially able to fulfil this requirement. An introduction briefly sketching the relevant features of quantum-based key distribution is given first before the description of a quantum dot diode under development.[12]

Cryptography *keys* are long strings of bits with which a confidential data stream is scrambled to form a *cryptogram*. After transmission from one party to another, the key is used to unscramble and recover the original data from the cryptogram. Provided the key itself can be transmitted between the parties involved secure from eavesdropping, any message it is used to encrypt can be equally secure. It turns out that the quantum nature of electromagnetic radiation can be exploited to make it certain that any attempt to eavesdrop during the exchange of a cryptography key will be detected. The BB84 protocol proposed by Bennett and Brossard[13] in 1984 will be used here to illustrate the principles of

[11]See for example J. Tatebayashi, M. Ishida, N. Hatori, H. Ebe, H. Sudou, A. Kuramat, M. Sugawara and Y. Arakawa; *IEEE Journal of Selected Topics in Quantum Electronics* 11, 1027 (2005).

[12]For a much fuller account of the techniques of quantum key distribution see 'Quantum Cryptography' by N. Gisin, G. Ribordy, W. Tittel and H. Zbinden: *Reviews of Modern Physics* 74, 145 (2002).

[13]C. H. Bennett and G. Brassard: *Proceeedings of the IEEE International Conference on Computer Systems and Signal Processing*, pages 175-9, published by IEEE

the quantum approach.

A setup for the BB84 protocol is shown in figure 20.14. Using the customary names, Alice sends Bob the key in the form of a string of single photons whose polarizations carry the bit information. Eve, the would–be eavesdropper, intercepts the photons. She also attempts to copy the photons and pass these copies on to Bob with the intention of concealing the interception. If the connection between Alice and Bob is made over an optical fibre link then Eve might break the fibre at some point and insert a repeater there.

Alice's photon source provides plane polarized photons. First Alice

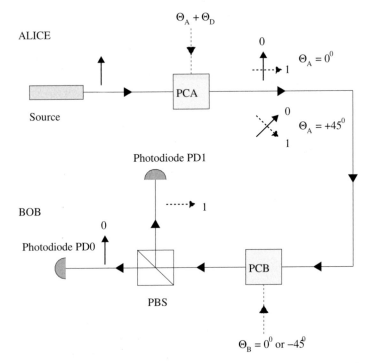

Fig. 20.14 An implementation of the BB84 protocol for secure quantum key exchange using photon polarization to carry the bit information.

rotates the polarization vector of each photon through an angle $\theta_A + \theta_D$ using the Pockels cell PCA. Alice sets θ_D to $0°(90°)$ when the current data bit in the key is $0(1)$. The way that Alice exploits the quantum property of light is by resetting θ_A randomly for each pulse to either $0°$ or $45°$. Alice records the value of θ_A used for each bit transmitted. At Bob's receiver the polarization is rotated through a further angle θ_B by the Pockels cell PCB. The choice of θ_B is also made randomly between the two values $0°$ or $-45°$. This choice is made quite independent of Al-

New York (1984).

Table 20.2 Possible combinations of Bob and Eve's analyser settings for a given polarization set by Alice. Bob's success rate during the interception is shown in the final column.

Alice	Eve	Bob	Bob's success rate
⊕	⊕	⊕	100%
⊕	⊕	⊗	not used
⊕	⊗	⊕	50%
⊕	⊗	⊗	not used

ice's choice of θ_A. Finally the photon enters the polarizing beam splitter PBS and travels to one or other photodetector. For each bit received Bob records which photodetector fired and the corresponding setting of θ_B.

When the transmission of the key is complete Alice and Bob openly exchange information on the random settings that they used for each bit, namely the values of θ_A and θ_B. Whenever they find that the analyser and polarizer settings are complementary, namely that $\theta_B = -\theta_A$, then the bit that Bob receives will be exactly that sent by Alice. If Alice sent a '0' bit the photodetector PD0 would have fired, and if a '1' bit PD1 would have fired. Bob and Alice accept these bits. In cases for which they find that $\theta_B \neq -\theta_A$ the photon will be polarized at 45° or 135° on entry to Bob's polarizing beam splitter PBS. The photon will be sent quite randomly to one or other photodetector. Bob and Alice reject these bits. The bits accepted, known as the *sieved bits*, are used to form the cryptographic key.

Now we can examine how quantum mechanics works to foil Eve when she intercepts the photons with a setup identical to Bob's and tries to transmit copies to Bob. We can suppose that Eve knows, or infers, that Bob sets his analyser θ_B randomly to 0° or −45°. However she has no idea which setting is required for any given bit, just because the choice is made randomly. The four possible combinations of Bob's and Eve's analyser settings are listed in table 20.2 for one of Alice's settings. These four combinations are equally probable. The essence of the matter is that when Bob and Alice make complementary choices and Eve makes the opposite choice to Bob for her analyser's alignment – for example $\theta_B = \theta_A = 0°$ while Eve has $\theta_E = -45°$ – Eve will randomly get a 0-bit or a 1-bit. In such cases the bit Eve then sends to Bob has a 50% chance of being the same as that sent by Alice. This is noted in row 3 in table 20.2. On the other hand when Eve uses the same analyser setting as Bob he will receive a correct bit from Eve, as noted in row 1 in the table.

Alice and Bob can detect Eve's presence by comparing a subset of the sieved bits, for which they ought to have identical bit values if there is no eavesdropper. The sieved bits come from the analyser combinations in rows 1 and 3 of table 20.2. Therefore Bob and Alice will find that they disagree over the bit value in one quarter of all the subset they compare, and this tells them that the transmission has been intercepted. Obviously all sources of error must be eliminated to a level well below 25% in order that Bob and Alice can reliably detect Eve's interception.

The effectiveness of the BB84 protocol vividly illustrates the principle that it is impossible to determine the general quantum state of a single photon. This principle ensures that photons cannot be cloned, which is Eve's undoing. Eve measures the polarization in one basis and in doing so collapses the photon polarization state into one in her basis. This is all the information that Eve can extract and in the face of the random choice of basis made by Alice it cannot fully characterise the incoming photon's polarization. When bits are coded in this way as quantum states of photons (or material particles) they are called *qubits*.

In order to implement the BB84 protocol Alice should send pulses each containing a single photon. She needs to have a source that provides a single photon whenever she issues an electronic command: that is to say a *single photon on demand*. A quantum dot source under development which already shows great promise is described in the following section. Proposals for quantum computing require in addition that the single photons on demand should also be indistinguishable.[14] The quantum dot source just mentioned approaches some way toward this ideal. An account of this source given below describes the methods used to determine whether single photons are emitted, and whether the successive single photons emitted are indistinguishable.

Sources now used for quantum key distribution are attenuated laser beams. The photon number distribution is Poissonian so that a fraction of the bits are carried by pulses containing several photons. It is therefore feasible for Eve to pick off a photon from a pulse and pass the remainder on to Bob. This is called a *photon number splitting* attack. Bob still receives photons direct from Alice and is unaware of any interception. However this form of interception can be foiled by introducing *decoy* pulses. The majority of pulses sent by Alice, the signal pulses, have a mean photon count of μ_s. Interspersed among these Alice sends decoy pulses with a much lower mean photon count, μ_d. Eve does not have any way of anticipating the decoy pulses and has to treat all pulses in the same way. After transmission is complete Bob can com-

---

[14]A clear and compact introduction to these proposals can be found in Chapters 12 and 13 of *Quantum Optics* by Mark Fox, published by Oxford University Press (2006).

pare the mean number of photons in the signal and decoy pulses. If the ratio departs from that expected then there is evidence of eavesdropping.

In circumstances in which Alice and Bob find they must use a key exchange that contains errors and that may have been intercepted they need to perform two further steps. These steps provide *error correction* and *privacy enhancement* and are used in classical key distribution also. In the error correction step Alice and Bob take pairs of sieved bits and sum them modulo 2 (the exclusive or operation XOR) and compare their results. If the XORs agree then there is no error in either bit and they retain the first bit of each pair. Otherwise Alice and Bob discard both bits. Alice selects these comparison pairs randomly and communicates her choices openly to Bob: for example the 657th bit and the 2345th bit. This process continues until all the bits have been checked. In privacy amplification Alice and Bob again make XORs with pairs of bits now using only the retained error-free bits. Another random sequence selected by Alice decides the pairings. This time Alice and Bob do not declare their results but instead they retain the XORs. These XOR bits form the final key. Eve's valid information is depleted by these steps and her residual information is calculable from the bit error rate.

The range of quantum key distribution over installed optical fibre links is $\sim$50 km at rates above $100\,\mathrm{kb\,s^{-1}}$. Transfering quantum keys through nodes in a network presents difficulties. One idea which avoids nodes is to use a satellite link just for the key transfer.

20.7.1   Microcavity diode

The objective of single photon production from quantum dots by electrical stimulation has been pursued by the Quantum Information Group at the Toshiba Laboratories in Cambridge.[15] A cross-section through their prototype single photon emitting diode is sketched in figure 20.15. Its electrical response is shown in the same figure. A layer of sparse InGaAs quantum dots is embedded in a GaAs slab having an optical thickness of one wavelength.This GaAs slab is sandwiched between distributed Bragg reflectors made up of pairs of GaAs/AlAs layers, each layer being a quarter wavelength in optical thickness. There are twelve layer pairs below, and two pairs above the GaAs. The reflectors increase the efficiency with which photons from the quantum dots are collected by an order of magnitude, but the Q value of the GaAs cavity between them is so small that there is negligible Purcell enhancement. If the cavity were made small enough, $\sim (\lambda/n)^3$, so that the Purcell enhancement became significant, then confining the electrical excitation to such a small region would be difficult. That is one reason why studies often employ photoluminescence rather than electroluminescence.

---

[15]See A. J. Bennett, R. B. Patel, A. J. Shields, K.Cooper, P. Atkinson, C. A. Nicoll and D. A. Ritchie, *Applied Physics Letters* 92, 193503 (2008).

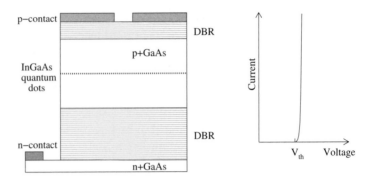

Fig. 20.15 Quantum dot diode. On the left the physical structure is shown; on the right the diode's characteristic. Courtesy Professor Shields.

Figure 20.16 shows the drive circuit and the setup for detecting the photons emitted. Radiation from a negatively charged exciton in an

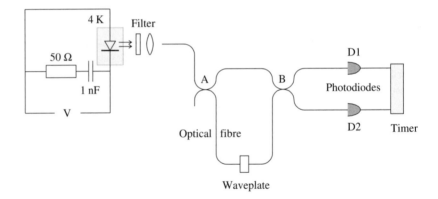

Fig. 20.16 Sketch of the apparatus used to study photon production from a quantum dot diode. Adapted from figure 1(a) from A. J. Bennett, R. B. Patel, A. J. Shields, K. Cooper, P. Atkinson, C. A. Nicoll and D. A. Ritchie, *Applied Physics Letters* 92, 193503 (2008). Courtesy Professor Shields and the American Institute of Physics.

isolated quantum dot is collected by a microscope objective and focused through a narrow bandwidth filter centred on 1.31475 eV in order to select radiation from the ground state exciton decay only. The radiation is imaged onto a single mode optical fibre. This takes the radiation through the 50%:50% beam splitter labelled A, and into an optical fibre Mach–Zehnder interferometer. The unequal length interferometer arms connect to a similar beam splitter labelled B. From there the outputs go to avalanche photodiodes that count individual photons. Finally the electrical signals generated by these photons are used as the start and stop pulses for circuitry which times and records the interval between photons arriving at the two photodiodes. If the pulse from photodiode D1 precedes that at photodiode D2 a positive time is recorded, and if

D2 precedes D1 a negative time interval is recorded.

The diode containing the dots is held off most of the time with a bias maintained below the threshold value V_{th} indicated in figure 20.15, and is briefly pulsed every 2 ns. These pulses have two components. There is an initial positive going component which injects charge into the quantum dots and so excites an exciton. This is followed by a negative going component which stops the injection of charge, and shifts the exciton energy via the Stark effect. It is arranged that for a short period, lasting around 30 ps at the pulse minimum, the exciton emission wavelength lies within the narrow acceptance bandwidth of the optical filter. This drastically restricts the interval during which any of the detected photons could be emitted. In this way the dephasing of the exciton due to interactions with its surroundings is effectively eliminated. Successive photons entering the detectors should therefore be coherent, that is to say indistinguishable. A disadvantage of this technique is that by turning off the emission before all the excitons have had time to decay the quantum efficiency of the source is reduced fivefold.

In a first series of measurements the avalanche photodiodes were connected directly to the output ports of the first beam splitter. The apparatus then has the configuration of the Hanbury Brown and Twiss interferometer met in Section 19.6, with pulsed rather than continuous illumination. Figure 20.17 shows the distribution of time intervals observed between pulses at the two photodiodes. Counts for intervals of duration one, two, three .... pulse periods are nearly equal, while that for a zero time interval is only a small fraction, 0.03 ± 0.01, of the others. This distribution shows that only rarely are photons detected simultaneously at the photodiodes. Most such examples can be attributed to noise. Evidently the quantum dot is emitting one single photon during any excitation, which is the desired photon on demand. Unfortunately the overall efficiency is low, with only a few percent of the photons reaching the detectors. This number would be an order of magnitude lower if the quantum dots were not located in a cavity.

In the quantum view the reason that there are no simultaneous photons is quite obvious. A second photon could only appear if the exciton were re-excited after being de-excited by the emission of the first photon; and there is effectively no time available for this process. Similar demonstrations that photons appear singly were used to definitively rule out a tenacious, semi-classical picture in which states of matter would be quantized but electromagnetic radiation would be purely wavelike. Without going into details here, it had proved straightforward to account in this semi-classical way for the outcomes of the standard experiments used in introducing quanta in Chapter 12. However such a semi-classical picture predicts that simultaneous detections would be as frequent as the others in figure 20.17.

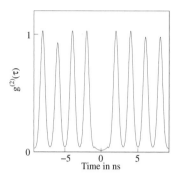

Fig. 20.17 Distribution of time intervals between pulses in the HBT configuration. Adapted from figure 3(a) from A. J. Bennett et al., *Applied Physics Letters* **92**, 193503 (2008). Courtesy Professor Shields and the American Institute of Physics.

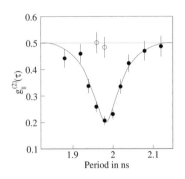

Fig. 20.18 Photon correlations as a function of repetition rate of the drive voltage with the HOM configuration. Adapted from figure 3(d) from A. J. Bennett, R. B. Patel, A. J. Shields, K.Cooper, P. Atkinson, C. A. Nicoll and D. A. Ritchie, *Applied Physics Letters* **92**, 193503 (2008). Courtesy Professor Shields and the American Institute of Physics.

In a second measurement phase the apparatus including the Mach–Zehnder interferometer was used to test whether the photons were also indistinguishable. This layout has the form of the Hong–Ou–Mandel interferometer described in Section 19.8. The pulse period was varied in small steps to cover a range of time intervals centred on 1.98 ps. This particular interval, 1.98 ps, is the difference between the travel times along the two arms of the interferometer. If the pulse period is exactly 1.98 ps photons from consecutive pulses can take opposite paths through the interferometer and arrive simultaneously at the two entry ports of beamsplitter B. Timing histograms like that shown in figure 20.17 were obtained in the Hong–Ou–Mandel configuration for each delay period selected. At each such step the area of the central peak for coincident photons at the detectors was divided by the area of the average of the other peaks. Two such data sets are shown in figure 20.18 plotted versus the time delay. One set was taken with the photons having identical polarization; the other with the photons orthogonally polarized. For these selections linear polarizers were placed in front of the two detectors.

When photons have orthogonal polarization they are distinguishable and the central peak should have an area half that of the other peaks. The corresponding measurements are indicated by the hollow circles in figure 20.18 and agree well with the prediction. Data taken with identical polarization are indicated by the full circles. For their interpretation we can refer to the description of the Hong–Ou–Mandel interferometer in Section 19.8. When indistinguishable photons arrive simultaneously at opposite input ports of the beam splitter both photons will emerge from the same exit port. There will never be any coincidences between the detectors. This will ideally be the situation in the experiment described here if consecutive photons are indistinguishable and if the interval between pulses is exactly equal to the time delay between the two arms of the Mach–Zehnder. There is certainly a dip in the data at the predicted pulse interval. When allowance is made for the photodiode dark counts the visibility at the dip is 64 %. Hence not only are the photons appearing on demand but they are also *generally* indistinguishable.

An improved collection efficiency of 14% has been attained by the Cambridge group with a micropillar device in which a $1.4\,\mu m^3$ GaAs cavity lies between 17 and 24 period Bragg mirrors.[16] Layers were partially oxidized to create a smooth cylindrical central cavity with low losses. As a result the measured Purcell factor of 2.5 is achieved, not very different from the predicted value of 2.8. It will be some time before sources of single indistinguishable photons on demand are available commercially. Unfortunately all the devices considered in this chapter require cryogenic cooling. Low temperatures reduce the interactions with the quantum dot's environment so that individual spectral lines are

---

[16]See D. J. P. Ellis, A. J. Bennett, S. J. Dewhurst, C. A. Nicoll, D.A. Ritchie and A. J. Shields; *New Journal of Physics* **10**, 043035 (2008).

well resolved. The lifetimes of order a nanosecond may be long enough for use in quantum computing.

20.8 Further reading

Quantum Optics: An Introduction by Mark Fox, published by Oxford University Press (2006) delves more deeply. In particular it contains three chapters on quantum information processing.

'Quantum Cryptography' by N. Gisin, G. Ribordy, W. Tittel and H. Zbinden: *Reviews of Modern Physics* 74, 145 (2002). This is a comprehensive review of this rapidly developing field by active experts.

Exercises

(20.1) An InGaAs quantum dot is embedded in a GaAs cavity of volume $0.3\,\mu\mathrm{m}^3$. The cavity resonant wavelength is 950 nm and the lifetime of photons in the cavity is 100 ps. Calculate the expected Purcell factor. The relative permittivity of GaAs is 3.5.

(20.2) A GaAs pillar microcavity of volume $0.20\,\mu\mathrm{m}^3$ contains an InGaAs quantum dot. The exciton wavelength is $0.95\,\mu\mathrm{m}$, and its lifetime is 0.1 ns. The dipole moment for the exciton transition is $1.6\,10^{-28}\,\mathrm{C\,m}$, and the refractive index of GaAs is 3.5. What is the vacuum Rabi splitting of the exciton? How large should the cavity Q value be if the Purcell effect is to be observable?

(20.3) If the cavity Q value is 1000 for the system described in the previous question, what is the Purcell factor?

(20.4) In the system described in the previous question what photon population in the mode will lead to strong coupling?

(20.5) In Section 20.7.1 it is stated that if the photons are distinguishable, then the central peak of the time interval distributions should be of half height. Prove this statement, and also that the height of peaks for ± 1 pulse interval are three-quarters the height of other peaks.

(20.6) InGaAs quantum dots are enclosed in a GaAs cavity of refractive index 3.5 and volume $0.1\,\mu\mathrm{m}^3$.

The angular frequency of the cavity mode is $2.0\,10^{15}\,\mathrm{rad\,s}^{-1}$ and its halfwidth is $5.0\,10^{12}\,\mathrm{rad\,s}^{-1}$. An exciton has a decay angular frequency $2.01\,10^{15}\,\mathrm{rad\,s}^{-1}$, and the corresponding electric dipole moment is $10^{-28}\,\mathrm{C\,m}$. What is the exciton decay rate? What would it be if the quantum dot were in an unbounded volume of GaAs?

(20.7) The two states involved in an exciton decay are $|\psi_1\rangle$ and $|\psi_2\rangle$. Write down the state vector after a $\pi/2$ pulse is used to excite the lower state. Suppose that α is the rate at which the phase of the upper state drifts relative to the phase of the lower state. Write down the state vector at a time T after the $\pi/2$-pulse. At that moment a π-pulse is applied. Write down the state vector of the quantum dot immediately after this π-pulse and again after a further second interval T.

Suppose that the same sequence is followed for a collection of quantum dots whose phase drifts vary from dot to dot. After the $\pi/2$-pulse the excited states initially radiate in phase, and this pulse of radiation dies away as their phases drift to give a random distribution. The time T is long compared to the time taken for this randomization. Explain why it is that after the second interval T the ensemble again radiates coherently. This is known as a *photon echo* and also occurs in other optical systems.

Appendix: Physical constants and parameters

Electron charge $e = 1.6022\,10^{-19}\,\mathrm{C}$
Electron mass $m_{\mathrm{e}} = 9.1094\,10^{-31}\,\mathrm{kg}$
Electron volt $\mathrm{eV} = 1.6022\,10^{-19}\,\mathrm{J}$
Proton mass $m_{\mathrm{p}} = 1.6726\,10^{-27}\,\mathrm{kg}$
Atomic mass unit $u = 1.6605\,10^{-27}\,\mathrm{kg}$

Planck's constant $h = 6.626\,10^{-34}\,\mathrm{J\,s}$
$\hbar = 1.0546\,10^{-34}\,\mathrm{J\,s} = 6.582\,10^{-16}\,\mathrm{eV\,s}$

Permittivity of free space $\varepsilon_0 = 8.854\,10^{-12}\,\mathrm{F\,m^{-1}}$
Permeability of free space $\mu_0 = 4\,\pi\,10^{-7}\,\mathrm{N\,A^{-2}}$

Fine structure constant $\alpha = e^2/4\pi\varepsilon_0\hbar c = 7.2974\,10^{-3}\,(\approx 1/137)$
Bohr radius $a_\infty = 4\pi\varepsilon_0\hbar^2/m_{\mathrm{e}}e^2 = 0.5292\,10^{-10}\,\mathrm{m}$

Avogadro's number $N_{\mathrm{A}} = 6.022\,10^{23}\,\mathrm{mol^{-1}}$
Boltzmann's constant $k_{\mathrm{B}} = 1.3807\,10^{-23}\,\mathrm{J\,K^{-1}}$

Wien's constant $\lambda_{\max}T = 2.898\,10^{-3}\,\mathrm{m\,K}$
Stefan–Boltzmann constant $\pi^2 k^4/60\hbar^3 c^2 = 5.671\,10^{-8}\,\mathrm{W\,m^{-2}\,K^{-4}}$

Appendix: Cardinal points and planes of lens systems

The matrix approach is used here to locate the *cardinal points* of a lens system: the focal, principal and nodal points. The initial and final media have the same refractive index n. Let the transfer matrix through the system between the first and last refracting surfaces be

$$\begin{pmatrix} a & b \\ c & d \end{pmatrix}.$$

Distances in the object and image spaces will be measured according to the Cartesian sign convention in order that the final results can be readily compared to the contents of Chapter 3. This requires care in the case of the object space. In the Cartesian convention distances in the object space are measured from the first refracting surface, whereas in the matrix formalism the first path element is measured forward from the object to the first refracting surface.

We suppose the system forms an image at a distance s_2 to the right of the last refracting surface of an object a distance s_1 measured *from* the first refracting surface to the object. Then the matrix equation relating an outgoing ray at the image at height y_2 and making an angle α_2 with the optic axis to the same ray from the object at height y_1 and at an angle α_1 is

$$\begin{pmatrix} n\alpha_2 \\ y_2 \end{pmatrix} = \begin{pmatrix} 1 & 0 \\ s_2 & 1 \end{pmatrix} \begin{pmatrix} a & b \\ c & d \end{pmatrix} \begin{pmatrix} 1 & 0 \\ -s_1 & 1 \end{pmatrix} \begin{pmatrix} n\alpha_1 \\ y_1 \end{pmatrix}.$$

The rightmost square matrix describes the straight line propagation from the object to the front surface of the first lens. s_1 has acquired the necessary minus sign to make it the distance from the object to the first refracting surface. The second square matrix describes the effect of the optical system, and the final square matrix describes the straight line propagation from the final lens surface to the image plane, over a positive distance s_2. Thus

$$\begin{pmatrix} n\alpha_2 \\ y_2 \end{pmatrix} = \begin{pmatrix} 1 & 0 \\ s_2 & 1 \end{pmatrix} \begin{pmatrix} a - bs_1 & b \\ c - ds_1 & d \end{pmatrix} \begin{pmatrix} n\alpha_1 \\ y_1 \end{pmatrix}$$

$$= \begin{pmatrix} a - bs_1 & b \\ as_2 - bs_1s_2 + c - ds_1 & bs_2 + d \end{pmatrix} \begin{pmatrix} n\alpha_1 \\ y_1 \end{pmatrix}. \quad \text{(B.1)}$$

Then:

$$y_2 = (as_2 - bs_1s_2 + c - ds_1)n\alpha_1 + (bs_2 + d)y_1. \quad \text{(B.2)}$$

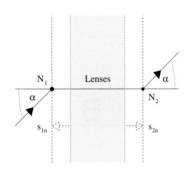

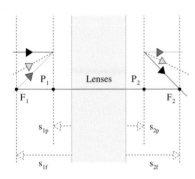

Fig. B.1 The upper panel shows the nodal planes of a lens system, and the lower panel the principal planes and the focal planes. The dotted lines at the edge of the shaded region indicate the outer surfaces of the final lenses. Note that the principal and nodal planes can lie inside the shaded region and also that the first principal/nodal plane could lie to the right of the second. In air the nodal and principal planes coincide.

The image distance off-axis y_2 will not depend on the direction α_1 that the ray leaves the object, so the coefficient

$$as_2 - bs_1s_2 + c - ds_1 = 0$$

$$\text{so} \quad s_2 = -\frac{c - ds_1}{a - bs_1} \tag{B.3}$$

$$\text{or} \quad s_1 = \frac{c + as_2}{d + bs_2}. \tag{B.4}$$

The distances of the focal points from the rear and front lens' surface respectively are obtained by making first s_1 and then s_2 equal to infinity

$$s_{2f} = -d/b, \quad \text{and} \quad s_{1f} = a/b. \tag{B.5}$$

Also from eqn B.2

$$y_2/y_1 = bs_2 + d.$$

The planes of unit magnification (principal planes) are found by setting $y_2/y_1 = 1$. This gives

$$s_{2p} = (1 - d)/b, \tag{B.6}$$

and

$$s_{1p} = (c + as_{2p}) = (-ad + a + cb)/b$$
$$= (a - 1)/b, \tag{B.7}$$

where the requirement that the determinant is unity, $ad - bc = 1$ was used. Collecting these results

$$s_{2p} = (1 - d)/b, \quad \text{and} \quad s_{1p} = (a - 1)/b. \tag{B.8}$$

The determination of the points of unit angular magnification (nodal points) starts from the imaging equation B.1

$$n\alpha_2 = (a - bs_1)n\alpha_1 + by_1.$$

On axis this reduces to

$$n\alpha_2 = (a - bs_1)n\alpha_1.$$

At the nodal points when $\alpha_2 = \alpha_1$

$$s_{1n} = (a - 1)/b = s_{1p}. \tag{B.9}$$

Similarly

$$s_{2n} = (1 - d)/b = s_{2p}. \tag{B.10}$$

These results confirm that where the object and image media are the same ($n_1 = n_2 = n$) the principal and nodal points coincide. We also see that the distances from the principal points to the focal points are the same in the object and image regions

$$s_{1p} - s_{1f} = -1/b, \tag{B.11}$$
$$s_{2f} - s_{2p} = -1/b. \tag{B.12}$$

Note that the directions of s_{1p}, etc. are given in figure B.1 according to the Cartesian convention. Clearly $-1/b$ is the focal length of the equivalent thin lens. Thus we have justified the use of the equivalent thin lens in treating any coaxial system of lenses in the paraxial approximation. For an explicit example see eqn. 3.36 where the optical system is a pair of lenses separated by a distance t.

Appendix: Kirchhoff's analysis of wave propagation

Kirchhoff, using a model for electromagnetic waves with scalar fields, was able to describe how electromagnetic disturbances propagate through an aperture. His results provide the formal basis for the Huygens–Fresnel use of secondary waves in predicting interference effects, including diffraction. Kirchhoff considered a monochromatic *scalar* field

$$E = \phi(\mathbf{r}_{\text{out}}) \exp(-i\omega t),$$

satisfying the wave equation

$$\nabla^2 E = [\partial^2 E/\partial t^2]/c^2, \tag{C.1}$$

which is the form the wave equation, eqn. 1.3, would take if the electric field were a scalar. Note that the radius vector is named $\mathbf{r}_{\text{out}}$ rather than $\mathbf{r}$ in order to avoid a change of variable later. The point at which the field is to be determined is taken to be the origin. Substituting the scalar field into the wave equation gives

$$\nabla^2 \phi = -k^2 \phi, \tag{C.2}$$

which is known as the *Helmholz equation* and where $k = \omega/c$. An outgoing spherical wave

$$\phi^r(\mathbf{r}_{\text{out}}) = \exp(ikr_{\text{out}})/r_{\text{out}}, \tag{C.3}$$

provides a solution that is valid at any point away from the origin itself but diverges at the origin. This will be used as the *reference solution*. Its derivative will also be required, and this is

$$\nabla \phi^r(\mathbf{r}_{\text{out}}) = ik\mathbf{r}_{\text{out}} \exp(ikr_{\text{out}})/r_{\text{out}}^2 - \mathbf{r}_{\text{out}} \exp(ikr_{\text{out}})/r_{\text{out}}^3$$
$$= (ikr_{\text{out}} - 1) \exp(ikr_{\text{out}}) \mathbf{r}_{\text{out}}/r_{\text{out}}^3. \tag{C.4}$$

Suppose the actual wave is $\phi(\mathbf{r}_{\text{out}})$, then applying eqn. C.2 twice gives

$$\phi\nabla^2\phi^r - \phi^r\nabla^2\phi = -k^2\phi\phi^r + k^2\phi^r\phi = 0,$$

at any point *away from the origin*. The next step is to integrate this function over a volume V of any form enclosing the origin but excluding a small sphere centred on the origin. This gives

$$\int_V [\phi \nabla^2 \phi^r - \phi^r \nabla^2 \phi] \, dV = 0.$$

Applying Gauss' theorem converts the LHS to an integral over the inner, S_i, and outer, S_o, surfaces enclosing V

$$\int_{S_i + S_o} [\phi \nabla \phi^r - \phi^r \nabla \phi] \cdot \mathbf{n} dS = 0, \tag{C.5}$$

where $\mathbf{n} dS$ is a vector element of either surface directed outward from V. Using eqns. C.3 and C.4 this becomes

$$\int_{S_i + S_o} [\exp(ikr_{out})/r_{out}^3] [(ikr_{out} - 1)\, \phi \, \mathbf{r}_{out} - r_{out}^2 \nabla \phi] \cdot \mathbf{n} \, dS = 0. \tag{C.6}$$

The contribution to the integral from the small sphere radius δr surrounding the origin is evaluated with $\mathbf{r}_{out} = -\delta r \mathbf{n}$, $dS = \delta r^2 d\Omega$, and $\phi = \phi(0)$. This gives

$$I_i = \int_{S_i} [-\exp(ik\delta r)/\delta r^3] [(ik\delta r - 1)\, \phi \, \delta r \mathbf{n} - \delta r^2 \nabla \phi] \cdot \mathbf{n} \, \delta r^2 d\Omega$$

$$= -\int_{S_i} \exp(ik\delta r) [(ik\delta r - 1)\phi(0) - \delta r \, \nabla \phi(0) \cdot \mathbf{n}] \, d\Omega.$$

Now δr can be made arbitrarily small and in the limit

$$\mathrm{Lt}_{\delta r \to 0}[I_i] \to 4\pi \phi(0).$$

Inserting this result in eqn. C.6 gives

$$\int_{S_o} [\phi \nabla \phi^r - \phi^r \nabla \phi] \cdot \mathbf{n} dS = -4\pi \phi(0), \tag{C.7}$$

which is the general result sought for any wave ϕ satisfying eqn. C.2. Thus if ϕ is known over a surface enclosing the point of interest the value of ϕ there can be calculated using eqn. C.7 where ϕ^r is the simple spherical wave reference function.

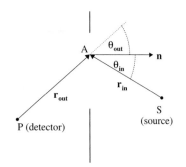

Fig. C.1 The point source S illuminates an aperture. P is the point of observation and A is a representative point on the aperture.

The simple optical arrangement pictured in figure C.1 will be used to obtain a general expression useful in calculating diffraction effects. A point source S illuminates an aperture, and the resulting wave intensity is observed at P. First we calculate the wave amplitude across the aperture and then insert that in eqn. C.7 to obtain the wave amplitude at P. A is any point in the aperture, and the unit normal at A to the surface through A spanning the aperture is $\mathbf{n}$. By definition

$$\mathbf{r}_{out} = \vec{PA}; \quad \mathbf{r}_{in} = \vec{SA}.$$

There are no obstructions between the source and A so the wave amplitude at A is

$$\phi(\mathbf{r}_{in}) = \exp(ikr_{in})/r_{in},$$

and its gradient at A is

$$\nabla\phi(\mathbf{r}_{\text{in}}) = \mathbf{r}_{\text{in}}(ikr_{\text{in}} - 1)\exp{(ikr_{\text{in}})}/(r_{\text{in}})^3.$$

Inserting $\phi(\mathbf{r}_{\text{in}})$ for ϕ and $\phi^r(\mathbf{r}_{\text{out}})$ for ϕ^r in eqn. C.7 gives the wave amplitude at P

$$\int_{S_o} \exp{[ik(r_{\text{in}} + r_{\text{out}})]}\{(ikr_{\text{out}} - 1)\mathbf{r}_{\text{out}} \cdot \mathbf{n}/(r_{\text{in}}r_{\text{out}}^3)$$
$$-(ikr_{\text{in}} - 1)\mathbf{r}_{\text{in}} \cdot \mathbf{n}/(r_{\text{in}}^3 r_{\text{out}})\}\mathrm{d}S = -4\pi\phi(0),$$

where the integral is restricted to the surface spanning the aperture, because this is the only region from which light can reach P. This presumes that the screen has no other effect than simply absorbing the wave incident on it. In the far field situation where the distance from the aperture to source S, and from the aperture to P are very much larger than the wavelength of the radiation, $kr_{\text{in}} \gg 1$ and $kr_{\text{out}} \gg 1$. With these restrictions the result is the Fresnel–Kirchhoff equation

$$\phi(0) = (-ik/2\pi)\int_{S}\{\exp{[ik(r_{\text{out}} + r_{\text{in}})]}/(r_{\text{out}}r_{\text{in}})\}$$
$$\times\{[\cos{(\theta_{\text{in}})} + \cos{(\theta_{\text{out}})}]/2\}\,\mathrm{d}S. \qquad (\text{C.8})$$

The term $[\cos{(\theta_{\text{in}})} + \cos{(\theta_{\text{out}})}]/2$ is called the *inclination factor* and in the backward direction this is exactly zero. This eliminates a perceived weakness of the Huygens–Fresnel theory that the secondary waves from an unobstructed plane wave would add up equally well in the backward as in the forward direction. In most cases met with in instrumental optics $\vec{SA}$ and $\vec{AP}$ are nearly collinear and the inclination factor can be taken to be unity. For such cases the above analysis is adequate.

The analysis presented here will not apply in the *near field* region close to the aperture. It is then no longer true that $kr_{\text{in}} \gg 1$ and $kr_{\text{out}} \gg 1$, and in addition the actual *vector* electric fields due to light from different points on the aperture are no longer all parallel. Hence a scalar field cannot be used. A calculation starting from Maxwell's equations with boundary conditions appropriate to the aperture is required rather than assuming the waves are absorbed totally over the surface of the screen without any other effect. At large distances, the *far field* region, such a calculation simply reproduces Kirchhoff's result. More details can be found in Born and Wolf.[1]

---

[1] Chapter 8 of the seventh edition of *Principles of Optics* by M. Born and E. Wolf published by Cambridge University Press (1999).

Appendix: The non-linear Schroedinger equation

The wave equation for the transmission of electromagnetic waves along an optical fibre is derived here, taking account of both dispersion and the optical Kerr effect. Then the soliton solution is presented. A direct derivation from Maxwell's equations is avoided because this is rather involved: instead a more general approach is used applicable to other systems. An important point to emphasize immediately is that the soliton waveform on a fibre is the envelope of the electromagnetic pulse.

The refractive index of the fibre for waves at angular frequency ω_0 is taken to be n_0 at low intensity, when non-linear effects are negligible. Then at a nearby angular frequency ω, equal to $\omega_0 + \Delta\omega$, and at intensities used in optical communications, the refractive index is

$$n = n_0 + \Delta\omega(\mathrm{d}n/\mathrm{d}\omega) + \Delta\omega^2(\mathrm{d}^2n/\mathrm{d}\omega^2)/2 + n_2P/A_{\mathrm{eff}}, \qquad \text{(D.1)}$$

where n_2 is defined by eqn. 14.55, P is the optical power and A_{eff} is the effective area of the optical mode. Multiplying this by $2\pi/\lambda$ where λ is the wavelength in free space, and re-expressing the result in terms of wave numbers gives

$$k = k_0 + k_1\Delta\omega + k_2(\Delta\omega)^2/2 + k_{\mathrm{NL}}P, \qquad \text{(D.2)}$$

where

$$k_0 = 2\pi n_0/\lambda, \;\; k_{\mathrm{NL}} = 2\pi n_2/(A_{\mathrm{eff}}\lambda), \;\; k_1 = \mathrm{d}k/\mathrm{d}\omega, \;\; k_2 = \mathrm{d}^2k/\mathrm{d}\omega^2.$$

A plane wave is chosen

$$E = u\exp\left[i(k_0z - \omega_0t)\right], \qquad \text{(D.3)}$$

where u is the envelope

$$u = u_0\exp\left[i(\Delta kz - \Delta\omega t)\right], \qquad \text{(D.4)}$$

and we put

$$P = P_0|u|^2. \qquad \text{(D.5)}$$

In the special circumstances of soliton propagation this envelope, u, will be invariant as the waves travel: it is therefore the component of interest. We now assert that if u is a solution of the wave equation

$$-i\partial u/\partial z = ik_1\partial u/\partial t - (k_2/2)\partial^2u/\partial t^2 + k_{\mathrm{NL}}Pu, \qquad \text{(D.6)}$$

then the dispersion relation, eqn. D.2, also holds true. In order to check this assertion the trial wave of eqn. D.4 is substituted into eqn. D.6. This gives

$$\Delta k \, u = k_1 \Delta \omega \, u + k_2 (\Delta \omega)^2 u / 2 + k_{\mathrm{NL}} P \, u. \qquad \text{(D.7)}$$

Dividing this by u then gives the expected dispersion relation. The wave equation is now simplified by transforming to the frame moving with the group velocity, k_1. This requires the coordinate substitutions

$$\tau = (t - k_1 z)/t_0, \quad \xi = z/z_c,$$

where z_c and t_0 are free parameters at this stage, but will acquire special significance later on when they are chosen appropriately. It follows that

$$t = t_0 \tau + k_1 z_c \xi, \quad z = z_c \xi,$$

and then

$$\partial/\partial\tau = (\partial t/\partial\tau)(\partial/\partial t) + (\partial z/\partial\tau)(\partial/\partial z) = t_0(\partial/\partial t);$$
$$\partial/\partial\xi = (\partial t/\partial\xi)(\partial/\partial t) + (\partial z/\partial\xi)(\partial/\partial z) = k_1 z_c(\partial/\partial t) + z_c(\partial/\partial z).$$

In this new frame the wave equation D.6 becomes

$$(-i/z_c)\partial u/\partial\xi = -(k_2/2t_0^2)\partial^2 u/\partial\tau^2 + k_{\mathrm{NL}} P_0 |u|^2 u. \qquad \text{(D.8)}$$

The parameters t_0 and z_c are now chosen in such a way as to link them to P_0, which anticipates the inherent properties of solitons. Setting

$$P_0 = (k_{\mathrm{NL}} z_c)^{-1} \qquad \text{(D.9)}$$
$$\text{and} \ \ t_0^2/z_c = -k_2, \qquad \text{(D.10)}$$

the wave equation, eqn. D.8 simplifies to

$$i\frac{\partial u}{\partial\xi} + \frac{1}{2}\frac{\partial^2 u}{\partial\tau^2} + |u|^2 u = 0, \qquad \text{(D.11)}$$

which is known as the *non-linear Schroedinger equation* (NLS).

The NLS has a simple solution of the form

$$u = \mathrm{sech}(\tau)\exp{(i\xi/2)}, \qquad \text{(D.12)}$$

which is the envelope exhibited in the second panel of figure 16.30. This is produced by an incident pulse whose envelope at $z = 0$ is $\mathrm{sech}(t/t_0)$, and is called the *fundamental* soliton. The requirement that the pulse shape remains invariant both constrains the shape to that given by eqn. D.12, and also limits the possible combinations of values that z_c, t_0 and P_0 can take. Starting from a fundamental soliton of given shape and energy others can be generated by scaling P_0, z_c and t_0 appropriately: P_0 is scaled by a factor m^2, z_c by a factor m^{-2} and t_0 by a factor m^{-1}; m being any arbitrary factor.

Other solitons can be produced if the initial pulse at $z = 0$ has the form $N \operatorname{sech}(t/t_0)$ where N is a positive integer. Solitons with values of N larger than unity undergo changes of shape as they travel, always returning to their initial shape after cycles of length $z_0 = \pi z_c/2$. The parameter z_0 is therefore called the *soliton period*, which can be expanded as

$$z_0 = -\pi t_0^2/(2k_2) = \pi^2 c t_0^2/(\lambda^2 D_{\mathrm{m}}), \qquad (\mathrm{D}.13)$$

where D_{m} is the group velocity dispersion defined in eqn. 16.17. If λ = 1550 nm, the time is expressed in picoseconds and the group velocity dispersion in ps/km/nm, then we have with these choices

$$z_0 \approx t_0^2/D_{\mathrm{m}} \text{ km}. \qquad (\mathrm{D}.14)$$

The half width of the soliton pulse at the half height of the pulse's power is given by

$$0.5 = \operatorname{sech}^2(\tau), \qquad (\mathrm{D}.15)$$

whence $\tau = 0.88$ and $t = 0.88 t_0$. Evidently t_0 is a measure of the soliton pulse width. In a similar way the power is constrained by

$$P_0 = 1/k_{\mathrm{NL}} z_c = A_{\mathrm{eff}} \lambda/(4 n_2 z_0). \qquad (\mathrm{D}.16)$$

A typical mode in telecom single mode fibre has an effective area $70 \, \mu \mathrm{m}^2$, while the Kerr coefficient for fused silica is $2.6 \, 10^{-20} \, \mathrm{m}^2 \, \mathrm{W}^{-1}$, so that

$$P_0 \approx 1000/z_0 \text{ mW}, \qquad (\mathrm{D}.17)$$

with z_0 again in kilometres.

Appendix: State vectors

The quantum state of photons (or other particles) is best described using state vectors when second quantization is being carried through. Consider a system whose eigenstates provide a set of orthonormal wavefunctions $\{\phi_i\}$. Suppose ψ is a normalized wavefunction describing a state of the system. This can be expanded as

$$\psi = \sum_i c_i \phi_i, \tag{E.1}$$

in which the c_is are complex coefficients. Then the integral

$$\int \psi^* \psi \, \mathrm{d}V = \sum_{ij} c_i^* c_j \int \phi_i^* \phi_j \, \mathrm{d}V = \sum_i c_i^* c_i. \tag{E.2}$$

The right hand side of this equation can be expressed in matrix notation as

$$\begin{pmatrix} c_1^* & c_2^* & c_3^* & \cdots \end{pmatrix} \begin{pmatrix} c_1 \\ c_2 \\ c_3 \\ \cdot \\ \cdot \end{pmatrix}, \tag{E.3}$$

which is identical to the scalar product of two vectors with coordinate lengths referred to the same set of orthogonal axes $(c_1^*, c_2^*, c_3^*, \ldots)$ and $(c_1, c_2, c_3, \ldots)$. The state of the system with wavefunction ψ can then be pictured as a *state vector* in a multi-dimensional space in which the unit length vectors along the axes are the state vectors of the eigenstates $\{\phi_i\}$: this space is known as a *Hilbert space*. Such state vectors are very useful in visualizing and making calculations in quantum mechanics. The following account uses the notation introduced by Dirac. A state vector for a column matrix is written $|\psi\rangle$ and called a *ket vector*. For an eigenstate with wavefunction ϕ_i the ket is $|\phi_i\rangle$. Then in vector form eqn. E.1 becomes

$$|\psi\rangle = \sum_i c_i |\phi_i\rangle. \tag{E.4}$$

Another type of state vector is needed to correspond to the row matrix in eqn. E.3. These are called *bra vectors*, and are written $\langle\psi|$ and $\langle\phi_i|$. Of course the vectors $|\psi\rangle$ and $\langle\psi|$ describe exactly the same state.[1] For

---

[1] The Hilbert spaces containing the bra and ket vectors are actually separate vector spaces known as dual spaces. See for example the second edition of *The Principles of Quantum Mechanics* by R. Shankar, published by Kluwer, Dordrecht (1994).

the bra vectors

$$\langle\psi| = \sum_i c_i^* \langle\phi_i|. \tag{E.5}$$

The scalar product $\langle\phi_i|\phi_j\rangle$ is defined by

$$\langle\phi_i|\phi_j\rangle = \int \phi_i^* \phi_j \, \mathrm{d}V = \delta_{ij}. \tag{E.6}$$

In addition we have

$$\langle\psi|\phi_i\rangle = c_i^*,$$
$$\langle\phi_i|\psi\rangle = c_i.$$

Then the state vectors $|\psi\rangle$ and $\langle\psi|$ can be expanded in this way

$$|\psi\rangle = \sum_i \langle\phi_i|\psi\rangle\,|\phi_i\rangle, \tag{E.7}$$

$$\langle\psi| = \sum_i \langle\psi|\phi_i\rangle\,\langle\phi_i|. \tag{E.8}$$

The scalar product of two different states, $|\psi_1\rangle$ and $|\psi_2\rangle$, is

$$\int \psi_2^* \psi_1 \, \mathrm{d}V = \langle\psi_2|\psi_1\rangle$$
$$= \sum_{ij} \langle\psi_2|\phi_i\rangle\,\langle\phi_i|\phi_j\rangle\,\langle\phi_j|\psi_1\rangle$$
$$= \sum_i \langle\psi_2|\phi_i\rangle\,\langle\phi_i|\psi_1\rangle$$
$$= \langle\psi_2|\,\sum_i[\,|\phi_i\rangle\,\langle\phi_i|\,]\,|\psi_1\rangle. \tag{E.9}$$

Comparing the first and fourth lines of this equation we see that the sum in the fourth line is the identity. Writing this in matrix notation gives

$$\sum_i [\,|\phi_i\rangle\,\langle\phi_i|\,] = \begin{pmatrix} 1 & 0 & 0 & . & . \\ 0 & 1 & 0 & . & . \\ 0 & 0 & 1 & . & . \\ . & . & . & . & . \\ . & . & . & . & . \end{pmatrix}, \tag{E.10}$$

where the right hand side is known as I the identity matrix. This useful result is known as a *closure relation*. The expectation value of an observable A in state described by $|\psi\rangle$ is given by

$$\langle\hat{A}\rangle = \int \psi^* \hat{A}\psi \, \mathrm{d}V$$
$$= \sum_{ij} c_i^* c_j \,\langle\phi_i|\hat{A}|\phi_j\rangle, \tag{E.11}$$

which can be re-expressed in this way

$$\langle\hat{A}\rangle = \sum_{ij} c_i^* A_{ij} c_j, \tag{E.12}$$

where the terms A_{ij} form a matrix with elements

$$A_{ij} = \langle \phi_i | \hat{A} | \phi_j \rangle. \tag{E.13}$$

Observables have real values so that $\sum_{ij} c_i^* c_j \langle \phi_i | \hat{A} | \phi_j \rangle$ must also be real. This can only be true if the combinations

$$c_i^* c_j A_{ij} + c_i c_j^* A_{ji}^*$$

are always real. Therefore it must be the case that

$$A_{ij}^* = A_{ji}. \tag{E.14}$$

In words, the *complex conjugate transpose* of any element of the matrix is equal to the element itself. Writing the complex conjugate transpose of the operator $\hat{A}$ as $\hat{A}^\dagger$ we have in turn that

$$\hat{A}^\dagger = \hat{A}, \tag{E.15}$$

which re-expresses the result noted in Chapter 13 that observables are associated with *Hermitean operators*.

In order to carry out the reverse process of obtaining a wavefunction referred to a point $(\mathbf{r}, t)$ in space time from a state vector we introduce a new state vector, $|\mathbf{r}, t\rangle$. This is a state vector whose wavefunction is a delta function at the space time point with coordinates $(\mathbf{r}, t)$. Then

$$\psi(\mathbf{r}, t) = \langle \mathbf{r}, t | \psi \rangle. \tag{E.16}$$

Rotations can be made in Hilbert space similar to rigid rotations in normal coordinate space and leave any product of state vectors $\langle \psi | \xi \rangle$ unaffected. However these transformations differ from rigid rotations in cordinate space because the length of a vector in Hilbert space is a complex number. Suppose U is such a *unitary transformation*, then

$$\langle \psi | \xi \rangle = \langle U\psi | U\xi \rangle. \tag{E.17}$$

Now $\langle U\psi |$ has components

$$(U\psi)_i^* = \sum_j U_{ij}^* \psi_j^* = \sum_j U_{ji}^\dagger \psi_j^* = \sum_j \psi_j^* U_{ji}^\dagger,$$

where the matrix $U^\dagger$ is the complex conjugate transpose of U. Then we have

$$\langle U\psi | = \langle \psi | U^\dagger. \tag{E.18}$$

Substituting this is in eqn. E.17

$$\langle \psi | \xi \rangle = \langle \psi | U^\dagger U \xi \rangle, \tag{E.19}$$

whence it follows that

$$U^\dagger U = I. \tag{E.20}$$

This means that the complex conjugate transpose of a unitary operator is also its inverse, U^{-1}, because $U^{-1}U = I$. Rigid rotations in coordinate space are real. Then not only are they unitary but they are also orthogonal, meaning that $\tilde{U}U = I$, where $\tilde{U}$ is the transpose of U, with $\tilde{U}_{ij} = U_{ji}$.

Appendix: Representations

There are alternative ways to assign the time dependence of a quantum system and the choice between these alternatives is made in a way that eases calculation and interpretation. The choice made in Chapter 13 when introducing quantum mechanics is called the *Schroedinger representation*, in which the wavefunction varies with time as described by eqn. 13.10

$$\Psi_S(t) = \exp\left(-i\hat{E}t/\hbar\right)\Psi_S(0), \tag{F.1}$$

while the operators are considered as not varying with time. The subscript S is being used here to distinguish quantities in the Schroedinger representation from those in the other representations described below. The expectation value of an observable A is given by eqn. 13.29

$$\langle \hat{A} \rangle = \int \Psi_S^*(t)\hat{A}_S\Psi_S(t)\,dV$$

$$= \int \Psi_S^*(0)\exp\left(i\hat{E}t/\hbar\right)\hat{A}_S\Psi_S(0)\exp\left(-i\hat{E}t/\hbar\right)dV. \tag{F.2}$$

The second representation is the *Heisenberg representation*, indicated by a subscript H, in which the time variation is shifted to the operators. It must remain true that

$$\langle \hat{A} \rangle = \int \Psi_H^*\hat{A}_H(t)\Psi_H\,dV, \tag{F.3}$$

but now

$$\Psi_H = \Psi_S(0), \tag{F.4}$$

$$\hat{A}_H(t) = \exp\left(i\hat{E}t/\hbar\right)\hat{A}_S\exp\left(-i\hat{E}t/\hbar\right). \tag{F.5}$$

In the Heisenberg representation the wavefunction is constant in time while the operators are time varying. The time variation is given by

$$i\hbar\left(d\hat{A}_H(t)/dt\right) = -\hat{E}\exp\left(i\hat{E}t/\hbar\right)\hat{A}_S\exp\left(-i\hat{E}t/\hbar\right)$$
$$+ \exp\left(i\hat{E}t/\hbar\right)\hat{A}_S\exp\left(-i\hat{E}t/\hbar\right)\hat{E}$$
$$= -\hat{E}\hat{A}_H + \hat{A}_H\hat{E}$$
$$= [\hat{A}_H(t), \hat{E}]. \tag{F.6}$$

A third common representation lies formally midway between the other two, and is called the *interaction representation*. This representation is useful if there is an interaction which weakly couples two systems so that the total energy operator can be written

$$\hat{E} = \hat{E}_0 + \hat{E}_I, \tag{F.7}$$

where E_0 is the energy of the two systems in the absence of any interaction, and E_I is the relatively small interaction energy. Then

$$\Psi_I(t) = \exp\left(i\hat{E}_0 t/\hbar\right)\Psi_S(t), \tag{F.8}$$
$$i\hbar\left(\mathrm{d}\Psi_I(t)/\mathrm{d}t\right) = \hat{E}_I\Psi_I(t), \tag{F.9}$$
$$\hat{A}_I = \exp\left(i\hat{E}_0 t/\hbar\right)\hat{A}_S \exp\left(-i\hat{E}_0 t/\hbar\right), \tag{F.10}$$
$$i\hbar\left(\mathrm{d}\hat{A}_I/\mathrm{d}t\right) = [\,\hat{A}_I(t), \hat{E}_0\,]. \tag{F.11}$$

The wavefunction evolves like a Schroedinger wavefunction under the action of the interaction energy alone. On the other hand the operators evolve like Heisenberg operators under the action of the uncoupled energies. In the absence of any interaction the interaction representation collapses to the Heisenberg representation. The interaction representation is the appropriate choice when the interaction energy is much smaller than energies of the non-interacting systems.

While the analysis in Chapters 12, 13 and 18 uses the Schroedinger representation, that presented in Chapter 19 uses the Heisenberg representation. In Chapter 18 the interaction of the em field and the atom is treated classically without operators. However in Chapter 19 the quantized electric fields are the time varying operators, while the photon states – and that includes the coherent laser-like states, the Fock states and the thermal states – are all time independent.

Appendix: Fermi's golden rule

Fermi's golden rule is a widely used method for calculating transition rates. Here it will be derived for atomic transitions involving electromagnetic radiation. Two assumptions are made: first that the interaction between the radiation and the atom is weak enough to be treated perturbatively; second that the spectral distribution of radiation is continuous across the frequencies involved in the transition. Results will be taken from the analysis presented in Section 18.2 rather than duplicating the steps here.

The interaction energy responsible for the transition is ΔW and is left unspecified. From Section 18.2, the fraction of atoms originally in the lower energy state 1 which enter the higher energy state 2 in time t is given by eqn. 18.26:

$$|c_2(t)|^2 = \Omega_0 \sin^2\left(\Delta\omega t/2\right)/\Delta\omega^2. \tag{G.1}$$

The term Ω_0^2 can be expanded to the form it took in Section 18.2 with the help of eqns. 18.19 and 18.24 giving

$$\Omega_0^2 = [4/\hbar^2]|\Delta W_{12}|^2.$$

Although the particular choice of an electric dipole interaction was used in Section 18.2 this expression will hold good for a perturbative process with interaction energy ΔW. Also in Section 18.2 the quantity $\hbar\Delta\omega$ was the difference between the photon's energy $\hbar\omega$ and the energy required for the atomic transition $E_2 - E_1$. Generalizing to include other perturbative processes $\hbar\Delta\omega$ becomes the overall energy imbalance in the reaction. Then more generally

$$\Omega_0^2 = [4/\hbar^2]\left| \int \psi_f^* \Delta W \psi_i \, dV \right|^2, \tag{G.2}$$

where ψ_f and ψ_i are now the waveforms describing the total final and total initial states with total energies $\hbar\omega_f$ and $\hbar\omega_i$ respectively. We define the transition rate for a general perturbative process by taking the limit

$$\gamma = \mathrm{Lt}_{t\to\infty}\{\,|c_2(t)|^2/t\,\}. \tag{G.3}$$

The essential component in this limit is

$$\mathrm{Lt}_{t\to\infty}\{\,\sin^2\left(\Delta\omega t/2\right)/\Delta\omega^2 t\,\} = (\pi/2)\delta(\Delta\omega). \tag{G.4}$$

The delta function emerging here ensures that when a rate is calculated the total energy will be conserved. Now reassembling the components of eqn. G.3

$$\gamma = [2\pi/\hbar^2] \left| \int \psi_f^* \Delta W \psi_i \, dV \right|^2 \delta(\omega_i - \omega_f), \tag{G.5}$$

which is known as Fermi's golden rule for calculating rates of reactions. If the initial and final states have state vectors $|i\rangle$ and $|f\rangle$ respectively, this can be rewritten

$$\gamma = [2\pi/\hbar^2] \sum_f | M_{if} |^2 \, \delta(\omega_i - \omega_f), \tag{G.6}$$

where

$$M_{if} = \langle f \mid \Delta W \mid i \rangle \tag{G.7}$$

is known as the *matrix element* for the transition. When the available final states form a continuum in energy we can proceed as follows. Define $N_f(\omega)$ to be the total number of states with energies from zero to $\hbar\omega$, then Fermi's golden rule becomes

$$\gamma = (2\pi/\hbar^2) \int | M_{if} |^2 \, \delta(\omega_i - \omega_f) \, (dN_f/d\omega_f) \, d\omega_f$$
$$= (2\pi/\hbar^2) \, | M_{if} |^2_{(\omega_f = \omega_i)} \, (dN_f/d\omega_f)_{(\omega_f = \omega_i)}. \tag{G.8}$$

In this prediction both the matrix element M_{if} and the density of the final states $(dN_f/d\omega_f)$ must be evaluated with ω_f equal to ω_i.

Appendix: Density matrix

The simplest states of the electromagnetic field are the eigenstates of Maxwell's equations which satisfy the boundary conditions: for example the TEM_{00} mode within a Fabry–Perot etalon. Then again the electromagnetic field could be a linear superposition of orthogonal eigenstates, which is still a coherent state of the field. Least simple are incoherent mixtures of coherent states of the electromagnetic field, and this is the structure of most beams of light.

A coherent state $|\psi\rangle$ can always be expanded as a linear superposition of orthogonal eigenstates $|n\rangle$

$$|\psi\rangle = \sum_n p_n |n\rangle, \tag{H.1}$$

where p_n is the amplitude of the eigenstate $|n\rangle$ in the superposition. Then the expectation value of an observable is given by

$$\langle \hat{A} \rangle = \sum_{m,n} p_m^* p_n \langle m|\hat{A}|n\rangle. \tag{H.2}$$

When there is an incoherent mixture of eigenstates $|n\rangle$ all the interference terms for which $m \neq n$ will average out to zero so that

$$\langle \hat{A} \rangle = \sum_n P_n \langle n|\hat{A}|n\rangle, \tag{H.3}$$

where P_n is the probability of the state $|n\rangle$ in the mixture and $\sum_n P_n = 1.0$.[1] Any coherence is signaled by the presence of non-zero off-diagonal elements in the density matrix.

The density matrix is a tool with which the analysis of incoherent, partially coherent or fully coherent states can be handled in a unified way. In the case of any state $|\psi\rangle$, mixed or pure, the *density matrix elements* are defined by

$$\rho_{nm} = \langle n|\psi\rangle\langle\psi|m\rangle, \tag{H.4}$$

and the expectation of the value of the observable A in the state $|\psi\rangle$ is

$$\langle \hat{A} \rangle = \langle\psi|\hat{A}|\psi\rangle$$
$$= \sum_{m,n} \langle\psi|m\rangle\langle m|\hat{A}|n\rangle\langle n|\psi\rangle, \tag{H.5}$$

[1] The eqns. H.2 and H.3 are repeated in Chapter 19 as eqns. 19.66 and 19.67 respectively.

where the closure relation, eqn. E.10, has been used twice. Then

$$\langle \hat{A} \rangle = \sum_{mn} \rho_{nm} A_{mn}, \tag{H.6}$$

where

$$A_{mn} = \langle m|\hat{A}|n \rangle. \tag{H.7}$$

When $|\psi\rangle$ is a coherent state, that is a linear superposition of eigenstates,

$$\rho_{nm} = p_m^* p_n, \tag{H.8}$$

hence it follows that

$$\rho_{mn} = \rho_{nm}^*. \tag{H.9}$$

On the other hand, if the state is totally incoherent

$$\rho_{mn} = P_m \text{ if } m = n,$$
$$\rho_{mn} = 0 \quad \text{if } m \neq n. \tag{H.10}$$

If there is partial coherence the values of the corresponding off-diagonal elements ρ_{mn} depart from zero and their presence will be signalled by the effects associated with coherence. For example interference fringes appear with low visibility if there is weak coherence. There can also be decoherence of an initially coherent system, as will happen when atoms excited by a laser into Rabi oscillations then make collisions which disturb their phases randomly. The off-diagonal components of the density matrix then decrease and will in general no longer satisfy eqn. H.9

Corresponding to the density matrix there is a *density matrix operator*

$$\hat{\rho} = |\psi\rangle\langle\psi| = \sum_{i,j} |i\rangle\langle j|\rho_{ij}. \tag{H.11}$$

This is unlike all operators met previously, because it is defined afresh for each mixture of states considered. Using eqn. H.6

$$\langle \hat{A} \rangle = \sum_{m,n} \rho_{nm} A_{mn}$$
$$= \sum_{m,n} \langle n|\hat{\rho}|m \rangle \langle m|\hat{A}|n \rangle$$
$$= \sum_{n} \langle n|\hat{\rho}\hat{A}|n \rangle. \tag{H.12}$$

An alternative way of presenting this result is

$$\langle \hat{A} \rangle = \text{Trace}[\hat{\rho}\hat{A}], \tag{H.13}$$

where the trace is the sum of the diagonal elements of the matrix for any selected eigenstates as the basis. In a pure eigenstate $|m\rangle$ we have

$$\hat{\rho} = |m\rangle\langle m|,$$
$$\text{and } \hat{\rho}^2 = |m\rangle\langle m|m\rangle\langle m| = \hat{\rho}, \tag{H.14}$$

which provides a useful test of the purity of a state.

Appendix: The Kramers–Kronig relations

The correlated behaviour of the absorption coefficient and refractive index close to an atomic or molecular transition frequency shown in figure 11.3 is quite distinctive. Figure 11.4 shows that the predictions of the model used in Section 11.5 accurately reproduce this behaviour around resonance. The relationship between dispersion and absorption is more general and has its origin in the physical principle known as *causality*.

Causality is the requirement that any physical effect should follow and not precede an action which causes the effect. Applying this principle to the polarization produced by an applied electric field will allow us to deduce the connection between absorption and dispersion. This connection takes the form of equations linking the real and imaginary parts of the electric susceptibility called the Kramers–Kronig relations.

The polarization $P(t)$ at time t can only depend on the values of the electric field up to the moment t. Thus

$$P(t) = \varepsilon_0 \int_{-\infty}^{\infty} G(t - t')E(t')\, \mathrm{d}t', \tag{I.1}$$

where causality requires that $G(t) = 0$ whenever $t < 0$. Also we can expect that for real materials $G(t)$ and its integral are finite. In addition because both $E(t)$ and $P(t)$ are real $G(t)$ must be real too. Taking the Fourier transform of eqn. I.1 gives

$$P(\omega) = \varepsilon_0 E(\omega) \int_{0}^{\infty} G(t)\exp\left(+i\omega t\right)\, \mathrm{d}t, \tag{I.2}$$

where the convolution theorem, eqn. 7.26, has been used. This equation can be compared to eqn. 9.3

$$P(\omega) = \varepsilon_0 \chi(\omega)E(\omega), \tag{I.3}$$

where $\chi(\omega)$ is the susceptibility of the material. We see immediately that

$$\chi(\omega) = \int_{0}^{\infty} G(t)\exp\left(+i\omega t\right)\, \mathrm{d}t. \tag{I.4}$$

Separating this equation into its real and imaginary parts

$$\chi_r(\omega) = \int_0^\infty G(t)\cos\omega t\,dt,$$

$$\chi_i(\omega) = \int_0^\infty G(t)\sin\omega t\,dt.$$

where the real part determines the refractive index and dispersion, while the imaginary part determines the absorption.[1] Hence

$$\chi_r(-\omega) = \chi_r(\omega) \text{ and } \chi_i(-\omega) = -\chi_i(\omega). \tag{I.5}$$

In order to proceed we need to extend eqn. I.4 to complex values, $\omega = \omega_r + i\omega_i$, for which the Argand diagram is shown in figure I.1. Within the upper half-plane of this Argand diagram

$$\exp(+i\omega t) = \exp(+i\omega_r t)\exp(-\omega_i t), \tag{I.6}$$

where $\omega_i = |\omega|\sin\theta$ is positive. It follows that the integrand appearing in eqn. I.4 converges to finite values everywhere in the upper half-plane. Thus $\chi(\omega)$ is analytic everywhere in the upper half-plane. This result is the direct consequence of the causality requirement: $G(t) = 0$ for $t < 0$.

It is also necessary to check that $\chi(\omega)$ vanishes as $|\omega| \to \infty$ in the upper half-plane. We have

$$\chi(\omega) \le \text{Max}\,G \int_0^\infty \exp(-|\omega|t\sin\theta)\,dt, \tag{I.7}$$

where $\text{Max}\,G$, the largest value taken by G, is bounded. Then

$$\chi(\omega) \le \text{Max}\,G/[|\omega|\sin\theta], \tag{I.8}$$

which tends to zero as $|\omega|$ tends to infinity except along the real axis where $\sin\theta = 0$. Along this axis

$$\chi(\omega_r) = \int_0^\infty G(t)\exp(+i\omega_r t)\,dt. \tag{I.9}$$

As $|\omega_r|$ tends to infinity the exponential term oscillates faster than any variation in $G(t)$, which means that contributions to the integral over each cycle cancel. Hence as $\omega_r \to \infty$, $\chi(\omega_r) \to 0$. It can also be proved that in the upper half plane as $|\omega| \to \infty$, $\omega\chi(\omega) \to 0$.

The final step is to obtain the relationships between $\chi_r(\omega)$ and $\chi_i(\omega)$. We define

$$u(\omega) = \chi(\omega)/(\omega - \omega_0), \tag{I.10}$$

where ω_0 is purely real. This function has a single pole at $\omega = \omega_0$ with residue $\chi(\omega_0)$, which is the function of interest. We have just proved that, due to causality, $\chi(\omega)$ is analytic in the upper half-plane; hence $u(\omega)$ has no poles in the upper half-plane. Referring to figure I.1 we

[1] If the choice is made to express complex sinusoidal plane waves by $\exp[i(\omega t - kz)]$ rather than the choice made throughout this book, $\exp[i(kz - \omega t)]$, then the effect is to replace χ_i by $-\chi_i$.

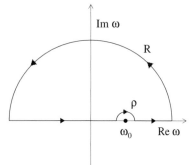

Fig. I.1 Contour used in the complex ω plane for the integral.

integrate $u(\omega)$ around the closed path indicated. There are no poles within this closed path so according to Cauchy's theorem the integral round the whole path is identically zero. The two semicircles have radii ρ and R, with R being allowed to go to infinity and ρ to zero. In the limit as $R \to \infty$ the contribution from the large half circle vanishes because $u(\omega) \to 0$ faster than $1/R$. Using Cauchy's theorem again the contribution from the small half circle is

$$\int u(\omega)\,\mathrm{d}\omega = -i\pi\chi(\omega_0).\tag{I.11}$$

The final part of the path along the real axis contributes

$$P\int_{-\infty}^{\infty} [\chi(\omega)/(\omega-\omega_0)]\,\mathrm{d}\omega,\tag{I.12}$$

where P specifies the *principal value* of the integral. This is its value with an infinitesimal segment deleted around $\omega = \omega_0$. Therefore because the total path integral is zero

$$P\int_{-\infty}^{\infty} [\chi(\omega)/(\omega-\omega_0)]\,\mathrm{d}\omega - i\pi\chi(\omega_0) = 0,\tag{I.13}$$

which when rearranged becomes

$$\chi(\omega_0) = \frac{-i}{\pi}P\int_{-\infty}^{\infty} [\chi(\omega)/(\omega-\omega_0)]\,\mathrm{d}\omega.\tag{I.14}$$

Next this equation is separated into its real and imaginary parts. First the real part

$$\chi_{\mathrm{r}}(\omega_0) = \frac{1}{\pi}P\int_{-\infty}^{\infty} [\chi_{\mathrm{i}}(\omega)/(\omega-\omega_0)]\,\mathrm{d}\omega$$
$$= \frac{1}{\pi}P\int_{0}^{\infty} [\chi_{\mathrm{i}}(\omega)/(\omega-\omega_0) - \chi_{\mathrm{i}}(-\omega)/(\omega+\omega_0)]\,\mathrm{d}\omega.$$

Making use of eqn. I.5 this becomes

$$\chi_{\mathrm{r}}(\omega_0) = \frac{2}{\pi}P\int_{0}^{\infty} \frac{\omega\chi_{\mathrm{i}}(\omega)}{(\omega^2-\omega_0^2)}\,\mathrm{d}\omega.\tag{I.15}$$

Next the imaginary part

$$\chi_{\mathrm{i}}(\omega_0) = \frac{-1}{\pi}P\int_{-\infty}^{\infty} [\chi_{\mathrm{r}}(\omega)/(\omega-\omega_0)]\,\mathrm{d}\omega$$
$$= \frac{-1}{\pi}P\int_{0}^{\infty} [\chi_{\mathrm{r}}(\omega)/(\omega-\omega_0) - \chi_{\mathrm{r}}(-\omega)/(\omega+\omega_0)]\,\mathrm{d}\omega.$$

Making use again of eqn. I.5 this becomes

$$\chi_{\mathrm{i}}(\omega_0) = \frac{2\omega_0}{\pi}P\int_{0}^{\infty} \frac{\chi_{\mathrm{r}}(\omega)}{(\omega_0^2-\omega^2)}\,\mathrm{d}\omega.\tag{I.16}$$

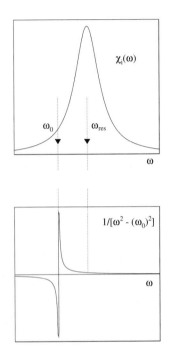

$\chi_i(\omega)$

ω_0 ω_{res}

ω

$1/[\omega^2 - (\omega_0)^2]$

ω

Fig. I.2 Components of the Kramers–Kronig integrand for χ_r.

These two equations I.15 and I.16, which are the consequences of the principle of causality, are known as the *Kramers–Kronig* relations.[2] The way these equations give rise to correlations between the dispersion and absorption near a resonance will be demonstrated next.

The functions whose product appears in the integral for $\chi_r(\omega_0)$ are depicted in figure I.2 near an absorptive peak at ω_{res}. The example shown is for $\omega_0 < \omega_{\text{res}}$. We see that the integral of the product is net positive and that it increases as ω_0 enters the region of the resonance from below. χ_r reaches a maximum value when ω_0 is below resonance because it evidently falls to zero at resonance. Above resonance it goes negative and reaches a minimum value when ω_0 is some way above resonance. Asymptotically the integral tends to zero far below or above resonance. This behaviour matches exactly the observed behaviour of χ_r near resonance.

A natural question to pose here is the following: how does $\chi(\omega)$ behave in the lower half-plane of ω? In this region $\chi(\omega)$ is not analytic; it is where the poles due to the atomic and molecular transitions (resonances) reside. The analysis, given in Section 11.5.1, of the electric susceptibility of a gas at low pressure provides a useful example. From eqn. 11.20 we see that a resonance at ω_{res} with width γ produces a corresponding pole in $\chi(\omega)$ at $\omega = \omega_{\text{res}} - i\gamma/2$. Thus the sharper the resonance the closer the pole lies to the real axis. The corollary of this statement is that a pole on the real axis corresponds to a stable state, one that never decays.

[2]For more mathematical details see *Mathematics of Classical and Quantum Physics* by F. W. Byron Jr. and R. W. Fuller, published by Dover Publications Inc., New York (1992).

Appendix: Solutions

J

Exercises

Chapter 1

(1.1) $1.06\,\mathrm{kVm^{-1}}$; $4.24\,10^{26}$ W.

(1.2) The light received at the comet's surface will be red-shifted, and there will be a further red-shift when the reflected light is received by the observer. Half of $4.74\,10^4\,\mathrm{m\,s^{-1}} = 2.37\,10^4\,\mathrm{m\,s^{-1}}$.

(1.3) The wavelength change by a factor of 4.62 which would give a recession velocity $0.91c$. In fact the waves have got longer because space has stretched by this factor since the radiation left the quasar.

(1.4) (a) Period = $0.33\,\mathrm{s}$, wavelength = 0.0667 m, velocity $0.2\,\mathrm{m\,s^{-1}}$ in the negative x direction. (b) Use $15x + 20y = k_x x + k_y y$ where k_x and k_y are the components of a vector **k** with $k = 1/\lambda$. Period = $0.2\,\mathrm{s}$, wavelength = 0.04 m, velocity = $0.02\,\mathrm{m\,s^{-1}}$. The velocity is rotated by $53.1°$ from the positive x-direction toward the positive y-direction.

(1.5) Using the pieces of data in the order given. $\phi = 0$, $\lambda = 2437$ m and $f = 1.23\,10^5$ Hz.

(1.6) $3.33\,\mu\mathrm{J}$.

(1.7)

(1.8) $\lambda = 30$ cm, $E = 137\,\mathrm{Vm^{-1}}$.

(1.9) $\lambda = 12.2$ cm.

(1.10) $x\Delta n/c = 10^{-9}$, so that the length is 0.6 m.

(1.11) α is approximately $1\,\mathrm{cm^{-1}}$, and the fraction penetrating is 0.37.

Chapter 2

(2.1) $465\ \mu$m.

(2.2) 1.77.

(2.3) $23.9°$, $32.6°$.

(2.4) NA = 0.123.

(2.5) Any imperfections on the outer surface of the glass thread would lead to light losses, and on a long run of many kilometres these losses would be intolerable. Any similar imperfections on the cladding are unimportant because the radiation does not reach its outer surface.

(2.6)

(2.7) Applying Snell's law to each pair of surfaces crossed in sequence gives $n_{\mathrm{air}} \sin\theta_{\mathrm{air}} = n_m \sin\theta_m$ for the mth layer. The final layer is air so the ray emerges at $45°$. By the same token it would require some sheet to have a refractive index less than air to cause TIR, which is impossible.

Chapter 3

(3.1) The image is at $+8.2$ cm beyond the postive lens, i.e. a virtual image. The magnification is $+0.58$ so the image is upright.

(3.2) Using eqn 3.37 the focal length is $+80$ cm, while eqn. 3.36 gives the matrix for the combination. Equation B.5 gives the distances between the focal planes and the front/rear lens surfaces. Equation B.8 gives the distances between the principal planes and the front/rear lens surfaces. The focal planes are 30 cm in front of the negative lens and 147 cm behind the positive lens. The first principal plane is 50 cm to the right of the negative lens and the second is 67 cm to the right of the positive lens. Compare this with the format of a telephoto lens.

(3.3) Figure 3.35 shows that reflected rays at the mirror edges. Thus the angular coverage is twice $\beta + \theta$, that is $2\alpha + 4\theta$, which in the case given is $\approx 74°$. Using eqn 3.5, $v = +0.47$ m, so the image is behind the mirror and 2.47 m from the shopkeper. The magnification is 0.067 and so the image of the hand is 0.67 cm across. This subtends an angle 0.3 mrad at the shopkeeper's eye.

(3.4) The image lies 100 cm in front of the mirror. The image size is 1.6 cm.

(3.5) 21 cm. Arrange them with curved faces inward which minimizes spherical aberration. A symmetric arrangement also minimizes coma and distortion so the stop is placed mid-way between the lenses.

(3.6) A focal length +19.49 cm lens from BK7 glass and focal length −34.37 cm lens from DF glass. Place the lens a distance apart equal to half the sum of their focal lengths.

(3.7) The distance separating object and image, $(v − u)$, can be rewritten using the thin lens equation in terms of u and f. Then setting to zero the differential with respect to u gives the minimum. There are two solutions, one being $u = 0$. Why can you reject this one?

(3.8) Substitute $u = −f + x_1$ and $v = f + x_2$ in the thin lens equation. This result can be useful in finding the principal planes of a lens system. The focal points can easily be located using parallel light. Then this equation can be used to determine the focal length from measurements of x_1 and x_2 for a few object and image points.

(3.9) −24.7 cm.

(3.10) An equation analogous to eqn. 3.36 results with F_1 and F_2 being the powers of the two surfaces, and F being the power of the lens. Substituting for the powers of the surfaces gives: $1/f = (n−1)(1/r_1 − 1/r_2) + T(n−1)^2/(nr_1r_2)$. Note that the reduced thickness $t = T/n$. Using eqn. B.8 the distances from the lens' surfaces to the principal planes are $TF_2/(nF)$ and $−TF_1/(nF)$.

(3.11) The proof in Section 3.4 can be repeated to give

$$1/v − 1/u = \frac{n − n_w}{n_w}(1/r_1 − 1/r_2).$$

Thus the focal length is multiplied by $(n − n_w)/[n_w(n − 1)]$ which in this case is 0.408, so that the focal length in water is 8.16 cm.

Chapter 4

(4.1) The image of something of angular size α formed by the objective is $f_o\alpha$ and this has angular size $f_o\alpha/f_e$ at the eyelens. The angular magnification is 8× and this is also the laser beam expansion factor.

The image of the objective formed by the eyelens is virtual: 5.5 mm diameter and 1.75 cm from the eyelens toward the objective. Evidently the eye cannot be placed at this point so there is a mismatch. The eye's image formed by the two lenses is 176.1 cm behind the objective and 40 mm diameter. The eyelens image formed by the objective is 112 cm behind the objective and 80 mm across. Thus the eye pupil is the aperture stop because seen from the object space in front of the objective its image subtends an angle smaller than the objective or the image of the eye lens.

(4.2) The objective is ×16 and the eyepiece ×10. The eyering is 28.7 mm from the eyelens and has diameter 0.30 mm. The small diameter of a single lens objective produces an eyering much smaller than the eye pupil; compound objectives produce a better match. The angular field of view is the eye lens' diameter divided by the lens separation, 15/195, or 4.6°. Projected onto the slide surface this defines a circle of diameter 150/195 or 0.77 mm.

(4.3) A lens to image ∞ at −3 m is required: f = −3 m.

(4.4) The lens needed will form an image at −1 m of an object at −25 cm: f = +33.3 cm.

(4.5) The mirrors are separated at their poles by a distance equal to the focal length of either. The upper one produces an image of the coin at infinity. The lower mirror images this at its focal point which is precisely in the aperture of the upper mirror.

(4.6) The first lens is placed so that the object is at its focus and the image is at infinity. The second lens images this at its focal plane inverted and at full size. Such a relay lens combination can be used to move an image longitudinally as required in a periscope.

(4.7) 25.7 m.

(4.8) From eqn 3.37 the focal length is $3f/4$. Either by imaging an object at infinity or by using eqn. B.5 the focal plane can be found to be a distance $f/4$ beyond the second lens. This puts the second principal plane $−f/2$ from the second lens, and $+f/6$ from the first lens. The first principal plane is $+f/2$ from the first lens and $−f/6$ from the second. Thus the first principal plane is to the right of the second principal plane. This is worth sketching in order to appreciate how ray tracing works in such cases. The image of the second lens formed by the first is $2f$ from the first lens, that is on the far side of the second lens from the first, and magnified 3×. Evidently this would subtend a larger angle at the objective than the first lens in the eyepiece, so the first lens is the field stop.

(4.9) In the figure referred to, the angle of incidence, α, at the surface of the spherical lens of the extreme ray is given by $\sin \alpha = \mathrm{PC}_1/R$ where R is the radius of this lens. Thus $\sin \alpha$ equals $1/($the refractive index of the sphere$)$. The angle of refraction is thus $90°$. Then the semi-angle of the cone of rays is $\alpha = \sin^{-1}(1/1.4)$ or $45.6°$.

(4.10) Using eqn. 4.24 the widest angle which a ray at the optical axis makes with this axis in a GRIN lens is given by

$$n(0)\cos\alpha(0) = n(R) = n(0)\sqrt{(1 - g^2 R^2)}.$$

Thus the numerical aperture is

$$\mathrm{NA} = n(0)\sin\alpha(0) = n(0)gR.$$

Chapter 5

(5.1) 9.5 mm.

(5.2) The angle subtended by the source aperture at the slits, θ_s, determines the lateral coherence length $d_c = \lambda/\theta_s$. With the given aperture and wavelength: $\theta_s < 6.33\,10^{-3}$ and the corresponding limit on the slit width is $w = \theta_s f < 1.9$ mm.

(5.3) The coherence length, l_c is that of the wavepackets. Then the coherence time, $\tau_c = l_c/c = 3.3\,10^{-9}$ s. In turn the frequency spread $\Delta\nu = 1/\tau_c = 0.3$ GHz. This is a fraction $5\,10^{-7}$ of the frequency. Note that $\Delta\lambda/\lambda = -\Delta\nu/\nu$ so this is also the wavelength fraction.

(5.4) Between consecutive fringes the path has changed by λ and the gap thickness has changed by $\lambda/2 = 366.5$ nm. Thus the angle required $\alpha = 366.5\,10^{-9}/1.5\,10^{-2} = 2.4\,10^{-5}$ rad.

(5.5) The angular frequency of the Earth's rotation is $\Omega = 2\pi/86400 = 7.27\,10^{-5}$ rad s^{-1}. Then using eqn. 5.41, $\Delta f = 27.27\,10^{-5}\cos 40°/6.33\,10^{-7} = 176$ Hz.

(5.6) The coherence time is $\tau_c = 10^{-14}$ s so the coherence length is $3\,\mu$m. The frequency spread is $\Delta f = 1/\tau_c = 10^{14}$ Hz. This compares to a mean frequency of $c/6.33\,10^{-7} = 4.7\,10^{14}$ Hz. Consequently the wavelength spread is $\Delta\lambda = \lambda\Delta f/f = 133$ nm.

(5.7) The change in the optical path length through the arm filled with gas is $(n-1)L$ so that the difference between the two paths changes by this length also. This is $(n-1)L/\lambda$ wavelengths so the number of fringes passing the centre of the field will be the same.

(5.8) At a 2 cm separation of what are effectively Young's slits the angular fringe separation is $\lambda/d = 2.5\,10^{-5}$ rad. The lens produces a fringe separation of $2.5\,10^{-6}$ m. These need to be viewed with a low power microscope.

(5.9) The finesse is 61.2 and the free spectral range $\Delta\lambda_{\mathrm{fsr}}$ is 0.0417 nm. In order to exploit the finesse the surfaces need to be flat to $\lambda/\mathcal{F} = 10$ nm. The chromatic resolving power is $7.34\,10^5$. The etendue is the product of the pinhole area and the solid angle of light arriving. The angular radius of central fringe, θ, is obtained using

$$2nd\cos\theta = 2nd(1 - \theta^2/2) = m(\lambda - \Delta\lambda)$$

and

$$2nd = m\lambda$$

so that $\theta = \sqrt{\Delta\lambda/\lambda}$. This requires a pinhole diameter $2f\theta = 117\,\mu$m. The area of this pinhole is $A_{\mathrm{pin}} = 1.07\,10^{-8}$ m^2. The area of the lens which is illuminated is just the clear area of the etalon which is of diameter 2.5 cm. The radius subtends an angle at the pinhole of $\theta_{\mathrm{ill}} = \tan^{-1}(1.25/5) = 14°$. Thus the solid angle subtended by the area of the lens illuminated is $\Omega_{\mathrm{pin}} = 2\pi(1 - \cos\theta_{\mathrm{ill}}) = 0.188$ sterad. Hence the etendue is $A_{\mathrm{pin}}\Omega_{\mathrm{pin}} = 202\,\mu$m^2 sterad.

(5.10) $2\Delta f_{1/2} = \Delta f_{\mathrm{fsr}}/\mathcal{F}$.

Chapter 6

(6.1) The zeroes occur when the phase delay between slits is $\pi/2$, π and $3\pi/2$; the subsidiary maxima at $2\pi/3$ and $4\pi/3$.

(6.2) CRP is 40 000. Blaze angle is $30°$. The nth principal maximum for wavelength λ/n will coincide with the first principal maximum for wavelength λ. Hence to exclude these the free spectral range is around 250 nm.

(6.3) 1.3 arcsec $= 6.30\,\mu$rad, and the diameter required is at least 9.7 cm.

(6.4) The velocity of the Earth's surface under the Moon's shadow is, at the equator, 466 m s^{-1}. Hence the fringe frequency is $466/12 = 39$ Hz. Taking into account the Moon's motion round the Earth changes this frequency by 3%.

(6.5) The required CRP is $\lambda/\Delta\lambda = 589.3/0.597 = 987$. To achieve this a grating of at least 5 mm width is required. The maxima are located at $\theta = \lambda/($line spacing$) = 0.12$ rad $= 6.8°$. Their angular separation is given by $\cos\theta\Delta\theta = \Delta\lambda/($line spacing$)$

= $1.19\,10^{-4}$ rad. Since the angles are small this translates at the focal plane to $f\Delta\theta = 59\,\mu$m. The angular resolution of the lens must be much less than $\Delta\theta$, and using eqn. 6.19 shows the diameter $D \gg 1.22\lambda/\Delta\theta = 6$ mm, so the $f/\#$ required ($\gg 0.01$) is easily obtained.

(6.6) The zeroes of the single slit pattern are at angles $\pm\lambda/d$ and the principal maxima are at angles $\pm n\lambda/a$ where n is integral. Thus the last maximum inside will be the mth for which $m < a/d$. Hence there are $a/d-1$ on either side plus the central one, in all $2a/d - 1$.

(6.7) Optical path lengths increase by a factor equal to the refractive index of μ. Hence phase differences increase by the same factor. Consequently the angular positions of the Fraunhofer single slit pattern zeroes previously given by eqn. 6.6 are now given by $\mu d\sin\theta = n\lambda$, with λ being the wavelength in free space as before. Thus the angular pattern shrinks by a factor around μ, or 1.33 for water. Evidently this shrinkage will occur for multi-slit and for Fresnel diffraction patterns also.

(6.8) The phase difference between successive slits changes from that given in 6.11 to $\beta = ka\sin\theta + \pi/\cos\theta$. Then close to the forward direction, where $\cos\theta$ is close to unity eqn. 6.14 for principal maxima yields $a\sin\theta = (p - 1/2)\lambda$, so we see that the maxima have been displaced by half their angular spacing, and this includes the central peak. Away from the forward direction the displacement will gradually increase. Evidently the effect is similar to that of blazing the grating, only in this case the multislit pattern moves relative to the single slit envelope.

(6.9) The phase difference between successive zones is converted from π to 2π so that all are in phase. Thus the intensity is increased by a factor 4.

(6.10) Substituting in eqn. 6.26 yields $\sin\alpha = 0.108$ and $\alpha = 6.2°$. The potential CRP is 60 000. The angular width of the maxima is given by $\cos\alpha\Delta\alpha/(\sin\theta - \sin\alpha) = 1/$CRP. Whence $\Delta\alpha = 1.26\,10^{-5}$ rad and the slit width required to match this is 12.6 μm. Using the expression for etendue eqn. 6.39 gives $\mathcal{T} = 0.0016$ mm^2 sterad. This result will be of interest when examining the relative performance of grating and Fourier transform interferometers in the following chapter.

(6.11) Using the notation of Section 6.15 the area of the mode at the waist is πw_0^2 and the solid angle spread there is $\pi\theta^2$. The product is the etendue, and using eqn. 6.63 this comes to λ^2.

(6.12) First put $\pi w_0^2/z\lambda = A$. Then using eqns 6.64 and 6.72 we get

$$\lambda R/\pi w^2 = \lambda z(1 + A^2)/[\pi w_0^2(1 + A^{-2})] = A,$$

as required. Replacing $\pi w_0^2/z\lambda$ by $\lambda R/\pi w^2$ converts eqns. 6.64 and 6.72 to eqns. 6.73 and 6.74.

Chapter 7

(7.1) $f(x) = \sum(-\lambda/m\pi)\sin(2\pi x/\lambda)\cos(m\pi)$.

(7.2) The count is $\lambda_{\mathrm{mean}}/\Delta\lambda = 987$.

(7.3) (a) In the integral

$$\int_{-\infty}^{\infty} f(x')g(x - x')\,\mathrm{d}x'$$

make the substitution $y = x - x'$. Alternatively you can use the convolution theorem. (b) The corresponding square pulse will be negative rather than positive and three times taller. One way to picture a convolution is to imagine one function, f, to be a sequence of contiguous delta functions, each with a multiplier equal to the value of f at its location, $f(x_m)$. When the second function g is convolved with the delta function at x_m it is displaced to lie centred at x_m and is scaled up by the value of the first function there.

(7.4) The microwave wavelength corresponding to a frequency of 50 MHz in lithium niobate is $1.32\,10^{-4}$ m. We can use eqn. 7.72 for the angular spread of the microwave beam, giving $\Delta\theta_{\mathrm{B}} = (6.32\,10^{-7})(2\,10^7)/2(6.6\,10^3)$, that is $0.958\,10^{-3}$ rad. The width of the laser beam can be obtained from eqn. 7.73 $w = (6.32\,10^{-7})100/2(0.958\,10^{-3})$ or 3.3 cm.

(7.5) 2.44 GHz frequency and 0.00337 nm in wavelength.

(7.6) The spacing across the surface is 632.8 nm/($2\sin 30°$) $= 0.6328\,\mu$m, or 1580 lines/mm.

(7.7) If the required $f/\#$ is fn then $1.22\,\lambda\,fn = 10^{-5}$ m, whence $fn = 12.9$.

(7.8) (a) The complex electric field is $E(t) = E(0)\exp(-\gamma t/2)\exp(-i\omega t)$. Thus

$$g^{(1)}(\tau) = \int E^*(t)E(t + \tau)\,\mathrm{d}t\Big/\int E^*(t)E(t)\,\mathrm{d}t$$

$$= \int \exp(-\gamma(t + \tau/2))\exp(-i\omega\tau)\,\mathrm{d}t$$

$$\Big/\int \exp(-\gamma t)\,\mathrm{d}t$$

$$= \exp(-i\omega\tau - \gamma\tau/2).$$

This result is independent of the sign of τ, so that $g^{(1)}(\tau) = \exp\left(-i\omega\tau - \gamma|\tau|/2\right)$. (b) The visibility is $\exp\left(-\gamma x/2c\right)$ where x is the path difference.

(7.9) (a) The complex electric field is $E(t) = E(0)\exp\left(-\sigma_\omega^2 t^2\right)\exp\left(-i\omega_0 t\right)$, so that

$$g^{(1)}(\tau) = \int \exp\left(-i\omega_0\tau\right)\exp\left(-\sigma_\omega^2 t^2\right)$$

$$\exp\left(-\sigma_\omega^2 (t+\tau)^2\right)\mathrm{d}t / \int \exp\left(-2\sigma_\omega^2 t^2\right)\mathrm{d}t$$

$$= \exp\left(-i\omega_0\tau - \sigma_\omega^2\tau^2/2\right)$$

$$\times \int \exp\left(-\sigma_\omega^2(\sqrt{2}t + \tau/\sqrt{(2)})^2\right)\mathrm{d}t$$

$$/ \int \exp\left(-2\sigma_\omega^2 t^2\right)\mathrm{d}t$$

$$= \exp\left(i\omega\tau - \sigma_\omega^2\tau^2/2\right).$$

(b) The visibility is $\exp\left(-\sigma_\omega^2\tau^2/2\right)$.

(7.10) The pulsed electric field is $E(t) = s(t)\,p(t)$, so the Fourier transform is the convolution $e(\omega) = \mathrm{FT}[s] \otimes \mathrm{FT}[p]$ where s is the sinusoidal waveform and p the pulse. Thus

$$\mathrm{FT}[s] = \int \exp\left(\omega_0 t\right)\exp\left(i\omega t\right)\mathrm{d}t$$

$$= \delta(\omega - \omega_0).$$

The repetitive pulse train has a similar expansion to that given in eqn. 7.4

$$p(t) = (2/\pi)\left[\cos\left(\omega_\mathrm{p} t\right) - \cos\left(3\omega_\mathrm{p} t\right)/3 + \dots\right].$$

Its Fourier transform is

$$\mathrm{FT}[p] = (1/\pi)\left[\delta(\omega - \omega_\mathrm{p}) + \delta(\omega + \omega_\mathrm{p})\right.$$
$$\left. -(1/3)(\delta(\omega - 3\omega_\mathrm{p}) + \delta(\omega + 3\omega_\mathrm{p})) + \dots\right].$$

Now for example the convolution of $\delta(\omega - \omega_0)$ with $\delta(\omega - \omega_\mathrm{p})$ gives

$$\int \delta(\omega' - \omega)\delta(\omega - \omega_\mathrm{p} - \omega')\,\mathrm{d}\omega'$$

$$= \delta(\omega - \omega_\mathrm{p} - \omega_0).$$

Then the electric field as a function of angular frequency is

$$e(\omega) =$$
$$(1/\pi)\left[\delta(\omega - \omega_0 - \omega_\mathrm{p}) + \delta(\omega - \omega_0 + \omega_\mathrm{p})\right]$$
$$-(1/3\pi)\left[\delta(\omega - \omega_0 - 3\omega_\mathrm{p}) + \delta(\omega - \omega_0 + 3\omega_\mathrm{p})\right]$$
$$+ \dots.$$

Thus the structure is a line spectrum which contains lines spaced at equal intervals $2\omega_\mathrm{p}$ from $\omega = \omega_0 + \omega_\mathrm{p}$.

(7.11) For the Lorentzian $P(\omega_0) = 2/\pi\gamma$, or $0.635/\gamma$. By inspection the FWHM, $\Delta\omega$, is γ. In the case of the Gaussian $P(\omega_0) = 1/(\sqrt{2\pi}\sigma_\omega)$, or $1/2.51\sigma_\omega$. At the half height point, $\omega + \Delta\omega/2$, we have $\exp\left(-\Delta\omega^2/8\sigma_\omega^2\right) = 1/2$, thus $\Delta\omega = \sqrt{(8\ln 2)}\sigma_\omega$, or $2.35\sigma_\omega$. Then $P(\omega_0)$ is approximately $1/\Delta\omega$ in both cases.

(7.12) The first differential is

$$\frac{\mathrm{d}f}{\mathrm{d}x} = \frac{\mathrm{d}}{\mathrm{d}x}\int F(k)\exp\left(ikx\right)\mathrm{d}k/(2\pi)$$

$$= \int (ik)F(k)\exp\left(ikx\right)\mathrm{d}k/(2\pi).$$

Thus the Fourier transform of $\mathrm{d}f/\mathrm{d}x$ is $ikF(k)$. Similarly the Fourier transform of $\mathrm{d}^2 f/\mathrm{d}x^2$ is $-k^2 F(k)$.

(7.13) The parallel analysis is made for the case that the electric field has been extended to the complex form described in eqn. 7.18. Then the auto-correlation

$$\overline{E^*(t)E(t+\tau)}$$
$$= \int_{-\infty}^{+\infty}\int_0^{+\infty}\int_0^{+\infty} e^*(\omega)e(\omega')$$
$$\exp\left(i\omega t\right)\exp\left[-i\omega'(t+\tau)\right]\mathrm{d}\omega\,\mathrm{d}\omega'\,\mathrm{d}t/(4\pi^2 T).$$

Proceeding as in Section 7.3.2, and noting that the delta function vanishes whenever $\omega < 0$, we get

$$\overline{E^*(t)E(t+\tau)}$$
$$= (4/T)\int_0^\infty |e(\omega)|^2 \exp\left(-i\omega\tau\right) \text{ for } \omega > 0,$$
$$= 0 \text{ for } \omega < 0.$$

Taking the Fourier transform gives

$$\int_{-\infty}^\infty \overline{E^*(t)E(t+\tau)}\exp\left(i\omega t\right)\mathrm{d}\tau = (4/T)|e(\omega)|^2.$$

Consequently the power spectrum given by eqn. 7.39 is

$$P(\omega) = FT\{\overline{E^*(t)E(t+\tau)}\}/(4\pi Z_0),$$

differing by a factor $1/4$ from eqn. 7.42.

Chapter 8

(8.1) The apparent luminosity is given by eqn. 8.5 $\ell = 2.52\,10^{-8}\,10^{-16/2.5}$ which gives $0.577\,10^{-8}\,\mathrm{W\,m^{-2}}$. Now the power received by the telescope is $\ell\pi d^2/4$ where d is the primary mirror diameter, which gives $2.61\,10^{-8}\,\mathrm{W}$. Over $1\,\mathrm{ms}$ the charge accumulated is $26.1\,\mathrm{pC}$.

(8.2) When the Fried prameter is $20\,\mathrm{cm}$ the root mean square wavefront distortion over $20\,\mathrm{cm}$ is $1\,\mathrm{rad}$. If this is all tip/tilt the rms waveform tilt is $\lambda/6$ or $100\,\mathrm{nm}$ in $20\,\mathrm{cm}$, making an angle of $5\,10^{-7}\,\mathrm{rad}$. Consequently the separation fluctuates by $30\,(5\,10^{-7}) = 15\,\mu\mathrm{m}$ in the image plane.

(8.3) If the baseline is aligned parallel to the line joining the stars the separation B must be such that $\lambda/B < 0.003\,\mathrm{arcsec}$, that is $1.5\,10^{-8}\,\mathrm{rad}$. Thus $B > 35\,\mathrm{m}$. If B is at an angle θ to this direction the resolution drops by a factor $\cos\theta$. When the baseline is perpendicular to the line joining the stars the resolution is simply that of either telescope alone.

(8.4) The complex visibility becomes

$$\int_{-M}^{M}\int_{-L}^{L} I(\ell,m)\exp\left(\left[i(lu+mv)\right]\right)\mathrm{d}\ell\mathrm{d}m.$$

Taking one integral

$$\int_{-L}^{L} I(\ell,m)\exp(ilu)\mathrm{d}\ell$$

$$= \left[\int_{-L}^{0} + \int_{0}^{L}\right] I(\ell,m)\exp(ilu)\mathrm{d}\ell.$$

Putting $p = -\ell$ in the first integral gives

$$-\int_{L}^{0} I(-p,m)\exp(-ipu)\mathrm{d}p$$

$$= \int_{0}^{L} I(p,m)\exp(-ipu)\mathrm{d}p,$$

because $I(p,m)$ is symmetric in both arguments. Thus the integral over the range $-L$ to $+L$ is

$$\int_{0}^{L} I(\ell,m)\cos(i\ell u)\mathrm{d}\ell,$$

which is real. By extension the visibility is real.

(8.5) Using eqn. 8.37, and taking a measurement time of $10\,\mathrm{ms}$, the power required is $(2.8\,10^{-19}/10^{-21})^2/10$ watts or $8\,\mathrm{kW}$. The time for one pass to and fro in a cavity is $26.7\,\mu\mathrm{s}$, so with a finesse of 500 the storage time roughly $13\,\mathrm{ms}$.

(8.6) The total source intensity is $[\delta(\ell-\theta)+\delta(\ell+\theta)]\,\delta(m)/2$ where the line joining the two stars is taken as the baseline axis. Inserting this in the expression for visibility, eqn. 8.21, gives

$$\int [\delta(\ell-\theta)+\delta(\ell+\theta)]\,[\exp(i\ell u)/2]\,\mathrm{d}\ell$$

$$= [\exp(i\theta u)+\exp(-i\theta u)]/2 = \cos(\theta u).$$

This gives unity when the stars merge so that θ is zero; which confirms the choice of the constant factor, 0.5, in the source intensity.

(8.7) On the sky the pixel angular size is p/f. This should equal half the angular resolution, that is $\lambda/2D$. Thus $f/D = 2p/\lambda$. With oversampling there is a possibility of detecting spurious finer spatial detail than the image actually contains. See Section 7.3.2.

(8.8) Use figure 8.2. The distance the secondary's first focus lies behind its pole is $a(e-1) = 0.8\,\mathrm{m}$. Thus the mirrors lie $8-0.8 = 7.2\,\mathrm{m}$ apart. The distance the second focus of the secondary lies in front of its pole is $a(e+1) = 8.8\,\mathrm{m}$ so that the final image lies $8.8-7.2 = 1.6\,\mathrm{m}$ behind the primary. The primary image size for a source of angular size θ is $8\theta\,\mathrm{m}$, and the magnification of the secondary is $(e+1)/(e-1) = 11$, Thus the final image size is $88\theta\,\mathrm{m}$. There are $206\,265$ arcsec per rad, so the plate scale is $88/206\,265 = 4.32\,10^{-4}\,\mathrm{m/arcsec}$, that is $432\,\mu\mathrm{m/arcsec}$.

(8.9) The f/# of the first telescope is 2.08. Using eqn. 8.2 gives its comatic flare to be $(\theta/17.4)$ arcsec at an angle θ off axis. Thus the field of view with less than 0.5 arcsec distortion has a semi-angle 8.7 arcsec. In the case of the HST eqn. 8.4 gives a field of view of $12\,\mathrm{arcmin}$ which is roughly 100 times larger.

(8.10) Eqn. 6.64 relates the width, w, at a distance z from the waist, to the width at the waist, w_0. In the case of the confocal Fabry–Perot cavity $z = L/2$, so we have

$$w^2 = w_0^2 + \lambda^2 L^2/(4\pi^2 w_0^2).$$

The first term is smaller the smaller w_0 is made, but the second, diffractive term grows as w_0 is made smaller. To get the minimum we calculate $\partial w^2/\partial w_0^2$ and set it to zero.

$$\partial w^2/\partial w_0^2 = 1 - \lambda^2 L^2/(4\pi^2 w_0^4),$$

which is zero when $w_0^2 = \lambda L/(2\pi)$. Substituting this value back into the original expression for w^2 gives $w^2 = \lambda L/\pi$. In the example given w_0 is

1.78 cm and the width at the mirrors is 2.52 cm. In practice the mirrors have to be made several times larger in radius in order to capture the tails of the Gaussian distribution. If the mirrors were 2.52 cm in radius the loss of light spilling over round a mirror's edge would be of order 1% per reflection with perfect alignment. However the stored radiation undergoes a very large number of reflections so that a loss even at this low level per reflection is prohibitive.

Chapter 9

(9.1) $1.28\,10^8\,\mathrm{m\,s^{-1}}$; $160\,\Omega$; $269\,\mathrm{nm}$.

(9.2) Equations 9.55 and 9.56 give the non-zero electric and magnetic field components: E_θ and B_ϕ. At $90°$ these are $394\,\mathrm{Vm^{-1}}$ and $1.31\,\mu\mathrm{T}$ respectively. The energy flow is instantaneously $E_\theta B_\phi \cos^2{(\omega t)}/\mu_0$ and time averaging gives $206\,\mathrm{Wm^{-2}}$. At $45°$ the fields are $279\,\mathrm{Vm^{-1}}$ and $0.93\,\mu\mathrm{T}$, while the energy flow is $1.73\,\mathrm{GW}$. Along the dipole axis the fields vanish and hence the energy flow is zero.

(9.3) The critical angle is $41.81°$. Using eqn. 9.67 gives for the case of $42°$ incidence $k_t = \mathrm{i}\,0.854\,\mu\mathrm{m}^{-1}$ so that the intensity falls by a factor e in a distance perpendicular to the surface of $0.585\,\mu\mathrm{m}$. At $41.82°$ $k_t = \mathrm{i}\,0.194\,\mu\mathrm{m}^{-1}$ and the penetration distance is $2.57\,\mu\mathrm{m}$. The penetration depth remains small even at an angle only a fraction of a degree above the critical angle.

(9.4) Recall that $\tan\theta_B = n_2/n_1$, so that $\tan\theta_B = n_1/n_2$. This makes $\tan\theta_B = \cot\theta_B$, and the required result follows.

(9.5) Fresnel's equation for the s-polarization transmission coefficient is

$$t_s = 2n_1\cos\theta_i/(n_1\cos\theta_i + n_2\cos\theta_t)$$

so the reverse path requires that the refractive indices are exchanged and the angles exchanged in the above equation giving

$$t'_s = 2n_2\cos\theta_t/(n_2\cos\theta_t + n_1\cos\theta_i).$$

The same exchanges can be used to obtain r'_s from r_s. Then Stokes' equations follow.

(9.6) $r_s = -0.240$, $t_s = 0.760$, $r_p = 0.159$ and $t_p = 0.772$. These values can be checked against figure 9.12.

(9.7) According to Stokes' relation $r' = -r$ so the reflectances at the air/glass and the glass/air interface are the same. Thus the fraction travelling through the stack is $(1 - R)^{2n}$.

(9.8) The reflection law is

$$\mathbf{k_i} \wedge \mathbf{n} = -\mathbf{k_r} \wedge \mathbf{n},$$

and Snell's law is

$$\mathbf{k_i} \wedge \mathbf{n} = +\mathbf{k_t} \wedge \mathbf{n}.$$

Notice that the vector forms are compact, and are more complete because they also contain the requirement that these vectors are coplanar.

(9.9) Using eqn. 9.109 and squaring the result gives $R_1 = (1.7029^2 - 1.5)/(1.7029^2 + 1.5) = 0.10$, also $R_4 = (1.7029^8 - 1.5)/(1.7029^8 + 1.5) = 0.92$. Using eqn. 5.56 the finesse for a cavity with mirrors on glass made of four double layers is $\pi\sqrt{R_4}/(1 - R_4)$ or 37.0.

(9.10) Using the expression for the critical angle $\cos{(90° - \theta_c)} = 0.9999$. This gives $(90° - \theta_c) = 48.6\,\mathrm{arcmin}$. Thus TIR will be possible at angles up to $48.6\,\mathrm{arcmin}$ from grazing incidence.

(9.11) Fresnel's laws only apply to a simple interface between two dielectrics with different refractive indices n_1 and n_2. The relationship $r' = r$ applies when the surface is a symmetric thin layer of one or more materials where the initial and final refractive indices are the same. If Fresnel's laws are applied in the limit that the two refractive indices become the same, then both r and r' go to zero, in which case indeed $r' = \pm r$.

Chapter 10

(10.1) (a)
$$R(45°)HR(-45°) = \frac{1}{2}\begin{pmatrix} 1 & 1 \\ 1 & 1 \end{pmatrix}.$$

The polarizer changes the overall amplitude by a factor $1/\sqrt{2}$ and this lies at $45°$ to the axes. Thus the amplitude along each axis is $1/2$. (b)
$$\begin{pmatrix} 1/t & 0 \\ 0 & 1/t \end{pmatrix}.$$

(10.2) The magnetic field is

$$\mathbf{B} = [E_0/(\sqrt{2}c)][\mathbf{e}_y\cos{(kz - \omega t)} + \mathbf{e}_x\sin{(kz - \omega t)}].$$

Thus $\mathbf{E} \wedge \mathbf{H} = [E_0^2/(2c\mu_0)]\mathbf{e}_z$.

(10.3) The delay is $(2\pi/\lambda)(d_1 - d_2)(n_e - n_o)$, where d_1 and d_2 are the thicknesses of the two prisms crossed by the light beam. The Babinet compensator permits the user to change the phase delay between the orthogonally polarized components of a wave by moving the compensator transverse to the optical path.

(10.4) At the prism interface the incident extraordinary ray goes from material with refractive index 1.486 to material in which it is an ordinary ray for which the refractive index is 1.658. Snell's law gives the angle of refraction as 39.33°. This ray is incident on the exit surface at an angle $(45° - 39.33°)$ or 5.67°. Snell's law gives the angle of refraction in air as 9.43°. Following the other ray gives its angle in air to be 10.57°. Thus the angle between the rays on exit is 20°.

(10.5) The angle between the polarization axes is $2\omega t$ so that the intensity is by Malus' law $I_0 \cos^2(2\omega t)$ or $(I_0/2)(\cos(4\omega t) + 1)$. The rotation angular frequency of the polarization is identical to that of the second polarizer ω.

(10.6) The frequency range is limited to well below the inverse of the transit time, c/nL, where n is the refractive index of LNO and L is the electrode length. In this case this cut-off frequency is 13.6 GHz. If the voltage swing about the bias point v is not large compared to V_π, then the optical response is proportional to $\sin(\pi v/V_\pi) \approx \pi v/V_\pi - (\pi v/V_\pi)^3/3 + ...$, which approximates to the desired linear response. When v is $0.1 V_\pi$ then the distortion from a linear response is 0.3%.

(10.7) Using eqn. 10.53 the requirement is that $B = \pi/4VL$, where V is the Verdet constant and L the length. With the values given the field is 0.43 T. The Verdet constant is approximately inversely proportional to the wavelength so that at 533 nm the rotation would not be 45° but $45(633/533) = 53.4°$. Assuming that the reflected beam has the same polarization as the second polarizer, the fraction of light leaking through would be $\sin^2(53.4° - 45°) = 0.02$.

(10.8) The wavelengths are $589/1.658 = 355$ nm and $589/1.486 = 396$ nm for the ordinary and extraordinary waves. The transmission coefficients are given by eqn. 9.80. For ordinary waves 0.94 and for extraordinary 0.96.

(10.9) The circularly polarized part is $E_a[\mathbf{e}_x \cos\chi + \mathbf{e}_y \sin\chi$ and the plane polarized part is $(E_b - E_a)\mathbf{e}_y \sin\chi$, where χ is $(kz - \omega t)$. In Jones vectors this is

$$\left[-i(E_b - E_a) \begin{pmatrix} 0 \\ 1 \end{pmatrix} + E_a \begin{pmatrix} 1 \\ -i \end{pmatrix} \right] \exp(i\chi).$$

(10.10)

(10.11) The extraordinary wave terms in eqn. 10.20 are

$$[n_o^2 - n^2 \cos^2\theta]E_x + n^2 \cos\theta \sin\theta E_z = 0,$$

and

$$n^2 \cos\theta \sin\theta E_x + [n_e^2 - n^2 \sin^2\theta]E_z = 0.$$

These can only be consistent if

$$[n_o^2 - n^2 \cos^2\theta][n_e^2 - n^2 \sin^2\theta] - n^4 \cos^2\theta \sin^2\theta = 0,$$

whence the required result follows.

(10.12)

Chapter 11

(11.1) Differentiating eqn. 11.44 gives $d\omega/dk = kc^2/\omega$, whence the result follows.

(11.2) Using eqn. 11.19 ω_p is $5.64\,10^7$ rad s^{-1} and f_p is 8.98 MHz. Substituting for ω and ω_p in eqn. 11.44 gives $k = 9.223\,10^{-2}$ m^{-1}. Then $v_p = \omega/k$ gives a phase velocity of $6.81\,10^8$ m s^{-1}. Using the result proved in the previous question $v_g = c^2/v_p$, the group velocity is $1.32\,10^8$ m s^{-1}.

(11.3) A distance of 1 km is 333.3 times 3 m, so the loss factor in 1 km is $\exp(-333.3)$ or $10^{-144.6}$. This is -1446 dB km^{-1}.

(11.4) In normal usage on paper a red ink is absorbed by the fibres in paper. Light incident on the paper enters the open structure formed by the fibres and is reflected many times before re-emerging. Any wavelengths that are absorbed strongly by the dye in the ink will be missing from the spectrum of the re-emerging light. Thus red ink absorbs the non-red, that is the green part of the spectrum. Hence a solid layer of such dried ink will therefore reflect green light strongly, while the small fraction of green light penetrating the ink will be absorbed in a short distance.

(11.5) Using eqn. 11.6 β is $7.81\,10^{-6}$ m^{-1}. Thus in 200 km the intensity is reduced by a factor $\exp(-1.562)$ or 0.210.

(11.6) Rewriting eqn. 11.20 in terms of wavelengths, dropping the damping term and rewriting the resonant wavelength as λ_i gives

$$n^2 = 1 + (1/\lambda_p^2)/[(1/\lambda_i^2) - (1/\lambda^2)]$$
$$= 1 + (\lambda\lambda_i/\lambda_p)^2/(\lambda^2 - \lambda_i^2). \quad (11.1)$$

It is seen that the empirical formula applies to narrow lines, that is with weak damping, and that A is $(\lambda_i/\lambda_p)^2$, while λ_i is resonant wavelength.

(11.7) Using eqn. 11.33 the reflectance is

$$R = rr^* = [(n_r - 1)^2 + n_i^2]/[(n_r + 1)^2 + n_i^2].$$

This gives a reflectance of 0.74.

(11.8)

(11.9) The dispersion relation of a surface plasma wave in this case would have a slope at the origin of $c/\sqrt{\varepsilon_p}$. This is to be compared to the slope of the dispersion relation for light incident in the prism, $\omega/k = c/(\sqrt{\varepsilon_p} \sin \theta_p)$. Evidently the latter is steeper so that the two curves never cross away from the origin, and excitation of such a SPW is not possible.

(11.10) From the definition of v_g in eqn. 11.41

$$\begin{aligned} 1/v_g &= dk/d\omega = d(n\omega/c)/d\omega \\ &= [(dn(\omega)/d\omega)\omega + n(\omega)]/c \\ &= 1/v_p + [\omega/c][dn(\omega)/d\omega]. \end{aligned}$$

Chapter 12

(12.1) Using the Einstein relation eqn. 12.12 the cut-off frequency f_0 is $1.53\,10^{15}$ Hz, the wavelength is c/f_0 = 196 nm. This excludes over 99% of the solar spectrum.

(12.2) The wavelength shift is given by eqn. 12.17 in which the constant factor has a value 0.0024 nm. Thus for scattering through $17°$, $\Delta\lambda = 0.0024(1 - \cos 17°)$, which gives $1.05\,10^{-4}$ nm. The left hand peaks are due to scattering off the nucleus. This has a mass, M, several thousand times that of the electron so that the constant h/Mc is correspondingly several thousand times smaller. The resulting wavelength shift was undetectable in Compton's experiment.

(12.3) In the case of the photon $\lambda = c/f = ch/E$, where E is the (total) energy. This gives $1.26\,\mu$m. In the case of the electron we could proceed the same way but need to recall that the total energy is the kinetic plus the rest mass energy, the latter being 511 keV. Then instead of c we must use the electron velocity. It is of course easier to obtain the electron's momentum, $\sqrt{2m\mathrm{KE}}$, where KE is its kinetic energy, and then use the de Broglie relation. With the electron mass of $9.11\,10^{-31}$ kg this gives a momentum $p = 5.40\,10^{-25}$ kg m s^{-1}, and $\lambda = h/p = 1.23\,10^{-9}$ m.

(12.4) Using the Boltzmann relation eqn. 12.36 the relative number of atoms in the first excited state with binding energy 3.39 eV divided by the number in the ground state

with binding energy 13.61 eV at 100 000 K is $\exp(10.22)(1.602\,10^{-19})/(10^5 \times 1.381\,10^{-23})$ or 0.305. At 1000 K the fraction is negligible. At 100 000 K both Balmer and Lyman absorption lines would be seen, but at 1000 K only the Lyman series, because so few atoms are in an excited state.

(12.5) Using equation eqn. 12.42 the natural line width $\Delta E = \hbar/(2\Delta t)$, where Δt is the time available for the measurement. Combining the effect of the lifetimes, and expressing $\hbar$ as $6.582\,10^{-16}$ eV s gives

$$\begin{aligned} \Delta E &= (\hbar/2)(1/\Delta t_{\text{parent}} + 1/\Delta t_{\text{daughter}}) \\ &= (6.582\,10^{-16}/2)(0.5833\,10^{-8}) = 1.92\,10^{-8}\,\text{eV}. \end{aligned}$$

Then $\Delta f = 4.64$ MHz.

(12.6) Using Wien's law eqn. 12.8 gives 483 nm. The irradiance is given by Stefan's law eqn. 12.9 $7.34\,10^7$ W m^{-2}. The peak wavelength at room temperature is $9.66\,\mu$m in the infrared, and 1.054 mm for the CBR, which is in the microwave region at 284 GHz.

(12.7) In the Balmer series the $n = 3$ to $n = 2$ level transition has the longest wavelength and lowest energy, while the $n = \infty$ to $n = 2$ level transition gives the shortest wavelength and highest energy. Using eqn. 12.22 the wavelengths are 656.5 nm and 364.7 nm. For the Paschen series the transitions are to the $n = 3$ level and give 1876 and 821 nm. For the Brackett series the wavelengths are 4052 nm and 1459 nm.

(12.8) The reduced mass of the electron in positronium is $m^2/(m + m)$ or half the electron mass. Thus the Rydberg constant for positronium $R_p = R_\infty/2$. The wavelengths are thus twice those of the hydrogen spectrum and the Lyman α-line has wavelength 243 nm.

(12.9) The slopes give close to $6.57\,10^{-34}$ J s which is the value that Millikan found.

(12.10) The angular size of the lens seen from the image plane is D/f where D is the lens diameter and f its focal length. The ratio is precisely $1/[f/\#]$. The momentum is h/λ so that the uncertainty in the transverse momentum is $\Delta p_y = h/[\lambda f/\#]$. Applying the uncertainty principle the resolution is at best $\Delta p_y = h/\Delta_y = \lambda[f/\#]$. This can be compared to eqn. 4.19.

(12.11) The longitudinal momentum spread is $\Delta p_z = h\Delta\lambda/\lambda^2$. The lateral spreads are $\Delta p_x = \Delta p_y = (h/\lambda)\Delta\theta$. Thus the product of the three uncertainties in the momentum components is $X =$

$\hbar^3 \Delta\lambda 10^{-4}/\lambda^4$. Consequently the product of the three spatial uncertainties is greater than or equal to $\hbar^3/X = 10^4\lambda^4/\Delta\lambda$ which is $6.25\,10^6\,\mu m^3$. This does not tell how wide or long the beam is, but simply the size of the coherence volume.

(12.12) Suppose the light is absorbed by atoms or molecues in their ground states, then after absorbing light they are in some excited state with excitation energy equal to the photon energy. If the excited state prefers to decay into a state which is also an excited state the energy of the photon emitted in this step is less than the energy of the incident photon. This will be true for any further decays that end up at the ground state. All the wavelengths will therefore be longer than that of the incident light.

(12.13) Let the amplitude of the ordinary and extraordinary polarized light at CCD A from slit a be a and $b\exp(i\alpha)$ respectively, while the corresponding amplitudes from slit b are $b\exp(i\gamma)$ and $a\exp i(\gamma + \alpha)$ respectively. Here α is the phase difference between the ordinary and extraordinary light after passing through the Wollaston prism, while γ is the phase lag due to the path difference from the two slits to the image point for the ordinary polarization. The intensity is thus

$$I = (a + b\exp(i\gamma))(a + b\exp(-i\gamma))$$
$$+[b\exp(i\alpha) + a\exp i(\alpha + \gamma)]$$
$$[b\exp(-i\alpha) + a\exp[-i(\alpha + \gamma)]]$$
$$= 2[a^2 + b^2 + 2ab\cos\gamma].$$

In the example $a^2 = 0.9$ and $b^2 = 0.1$ so that $I = 2(1 + 0.6\cos\gamma)$, where γ varies across the image. The visibility is $(I_{max} - I_{min})/(I_{max} + I_{min})$ giving 0.6.

(12.14) Using Wien's law the temperature at decoupling was $2.9\,10^{-3}/\lambda_{max}$ which is $2.9\,10^{-3}E_{photon}/hc$. This gives 3500 K. Comparing the peak wavelengths the universe has expanded by a factor $3500/2.75$ or around 1270.

Chapter 13

(13.1) At 3000 K, and using the value $8.6174\,10^{-5}\,eV\,K^{-1}$ for the Boltzmann constant gives $k_BT = 0.259\,eV$. Then using the eqns. 13.79, 13.80 and 13.81 gives for the respective cases gives

$$\langle n_{photon}\rangle = 1/[\exp(0.4/0.259) - 1] = 0.27,$$
$$\langle n_{electron}\rangle = 1/[\exp(0.4/0.259) + 1] = 0.18,$$
$$\langle n_{classical}\rangle = \exp(-0.4/0.259) = 0.21.$$

(13.2) At 3 K, and using the value $8.6174\,10^{-5}\,eV\,K^{-1}$ for the Boltzmann constant gives $k_BT = 2.59\,10^{-4}\,eV$. Then eqns. 13.79, 13.80 and 13.81 give almost identical tiny values: $\exp(-1540)$.

(13.3) Let $\{\phi_i\}$ be the set of eigenfunctions of these states and let $\hat{A}$ be the operator. Then

$$A_{ii} = \langle\phi_i|\hat{A}|\phi_i\rangle = a_i,$$
$$A_{ij} = \langle\phi_i|\hat{A}|\phi_j\rangle = a_j\langle\phi_i|\phi_j\rangle = 0,$$

where all the a_i are real. Thus for all i, j $A_{ij}^* = A_{ji}$ so that $\hat{A}$ is hermitean.

(13.4) For the momenta

$$[p_x, p_y]\psi = -\hbar^2\left[\frac{\partial^2}{\partial x\partial y} - \frac{\partial^2}{\partial y\partial x}\right]\psi.$$

Now the partial differentials act only on the variable specified treating the other as a constant, so the order of differentiation is unimportant. Thus the commutator vanishes. In the case of the coordinates neither is an operator acting on the other so their commutator vanishes. Finally

$$[x, p_y]\psi = -i\hbar\{x(\partial/\partial y)\psi - (\partial/\partial y)(x\psi)\}$$
$$= -i\hbar\{x(\partial/\partial y)\psi - x(\partial/\partial y)\psi\} = 0.$$

(13.5)

(13.6) The photon energy is 12.1 eV, and its momentum is 12.1 eV/c. The atom will have a momentum equal and oppositely directed. Its mass is $1.67\,10^{-27}$ kg or 938 MeV/c^2. Thus its velocity is $1.29\,10^{-8}c$ or $3.87\,ms^{-1}$. Its kinetic energy is $1.25\,10^{-26}$ J or $7.81\,10^{-8}$ eV. This is a tiny fraction of the photon's energy, but must come from the excess energy in the excited state. In calculating the decay kinematics one should apply the laws of conservation of energy and momentum incuding both the photon and recoil atom. This clearly depresses the energy of the photon slightly below the energy difference between the excited state and ground state. When a photon is absorbed it has to supply both the energy difference between the ground and excited state and the kinetic energy of the atomic recoil. It looks as though an emitted photon is not energetic enough to cause the upward transition! In practice the energy uncertainty in the excited state is enough to mask the tiny energy deficit.

(13.7) The angular momentum vector has length $\sqrt{12}\hbar$ and its component is $3\hbar$. Thus the angle is $\cos^{-1}(3/\sqrt{12}) = 0.866$, that is $30°$.

(13.8) The angular distribution is

$$I \propto |Y_{1,-1}(\theta,\phi)|^2 + |Y_{1,0}(\theta,\phi)|^2 + |Y_{1,+1}(\theta,\phi)|^2$$
$$= (3/8\pi)\sin^2\theta + (3/4\pi)\cos^2\theta + (3/8\pi)\sin^2\theta$$
$$= 3/4\pi,$$

which is independent of the angular coordinates, that is it is isotropic.

(13.9)

$$[L_z, L_x]\psi = -\hbar^2\left[x\frac{\partial}{\partial y} - y\frac{\partial}{\partial x}\right]\left[y\frac{\partial}{\partial z} - z\frac{\partial}{\partial y}\right]\psi$$
$$+\hbar^2\left[y\frac{\partial}{\partial z} - z\frac{\partial}{\partial y}\right]\left[x\frac{\partial}{\partial y} - y\frac{\partial}{\partial x}\right]\psi$$
$$= -\hbar^2\left[x\frac{\partial}{\partial y}y\frac{\partial}{\partial z} + y\frac{\partial}{\partial x}z\frac{\partial}{\partial y}\right.$$
$$\left. -y\frac{\partial}{\partial z}x\frac{\partial}{\partial y} - z\frac{\partial}{\partial y}y\frac{\partial}{\partial x}\right]\psi$$
$$= -\hbar^2\left[x\frac{\partial}{\partial z} - z\frac{\partial}{\partial x}\right]\psi$$
$$= i\hbar L_y\psi.$$

(13.10)

$$\langle p^m\rangle = (1/L)\int_{-L/2}^{L/2}\exp\left[-i(kx-\omega t)\right]$$
$$\left(-i\hbar\frac{\partial}{\partial x}\right)^m\exp\left[i(kx-\omega t)\right]\mathrm{d}x$$
$$= (\hbar k)^m\int_{-L/2}^{L/2}\mathrm{d}x/L = (\hbar k)^m.$$

(13.11)

Chapter 14

(14.1) The finesse, given by eqn. 5.56 is 313. The mode spacing is the free spectral range given by eq. 5.59 and is 0.3 GHz, so there are five modes lying within the FWHM.

(14.2) The sheeting should be formed as a cylinder, somewhat longer than the rod, having an elliptic cross-section. The lamp and rod are mounted with their axes along the two focal lines of the cylinder. In this way the light leaving the lamp perpendicular to the flash lamp gets reflected onto the laser rod. If the ellipse is made very narrow the lamp will shield the rod from some of the radiation it emitted, and if too large light will escape from the ends.

(14.3) The ring-down time for the empty cavity given by eqn. 14.30 is 3.3 ms. Using eqn. 14.29 in the case the cavity is full of gas, the ring down time is 1.7 ms.

(14.4) Using eq. 14.70 the angle between the common direction of the wave vectors and the crystal optic axis is given by $\sin^2\theta = 0.15173$ and θ is $22.9°$. The walk-off angle is that between the Poynting vectors, and this is the same as the angle between the **E** and **D** fields of the frequency doubled radiation. Formulae for these appear in eqns. 10.24 and 10.25. Then

$$E = \sqrt{n_e^4(532)\cos^2\theta + n_o^4(532)\sin^2\theta}$$
$$= 2.4807,$$
$$D/\varepsilon_0 = n_o^2(532)n_e^2(532) = 6.784,$$
$$\mathbf{E}\cdot\mathbf{D}/\varepsilon_0 = n_o^2(532)n_e^2(532)$$
$$\times[n_e^2(532)\cos^2\theta + n_o^2\sin^2\theta]$$
$$= 16.803.$$

Then the walk-off angle is given by

$$\cos\phi = \mathbf{E}\cdot\mathbf{D}/(ED) = 0.998$$

so that the walk off angle is only $3.2°$.

(14.5) From eqns. 14.21 and 14.29 we have $\gamma_{\text{th}} = 1/(c\tau)$.

(14.6) From eqn. 14.21 the threshold gain is $0.025\,\text{m}^{-1}$. Then using eqn. 14.16

$$\Delta N = 8\pi\Delta f\tau_{\text{sp}}\gamma_{\text{th}}/\lambda^2,$$

where Δf is the FWHM. This gives $3.14\,10^{14}\,\text{m}^{-3}$.

(14.7) From eqn. 14.16 the required population inversion density is

$$\Delta N = 8\pi\Delta f\tau_{\text{sp}}n^2/(c\lambda^2\tau_c)$$

which gives $3.83\,10^{20}\,\text{m}^{-3}$.

(14.8) Using the result from eqn. 7.56

$$\sigma_\omega = \sqrt{(\omega_0^2 k_B T/Mc^2)},$$

where ω_0 is the laser angular frequency, and M is the atomic mass of ^{20}Ne. This gives σ_ω equal to $4.06\,10^9\,\text{rad}^{-1}$. Then the FWHM is

$$\Delta\omega = 2.35\sigma_\omega = 9.53\,10^9\,\text{rad s}^{-1},$$

so that in frequency the FWHM is 1.52 GHz.

(14.9) In YAG the Nd ions are located in a regular lattice and consequently the environments of all of them are closely similar. Therefore the broadening is common to all. On the other hand glass is amorphous so that the environments vary a great deal. This makes for inhomogeneous broadening. The gain bandwidth is as a result sixty times wider for Nd:glass lasers at 30 nm compared to 0.5 nm for Nd:YAG.

(14.10) When the degeneracy of the lower level is large it can happen that the population inversion ($N_2 - (g_2/g_1)N_1$) is positive. Thus in principle it would be possible.

(14.11) The scattering is from the acoustic waves giving a Doppler shift of

$$\Delta f/f = 2v_{\text{acoustic}}/v_{\text{light}} = 3\,10^{-5}.$$

At $1\,\mu\text{m}$ wavelength the frequency shift is then $9\,\text{GHz}$.

(14.12) The separation of the teeth is fixed by the repetition rate at $100\,\text{MHz}$ and the envelope is of order $1/20\,\text{ps}$ or $50\,\text{GHz}$.

(14.13) The thickness of the well will equal a whole number of half waves, $L_z = n\lambda/2$. Then the kinetic energy of the electron in the z-direction is given by eqn. 14.40

$$E = \hbar^2 k^2/2m^* = h^2/(2m^*\lambda^2),$$

where k is the wave vector. Then replacing λ gives $E = n^2 h^2/(8m^* L_z^2)$.

(14.14) The kinetic energy of the lowest bound state in the quantum well is $h^2/(8m^* L)$, where L is the well width and m^* the effective mass of the bound particle. This gives $19\,\text{meV}$ for the electron in the conduction band and $3\,\text{meV}$ for the hole in the valence band. In all $22\,\text{meV}$ which translates into a frequency change of $5.3\,10^{12}\,\text{Hz}$.

Chapter 15

(15.1) The absorption length at wavelength $1\,\mu\text{m}$ obtained from figure 15.1 is $100\,\mu\text{m}$ so that the depth in which 90% is absorbed is d where $\exp(-d/100) = 0.1$, that is $230\,\mu\text{m}$. The resistance, R, of the $1\,\text{mm}$ long $230\,\mu\text{m}$ by $2\,\text{mm}$ area photoconductor is $6.52\,\text{M}\Omega$. The capacitance, C, is $\varepsilon_0\varepsilon_r \times \text{area/length} = 4.80\,10^{-14}\text{F}$. The time constant, RC, is $0.313\,\mu\text{s}$.

(15.2) The electric field, E, across the photoconductor is $200\,\text{Vm}^{-1}$ so the electron velocity $E\mu$ is $27\,\text{ms}^{-1}$, and the transit time is $37\,\mu\text{s}$. This is much longer than the electrical time constant. The gain is given by eqn. 15.4 and is 2.7.

(15.3) The reflectance at normal incidence is obtained from eqn. 9.79, with the silicon refractive index being $\sqrt{11.8} = 3.435$. Thus the required reflectance, R, is 0.301. In turn the quantum efficiency is $0.9 \times (1 - R)$ or 0.63. Finally the responsivity given by eqn 15.6 is $1.37\,\text{A W}^{-1}$.

(15.4) No, because the recombination time is much longer than the period of oscillation.

(15.5) The shot noise is given by eqn. 15.18 $\sigma_s^2 = 2eB(i_s + i_d)$, where i_s is the photocurrent $0.6 \times 2 = 1.2\,\mu\text{A}$. This gives $3.84\,10^{-17}\text{A}^2$. The Johnson noise is given by eqn. 15.22, $\sigma_j^2 = 4k_{\text{B}}TB/R$, which gives $3.31\,10^{-14}\text{A}^2$. Thus the signal to noise ratio $i_s^2/(\sigma_s^2 + \sigma_j^2)$ is 43.4. The dark current shot noise σ_d^2 is $1.6\,10^{-19}$, negligible compared to the Johnson noise. Using eqn. 15.25 the NEP is $3.03\,10^{-11}\,\text{W}/\sqrt{\text{Hz}}$.

(15.6) The shot noise is now $3.846\,10^{-14}\text{A W}^{-1}$, and the Johnson noise is unchanged. Thus the SNR is $2.01\,10^7$.

(15.7) The gain $G = 5^{10} = 10^7$. If n photons are incident per second the power incident, P, is $nhf = nhc/\lambda$, where f is the frequency, λ the wavelength. This gives a power $2.48n\,\text{eV}$. The anode current is $Gn\eta e$ where η is the quantum efficiency. Thus the current, i, is $2\,10^6 ne$. Then the responsivity i/P is $0.81\,\text{A}\,\mu\text{W}^{-1}$. The dark current shot noise is $2eGBi_d$, where i_d is the dark current and B is the bandwidth, giving $(10^{-20}B)\,\text{A}^2$. The Johnson noise $4k_{\text{B}}TB/R$ is $(3.31\,10^{-22}B)\,\text{A}^2$. Thus the NEP, using eqn. 15.25, is $0.122\,\text{fW}/\sqrt{\text{Hz}}$. Then the detectivity D^* given by eqn. 15.27 is $1.16\,10^{-14}\,\text{m}\sqrt{\text{Hz}}\text{W}^{-1}$.

(15.8) The APD in the Geiger mode gives uniform large pulses however many photons arrive. Photomultipliers would give larger pulses when two electrons arrive, but there is still a spread in the distribution of pulse height so that there is some overlap between the distributions for single and double photons. Partial separation of two and one photon events is feasible.

(15.9) In a photodiode the hole and electron pair produced by photon absorption travel under the applied electric field out of the depletion layer to the p-doped and n-doped regions respectively. They immediately recombine with the excess minority carrier charges that exist at the edge of the depletion layer. Hence the recombination time is exactly the same as the transit time, giving unity gain. In the case of photoconductors a high gain implies a long recombination time and in turn this implies a slow response.

(15.10) The responsivity is given by eqn. 15.6. Generally the recombination tim, τ_r, is longer than the transit time, τ_x, in order to achieve gain. Thus the recombination time determines the bandwidth, $B = 1/\tau_r$, from which the result given follows.

(15.11) Using eqn. 15.6, with unity gain, the responsivity is $0.515\,\mathrm{A\,W^{-1}}$. The photocurrent for unit gain would then be this responsivity times the input power, or $5.15\,10^{-8}\,\mathrm{A}$. The ratio between the actual current and this current is the gain, 970.

Chapter 16

(16.1) The loss in 10 km is 200 dB, an attenuation by a factor 10^{-20}. The number of photons is thus $(60)10^{-20}(\lambda/hc)$, that is 4.7.

(16.2) Using the equations given in Section 16.4 the dispersion over 5 km in step index fibre is $0.34\,\mu\mathrm{s}$. For step index fibre at a data rate of $3.3\,\mathrm{Mb\,s^{-1}}$ each bit would overlap its neighbour. Thus a rate of about $300\,\mathrm{kB\,s^{-1}}$ could be maintained. The bandwidth-distance product is then $1.5\,\mathrm{Mb\,s^{-1}km}$. For graded index fibre the intermodal dispersion is 0.58 ns. The chromatic dispersion is 40 ns. The data rate is limited by the chromatic dispersion to about $2.5\,\mathrm{Mb\,s^{-1}}$ and so the fibre capability is $12.5\,\mathrm{Mb\,s^{-1}\,km}$.

(16.3) The material dispersion over a distance L when the wavelength spread is $\Delta\lambda$ is given using eqn. 16.16,

$$\Delta t = \mathrm{d}(L/v_{\mathrm{g}})/\mathrm{d}\lambda\,\Delta\lambda = (L/c)\,\lambda\,\Delta\lambda\,\mathrm{d}^2 n/\mathrm{d}\lambda^2),$$

where v_{g} is the group velocity and n the refractive index. Inserting the parameters given yields Δt is 60 ns.

(16.4) Over 50 km the attenuation is $-10\,\mathrm{dB}$ or by a factor 10. Using eqn 16.30 the shot noise contribution to the mean square deviation in the current is

$$[2\lambda\eta\,e^2/(hc)]\,P_0\exp(-\alpha L)\Delta f$$
$$= 3.60\,10^{-14}\,\mathrm{amp}^2.$$

The RIN contribution is similarly

$$[(\lambda\eta\,e/(hc))]^2\,RIN\,P_0^2\exp(-2\alpha L)\,\Delta f$$
$$= 2.67\,10^{-14}\,\mathrm{amp}^2.$$

The noise contribution from the detector is

$$(4k_{\mathrm{B}}T/R_L)\Delta f = 2.48\,10^{-14}\,\mathrm{amp}^2.$$

Adding these three contributions gives $8.75\,10^{-14}\,\mathrm{A}^2$. Thus the rms noise is $0.296\,\mu\mathrm{A}$. The signal strength is

$$[\lambda\eta e/hc]\,P_0\exp(-\alpha L) = 74.95\,\mu\mathrm{A}.$$

The SNR in the current is thus 256, and in terms of power the SNR is 48 dB. The poor connection near the detector reduces the signal by a factor 10, and the mean square shot noise current by a factor 10, the RIN by a factor 100, but leaves the detector noise unaffected. Thus the rms noise current becomes $0.169\,\mu\mathrm{A}$, and the signal current falls to $7.495\,\mu\mathrm{A}$. The new current SNR is 44.25 and the power SNR falls to 33 dB.

(16.5) Suppose Λ_{s} is the pitch at the start of the grating and Λ_e the pitch at the end of the grating, and that the grating length is L. Then light of wavelength $\lambda_{\mathrm{s}} = 2\Lambda_{\mathrm{s}}n$ will be reflected at the start of the grating and light of wavelength $\lambda_e = 2\Lambda_e n$ will be reflected at the end of the grating, where n is the mean refractive index. Consequently light of wavelength λ_e will be delayed by a time $2Ln/c$ compared to light of wavelength λ_{s}.

(16.6) The reflection coefficient at each air/glass interface is given by eqn. 9.79 $R = [(n-1)/(n+1)]^2 \approx 0.04$. Thus the loss at each surface in dB is $10\log_{10}(1-R) = -0.177$. Overall the loss is thus 0.35 dB.

(16.7) Suppose that light of wavelength λ_0 whose mth principal maximum falls on the far left hand output port produces an $(m+1)$th principal maximum which falls on the far right hand output port. If the mth principal maximum on the far right hand port is formed by light of wavelength $\lambda_0 + \Delta\lambda$, then the free spectral range is just $\Delta\lambda$. Then we have

$$m(\lambda_0 + \Delta\lambda) = (m+1)\lambda_0.$$

Thus $\Delta\lambda \approx \lambda_0/m$. The optical path difference between successive waveguides is $n\Delta L = 15\,\mu\mathrm{m}$. Thus the diffraction order is $n\Delta L/\lambda_0 = 15/1.55 \approx 9$. Consequently $\Delta\lambda = 1550/9 = 172\,\mathrm{nm}$.

(16.8) Using eqn 16.41 the spacing is

$$\Delta\lambda = xd\lambda_0/(f\Delta L),$$

where x and d are the port spacings, f is the slab length and ΔL is the difference in length between successive guides. Inserting the values given $\Delta\lambda$ is 0.895 nm. The diffraction order is the optical path difference between successive guides divided by the wavelength, that is 61. The free spectral range is then $1550/61 = 25.4\,\mathrm{nm}$. The useful number of channels is thus $25.4/0.895$, that is 28.

(16.9) Using eqn. 16.44 the soliton period is 33 km. Using eqn. 16.45 the pulse power is 30 mW. Then the pulse energy, E, is the pulse power times its duration, or 300 fJ. This contains $E\lambda/hc$ photons. There are 2.3 million photons in the pulse.

(16.10) The numerical aperture is given by eqn. 2.41, NA $= \sqrt{(n_1^2 - n_2^2)}$, giving 0.12. Using eqn. 16.11 the cut-off wavelegth, λ is given by

$$\sqrt{(n_1^2 - n_2^2)}(2\pi a/\lambda) = 2.405,$$

where a is the core radius. After substituting the values given the cut-off wavelength is 1590 nm.

(16.11) The notation of Section 16.7.3 is used here. After a path length $3L_{12}/2$ the phase of the jth mode relative to the first mode is $(\beta_j - \beta_1)(3L_{12}/2) = \pi(j^2 - 1)$. Therefore the phases of modes with j odd have changed relative to the phase of the first mode by an integral multiple of 2π. The phases of the even modes relative to the first mode have all moved by an odd multiple of π. Any odd order mode has a transverse distribution across the slab of the form

$$A \cos{(k_t x + \alpha)},$$

where x is the transverse coordinate measured from the midline down the slab. Each even mode is of the form

$$A' \sin{(k_t' x)}.$$

After a distance $3L_{12}/2$ these become

$$A \cos{(k_t x + 2\pi)} = A \cos{(k_t x)},$$
$$A' \sin{(k_t' x + \pi)} = -A' \sin{(k_t x)}.$$

Making the exchange $x \to -x$ then reproduces the entry surface.

(16.12) Using the expression given in eqn. 16.14 gives 1825 modes.

Chapter 17

(17.1) Substituting in eqn. 17.16 the central wavelength is 495 nm. Then using eqn. 17.17 the width is approximately 60 nm. The Bragg wavelength is 500 nm.

(17.2) The product

$$(\mathbf{a}_1 \cdot \mathbf{a}_2 \wedge \mathbf{a}_2) = (\mathbf{i} + \mathbf{j} - \mathbf{k}) \cdot (2\mathbf{j} + 2\mathbf{i})/8 = 1/2.$$

Then

$$\mathbf{b}_1 = 2\pi(-\mathbf{i} + \mathbf{j} + \mathbf{k}) \wedge (\mathbf{i} - \mathbf{j} + \mathbf{k})/2 = 2\pi(\mathbf{i} + \mathbf{j}).$$

Similarly $\mathbf{b}_2 = 2\pi(\mathbf{j} + \mathbf{k})$, and $\mathbf{b}_3 = 2\pi(\mathbf{k} + \mathbf{i})$. These vectors join one corner of the face-centred cubic lattice to the nearest face centres, the cube having side length $1/\pi$.

(17.3) From the equation for the M-matrix we have

$$U_2 = AU_1 + BV_1 \text{ and } V_2 = CU_1 + DV_1.$$

Then

$$V_1 = (V_2 - CU_1)/D, \text{ and}$$
$$U_2 = (A - BC/D)U_1 + (B/D)V_2.$$

Hence

$$S = \begin{bmatrix} AD - BC & B \\ -C & 1 \end{bmatrix}/D.$$

Equally: $r_{12} = -C/D$, $r_{21} = B/D$, $t_{12} = (AD - BC)/D$ and $t_{21} = 1/D$.

(17.4) There are a variety of methods. One is to build a waveguide in the form of a linear defect alongside the cavity at a distance such that the cavity and waveguide modes overlap. Alternatively a stripped and tapered optical fibre can be attached similarly close to the cavity.

(17.5) In the case of the unpatterned Mach–Zehnder the shortest length required to give destructive interference is given by

$$\Delta n L_{\text{none}} = \lambda/2,$$

whence

$$L_{\text{none}} = (1/\Delta n)/\lambda/2.$$

Comparing with eqn. 17.34 gives

$$L_{\text{pattern}}/L_{\text{none}} = n v_{\text{g}}/c.$$

Inserting the values from the text, $n = 1.5$ and $v_{\text{g}}/c = 0.1$, this ratio is 0.15.

(17.6) The modulus of the amplitude of the scattered wave from a segment of the crystal is $f(\mathbf{r}) \, d\mathbf{r}$, where

$$f(\mathbf{r}) = \int \sum_i \delta(\mathbf{s} - \boldsymbol{\rho}_i) \, h(\mathbf{r} - \mathbf{s}) \, d\mathbf{s}.$$

Now the phase of the incident wave at $\mathbf{r}$ at time t is $(\mathbf{k}_0 \cdot \mathbf{r} - \omega t)$. Thus the phase of the wave scattered from this segment and arriving at some later time t' at a point $\mathbf{r}'$ is

$$[\mathbf{k}_0 \cdot \mathbf{r} - \omega t] + [\mathbf{k} \cdot (\mathbf{r}' - \mathbf{r}) - \omega(t' - t)]$$
$$= -\Delta \mathbf{k} \cdot \mathbf{r} + [\mathbf{k} \cdot \mathbf{r}' - \omega t'].$$

The corresponding amplitude contributed by scattering from the same element $d\mathbf{r}$ of the crystal is therefore

$$f(\mathbf{r})[\exp{(-i\Delta \mathbf{k} \cdot \mathbf{r})}] \, d\mathbf{r}.$$

Thus the total scattered amplitude is the Fourier transform of $f(\mathbf{r})$. Using the convolution theorem the Fourier transform of $f(\mathbf{r})$ is

$$F(\Delta \mathbf{k}) = \sum_i H(\Delta \mathbf{k}) \exp{(-i\Delta \mathbf{k} \cdot \boldsymbol{\rho}_i)}.$$

This must remain the same when the coordinates are displaced by any lattice vector so that

$$\Delta \mathbf{k} \cdot (\boldsymbol{\rho}_i - \boldsymbol{\rho}_j) = 2N\pi,$$

where N is an integer. For all other values there is cancellation of the scattered radiation. We can recognize that $(\boldsymbol{\rho}_i - \boldsymbol{\rho}_j)$ are lattice vectors. Hence eqn. 17.9 follows, which proves the required equality.

Chapter 18

(18.1) If the displacement is $x = x_0 \sin(\omega_s t)$ then the velocity $v = \omega_s x_0 \cos(\omega_s t)$, and the kinetic energy is $(m\omega_s^2 x_0^2/2) \cos^2(\omega_s t)$. At the midpoint of the motion the potential energy is zero so the total energy is $m\omega_s^2 x_0^2/2$. Equating this to the quantum of vibrational energy $\hbar\omega_s$ gives $x_0^2 = 2\hbar/m\omega_s$. Then inserting for the atomic mass 200 times the atomic mass unit gives x_0 approximately 10 nm.

(18.2) $\chi = (N\mu^2/\varepsilon_0\hbar)(2i/\gamma)$. Inserting the values given $\chi = 0.000615i$. Thus the relative permittivity is $1 + 0.000615i$, and the refractive index $1 + 0.0003075i$. The dependence of the electric field on the distance z through the cloud is $E(z) = E(0)\exp(-\omega n_i z/c)$ or $E(z) = E(0)\exp(-2\pi n_i z/\lambda)$, where λ is the wavelength in free space. Thus the intensity $|E(z)|^2$ falls by a factor e in a distance $z = \lambda/(4\pi n_i)$ which is 152 μm.

(18.3) The lifetime $\tau = 1.59 \, 10^{-8}$ s. The saturation electric field is given by substitution in eqn. 18.18 with $\Omega_0 = \gamma/\sqrt{2}$. From this the saturation value of the electric field is determined. Finally this can be inserted into eqn. 18.37 to obtain the Einstein coefficient. The saturation intensity is 63.6 W m^{-2}.

(18.4) The atom is stretched by the permanent field so that the centre of charge of the electron cloud no longer coincides with the centre of charge of the nucleus. The wavefunctions are modified so that they are solutions of eqn. 18.9, with both W and ΔW being time independent. As a result there will not be any pumping between states.

(18.5) If the electric field in the laser beam is of amplitude E_0, then the power density is $P = c\varepsilon_0 E_0^2/2$. Using eqn. 18.18 gives $\Omega_0 = (D_{12}/\hbar)\sqrt{(2P/\varepsilon_0 c)}$. Putting in the values given yields $\Omega_0 = 4.41 \, 10^6 \sqrt{P}$. Hence a Rabi frequency of 5 MHz requires a power density of 50.7 W m^{-2}. Over the area quoted the power is 0.507 μW, which would be an appropriate laser power because the area matches well to the cross-section of a laser beam.

(18.6) The Rabi angular frequency for no detuning is given by eqn. 18.18, Ω_0 is $2.01 \, 10^9$ rad s^{-1}. The angular frequency offset is $\Delta\omega = 2\pi c \Delta\lambda/\lambda^2$, which gives $5.43 \, 10^8$ rad s^{-1}. Then the generalized Rabi frequency $\sqrt{(\Omega_0)^2 + \Delta\omega^2}$ is $5.79 \, 10^9$ rad s^{-1}. Finally using eqn. 18.86 $\sin\theta = 0.608$, so that θ is $37.4°$.

(18.7) We use eqn. 18.97 giving $\omega = 3.2 \, 10^{15}$ rad s^{-1}. The electric field amplitude is 614 Vm^{-1}; the Rabi frequency of the coupling laser Ω_c is 123 MHz. Finally v_g is 450 m s^{-1}.

(18.8) The electric field amplitude E_0 is 1735 Vm^{-1}. Using eqn. 18.18 the dipole moment is $1.90 \, 10^{-29}$ C m. Then using eqn. 18.37 the Einstein coefficient for spontaneous decay is $2.37 \, 10^5$ s^{-1}.

(18.9) The electric field amplitude E_0 is 43.4 kVm^{-1}. Using eqn. 18.18 the dipole moment is $3.08 \, 10^{-30}$ C m. Then using eqn. 18.37 the Einstein coefficient for spontaneous decay is $1.2 \, 10^6$ s^{-1}.

(18.10) Using eqn. 18.97 and replacing Ω_c by $\mu_c E_c/\hbar$,

$$v_g = \mu_c^2 E_c^2 \varepsilon_0 c/(2\hbar\omega N \mu_p^2)$$
$$= (\mu_c/\mu_p)^2 P_c/(N\hbar\omega).$$

For simplicity take the dipole moments to be equal, then

$$v_g = P_c/(N\hbar\omega).$$

Now $\hbar\omega$ is the excitation energy per atom and N is the density of atoms, which supports the interpretation offered.

Chapter 19

(19.1) Suppose that the count for mode i is n_i, then using eqn. 19.91

$$\langle n_i^2 \rangle = \langle n_i \rangle + 2\langle n_i \rangle^2$$
$$= \langle N \rangle/m + 2\langle N \rangle^2/m^2,$$

where N is the total count. Then the variance is

$$\left\langle \left(\sum_i n_i \right)^2 \right\rangle - \left\langle \sum_i n_i \right\rangle^2$$
$$= \left\langle \sum_{i \neq j} n_i n_j \right\rangle + \left\langle \sum_i n_i^2 \right\rangle - \left\langle \sum_i n_i \right\rangle^2$$
$$= m(m-1)\langle N \rangle^2/m^2 + \langle N \rangle + 2\langle N \rangle^2/m - \langle N \rangle^2$$
$$= \langle N \rangle + \langle N \rangle^2/m.$$

(19.2) Suppose that M is the number of photons incident and m is the number detected. The variance on m is given by

$$\Delta m^2 = \langle m^2 \rangle - \langle m \rangle^2.$$

Now although $\langle m \rangle^2 = \eta^2 \langle M \rangle^2$, $\langle m^2 \rangle \neq \eta^2 \langle M^2 \rangle$: when one photon has been detected there are only $M - 1$ remaining which are potentially detectable. Instead $\langle m(m-1) \rangle = \eta \langle M(M-1) \rangle$. Then

$$\begin{aligned}\Delta m^2 &= \langle m(m-1) \rangle + \langle m \rangle - \langle m \rangle^2 \\ &= \eta^2 \langle M(M-1) \rangle + \eta \langle M \rangle - \eta^2 \langle M \rangle^2 \\ &= \eta^2 \langle M^2 \rangle + (\eta - \eta^2) \langle M \rangle - \eta^2 \langle M \rangle^2 \\ &= \eta^2 (\langle M^2 \rangle - \langle M \rangle^2) + \eta(1 - \eta) \langle M \rangle \\ &= \eta^2 \Delta M^2 + \eta(1 - \eta) \langle M \rangle.\end{aligned}$$

(19.3)

$$\langle E^2 \rangle = \zeta_\omega^2 \langle \alpha | (\exp(i\xi)\hat{a} + \exp(-i\xi)\hat{a}^\dagger) \\ (\exp(i\xi)\hat{a} + \exp(-i\xi)\hat{a}^\dagger)|\alpha\rangle.$$

Then using eqns. 19.15, 19.72 and 19.73,

$$\begin{aligned}\langle E^2 \rangle &= \zeta_\omega^2 \left[\exp(2i\xi)\alpha^2 \right. \\ &\quad \left. + \exp(-2i\xi)(\alpha^*)^2 + 1 + 2|\alpha|^2 \right] \\ &= 4\zeta_\omega^2 \cos^2(\xi + \theta)|\alpha|^2 + \zeta_\omega^2 \\ &= \langle E \rangle^2 + \zeta_\omega^2.\end{aligned}$$

(19.4) Use eqns. 19.27 and 19.28 n times.

(19.5)

(19.6) In free space $\mathbf{B}$ is perpendicular to $\mathbf{E}$, in phase with $\mathbf{E}$ and $B = Ek/\omega$. Then the result follows.

(19.7) First show that $|\{\alpha\}\rangle$ is an eigenstate of $\hat{E}^+$, $\hat{E}^+(t)|\{\alpha\}\rangle = E_\alpha(t)|\{\alpha\}\rangle$. Then substitute for $\hat{E}^+(t)$, $\hat{E}^+(t + \tau)$ etc. in eqn. 19.62.

(19.8) Form the expectation value of this quantity using pure and incoherent states using eqns. 19.66 and 19.67.

(19.9) The initial state with a photon located with equal probability at either slit is

$$|\text{in}\rangle = (1/\sqrt{2})(\hat{a}_1^\dagger + \hat{a}_2^\dagger)|0\rangle.$$

In addition

$$\hat{E}^+ = \zeta \left[\hat{a}_1 \exp(-ixi_1) + \hat{a}_2 \exp(-ixi_2) \right],$$

where $\xi_1 - \xi_2 = \mathbf{k} \cdot \mathbf{\Delta r}$. Also

$$\hat{E}^- = \zeta \left[\hat{a}_1^\dagger \exp(+ixi_1) + \hat{a}_2^\dagger \exp(+ixi_2) \right].$$

Then

$$\begin{aligned}&\langle \text{in} | \hat{E}^- \hat{E}^+ | \text{in} \rangle \\ &= (\zeta^2/2) \left[\langle 1| + \langle 2| \right] \hat{E}^- \hat{E}^+ \left[|1\rangle + |2\rangle \right] \\ &= (\zeta^2/2) \left[2 + \exp[i(\xi_1 - \xi_2)] + \exp[i(\xi_2 - \xi_1)] \right] \\ &= \zeta^2 \left[1 + \cos(\mathbf{k} \cdot \mathbf{\Delta r}) \right].\end{aligned}$$

(19.10) $\langle n_1 n_2 \rangle = \langle \psi | \hat{a}_1^\dagger \hat{a}_2^\dagger \hat{a}_2 \hat{a}_1 | \psi \rangle$ where $\hat{a}_1$ and $\hat{a}_2$ are the annihilation operators for the two ports. The state vector is $|\psi\rangle = \{(\hat{a}_1^\dagger + \hat{a}_2^\dagger)/\sqrt{2}\}^n |0\rangle$. Then the expectation value is

$$\begin{aligned}&\langle 0|(\hat{a}_1^n + n\hat{a}_1^{n-1}\hat{a}_2 + \dots \hat{a}_2^n) \\ &\hat{a}_1^\dagger \hat{a}_2^\dagger \hat{a}_2 \hat{a}_1 ((\hat{a}_1^\dagger)^n + n(\hat{a}_1^\dagger)^{n-1}\hat{a}_2^\dagger + \dots (\hat{a}_2^\dagger)^n))|0\rangle/(2^n n!) \\ &= \{\langle 0|\hat{a}_1^n \hat{a}_1^\dagger \hat{a}_2^\dagger \hat{a}_2 \hat{a}_1 (\hat{a}_1^\dagger)^n|0\rangle \\ &+ n^2 \langle 0|\hat{a}_1^{n-1}\hat{a}_2 \hat{a}_1^\dagger \hat{a}_2^\dagger \hat{a}_2 \hat{a}_1 (\hat{a}_1^\dagger)^{n-1}\hat{a}_2^\dagger|0\rangle + \dots\}/(2^n n!) \\ &= \{0 + n^2(n-1)(1)(n-1)!(1!) + \\ &(n(n-1)/(2!))^2(n-2)(2)(n-2)!(2!) + \dots\} \\ &= n(n-1)Lt_{x \to 1}(1+x)^{n-2}/2^n \\ &= n(n-1)/4.\end{aligned}$$

(19.11) In the case of peak amplitude (phase) squeezing the uncertainty area is an ellipse with its short axis radial (tangential).

(19.12) The phase mismatch is

$$\begin{aligned}\Delta k dx &= \left[k_{(2\omega)} - 2k_{(\omega)} \right] dx \\ &= (2\omega/c) \left[n_{(2\omega)} - n_\omega \right] dx.\end{aligned}$$

Then the amplitude of the outgoing waves generated at a distance x from the end of the crystal is $\exp(i\Delta kx)$ at exit. Summing and averaging the contributions along the crystal the resulting amplitude is

$$\begin{aligned}E &= \int_0^L \exp(i\Delta kx)dx/L \\ &= \exp(i\Delta kL/2)2\sin(\Delta kL/2)/(\Delta kL) \\ &= \exp(i\Delta kL/2)\text{sinc}(\Delta kL/2),\end{aligned}$$

whence the intensity varies like $\text{sinc}^2(\Delta kL/2)$.

Chapter 20

(20.1) If the lifetime is $100\,\mathrm{ps}$ the frequency halfwidth $\Delta f = 1/(2\pi 10^{-10}) = 1.59\,10^9\mathrm{Hz}$. The wavelength halfwidth is

$$\Delta\lambda = \Delta f \lambda^2/c$$
$$= (0.950\,10^{-6})^2 1.59\,10^9/3\,10^8$$
$$= 4.78\,10^{-12}\,\mathrm{m}.$$

Thus $Q = \lambda/\Delta\lambda$ is $1.99\,10^5$. Then using eqn. 20.16 the Purcell factor is $3\,(1.99\,10^5)\,(0.95\,10^{-6}/3.5)^3/(4\pi^2 3\,10^{-19})$, which reduces to 1000.

(20.2) The vacuum Rabi splitting is given by eqn. 20.8. With the values given $\zeta_\omega = 6.95\,10^4\mathrm{Vm}^{-1}$, and ω is $1.984\,10^{15}$, so that the splitting ΔE is $139\,\mu\mathrm{eV}$. If the Purcell effect is to be observable the cavity linewidth must be much greater than the exciton line width. Now for the latter

$$\Delta\lambda/\lambda = \Delta f/f = 1/(\omega\tau),$$

where f is the frequency and τ the exciton lifetime. Then
$$Q < \lambda/\Delta\lambda = \omega\tau = 1.98\,10^5.$$

(20.3) The Purcell factor is given by eqn. 20.16. With the values given V is $1.0 \times (\lambda/n)^3$ and substituting further $F_\mathrm{P} = 3\,10^3/4\pi^2$, that is 7.6.

(20.4) The vacuum Rabi angular frequency is $\Delta E/\hbar$, and using the values given this is $2.11\,10^{11}\,\mathrm{rads}^{-1}$. The cavity decay time is shorter than the exciton lifetime and is given by

$$\tau_\mathrm{cav} = 1/(2\pi\Delta f)$$
$$= (1/2\pi)(\lambda/\Delta\lambda f)$$
$$= Q/\omega = 5.04\,10^{-13}\,\mathrm{s}.$$

Therefore the Rabi angular frequency must exceed $1/\tau_\mathrm{cav}$ or $1.98\,10^{12}\,\mathrm{rads}^{-1}$ comfortably if there is to be strong coupling. This is ten times the vacuum Rabi angular frequency. Now the Rabi frequency is proportional to the square root of the photon population of the mode, hence it requires a population of well above 100 photons.

(20.5) Consider the general case that the photons are n pulse intervals apart. The possibilities for their paths through the Mach–Zehnder are these: 1) short–short; 2) short–long; 3) long–short; 4) long–long. Each may then enter either detector. Only when they enter different detectors will a time interval be registered. The possibilities for these interval are respectively: 1) n, n; 2) $(n-1)$, $(n+1)$;

3) $(n+1)$, $(n-1)$; 4) n, n. We can summarize the frequencies of registering intervals when the pulses start out n intervals apart in this way

$$[n : 4 \times n; 2 \times (n-1); 2 \times (n+1)].$$

Around $n = 0$, which is missing because only single photons are emitted in each pulse, we have

$$[-1 : 4 \times (-1);\ 2 \times (-2);\ 2 \times (0)],$$
$$[+1 : 4 \times (+1);\ 2 \times (0);\ 2 \times (+2)],$$
$$[+2 : 4 \times (+2);\ 2 \times (+1);\ 2 \times (+3)],$$
$$[+2 : 4 \times (+3);\ 2 \times (+2);\ 2 \times (+4)],$$

and so on. Adding the number of occurences gives: $4 \times (0)$; $6 \times (+1)$; $8 \times (n \geq 2)$, with identical rates for the negative as for the positive intervals. Consequently the relative heights of the central peak and its immediate neighbours are 0.5 and 0.75 of the other peaks.

(20.6) The exciton is far off resonance with the cavity so that eqn. 20.17 is applicable. Substituting the values given $[\gamma_c/(\omega_c - \omega)]^2 = 1/4$. Then the decay rate of the exciton $\gamma_\mathrm{off} = 4.38\,10^8\,\mathrm{s}^{-1}$. In the absence of a cavity eqn. 19.146 is applicable, and $\gamma_\mathrm{open} = 3.54\,10^9$.

(20.7) Immediately after the $\pi/2$ pulse the state vector is

$$|\psi_1\rangle + |\psi_2\rangle]/\sqrt{2},$$

after a time T later it is

$$[|\psi_1\rangle + \exp{(i\alpha T)}|\psi_2\rangle]/\sqrt{2}.$$

The π pulse interchanges the states so that the vector is

$$[|\psi_2\rangle + \exp{(i\alpha T)}|\psi_1\rangle]/\sqrt{2}.$$

Then after a further interval T we have

$$\exp{(i\alpha T)}[|\psi_2\rangle + |\psi_1\rangle]/\sqrt{2}.$$

Thus the state becomes identical to that after the $\pi/2$ pulse. In the case of an ensemble of quantum dots their phase drifts will be different because of their different environments. Thus following the $\pi/2$ pulse the excitons initially radiate in phase and then drift out of phase with one another. So there is a pulse which fades rapidly. After an interval T following the π pulse coherence is restored and the excitons radiate in phase again. This gives a second pulse of radiation, the photon echo pulse.

Index